T0206417

Fourth Edition

Plant Systematics

An Integrated Approach

Gurcharan Singh

Retired Associate Professor

University of Delhi, New Delhi, India

CRC Press is an imprint of the

Taylor & Francis Group, an **informa** business

A SCIENCE PUBLISHERS BOOK

CRC Press
Taylor & Francis Group
6000 Broken Sound Parkway NW, Suite 300
Boca Raton, FL 33487-2742

First issued in paperback 2021

© 2019 by Taylor & Francis Group, LLC
CRC Press is an imprint of Taylor & Francis Group, an Informa business

No claim to original U.S. Government works

Version Date: 20190302

ISBN 13: 978-0-367-77966-5 (pbk)
ISBN 13: 978-0-367-25088-1 (hbk)

This book contains information obtained from authentic and highly regarded sources. Reasonable efforts have been made to publish reliable data and information, but the author and publisher cannot assume responsibility for the validity of all materials or the consequences of their use. The authors and publishers have attempted to trace the copyright holders of all material reproduced in this publication and apologize to copyright holders if permission to publish in this form has not been obtained. If any copyright material has not been acknowledged please write and let us know so we may rectify in any future reprint.

Except as permitted under U.S. Copyright Law, no part of this book may be reprinted, reproduced, transmitted, or utilized in any form by any electronic, mechanical, or other means, now known or hereafter invented, including photocopying, microfilming, and recording, or in any information storage or retrieval system, without written permission from the publishers.

For permission to photocopy or use material electronically from this work, please access www.copyright.com (http://www.copyright.com/) or contact the Copyright Clearance Center, Inc. (CCC), 222 Rosewood Drive, Danvers, MA 01923, 978-750-8400. CCC is a not-for-profit organization that provides licenses and registration for a variety of users. For organizations that have been granted a photocopy license by the CCC, a separate system of payment has been arranged.

Trademark Notice: Product or corporate names may be trademarks or registered trademarks, and are used only for identification and explanation without intent to infringe.

Library of Congress Cataloging-in-Publication Data

Names: Singh, Gurcharan, 1945- author.
Title: Plant systematics : an integrated approach / Gurcharan Singh.
Description: First edition. | Boca Raton, FL : CRC Press/Taylor & Francis Group, [2019] | Includes bibliographical references and index.
Identifiers: LCCN 2019007826 | ISBN 9780367250881 (hardback)
Subjects: LCSH: Plants--Classification.
Classification: LCC QK95 .S572 2019 | DDC 580.1/2--dc23
LC record available at https://lccn.loc.gov/2019007826

Visit the Taylor & Francis Web site at
http://www.taylorandfrancis.com

and the CRC Press Web site at
http://www.crcpress.com

Preface

The fourth edition of "Plant Systematics: An Integrated Approach" has been revised and reorganised thoroughly, and significantly updated to incorporate the developments of the last few years. The publication of current edition was forced mainly because of publication of revision of International Code of Nomenclature for Algae, Fungi and Plants (Shenzhen Code, 2018), new version of PhyloCode (Beta version of Phylocode 5, 2014), updated APweb version 14 (September, 2018 update), revised Angiosperm Phylogeny Group classification APG IV (2016), new Pteridophyte Phylogeny Group Classification PPG I (2016) and accumulation of large amount of information subsequent to the publication of third edition. There has been development of new tools of biotechnology, vigorous utilization of molecular data in understanding phylogeny, and redefining of affinities and arrangements of plant groups. Recent years have also seen disappearance of gaps between numerical and cladistical methodologies, and integration of former into latter for complete understanding of phylogenetic relationships. These trends have largely influenced the combination of numerical and cladistic methods under one chapter, and enlarged discussion on Molecular Systematics, discussing new concepts, tools and recent achievements and need to align these with morphological traits so that plants can be conveniently identified in the field. New chapters on Pteridophytes and Gymnosperms added in third edition have been retained and updated for complete understanding of systematics of vascular plants.

Actual photographs of plants and plant parts enable better understanding of taxonomic information. This practice (of using actual photographs) has been followed in several other recent publications too. The present edition incorporates numerous photographs of plants from diverse families of plants (color in the ebook, B&W in the print version). This has largely been possible with the assistance of my son Manpreet Singh who sponsored my recent visit to California and provided me the opportunity to visit and photograph temperate plants in and around California. The book, as such, contains images of both tropical plants, Himalayan plants, temperate American plants and plants from other parts of the world growing in the Botanical Gardens of University of California and San Francisco Botanical Garden, as also natural forest regions in California. In addition, 305 black & white illustrations help in a better understanding of the plants covered in the book.

The recent decades have seen major changes in schemes of classification of angiosperms. The traditional distinction between dicots and monocots seems to be disappearing, dicots are being split into the basal families and more advanced eudicots, and monocot families interpolated between these two groups of dicots. Whereas Angiosperm Phylogeny Group is attempting to establish monophyletic groups at the most up to the order level, grouped into informal groups, Thorne aims at blending these developments with the traditional Linnaean hierarchy. Both sets of classification have undergone major modifications in the recent years, such as the latest version of Thorne's classification (2007), Takhtajan's classification (revised in 2009), APG III (2009), APG IV (2016) and APweb Version 14 (2017). Attempts are also being made to provide unified Code of nomenclature for all living organisms.

The author has attempted to strike a balance between classical fundamental information and the recent developments in plant systematics. Special attention has been devoted to the information on botanical nomenclature, identification and phylogeny of angiosperms with numerous relevant examples and detailed explanation of important nomenclatural problems. An attempt has been made to present a continuity between orthodox and contemporary identification methods by working on a common example. The methods of identification using computers have been further explored to help better on-line identification. Outputs of computer programs especially used in molecular studies and construction of phylogenetic trees have been included based on actual or hypothetical data. This will acquaint readers with the handling of raw data and working of computer programs.

Internet highways are revolutionizing the exchange of scientific information. Botanical organizations have plunged into this revolution in a big way. Instant information on major classification systems, databases, herbaria, gardens, indices and thousands of illustrations are available to users Worldwide at the touch of a button. Discussion on important aspects of this information highway and useful links have been provided in this book. My interaction with colleagues especially the members of the list 'Taxacom', 'efloraofindia' Facebook groups 'Plant Identification', 'California Native Plant Society' and 'Indian Flora' has been very helpful in resolving some doubts. I am greatful to all members whose information has been valuable on several occasions.

For providing me inspiration for this book, I am indebted to my students, who helped me to improve the material through frequent interactions. I am also indebted to my wife Mrs. K. G. Singh for constantly bearing with my overindulgence with this book and also providing ready help in computer-related activities during the processing of this work. I am also thankful to my son Manpreet Singh and daughter-in-law Komal for our visit to California, where many plants were photographed. My younger son Kanwarpreet Singh & daughter-in-law Harpreet provided help on various occasions. Manpreet and Kanwarpreet also provided important photographic tips.

I wish to record my thanks to all colleagues whose inputs have helped me to improve the information presented here. I also wish to extend my thanks to Dr. Jef Veldkamp for valuable information on nomenclature, Dr. Gertrud Dahlgren for photographs and literature, Dr. P. F. Stevens for literature on APG IV, APweb Version 14 (updated September 2018) and trees from his APweb, Dr. Robert Thorne for making available his 2007 classification, Dr. James Reveal for his help on nomenclatural problems, Dr. Patricia Holmgren for information about major world herbaria, Dr. D. L. Dilcher for photograph, Dr. Julie Barcelona and Harry Wiriadinata for photographs of *Rafflesia*, the authorities of New York Botanical Garden, Missouri Botanical Garden, USA, Royal Botanic Gardens Kew and University of California, Santa Cruz, for photographs used in the book. Special thanks to Philip Cantino of Ohio University for making available the Beta Version of PhyloCode 5. Thanks, are also due to all whose contributions are cited in this book.

New Delhi
January 2019

Gurcharan Singh

Contents

List of Plates

Chapter 1

Plants, Taxonomy and Systematics

Taxonomy (or systematics) is basically concerned with the classification of organisms. Living organisms are placed in groups based on similarities and differences at the organismic, cellular, and molecular levels. Described species of organisms on earth were estimated between 1.4 million and 1.8 million (Stork, 1988; Barnes, 1989; Hammond, 1992). Wilson (1992) estimated that roughly 1.5 million species of described species comprised 73.1% Animals, 17.6% Plants, 4.9% Fungi, 4.1% protists and only 0.4% Bacteria. Hammond (1992) proposed a working estimate of 12.5 million species organisms on earth consisting of 9.6 m animals (1.2 m described), 300,000 plants (250,000 described), 990,000 Fungi (80,000 described), 400,000 protists (90,000 described) and nearly 400,000 bacteria (only 6000 described).

The United Nations Environment Programme's *Global Biodiversity Assessment* published by Convention on Biological Diversity (2001) estimates the number of described species of living organisms as approximately 1.75 million, slightly more than ten percent of estimated 14 million species. Whereas nearly 90 percent of estimated plant species have already been described (270,000 of 300,000), only 12.5 percent of animals (1.3 m of 10.6 m), 4.7 percent of Fungi (70,000 of 1.5 m), 13 percent of Protoctists (80,000 of 600,00) and only 0.4 percent of Bacteria (4,000 of 1 m) have been described so far. The list grows every year. Subsequent publication by Chapman (2009) puts figures of described plant species as 310,129 (of 390,800 estimated) that includes 268,600 described angiosperms (estimated 352,000). The more recent estimates, however, put figures of flowering plants only as around 400,000 (Edwards, 2010 on Physorg.com). Best estimate by Census of Marine life, published by Mora et al. (2011) in PLoS Biology puts total number of Eukaryotic species on earth as 8.7 million, with 6.5 million species on land and 2.2 million in oceans, as against the earlier rough estimates between 3 million and 100 million. These 8.7 m species comprised 7.77 m animals (954,434 described), 611,000 Fungi (43,271 described), 298,000 plants (215,644 described), 36,400 Protozoa or protists (8,118 described) and 10,358 bacteria (described, predicted 9680). Only 1.2 m species of 8.7 m estimated have been described and catalogued. Furthermore, the study says a staggering 86% of all species on land and 91% of those in the seas have yet to be discovered, described and catalogued.

Pappas (2016), concluded that there may be 100 million to even 1 trillion species on this planet, based on scaling rules that linked the number of individual organisms to the number of total species. The list grows longer every year.

Christenhusz and Byng (2016) who have compiled a list of estimated number of genera and species in families recognized mainly in APG IV (2016) and other recent classifications estimate described and accepted number of plant species as ca 374,000, of which approximately 308,312 are vascular plants, with 295,383 flowering plants. Global numbers of smaller plant groups are as follows: algae ca 44,000, liverworts ca 9,000, hornworts ca 225, mosses 12,700, lycopods 1,290, ferns 10,560 and gymnosperms 1,079.

Most recent study by Brenden et al. (2017) estimates suggest that there are likely to be at least 1 to 6 billion species on Earth, and in contrast to previous estimates, rather than being dominated by insects, the new Pie of Life is dominated by bacteria (approximately 78%), protists 7.3%, plants 0.02%, Fungi 7.4%, and animals 7.3% (Figure 1.1). Classifying these organisms has been a major challenge, and the last few decades have seen a lot of realignments as additional ultrastructural and molecular information piles up. These realignments have primarily been the result of realization that the branches of the phylogenetic tree must be based on the concept of monophyly, and each taxonomic group, kingdoms included, should be monophyletic. Unfortunately, only a fraction of these estimates, just 1.5 million species have been described.

Before attempting to classify the various organisms, it is necessary to identify and name them. A group of individuals, unique in several respects, is given a unique binomial, and is recognized as a species. These species are grouped into taxonomic groups, which are successively assigned the ranks of genera, families, orders, and the process continues till all the species have been arranged (classified) under a single largest, most inclusive group. Classifying organisms and diverse forms of life is challenging task before the biologists.

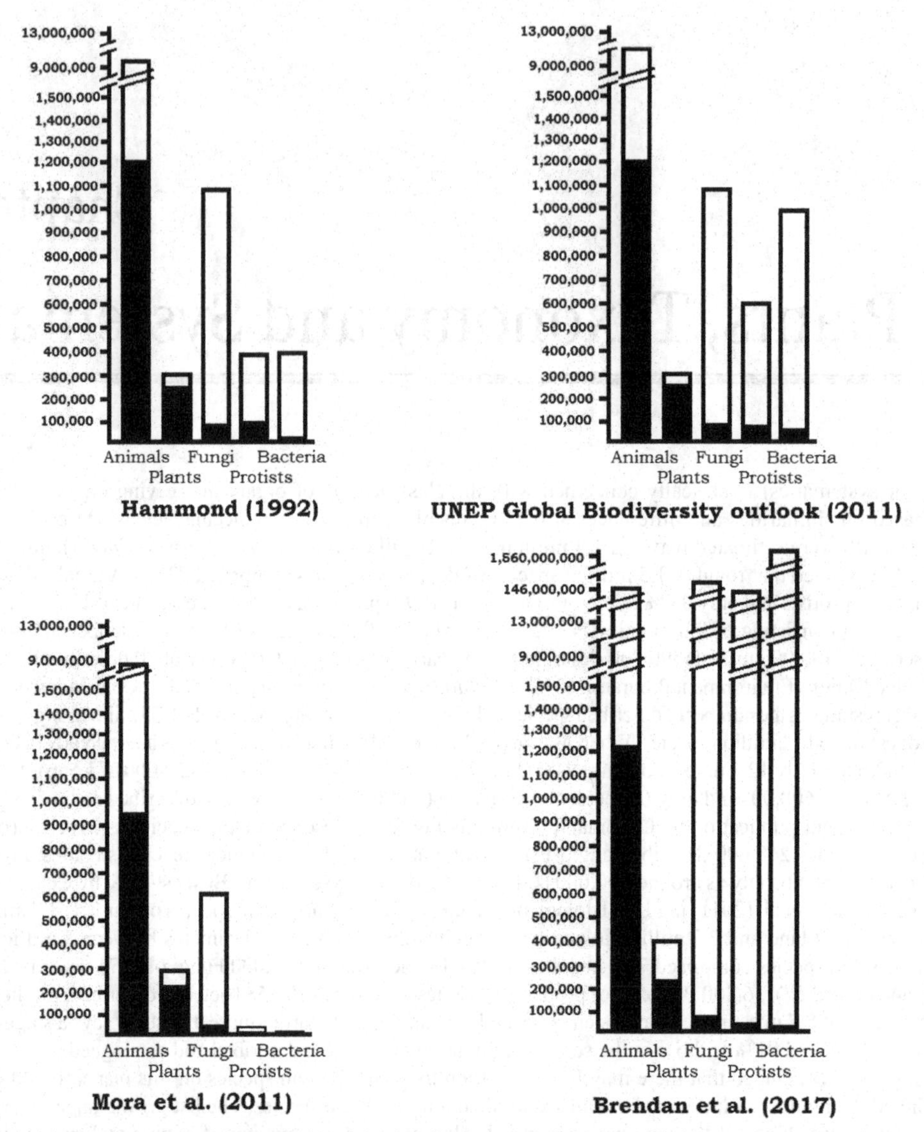

Figure 1.1: Estimated number of species on earth and the described number of species (solid lower portion of bar) in major groups of organisms. The estimates in recent years have shifted from Animals to Bacteria, Fungi and Protists as dominant groups.

PLANTS AND KINGDOMS OF LIFE

Plants are man's prime companions in this universe, being the source of food and energy, shelter and clothing, drugs and beverages, oxygen and aesthetic environment, and as such they have been the dominant component of his taxonomic activity through the ages. Before attempting to explore the diversity of plant life it is essential to understand as to what our understanding of the term **Plant** is, and the position of plants in the web of life. Traditionally the plants are delimited as organisms possessing cell wall, capable of photosynthesis, producing spores and having sedentary life. A lot of rethinking has resulted in several different interpretations of the term plant.

Two Kingdom System

The living organisms were originally grouped into **two kingdoms**. Aristotle divided all living things between plants, which generally do not move or have sensory organs, and animals. Linnaeus in his *Systema naturae* published in 1735 placed them under **Animalia** (Animals) and **Vegetabilia** (Plants) as two distinct kingdoms (Linnaeus placed minerals in the third kingdom Mineralia). Linnaeus divided each kingdom into classes, later grouped into phyla for animals and divisions for plants. When single-celled organisms were first discovered, they were split between the two kingdoms: mobile forms in the animal phylum Protozoa, and colored algae and bacteria in the plant division Thallophyta or Protophyta. As a result, Ernst Haeckel (1866) suggested creating a third kingdom **Protista** for them, although this was not very popular until relatively recently (sometimes also known as **Protoctista**). Haeckel recognized **three kingdoms**: **Protista**, **Plantae** and **Animalia**.

Two Empires Three Kingdoms

The subsequent discovery that bacteria are radically different from other organisms in lacking a nucleus, led Chatton (1937) to propose a division of life into **two empires**: organisms with a nucleus in **Eukaryota** and organisms without in **Prokaryota**. **Prokaryotes** do not have a nucleus, mitochondria or any other membrane bound organelles. In other words, neither their DNA nor any other of their metabolic functions are collected together in a discrete membrane enclosed area. Instead everything is openly accessible within the cell, though some bacteria have internal membranes as sites of metabolic activity these membranes do not enclose a separate area of the cytoplasm. **Eukaryotes** have a separate membrane bound nucleus, numerous mitochondria and other organelles such as the Golgi Body within each of their cells. These areas are separated off from the main mass of the cell's cytoplasm by their own membrane in order to allow them to be more specialized. The nucleus contains all the Eukaryote cell DNA, which gets organized into distinct chromosomes during the process of mitosis and meiosis. The energy is generated in mitochondria. The exception to this rule are red blood cells which have no nucleus and do not live very long. Haeckel (1966) proposed a three kingdom classification recognizing Protista, Plantae and Animalis, the additional kingdom Protista including all single-celled organisms that are intermediate between animals and plants. Herbert Copeland (1938), who gave the prokaryotes a separate kingdom, originally called Mycota but later referred to as **Monera** or **Bacteria**. Copeland later on (1956) proposed a **four-kingdom system** placing all eukaryotes other than animals and plants in the kingdom **Protoctista,** thus recognizing four kingdoms **Monera**, **Protoctista**, **Plantae** and **Animalia**. The importance of grouping these kingdoms in two empires, as suggested earlier by Chatton was popularized by Stanier and van Niel (1962), and soon became widely accepted.

Five Kingdom System

American biologist Robert H. Whittaker (1969) proposed the removal of fungi into a separate kingdom thus establishing a **five kingdom system** recognizing **Monera**, **Protista**, **Fungi**, **Plantae** and **Animalia** as distinct kingdoms. The fungi like plants have a distinct cell wall but like animals lack autotrophic mode of nutrition. They, however, unlike animals draw nutrition from decomposition of organic matter, have cell wall reinforced with chitin, cell membranes containing ergosterol instead of cholesterol and have a unique biosynthetic pathway for lysine. The classification was followed widely in textbooks.

Six or Seven Kingdoms?

Subsequent research concerning the organisms previously known as archebacteria has led to the recognition that these creatures form an entirely distinct kingdom **Archaea**. These include anaerobic bacteria found in harsh oxygen-free conditions and are genetically and metabolically completely different from other, oxygen-breathing organisms. These bacteria, called **Archaebacteria**, or simply Archaea, are said to be "living fossils" that have survived since the planet's very early ages, before the Earth's atmosphere even had free oxygen. This together with the emphasis on phylogeny requiring groups to be monophyletic resulted in a **six kingdom system** proposed by Carl Woese et al. (1977). They grouped Archaebacteria and Eubacteria under Prokaryotes and rest of the four kingdoms Protista, Fungi, Plantae and Animalia under Eukaryotes. They subsequently (1990) grouped these kingdoms into three **domains Bacteria** (containing Eubacteria), **Archaea** (containing Archaebacteria) and **Eukarya** (containing Protista, Fungi, Plantae and Animalia).

Margulis and Schwartz (1998) proposed term **superkingdom** for domains and recognized two superkingdoms: **Prokarya** (Prokaryotae) and **Eukarya** (Eukaryotae). Former included single kingdom **Bacteria** (Monera) divided into two subkingdoms **Archaea** and **Eubacteria**. **Eukarya** was divided into four kingdoms: **Protoctista (Protista), Animalia, Plantae** and **Fungi**.

Several recent authors have attempted to recognize seventh kingdom of living organisms, but they differ in their treatment.

Ross (2002, 2005) recognized Archaebacteria and Eubacteria as separate kingdoms, named as **Protomonera** and **Monera**, respectively again under separate superkingdoms (domains of earlier authors) **Archaebacteria** and **Eubacteria**. He added seventh kingdom **Myxomycophyta** of slime molds under superkingdom Eukaryotes. Two additional superkingdoms of extinct organisms Progenotes (first cells) and Urokaryotes (prokaryotic cells that became eukaryotes):

Superkingdom Progenotes*........first cells now extinct
Superkingdom Archaebacteria
Kingdom Protomonera.....archaic bacteria
Superkingdom Eubacteria
Kingdom Monera........bacteria
Superkingdom Urkaryotes* ...prokaryotic cells that became eukaryotes
Superkingdom Eukaryotes ...cells with nuclei
Kingdom Protista..........protozoans
Kingdom Myxomycophyta.....slime molds
Kingdom Plantae...........plants
Kingdom Fungi.............fungi
Kingdom Animalia...........animals

Patterson and Sogin (1992; Figure 1.2) recognized seven kingdoms but include slime molds under Protozoa (Protista) and instead establish **Chromista** (diatoms) as seventh kingdom. Interestingly the traditional algae now find themselves distributed in three different kingdoms: eubacterial prokaryotes (the blue-green cyanobacteria), chromistans (diatoms, kelps), and protozoans (green algae, red algae, dinoflagellates, euglenids).

Cavalier-Smith (1981) suggested that Eukaryotes can be classified into nine kingdoms each defined in terms of a unique constellation of cell structures. Five kingdoms have plate-like mitochondrial cristae: (1) **Eufungi** (the non-ciliated fungi, which unlike the other eight kingdoms have unstacked Golgi cisternae), (2) **Ciliofungi** (the posteriorly ciliated fungi), (3) **Animalia** (Animals, sponges, mesozoa, and choanociliates; phagotrophs with basically posterior ciliation), (4) **Biliphyta** (Non-phagotropic, phycobilisome-containing, algae; i.e., the Glaucophyceae and Rhodophyceae), (5) **Viridiplantae** (Non-phagotrophic green plants, with starch-containing plastids). Kingdom (6), the **Euglenozoa**, has disc-shaped cristae and an intraciliary dense rod and may be phagotrophic and/or phototrophic with plastids with three-membraned envelopes. Kingdom (7), the **Cryptophyta**, has flattened tubular cristae, tubular mastigonemes on both cilia, and starch in the compartment between the plastid endoplasmic reticulum and the plastid envelope; their plastids, if present, have phycobilins inside the paired thylakoids and chlorophyll c2. Kingdom (8), the **Chromophyta**, has tubular cristae, together with tubular mastigonemes on one anterior cilium and/or a plastid endoplasmic reticulum and chlorophyll c1 + c2. Members of the ninth kingdom, the **Protozoa**, are mainly phagotrophic, and have tubular or vesicular cristae (or lack mitochondria altogether), and lack tubular mastigonemes on their (primitively anterior) cilia; plastids if present have three-envelop membranes, chlorophyll c2, and no internal starch, and a plastid endoplasmic reticulum is absent. Kingdoms 4–9 are primitively anteriorly biciliate. A simpler system of five kingdoms suitable for very elementary teaching is possible by grouping the photosynthetic and fungal kingdoms in pairs. It was suggested that Various compromises are possible between the nine and five kingdoms systems; it is suggested that the best one for general scientific use is a system of seven kingdoms in which the Eufungi and Ciliofungi become subkingdoms of the Kingdom Fungi, and the Cryptophyta and Chromophyta subkingdoms of the Kingdom Chromista; the Fungi, Viridiplantae, Biliphyta, and Chromista can be

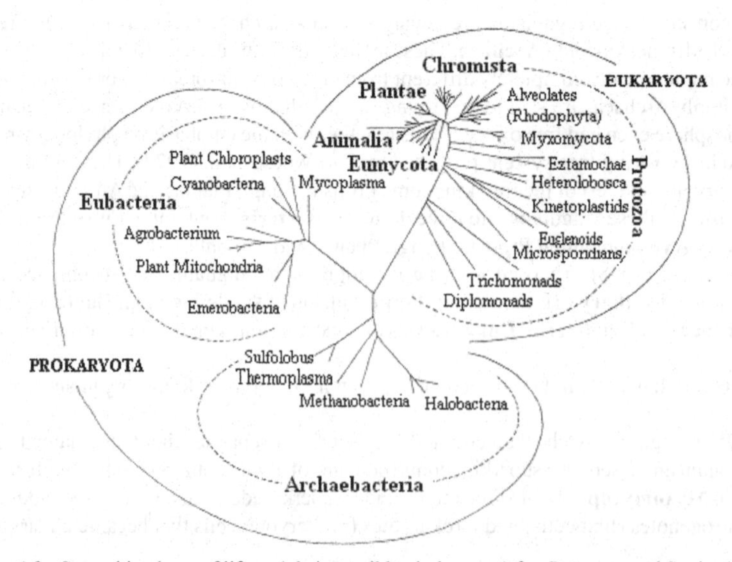

Figure 1.2: Seven kingdoms of life and their possible phylogeny (after Patterson and Sogin, 1992).

subject to the Botanical Code of Nomenclature, while the Zoological Code can govern the Kingdoms Animalia, Protozoa and Euglenozoa.

These 9 kingdoms together with two or one kingdom of prokaryotes total eleven or ten kingdoms of life. Subsequently, however, Cavalier-Smith (1998, 2000, 2004) reverted to six kingdom classification recognizing **Bacteria, Protozoa, Animalia, Fungi, Plantae and Chromista** under two empires Prokaryota and Eukaryota. Prokaryotes constitute a single kingdom, Bacteria, here divided into two new subkingdoms: Negibacteria, with a cell envelope of two distinct genetic membranes, and Unibacteria, comprising the phyla Archaebacteria and Posibacteria. Outline of the classification is as under:

Empire Prokaryota

Kingdom **Bacteria**

Subkingdom **Negibacteria** (phyla Eobacteria, Sphingobacteria, Spirochaetae, Proteobacteria, Planctobacteria, Cyanobacteria)

Subkingdom **Unibacteria** (phyla Posibacteria, Archaebacteria)

Empire Eukaryota

Kingdom **Protozoa**

Subkingdom **Sarcomastigota** (phyla Amoebozoa, Choanozoa)

Subkingdom **Biciliata**

Kingdom **Animalia** (Myxozoa and 21 other phyla)

Kingdom **Fungi** (phyla Archemycota, Microsporidia, Ascomycota, Basidiomycota)

Kingdom **Plantae**

Subkingdom **Biliphyta** (phyla Glaucophyta, Rhodophyta)

Subkingdom **Viridaeplantae** (phyla Chlorophyta, Bryophyta, Tracheophyta)

Kingdom **Chromista**

Subkingdom **Cryptista** (phylum Cryptista: cryptophytes, goniomonads, katablepharids)

Subkingdom **Chromobiota**

The name archaebacteria seems to be confusing. They were so named because they were thought to be the most ancient (Greek '*archaio*' meaning ancient) and sometimes labelled as living fossils, since they can survive in anaerobic conditions (methanogens—which use hydrogen gas to reduce carbon dioxide to methane gas), high temperatures (thermophiles, which can survive in temperatures of up to 80 degree C), or salty places (halophiles). They differ from bacteria in having methionine as aminoacid that initiates protein synthesis as against formyl-methionine in bacteria, presence of introns in some genes, having several different RNA polymerases as against one in bacteria, absence of peptidoglycan in cell wall, and growth not inhibited by antibiotics like streptomycin and chloramphenicol. In several of these respects archaebacteria are more similar to eukaryotes. Bacteria are thought to have diverged early from the evolutionary line (the clade neomura, with many common characters, notably obligately co-translational secretion of N-linked glycoproteins, signal recognition particle with 7S RNA and translation-arrest domain, protein-spliced tRNA introns, eight-subunit chaperonin, prefoldin, core histones, small nucleolar ribonucleoproteins (snoRNPs), exosomes and similar replication, repair, transcription and translation machinery) that gave rise to archaebacteria and eukaryotes. It is, as such more appropriate to call archaebacteria as **metabacteria**.

The eukaryotic host cell (Figure 1.3) evolved from something intermediate between posibacteria and metabacteria ("archaebacteria"), which had evolved many metabacterial features but not yet switched to ether-linked lipid membranes in a major way. They would no doubt cladistically fall out as primitive metabacteria, but whether such forms are still extant is uncertain. There are lots of metabacteria out there which are uncultured (only known from environmental sequences) or just undiscovered, so who knows.

The further shift from archaebacteria to Eukaryotes involved the transformation of circular DNA into a **linear DNA** bound with histones, formation of **membrane bound nucleus** enclosing chromosomes, development of **mitosis**, occurrence of **meiosis** in sexually reproducing organisms, appearance of membrane bound organelles such as **endoplasmic reticulum, golgi bodies** and **lysosomes**, appearance of **cytoskeletal elements** like actin, myosin and tubulin, and the formation of **mitochondria** through endosymbiosis.

A major shift in this eukaryotic line which excluded animal and fungi, involved the development of **chloroplast** by an eukaryotic cell engulfing a photosynthetic bacterial cell (probably a cyanobacterium). The bacterial cell continued to live and multiply inside the eukaryotic cell, provided high energy products, and in turn received a suitable environment to live in. The two thus shared **endosymbiosis**. Over a period of time the bacterial cell lost ability to live independently, some of the bacterial genes getting transferred to eukaryotic host cell, making the two biochemically interdependent. Chloroplast evolution in Euglenoids and Dinoflagellates occurred through **secondary endosymbiosis**, wherein eukaryotic cell engulfed an eukaryotic cell containing a chloroplast. This common evolutionary sequence is shared by green plants (including green algae; green chloroplast), red algae (red chloroplast) and brown algae and their relatives (commonly

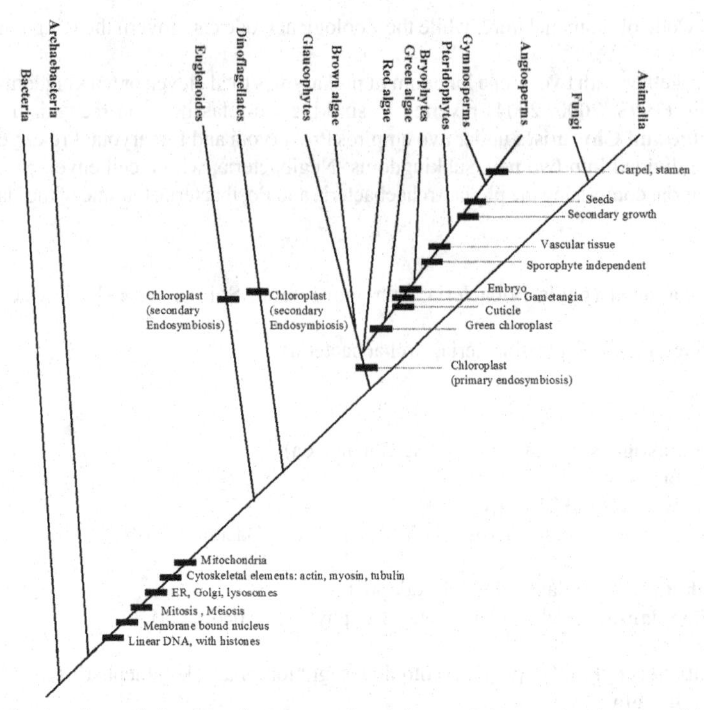

Figure 1.3: Cladogram showing the evolution of major groups of organisms and the associated apomorphies. Chloroplast evolution has occurred twice, once (primary endosymbiosis) eukaryote cell engulfing a photosynthetic bacterial cell, and elsewhere (secondary endosymbiosis) eukaryotic cell engulfing an eukaryotic cell containing chloroplast.

known as stramenopiles; brown chloroplast), in which diversification of chloroplast pigments occurred, along with the thylakoid structure and a variety of storage products.

The Plant Kingdom

It is now universally agreed that members of the plant kingdom include, without doubt the green algae, liverworts and mosses, pteridophytes, gymnosperms and finally the angiosperms, the largest group of plants. All these plants share a green chloroplast. Red algae, Brown algae and Glaucophytes, latter two together known as stramenophiles, also belong to this kingdom. All these groups share the presence of a chloroplast. All green plants share a green chloroplast with chlorophyll b, chlorophyll a, thylakoids and grana, and starch as storage food. Evolution of cuticle combined with gametangia and embryo characterizes embryophytes, including bryophytes, pteridophytes and seed plants. The development of vascular tissue of phloem and xylem, and independent sporophyte characterize tracheophytes including pteridophytes and seed plants. Secondary growth resulting in the formation of wood and seed habit differentiates seed plants. The final evolution of a distinct flower, carpels and stamens, together with vessels and sieve tubes set apart the angiosperms, the most highly evolved group of plants.

Out of nearly 3,74,000 species of plants known nearly 44,000 belong to Algae, 22,500 species belong to usually overlooked mosses and liverworts, 12,000 ferns and their allies, 1,050 to gymnosperms and 295,000 to angiosperms (belonging to about 485 families and 13,372 genera), considered to be the most recent and vigorous group of plants that have occupied earth. Angiosperms occupy the majority of the terrestrial space on earth and are the major components of the world's vegetation.

Brazil and Colombia, both located in the tropics, are countries with the most diverse angiosperms floras and which rank first and second. China, even though the main part of her land is not located in the tropics, the number of her angiosperms still occupies the third place in the world, and has approximately 300 families, 3,100 genera and 30,000 species.

TAXONOMY AND SYSTEMATICS

There are slightly more than one third of a million species of plants known to man today, the information having been accumulated through efforts of several millenniums. Although man has been classifying plants since the advent of civilization, **taxonomy** was recognized as a formal subject only in 1813 by A. P. de Candolle as a combination of Greek words **taxis** (arrangement) and **nomos** (rules or laws) in his famous work *Theorie elementaire de la botanique*. For a long time, plant taxonomy was considered as 'the science of identifying, naming, and classifying plants' (Lawrence, 1951). Since identification and nomenclature are important prerequisites for any classification, taxonomy is often defined as the '***science dealing with the study of classification, including its bases, principles, rules and procedures***' (Davis and Heywood, 1963).

Although **Systematics** was recognized as a formal major field of study only during the latter half of twentieth century, the term had been in use for a considerable period. Derived from the Latin word *systema* (organized whole), forming the title of the famous work of Linnaeus *Systema naturae* (1735), the term Systematics first appeared in his *Genera Plantarum* (1737), though Huxley (1888) is often credited to have made the first use of the term in his article in *Nature* on the systematics of birds. Simpson (1961) defined **systematics** as a '***scientific study of the kinds and diversity of organisms, and of any and all relationships between them***'. It was recognized as a more inclusive field of study concerned with the study of diversity of plants and their naming, classification and evolution. The scope of taxonomy has, however, been enlarged in recent years to make taxonomy and systematics synonymous. A broader definition (Stace, 1980) of **taxonomy,** to coincide with **systematics** recognized it as '***the study and description of variation in organisms, the investigation of causes and consequences of this variation, and the manipulation of the data obtained to produce a system of classification***'.

Realization of the fact that a good number of authors still consider taxonomy to be a more restricted term and systematics a more inclusive one has led recent authors to prefer the term **systematics** to include discussion about all recent developments in their works. Modern approach to systematics aims at reconstructing the entire chronicle of evolutionary events, including the formation of separate lineages and evolutionary modifications in characteristics of the organisms. It ultimately aims at discovering all the branches of the evolutionary tree of life; and to document all the changes and to describe all the species which form the tips of these branches. This won't be possible unless information is consolidated in the form of an unambiguous system of classification. This, however, is again impossible without a clear understanding of the basic identification and nomenclatural methods. Equally important is the understanding of the recent tools of data handling, newer concepts of phylogenetics, expertise in the judicious utilization of fast accumulating molecular data in understanding of affinities between taxa.

Prior to the evolutionary theory of Darwin, relationships were expressed as **natural affinities** on the basis of an overall similarity in morphological features. Darwin ushered in an era of assessing **phylogenetic relationships** based on the course of evolutionary descent. With the introduction of computers and refined statistical procedures, overall similarity is represented as phenetic **relationship**, which takes into account every available feature, derived from such diverse fields as anatomy, embryology, morphology, palynology, cytology, phytochemistry, physiology, ecology, phytogeography and ultrastructure.

With the advancement of biological fields, new information flows continuously, and the taxonomists are faced with the challenge of integrating and providing a synthesis of all the available data. Systematics now is, thus, an **unending synthesis,** a dynamic science with never-ending duties. The continuous flow of data necessitates rendering descriptive information, revising schemes of identification, revaluating and improving systems of classification and perceiving new relationships for a better understanding of the plants. The discipline as such includes all activities that are a part of the effort to organize and record the diversity of plants and appreciate the fascinating differences among the species of plants. Systematic activities are basic to all other biological sciences, but also depend, in turn, on other disciplines for data and information useful in constructing classification. Certain disciplines of biology such as cytology, genetics, ecology, palynology, paleobotany and phytogeography are so closely tied up with systematics that they cannot be practiced without basic systematic information. Experiments cannot be carried out unless the organisms are correctly identified and some information regarding their relationship is available. The understanding of relationships is particularly useful in the applied fields of plant breeding, horticulture, forestry and pharmacology for exploring the usefulness of related species. Knowledge of systematics often guides the search for plants of potential commercial importance.

Basic Components (Principles) of Systematics

Various systematic activities are directed towards the singular goal of constructing an ideal system of classification that necessitates the procedures of identification, description, nomenclature and constructing affinities. This enables a better management of information to be utilized by different workers, investigating different aspects structure and functioning of different species of plants.

Identification

Identification or determination is recognizing an unknown specimen with an already known taxon and assigning a correct rank and position in an extant classification. In practice, it involves finding a name for an unknown specimen. This may be achieved by visiting a herbarium and comparing unknown specimen with duly identified specimens stored in the herbarium. Alternately, the specimen may also be sent to an expert in the field who can help in the identification.

Identification can also be achieved using various types of literature such as Floras, Monographs or Manuals and making use of identification keys provided in these sources of literature. After the unknown specimen has been provisionally identified with the help of a key, the identification can be further confirmed by comparison with the detailed description of the taxon provided in the literature source.

A method that is becoming popular over the recent years involves taking a photograph of the plant and its parts, uploading this picture on the website and informing the members of appropriate electronic Lists or Newsgroups, who can see the photograph at the website and send their comments to the enquirer. Members of the fraternity could thus help each other in identification in a much efficient manner.

Description

The description of a taxon involves listing its features by recording the appropriate character states. A shortened description consisting of only those taxonomic characters which help in separating a taxon from other closely related taxa, forms the **diagnosis**, and the characters are termed as **diagnostic characters.** The diagnostic characters for a taxon determine its **circumscription**. The description is recorded in a set pattern (habit, stem, leaves, flower, sepals, petals, stamens, carpels, fruit, etc.). For each **character**, an appropriate **character-state** is listed. Flower color (character) may thus be red, yellow, white, etc. (states). The description is recorded in semi-technical language using specific terms for each character state to enable a proper documentation of data.

Whereas the fresh specimens can be described conveniently, the dry specimens need to be softened in boiling water or in a **wetting agent** before these could be described. Softening is often essential for dissection of flowers in order to study their details.

Nomenclature

Nomenclature deals with the determination of a **correct name** for a taxon. There are different sets of rules for different groups of living organisms. Nomenclature of plants (including fungi) is governed by the International **Code of Nomenclature for Algae, Fungi and Plants (ICN)** since 2011 (earlier known as International Code of Botanical Nomenclature, ICBN) through its rules and recommendations. Latest Shenzhen Code adopted in 2017 has been published in June 2018, containing several changes in rules and recommendations for these groups. Updated every six years or so, the **Code** helps in picking up a single correct name out of numerous scientific names available for a taxon, with a particular circumscription, position and rank. To avoid inconvenient name changes for certain taxa, a list of conserved names is provided in the Code. Cultivated plants are governed by the International Code of Nomenclature for Cultivated Plants (**ICNCP**), slightly modified from and largely based on the Botanical Code.

Names of animals are governed by the International Code of Zoological Nomenclature (**ICZN**); those of bacteria by International Code for the Nomenclature of Bacteria (**ICNB**), now called Bacteriological Code (**BC**). A separate Code exists for viruses, named the International Code of Virus Classification and Nomenclature (**ICVCN**).

With the onset of electronic revolution and the need to have a common database for living organisms for global communication a common uniform code is being attempted. The **Draft BioCode** is the first public expression of these objectives. The first draft was prepared in 1995. After successive reviews the fourth draft, named **Draft BioCode (1997)** prepared by the International Committee for Bionomenclature was published by Greuter et al. (1998), latest version **Draft BioCode 2011** (Hawksworth et al., 2011) is available on the web. The last decade of twentieth century also saw the development of rankless **PhyloCode** based on the concepts of phylogenetic systematics. It omits all ranks except species and 'clades' based on the concept of recognition of monophyletic groups. The latest version of PhyloCode (PhyloCode5, 2014) is also available on the web.

Phylogeny

Phylogeny is the study of the genealogy and evolutionary history of a taxonomic group. Genealogy is the study of ancestral relationships and lineages. Relationships are depicted through a diagram better known as a **phylogram** (Stace, 1989), since the commonly used term **cladogram** is more appropriately used for a diagram constructed through cladistic methodology. A phylogram is a branching diagram based on the degree of advancement (**apomorphy**) in the descendants, the longest branch representing the most advanced group. This is distinct from **a phylogenetic tree** in which the vertical scale represents a geological time-scale and all living groups reach the top, with primitive ones near the centre and advanced ones near the periphery. Monophyletic groups, including all the descendants of a common ancestor, are recognized and form entities in a classification system. Paraphyletic groups, wherein some descendants of a common ancestor are left out, are reunited. Polyphyletic groups, with more than one common ancestor, are split to form monophyletic groups. Phenetic information may often help in determining a phylogenetic relationship.

Classification

Classification is an arrangement of organisms into groups on the basis of similarities. The groups are, in turn, assembled into more inclusive groups, until all the organisms have been assembled into a single most inclusive group. In sequence of increasing inclusiveness, the groups are assigned to a fixed hierarchy of categories such as species, genus, family, order, class and division, the final arrangement constituting a system of classification. The process of classification includes assigning appropriate **position** and **rank** to a new **taxon** (a taxonomic group assigned to any rank; pl. **taxa**), **dividing** a taxon into smaller units, **uniting** two or more taxa into one, **transferring** its position from one group to another and **altering** its rank. Once established, a classification provides an important mechanism of information storage, retrieval and usage. This ranked system of classification is popularly known as the **Linnaean system**. Taxonomic entities are classified in different fashions:

1. **Artificial classification** is utilitarian, based on arbitrary, easily observable characters such as habit, color, number, form or similar features. The **sexual system** of Linnaeus, which fits in this category, utilized the number of stamens for primary classification of the flowering plants.
2. **Natural classification** uses overall similarity in grouping taxa, a concept initiated by M. Adanson and culminating in the extensively used classification of Bentham and Hooker. Natural systems of the eighteenth and nineteenth centuries used morphology in delimiting the overall similarity. The concept of overall similarity has undergone considerable refinement in recent years. As against the sole morphological features as indicators of similarity in natural systems, overall similarity is now judged on the basis of features derived from all the available fields of taxonomic information (phenetic relationship).
3. **Phenetic classification** makes the use of overall similarity in terms of a phenetic relationship based on data from all available sources such as morphology, anatomy, embryology, phytochemistry, ultrastructure and, in fact, all other fields of study. Phenetic classifications were strongly advocated by Sneath and Sokal (1973) but did not find much favor with major systems of classification of higher plants. Phenetic relationship has, however, been very prominently used in modern phylogenetic systems to decide the realignments within the system of classification.
4. **Phylogenetic classification** is based on the evolutionary descent of a group of organisms, the relationship depicted either through a **phylogram, phylogenetic tree** or a **cladogram**. Classification is constructed with this premise in mind, that all the descendants of a common ancestor should be placed in the same group (i.e., group should be **monophyletic**). If some descendents have been left out, rendering the group **paraphyletic**, these are brought back into the group to make it monophyletic (merger of Asclepiadaceae with Apocynaceae, and the merger of Capparaceae with Brassicaceae in recent classifications). Similarly, if the group is polyphyletic (with members from more than one phyletic lines), it is split to create monophyletic taxa (Genus *Arenaria* split into *Arenaria* and *Minuartia*). This approach, known as **cladistics**, is practiced by **cladists**.
5. **Evolutionary taxonomic classification** differs from a phylogenetic classification in that the gaps in the variation pattern of phylogenetically adjacent groups are regarded as more important in recognizing groups. It accepts leaving out certain descendants of a common ancestor (i.e., recognizing **paraphyletic** groups) if the gaps are not significant, thus failing to provide a true picture of the genealogical history. The characters considered to be of significance in the evolution (and the classification based on these) are dependent on expertise, authority and intuition of systematists. Such classifications have been advocated by Simpson (1961), Ashlock (1979), Mayr and Ashlock (1991) and Stuessy (1990). The approach, known as **eclecticism**, is practiced by **eclecticists**.

The contemporary phylogenetic systems of classification, including those of Takhtajan, Cronquist, Thorne and Dahlgren, are largely based on decisions in which **phenetic information** is liberally used in deciding the phylogenetic relationship between groups, differing largely on the weightage given to the cladistic or phenetic relationship.

There have been suggestions to abandon the hierarchical contemporary classifications based on the **Linnaean system**, which employs various fixed ranks in an established conventional sequence with a '**phylogenetic taxonomy**' in which monophyletic groups would be unranked names, defined in terms of a common ancestry, and diagnosed by reference to synapomorphies (de Queiroz and Gauthier, 1990; Hibbett and Donoghue, 1998). Angiosperm Phylogeny Group is continuously evolving APG system of classification, latest APG IV (2016) and its variant APWeb IV developed in 2017 are also available on the web.

Classification not only helps in the placement of an entity in a logically organized scheme of relationships, it also has a great predictive value. The presence of a valuable chemical component in one species of a particular genus may prompt its search in other related species. The more a classification reflects phylogenetic relationships, the more predictive it is supposed to be. The meaning of a natural classification is gradually losing its traditional sense. A 'natural classification' today is one visualized as truly phylogenetic, establishing monophyletic groups making fair use of the phenetic information so that such groups also reflect a phenetic relationship (overall similarity) and the classification represents a reconstruction of the evolutionary descent.

Aims of Systematics

The activities of plant systematics are basic to all other biological sciences and, in turn, depend on the same for any additional information that might prove useful in constructing a classification. These activities are directed towards achieving the undermentioned aims:

1. To provide a convenient method of identification and communication. A workable classification having the taxa arranged in hierarchy, detailed and diagnostic descriptions are essential for identification. Properly identified and arranged herbarium specimens, dichotomous keys, polyclaves and computer-aided identification are important aids for identification. The **Code** (ICN), written and documented through the efforts of IAPT (International Association of Plant Taxonomy), helps in deciding the single correct name acceptable to the whole botanical community.
2. To provide an inventory of the world's flora. Although a single world Flora is difficult to come by, floristic records of continents (**Continental Floras;** *cf. Flora Europaea* by Tutin et al.), regions or countries (**Regional Floras;** *cf. Flora of British India* by J. D. Hooker) and states or even counties (**Local Floras;** *cf. Flora of Delhi* by J. K. Maheshwari) are well documented. In addition, **World Monographs** for selected genera (e.g., *The genus Crepis* by Babcock) and families (e.g., *Das pflanzenreich* ed. by A. Engler) are also available.
3. To detect evolution at work; to reconstruct the evolutionary history of the plant kingdom, determining the sequence of evolutionary change and character modification.
4. To provide a system of classification which depicts the evolution within the group. The phylogenetic relationship between the groups is commonly depicted with the help of a phylogram, wherein the longest branches represent more advanced groups and the shorter, nearer the base, primitive ones. In addition, the groups are represented by balloons of different sizes that are proportional to the number of species in the respective groups. Such a phylogram is popularly known as a **bubble diagram**. The phylogenetic relationship could also be presented in the form of a phylogenetic tree (with vertical axis representing the geological time scale), where existing species reach the top and the bubble diagram may be a cross-section of the top with primitive groups towards the center and the advanced ones towards the periphery.
5. To provide an integration of all available information. To gather information from all the fields of study, analyzing this information using statistical procedures with the help of computers, providing a synthesis of this information and developing a classification based on overall similarity. This synthesis is unending, however, since scientific progress will continue, and new information will continue to pour and pose new challenges for taxonomists.
6. To provide an information reference, supplying the methodology for information storage, retrieval, exchange and utilization. To provide significantly valuable information concerning endangered species, unique elements, genetic and ecological diversity.
7. To provide new concepts, reinterpret the old, and develop new procedures for correct determination of taxonomic affinities, in terms of phylogeny and phenetics.
8. To provide integrated databases including all species of plants (and possibly all organisms) across the globe. Several big organizations have come together to establish online searchable databases of taxon names, images, descriptions, synonyms and molecular information.

Advancement Levels in Systematics

Plant systematics has made considerable strides from herbarium records to databanks, recording information on every possible attribute of a plant. Because of extreme climatic diversity, floristic variability, inaccessibility of certain regions and economic disparity of different regions, the present-day systematics finds itself in different stages of advancement in different parts of the world. Tropical Asia and tropical Africa are amongst the richest areas of the world in terms of floristic diversity but amongst the poorest as far as the economic resources to pursue complete documentation of systematic information. The whole of Europe, with more than 30 m square kilometers of landscape and numerous rich nations with their vast economic resources, must account for slightly more than 12 thousand species of vascular plants. India, on the other hand, with meager resources, less than one tenth of landscape, has to account for the study of twice the number vascular plants. A small country like Colombia, similarly, has estimated 4,500 different species, with only a few botanists to study the flora. Great Britain, on the other hand, has approximately 1370 taxa (Woodland, 1991), with thousands of professional and amateur botanists available to document the information. It is not strange, as such, that there is lot of disparity in the level of advancement concerning knowledge about respective floras. Taxonomic advancement today can be conveniently divided into four distinct **phases** encountered in different parts of the world:

Exploratory or Pioneer Phase

This phase marks the beginning of plant taxonomy, collecting specimens and building herbarium records. The few specimens of a species in the herbarium are the only record of its variation. These specimens are, however, useful in a preliminary inventory of flora through discovery, description, naming and identification of plants. Here, morphology and distribution provide the data on which the systematists must rely. Taxonomic experience and judgement are particularly important in this phase. Most areas of tropical Africa and tropical Asia are passing through this phase.

Consolidation or Systematic Phase

During this phase, herbarium records are ample, and enough information is available concerning variation from field studies. This development is helpful in the preparation of Floras and Monographs. It also aids in better understanding of the degree of variation within a species. Two or more herbarium specimens may appear to be sufficiently different and regarded as belonging to different species on the basis of a few available herbarium records, but only a field study of populations involving thousands of specimens can help in reaching at a better understanding of their status. If there are enough field specimens to fill in the gaps in variation pattern, there is no justification in regarding them as separate species. On the other hand, if there are distinct gaps in the variation pattern, it strengthens their separate identity. In fact, many plants, described as species on the basis of limited material in the pioneer phase, are found to be variants of other species in the consolidation phase. Most parts of central Europe, North America and Japan are experiencing this phase.

Experimental or Biosystematic Phase

During this phase, the herbarium records and variation studies are complete. In addition, information on **biosystematics** (studies on transplant experiments, breeding behavior and chromosomes) is also available. Transplant experiments involving collecting seeds, saplings or other propagules from morphologically distinct populations from different habitats and growing them under common environmental conditions. If the differences between the original populations were purely ecological, the differences would disappear under a common environment, and there is no justification in regarding them as distinct taxonomic entities. On the other hand, if the differences still persist, these are evidently genetically fixed. If these populations are allowed to grow together for several years, their breeding behaviors would further establish their status. If there are complete reproductive barriers between the populations, they will fail to interbreed, and maintain their separate identity. These evidently belong to different species. On the other hand, if there is no reproductive isolation between them, over the years, they would interbreed, form intermediate hybrids, which will soon fill the gaps in their variation. Such populations evidently belong to the same species and better distinguished as ecotypes, subspecies or varieties. Further chromosomal studies can throw more light on their affinities and status. Central Europe has reached this phase of plant systematics.

Encyclopaedic or Holotaxonomic Phase

Here, not only the previous three phases are attained, but information on all the botanical fields is also available. This information is assembled, analyzed, and a meaningful synthesis of analysis is provided for understanding phylogeny. Collection of data, analysis and synthesis are the jobs of an independent discipline of systematics, referred to as **numerical taxonomy**.

The first two phases of systematics are often considered under **alpha-taxonomy** and the last phase under **omega-taxonomy**. At present, only a few persons are involved in encyclopedic work and that too, in a few isolated taxa. It may thus be safe to conclude that though in a few groups omega-taxonomy is within reach, for the great majority of plants, mainly in the tropics, even the 'alpha' stage has not been crossed. The total integration of available information for the plant kingdom is, thus, only a distant dream at present. Concerted efforts are essential to focus on unexplored regions and little-known taxa.

Chapter 2

Descriptive Terminology

For better understanding of any plant species, it is essential to know features based on which it can identified in the field. A botanical analysis necessitates the availability of information about its characteristics. The descriptive information about the morphology of a plant (**phytography**) is suitably expressed in semi-technical language through a set of terms, which provide an unambiguous representation of the plant. The descriptive terminology thus precedes any taxonomic or phylogenetic analysis of a taxon. Whereas the vegetative morphology of vascular plants (**Tracheophyes**) uniformly includes information about the organs such as **root**, **stem** as **leaves**, the reproductive morphology may differ in different groups. The Pteridophytes are represented by strobili, cones, sporophylls, microsporophylls, megasporophylls and spores, Gymnosperms by cones, megasporophylls, microsporophylls and seeds. The flowering plants have distinct **inflorescences**, **flowers**, **seeds** and **fruits**. All these organs show considerable variability, amply depicted through a large vocabulary of descriptive terms.

Morphological terminology has been in use for description of species for several centuries and continues to be the principal source of taxonomic evidence. The descriptive terminology is very exhaustive, and as such only the most commonly used terms are illustrated here.

HABIT AND LIFE SPAN

Annual: A plant living and completing its life cycle in one growing season. **Ephemerals** are annuals surviving for one or two weeks (*Boerhaavia repens*).

Biennial: A plant living for two seasons, growing vegetatively during the first and flowering during the second.

Perennial: A plant living for more than two years and flowering several times during the life span (except in monocarpic plants which live for several years but perish after flowering, as in several species of *Agave* and bamboos). In **herbaceous perennials,** the aerial shoot dies back each winter, and the annual shoots are produced from subaerial stock every year, those with a rhizome, tuber, corm or bulb better known as **geophytes**. A **woody perennial,** on the other hand has woody aerial shoots which live for several years. A woody perennial may be a **tree** (with a distinct trunk or bole from the top of which the branches arise—**deliquescent tree** as in banyan, a totally unbranched **caudex** with a crown of leaves at top as in palms, or the main stem continues to grow gradually narrowing and producing branches in acropetal order—**excurrent tree** as in *Polyalthia*) or a **shrub** (with several distinct branches arising from the ground level). A **suffrutescent plant** is intermediate between woody and herbaceous plants, with the basal woody portion persisting year after year whereas the upper portion dies back every year. A weak climbing plant may be woody (**liana**) or herbaceous (**vine**).

It should be noted that the terms herb, shrub, suffrutescent plant and tree represent different forms of habit. Annual, biennial and perennial denote the life span or duration of the plant.

HABITAT

Plants grow in a variety of habitats. **Terrestrial** plants grow on land, **aquatic** plants in water and those on other plants as **epiphytes**. Terrestrial plant may be a **mesophyte** (growing in normal soil), **xerophyte** (growing on dry habitats: **psammophyte** on sand, **lithophyte** on rock). An aquatic plant may be **free-floating** (occurring on water surface), **submerged** or **emersed** (wholly under water), **emergent** (Anchored at bottom but with shoots exposed above water), **floating-leaved** (anchored at bottom but with floating leaves), or a **helophyte** (emergent marsh plant in very shallow waters). A plant growing in saline habitats (terrestrial or aquatic) is known as halophyte, whereas one in acidic soils as **oxylophyte** or **oxyphyte**. **Saprophyte** grow on decaying organic matter, **parasite** lives and depends on another organism. It may be a

partial parasite (Hemiparasite) when it is green, contains chlorophyll and can perform its photosynthesis deriving only water and nutrients from the host plant (*Viscum album*). A total parasite (holoparasite), on the other hand, is generally nongreen, lacks chlorophyll and depends on the host plant for both food and nutrients (*Cuscuta*).

ROOTS

Roots unlike stems lack nodes and internodes, have irregular branching and produce endogenous lateral roots. Upon seed germination, usually the radicle elongates into a primary root, forming a **taproot**, but several other variations may be encountered:

Adventitious: Developing from any part other than radicle or another root.

Aerial: Grows in air. In epiphytes, the aerial roots termed **epiphytic roots** are found hanging from the orchids and are covered with a spongy **velamen** tissue. Orchids also carry some clinging roots which penetrate crevices and help in anchorage.

Assimilatory: Green chlorophyll-containing roots capable of carbon assimilation as in *Tinospora cordifolia*, and many species of Podostemaceae.

Fibrous: Thread like tough roots common in monocots, especially grasses, usually adventitious in nature.

Buttressed: Enlarged, horizontally spread and vertically thickened roots at the base of certain trees of marshy areas.

Fleshy: Thick and soft with a lot of storage tissue. Storage roots may be the modification of taproot:

(i) **Fusiform:** Swollen in the middle and tapering on sides, as in radish (*Raphanus sativus*).

(ii) **Conical:** Broadest on top and gradually narrowed below, as in carrot (*Daucus carota*).

(iii) **Napiform:** Highly swollen and almost globose and abruptly narrowed below, as in turnip (*Brassica rapa*).

Modifications of the storage adventitious roots include:

(i) **Tuberous:** Clusters of tubers growing out from stem nodes, as in sweet potato (*Ipomoea batatas*) and tapioca (*Manihot esculenta*).

(ii) **Fasciculated:** Swollen roots occurring in clusters, as in *Asparagus* and some species of *Dahlia*.

(iii) **Nodulose:** Only the apices of adventitious roots becoming swollen like beads, as in *Curcuma amada* and *Costus speciosus*.

(iv) **Moniliform:** Portions of a root are alternately swollen and constricted giving beaded appearance, as in *Dioscorea alata*.

Haustorial (sucking): Small roots penetrating the host xylem tissue for absorbing water and nutrients as in partial parasites (*Viscum*) or also the photosynthetic materials by penetrating the phloem tissue as well, as in total parasites (*Cuscuta*).

Mycorrhizal: Roots infested with fungal mycelium which helps in root absorption. The fungal mycelium may penetrate cortical cells (**endotrophic mycorrhizae** found in orchids) or may largely form a mantle over the root with a few hyphae

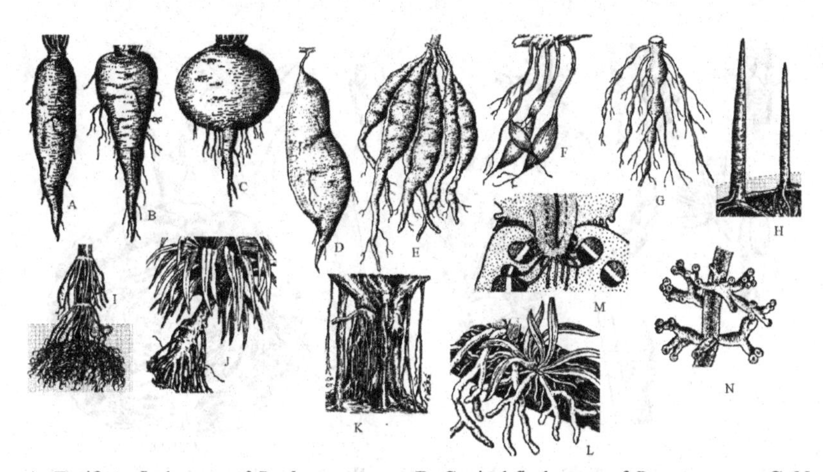

Figure 2.1: Roots. A: Fusiform fleshy root of *Raphanus sativus*; **B:** Conical fleshy root of *Daucus carota*; **C:** Napiform fleshy root of *Brassica rapa*; **D:** Root-tuber of *Ipomoea batatas*; **E:** Fasciculated tuberous roots of *Dahlia*; **F:** Nodulose roots of *Curcuma amada*; **G:** Moniliform roots; **H:** Pneumatophores of *Avicennia*; **I:** Stilt roots of *Zea mays*; **J:** Stilt roots of *Pandanus*; **K:** Prop roots of *Ficus benghalensis*; **L:** Aerial roots of *Dendrobium*; **M:** Haustorial roots of *Viscum*, sending haustoria only into the host xylem; **N:** Mycorrhizal roots of *Pinus*.

penetrating between the outer cells (**ectotrophic mycorrhizae** found in conifers). In specialized **VAM (vesicular arbuscular mycorrhizae)** found in grasses, the fungal hyphae penetrate cortical cells, forming a hyphal mass called **arbusculum**.

Respiratory: Negatively geotropic roots of some mangroves (e.g., *Avicennia*) which grow vertically up and carry specialized lenticels (**pneumathodes**) with pores for gaseous exchange. Such roots are also known as **pneumatophores**.

Prop: Elongated aerial roots arising from horizontal branches of a tree, striking the ground and providing increased anchorage and often replacing the main trunk as in several species of *Ficus* (e.g., the great banyan tree *F. benghalensis* in the Indian Botanical Garden at Sibpur, Kolkata). The large hanging prop roots of *Ficus* species are often used in bungee jumping sport.

Stilt: Adventitious roots arising from the lower nodes of the plant and penetrating the soil in order to give increased anchorage as in maize (*Zea mays*), screw-pines (*Pandanus*) and *Rhizophora*.

STEMS

Stems represent the main axes of plants, being distinguished into **nodes** and **internodes**, and bearing **leaves** and axillary **buds** at the nodes. The buds grow out into lateral shoots, inflorescences or flowers.

A plant may lack stem (acaulescent) or have a distinct stem (caulescent). The latter may be aerial (erect or weak) or even underground.

Acaulescent: Apparently a stemless plant having very inconspicuous reduced stem. The reduced stem may often elongate at the time of flowering into a leafless flowering axis, known as **scape** as found in onion.

Arborescent: Becoming tree-like and woody, usually with a single main trunk.

Ascending: Stem growing upward at about 45–60° angle from the horizontal.

Bark: Outside covering of stem, mainly the trunk. Bark may be **smooth**, **exfoliating** (splitting in large sheets), **fissured** (split or cracked), or **ringed** (with circular fissures).

Bud: Short embryonic stem covered with bud scales and developing leaves and often found in leaf axils. Buds are frequently helpful in identification and may present considerable diversity:

(i) **Accessary bud:** An extra bud on either side (**collateral** bud) or above (**superposed bud** or **serial bud**) the axillary bud.

(ii) **Adventitious bud:** Bud developing from any place other than the node.

(iii) **Axillary (lateral) bud:** Bud located in the axil of a leaf.

(iv) **Bulbil:** Modified and commonly enlarged bud meant for propagation. In *Agave* and top onion (*Allium* x *proliferum*) flower buds get modified into bulbils.

(v) **Dormant (winter) bud:** Inactive well protected bud usually to survive winter in cold climates.

(vi) **Flower bud:** Bud developing into flower.

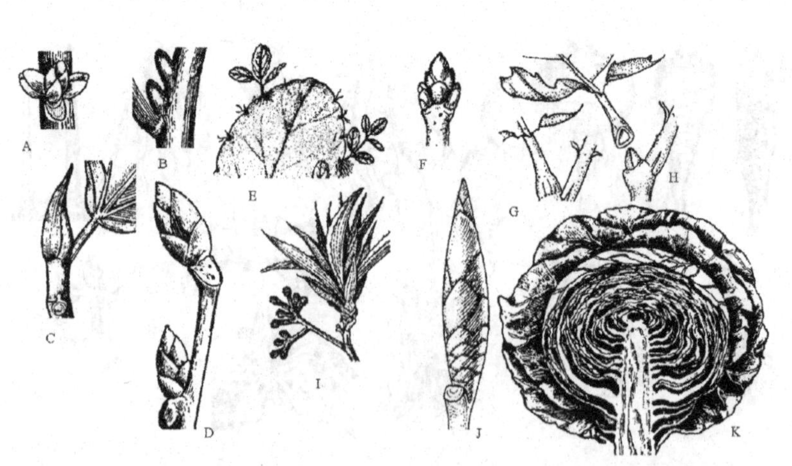

Figure 2.2: Buds. A: Axillary bud with 2 collateral buds in *Acer*; **B:** Axillary bud and a superposed bud in *Juglans regia*; **C:** Scaly bud of *Ficus* covered with bud-scale; **D:** Winter buds in *Salix*; **E:** Vegetative bud with embryonic leaves; **F:** Terminal bud with two collateral buds; **G:** Intrapetiolar bud hidden by petiole base; **H:** Same with petiole removed; **I:** Bulbil developing from one flower of *Agave*; **J:** Pseudoterminal bud, taking terminal position due to death or non-development of terminal bud; **K:** Vegetative bud of *Brassica oleracea* var. *capitata* (cabbage).

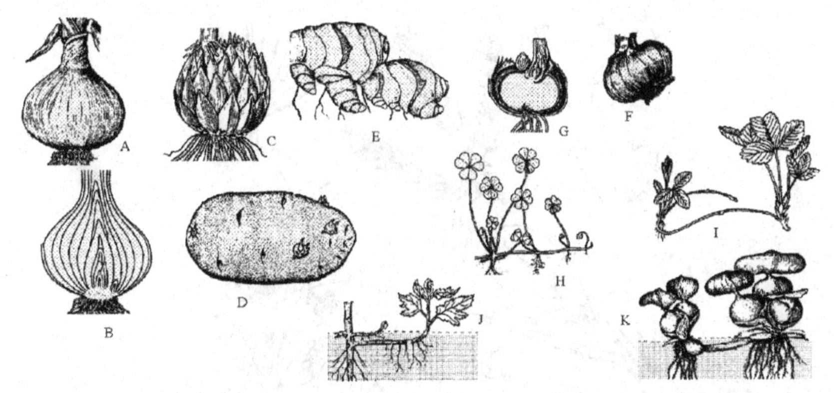

Figure 2.3: Stem, subaerial and underground modifications. A: Tunicated bulb of *Allium cepa*; **B:** Same in vertical section, showing concentric layers of leaf sheaths; **C:** Scaly bulb of *Lilium* with separate fleshy leaf sheaths; **D:** Stem tuber of *Solanum tuberosum* with eye buds; **E:** Rhizome of *Zingiber officinale* with fleshy branched horizontal stem; **F:** Corm of *Crocus sativus* covered with scale leaves; **G:** Same in longitudinal section showing the solid inside as opposed to the bulb; **H:** Runner of *Oxalis,* rooting at nodes; **I:** Stolon of *Fragaria vesca,* arching down to strike roots at nodes; **J:** Sucker in *Chrysanthemum,* underground and rising up to produce shoot; **K:** Offset in *Eichhornia crassipes,* like runner but shorter and thicker.

(vii) **Mixed bud:** A bud bearing both embryonic leaves and flowers.

(viii) **Naked bud:** Not covered by bud scales.

(ix) **Pseudoterminal bud:** Lateral bud near the apex appearing terminal due to death or non-development of terminal bud.

(x) **Scaly (covered) bud:** Covered by bud scales.

(xi) **Terminal bud:** Located at stem tip.

(xii) **Vegetative bud:** Bearing embryonic leaves.

Caulescent: With a distinct stem.

Caudiciform: Low swollen storage stem at ground level, from which annual shoots arise as in *Calibanus* and some species of *Dioscorea.*

Culm: Flowering and fruiting stem of grasses and sedges.

Erect: Growing erect as an herb, shrub or a tree.

Lignotuber: Swollen woody stem at or below ground level, from which persistent woody aerial branches arise, as in *Manzanita.*

Pachycaul: Woody trunk-like stem swollen at base functioning for storage as in bottle tree *Brachychiton.*

Phylloclade (cladophyll): Stem flattened and green like leaves bearing scale leaves as in *Opuntia.* A phylloclade of one internode length found in *Asparagus* in known as **Cladode**.

Pseudobulb: Short erect aerial storage or propagating stem of certain epiphytic orchids.

Subaerial: Generally perennial partially hidden stems:

(i) **Runner:** Elongated internodes trailing along the ground and generally producing a daughter plant at its end as in *Cynodon* and *Oxalis.*

(ii) **Sobol:** Like runner but partially underground as in *Saccharum spontaneum*, and unlike rhizome, not a storage organ.

(iii) **Stolon:** Like runner but initially growing up and then arching down and striking roots in soil as in strawberry.

(iv) **Sucker:** Like runner but underground and growing up and striking roots to form new plant as in *Chrysanthemum* and *Mentha arvensis.*

(v) **Offset:** Shorter than runner and found in aquatic plants like *Eichhornia crassipes.*

Subterranean (underground): Growing below the soil surface and often specially modified:

(i) **Bulb:** A reduced stem surrounded by thick fleshy scale leaves. The leaves may be arranged in a concentric manner surrounded by a thin membranous scale leaf (**tunicated bulb** of onion—*Allium cepa*) or leaves only overlapping along margins (**scaly** or **imbricate** bulb of garlic—*Allium sativum*).

(ii) **Corm:** A vertical fleshy underground stem covered with some scale leaves and with a terminal bud, as in *Gladiolus.*

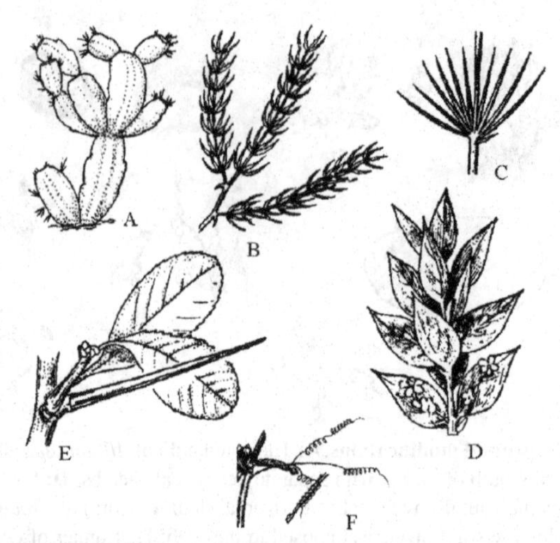

Figure 2.4: Stem, aerial modifications. **A:** Phylloclade of *Opuntia*; **B:** Cladodes in *Asparagus*; **C:** Portion enlarged to show whorl of cladodes in axil of scale-leaf; **D:** Phylloclades of *Ruscus*, leaf-like and bearing flowers; **E:** Thorn of *Prunus*; **F:** Tendril of *Luffa*.

(iii) **Rhizome:** A horizontal dorsiventral fleshy underground stem with nodes and internodes and covered with scale leaves, as in Ginger.

(iv) **Stem tuber:** Underground portions of stem modifies into tubers as in potato.

Thorn: Branch or axillary bud modified into a hard, sharp structure, being deep-seated and having vascular connections as opposed to prickles which are mere superficial outgrowths without vascular connections. Spine is like a thorn but generally weaker and developing from the leaf or stipule. Thorns may bear leaves (*Duranta*), flowers (*Prunus*), or may be branched (*Carissa*).

Weak: Plant not strong enough to grow erect:

(i) **Creeper:** Growing closer to ground and often rooting at the nodes, as in *Oxalis*.

(ii) **Trailer:** Trailing along the surface and often quite long. They are usually **prostrate** or **procumbent,** lying flat on ground as in *Basella*, but sometimes **decumbent** when the tips start growing erect or ascending, as in *Portulaca*.

(iii) **Climber:** Weak plant which uses a support to grow up and display leaves towards sunlight. This may be achieved in several ways:

(a) **Twiner (stem climber):** Stem coiling round the support due to special type of growth habit, as in *Ipomoea* and *Convolvulus*.

(b) **Root climber:** Climbing with the help of adventitious roots which cling to the support, as in species of *Piper*.

(c) **Tendril climber:** Climbing with the help of tendrils which may be modified stem (*Passiflora*, *Vitis*), modified inflorescence axis (*Antigonon*), modified leaf (*Lathyrus aphaca*), modified leaflets (*Pisum sativum*), modified petiole (*Clematis*), modified leaf tip (*Gloriosa*), modified stipules (*Smilax*) or even modified root (*Parthenocissus*).

(d) **Scrambler:** Spreading by leaning or resting on support, as in Rose.

(e) **Thorn climber:** Climbing or reclining on the support with the help of thorns, as in *Bougainvillea*.

(f) **Hook climber:** Climbing with the help of hooked structures (*Galium*).

LEAVES

Leaves are green photosynthetic organs of a plant arising from the nodes. Leaves are usually flattened, either **bifacial** (**dorsiventral**) with **adaxial** side (upper surface facing stem axis) different from **abaxial** side (lower surface facing away from stem axis) or may be **unifacial** (**isobilateral**) with similar adaxial and abaxial surfaces. A leaf is generally differentiated into a **leaf blade** (**lamina**) and a **petiole**. A leaf with a distinct petiole is termed **petiolate**, whereas one lacking a petiole is **sessile**. A petiole A leaf with a distinct petiole is termed **petiolate**, whereas one lacking a petiole is **sessile**. A petiole may be winged (*Citrus*), swollen (*Eichhornia*), modified into tendril (*Clematis*), spine (*Quisqualis*) or become modified into a flattened photosynthetic **phyllode** (Australian Acacia). Two small stipules may be borne at the base of the petiole. The leaf terminology affords a wide diversity. The leaf base may sometimes be sheathing or **pulvinate** (swollen).

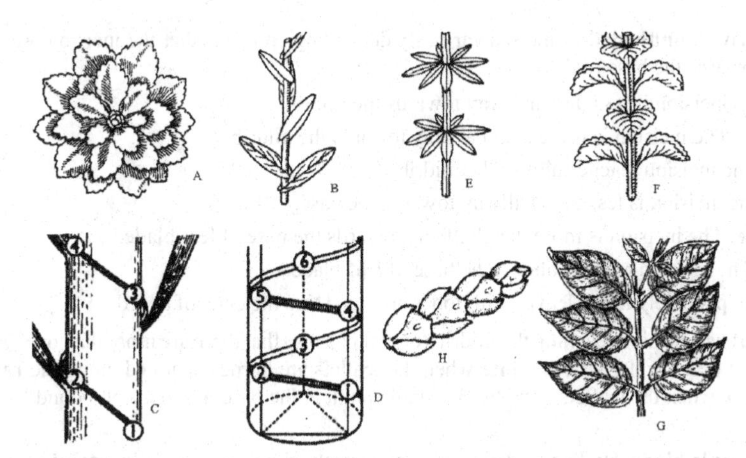

Figure 2.5: Phyllotaxy of leaves. A: Rosulate; **B:** Alternate; **C:** Diagramatic representation of distchous (2-ranked) arrangement; **D:** Diagramatic representation of tristichous (3-ranked) arrangement; **E:** Whorled leaves in *Galium*; **F:** Opposite and decussate leaves of *Lamium*; **G:** Opposite and superposed leaves of *Quisqualis*; **H:** Imbricated leaves.

Leaf Arrangement (Phyllotaxy)

Alternate: Bearing one leaf at each node. The successive leaves usually form a **spiral** pattern, in mathematical regularity so that all leaves are found to lie in a fixed number of vertical rows or **orthostichies**. The arrangement commonly agrees with the **Fibonacci series (Schimper-Brown series)**, wherein numerator and denominator in each case are obtained by adding up the preceding two (1/2, 1/3, 1+1/2+3 = 2/5, 1+2/3+5 = 3/8, and so on). In grasses the leaves are in two rows (2-ranked, **distichous** or **½ phyllotaxy**), so that the third leaf is above the first leaf. Sedges have three rows of leaves (3-ranked, **tristichous,** or **1/3 phyllotaxy**), the fourth leaf above the first leaf. China rose, and banyan show **pentastichous** arrangement, where the sixth leaf lies above the first one, but in doing so leaves complete two spirals and the phyllotaxy is known as **2/5 phyllotaxy**. *Carica papaya* depicts **octastichous** arrangement, wherein the ninth leaf lies above the first one and three spirals are completed in doing so, thus a **3/8 phyllotaxy**. Leaf bases of date palm and sporophylls of pinecone are closely packed and internodes are extremely short making it difficult to count the number of rows (**orthostichies**). Such an arrangement is known as **parastichous**.

Imbricated: The leaves closely overlapping one another, as in *Cassiope*.

Opposite: Bearing pairs of leaves at each node. The pairs of successive leaves may be parallel (**superposed**) as in *Quisqualis* or at right angles (**decussate**) as in *Calotropis* and *Stellaria.*

Whorled (verticillate): More than three leaves at each node as in *Galium*, *Rubia* and *Nerium*.

Radical: Leaves borne at the stem base often forming a rosette (**rosulate**) in reduced stems, as in *Primula* and *Bellis.*

Cauline: Leaves borne on the stem.

Ramal: Leaves borne on the branches.

Leaf Duration

Leaves may stay and function for few days to many years, largely determined by the adaptation to climatic conditions:

Caducuous (fugacious): Falling off soon after formation, as in *Opuntia.*

Deciduous: Falling at the end of growing season so that the plant (tree or shrub) is leafless in winter/dormant season. In tropical climate, the tree may be leafless for only a few days. *Salix* and *Populus* are common examples.

Evergreen (persistent): Leaves persisting throughout the year, falling regularly so that tree is never leafless, as in mango, pines and palms. It must be noted that whereas the term persistent is used for the leaves, the term evergreen is commonly associated with trees with such leaves.

Marcescent: Leaves not falling but withering on the plant, as in several members of Fagaceae.

Leaf Incision/Type of Leaves

A leaf with a single blade (divided or not) is termed **simple**, whereas one with two or more distinct blades (**leaflets**) is said to be compound.

A **Simple leaf** may be **undivided** or incised variously depending upon whether the incision progresses down to the midrib (pinnate) or towards the base (palmate):

(i) **Pinnatifid:** The incision is less than halfway towards the midrib.
(ii) **Pinnatipartite:** The incision is more than halfway towards the midrib.
(iii) **Pinnatisect:** The incision reaches almost the midrib.
(iv) **Palmatifid:** The incision is less than halfway towards the base.
(v) **Palmatipartite:** The incision is more than halfway towards the base of leaf blade.
(vi) **Palmatisect:** The incision reaches almost the base of leaf blade.
(vii) **Pedate:** Deeply palmately lobed leaves with lobes arranged like the claw of a bird.

A **compound leaf** has incision reaching the midrib (or leaf base) so that there are more than one distinct blades called as leaflets or pinnae. It may similarly be **pinnate** when the leaflets are borne separated along the **rachis** (cf. midrib of simple leaf) or **palmate** when the leaflets arise from a single point at the base. Pinnate compound leaves may be further differentiated:

(i) **Unipinnate (simple pinnate):** The leaflets are borne directly along the rachis. In **paripinnate** leaf (*Cassia*), the leaflets occur in pairs and as such the terminal leaflet is missing and there are even numbers of leaflets. In an **imparipinnate** (*Rosa*) leaf, on the other hand, there is a terminal leaflet, resulting in odd number of leaflets.
(ii) **Bipinnate (twice pinnate):** The **pinnae** (primary leaflets) are again divided into **pinnules**, so that the leaflets (**pinnules**) are borne on the primary branches of the rachis as in *Mimosa pudica.*
(iii) **Tripinnate (thrice pinnate):** The dissection goes to the third order so that the leaflets are borne on secondary branches of the rachis as in *Moringa*.
(iv) **Decompound:** Here the dissections go beyond the third order, as in Fennel. The term is sometimes used for leaves more than once compound.
(v) **Ternate:** The leaflets are present in groups of three. Leaf may be ternate (pinnate with three leaflets, i.e., trifoliate), biternate (twice pinnate with three pinnae and three pinnules) triternate or decompound ternate.

Palmate compound leaf does not have a rachis and the leaflets arise from the top of the petiole:

(i) **Unifoliate:** A modified situation in commonly a trifoliate leaf when the lower two leaflets are reduced, and the terminal leaflet looks like a simple leaf but has a distinct joint at base, as seen in *Citrus* plants.
(ii) **Bifoliate (binnate):** A leaf with two leaflets, as found in *Hardwickia*.

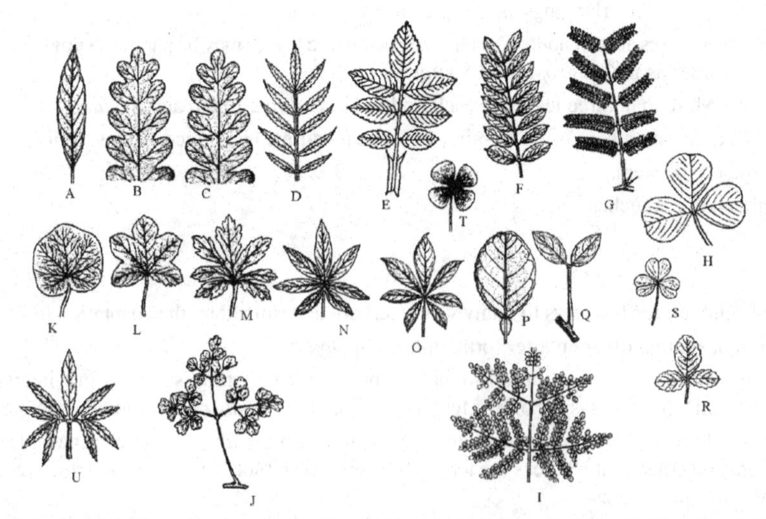

Figure 2.6: Leaf incision. A: Undivided with pinnate venation; **B:** Pinnatifid; **C:** Pinnatipartite; **D:** Pinnatisect; **E:** Pinnate compound-imparipinnate leaf of *Rosa*; **F:** Pinnate compound-paripinnate leaf of *Cassia*; **G:** Bipinnate leaf of *Acacia nilotica*; **H:** Pinnate-trifoliate leaf of *Medicago*, note middle leaflet with longer petiolule; **I:** Tripinnate leaf of *Moringa*; **J:** Triternate leaf of *Thalictrum*; **K:** Undivided with palmate venation; **L:** Palmatifid; **M:** Palmatipartite; **N:** Palmatisect; **O:** Palmate compound-digitate; **P:** Unifoliate leaf of *Citrus*; **Q:** Bifoliate; **R:** Trifoliate leaf of *Trifolium*, note all leaflets with equal petiolules as opposed to pinnate trifoliate leaf; **S:** Trifoliate leaf of *Oxalis*; **T:** Quadrifoliate leaf of *Marsilea*; **U:** pedate leaf of *Vitis pedata.*

(iii) **Trifoliate (ternate):** A leaf with three leaflets, as in *Trifolium*. The trifoliate leaf of *Medicago* and *Melilotus* has terminal leaflet with a longer **petiolule** (stalk of leaflet) than basal leaflets and is accordingly a pinnate trifoliate leaf.

(iv) **Quadrifoliate:** A leaf with four leaflets, as in *Paris* and aquatic pteridophyle *Marsilea.*

(v) **Multifoliate (Digitate):** A leaf with more than four leaflets, as in *Bombax*.

Stipules

The leaves of several species bear two small stipules as outgrowths from the leaf base. Leaves with stipules are termed **stipulate** and those without stipules as **exstipulate**. They show a lot of structural diversity:

Free-lateral: Free and lying on either side of the petiole base, as in china-rose (*Hibiscus rosa-sinensis*).

Adnate: Attached to the base of petiole for some distant, as in Rose.

Intrapetiolar: The two stipules are coherent to form one, which lies in the axil of a leaf as in *Gardenia.*

Interpetiolar: A stipule lying between the petioles of two adjacent leaves, commonly due to fusion and enlargement of two adjacent stipules of different leaves as found in several members of Rubiaceae like *Ixora.*

Ochreate: The two stipules united and forming a tubular structure **ochrea**, found in family Polygonaceae.

Foliaceous: Modified and enlarged to function like leaves as in *Lathyrus aphaca*, where the whole leaf blade is modified into tendril and stipules are foliaceous.

Tendrillar: Stipules modified into tendrils as in *Smilax*.

Spiny: Stipules modified into spines as in *Acacia.*

Leaf Shape (Outline of Lamina)

The shape of leaf/leaflet blade shows considerable variability and is of major taxonomic value.

Acicular: Needle shaped, as in pine.

Cordate: Heart shaped, with a deep notch at base, as in *Piper betle*.

Cuneate: Wedge-shaped, tapering towards the base, as in *Pistia.*

Deltoid: Triangular in shape.

Elliptical: Shaped like an ellipse, a flattened circle usually more than twice as long as broad, as in *Catharanthus roseus.*

Hastate: Shaped like an arrow head with two basal lobes directed outwards, as in *Typhonium*; also referring to hastate leaf base.

Lanceolate: Shaped like a lance, much longer than broad and tapering from a broad base towards the apex, as in bottle-brush plant (*Callistemon citrinus syn: C. lanceolatus*).

Linear: Long and narrow with nearly parallel sides as in grasses and onion.

Lunate: Shaped like half-moon, as in *Passiflora lunata*.

Lyrate: Lyre-shaped; pinnatifid with large terminal lobe and smaller lower lobes, as in *Brassica campestris.*

Oblanceolate: Like lanceolate but with broadest part near apex.

Obcordate: Like cordate but with broadest part and notch at apex, as in *Bauhinia.*

Oblong: Uniformly broad along the whole length as in banana.

Obovate: Ovate, but with broadest part near the apex, as in *Terminalia catappa.*

Ovate: Egg-shaped, with broadest part near the base, as in *Sida ovata.*

Orbicular (rotund): Circular in outline. The peltate leaf of *Nelumbo* is orbicular in outline.

Pandurate: Fiddle shaped; obovate with sinus or indentation on each side near the base and with two small basal lobes, as in *Jatropha panduraefolia.*

Peltate: Shield shaped with petiole attached to the lower surface of leaf (and not the margin), as in *Nelumbo.*

Reniform: Kidney-shaped, as *Centella asiatica.*

Runcinate: Oblanceolate with lacerate or parted margin, as in *Taraxacum.*

Sagittate: Shaped like an arrowhead with two basal lobes pointed downwards, as in *Sagittaria* and *Arum*; also referring to sagittate leaf base.

Spathulate (spatulate): Shaped like a spatula, broadest and rounded near the apex, gradually narrowed towards the base, as in *Euphorbia neriifolia.*

Subulate: Awl-shaped, tapering from a broad base to a sharp point.

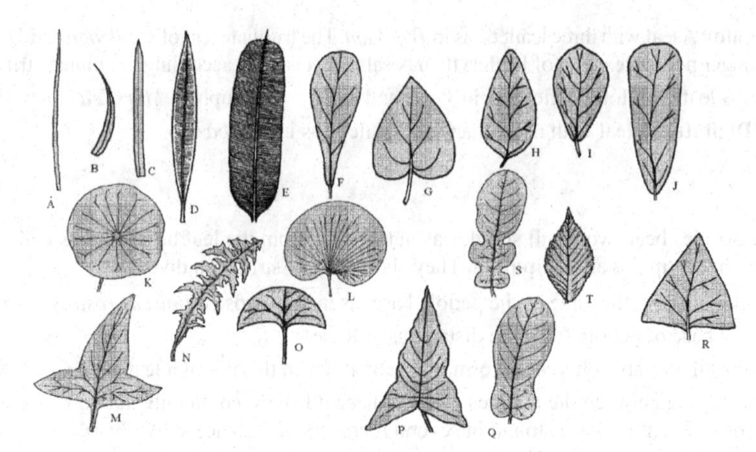

Figure 2.7: Leaf outline. A: Acicular; **B:** Subulate; **C:** Linear, common in grasses; **D:** Lanceolate; **E:** Oblong; **F:** Spathulate; **G:** Cordate; **H:** Ovate; **I:** Obovate; **J:** Oblanceolate; **K:** Peltate; **L:** Reniform; **M:** Hastate; **N:** Runcinate; **O:** Lunate; **P:** Sagittate; **Q:** Pandurate; **R:** Deltoid; **S:** Lyrate; **T:** Elliptic.

Leaf Margin

The edge of a leaf blade is known as margin and may show any of the following conditions:

Crenate: With low rounded or blunt teeth, as in *Kalanchoe.*

Crisped: Margin strongly winding in vertical plane giving ruffled appearance to leaf.

Dentate: With sharp teeth pointing outwards.

Denticulate: Minutely or finely dentate.

Double crenate (bi-crenate): Rounded or blunt teeth are again crenate.

Double dentate: Sharp outward teeth are again dentate. The term **bi-dentate,** though sometimes used here, is inappropriate, as it more correctly refers to a structure bearing two teeth.

Double serrate (bi-serrate): The serrations are again serrate similarly as in *Ulmus.*

Entire: Smooth, without any indentation, as in Mango.

Retroserrate: Teeth pointed downwards.

Revolute: Margin rolled down.

Serrate: With sharp teeth pointing upward like saw, as seen in rose.

Serrulate: Minutely or finely serrate.

Sinuate: Margin winding strongly inward as well as outward.

Undulate (repand, wavy): Margin winding gradually up and down and wavy, as in *Polyalthia.*

Leaf Base

In addition to the terms cordate, cuneate, hastate, sagittate already described above when referring to the leaf base, the following additional terms are frequently used:

Amplexicaul: The auriculate leaf base completely clasps the stem.

Attenuate: Showing a long gradual taper towards the base.

Auriculate: With ear like appendages at the base, as in *Calotropis.*

Cuneate: Wedge shaped, with narrow end at the point of attachment.

Decurrent: Extending down the stem and adnate to the petiole.

Oblique: Asymmetrical with one side of the blade lower on petiole than other.

Perfoliate: The basal lobes of leaf fusing so that the stem appears to pass through the leaf, as in *Swertia.* When the bases of two opposite leaves fuse and the stem passes through them, it is termed **connate perfoliate** as seen in *Canscora.*

Rounded: With a broad arch at the base.

Truncate: Appearing as if cut straight across.

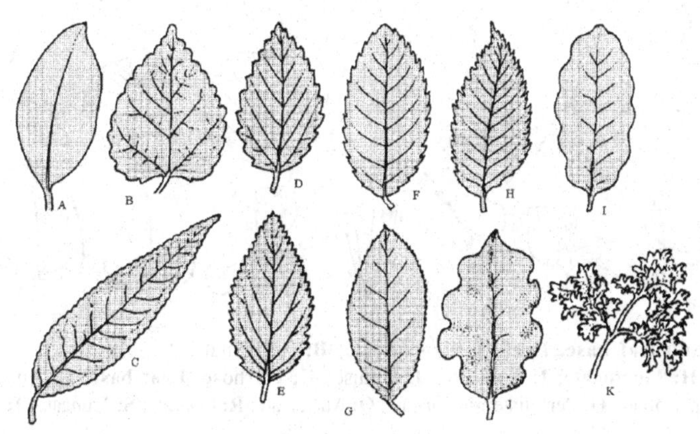

Figure 2.8: Leaf margin. A: Entire; **B:** Crenate; **C:** Crenulate; **D:** Dentate; **E:** Denticulate; **F:** Serrate; **G:** Serrulate; **H:** Bi-serrate; **I:** Undulate; **J:** Sinuate; **K:** Crispate.

Leaf Apex

Leaf apex may similarly present a number of diverse terms:

Acute: Pointed tip with sides forming acute angle, as in mango.

Acuminate: Tapering gradually into a protracted point, as in *Ficus religiosa.*

Aristate: With a long bristle at the tip.

Attenuate: Tip drawn out into a long tapering point.

Caudate: Apex elongated and tail-like.

Cirrhose: With slender coiled apex, as in banana.

Cuspidate: Abruptly narrowed into sharp spiny tip, as in pineapple.

Emarginate: With a shallow broad notch at tip, as in *Bauhinia.*

Mucronate: Broad apex with a small point, as in *Catharanthus.*

Obtuse: Broad apex with two sides forming an obtuse angle, as in banyan.

Retuse: With a slight notch generally from an obtuse apex, as in *Crotalaria retusa.*

Rounded: With a broad arch at the base.

Truncate: Appearing as if cut straight across.

Leaf Surface

The surface of leaves, stems and other organs may present a variety of surface indumentation, whose characteristics are highly diagnostic in several taxa. The surface may be covered by trichomes (hairs, glands, scales, etc.) arranged variously:

Arachnoid: Covered with entangled hairs giving a cobwebby appearance.

Canescent: Covered with grey hairs.

Ciliate: With marginal fringe of hairs.

Floccose: Covered with irregular tufts of loosely tangled hairs.

Glabrate: Nearly glabrous or becoming glabrous with age.

Glabrous: Not covered with any hairs. Sometimes but not always synonymous with **smooth** surface.

Glacous: Surface covered with a waxy coating, which easily rubs off.

Glandular: Covered with glands or small secretory structures.

Glandular-punctate (gland-dotted): Surface dotted with immersed glands, as in *Citrus.*

Hirsute: Covered with long stiff hairs.

Hispid: Covered with stiff and rough hairs.

Lanate: Wooly, with long intertwined hairs.

Pilose: Covered with long distinct and scattered hairs.

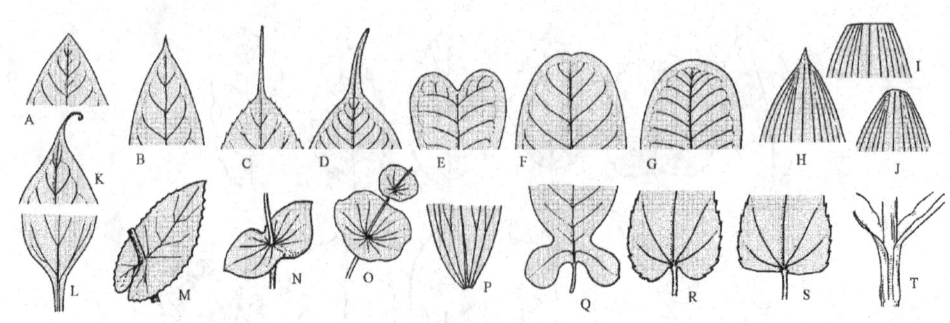

Figure 2.9: Leaf apex and leaf base. Leaf apex. A: Acute; **B:** Acuminate; **C:** Aristate; **D:** Caudate; **E:** Emarginate; **F:** Retuse; **G:** Rounded; **H:** Mucronate; **I:** Truncate; **J:** Obtuse; **K:** Cirrhose. **Leaf base. L:** Attenuate; **M:** Amplexicaul; **N:** Connate-perfoliate; **O:** Perfoliate; **P:** Cuneate; **Q:** Auriculate; **R:** Cordate; **S:** Truncate; **T:** Decurrent.

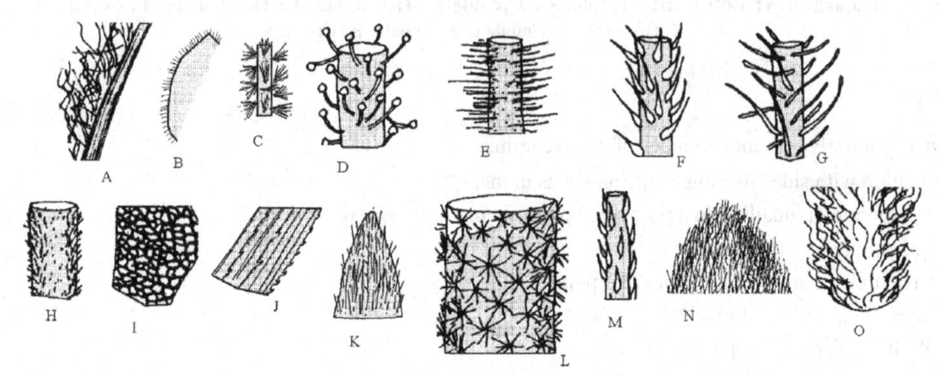

Figure 2.10: Surface coverings. A: Arachnoid; **B:** Ciliate; **C:** Floccose; **D:** Glandular; **E:** Hirsute; **F:** Hispid; **G:** Pilose; **H:** Puberulent; **I:** Rugose; **J:** Scabrous; **K:** Sericeous; **L:** Stellate; **M:** Strigose; **N:** Tomentose; **O:** Villous.

Puberulent: Minutely pubescent.
Pubescent: Covered with soft short hairs.
Rugose: With wrinkled surface.
Scabrous: Surface rough due to short rough points.
Scurfy: Covered with scales.
Sericeous: Covered with soft silky hairs, all directed towards one side.
Stellate: Covered with branched star-shaped hairs.
Strigose: Covered with stiff appressed hairs pointing in one direction.
Tomentose: Covered with densely matted soft hairs, wooly in appearance.
Velutinous: Covered with short velvety hairs.
Villous: Covered with long, fine soft hairs, shaggy in appearance.

The hairs covering the surface may be **unicellular** or **multicellular**, **glandular** or **nonglandular**. The hairs may be unbranched or branched variously. They may bear one row of cells (**uniseriate**), two rows (**biseriate**) or several rows (**multiseriate**). Some species of plants, especially some acacias bear specialized glands **domatia** at the leaf base, which house ants which protect plants from herbivores.

Venation

The distribution of vascular bundles that are visible on the leaf surface as veins constitutes venation. Dicots exhibit a network of veins (**reticulate venation**); whereas monocots usually have non-intersecting parallel veins (**parallel venation**). Each type of venation may encounter a single **midrib** from which the secondary veins arise (**Unicostate** or **pinnate**), or more than one equally strong veins entering the leaf blade (**multicostate** or **palmate**). In ferns and *Ginkgo*, the venation is **dichotomous** with forked veins.

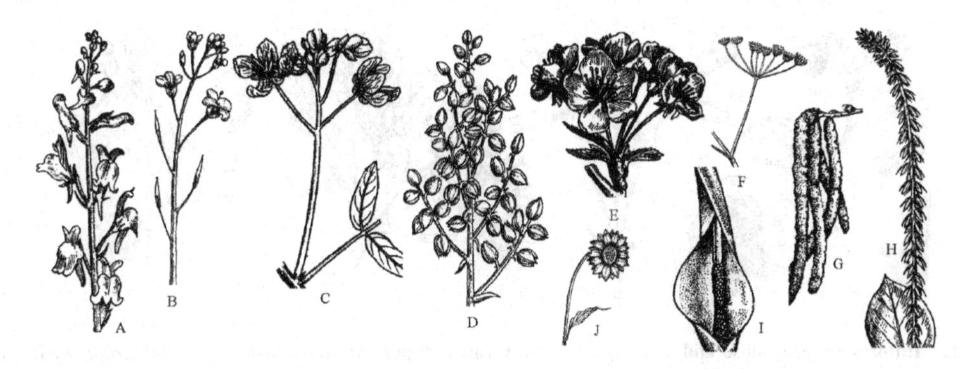

Figure 2.11: Inflorescence: racemose types. A: Raceme of *Linaria*; **B:** Corymbose raceme of *Brassica*; **C:** Corymb of *Cassia*; **D:** Panicle of *Yucca*; **E:** Umbel of *Prunus*; **F:** Compound umbel of *Foeniculum*; **G:** Catkins of *Betula*; **H:** Spike of *Achyranthes*; **I:** Spadix of *Colocasia*; **J:** Capitulum of *Helianthus*.

INFLORESCENCE

Inflorescence is a modified shoot system bearing flowers (modified shoots). The term inflorescence appropriately refers to the arrangement of flowers on the plant. The flowers may either occur singly (in leaf axils—**solitary axillary** or terminal on the stem—**solitary terminal**) or may be organized into distinct inflorescences. Two principal types of inflorescences are differentiated. In **racemose (indeterminate** or **polytelic)**, inflorescence the axis is of unlimited growth, apical bud continuing to grow, thus bearing oldest flower towards the base and youngest towards the top. In **cymose (determinate** or **monotelic)** inflorescence, on the other hand, the main axis has limited growth, being terminated by the formation of a flower, and as each level of branching bears one flower, there are generally a limited number of flowers, and the oldest flower is either in the centre, or flowers of different ages are mixed up. An inflorescence is sometimes carried on a leafless axis. Such a leafless axis arising from aerial stems is termed a **peduncle** (inflorescence pedunculate) and the one arising from basal rosette of leaves as **scape** (inflorescence scapigerous).

Racemose Types

The following variations of the **racemose** type are commonly encountered:

Raceme: A typical racemose inflorescence with single (unbranched) axis bearing flowers on distinct pedicels, as in *Delphinium*.

Panicle: Branched raceme, the flowers being borne on the branches of the main axis, as in *Yucca*.

Spike: Similar to raceme but with sessile flowers, as in *Adhatoda*.

Spadix: Variation of a spike where the axis is fleshy and the flowers are covered by a large bract known as **spathe**, as found in *Alocasia* and *Arum*.

Corymb: Flat-topped racemose inflorescence with longer lower pedicels and shorter upper pedicels so that all flowers reach the same level, as in *Iberis amara*.

Corymbose-raceme: Intermediate between a typical raceme and a typical corymb, all flowers not managing to reach the same height, as in *Brassica campestris*.

Catkin (ament): A spike-like inflorescence of reduced unisexual flowers, as in *Morus*.

Umbel: Flowers arising from one point due to condensation of axis, with oldest flowers towards the periphery and youngest towards the center as in the family Apiaceae (Umbelliferae). **Compound umbel** has branches bearing the umbels also borne in umbellate manner.

Head: Flat-topped axis bearing crowded sessile flowers as in *Acacia* and *Mimosa*.

Capitulum: Flat-topped inflorescence like head (and often known as head) but with distinct ray florets and disc florets (one or both types), surrounded by involucre bracts (phyllaries), as found in the family Asteraceae (Compositae).

Cymose Types

A cymose inflorescence may be primarily differentiated on account of bearing one or more determinate branches arising below the terminal flower at each level:

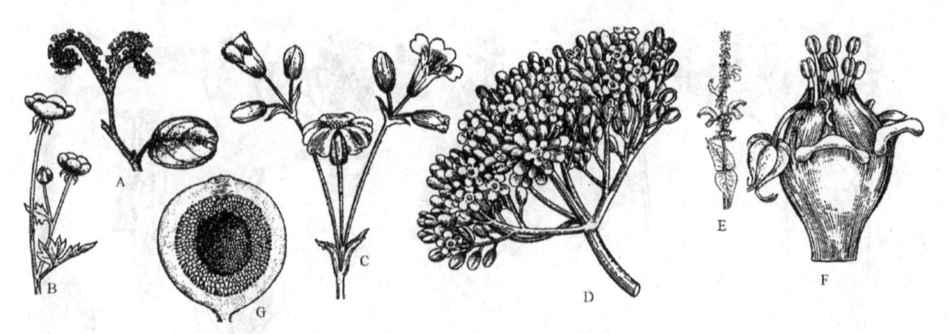

Figure 2.12: Inflorescence: cymose and specialized types. **Cymose types. A:** Helicoid cyme of *Heliotropium*; **B:** Scorpioid cyme of *Ranunculus bulbosus*; **C:** Biparous cyme of *Dianthus*; **D:** Multiparous cyme of *Viburnum*. **Specialized types. E:** Verticillaster of *Salvia*; **F:** Cyathium of *Euphorbia*; **G:** Hypanthodium of *Ficus cunia*.

Monochasial (Uniparous) cyme: One branch arising at each node so that when the **sympodial** (false) axis differentiates, a limited number of bract-opposed flowers (instead of many and axillary in raceme) are formed. Two types of monochasia are found:

(i) **Helicoid cyme:** Successive branches (each forming one flower) are borne on same side so that the inflorescence is often coiled, as in the family Boraginaceae (e.g., *Myosotis*).

(ii) **Scorpioid cyme:** Successive branches (each forming one flower) are borne on alternate sides. In **rhipidium** found in *Solanum nigrum,* all the flowers lie in same plane as the main axis.

Dichasial (Biparous) cyme: Two branches arising below the apical flower at each level so that the flower is between the fork of two branches, as in *Stellaria* and *Dianthus*.

Polychasial (multiparous) cyme: More than two branches arising at each node below the terminal flower so that a broad inflorescence of several flowers is formed, as in *Viburnum*.

Cymose cluster: Cymose group of flowers arising from a point due to reduction of axis.

Cymose umbel: Looking like an umbel but formed by grouping together of numerous cymes so that the flowers of different ages are mixed up, as found in *Allium*.

Specialized Types

In addition to the typical determinate and indeterminate types, some mixed and specialized types are also encountered:

Cyathium: Complex type of inflorescence met in genus *Euphorbia*, having a cup-shaped involucre (formed by fused bracts) usually carrying five nectaries along the rim and enclosing numerous male flowers (in scorpioid cymes, without perianth and bearing a single stamen) in axils of bracts and single female flower in the centre.

Verticillaster: Characteristic inflorescence of family Lamiaceae. Each node of the inflorescence bears two opposite clusters of dichasial cymes, subsequently becoming monochasial as the number of flowers in each cluster exceeds three. Due to the condensation of the axis, flowers of different ages appear to form a false whorl or **verticel**.

Hypanthodium: Typical inflorescence of figs having vessel like receptacle with a small opening at the top and bearing flowers along the inner wall.

Thyrse: A mixed inflorescence with racemose main axis but with cymose lateral clusters as seen in grape vine.

FLOWER

A flower is a highly modified shoot bearing specialized floral leaves. The axis of the flower is condensed to form **thalamus** (**torus** or **receptacle**) commonly bearing four whorls of floral parts: **calyx** (individual parts sepals), **corolla** (individual parts petals), **Androecium** (individual parts stamens) and **Gynoecium** (individual parts carpels). In some plants, the calyx and corolla may not be differentiated and represented by a single or two similar whorls of **perianth** (individual members **tepals**: a term formerly restricted to petal like perianth of monocots). The flower is usually carried on a **pedicel** and may or may not be subtended by a reduced leaf known as **bract**. The pedicel may sometimes carry small **bracteoles** (if present usually two in dicots, one in monocots). As a general rule, members of different whorls alternate each other. The terms associated with the general description of flower in usual sequence includes:

Bract

Bracteate: Flower in the axil of a bract.

Ebracteate: Bract absent.

Bracteolate: Bracteoles present on pedicel.

Pedicel

Pedicellate: Pedicel distinct, often longer than flower.

Subsessile: Pedicel much shorter, often shorter than flower.

Sessile: Pedicel absent.

Complete: All the four floral whorls present.

Incomplete: One or more floral whorl lacking.

Symmetry: Symmetry of a flower is largely based on relative shapes and sizes of sepals (or calyx lobes) in calyx whorl and/or relative shapes and sizes of petals (or corolla lobes) in the corolla whorl.

Actinomorphic: Symmetrical flower which can be divided into equal halves when cut along any vertical plane. In practice an actinomorphic flower has all parts of the calyx and all parts of the corolla (or all parts of perianth) more or less of the same shape and size.

Zygomorphic: Asymmetrical flower, which may be divided into equal halves by one or more but not all vertical planes. In practice such flower has parts of calyx and/or corolla (or perianth) of different shapes and sizes.

Sexuality

Bisexual (perfect): Bearing both stamens and carpels.

Unisexual (imperfect): Bearing either stamens or carpels.

Staminate (male)**:** Bearing stamens only.

Pistillate (female)**:** Bearing carpels only.

Dioecious: With male and female flowers on the same plant.

Monoecious: With male and female flowers on different plants.

Polygamous: With male, female and bisexual flowers on the same plant.

Insertion: Insertion of floral parts on the thalamus not only determines the shape of the thalamus, it also reflects on the relative position of floral whorls, as also whether the ovary is superior (and, consequently, other whorls inferior) or inferior (and, consequently, other whorls superior):

Hypogynous: The thalamus is convex so that the other floral parts are inserted below the ovary. The ovary in this case is **superior** and other floral whorls inferior. There is no hypanthium.

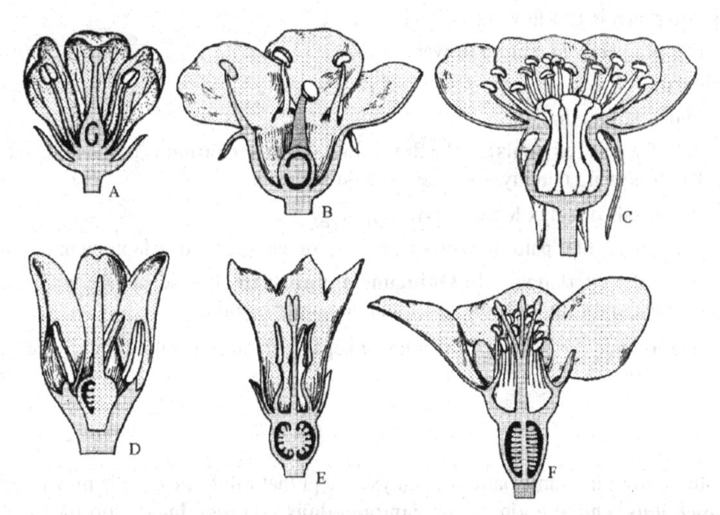

Figure 2.13: Insertion of floral parts. A: Hypogynous with superior ovary; **B:** Perigynous with cup-shaped hypanthium and superior ovary; **C:** Perigynous with flask-shaped hypanthium, ovary superior; **D:** Perigynous with partially immersed semi-inferior ovary; **E:** Epigynous with inferior ovary, without free hypanthium above the ovary; **F:** Epigynous with inferior ovary and with free hypanthium above the ovary.

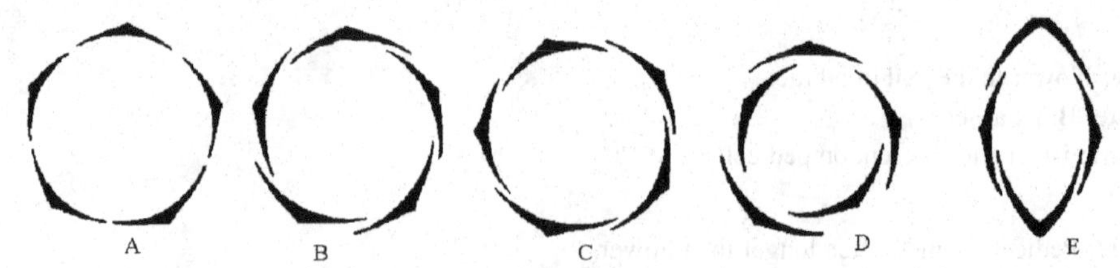

Figure 2.14: Aestivation of calyx and corolla parts. A: Valvate; **B:** Twisted; **C:** Imbricate; **D:** Quincuncial imbricate; **E:** Vexillary.

Perigynous: The thalamus is depressed to the extent that the level of ovary is lower than the other whorls and the thalamus forms either a saucer-shaped, cup-shaped or flask-shaped **hypanthium**. It must be noted that although hypanthium surrounds the ovary, it is free from the ovary, the other floral whorls are borne along the rim of the hypanthium, yet the ovary is morphologically still superior and other floral whorls inferior. The ovary may sometimes be partially immersed and thus **semi-inferior**.

Epigynous: The hypanthium is fused with the ovary, so that the other floral whorls appear to arise from the top of the ovary. The ovary is obviously **inferior** and other floral whorls superior. There may or may not be a free hypanthium above the ovary; in the former case, other floral parts appear to arise from the top of ovary.

Pentamerous: Five members in each floral whorl (excluding stamens and carpels), typical of dicots.

Tetramerous: Four members in each floral whorl, as in crucifers.

Trimerous: Three members in each floral whorl, as in monocots.

Cyclic (tetracyclic): Calyx, corolla, androecium and gynoecium in four separate whorls.

Spirocyclic: Calyx and corolla cyclic but stamens and carpels spirally arranged, as in Ranunculaceae.

Calyx

Description of the calyx starts with the number of sepals in same whorl (5—typical on dicots, 3—typical of monocots), in two whorls (2+2, as in crucifers) or forming two lips (1/4 in *Ocimum*, 3/2 in *Salvia*):

Polysepalous (aposepalous, chorisepalous): Sepals free, and consequently more than one units (poly—many).

Gamosepalous: Sepals fused. Once the calyx is gamosepalous, it commonly gets differentiated two parts: calyx tube, the fused part and calyx lobes (no longer sepals), the free part. The shape of the calyx tube should be described. It may be **campanulate** (bell-shaped as in *Hibiscus*), **urceolate** (urn-shaped as in fruiting calyx of *Withania*), **tubular** (tube-like as in *Datura*), or **bilabiate** (two-lipped as in *Ocimum*).

Caducous: Falling just after opening of flowers.

Deciduous: Falling along with petals in mature flower.

Persistent: Persisting in fruit.

Accrescent: Persisting and enlarging in fruit.

Aestivation: Arrangement of sepals (or petals) in the flower bud. Term **vernation** is used exclusively for arrangement of young leaves in a bud. The following main types of aestivation are met:

(i) **Valvate:** Margins of sepals or calyx lobes not overlapping.

(ii) **Twisted:** Overlapping in regular pattern, with one margin of each sepal overlapping and other being overlapped.

(iii) **Imbricate:** With irregular overlapping. In **Quincuncial imbricate,** two sepals are with both margins outer, two with both margins inner, and fifth with one outer and one inner margin.

Description of aestivation may be followed by color of sepals (green or petaloid), and whether they are inferior or superior.

Corolla

Description of the corolla follows the same pattern as calyx except that **bilabiate** corolla may be 4/1 or 3/2, corolla may be **polypetalous (apopetalous, choripetalous)**, or **gamopetalous (sympetalous)**, corolla tube may be additionally **infundibuliform** (funnel-shaped) as in *Datura*, **rotate** (tube very short with large lobes spreading out at right angle to the tube like spokes of a wheel), as in *Solanum*, or **salverform (salver-shaped, hypocrateriform)**, as in *Catharanthus*. The junction of corolla tube and lobes (constituting **limb**) is known as **throat**. Petals may sometimes be narrowed into a stalk

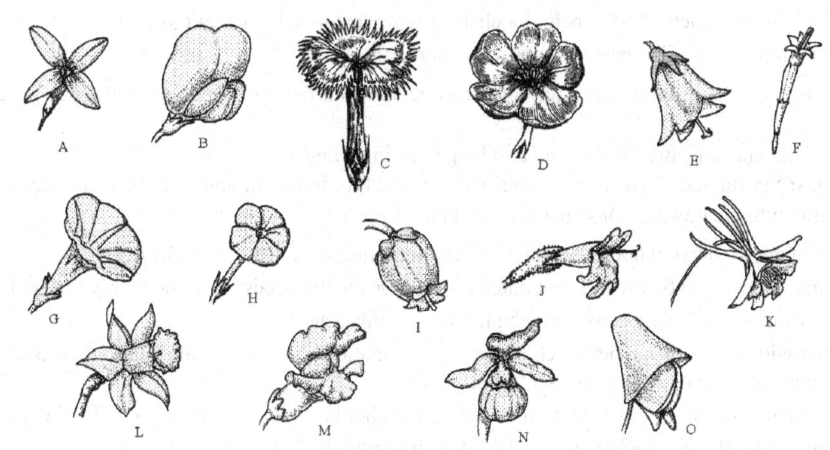

Figure 2.15: Corolla types. A: Cruciform; **B:** Papilionaceous; **C:** Caryophyllaceous; **D:** Rosaceous; **E:** Campanulate; **F:** Tubular; **G:** Infundibuliform (Funnel-shaped); **H:** Hypocrateriform; **I:** Urceolate; **J:** Bilabiate; **K:** Spurred (Calcarate); **L:** Coronate; **M:** Personate; **N:** Calceolate; **O:** Galeate.

termed as **claw**, the broader part then constituting the limb. Specialized types of corolla are encountered in Brassicaceae (**cruciform**—four free petals arranged in the form of a cross), Caryophyllaceae (**caryophyllaceous**—five free clawed petals with limb at right angles to the claw), Rosaceae (**rosaceous**—five sessile petals with limbs spreading outwards) and Fabaceae (**Papilionaceous**—resembling a butterfly with single large posterior petal **vexillum** or **standard**, two lateral petals **alae** or **wings**, and two anterior petals slightly united to form **keel** or **carina;** the aestivation is **vexillary** or descending imbricate, with the standard being the outermost, overlapping two wings, which in turn overlap keel). The petals may similarly be variously colored. In some cases, sepals or petals may bear a small pouch a condition known as **saccate** (lateral sepals of *Brassica* or corolla of *Cypripedium*—is more like slipper and called **calceolate**). Sometimes the base may be produced into a tube like structure known as **spur** (corolla as **calcarate**) as in *Delphinium* and *Aquilegia.* In some flowers (*Aconitum*), the corolla may be shaped like a helmet, when it is termed as **galeate**.

Present inner to corolla in some cases is an additional whorl generally attached to the throat of the corolla (or inner whorl of the perianth). Such a whorl is known as **corona** and may be consisting of appendages from perianth (*Narcissus*), corolla (corolline corona as in *Nerium*) or from stamens (staminal corona as in *Hymenocallis*). The flower is known as **coronate**.

Perianth

The description of **perianth** in the flowers lacking distinct calyx and corolla follows the same pattern specifying the number, number of whorls, perianth being **polyphyllous (apotepalous)** or **gamophyllous (syntepalous)**, aestivation, and the color of the perianth. The parts when free are called **tepals** in place of sepals or petals.

Androecium

Stamens representing the androecium present a more complicated architecture as compared to sepals and petals. Each stamen has an **anther**—typically tetrasporangiate with two anther sacs (microsporangia) in each of the two anther lobes—carried on a **filament**. The two anther lobes are often joined with the help of a **connective**, which in some primitive families, is a continuation of the filament. The description of androecium, likewise, starts with the number of stamens in a single or more whorls. Major descriptive terms include:

Fusion: Stamens may generally be free, but if fused it can take a variety of forms:

Polyandrous: Stamens free throughout.

Monadelphous: Filaments of all stamens united in a single group, as in family Malvaceae.

Diadelphous: Filaments of stamens united in two groups, as in *Lathyrus.*

Polyadelphous: Filaments united in more than two groups, as in *Citrus.*

Syngenesious (synantherous): Filaments free but anthers connate into a tube, as in family Asteraceae.

Synandrous: Stamens fused completely through filaments as well as anthers, as in *Cucurbita.*

Epipetalous: Filaments attached to the petals, a characteristic feature of sympetalous families.

Epiphyllous (epitepalous): Filaments attached to the perianth.

Relative size: Stamens in a flower are generally of the same size, but the following variations may be encountered in some flowers:

Didynamous: Four stamens, two shorter and two longer, as in *Ocimum*.

Tetradynamous: Six stamens, two shorter in outer whorl and four longer in inner whorl, as in crucifers.

Heterostemonous: Same flower with stamens of different sizes, as in *Cassia*.

Diplostemonous: Stamens in two whorls, the outer whorl alternating with petals as in *Murraya*.

Obdiplostemonous: Stamens in two whorls but outer whorl opposite the petals, as in the family Caryophyllaceae.

Antipetalous: Stamens opposite the petals, as in the family Primulaceae.

Bithecous: Stamen with two anther lobes (each anther lobe at maturity becomes unilocular due to coalescence of two adjacent microsporangia) so that anther is two-celled at maturity.

Monothecous: Stamen with single anther lobe so that mature anther is single-celled, as in family Malvaceae.

Attachment: Common modes of attachment of filament to the anther include:

(i) **Adnate:** Filament continues into connective which is almost as broad, as found in *Ranunculus*.

(ii) **Basifixed:** The filament ends at the base of anther (when connective extends up to base of anther)or at least base of connective (when anther lobes extend freely below the connective). The resultant anther is erect, as in *Brassica*.

(iii) **Dorsifixed:** Filament attached on the connective above the base. The resultant anther is somewhat inclined, as in *Sesbania*.

(iv) **Versatile:** Filament attached nearly at the middle of connective so that anther can swing freely as, in *Lilium* and grasses.

Dehiscence: Anther dehiscence commonly occurs by the formation of sutures along the point of contact of two anther sacs, but considerable variation in their location may be found:

Longitudinal: The two sutures extend longitudinally, one on each anther lobe as in *Datura*.

Transverse: Suture placed transversely, as in monothecous anthers of family Malvaceae.

Poricidal (apical pores): Anther opening by pores at the tip of anther, as in *Solanum nigrum*.

Valvular: Portions of anther wall opening through flaps or valves, as in *Laurus*.

Centripetal: Developing from the outside to the inside so that the oldest stamens are towards the periphery.

Centrifugal: Developing from centre towards the periphery, so that the oldest flowers are towards the centre.

Included: Stamens are shorter than the corolla.

Exserted: Stamens protruding far beyond the petals as in Umbellifers.

Introrse: Slits of the anther facing towards the centre.

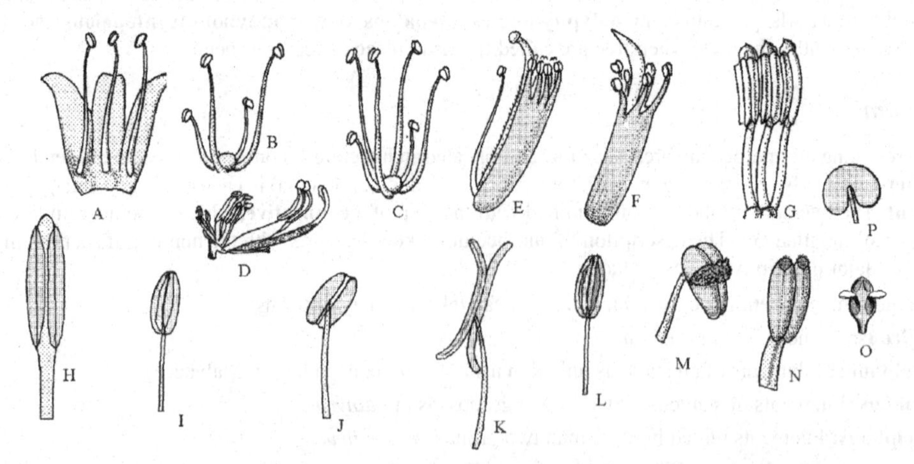

Figure 2.16: Androecium types. A: Epipetalous staments. **Length. B:** Didynamous; **C:** Tetradynamous; **D:** Heterostemonous. **Fusion. E:** Diadelphous; **F:** Monadelphous; **G:** Syngenesious. **Attachment. H:** Adnate; **I:** Basifixed; **J:** Dorsifixed; **K:** Versatile. **Dehiscence. L:** Longitudinal; **M:** Transverse; **N:** Poricidal; **O:** Valvular; **P:** Monothecous reniform anther.

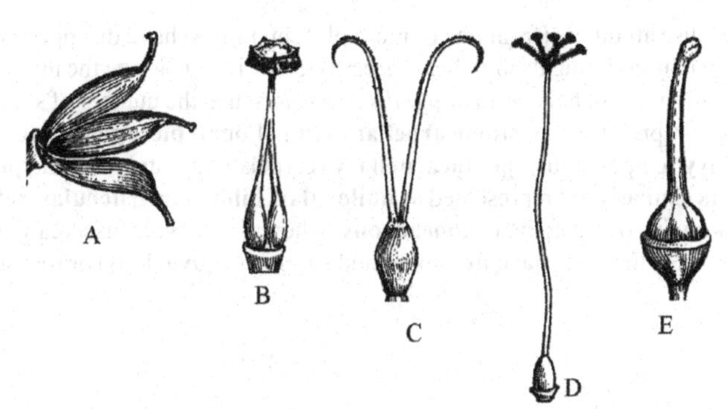

Figure 2.17: Carpel fusion. A: Apocarpous; **B:** Apocarpous with fused styles and stigmas (which, in turn, also fused with anthers to form gynostegium); **C:** Syncarpous with free styles and stigmas (synovarious); **D:** Syncarpous with free stigmas (synstylovarious); **E:** Syncarpous.

Extrorse: Slits of the anther facing towards outside.
Androphore: Extension of thalamus bearing stamens.
Gynostegium: Structure formed by the fusion of stamens with the stigmatic disc, as in family Asclepiadaceae.
Gynostemium: Structure formed by fusion of stamens with gynoecium, as in family Orchidaceae.

Gynoecium

Gynoecium represents a collection of carpels in a flower. The distinction between **carpel** and **pistil** is often ambiguous. In reality the carpels are components of a gynoecium whereas the pistils represent visible units. Thus, if carpels are free, there would be as many pistils (**simple pistils**). On the other hand, if the carpels are united (and obviously more than one), the flower would have only one pistil (compound pistil). Each carpel is differentiated into a broad basal **ovary** containing **ovules**, an elongated **style**, and pollen-receptive apical part **stigma**. Any attempt to describe gynoecium requires a transverse section through the ovary. An additional longitudinal section is always helpful.

Carpel Number and Fusion

A flower having more than one separate pistils would have as many carpels, which are free. On the other hand, if the pistil is one, there could either be one carpel, or more than one fused carpels. A section through the ovary helps to resolve the matter in most cases. If the ovary is single chambered, the number of rows of ovules (placental lines) would equal the number of united carpels. A solitary carpel would obviously have a single chamber with a single ovule or a single row of ovules. On the other hand, if ovary is more than one chambered, it obviously has more than one carpel, and the number of chambers would indicate the number of carpels. There are, however, atypical cases. Single chambered ovary may have a central column bearing ovules (since septa disappeared), or in a single chambered ovary there

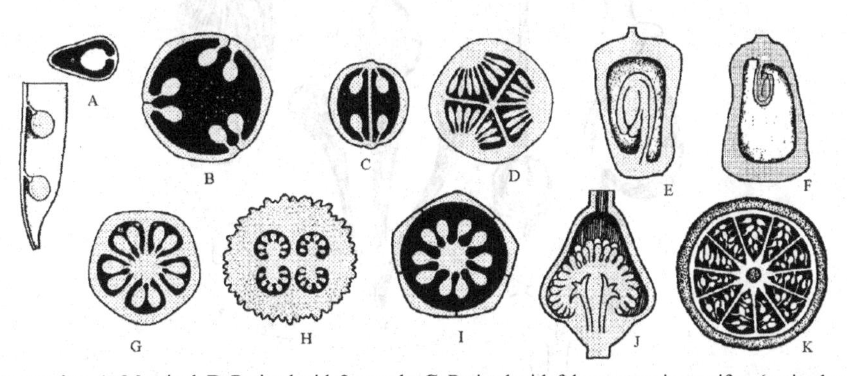

Figure 2.18: Placentation. A: Marginal; **B:** Parietal with 3 carpels; **C:** Parietal with false septum in crucifers (parietal-axile); **D:** Parietal with false septa in cucurbits; **E:** Basal; **F:** Apical; **G:** Axile; **H:** Axile with false septa in *Datura*; **I:** Free central with usual central column attached at the base and top of the ovary; **J:** Free central in Primulaceae in Longitudinal section showing placental column projecting from the base; **K:** Superficial in *Nymphaea*.

may be single large ovule because all others (from one or more placental lines) have disappeared. In both these cases, the number of carpels can be known by counting the number of free styles, or if style is one the number of stigmas or stigmatic lobes. In extreme cases, even this may not help, as in *Anagallis arvensis*, when the number of suture lines on the fruit would help. The number of carpels is represented as **monocarpellary** (carpel one), **bicarpellary** (carpels two), **tricarpellary** (carpels three), **tetracarpellary** (carpels four), **pentacarpellary** (carpels five), and **multicarpellary** (carpels more than five). The number of chambers similarly are represented as **unilocular, bilocular, trilocular, tetralocular, pentalocular** and **multilocular**. Gynoecium with free carpels is **apocarpous**, whereas one with fused carpels (at least ovaries fused) as **syncarpous.** Syncarpous gynoecium may have free styles and stigma (**synovarious**) or free stigmas (**synstylovarious**) or all fused.

Placentation

Placentation refers to the distribution of placentae on the ovary wall and, consequently, the arrangement of ovules. The following major types are found:

(i) **Marginal:** Single chambered ovary with single placental line commonly with single row of ovules, as in *Lathyrus*.

(ii) **Parietal:** Single chambered ovary with more than one discrete placental lines as, in family Capparaceae. In family Brassicaceae, the ovary later becomes bilocular due to the formation of a false septum, the ovules present at the junction of septum and ovary wall, a condition often known as **parietal-axile**. In some members of Aizoaceae, the ovules arise from inner ovary walls of septate ovary, a condition known as **parietal-septate**. In family Cucurbitaceae, the three parietal placentae intrude into ovary cavity and often meet in the centre making **false-axile** placentation.

(iii) **Axile:** Ovary more than one chambered and placentae along the axis as in *Hibiscus*.

(iv) **Free-central:** Ovary single chambered, ovules borne along the central column, as in family Caryophyllaceae.

(v) **Basal:** ovary single chambered, with single ovule at the base, as found in family Asteraceae (Compositae).

(vi) **Superficial:** Multilocular ovary with whole inner wall of ovary lined with placentae as in *Nymphaea*. In **laminar** placentation, the ovules arise from surface of septa.

Style and Stigma

Simple: Single style or stigma resulting from single carpel or fused styles or stigmas.

Bifid: Style or stigma divided into two as in family Asteraceae.

Terminal style: Arising from the tip of ovary, the most common type.

Gynobasic style: Arising from central base of the ovary, as in family Lamiaceae.

Capitate: Stigma appearing like a head.

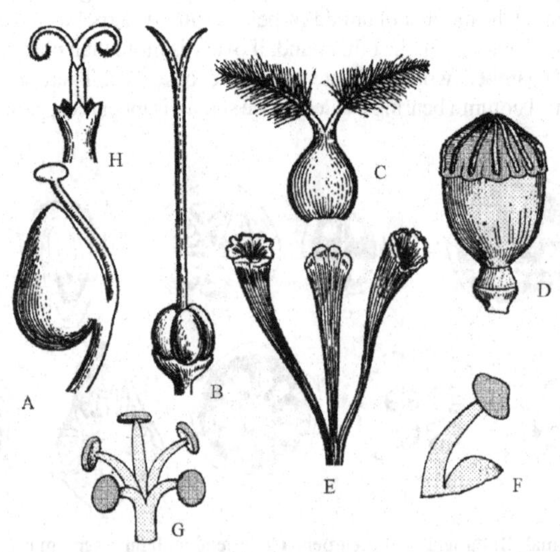

Figure 2.19: Style and stigma. A: Lateral style; **B:** Gynobasic style; **C:** Bifid feathery stigma in Poaceae; **D:** Sessile and radiate stigma of *Papaver*; **E:** Tripartite funnel-shaped stigma of *Crocus*; **F:** Capitate stigma of *Alchemilla*; **G:** Discoid stigmas of *Hibiscus*; **H:** Bifid stigma in Asteraceae.

Lateral style: Style arising from the side of the ovary, as in *Mangifera* and *Alchemilla*.
Stylar beak: Persistent style, extended into a long beak.
Pistillode: Sterile pistil, devoid of any fertile ovules, as in ray floret of radiate head of *Helianthus*.
Radiate stigma: Sessile disc like with radiating branches, as in *Papaver*.
Stylopodium: Swollen basal part of style surrounded by nectary persisting in fruit of umbellifers.
Sessile stigma: Seated directly on ovary, style being reduced as in *Sambucus*.
Discoid stigma: Disc-shaped stigma.
Globose stigma: Stigma spherical in shape.
Plumose stigma: Feathery stigma with trichome-like branches as in Poaceae and Cyperaceae.

Ovule

Ovule represents megasporangium, attached to the placenta by **funiculus**, which joins the ovule at the **hilum**. Base of the ovule is known as **chalaza**, and the tip as **micropyle**. Ovule has a female gametophyte (**embryo sac**) surrounded by nucellus, in turn, enveloped by two **integuments**. The following terms are commonly associated with ovules:
Orthotropous (atropous): Straight erect ovule with funiculus, chalaza and micropyle in one line, as in family Polygonaceae.
Anatropous: Inverted ovule with micropyle facing and closer to funiculus, as in *Ricinus*.
Amphitropous: Ovule placed at right angles to the funiculus, as in *Ranunculus*.
Campylotropous: Curved ovule so that micropyle is closer to chalaza, as in Brassicaceae.
Circinotropous: Funiculus very long and surrounding the ovule, as in *Opuntia*.
Hemianatropous (hemitropous): Body half-inverted so that funiculus is attached near middle with micropyle terminal and at right angles.
Bitegmic: Ovule with two integuments, common in polypetalous dicots.
Unitegmic: Ovule with single integument, common in sympetalous dicots.
Crassinucellate: Ovule with massive nucellus, found in primitive polypetalous dicots.
Tenuinucellate: Ovule with thin layer of nucellus, as in sympetalous dicots.

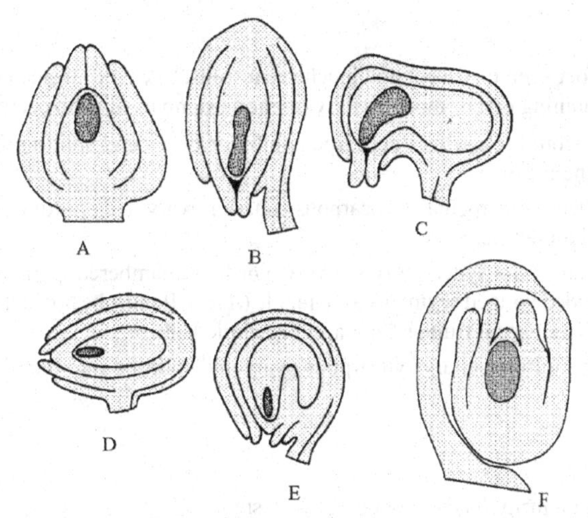

Figure 2.20: Ovules. A: Orthotropous; **B:** Anatropous; **C:** Campylotropous; **D:** Hemianatropous; **E:** Amphitropous; **F:** Circinotropous.

FRUITS

A fruit is a matured and ripened ovary, wherein the ovary wall gets converted into the fruit wall **pericarp** (differentiated into outer **epicarp**, middle **mesocarp** and inner **endocarp**), and the ovules into seeds. Three main categories of fruits are recognized: **simple fruits** developing from a single ovary of the flower, **aggregate fruits** developing from several free carpels within the flower, and **composite fruits** involving several flowers or the whole inflorescence.

Simple Fruits

A simple fruit develops from a flower having a single carpel or several united carpels so that the flower has a single ovary. Such a fruit may be dehiscent opening by a suture exposing seeds or remain indehiscent.

Dehiscent Fruits

Such fruits are generally dry and burst along the suture to release their seeds. Common types are enumerated below:

Follicle: Fruit developing from superior monocarpellary ovary and dehiscing along one suture, as in *Consolida*.

Legume or pod: Fruit developing like follicle from monocarpellary superior ovary but dehiscing along two sutures, as in legumes.

Lomentum: Modified legume, which splits transversely at constrictions into one- or many-seeded segments, as in *Mimosa*. Sometimes considered as a type of schizocarpic fruit.

Siliqua: Fruit developing from bicarpellary syncarpous superior ovary, which is initially one chambered but subsequently becomes two chambered due to the formation of a **false septum**, visible on the outside in the form of a rim known as **replum**. The fruit dehisces along both sutures from the base upwards, valves separating from septum and seeds remaining attached to the rim (replum), characteristic of the family Brassicaceae. The fruit is narrower and longer, at least three times longer than broad, as in *Brassica* and *Sisymbrium*.

Silicula: Fruit similar to siliqua but shorter and broader, less than three times longer than broad as seen in *Capsella*, *Lepidium* and *Alyssum*. Silicula is commonly flattened at right angles to the false septum (*Capsella*, *Lepidium*) or parallel to the false septum (*Alyssum*).

Capsule: Fruit developing from syncarpous ovary and dehiscing in a variety of ways:

- **Circumscissile (pyxis):** Dehiscence transverse so that top comes off as a lid or operculum, as in *Anagallis arvensis*.
- **Poricidal:** Dehiscence through terminal pores as in poppy (*Papaver*).
- **Denticidal:** Capsule opening at top exposing a number of teeth as in *Primula* and *Cerastium*.
- **Septicidal:** Capsule splitting along septa and valves remaining attached to septa as in *Linum*.
- **Loculicidal:** Capsule splitting along locules and valves remaining attached to septa, as in family Malvaceae.
- **Septifragal:** Capsule splitting so that valves fall off leaving seeds attached to central axis as in *Datura*.

Schizocarpic Fruits

This fruit type is intermediate between dehiscent and indehiscent fruits. The fruit, instead of dehiscing, rather splits into number of segments, each containing one or more seeds. Common examples of schizocarpic fruits are:

Cremocarp: Fruit developing from bicarpellary syncarpous inferior ovary and splitting into two one seeded segments known as **mericarps,** as in umbellifers.

Carcerulus: Fruit developing from bicarpellary syncarpous superior ovary and splitting into four one seeded segments known as nutlets, as in family Lamiaceae.

Double samara: Fruit developing from syncarpous ovary, two or four chambered, pericarp of each chamber forming a wing, fruit splitting into one-seeded winged segments as in maple (*Acer*). It must be noted that **single samara** of *Fraxinus*, is a single-seeded dry winged indehiscent fruit and not a schizocarpic fruit.

Regma: Fruit developing from multicarpellary syncarpous ovary and splitting into one-seeded **cocci,** as in *Ricinus* and *Geranium*.

Indehiscent Fruits

Such fruits do not split open at maturity. They may be dry or fleshy:

Dry indehiscent fruits: Such fruits have dry pericarp at maturity, and are represented by:

- **Achene:** Single seeded dry fruit developing from a single carpel with superior ovary. Fruit wall is free from seed coat. Achenes are often aggregated, as in family Ranunculaceae.
- **Cypsela:** Single seeded dry fruit, similar to (and often named achene) but developing from bicarpellary syncarpous inferior ovary, as in family Asteraceae.
- **Caryopsis:** Fruit similar to above two but fruit wall fused with seed coat as seen in grasses.
- **Nut:** One-seeded, generally large fruit developing from multicarpellary ovary and with hard woody or bony pericarp, as seen in *Quercus* and Litchi.

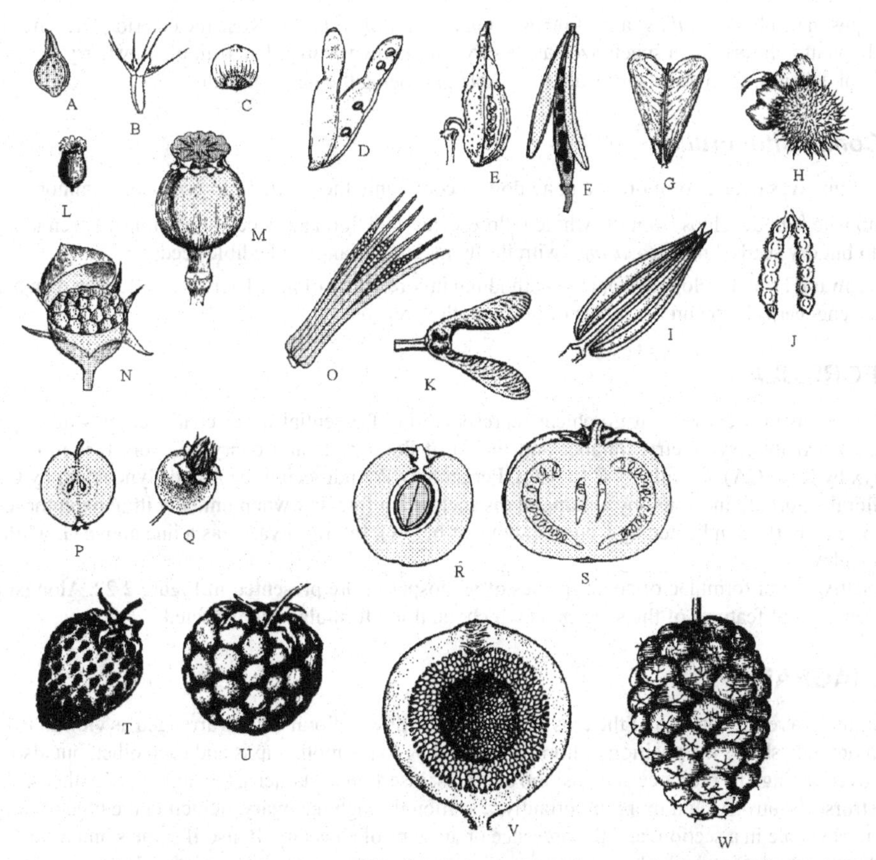

Figure 2.21: Fruits. A: Achene of *Ranunculus*; **B:** Cypsela of *Ageratum* with scaly pappus; **C:** Nut of *Castanea*; **D:** Pod of *Pisum*; **E:** Single follicle of *Calotropis*; **F:** Siliqua of *Brassica*; **G:** Silicula of *Capsella bursa-pastoris*; **H:** Capsule of *Datura*; **I:** Cremocarp in umbellifers; **J:** A pair of lomentum fruits in *Mimosa*; **K:** Double samara of *Acer*; **L:** Capsule of *Primula* dehiscing by apical teeth (denticidal); **M:** Operculate capsule of *Papaver* with poricidal dehiscence; **N:** Pyxis of *Celosia* with circumscissile dehiscence; **O:** Capsule of *Abelmoschus esculentus* with loculicidal dehiscence; **P:** Pome of *Malus pumila*; **Q:** Hip of *Rosa* with etaerio of achenes inside; **R:** Drupe of *Prunus*; **S:** Berry of *Solanum lycopersicum*; **T:** Pseudocarp of *Fragaria*, an accessary fruit with etaerio of achenes; **U:** Etaerio of drupes in *Rubus*; **V:** Syconium of *Ficus* developing from hypanthodium inflorescence; **W:** Sorosis of *Morus*.

Utricle: Like nut but with papery often inflated pericarp as in *Chenopodium*.

Fleshy indehiscent fruits: Such fruits have fleshy and juicy pericarp even at maturity. Common examples are:

Drupe: Fruit with usually skinny epicarp, fibrous or juicy mesocarp and hard stony endocarp, enclosing single seed, as seen in mango, plums and coconut.

Berry: Fruit with uniformly fleshy pericarp with numerous seeds inside, as seen in *Solanum*, tomato and brinjal.

Pepo: Fruit formed from inferior ovary of cucurbits with epicarp forming tough rind.

Hesperidium: Fruit developing from superior ovary with axile placentation, epicarp and mesocarp forming common rind and endocarp produced inside into juice vesicles, as seen in citrus fruits.

Pome: Fruit developing from inferior ovary, an example of **accessory (false) fruit,** wherein fleshy part is formed by thalamus and cartilaginous pericarp is inside, as seen in apple.

Balausta: Fruit developing from inferior ovary, pericarp tough and leathery, seeds attached irregularly, succulent testa being edible, as seen in pomegranate (*Punica granatum*).

Aggregate Fruits

Aggregate fruits develop from multicarpellary apocarpous ovary. Each ovary forms a fruitlet, and the collection of fruitlets is known as **etaerio**. Common examples are **etaerio of achenes** in Ranunculaceae, **etaerio of follicles** in *Calotropis*,

etaerio of drupes in raspberry (*Rubus*) and **etaerio of berries** in *Polyalthia*. In Rose the etaerio of achenes is surrounded by a cup like hypanthium forming a specialized accessory fruit known as **hip**. The fruit of strawberry (*Fragaria*), though also an etaerio of achenes, is an accessory fruit, the edible part being the fleshy thalamus.

Multiple (Composite) Fruits

A multiple fruit involves ovaries of more than one flower, commonly the whole inflorescence. Common examples are:

Sorosis: Composite fruit develops from the whole inflorescence and floral parts become edible, as seen in *Morus* (having fleshy perianth but dry seeds) and *Artocarpus* (with fleshy rachis, perianth and edible seeds).

Syconium (syconus): Fruit developing from hypanthodium inflorescence of figs. There is a collection of achenes or drupes (surrounded by endocarp) borne on the inside of fleshy hollow receptacle.

FLORAL FORMULA

The floral formula enables convenient graphical representation of essential floral characteristics of a species, mainly incorporating its sexuality, symmetry, number and fusion of floral parts and ovary position. It is more convenient to represent Calyx by K (or CA), Corolla by C (or CO), Perianth by P, Androecium by A and Gynoecium by G. The number of parts in a floral whorl are indicated by a numeral (as such when free, but when united within parentheses or a circle). Adnation between whorls is indicated by a curve (above or below). Inferior ovary has a line above G, while the superior ovary has one below.

Representative floral formulae of some species of angiosperms are presented in Figure 2.22. Alongside each floral formula is given a list of features of the species on which the floral formula is constructed.

FLORAL DIAGRAM

The floral diagram is a representation of the cross-section of the flower, floral whorls arranged as viewed from above. The floral diagram not only shows the position of floral parts relative to the mother axis and each other, but also their number, fusion or not, overlapping, the presence and position of bracts, insertion of stamens, the number of anther sacs, whether the anthers are extrorse or introrse, and more importantly, a section through the ovary, depicting the type of placentation, the number of ovules visible in a section, and the presence or absence of a nectary. It also if some stamens are nonfunctional (represented by staminodes) and whether the ovary is functional or represented by a pistillode.

The branch (or the inflorescence axis) bearing the flower is known as **mother axis**, and the side of flower facing it as **posterior side**. The bract, if present is opposite the mother axis, and the side of flower facing it is the anterior side. The remaining components of the flower—depending upon whether they are closer to the mother axis or the bract—occupy postero-lateral and antero-lateral positions, respectively. The members of different floral whorls are shown arranged in concentric rings, calyx being the outermost and the gynoecium the innermost. A large majority of dicot flowers are pentamerous, and as such the five members of each whorl (excluding gynoecium in the centre) are arranged such a way that four of them occur in pairs (members of each pair occupying complementary position) the fifth one is the odd member. It is also to be remembered that in large majority of dicots (except Fabaceae and few others), the odd sepal occupies posterior position (of the remaining four, two form antero-lateral pair, and the remaining two the postero-lateral pair). The different whorls usually alternate each other, and accordingly the odd petal occupies anterior position, the petals alternate with sepals. The stamens accordingly alternate with petals and are opposite the sepals. In flowers with two whorls of stamens, the outer whorl alternates with petals, whereas the inner is opposite the petals (because it alternates with the outer whorl of stamens). The stamens are represented in the floral diagram by anthers, each with two anther lobes (shown by a deep fissure) and latter, in turn, with two anther sacs (with a less deep cleft). The lobes face towards the outside in extrorse anthers and towards the ovary in introrse anthers. Epipetalous stamens are shown by a line joining the anthers with the petals. A few representative types of floral diagram are shown in Figure 2.23.

The floral diagram summarizes the information about the presence or absence of bracts and bracteoles, number, fusion and aestivation of sepals and petals (or tepals if there are no separate sepals and petals, as shown in Moraceae). The calyx and corolla forming bilabiate arrangement are appropriately shown with the number of lobes in upper and lower lip (as seen in Lamiaceae). The stamens with united filaments are depicted by joining anthers via lines (diadelphous condition in Fabaceae-Faboideae), whereas the united anthers are shown by physically touching anther margins. In families with complex floral arrangement such as the cyathium in *Euphorbia*, floral diagram for the entire cyathium may be drawn, supplemented by floral diagrams of male and female flowers. In family Poaceae also, it is helpful to make a floral diagram for the whole spikelet (shown in *Avena sativa*), or separate diagrams for male and female spikelets if the male and female flowers occur in separate inflorescences or at least separate spikelets (shown in *Zea mays*).

Family	*Species*	*Floral Formula*	
Solanaceae	*Solanum nigrum*	$\oplus$ ⚥ $K_{(5)}\widehat{C_{(5)}A_{5}}G_{(\underline{2})}$	Flowers actinomorphic, bisexual, sepals 5 united, petals 5 united, stamens 5 free epipetalous, carpels 2 united, ovary superior.
Lamiaceae	*Ocimum basilicum*	% ⚥ $K_{(1/4)}\widehat{C_{(4/1)}A_{2+2}}G_{(\underline{2})}$	Flowers zygomorphic, bisexual, calyx bolabiate, upper lip 1 lobed lower 4 lobed, corolla bilabiate, upper lip 4 lobed lower 1 lobed, stamens 4 didynamous epipetalous, carpels 2 united, ovary superior.
Brassicaceae	*Brassica campestris*	$\oplus$ ⚥ $K_{2+2}C_{4\times}A_{2+4}G_{(\underline{2})}$	Flowers actinomorphic, bisexual, sepals 4 fee in 2 whorls, petals 4 cruciform, stamens 6 tetradynamous, carpels 2 united, ovay superior.
Fabaceae	*Lathyrus odoratus*	% ⚥ $K_{(5)}C_{1+2+(2)}A_{1+(9)}G_{\underline{1}}$	Flowers zygomorphic, bisexual, sepals 5 united, petals 5 free papilionaceous, stamens 10, diadelphous 9 united 1 free, carpel 1, ovary superior.
Malvaceae	*Hibiscus rosa-sinensis*	$\oplus$ ⚥ Epi $K_{5\text{-}7}K_{(5)}C_{5}A_{(\infty)}G_{(\underline{5})}$	Flowers actinomorphic, bisexual, epicalyx 5-7 free, sepals 5 united, petals 5 free, stamens many monadelphous epipetalous, carpels 5 united, ovary superior.
Asteraceae	*Helianthus annuus*	**Ray floret** % ♀ $K_{pappus}C_{(5)}A_{0}G_{(\overline{2})}$ **Disc floret** $\oplus$ ⚥ $K_{pappus}\widehat{C_{(5)}A_{(5)}}G_{(\overline{2})}$	Flowers of 2 types. Ray florets zygomorphic, pistillate, calyx represented by pappus, petals 5 united, stamens absent, carpels 2 united, ovary inferior. Disc floret actinomorphic, bisexual,calyx represented by pappus, corolla 5 united, stamens 5 epipetalous united (syngenesious), carpels 2, united, ovary inferior.
Chenopodiaceae	*Chenopodium album*	$\oplus$ ⚥ $P_{(5)}A_{5}G_{(\underline{2})}$	Flowers actinomorphic, besexual, tepals 5 united, stamens 5 free, carpels 2 united, ovary superior.
Caryophyllaceae	*Stellaria media*	$\oplus$ ⚥ $K_{5}C_{5}A_{5+5}G_{(\underline{3})}$	Flowers actinomorphic, bisexual, sepals 5 free, petals 5 free, stamens 10 in two whorls, carpels 3 united, ovary superior.

Figure 2.22: Floral formulae of some representative species of few families of angiosperms depicting diversity of features depicted. The important features on which each formula is based are shown in the right column.

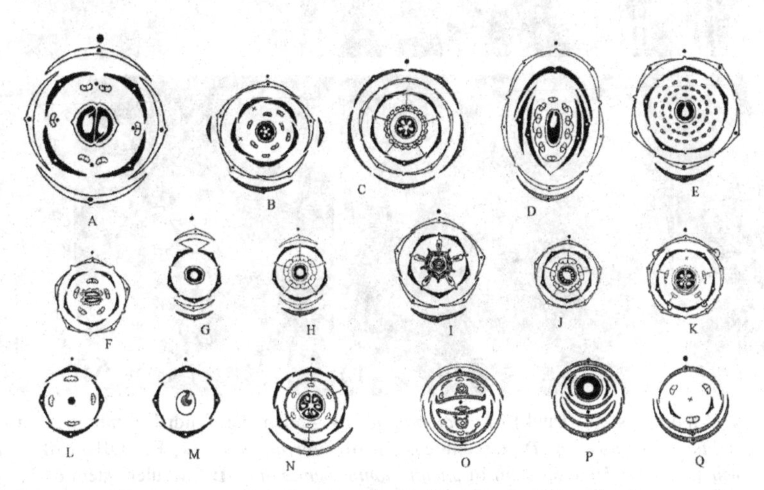

Figure 2.23: Floral diagrams of some representative members of major families. A: *Brassica campestris* (Brassicaceae); **B:** *Stellaria media* (Caryophyllaceae); **C:** *Hibiscus rosa-sinensis* (Malvaceae); **D:** *Lathyrus odoratus* (Fabaceae-Faboideae); **E:** *Acacia nilotica* (Fabaceae-Mimosoideae); **F:** *Foeniculum vulgare* (Apiaceae); **G:** Ray floret of *Helianthus annuus* (Asteraceae); **H:** Disc floret of *H. annuus*; **I:** *Calotropis procera* (Apocynaceae-Asclepiadoideae); **J:** *Withania somnifera* (Solanaceae); **K:** *Ocimum basilicum* (Lamiaceae); **L:** Male flower of *Morus alba* (Moraceae); **M:** Female flower of *M. alba*; **N:** *Narcissus pseudo-narcissus* (Amaryllidaceae); **O:** *Avena sativa* (Poaceae), floral diagram of spikelet; **P:** *Zea mays* (Poaceae), floral diagram of female spikelet; **Q:** *Z. mays*, floral diagram of male spikelet.

Figure 2.24: Stems. A: Arboreus stem (trunk) of *Cyclobalanopsis glauca*; **B:** Tendril climbing stem of *Luffa cylindrica*; **C:** Scandent stem of *Allamanda violacea*; **D:** Creeping stem of *Zebrina pendula*; **E:** Offset of *Eichhornia crassipes*; **F:** Runner of *Oxalis corniculata*; **G:** Twining stem of *Jacquemontia pentantha*; **H:** Succulent stem of *Echinopsis terescheckii*; **I:** Rhizome of *Zingiber officinale*; **J:** Phylloclade of *Ruscus aculeatus*; **K:** Bulb of *Allium cepa*; **L:** Tuber of *Solanum tuberosum*; **M:** Corm of *Alocasia*; **N:** Phylloclade of *Opuntia elatior*.

Figure 2.25: Leaves. A: Alternate phyllotaxy in *Citrus*; **B:** Opposite decussate phyllotaxy in *Calotropis procera*; **C:** Whorled phyllotaxy in *Alstonia scholaris*; **D:** Ovate long acuminate leaf of *Ficus religiosa*; **E:** Sagittate leaf of *Sagittaria sagitifolia*; **F:** Palmately lobed leaf of *Rubus trifidus*; **G:** Palmately lobed leaf-lobes further pinnately lobed in *Carica papaya*; **H:** Palmate leaf of *Acer palmatum*; **I:** Unifoliate compound leaf of *Citrus medica*; **J:** Palmately trifoliate compound leaf of *Oxalis corniculata*; **K:** Palmate compound leaf of *Cannabis sativa*; **L:** Pinnate compound leaf of *Rosa*; **M:** Peltate orbicular leaf of *Tropaeolum majus*; **N:** Bipinnate compound leaf of *Leucaena leucocephala*; **O:** Panduraeform leaf of *Jatropha panduraefolia*; **P:** Grass leaf with leaf sheath and free lamina of *Zea mays*; **Q:** Pitcher leaf of *Sarracenia flava*.

Figure 2.26: Inflorescences. A: Solitary flower of *Malvaviscus arboreus*; **B:** Corymbose-raceme of *Brassica campestris*; **C:** Corymb of *Iberis amara*; **D:** Rhipidium of *Solanum nigrum*; **E:** Cyathium of *Euphorbia milii*; **F:** Spike of *Adhatoda vasica*; **G:** Panicle of spikelets of *Zea mays*; **H:** Cob (spike of spikelets) of *Zea mays*; **I:** Spike of spikelets of *Triticum aestivum*; **J:** Raceme of verticillasters in *Salvia splendens*; **K:** Spike of *Acanthus spinosus*; **L:** Raceme of *Delphinium ajacis*; **M:** Hypanthodium of *Ficus religiosa*; **N:** Cymose cluster with spathaceous bracts of *Tradescatia spathacea (syn: Rhoeo discolor)*; **O:** Umbel of *Astrantia major*; **P:** Radiate capitulum of *Helianthus debilis*; **Q:** Discoid capitulum of *Ageratum houstonianum*; **R:** Spadix of *Amorphophalus titanum*; **S:** Cymose umbel of *Agapanthus umbellatus*.

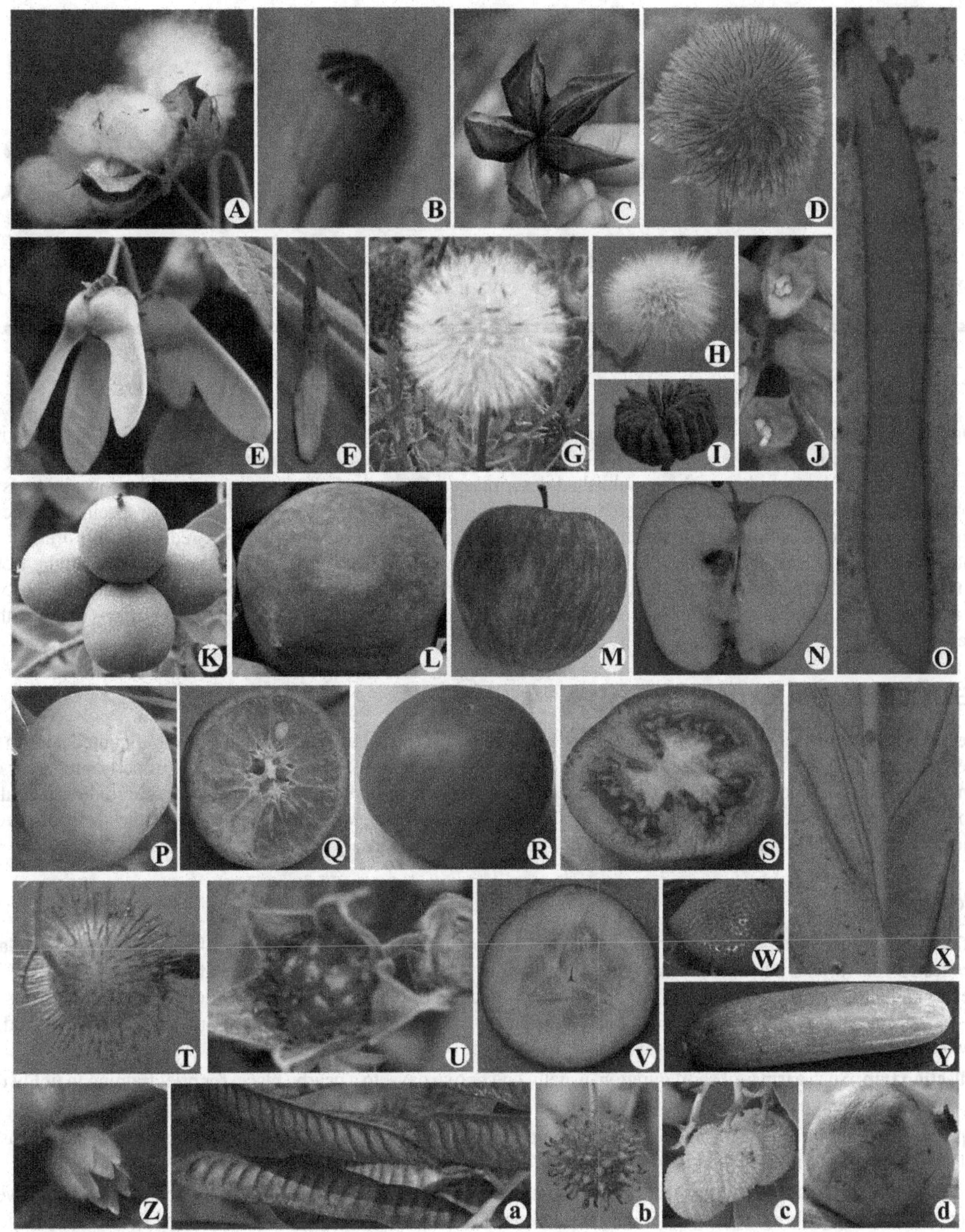

Figure 2.27: Fruits. A: Dehisced capsule of *Gossypium hirsutum* with exposed hairy seeds; **B:** Capsule of *Papaver orientale*; **C:** Dehisced capsule of *Chiranthodendron pentadactylon*; **D:** etaerio of achenes of *Anemone occidentalis*; **E:** Double samara of *Acer griseum*; **F:** Pod of D*albergia sissoo*; **G:** Cypsela of *Haplopappus macrocephalus*; **H:** Cypsela of *Sonchus oleraceous*; **I:** Schizocarp of *Abutilon indicum*; **J:** Carcerulus of *Salvia splendens*; **K:** Drupe of *Juglans nigra*; **L:** Drupe of *Prunus persica*; **M:** Pome of *Malus pumila*; **N:** Same in Longitudinal section; **O:** Pod of *Clitoria ternatea*; **P:** Hesperidium of *Citrus sinensis*; **Q:** Same in Transverse section; **R:** Berry of *Lycopersicon esculentum*; **S:** Same in Transverse section; **T:** Berry of *Ribes menziesii*; **U:** Etaerio of drupes of *Rubus nepalensis*; **V:** Pepo of *Cucumis sativus* in Transverse section; **W:** Whole pepo; **X:** Accessory fruit of *Fragaria vesca*; **Y:** Siliqua of *Brassica campestris*; **Z:** Dehisced capsule of *Stellaria media*; **a:** Pod of *Leucaena leucocephala*; **b:** Multiple fruit of *Liquidambar styraciflua*; **c:** Multiple fruit of *Arbutus unedo*; **d:** Balausta of *Punica granatum*.

Chapter 3

Process of Identification

Recognizing an unknown plant is an important constituent taxonomic activity. A plant specimen is identified by comparison with already known herbarium specimens in a herbarium, and by utilizing the available literature and comparing the description of the unknown plant with the published description/s. Since the bulk of our plant wealth grows in areas far removed from the centers of botanical research and training, it becomes imperative to collect many specimens on each outing. For proper description and documentation, these specimens must be suitably prepared for incorporation and permanent storage in herbarium. This goes a long way in compiling floristic accounts of the different regions of the world. The availability of the specimens in the herbaria often provides reasonable information about the abundance or rarity of a species and helps in preparing lists of rare or endangered species and provides sufficient inputs for efforts towards their conservation.

SPECIMEN PREPARATION

A specimen meant for incorporation in a herbarium needs to be carefully collected, pressed, dried, mounted and finally properly labelled, so that it can meet the demands of rigorous taxonomic activity. Specimens, properly prepared, can retain their essential features for a very long period, proving to be immensely useful for future scientific studies, including compilation of floras, taxonomic monographs and, in some cases, even experimental studies, since the seeds of several species can remain viable for many years even in dry herbarium specimens.

Fieldwork

The fieldwork of specimen preparation involves plant collection, pressing and partial drying of the specimens. The plants are collected for various purposes: building new herbaria or enriching older ones, compilation of floras, material for museums and class work, ethnobotanical studies, and introduction of plants in gardens. In addition, bulk collections are done for trade and drug manufacture. Depending on the purpose, resources, proximity of the area and duration of studies, fieldwork may be undertaken in different ways:

Collection trip: Such a trip is of short duration, usually one or two days, to a nearby place, for brief training in fieldwork, vegetation study and plant collection by groups of students.

Exploration: This includes repeated visits to an area in different seasons, for a period of a few years, for intensive collection and study, aimed at compilation of floristic accounts.

Expedition: Such a visit is undertaken to remote and difficult area, to study the flora and fauna, and usually takes several months. Most of our early information on Himalayan flora and fauna has been the result of European and Japanese expeditions.

Equipment

The equipment for fieldwork may involve a long list, but the items essential for collection include plant press, field notebook, bags, vasculum, pencil, cutter, pruning shears, knife and a digging tool (Figure 3.1).

Plant Press

A plant press consists of two wooden, plywood or wire mesh planks, each 12 inches × 18 inches (30 cm × 45 cm), between which are placed corrugated sheets, blotters and newspaper sheets (Figure 3.2). Two straps, chains or belts are used to

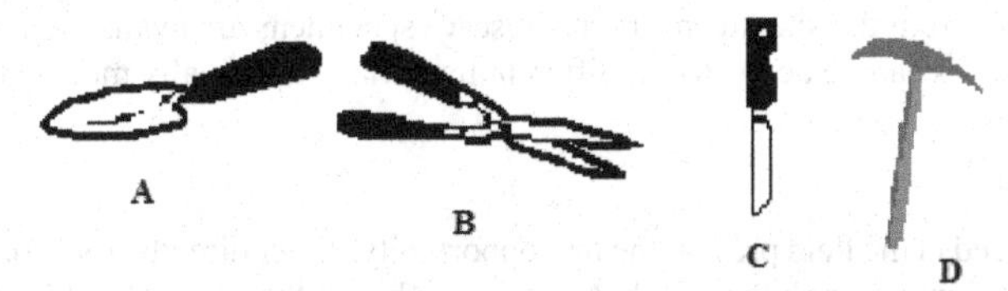

Figure 3.1: Common implements helpful in collection. **A:** Trowel; **B:** Pruning shears; **C:** Knife; **D:** Pickaxe.

Figure 3.2: Plant press containing pressed specimens. Vasculum placed alongside. (Photograph courtesy Mr. S. L. Kochhar.)

tighten the press. **Corrugated sheets** or ventilators are made of cardboard and help ventilation and the consequent drying of specimens. The ducts of the corrugated sheet run across and not lengthwise to in order to afford shorter distance and larger number of ducts.

The plant press carried in the field and called a **field press** is light weight and generally has one corrugated sheet alternating with one folded blotter containing ten newspaper sheets, one meant for each specimen. The plant press used for subsequent pressing and drying of specimens, kept at the base camp or the organization, is called the **drying press**. It is much heavier and has an increased number of corrugated sheets, one alternating each folded blotter containing one folded newspaper. In countries such as India which use thick coarse paper for newsprint, blotters can be dispensed with, in at least subsequent changes, as the paper soaks enough moisture and serves the purpose of blotters as well.

Field Notebook

A field notebook or **field diary** is an important item for a collector. A well-designed field notebook (Figure 3.3) has numbered sheets with printed proforma for entering field notes such as scientific name, family, vernacular name, locality, altitude, date of collection and for recording any additional data collected in the field. The multiple detachable slips at the lower end of the sheet, separated by perforated lines and bearing the serial number of the sheet, can be used as tags for multiple specimens of a species collected from a site, and serve as ready reference to the information recorded in the field notebook. The number also serves as the collection number for the collector.

Vasculum

A vasculum is a metallic box with a tightly-fitted lid and a shoulder sling. It is used to store specimens temporarily before pressing, and to store bulky parts and fruits. It is generally painted white to deflect heat and affords easy detection when left in the field. Being bulky, the vasculum is commonly substituted by a **polythene bag**, which is almost weightless. Several polythene bags can be carried for easy storage, as these can be readily made airtight using a rubber band and, as such, the plants retain their freshness for many hours.

Collection

The specimen collected should be as complete as possible. Herbs, very small shrubs, as far as possible, should be collected complete, in flowering condition, along with leaves and roots. Trees and shrubs should be collected with both vegetative and flowering shoots, to enable the representation of both leaves and flowers. All information concerning the plant should be recorded in the field notebook and a tag from the sheet attached to the concerned specimen. It is advisable to collect a few specimens of each species from the site, to ensure that reserve specimens are available if one or more get destroyed, and also to ensure that duplicates can be deposited in different herbaria, when finally mounted on sheets.

Pressing

The specimens should be placed in the field press at the first opportunity, either directly after collection, or sometimes after a temporary storage in a vasculum or a polythene bag. A specimen shorter than 15 inches (38 cm) should be kept directly in the folded newspaper after loosely spreading the leaves and branches. Herbs, which are generally collected along with the roots, if longer than 15 inches, can be folded in the form of a **V**, **N** or **W** (Figure 3.4A–C), always ensuring that the terminal part of the plant with leaves, flowers and fruits, is erect, and when finally mounted, the specimens can be easily studied, without having to invert the herbarium sheet. Specimens of grasses and some other groups, which show considerable elasticity, are difficult to hold in a folded condition. These specimens can be managed by using **flexostat** (a strip of stiff paper or card with 2.5 cm long slit). One flexostat inserted at each corner (Figure 3.4D) holds the specimen in place.

To press bulky fruits, these may be thinly sliced. Large leaves can be trimmed to retain any lateral half. It is useful to invert some leaves so that the under surface of the leaves can also be studied from a pressed leaf.

Handling Special Groups

A few groups of plants such as conifers, water plants, succulents and mucilaginous plants pose problems during collection and need special methods.

Conifers, although easy to collect and press, pose problems during drying. The tissues of conifers remain living for a long time and progressive desiccation during pressing and drying initiates an abscission layer at the base of leaves and sporophylls. As such a dry twig readily disintegrates, losing its leaves with a slight touch, a problem occasionally encountered in *Abies, Picea, Cedrus,* and several other genera. Before pressing, such twigs should be immersed in boiling water for one minute, a pretreatment that kills the tissues and prevents the abscission formation during drying. Page (1979) has suggested pretreatment method involving immersion in 70% ethyl alcohol for 10 minutes, followed by immersion in 50% aqueous glycerine solution for four days. Since the pretreatment removes the bloom and waxes, and results in a slight color change, an untreated portion of the plant should also be preserved, kept in a small pouch and attached to the herbarium sheet along with the pretreated specimen, for reference.

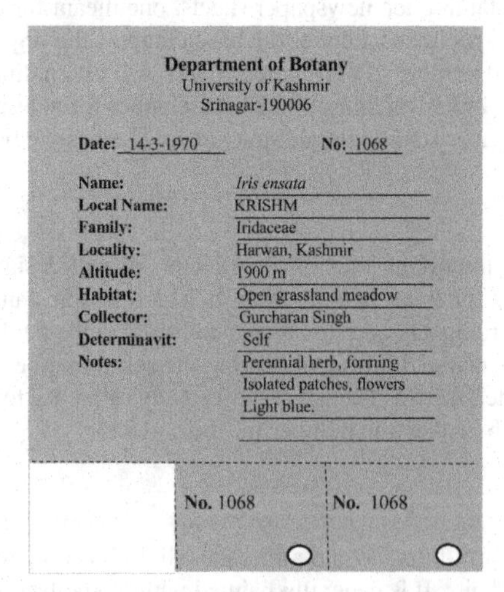

Department of Botany
University of Kashmir
Srinagar-190006

Date: 14-3-1970 **No:** 1068

Name: *Iris ensata*
Local Name: KRISHM
Family: Iridaceae
Locality: Harwan, Kashmir
Altitude: 1900 m
Habitat: Open grassland meadow
Collector: Gurcharan Singh
Determinavit: Self
Notes: Perennial herb, forming Isolated patches, flowers Light blue.

No. 1068 **No.** 1068

Figure 3.3: **A sheet from field notebook** with relevant entries.

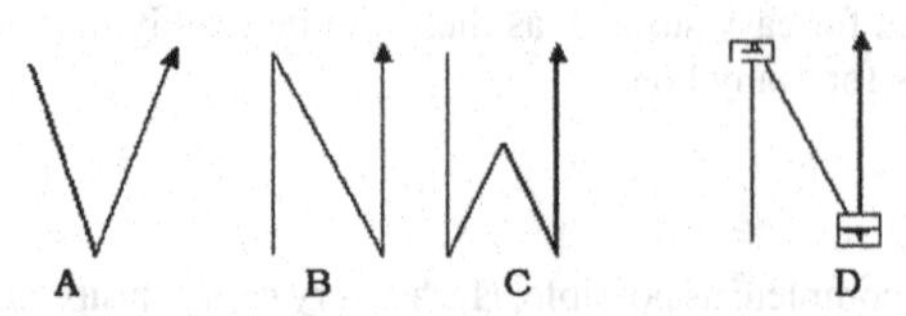

Figure 3.4: Folding in plant press A–C: Different methods of folding longer herbaceous plants; **D:** Use of flexostat slips for holding plants in folded condition. Note that the tip of the plant (arrow) would always be erect for convenient study of this important portion with leaves and flowers.

Water plants, especially with submerged leaves, readily collapse due to the absence of cuticle and are difficult to press normally. Such specimens are collected in bags and made to float in a tray filled with water, at the bottom of which a white sheet of paper is placed. The paper is lifted gently, carrying the specimen along and placed in a blotter and pressed. As the slender water plant sticks to the paper, the sheet along with the specimen is shifted from one blotter to another during the process of drying, and finally pasted on the herbarium sheet as such.

Succulents and cacti have a large amount of proliferated parenchyma storing water and, unless special care is taken, these plants readily rot and fungal infection sets in. Such plants are handled by giving slits on thick organs and scooping out the succulent tissue or, alternately, salt is sprinkled on the slits to drive out the moisture. The plants may also be killed by pretreatment with ethyl alcohol or formaldehyde.

Mucilaginous plants such as members of the family Malvaceae stick to the blotters and are difficult to process. These plants should be placed between waxed or tissue paper or else folds of muslin cloth. Only the blotter should be changed every time the press is opened and the specimen separated from the tissue paper or muslin only when fully dry.

Aroids and bulbous plants continue to grow even in a press even after they have presumably been properly pressed and dried. These should be killed with ethyl alcohol and formaldehyde prior to pressing.

Drying

Drying of pressed plant specimens is a slow process if no artificial heat is used. Natural drying of specimens is a slow process, which may take up to one month for complete drying. The plants, freshly collected, are placed in a press without corrugated sheets and the press is locked for 24 hours. During this **sweating period**, plants lose some moisture, become flaccid and can be easily rearranged. The folded sheet containing the specimen is lifted and placed in a fresh dry folded blotter. In countries using thick coarse newsprint, changing the newspaper is also necessary, and the plant should be carefully transferred from one newspaper to another. The use of a blotter in such a case can be dispensed with, especially after one or two changes. The change of blotters or newspaper sheets is repeated every few days, increasing the interval between the changes successively until the specimens are fully dry. The whole process of drying may take about 10 days to one month, depending on the specimens and the climate of the area.

Drying with the help of artificial heat takes 12 hours to two days. The specimens, after the initial sweating period in the field press, are transferred to a drying press, with an ample number of corrugated sheets, usually one alternating every folded blotter containing one specimen. The press is kept in a **drier**, a cabinet in which a kerosene lamp or electric bulb warms the air, drying the specimens by movement through the corrugates. Use of a hot air blower in the cabinet speeds up circulation of the hot air and, consequently, faster drying is achieved. Sinnott (1983) developed a **solar powered drier** capable of drying 100 specimens on a sunny day and attaining a temperature of up to 60° Celsius in the center of the press. The unit consists of a flat plate collector and a drying box to hold the press. The collector is composed of a wooden frame, a blackened aluminium absorber plate, insulation and a glass or Plexiglas glazing to retain and channel heat into the drying box. One-inch space is provided between the glazing and the absorber plate. The air enters the collector at the open bottom of the collector panel, is heated by conduction from the absorber, rises by convection into the drying box, moves through the corrugates and finally exits from the uncovered top of the drying box, taking with it moisture from the plant specimens. Drying is accomplished in a single day, occasionally two days for complete drying. This solar drier, with practically no operational cost, should provide a right step towards energy conservation.

The rapid drying of specimens using artificial heat has, however, inherent limitations of rendering plants brittle, loss of bloom and some color change in leaves. In arid regions, plants can be dried partially during travel, by placing the press horizontally on the luggage rack of the vehicle, with the corrugate ducts facing front, forcing the dry wind through the corrugates as the vehicle moves forward. Specimens pressed and dried are next mounted on herbarium sheets, and properly labelled before these can be incorporated in herbarium.

HERBARIUM METHODS

Herbarium is a collection of pressed and dried plant specimens, mounted on sheets bearing a label, arranged according to a sequence and available for reference or study. In practice, it is a name given to a place owned by an institution, which maintains this orderly collection of plant specimens. Most of the well-known herbaria of the world made their beginning from **botanical gardens**.

Botanical Gardens

Although gardens existed in ancient China, India, Egypt and Mesopotamia, these gardens were not botanical gardens in the true sense. They existed for growing food plants, herbs, and ornamentals for aesthetic, religious and status reasons. The famous 'hanging gardens' of Babylon in Mesopotamia is a typical example. The first garden for the purpose of science and education was maintained by Theophrastus in his Lyceum at Athens, probably bequeathed to him by his teacher, Aristotle.

Credit for establishment of the first modern botanical garden belongs to Luca Ghini (ca 1490–1556), a professor of botany who developed it at Pisa, Italy in 1544. These were followed by botanical gardens at Padua and Florence in 1545.

Roles of a Botanical Garden

Botanical gardens have been instrumental in motivating several well-known authors to develop their own **systems** of classification while trying to fit the plants grown in the garden, into some previous system of classification, e.g., Linnaeus, while working at Uppsala and Bernard de Jussieu at Versailles. Although the majority of the botanical gardens house plant species which the climate of the area can support, several well-known botanical gardens have controlled enclosures to support specific plants. Tropical gardens often need indoor growing space, **Screen houses** for most plants and **glasshouses** for most cacti and succulents in wet tropical and temperate gardens. Glasshouses in temperate gardens often require winter heating. Botanical gardens play the following important roles:

1. **Aesthetic appeal:** Botanical gardens have an aesthetic appeal and attract a large number of visitors for observation of general plant diversity as also the curious plants, as for example, the Great Banyan Tree (*Ficus benghalensis*) in the Indian Botanical Garden at Kolkatta.
2. **Material for botanical research:** Botanical gardens generally have a wide range of species growing together and offer ready material for botanical research, which can go a long way in understanding taxonomic affinities.
3. **On-site teaching:** Collection of plants is often displayed according to families, genera or habitats, and can be used for self-instruction or demonstration purposes.
4. **Integrated research projects:** Botanical gardens with rich living material can support broad-based research projects which can integrate information from such diverse fields as anatomy, embryology, phytochemistry, cytology, physiology and ecology.
5. **Conservation:** Botanical gardens are now gaining increased importance for their role in conserving genetic diversity, as also in conserving rare and endangered species. The Proceedings of the Symposium on Threatened and Endangered species, sponsored by New York Botanical Garden in 1976, published as *Extinction is Forever*, and the conference on practical role of botanical gardens in conservation of rare and threatened species sponsored by the Royal Botanical Garden, Kew and published as *Survival and Extinction*, are among the major examples of the role of botanical gardens in conservation.
6. **Seed exchange:** More than 500 botanical gardens across the world operate an informal seed exchange scheme, offering annual lists of available species and a free exchange of seeds.
7. **Herbarium and library:** Several major botanical gardens of the world have herbaria and libraries as an integral part of their facilities and offer taxonomic material for research at a single venue.
8. **Public services:** Botanical gardens provide information to the general public on identification of native and exotic species, methods of propagation and supply plant material through sale or exchange.

Major Botanical Gardens

Thousands of botanical gardens located worldwide are maintained by various institutes. Of these, nearly 800 important gardens are documented in the *International Directory of Botanical Gardens* published by Henderson (1983). A botanical garden today is an area set aside and maintained by an organization for growing various groups of plants for study, aesthetic, conservation, economic, educational, recreational and scientific purposes. Some of the major botanical gardens are discussed below:

New York Botanical Garden, USA: This garden was christened the New York Botanical Garden in 1891, when the Torrey Botanical Club adopted its foundation as a corporation chartered by the State. David Hosak founded the garden in 1801 as Algin Botanic Garden.

Professor N. L. Britton, the most productive taxonomist of his time, directed the idea of advancement of botanical knowledge through research at this botanical garden. The garden (Figure 3.5) today covers 100 ha in the heart of New York City along the Bronx River. In addition, 778 ha Mary Flager Cary Arboretum at Millbrook has been added to the jurisdiction of the garden. There are 15,000 species distributed in the demonstration gardens, Montgomery conifer collection, Stout day lily garden, Havemeyer lilac collection, *Rhododendron* and *Azalea* collection, Everett rock garden, herb garden, rose garden, arboretum and conservatory complex. The garden has a systematic arrangement of trees and shrubs that make it a place of interest for the general public as well as botanists and horticulturists. The garden plays a major role in conservation of rare and endangered species. The garden has a well-maintained herbarium of over 7.8 million specimens from all over the world, but mainly from the New World. The library houses over 550,000 volumes, 9600 pre-1860 books and 4,000 popular books. It also maintains a huge botanical database.

Figure 3.5: Haupt conservatory complex of New York Botanical Garden.

Royal Botanic Gardens, Kew: More popularly known as 'Kew Gardens', this historical garden is undoubtedly the finest botanical garden and botanical research and resource center in the world. The garden was developed in 1759 and William Aiton took over as its superintendent. Sir Joseph Banks introduced large collections from different parts of the world. In 1841, the management of the garden was transferred from the crown to the parliament and Sir William Hooker became its first official director. He was mainly responsible for the advancement of the garden, enlarging it from a mere 6 ha to more than 100 ha and building a palm house. Sir J. D. Hooker, who succeeded his father as its Director, added rhododendrons, and authored several important publications. John Hutchinson worked and developed his famous system of classification here.

The garden (Figure 3.6) has since grown into a premier Research and Educational Institute with excellent herbarium and library. Originally the garden covered an area of 120 ha. The outstation of the Royal Botanic Gardens, Kew at Wakehurst Place near Ardingly in West Sussex is a rural estate of 202 ha with an Elizabethan mansion and was acquired in 1965. The Royal Botanic Gardens Kew has directed and financed its development so that Wakehurst Place now makes a vital contribution in maintaining the international reputation of the Living Collections Department (LCD). In particular the practical *in situ* conservation policies pursued, and the rich and diverse plant collections, which are maintained, add greatly to the LCD's activities. The environmental conditions of the High Weald of Sussex contrast with those of Kew by offering varied topography, higher rainfall and more diverse and moisture retentive soils. These combine together to provide a range of microclimates, which make possible the successful cultivation of a great diversity of plants, many of which do not thrive at Kew. There are substantive differences in the layout and content of the collections at Wakehurst Place which act to complement those at Kew. In particular the botanical collections are laid out in a floristic manner reflecting the way that temperate plant communities have evolved. The botanical collections are supported by extensive ornamental displays exploiting the wide range of available biotopes and acting as primary visitor attractants. A final element of the woodland cover is forestry plots comprising high forest and Christmas tree plantations. Jodrell Laboratory at Kew has established itself as the world centre in the study of plant anatomy, cytogenetics and plant biochemistry.

The Royal Botanic Gardens' Living collections at Kew and Wakehurst Place are a multilevel encyclopaedic reference collection reflecting global plant diversity and providing a reference source which serves all the aspects of botanical and horticultural science within Kew, Great Britain and throughout the world. It is probably the largest and most diverse living collection in the world. The two sites provide quite different environments, allowing the development of two differing but complementary collections. The living collections at Kew are most diverse with 351 families, 5465 genera and over 28,000 species growing successfully. The arboretum covers the greatest area with large mature temperate trees. Tropical plants are maintained indoors, including Aroid House, Palm House, Filmy Fern House, etc. Several interesting plants such as *Victoria amazonica* from South America and *Welwitschia mirabilis* from Angola are also growing here. Kew Herbarium, undoubtedly the most famous herbarium of the world maintains over 7 million specimens of vascular plants and fungi from every country in the world. There are over 350,000 type specimens as well.

The library at Kew is very extensive with over 300,000 printed volumes, 5,000 journal titles and 20,000 maps. *Kew Bulletin* and *Index Kewensis* (now incorporated in IPNI) are its premier publications.

Kew maintains databases on plant names, taxonomic literature, economic botany, plants for arid lands and on plant groups of special economic and conservation value. Kew also makes about 10,000 identifications a year through its Herbarium service and provides specialist advice on taxonomy and nomenclature in difficult cases. Kew is involved in major biodiversity research programmes in many parts of the world, including tropical and West Asia, SE Asia, Africa, Madagascar, South America, and the Pacific and Indian Oceanic islands. The Herbarium runs an international Diploma Course in Herbarium Techniques. The General Catalogue now contains over 122,000 records and is available throughout RBG Kew on the network.

Figure 3.6: Princess of Wales House at Royal Botanic Gardens Kew.

Missouri Botanical Garden, USA: Considered one of the top three botanical gardens in the world, the Missouri Botanical Garden is a National Historical Landmark and a centre for botanical research, education and horticultural display. The garden was founded by an Englishman Henry Shaw and opened to public in 1859 with active help from Asa Gray and Sir William Hooker and Enelmann. Today, the garden covers 79 acres and operates the world's most active tropical botany research programme. Under the leadership of Dr Peter Raven, its former Director, the Garden played a leading role in strategies of conservation and sustainable living. The garden is known for its **Climatron®** conservatory, a geodesic greenhouse dome with climatic control, supporting a vibrant tropical rainforest, under a 0.5 acre roof (Figure 3.7). It also has a **Japanese Garden** (*Seiwa-en*) covering 14 acres, the largest Japanese strolling Garden in North America, with a proud collection of *Hamerocallis*, *Iris*, roses, *Hosta*, and several economic plants (Figure 3.8). There are also Chinese, English German and Victorian Gardens. Over 4,000 trees thrive on the grounds, including some rare and unusual varieties.

The Missouri Botanic Garden is one of the world's leading research centres for botanical exploration and research, with nearly 25 major flora projects. The information is shared via website **TROPICOS,** the world's largest database, containing more than 920,000 scientific plant names and over 1,800,000 specimen records. The garden's highly regarded education programme seeks to improve science instruction in the St. Louis region, reaching more than 137,000 students each year.

With more than 6.6 million specimens (mosses, ferns, gymnosperms, and angiosperms), the herbarium ranks second in the USA and 5th in the world. It has collections dating back to mid-1700s. The herbarium specializes in having collections of G. Boehmer, Joseph Banks, D. Solander (who accompanied Captain James Cook in his first voyage around the world), and Charles Darwin. During the last five years, the herbarium has added an average of 120,000 mounted specimens per year to its collection. In addition to the many gift specimens sent to the specialists, this herbarium loans an average of 34,000 specimens annually, and borrows about 27,000 specimens. The herbarium staff also provides identifications from their area of expertise. The pace of development of the herbarium can be judged from the fact from being number 13th in the world in 1990 (Woodland, 1991), the herbarium today has risen to number five. The reference library of the garden has over 220,000 volumes, including many rare books.

Among its major research activities include *Flora of North America* project, 17 volumes having already been published, covering the plants of USA, Canada and Greenland. The garden also coordinates *Flora of China* project, 25 volume publication completed in 2009.

Pisa Botanical Garden, Italy: The Pisa Garden, developed by Luca Ghini in 1544, is credited as the first modern botanical garden. The garden was known for the finest specimens of *Aesculus hippocastanum, Magnolia grandiflora*, and several other species. Though the garden does not exist today, the records of its design demonstrate geometric outlay of plantings that are characteristic of several continental gardens even today.

Padua Botanical Garden, Italy: The garden is a contemporary of Pisa Botanical Garden, established in 1545. The specialty of this garden is the elegance and Halian taste, which has been wedded to the service of science. The elegance and beauty of Padua Botanical Garden are equalled by Kew Gardens only.

Berlin Botanic Garden and Museum, Berlin-Dahlem: The Berlin Botanic Garden was set up in 1679 when the Grand Duke of Berlin gave instructions to open an agricultural model garden in Schoneberg, a village near Berlin. Due to lack of space, it was later relocated to Dahlem. The garden developed largely due to the efforts of C. L. Wildenow, who built it up from an old rundown royal garden. Adolph Engler and L. Diels, who were its subsequent directors, improved its quality and content. Much of the garden was destroyed during World War II. It was rebuilt largely through the efforts of Robert Pilger, its then director.

Figure 3.7: Climatron® at the Missouri Botanical Garden, a Geodesic dome with Climatic control and supporting tropical rainforest. (Photograph by Jack Jennings/Courtesy the Missouri Botanical Garden.)

Figure 3.8: Japanese Garden at the Missouri Botanical Garden. (Photograph by Jack Jennings/Courtesy the Missouri Botanical Garden.)

The botanical garden today comprises an area of 126 acres. About 20,000 different species of plants are cultivated here. The section on plant geography covering 39 acres, one of the biggest of its kind in the world, depicts the whole of the Northern Hemisphere. The arboretum and taxonomy section covers 42 acres and includes around 1800 species of trees and shrubs and nearly 1000 species of herbaceous plants, the latter arranged according to the classification system of Adolph Engler. The botanical museum specializes in the display of botanical exhibits, being the only museum of its kind in Central Europe with models of various life-forms.

The main tropical greenhouse (Figure 3.9), with its length of 60 m and height of 23 m, is one of the largest in the world, featuring tall trees with epiphytes, rich ground vegetation and lianas, which give an idea of the vast variety of tropical vegetation.

Cambridge University Botanical Garden: The Cambridge University Botanical Garden was founded in 1762 as a small garden on 5 acres of land in the centre of Cambridge. It was moved to the present location in 1831 when Prof. J. S. Henslow established it on newly-acquired land of the University covering 40 acres. The garden is artistically landscaped with systematic plantings, winter-hardy trails, an alpine garden and a chronological bed. The latter is in the form of a

Figure 3.9: Tropical greenhouse of the Berlin Botanic Garden at Dahlem.

narrow bed (300 × 7 feet) divided into 24 sections, each containing plants introduced during a 20-year period. Tropical houses are one of the major attractions of the garden and contain palms and other tropical plants.

Herbaria

It was again Luca Ghini who initiated the art of herbarium making by pressing and sewing specimens on sheets of paper. This art was disseminated throughout Europe by his students who mounted sheets and bound them into book volumes.

Although the herbarium technique was a well-known botanical practice at the time of Linnaeus, he departed from the convention of mounting and binding the specimens into volumes. He mounted specimens on single sheets, storing them horizontally, a practice followed even today.

From isolated personal collections, herbaria have grown into large institutions of national and international stature with millions of specimens from different parts of the world. *Index Herbariorum,* edited by Patricia Holmgren (Figure 3.10) (Holmgren et al., 1990) lists the world's important herbaria. Each herbarium is identified by an abbreviation that is valuable in locating the type specimens of various species. Records of *Index Herbariorum* estimate 3,001 active herbaria in the world with an estimated 387,007,790 specimens. The major herbaria of the world with approximate number of specimens in the order of importance are listed in the Table 3.1.

In India Central National Herbarium (CAL) of the Indian Botanic Garden, Botanical Survey of India, Kolkatta has over 1.3 million specimens. The herbarium of Forest Research Institute, Dehradun (DD) and National Botanical Research Institute, Lucknow (LUCK) are other major herbaria in India, with collections from all over the world.

Roles of a Herbarium

From a safe place for storing pressed specimens, especially type material, herbaria have gone a long way in becoming major centres of taxonomic research. Additionally, herbaria also form an important link for research in other fields of study. The classification of the world flora is primarily based on herbarium material and associated literature. More recently, the herbaria have gained importance for sources of information on endangered species and are of primary interest to conservation groups. The major roles played by a herbarium include:

1. **Repository of plant specimens:** Primary role of a herbarium is to store dried plant specimens, safeguard these against loss and destruction by insects, and make them available for study.
2. **Safe custody of type specimens:** Type specimens are the principal proof of the existence of a species or an infraspecific taxon. These are kept in safe custody, often in rooms with restricted access, in several major herbaria.
3. **Compilation of floras, manuals and monographs:** Herbarium specimens are the 'original documents' upon which the knowledge of taxonomy, evolution and plant distribution rests. Floras, manuals and monographs are largely based on herbarium resources.
4. **Training in herbarium methods:** Many herbaria carry facilities for training graduates and undergraduates in herbarium practices, organizing field trips and even expeditions to remote areas.

Figure 3.10: Patricia K. Holmgren Director Emerita of the Herbarium, New York Botanical Garden, the editor of *Index Herbariorum* and 2 volumes of *Intermountain Flora.* (Courtesy New York Botanical Garden, Bronx.)

Figure 3.11: Herbarium cabinet with filed specimens used in New York Botanical Garden Herbarium. (Photograph courtesy New York Botanical Garden.)

Table 3.1: Major herbaria of the world listed in the order of number of specimens.

Herbarium	Abbreviation	Number of specimens
1. Museum National d'Histoire Naturelle (Museum of Natural History), Paris, France*	P, PC	10,000,000
2. New York Botanical Garden, New York, USA.	NY	7,800,000
3. Komarov Botanic Institute, Saint Petersburg (Formerly Leningrad), Russia.	LE	7,180,000
4. Royal Botanic Gardens, Kew, Surrey, UK.	K	7,000,000
5. Missouri Botanical Garden, Saint Louis, Missouri, USA.	MO	6,600,000
6. Conservatoire et Jardin Botaniques (Conservatory and Botanical Garden), Geneva, Switzerland.	G	6,000,000
7. British Museum of Natural History, London, UK.	BM	5,200,000
8. Combined Herbaria, Harvard University, Cambridge, Massachusetts, USA**	A, FH, GH, ECON, AMES	5,000,000
9. Natural History Museum, Vienna, Austria.	W	5,000,000
10. US National Herbarium (Smithsonian), Washington, USA.	USA	5,000,000

Information updated based on Index Herbariorum records December 2017.

* Specimens are located at Laboratoire de Phanérogamie (P) and Laboratoire de Cryptogamie (PC).

** Consists of Arnold Arboretum (A), Farlow Herbarium (FH), Gray's Herbarium (GH), The Economic Herbarium of Oakes Ames (ECON) and Oaks Ames Orchid Herbarium (AMES).

5. **Identification of specimens:** The majority of herbaria have a wide-ranging collection of specimens and offer facilities for on-site identification or having the specimens sent to the herbarium identified by experts. Researchers can personally identify their collection by comparison with the duly identified herbarium specimens.
6. **Information on geographical distribution:** Major herbaria have collections from different parts of the world and, thus, scrutiny of the specimens can provide information on the geographical distribution of a taxon.
7. **Preservation of voucher specimens:** Voucher specimens preserved in various herbaria provide an index of specimens on which a chromosomal, phytochemical, ultrastructural, micromorphological or any specialized study has been undertaken. In the case of a contradictory or doubtful report, the voucher specimens can be critically examined in order to arrive at a more satisfactory conclusion.

Mounting of Specimens

Pressed and dried specimens are finally mounted on herbarium sheets. A standard herbarium sheet is **29** by **41.5** cm (**11½** by **16½ inches**), made of thick handmade paper or a card sheet. The sheet should be relatively stiff to prevent damage during handling of specimens. It should have a high rag content (preferable 100 percent) with fibres running lengthwise.

The specimens are attached to the sheet in a number of ways. Many older specimens in the herbaria are frequently found to have been sewn on the sheets. Use of adhesive linen, paper or cellophane strips is an easier and faster method of fixing specimens. **Archer method** involves the use of small strips of liquid plastic extruded from a container with a narrow nozzle. Most of the contemporary specimens are fixed using liquid paste or glue in one of the two ways, however:

(i) Paste or glue is applied to the backside (if distinguishable) of the specimen, which is later pressed onto the mounting sheet and allowed to dry in the pressed condition for a few hours. This method is slower but more economical.

(ii) Paste or glue is smeared on a glass or plastic sheet, the specimen placed on the sheet and the glued specimen transferred to a mounting sheet. This method is more efficient but expensive.

The use of methylcellulose as adhesive mixed in a solution of 40% alcohol, instead of pure water was suggested by Tillet (1989) for fixing herbarium specimens. It decreases the drying time and prevents growth of micro-organisms. The stem and bulky parts may often require adhesive strips or even sewing for secure fixing of specimens. Small paper envelops called **fragment packets** are often attached to the herbarium sheet to hold seeds, extra flowers or loose plant parts.

Labelling

An **herbarium label** is an essential part of a permanent plant specimen. It primarily contains the information recorded in the **field diary** (**Field notebook**) at the time of collection, as also the results of any subsequent identification process. The label is located on the lower right corner of the herbarium sheet (Figure 3.12), with the necessary information recorded on the pre-printed proforma, printed directly on the sheet or on the paper slips which are pasted on the sheets. It is ideal to type the information. If handwritten, it should be in permanent ink. Ball pens should never be used, as the ink often spreads after some years.

There is no agreement as to the size of a herbarium label, the recommendations being as diverse as **2¾** by **4¼** inches (Jones and Luchsinger, 1986) and **4** by **6** inches (Woodland, 1991). The information commonly recorded on the herbarium label includes:

Name of the institution
Scientific name
Common or vernacular name
Family
Locality
Date of collection
Collection number
Name of the collector
Habit and habitat including field notes

An expert visiting a herbarium may want to correct an identification or record a name change. Such correction is never done on the original label but on a small **annotation label** or **determination label**, usually **2 by 11** cm and appended left of the original label. This label, in addition to the correction, records the name of the person and the date on which the change was recorded. Such information is useful, especially when more than one annotation label is appended to a herbarium sheet. The last label is likely to be the correct one.

Voucher herbarium specimens of a research study often have authentic information about the specimens recorded in the form of a **voucher label**.

Figure 3.12: A **sample herbarium sheet** with mounted specimen and a label.

Filing of Specimens

Mounted, labelled and treated (to kill insect pests) specimens are finally incorporated in a herbarium, where they are properly stored and looked after. Small herbaria arrange specimens alphabetically according to family, genus and species. Larger herbaria, however, follow a particular system of classification. Most herbaria usually follow Bentham and Hooker (British herbaria and most commonwealth countries) or Engler and Prantl (Europe and North America). Many herbaria of the latter category follow the number **code** of families and genera given by Dalla Torre and Harms (1900–1907).

The specimens belonging to a species are placed in a folder made of thin strong paper, termed **species cover**. The species covers belonging to a particular genus are often arranged alphabetically and placed inside a **genus cover,** a heavy manilla folder made of a thicker paper. More than one genus cover may be used if the number of species are more, or if the specimens are to be arranged geographically, and often differently colored for different geographical regions.

The genus covers of a family are arranged according to the system of classification being followed. The demarcation between the two families (last genus of a family and first genus of the next family) is done using a sheet of paper with a **front-hanging label,** indicating the name of the next family. The folders are stacked in pigeonholes of the herbarium cases and the arrangement is suitable for shifting of folders as the number of specimens increase with time.

Unknown specimens are kept in separate folders marked **dubia,** placed towards the end of a genus (when the genus is identified) or a family (when the family is identified but not the genus) so that an expert can examine them conveniently. Standard **herbarium cases** are insect- and dust-proof with two or more tiers of pigeonholes, each 19 in deep, 13 in wide and 8 in high (Figure 3.11).

Type specimens are usually kept separately in distinct folders or often in separate herbarium cases, sometimes even separate rooms, for better care and safety.

Herbarium commonly maintains an **index register** in which all the genera in the herbarium are listed alphabetically and against each genus is indicated family number and the genus number, the two help in convenient incorporation and retrieval of specimens in a herbarium.

Pest Control

Herbarium specimens are generally sufficiently dry, and as such not attacked by bacteria or fungi. They are, however, easily attacked by pests such as **silverfish**, **dermestid beetles** (cigarette beetle, drugstore beetle and black carpet beetle). Control measures include:

1. **Treating incoming specimens:** Specimens must be pest free before they can be incorporated into herbarium. This is achieved in three ways:

   (i) ***Heating*** at temperatures up to 60ºC for 4–8 hours in a heating cabinet. The method is effective but the specimens become brittle.

(ii) ***Deep-freezers*** have now replaced heating cabinets in most herbaria of the world. A temperature of minus 20 to minus 60°C is maintained in most herbaria.

(iii) ***Microwave ovens*** have been used by some herbaria, but as indicated by Hill (1983), the use of microwave ovens has some serious shortcomings including:

(a) Stems containing moisture burst due to sudden vaporization of the water inside.

(b) Metal clips, staples on the sheets get overheated and may char the sheet.

(c) The embryo in the seed gets killed, thus destroying a valuable source of experimental research, as seeds from herbarium specimens are often used for growing new plants for research projects.

2. **Use of repellents:** Chemicals with an offensive odor or taste are kept in herbarium cases to keep pests away from specimens. **Naphthalene** and **Paradichlorobenzene** (PDB) are commonly used repellents, usually powdered and put in small muslin bags kept in pigeonholes. PDB is more toxic and as such prolonged exposure of workers should be avoided. For people working 8 hours a day in a 5 day per week schedule, the upper exposure level for naphthalene is 75 PPM and for PDB 10 PPM.

3. **Fumigation:** Despite pre-treatment of specimens and the use of repellents, fumigation is necessary for proper herbarium management. Fumigation involves exposing specimens to the vapours of certain volatile toxic substances. A mixture of **ethylene dichloride** (3 parts) and **carbon tetrachloride** (1 part) was once commonly used for fumigation. Ethylene dichloride is explosive without carbon tetrachloride, but the latter is extremely toxic to humans, causing liver damage, and as such the use of this fumigant has been banned. Some herbaria also use Ethylene bromide, Ethylene oxide, Lauryl pentachlorophenate (Mystox), Methyl bromide (often used synergistic with carbon dioxide) or Malathion for fumigation, but these are also toxic for humans.

   Integrated Pest Management program (IPMP) for safer herbarium use emphasises prevention of insect entering the herbarium, rather than relying on toxic chemicals to kill them once inside the herbarium. It also encourages the use of:

   (a) Deep freezing not only to treat incoming specimens but also to kill insects in the herbarium. The bundles of specimens are placed in plastic bag, the bag sealed, and the bundle placed in freezer for several days. The bag (sealed) is next left at room temperature for 7–10 days to allow any eggs to hatch and then refrozen for 3–4 days.

   (b) **Anoxic treatment**, involves the usage of bag impermeable to oxygen and containing specimens and a oxygen scavenger. The pest is killed by depriving it of oxygen.

   (c) Small **sticky trap** placed in hidden areas of the herbarium and herbarium cases to trap insects. Such traps should be checked regularly for insects trapped.

   (d) Pheromone traps involve the use of natural scents which insects use to communicate with each other. Certain insects are attracted to these traps from the surrounding area and are very effective. Specific traps are available for drug store beetles, Indian meal moth, cigarette beetles and warehouse beetles.

   (e) Insect electrocuters are useful for detecting and controlling flying insects. These emit ultraviolet light that attracts flying insects particularly flies and moths.

Where fumigation is essential use of safer fumigants like **Pyrethrin**, **sulphuryl fluoride** (Vikane), **dichlorovos** (no pest, Vapona resin strips or Raid strips are suitable for herbarium cases; one-third of a strip is placed in each herbarium case for seven to ten days twice a year), **carbon dioxide** and **cyanogen** (often used synergistic with carbon dioxide). **Dowfume-75** has been cleared by the Environmental Protection Agency for use in herbaria.

Virtual Herbarium

Virtual herbarium is a database of consisting of images of Herbarium specimens and the supporting text, available over the internet. It is a huge advancement in herbarium use and design, coupling physical specimens directly with internet and integrating complete specimen data, with resources or information generation and retrieval. Although a virtual herbarium cannot exist without a physical herbarium, it enjoys several advantages over a physical herbarium:

1. Images being available electronically, user may not have to handle physical specimens, thus reducing the damage substantially.
2. Whereas it may take months to sort out specimens of a collector or a country in a physical herbarium, the same can be done in few seconds through a virtual herbarium.
3. Virtual herbarium greatly increases the user interaction. Only few hundred visitors may visit a physical herbarium in a month, but during the same period thousands of users can access the virtual herbarium sitting in the comfort of their offices.

4. Physical herbarium usually stores only specimens, and the user has to spend considerable time in the library to collect relevant information. A virtual herbarium on the other hand provides information on descriptive details, geographical distribution, photographs, illustrations, manuscripts, published work, microscopic preparations, gene sequences and nomenclature through hyperlinks.
5. Physical herbarium can offer only own specimens for study, whereas portals of major virtual herbaria offer facility of searching several virtual herbaria simultaneously.
6. Physical specimens are prone to damage through handling or during hazardous situation. Thousands of specimens and valuable holotypes preserved in Berlin herbarium were destroyed during second the World War. Digitized images, on the other hand can be saved on several computers, at different locations.
7. Majority of research projects don't need physical specimens, and as such electronic images can be utilized, saving the time and cost for transportation of actual specimens.

Virtual herbaria with searchable database have been developed by many major organisations like New York Botanical Garden (KE EMu), Royal Botanic Gardens Melbourne (AVH), Fairchild Tropical Garden (E-FTG), Australian Virtual Herbarium (AVH) and Royal Botanic Gardens, Kew (ePIC).

Though initiated in 1990, the NYpc project of New York Botanical Garden became operational in 1995. The data was transferred to new platform KE EMu in 2004 with additional search and display capabilities. The Virtual herbarium presently consists of digital collection of 850,000 herbarium specimens and 120,000 high resolution images, updated daily. Garden persues the goal of digitizing all of its 7 million plant and fungi specimens.

Fairchild Tropical Garden Virtual Herbarium (eFTG) has record of more than 100,000 specimens, more than 200,000 photographs (including data labels). There are more than 20,000 high resolution photographs of specimens, that can be zoomed in or out of the browser. Nearly 60,000 records are searchable online by family, genus, collector and other fields. eFTG is the first truly Virtual Herbarium as Web portal of the herbarium allows simultaneous search through Virtual Herbaria of FTG (Fairchild Tropical Garden), FLAS (Florida Museum of Natural History), MO (TROPICOS-Missouri Botanical Garden), NY (Cassia- New York Botanical Garden), S (Linnean Herbarium, Swedish Museum of Natural History, Leiden), BM (British Museum of Natural History-including Clifford Herbarium), CAYM (National Trust for Cayman Islands), INB (Instituto Nacional de Biodiversidad, Costa Rica), TAMU (Texas A & M University). The virtual herbarium of FTG thus affectively includes not only specimens from Fairchild Herbarium, but also from other herbaria. It also provides species lists, interactive keys and photographs of living specimens in various databases and indices.

Australia's Virtual Herbarium (AVH) is a collaborative project of the State, Commonwealth and Territory herbaria, being developed under the auspices of the Council of Heads of Australian Herbaria (CHAH), representing the major Australian collections. It is an on-line botanical information resource accessible via the web, providing immediate access to the wealth of data associated with scientific plant specimens in each Australian herbarium.

Australian herbaria house over six million specimens that are a primary source of information on the classification and distribution of plants, algae and fungi. These specimens are the working tools of scientists who contribute to our knowledge and understanding of biodiversity and conservation through the discovery, classification and description of new species. These will be enhanced by images, descriptive text and identification tools.

The AVH is accessed via the website of any participating herbarium. A gateway at each of these herbaria links to the databases of all the other herbaria, consolidating the combined data into a nation-wide view of the botanical information. Most data related to specimens will be stored by the custodial institution, and there will be some resources, such as the scientific names database (Australian Plant Names Index, APNI) which will be common to all. More than 70% of the specimens housed in Australian herbaria have been databased, providing a comprehensive resource for accurate depiction of geographic distribution and occurrence, historical mapping, information valuable for understanding the threatening processes of vegetation clearance and weed invasion. Flexible on-line search options allow you to customize the data you generate to suit your requirements.

Australia's Virtual Herbarium provides the opportunity to deliver descriptions of the flora dynamically linked to data and information from across the continent and distributed on-line as an electronic Australian Flora—a one-stop source of current information on the plants, algae and fungi of the entire Australian continent. New observations can be released with minimal delay as they are confirmed and recorded in the database.

The Strong ePIC database software of Royal Botanic Gardens, Kew also provides a window for digitized herbarium specimens. The Herbarium's core digital collection programme was initiated in 2002 and since then digital resources have grown at an increasing rate, as well as central Herbarium Catalogue, have an image server and many project databases with information about specimens that were built before the Catalogue became available. These are being moved into the Catalogue as resources permit. Label data from dry and spirit specimens of flowering plants, ferns and gymnosperms held in Kew's herbarium are being uploaded. Information recorded includes the plant name, collection and determination data, locality and type status. Digitization is proceeding and as of February 2019, digitized specimens number 577,156 including 170,760 types. A potential c7,000,000 specimens (c275,000 types) may eventually be digitized.

IDENTIFICATION METHODS

Identification of an unknown specimen is a common taxonomic activity, and often combined with determination of a correct name. The combined activity is appropriately referred as **specimen determination**. Before the specimen can be identified, it is desirable to describe it and prepare a list of character-states, mainly pertaining to floral structure. Whereas fresh specimens may be described more conveniently, the dried specimens may be softened by immersing in water or a **wetting agent** such as **aerosol OT** (dicotyl sodium sulfosuccinate 1 per cent, distilled water 74 per cent and Methanol 25 per cent). **Pohl's softening agent** is an excellent detergent solution for softening flowers and fruits for dissection. The identification of an unknown plant may be achieved by comparison with identified herbarium specimens or through the help of taxonomic literature. Both methods may be combined for a more reliable identification.

The unknown specimen meant for identification is sent to a herbarium, where an expert on the plant group examines and identifies it by comparison with duly identified specimens (Figure 3.13). The user can also visit a herbarium and personally compare and identify his specimens.

Computers have entered in a big way into solving identification problems. Electronic revolution in recent years has opened up a new, faster and more reliable method of identification. The photograph, description or illustration of parts can be put up on a website, with information to a relevant **e-mail list**, whose members can help in achieving identification within hours.

Figure 3.13: The **researchers comparing specimens** inside the herbarium at the Missouri Botanical Garden. (Photograph by Jack Jennings/Courtesy the Missouri Botanical Garden.)

Taxonomic Literature

Various forms of literature incorporating description, illustrations and identification keys are useful for proper identification of unknown plants. The library is, therefore, as important in taxonomic work as a herbarium, and knowledge of taxonomic literature is vital to the practicing taxonomist. The literature of taxonomy is one of the oldest and most complicated literatures of science. Several **bibliographic references**, **indexes** and **guides** are available to help taxonomists to locate relevant literature concerning a taxonomic group or a geographical region. The major forms of literature helpful in identification are described below.

Floras

A **Flora** is an inventory of the plants of a defined geographical region. A Flora may be fairly exhaustive or simply synoptic. Lists of the Floras may be found in the *Geographical Guide to the Floras of the World* by S. F. Blake (Part I, 1941; Part II, 1961) and *Guide to the Standard Floras of the World* by Frodin (1984, 2nd edition 2009). Depending on the scope and the area covered, the Floras are categorized as:

1. **Local Flora** covers a limited geographical area, usually a state, county, city, a valley or a small mountain range. Examples: *Flora of Delhi* by J. K. Maheshwari (1963), *Flora Simlensis* by H. Collet (1921), *Flora of Tamil Nadu* by K. M. Mathew (1983), *Flora of Missouri* by J. A. Steyermark (1963) and *Flora of Central Texas* by R. G. Reeves (1972).

2. **Regional Flora** includes a larger geographical area, usually a large country or a botanical region. Examples: *Flora of British India* by Sir J. D. Hooker (1872–97), *Flora Malesiana* by C. G. Steenis (1948), *Flora Iranica* by K. H. Rechinger (1963), *Flora of Turkey and East Aegean Islands* by P. H. Davis and *Flora SSSR* by V. L. Komarov and B. K. Shishkin (1934–64). A Flora covering a country is more appropriately known as a **National Flora**.
3. **Continental Flora** covers the entire continent. Examples: *Flora Europaea* by T. G. Tutin et al. (1964–80) and *Flora Australiensis* by G. Bentham (1863–78).
4. **Comprehensive treatments** have a much broader scope. Although no world Flora has ever been written, several important works have attempted a worldwide view. Examples: *Genera plantarum* of G. Bentham and J. D. Hooker (1862–83), *Die Naturlichen pflanzenfamilien* of A. Engler and K. A. Prantl (1887–1915) and *Das Pflanzenreich* of A. Engler (1900–54).

Electronic Floras (eFloras)

Last few years have seen the online availability of digitized form of many popular floras. These Online Floras known as Electronic Floras (eFloras) provide opportunity for users to work dynamically on floristic treatments, and to browse and search these treatments. One such effort by Missouri Botanical Gardens has resulted in the publication of www.eFloras.org/, combining together the information from several Floras including Flora of Chile, Flora of China, Flora of Missouri, Flora of North America, Flora of Pakistan, Moss Flora of China, Trees and Shrubs of Andes and Ecuador, as also the Annotated Checklist of Flowering Plants of Nepal. These Floras can be searched through common search engine to obtain relevant information. The hyperlinks to families, genera and species are very handy in identification and retrieving information. The website also hosts the interactive **Actkey** provided by the Harvard University Herbarium, allowing visitors to locate and use a key for identifying an unknown specimen. The keys for Families of Angiosperms by Bertel Hansen & Knud Rahn, Families of Dicotyledons of the Western Hemisphere South of the United States, Generic Tree Flora of Madagascar, Key to Taxa of China in Ackey, Trees and Shrubs of Borneo, and Weeds of Rain Fed Lowland Rice Fields of Laos and Cambodia are already incorporated in a user friendly interface.

Royal Botanical Gardens Kew has hosted eFlora *Flora Zambesiaca* providing not only an easy way of searching the information but also an identification tool. This web site allows you to search for a plant name across the whole Flora, which would otherwise entail looking up separate indexes. It also allows the creation of lists of: endemics, species from a particular division or country, species that match a particular habit and of species that occur at a specific altitude. As far as possible no changes have been made to the information existing in the original text and the information is presented in the same way as in the original.

Manuals

A **manual** is a more exhaustive treatment than a Flora, always having keys for identification, description and glossary but generally covering specialized groups of plants. Examples: *Manual of Cultivated Plants* by L. H. Bailey (1949), *Manual of Cultivated Trees and Shrubs Hardy in North America by* A. Rehder (1940) and *Manual of Aquatic Plants* by N. C. Fassett (1957).

A manual differs from a monograph in the sense that the latter is a detailed taxonomic treatment of a taxonomic group.

Monographs

A **monograph** is a comprehensive taxonomic treatment of a taxonomic group, generally a genus or a family, providing all taxonomic data relating to that group. Usually the geographical scope is worldwide since it is impossible to discuss a taxon without including all its members, and often all its species, subspecies, varieties and forms are discussed. The monograph also includes an exhaustive review of literature, as also a report on author's research work. A monograph includes all information related to nomenclature, designated types, keys, exhaustive description, full synonymy and citation of specimens examined. Examples: *The Genus Pinus* by N. T. Mirov (1967), *The Genus Crepis* by E. B. Babcock (1947), *A Monograph of the Genus Avena* by B. R. Baum (1977), *The Genus Datura* by A. F. Blakeslee et al. (1959) and *The Genus Iris* by W. R. Dykes (1913).

A **revision** is less comprehensive than a monograph, incorporating less introductory material and including a synoptic literature review. A revision includes a complete synonymy, but the descriptions are shorter and often confined to diagnostic characters. The geographical scope is usually worldwide.

A **conspectus** is an effective outline of a revision, listing all the taxa, with all or major synonyms, with or without short diagnosis and with a brief mention of the geographical range. *Species plantarum* of C. Linnaeus (1753) is an ideal example. A **synopsis** is a list of taxa with much abbreviated diagnostic distinguishing statements, often in the form of keys.

Icones (Illustrations)

Illustrations, often with detailed analysis of the parts are usually published along with the text in Floras and Monographs but may sometimes be compiled exclusively and often serve as useful tools for identification. In fact, many species of plants based on published illustrations only, without any accompanying description or diagnosis before 1 January 1908 have been accepted as validly published. Two principal compilations of Icones are *Hooker's Icones* and *Wight's Icones*. Others of interest include Illustrations of plants from Europe (Hegi, 1906–1931), North America (Gleason, 1963), Pacific states (Abrams, 1923–1960), Pacific coast trees (McMinn and Maino, 1946), Germany (Garcke, 1972), Korea (Lee, 1979).

Journals

Whereas Floras, manuals and monographs are published after a lot of taxonomic input and it may take several decades before they are revised, if at all, taxonomic journals provide information on the results of ongoing research. A continuous update on additional taxa described or reported from a region, nomenclatural changes and other taxonomic information is essential for continuance of taxonomic activity. Reference to a publication in a journal includes volume number (all issues within a year bear the same volume number; trend not followed by a few journals), issue number (numbered within a volume, a monthly journal would have 12 issues, quarterly 4 issues and so on) and page numbers on which a particular article appears. Common journals devoted largely to taxonomic research include:

Taxon—The journal of the International Association for Plant Taxonomy devoted to systematic and evolutionary biology with emphasis on botany; published quarterly by the International Bureau for Plant Taxonomy and Nomenclature, Botanisches Institut der Universitaet Wien, Austria.

Kew Bulletin—International peer-reviewed Journal of Plant Taxonomy; published in four parts in one year by Royal Botanic Gardens, Kew; containing original articles of interest mainly to vascular plant and mycological systematists; each part illustrated with line drawings and photographs and also features a Book Review and Notices section.

Plant Systematics and Evolution—Published by Springer, Wien from 1974 onwards; originally started in 1851 under the name *Österreichisches Botanisches Wochenblatt* was published between 1958 to 1973 as *Österreichische Botanische Zeitschrift*; devoted to publishing original papers and reviews on plant systematics in the broadest sense, encompassing evolutionary, phylogenetic and biogeographical studies at the populational, specific, higher taxonomic levels; taxonomic emphasis is on green plants; volumes each with four numbers published randomly, usually 6–7 volumes in one year.

Botanical Journal of Linnaean Society—Published on behalf of Linnean Society by Blackwell Synergy, London; three volumes with four monthly issues each published in one year; publishes original research papers in the plant sciences.

Adansonia—Published by Muséum national d'Histoire naturelle (Museum of Natural History), Paris; a peer-reviewed journal of plant biology, devoted to the inventory, analysis and interpretation of vascular plants biodiversity; publishes original results, in French or English, of botanical research, particularly in systematics and related fields; two issues appear each year. **Adansonia** continues as from 1997 the *Bulletin du Muséum national d'Histoire naturelle, section B, Adansonia, Botanique, phytochimie*.

Other important taxonomic journals include *Journal of the Arnold Arboretum* (Harvard), *Bulletin Botanical Survey of India* (Calcutta), *Botanical Magazine* (Tokyo) and *Systematic Botany* (New York).

Supporting Literature

With a large amount of research material being published throughout the world, there is always need for supporting literature to give consolidated information about the works published the world over. They also help in tracking down material concerning a particular taxon covering a certain period. *Taxonomic Literature,* an exhaustive series of *Regnum vegetabile,* covers full bibliographical details of literature extremely helpful in searching type material, priority of names, dates of publication and biographic data on authors. Originally published in 1967, it is under constant revision with 3 supplements of the 2nd edition published between 1992–97 (Stafleu and Mennega).

Abstracts or **Abstracting journals:** These provide a summary of different articles published in various journals throughout the world. *Biological Abstracts* and *Current Advances in Plant Science* are more general in approach. The *Kew Record of Taxonomic Literature* covers all articles relevant to taxonomy.

Index: An Index provides an alphabetic listing of taxa with reference to their publication. *Index Kewensis* is by far the most important reference tool, first published in 2 volumes from Royal Botanic Gardens, Kew (1893–1895), covering names of species and genera of seed plants published between 1753 and 1885. Regular *Supplements* used to be published every 5 years and 18 *Supplements* appeared up to 1985. Supplement 19 was published in 1991 covering the years 1986 to 1990. Since then the listing has been published annually under the title *Kew Index*.

Index Kewensis (Figure 3.14) is a list of new and changed names of seed-bearing plants with bibliographic references to the place of first publication. At the beginning of the nineteen eighties the data was transferred to a computer database which continues to expand at the rate of approximately 6000 records per year. To make this data generally available, it was decided to publish the whole *Index.*

Kewensis as a CD-ROM in 1993. This contains almost 968,000 records. Illustrations of vascular plants can be located through *Index Londinensis*, which contains information up to 1935. More recent information can be found in the 2-volume work *Flowering Plant Index of Illustrations and Information* compiled by R. T. Isaacson (1979).

A listing of all generic names can be found in *Index Nominum Genericorum* (ING) a 3-volume work published in 1979 under the series *Regnum Vegetabile*. The first supplement appeared in 1986. It has now been put on the database and can be directly accessed through the Internet.

Index Holmiensis (earlier *Index Holmensis*) is an alphabetic listing of distribution maps found in taxonomic literature of vascular plants. It commenced publication in 1969.

Gray Herbarium Card Index is information on cards, which has now been set up on a database. Usually on the same pattern as *Index Kewensis*, the Index has been published in 10 volumes between 1893 and 1967. A 2-volume *supplement* was published by G. K. Hall in 1978. The *Gray Herbarium Index* Database currently includes 350,000 records of New World vascular plant taxa at the level of species and below. The Index includes from its 1886 starting point, the names of plant genera, species and all taxa of infraspecific rank. The Gray Index has in common with *Index Kewensis* its involvement with taxon names, although they differ in biological and geographical coverage. The *Gray Index* covers vascular plants of the Americas; *Index Kewensis* includes seed plants worldwide. Only the *Gray Index* has nomenclatural synonyms cross-referenced to basionyms. The information is now accessible over the Internet via keyword searches from the E-mail Data Server and through the Biodiversity and Biological Collections Gopher. Indices covering other groups of plants have also been published: *Index Filicum* for Pteridophytes, and *Index Muscorum* for Bryophytes.

The *Hu Card Index* is a file of 158,844 cards for Chinese plant names, now housed in the Harvard University Herbaria building where it is available for use in person. The Index was produced by Dr. Hu Shiu-ying (Arnold Arboretum of Harvard University) and his staff. The Hu Card Index was prepared in the early 1950s when the Arnold Arboretum undertook a project to prepare a flora of China.

Royal Botanic Gardens Kew, The Harvard University Herbaria, and the Australian National Herbarium, under the collaborative project, have developed *International Plant Names Index* (IPNI), a single web database which combines citation data for seed plants from *Index Kewensis*, the *Gray Herbarium Card Index,* and the *Australian Plant Names Index* (APNI). It provides information on names and associated basic bibliographical details of all seed plants. Its goal is to eliminate the need for repeated reference to primary sources for basic bibliographic information about plant names. The

AMEBIA, Regel, Pl. Nov. Fedsch. 58 (1882) err. typ = **Arnebia**, Forsk. (Boragin.).

AMERCARPUS, Benth in Lindl. Veg. Kingd. 554 (1847) = **Indigofera**, Linn. (Legumin.).

AMECHANIA, DC. Prodr. vii. 578 (1839)= **Agarista**, D. Don (Ericaceae).
hispidula, DC. l. c. 579 (=*Leucothoe hispidula*).
subcanescens, DC. l. c. (=*Leucothoe subcanescens*).
AMELANCHIER, Medic. Phil. Bot. i. 135 (1789).
ROSACEAE, Benth. & Hook.f. i. 628.
ARONIA, Pers. Syn. ii. 39 (1807).
PERAPHYLLUM, Nutt. in Torr. & Gray, Fl. N. Am. i. 474 (1840).
XEROMALON, Rafin. New Fl. Am. iii. 11 (1836).
alnifolia, *Nutt. in Journ. Acad. Phil.* vii. (1834)22.__ Amer. bor.
asiatica, Endl. in Walp. Rep. ii. 55= canadensis
Bartramiana, M. Roem. Syn. Rosifl. 145= canadensis.
Botryapium, DC. Prodr. ii. 632= canadensis.
canadensis, *Medic. Gesch.* 79; *Torr. & Gray, Fl. N. Am. i.* 473. __Am. bor.; As. or.
chinensis, Hort. ex Koch, Dendrol. i. 186=Sorbus arbutifolia.

Figure 3.14: Portion of a page from *Index Kewensis*. Generic name Amebia (Normal caps) Regel is synonym of genus *Arnebia* (Bold small case) Forsk. of family Boraginaceae. Generic name *Amelanchier* (Bold caps) Medic. is correct name with generic names *Aronia* Pers., *Peraphyllum* Nutt. and *Xeromalon* Rafin. as synonyms. Species names *Amelanchier alnifolia* (Normal small case) Nutt. and *A. canadensis* Medic. are correct, whereas the names *A. asiatica* (italics small case) Endl., *A. Batramiana* M. Roem. and *A. Botryapium* D.C. are synonyms of *A. canadensis*. *A. chinensis* Hort. is similarly synonym of *Sorbus arbutifolia*.

data are freely available and are gradually being standardized and checked. IPNI is intended to be a dynamic resource, depending on direct contributions by all members of the botanical community.

Numerous valuable **Dictionaries** have been published but by far the most useful is *Dictionary of Flowering Plants and Ferns* published by J. C. Willis. The 8th edition revised by Airy Shaw appeared in 1973. The book contains valuable information concerning genera and families providing name of the author, distribution, family and the number of species in the genus.

Taxonomic Keys

Taxonomic keys are ***aids for rapid identification of unknown plants***. They constitute important component of Floras, manuals, monographs and other forms of literature meant for the identifying plants. In addition, identification methods in recent years have incorporated the usage of keys based on cards, tables and computer programs. The latter are primarily designed for identification by non-professionals. These keys are fundamentally based on characters, which are stable and reliable. The keys are helpful in a faster preliminary identification, which can be backed up by confirmation through comparison with the detailed description of the taxon provisionally identified with. Before identification is attempted, however, it is necessary that the unknown plant is carefully studied, described and a list of its character states prepared. Based on the arrangement of characters and their utilization, two types of identification keys are differentiated:

1. Single-access or sequential keys; and
2. Multi-access or multientry keys (polyclaves).

Single Access or Sequential Keys

Single-access keys are usual components of Floras, manuals, monographs and other books meant for identification. The keys are based on **diagnostic** (important and conspicuous) characters **(key characters)** and as such the keys are known as **diagnostic keys**. Most of the keys in use are based on pairs of contrasting choices and as such are **dichotomous keys**. They were first introduced by J. P. Lamarck in his *Flore Francaise* in 1778. The construction of a dichotomous key starts with the preparation of a list of reliable characters for the taxon for which the key is to be constructed. For each character the two contrasting choices are determined (e.g., habit woody or herbaceous). Each choice constitutes a **lead** and the two contrasting choices form a **couplet**. For characters having more than two available choices the character can be split to make it dichotomous. Thus, if flowers in a taxon could be red, yellow or white the first couplet would constitute flowers red vs non-red and the second couplet flowers yellow vs white. We shall illustrate the construction of keys taking an example from family Ranunculaceae. The diagnostic characters of some representative genera are listed below:

1. *Ranunculus*: Plants herbaceous, fruit achene, distinct calyx and corolla, spur absent, petal with nectary at base.
2. *Adonis*: Plants herbaceous, fruit achene, calyx and corolla differentiated, spur absent, petals without nectary.
3. *Anemone*: Plants herbaceous, fruit achene, calyx not differentiated, perianth petaloid, spur absent.
4. *Clematis*: Plants woody, fruit achene, calyx not differentiated, perianth petaloid, spur absent.
5. *Caltha*: Plants herbaceous, fruit follicle, calyx not differentiated, perianth petaloid, spur absent.
6. *Delphinium*: Plants herbaceous, fruit follicle, calyx not differentiated, perianth petaloid, spur one in number.
7. *Aquilegia*: Plants herbaceous, fruit follicle, calyx petaloid, not differentiated from corolla, spurs five in number.

Based upon the above information the following couplets and leads can be identified:

1. Plants woody
Plants herbaceous
2. Fruit achene
Fruit follicle
3. Calyx and corolla differentiated
Calyx and corolla not differentiated
4. Spur present
Spur absent
5. Number of spurs 1
Number of spurs 5
6. Petal with nectary at base
Petal without nectary at base

It must be noted that three choices are available for spur (absent, one, five). It has been broken into two couplets to maintain the dichotomy. Based on the arrangement of couplets and their leads, three main types of dichotomous keys are in use: **Yoked** or **Indented key, Bracketed** or **parallel key,** and **Serial** or **numbered key**.

1. **Yoked** or **Indented key:** This is one of the most commonly used keys in Floras and manuals especially when the keys are smaller in size. In this type of key, the statements (leads) and the taxa identified from them are arranged in visual groups or yokes and additionally the subordinate couplets are indented below the primary one at a fixed distance from the margin, the distance increasing with each subordinate couplet. We shall select the fruit type as the first couplet, as it divides the group into two almost equal halves and the taxa excluded would be almost equal whether the fruit in the unknown plant is an achene or a follicle. The yoked or indented key for the taxa under consideration is shown below:

1. Fruit achene.
  2. Calyx differentiated from corolla.
    3. Petal with basal nectary..............1. *Ranunculus*
    3. Petal without basal nectary.........2. *Adonis*
  2. Calyx not differentiated from corolla.
    4. Plants woody…..…….................4. *Clematis*
    4. Plants herbaceous…...................3. *Anemone*
1. Fruit follicle.
  5. Spur present.
    6. Number of spurs 1......................6. *Delphinium*
    6. Number of spurs 5......................7. *Aquilegia*
  5. Spur absent…………….............5. *Caltha*

It is important to note that all genera with achene fruit appear together and form visual groups; leads of subordinate couplets are at increasing distance from the margin and the leads of initial couplets are far separated, whereas those of subsequent subordinate couplets are closer. Such an arrangement is very useful in shorter keys, especially those appearing on a single page, but if the key is very long running into several pages, an Indented key exhibits important drawbacks. Firstly, it becomes difficult to locate the alternate leads of initial couplets, as they may appear on any page. Secondly, with the number of subordinate couplets increasing substantially, the key becomes more and more sloping, thus reducing the space available for writing leads. This may result in wastage of a substantial page space. The problem is clearly visible in *Flora Europaea* where attempts to reduce the indentation distance in longer keys has further complicated the usage of keys. These two disadvantages are taken care of in the Parallel or Bracketed key.

2. **Bracketed** or **Parallel key:** This type of key has been used in larger floras such as *Flora of USSR, Plants of Central Asia*, and *Flora of British Isles*. The two leads of a couplet are always together and the distance from the margin is always the same. Several variations of this are used wherein the second lead of the couplet is not numbered, as in *Flora of British Isles* or else the second lead is prefixed with a + sign as in *Plants of Central Asia.* The arrangement of couplets in this type of key is useful for longer keys as the location of alternate leads is no problem (two are always together) and there is no wastage of page space. There is, however, one associated drawback; the statements are no longer in visual groups. The reference to primary lead is often difficult, but this problem is usually solved by indicating the number of primary lead within parenthesis as done in several Russian Floras such as *Flora Siberia* and *Plants of Central Asia.* A typical bracketed key is illustrated below:

1. Fruit achene…………………………….....1
1. Fruit follicle…………………………….....5
2. Calyx differentiated from corolla……....3
2. Calyx not differentiated from corolla…..4
3. Petal with basal nectary…………...1. *Ranunculus*
3. Petal without basal nectary............2. *Adonis*
4. Plants woody….......……………......4. *Clematis*
4. Plants herbaceous…………..............3. *Anemone*
5. Spur present………………………….....6
5. Spur absent…………................…....5. *Caltha*
6. Number of spurs 1…….................6. *Delphinium*
6. Number of spurs 5….....................7. *Aquilegia*

Retention of positive features of the Parallel key and visual groups of the Yoked key is achieved in the Serial key.

3. **Serial** or **numbered key:** Such a key has been used for the identification of animals and adopted in some botanical works. This key retains the arrangement of Yoked key, but with no indentation so that distance from the margin remains the same. The location of alternate leads is made possible by serial numbering of couplets (or leads when separated) and indicating the serial number of the alternate lead within parentheses. A serial key for the taxa in question would appear as under:

1. (6) Fruit achene.
2. (4) Calyx differentiated from corolla.
3. Petal with basal nectary.........1. *Ranunculus*
3. Petal without basal nectary...2. *Adonis*
4. (2) Calyx not differentiated from corolla.
5. Plants woody.......................4. *Clematis*
5. Plants herbaceous............... 3. *Anemone*
6. (1) Fruit follicle.
7. (9) Spur present.
8. Number of spurs 1...............6. *Delphinium*
8. Number of spurs 5...............7. *Aquilegia*
9. (7) Spur absent...................5. *Caltha*

Such a key retains the visual groups of statements and taxa, alternate leads, even though separated, are easily located and the there is no wastage of page space.

An inherent drawback of dichotomous keys is that the user has a single fixed choice of the sequence of characters decided by the person who constructs the key. In the said example if information about the fruit is not available, it is not possible to go beyond the first couplet.

Guidelines for Dichotomous Keys

Certain basic considerations are important for the construction of dichotomous keys. These include:

1. The keys should be strictly dichotomous, consisting of couplets with only two possible choices.
2. The two leads of a couplet should be mutually exclusive, so that the acceptance of one should automatically lead to the rejection of another.
3. The statements of the leads should not be overlapping. Thus, the two leads 'leaves 5–25 cm long' and 'leaves 20–40 cm long' would find it difficult to place taxa with leaves that are between 20 and 25 cm in length.
4. The two leads of a couplet should start with the same initial word. In our example, both leads of the first couplet start with 'Fruit'.
5. The leads of two successive couplets should not start with the same initial word. In our example the word 'spur' appears in two successive couplets and as such in the second one the language has been changed to start with 'Number'. If such a change were not possible it would be convenient to prefix the second couplet with 'The'. Thus, the other alternative for the second couplet would have the two leads worded as 'The spur 1' and 'The spurs 5'.
6. For identification of trees, two keys should be constructed based on vegetative and reproductive characters separately. As trees commonly have leaves throughout the major part of the year, and flowers appear briefly when in many trees leaves are not yet developed, such separate keys are essential for identification round the year.
7. Avoid usage of vague statements. Statements such as 'Flowers large' vs 'Flowers small' may often be confusing during actual identification.
8. An initial couplet should be selected in such a way that it divides the group into more or less equal halves, and the character is easily available for study. Such a selection would make the process of exclusion faster, whichever lead is selected.
9. For dioecious plants, it is important to have two keys based on male and female flowers separately.
10. The leads should be prefixed by numbers or letters. This makes location of leads easier. If left blank, the location of leads is very difficult, especially in longer keys.

The keys described above have a single character included in a couplet, with two contrasting statements about the character in the two leads. Such keys are known as **monothetic sequential keys**. The commonest forms of keys used in floras, however, have at least some couplets (Figure 3.15) with several statements about the different characters in each lead. These keys are known as **polythetic sequential keys**. Such polythetic keys, also known as **synoptic keys** are especially useful for constructing keys for higher categories. Such keys have three basic advantages over the monothetic keys:

1 Stem woody at base; achenes 3.5-5 mm **8. pustulatus**
1 Stem not woody; achenes 2-3.75 mm
 2 Annual or biennial
  3 Achenes smooth at least between the ribs; strongly compressed and ± winged **1. asper**
  3 Achenes rugose or tuberculate between the ribs, neither strongly compressed nor winged
   4 Leaf-lobes strongly constricted at base, or narrowly linear; terminal lobe usually about as large as lateral lobes; ligules longer than corolla-tube; achenes abruptly contracted at base **2. tenerrimus**
   4 Leaf-lobes (if present) not constricted at base; terminal lobe usually much larger than lateral lobes; ligules about as long as corolla-tube; achenes gradually narrowed at base **3. oleraceous**
 2 Perennial

Figure 3.15: Portion of a **polythetic key of the yoked type** used in *Flora Europaea* for genus *Sonchus* (vol. 4, p. 327).

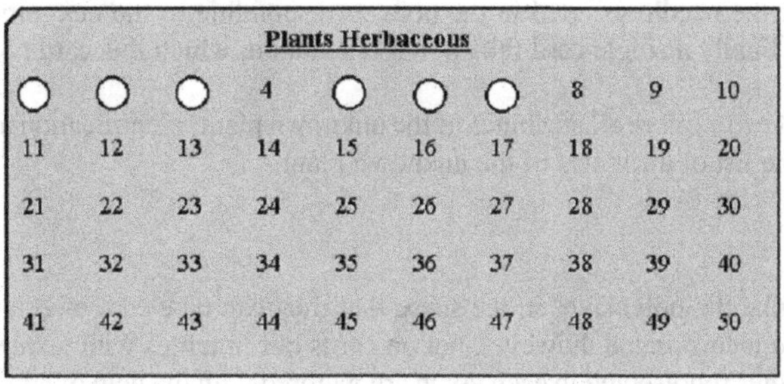

Plants Herbaceous

O O O 4 O O O 8 9 10
11 12 13 14 15 16 17 18 19 20
21 22 23 24 25 26 27 28 29 30
31 32 33 34 35 36 37 38 39 40
41 42 43 44 45 46 47 48 49 50

Figure 3.16: Body-punched card for herbaceous habit for the seven representative genera of Ranunculaceae: **1**—*Ranunculus*, **2**—*Adonis*, **3**—*Anemone*, **4**—*Clematis*, **5**—*Caltha*, **6**—*Delphinium*, **7**—*Aquilegia*. Note the diagonal trim on upper left corner of card for proper alignment of cards.

1. One or more characters may be unobservable due to damage or non-occurrence of requisite stage in the specimen. In such cases, a monothetic key becomes useless.
2. User can make a mistake in deciding about a single character. This error gets minimized if more than one character is used.
3. The single character used in the couplet may be exceptional. Such likelihood is not possible when more than one character is used.

Multi-Access Keys (Polyclaves)

Such multientry order-free keys are user-oriented. Many choices of the sequence of characters are available. Eventually, it is the user who decides the sequence in which to use the characters, and even if the information about a few characters is not available, the user can go ahead with identification. Interestingly, identification may often be achieved without having to use all the characters available to the user. Such identification methods often make use of cards. Two basic types of cards are in use:

Body-punched cards

These cards are also named **window cards** or **peek-a-boo cards** and make use of cards with appropriate holes in the body of the card (Figure 3.16). The process involves using one card for one attribute (character-state). In our example we shall need 11 cards (we have chosen only diagnostic characters above, whereas our list in polyclaves could include more characters, and thus more cards to make it more flexible).

It should be noted that we selected 12 leads and 6 couplets, with 4 leads for spur. Now we shall need only three actual attributes: 'spur absent', 'spur 1' and 'spurs 5'. Numbers are printed on the cards corresponding to the taxa for which the identification key is meant. In our example, we use only 7 of these numbers corresponding to our 7 genera. On each card, holes are punched corresponding to the taxa in which that attribute is present.

In our example card 'Habit woody' will have only one hole at number 4 (genus *Clematis*), and the card 'Habit herbaceous' will have holes at 1,2,3,5,6,7 (all seven except number 4). Once the holes are punched at appropriate positions in all the cards, we are ready for identification. The user studies the unknown plant and makes a list of characters, according to the sequence he wishes and the characters that are available to him. The user starts the identification process by picking up the first card concerning the first attribute in his list of attributes of the unknown plant. He next picks up the second card concerning the second attribute from his list and places it over the first card. This will close some holes of the first card and some of the second card. Only those holes will remain open which correspond to the taxa, which contain both the attributes. The third card is subsequently placed over the first two and the process is repeated with additional cards until finally only one hole is visible through the pack of selected cards. The taxon to which this hole corresponds is the identification of the unknown plant.

Edge-punched cards

An edge-punched card differs from the body punched card in that there is one card for each taxon and holes are punched all along the edge of the card, one for each attribute. In our example here, we shall need seven cards, one for each genus. These holes are normally closed along the edge (Figure 3.17). For each attribute, present in the taxon the hole is clipped out to form an open notch instead of a circular hole along the edge.

For actual identification, all the cards are held together as a pack. A needle is inserted in the hole corresponding to the first attribute of the unknown plant. As this needle is lifted up the taxa containing this attribute would fall down, and those lacking that attribute would remain in the pack lifted by the needle. The latter are rejected. The cards falling down are again arranged in a pack, the needle inserted in the hole corresponding to the next attribute of the unknown plant. The process is repeated until finally a single card falls down. The taxon, which this card represents, is the identification of the unknown plant.

Note that we may not have to explore all attributes of the unknown plant; identification may be achieved much before we have reached the end of the list of attributes of the unknown plant.

Tabular Keys

Tabular keys are essentially like the polyclaves in the sense that they can take care of exhaustive lists of attributes and are easier to use. The data are incorporated, however, not on cards but in tables with taxa along the rows and attributes along the columns. The attributes represented in each taxon are pictured with the help of appropriate symbols or drawings (Figure 3.18). The attributes not represented in a taxon show a blank space in the column. Thus, the table will have as many rows as taxa and as many columns as the number of attributes for which information is available.

The identification process begins with a strip of paper whose width is equal to each row and vertical lines separated by the width of the columns. The attributes present in the unknown plant are pictured on this strip of paper. The strip of paper is next placed towards the top of the table and slowly lowered and compared with each row. The row with which the entries match represents the identification of the unknown plant.

Taxonomic Formulae

A taxonomic formula is really an alphabetic formula based on a specific combination of alphabets. The various attributes in this method are coded with alphabets. Each taxon thus gets a unique alphabetic formula. These formulae are arranged in alphabetic order in the same manner as words in a dictionary. Based on the attributes of the unknown plant, its taxonomic formula is constructed. The next step is as simple as locating a word in the dictionary. The formula is located in the alphabetic list and its identification read against the formula.

The above example of Ranunculaceae could be extended here by assigning alphabets to the attributes: **A:** Woody; **B:** Herbaceous; **C:** Achene; **D:** Follicle; **E:** Spur absent; **F:** Spur 1; **G:** Spurs 5; **H:** Calyx differentiated from corolla; **I:** Calyx not differentiated from corolla, only perianth present; **J:** Nectary present; **K:** Nectary absent.

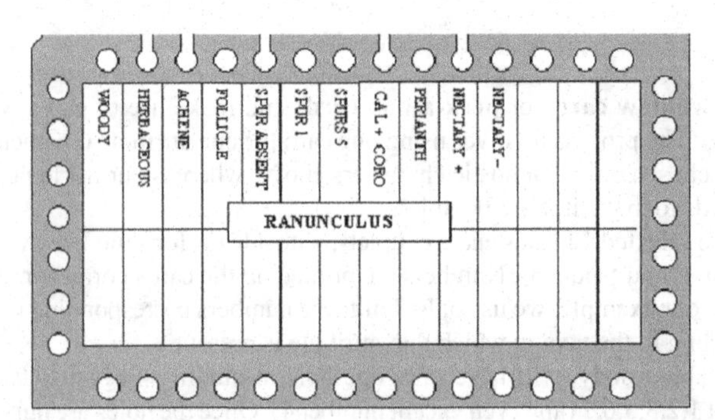

Figure 3.17: Edge-punched card for genus *Ranunculus*. Only the attributes represented in the example above are pictured. Many more attributes could be added along the vacant holes to make the identification process more versatile.

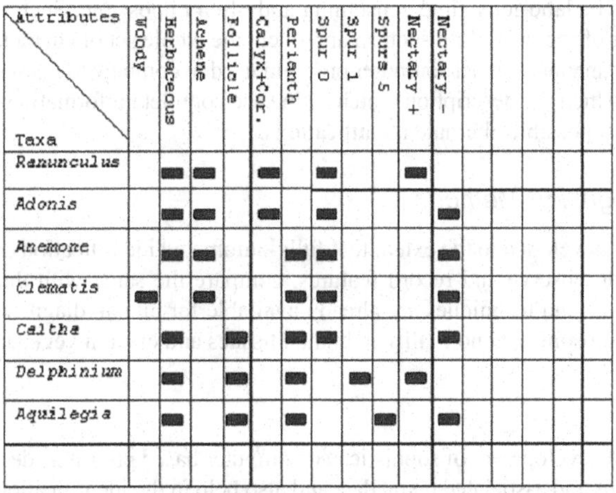

Attributes / Taxa	Woody	Herbaceous	Achene	Follicle	Calyx+Cor.	Perianth	Spur 0	Spur 1	Spurs 5	Nectary +	Nectary -
Ranunculus		■	■		■		■			■	
Adonis		■	■		■		■				■
Anemone		■	■			■	■				■
Clematis	■		■			■	■				■
Caltha		■		■		■	■				■
Delphinium		■		■		■		■		■	
Aquilegia		■		■		■			■		■

Figure 3.18: **Tabular key** for the identification of representative genera of family Ranunculaceae. Only selected attributes as in the above example are pictured. More attributes could be added in additional columns to make the identification process more versatile.

The seven representative genera would thus have the formulae as given below:

ACEIK ....*Clematis*
BCEHJ ....*Ranunculus*
BCEHK ....*Adonis*
BCEIK ....*Anemone*
BDEIK ....*Caltha*
BDFIJ ....*Delphinium*
BDGIK ....*Aquilegia*

Such formulae are useful in the identification process and have been incorporated in the written version of the multi-access key to the Genera of Apiaceae in the *Flora of Turkey* (Hedge and Lamond, 1972).

Computers in Identification

Over the years, computers have been increasingly used in data collection, processing and integration. They have also found use in a big way in scanning and identifying human ailments, which has greatly helped health management programmes. Computers have also found use in plant identification, whereby we no longer need trained botanists for this task. The following main approaches are used in computer identification:

Computer-Stored Keys

Dichotomous keys are constructed in the usual manner, fed into a computer and run using an appropriate program, which may be appropriately designed for step-wise processing of the key through a dialogue between the user and the computer. The computer program starts with the first couplet of the key, enquires about the attribute in the unknown plant and on the information provided, and handles the key asking relevant questions until finally the actual identification is achieved.

Computer-Constructed Keys

Appropriate programs may be developed which can construct a taxonomic key based on the taxonomic information about the taxa, in the same way and based on the same logic which is used by man to construct keys manually. Such keys permanently stored in a computer can be handled as above for the step-wise process of identification.

Simultaneous Character-Set Identification

Taxonomic keys are an aid to rapid identification and always provide only a provisional identification, confirmation being achieved only after comparison with a detailed description of the specific taxon. This comparison with the detailed description is not done in the first place, as comparing the description of the unknown plant with the description of all taxa of the group or the area would be laborious, time consuming and often impossible. Such a comparison can be achieved through a computer in a matter of seconds. With such an approach, the whole set of characters of the unknown plant may be fed into the computer simultaneously, and a computer program used to compare the description with the specific group and to suggest the taxon with which the description matches. In case complete information is not available, the computer program may be able to suggest possible alternate identifications.

Automated Pattern Recognition Methods

Computer technology has now developed to the extent that fully-automated identification can be achieved. The computer fitted with optical scanners can observe and record features, compare the same with those already known and make important conclusions. Programs and techniques are already available for human diagnosis, including chemical spectra and photomicrographs of chromosomes, abnormality in human tissues and even in vegetation and agricultural surveys.

Interactive Keys

Last two decades have seen the development of sophisticated computer based programs designed to collect, integrate and use it for organising descriptions and associated taxon data and also help in the identification of taxa through user friendly interfaces. Some of the Major ones are briefly described here.

DELTA System

The DELTA System is an integrated set of programs based on the DELTA format (**DEscription Language for TAxonomy**), which is a flexible and powerful method of recording taxonomic descriptions for processing by a computer. DELTA, a shareware program, has been adopted as a standard for data exchange by the International Taxonomic Databases Working Group. It enables the generation and typesetting of descriptions and **conventional keys**, conversion of DELTA data for use by **classification programs**, and the construction of **Intkey packages** for interactive identification and **information retrieval**. The System developed in the Natural Resources and Biodiversity Program of the CSIRO (Commonwealth Scientific and Industrial Research Organisation, Australia) Division of Entomology over a period of 30 years between 1971 and 2000 M. J. Dallwitz, T. A. Paine and E. J. Zurcher, is in use world-wide for diverse kinds of organisms, including fungi, plants, and wood. The programs are continually refined and enhanced in response to feedback from users.

The **DELTA program Key** generates conventional identification keys. Characters are selected by the program for inclusion in the key based on how well the characters divide the remaining taxa. This information is then balanced against subjectively determined weights, which specify the ease of use and reliability of the characters.

DELTA data can be readily converted to the forms required by programs for phylogenetic analysis, e.g., Paup, Hennig86 and MacClade. The characters and taxa for these analyses can be selected from the full dataset. Numeric characters are converted into multistate characters, as numeric characters cannot be handled by these programs. Printed descriptions can be generated to facilitate checking of the data.

Setting up a simple Delta identification: Although the DELTA system has capabilities of setting up of strong and sophisticated identification procedures, a simple one can be built with basic knowledge in computers. The first step in the process is to create a new data set (the existing one can also be used, some even downloaded from the internet). Create a new folder under Delta directory and give it an appropriate name. Open Delta Editor and click New Dataset from menu. This will open Attribute editor with 4 panels. A click in the upper left panel will open Item editor for the first taxon. Enter its name (images, comments and change of settings can be added later on) and click Done to come back to the Attribute editor (else add image, sound, change settings and then click done). Now click in the upper right panel will open character editor. Give appropriate name or description to the character. Select the character type from the list of **Unordered multistate** (Say for flower color with character states yellow, red, white, etc.; multistate includes binary characters also such as woody and herbaceous habit), **Ordered multistate** (height range such as 1–10 cm, 11–20 cm, 21–30 cm, etc.; similarly two states with plants up to 20 cm tall and more than 20 cm tall), **Integer numeric** (say leaves per node), **Real numeric** (seed size say 2.4 cm) or **text** information (say about habitats). If Multistate character has been selected (ordered or unordered), click states tab (if not already done), enter first Character state in the lower right panel, it will automatically be defined in the left panel. Now click below this entry in the left panel and enter second Character state in the right panel. Repeat this till all states (possible in the other taxa included in the identification process, but yet to be entered) have been entered. Next select character Number 2, give it a name, and select type. If the numeric (real or integer) has been selected, states tab won't appear. You will select the unit (cm, mm, leaves per node). For text character, add appropriate notes in the notes tab. After all the characters have been selected, click Done to go back to the Attribute editor.

Now click in the left upper panel to add the second taxon and repeat this till all the taxa have been added. The Attribute editor will now show a list of taxa in the upper left panel and the characters (identified as highlighted U-unordered multistate,

O-ordered multistate, I-integer numeric, R-real numeric and T-text) in the upper right panel. Any character missing from the list can be added and appropriately defined. Now select taxa one by one. For each taxon, enter (verify) the state in the right panel after expanding the character icon (+ not expanded, – expanded) till information for all the taxa has been entered. Save the dataset under the folder already created in the beginning. You can open the dataset now to add any images, comments or change settings if desired. The identification program needs a large number of files in the folder created for a particular dataset. The following procedure will create these files automatically.

Open Dataset in the Delta editor and click File-->Export directive. Delta Files to export dialogue box appears. Click OK, subsequently Done and then Close (if necessary, from X on top). Open Delta editor (if closed). Click view-->Action sets. Print character list appears. See that the Confor Tab is active. Select 'Print character list-RTF'. Click Run. In the next dialogue Box click Yes. Go to Action sets again, select 'Translate into Natural Language-RTF-Single file for all taxa' and click Run, subsequently click Yes. Go to Action sets again, select 'Translate into Key Format' and click Run and proceed similarly. Now open Action sets again, change Tab from Confor to Key, select 'Confirmatory character RTF' and click Run. In the Action sets again now change the Tab back to Confor, select 'Translate into Intkey format' and click Run. Go to the Action sets for the last time, change to Intkey Tab, select 'Intkey initializing File' and click Run. The process will complete and Intkey program window will open (don't forget to add dataset to the Intkey Index when prompted when you close Intkey program; or else add dataset when you open Intkey program window next time) with four panes with the list of characters in the upper left pane and the list of taxa in the upper right pane, both lower panes being empty. Using Intkey program window, one can identify an unknown plant from this group of taxa by reading the first character in the unknown plant, clicking the appropriate character in the left upper pane and clicking or entering the right choice of the character state when prompted. This will eliminate and show certain taxa in the lower right pane and the used character in the lower left pane (Figure 3.19). As you use more and more characters, some more taxa will be rejected, and the process will end when a single identified taxon is remains in the upper right pane. You can click i (information) icon to view image (if added) or read full description of taxon. Intkey can also be used to access Delta data and images over the internet. For this data files (such as iitems, ichars), intkey.ini, contents.ind (together with rtf files), and image files (optional) are put in a zip file (or self-extracting zip file) and uploaded to the website along with startup file (*.ink; which contains the information and the path of uploaded files of the project), intkey.ini, imagePath (optional) and InfoPath(optional). A data-set index file or link in WWW page must point to the special startup file (*.ink; not intkey.ini or intkey.ink). The startup file tells Intkey where the data set and its associated images are found on the website. When a person using an internet browser clicks on a link to an intkey startup file, the browser activates Intkey and passes it a copy of startup file. Intkey itself then retrieves the actual data set from the web, extracts its contents, and begins identification. For this web applicability, however, Intkey must be installed on both Web server and each client PC, and an association of files must be developed by the manager of Web server, where the project files are located.

Intkey based web applications are available for several families and Genera from Flora of China, Families of the World (Watson and Dallwitz), Grass Genera of the World (Watson and Dallwitz), Grass Species of the World (RBG, Kew), Tree and Shrub Genera of Borneo (J. K. Jarvie & Ermayanti), identification facility for the vascular flora of Western Australia, and is available in FloraBase. Additionally, interactive keys (using Intkey) to the families and genera of flowering plants in Western Australia are soon to be added to **FloraBase**, with specialist keys for certain significant genera also well advanced.

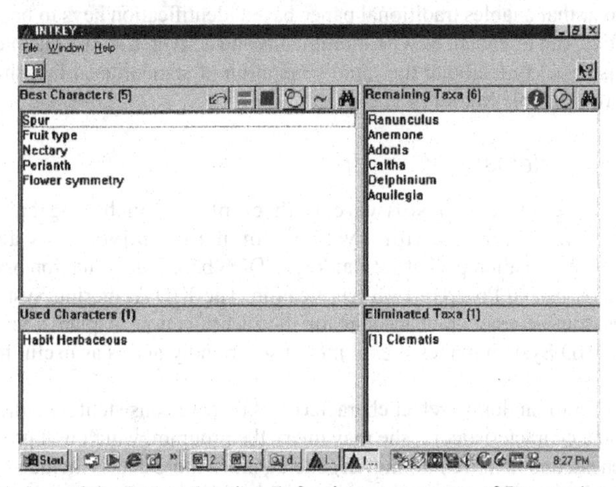

Figure 3.19: Identification window of the Intkey (version 5) for the seven genera of Ranunculaceae. Selection of character state herbaceous (shown in the panel of used characters) leads to the rejection of one taxon (Clematis, which has woody habit) shown in the panel of eliminated taxa. Selection further character states of the known plant would eliminate further taxa till only one identified taxon remains.

NaviKey

Navikey is a simple Java based interactive identification key, a free program, which works on Delta flat files (chars, items and specs-present in your folder if you have developed a database ready for identification through Intkey, as detailed in preceding paragraphs). NaviKey v. 4 is developed in the frame of BIOTa Africa project (An International Research Network on biodiversity, sustainable use and conservation) by Dieter Neubacher and Gerhard Rambold sustainable use and conservation) by Dieter Neubacher and Gerhard Rambold (University of Bayreuth, Germany), based on an earlier version (NaviKey v. 2.3 by Michael Bartley and Noel Cross, Harvard University Herbarium, Boston, USA). The program can be downloaded from www.NaviKey.net and can be used both as standalone application or as web application. After downloading the and unzipping the file, the folder will have several files on your computer. Simply add the three flat files of your project to this folder. For using it as standalone application simply click NaviKey.jar, and it will open up the identification window with four panels, like Intkey. Character panel is upper left window, but right upper panel shows character states, lower right panel the matching items panel showing matching or remaining taxa (click any taxon to get its full description) and the lower left panel the query criteria panel: display of previous (used) character state selections. NaviKey also allows checkbox matching options to: (a) Restrict view on used characters and character states of remaining items. (b) Retain items unrecorded for the selected characters. (c) Retain items matching at least one selected state of resp. characters. (d) Use extreme interval validation, and (e) Use overlapping interval validation. NaviKey does not display the list of excluded taxa but the total number of taxa and number remaining are displayed. The use of software as web application is very convenient. Just fill in the title and subtitle of the project being developed in NaviKeyAppletWebpageTemplate.html using html editor (say Frontpage), upload the whole folder to your website, and provide a link to NaviKey.html page. As this page opens, the java application gets loaded and the program is ready for interactive identification.

NaviKey identifications are available for several families and genera of Flora of China and genus *Arisaema* (Guy Gusman & Eric Gouda) and Flowering Plant Families of Jamaica (Gerald Guala & Jimi Sadle).

Lucid Systems

Lucid software (Lucid3) is a commercial powerful and widely acclaimed Lucid Professional identification and diagnostic software developed by Centre for Biological Information Technology, The University of Queensland, Brisbane Australia. The Lucid3 system comprises a Builder and Player for creating and deploying effective and powerful identification and diagnostic keys. It allows creation of interactive, random-access keys that can be deployed over the World Wide Web or CD. The key when used for the identification of an unknown specimen progressively eliminates entities that do not match the chosen features until only one or a few possible entities remain. Further information and images can be accessed to confirm the identification.

The basic elements of a Lucid3 key are: a list of entities; a list of features and states that may be used to describe those entities; a matrix of score data for the features associated with each of the entities for the features; and various attachments (images, web pages, etc.) for the entities and features, to provide extra information to users.

The Lucid3 Builder provides all the tools necessary to create the entity and feature lists, encode the score data, and attach information files to items. The package in addition includes Lucid Phoenix, a computer based dichotomous or pathway key Builder and Player that enables traditional paper-based identification keys to be published on the Internet or CD. Phoenix keys are interactive, can be enhanced with multimedia, and delivered across the Internet seamlessly. Additional Fact Sheet Fusion software is a tool to facilitate the rapid generation of standardised fact sheets in HTML (Hyper Text Markup Language) or XML (eXtensible Markup Language).

XID (Expert Identification Systems)

XID Services Inc. produces commercial software with emphasis on biological sciences and is one of the leading providers of expert identification systems for major universities and botanical gardens in United States. XID offers two identification packages: **Pankey**, a DOS based identification program, and ***XID Authoring Systems***, Windows based databases and Program for Identification. The XID Authoring System allows authors to create their own "smart key" or random-access expert system for the identification of plants, animals, or any other object. The elegant simplicity of the XID System makes it extremely user friendly and is as useful for school teacher as for the professional scientist.

XID System allows the user to randomly select characteristics that are consistent with their specimen and skill level. If the user cannot decide upon a characteristic, he/she may query the program, which will provide a list of suggestions in order of ease of use, effectiveness, and items remaining. In general, much more data is included on each item/species than is necessary to identify it. With this abundance of data, the user can identify any of the items/species using the characteristics most obvious and easy to describe. With each characteristic entered by the user, the program eliminates all species that do not have the combination of features entered.

XID also offers 1000 **Weeds of North America** CD ROM. This is the most comprehensive weed identification reference ever published in North America. Contains 140 grass-like and 860 broadleaf weeds, features include Interactive key, color photos of all species, illustrated glossary of terms, page number references to over 40 weed reference books, searchable geographic data, and State level distribution maps.

ActKey

ActKey is a web based interactive identification program developed by Hong Song of the Missouri Botanical Garden. This Java-based program uses MySQL as the database server, and can handle data sets in DELTA, MS Excel, MS Access and Lucid formats. ActKey identification is available for the floras of China, North America, Madagascar, Borneo, at the Harvard University Herbaria Editorial Center, and hosted at **eFlora website**. Examples include several keys to the large and medium-sized genera of China (also in Chinese); the genera of Brassicaceae of the world by Ihsan Al-Shehbaz; *Salix* (Salicaceae) of North America by George W. Argus (also in Chinese); angiosperm families by B. Hansen and K. Rahn (also in Chinese and Spanish); Trilliaceae (*Trillium* and *Paris*) of the world by Susan B. Farmer, the generic tree flora of Madagascar by George Schatz, and the trees & shrubs of Borneo by James K. Jarvie & Ermayanti, respectively.

Meka

MEKA (pronounced "mecca") is an interactive Multiple-Entry Key Algorithm to enable rapid identification of biological specimens, now designed to run under Windows. The program, distributed free, is developed by Christopher Meacham, Jepson Herbarium, Berkeley, CA. The user picks character states that are present in the specimen from a list of possibilities. As the character states are scored by picking them, MEKA eliminates taxa that no longer match the list of scored character states. Different windows display different aspects of the underlying database. As the identification progresses the windows are updated automatically. An index screen makes it easy to find and score particular classes of character states. MEKA does not lead the user in a fixed stepwise progression through a series of questions. Instead, the user can perform identifications by scoring character states in any order. This makes it possible to identify specimens that are much more fragmentary than is possible with dichotomous keys. New Windows version includes a conversion function that can convert any MEKA key to the SLIKS (Stinger's Light Weight Interactive Key Software) format developed by Gerald Guala for Web-based identification. Thomas J. Rosatti has developed many Meka keys to California plants, and Prof. Knud Ib Christensen of the Botanic Garden of the University of Copenhagen key to Old World *Crataegus*.

SLIKS software is a small free Javascript program developed to facilitate the use of interactive keys. SLIKS is written in simple Javascript and runs over the web or locally on your machine. Users can download their own copy or use it from your web site. It runs through the web browser, so it is essentially platform independent.

IdentifyIt

IdentifyIt is identification software of comprehensive commercial **Linnaeus II** multifunctional research tool developed by ETI BioInformatics, for systematists and biodiversity researchers. It facilitates biodiversity documentation and species identification. Linnaeus II supports the creation of taxonomic databases, optimizes the construction of easy-to-use identification keys, expedites the display and comparison of distribution patterns, and promotes the use of taxonomic data for biodiversity studies. There are three 'modules' of Linnaeus II: the 'Builder' to manage your data and to create an information system, the 'Runtime' engine to publish completed information systems on CD-ROM/DVD-ROM, and the 'Web Publisher' to publish your completed project as a Web site.

The package offers three identification modules: **Text Key™**—an electronic version of written dichotomous keys, The **Picture Key™**—similar to the Text Key but picture-based, and **IdentifyIt™**, the most powerful identification tool. It is a multiple-entry key based on a matrix of taxa, characters, and character states. Unlike the Species and Higher Taxa, which hold text descriptions of the taxa, in IdentifyIt taxa are described in a more structured format: as a series of character states. This allows you to easily obtain answers to specific questions like, "Which species are red and/or white".

Pl@nt Net

Pl@nt Net is a tool to help to identify plants with pictures. It is developed by scientists from four French research organisations: Cirad, INRA, Inria and IRD and the Tela Botanica Network, with the financial support of Agropolis fondation. It is organised into different databases for ornamental and wild plants Major databases include Useful Plants, useful plants of Tropical Africa, Useful Plants of Asia, Western Europe, Canada, USA, and several other region. Among other features, this free app helps identifying plant species from photographs, through a visual recognition software. Plant species that are well enough illustrated in the botanical reference database can be easily recognized.

Electronic Lists, Pictorial Online Identification

In addition to these interactive Keys Illustrations of plants from various parts of the World as also the illustrations of economic plants are put up at various websites hosted by different institutions, particularly one supporting Virtual herbaria and eFloras. These illustrations are available for help in identification.

Several **electronic lists** are maintained by listservers. **Taxacom** is one such list very active on taxonomic matters, subscribed to by numerous active taxonomists all over the world. There is a regular exchange on matters of taxonomic interest. Any member with a problem can seek opinions from all members simultaneously. An unknown plant can be identified by sending its description to the list. Still better, a photograph or illustration of the unknown plant can be put up on a website with information to the members. The members may go to the website, observe the photograph or illustration and send their comments to the member concerned or the list itself. Many users are being benefitted through this web-based interaction.

Last few years have seen the spurt of internet-based exchange of information. **Efloraofindia** (formerly **Indiantreepix**) https://sites.google.com/site/efloraofindia/ is one of the biggest non-commercial website, based on photographic collection of plants. It is documenting flora of India that is being discussed on efloraofindia google e-group https://groups.google.com/forum/indiantreepix, devoted to creating awareness, helping in identification, discussion and documentation of Indian Flora. It also has the largest database on net on Indian Flora with more than 12,000 species (along with more than 3,00,000 pictures). It also includes species from other parts of world.

Flowers of India, another website at http://www.flowersofindia.net/, devoted to Indian Flowering plants has separate databases with links to different species having photographs, description, common names and regional names, arranged according to botanical names (sorted alphabetically or familywise) or common names. Plants of different categories such as Flowering trees, Orchids, Medicinal Plants, Garden Flowers, Bulbous plants, Himalayan Flowers can be accessed through separate links. New images are being continuously added after confirmation by experts. **Vascular plant image library** (http//botany.csdl. tamu.edu/FLORA/gallery.htm) was developed originally with support from Texas Higher Education coordinating Board as a part of Digital Flora of Texas. Links are provided family wise to the images of plant species in databases including Flowers of India, CalPhotos, Flora of Chile, Missouri plants, Floral images, Plants of Hawaii, Oregon Flora image project, and several individual image collections. **CalPhotos** is a huge database developed under a project of BSCIT of University of California, Berkeley, and contains more than 640,793 images of plants, animals, fossils, peoples and landscapes around the world. Nearly 360,414 images of plants can be browsed alphabetically and searched through easy criteria such as scientific name, common name, location, country and photographer.

DNA Barcoding

DNA Barcoding is the most recent approach to fix the identity of different species, to ultimately facilitate a common database for living organisms. Consortium for the Barcode of Life (CBOL) is an international collaboration of natural history museums, herbaria, biological repositories, and biodiversity inventory sites, together with academic and commercial experts in genomics, taxonomy, electronics, and computer science. The mission of CBOL is to rapidly accelerate compiling of DNA barcodes of known and newly discovered plant and animal species, establish a public library of sequences linked to named specimens, and promote development of portable devices for DNA barcoding. **DNA barcoding** is a technique for characterizing species of organisms using a short DNA sequence from a standard and agreed-upon position in the genome. **DNA barcode sequences** are very short relative to the entire genome and they can be obtained reasonably quickly and cheaply. The cytochrome c oxidase subunit 1 mitochondrial region (COI) is emerging as the standard barcode region for higher animals. Because of its slow rate of evolution in higher plants, however, is not suitable for barcoding, and after experimenting with chloroplast plastid trn*H*-psb*A* intergeneric spacer gene, botanists at the Proceedings of National Academy of Sciences, Cameroon and Plant Working Group of CBOL in 2009, have decided to use two genes rbc*L* and mat*K* for DNA barcoding of plants. Once the barcodes of all species of plants are established, identification of plants may be possible through a handheld scanner. It may be useful for detecting illegal plants at check points and make the process of identification much simpler. However, at this point, detection of closely related species may not be possible, and traditional methods may be used before more defined methods of DNA barcoding are developed.

Chapter 4

Systematic Evidence Interdisciplinary Approach

Last few years have seen the utilization molecular data in redefining affinities between plant groups, resulting in major changes in increasingly popular APG classification (latest APG IV, 2016) and continuously updated APWeb (version 14, 2017, latest update December 2018). In addition, more information is being accumulated from different fields, giving strength to the realization that ultimately morphological features will be decisive in recognizing organisms in the field. Newer approaches in recent years include (a) increasing reliance on phytochemical information (**Chemotaxonomy**) mainly molecular data; (b) studies on ultrastructure and micromorphology; and (c) statistical analysis of the available data without much *a priori* weighting and providing a synthesis of all the available information using newer methods of analysis (**Taxometrics**); and (d) analysis of phylogenetic data to construct phylogenetic relationship diagrams (**Cladistics**). The aforesaid disciplines constitute the major **modern trends in taxonomy**. Data continues to flow from different disciplines, so that the process of analysis and synthesis is an ongoing activity. **Taxonomy (Systematics) is as such a field of unending synthesis**. The following disciplines have contributed to a greater or lesser extent to a better understanding of taxonomic affinities between plants.

MORPHOLOGY

Morphology has been the major criterion for classification over the last many centuries. The initial classifications were based on gross morphological characters. During the last two centuries, more and more microscopic characters of morphology were incorporated. Although floral morphology has been the major material for classifications, other morphological characters have also contributed in specific groups of plants. The diversity of morphological features has already been discussed in detail under Descriptive terminology in Chapter Two.

Habit

Life-forms—though of little significance to taxonomy—allow a means of estimating adaptiveness and ecological adjustment to the habitat. In *Pinus,* bark characters are used for identification of species. Woody and herbaceous characters have been the primary basis of recognition of Lignosae and Herbaceae series within dicots by Hutchinson (1926, 1973).

For several decades it was believed that trees or shrubs with simple leaves represented the most primitive condition within angiosperms. Increased evidence over the last decade, however, is pointing towards the assumption that the perennial herbaceous condition in **paleoherbs** such as Ceratophyllaceae, Nymphaeaceae and Piperaceae represents the archetype of the most primitive angiosperms.

Underground Parts

Rhizome characteristics are important for identification of various species of the genus *Iris*. Similarly, bulb structure (whether bulbs are clustered on rootstock or not) is an important taxonomic criterion in the genus *Allium*. Davis (1960) has divided Turkish species of the subgenus *Ranunculus* of genus *Ranunculus* based on rootstock and habit.

Leaves

Leaves are important for identification in palms, *Salix* and *Populus*. The genus *Azadirachta* has been separated from *Melia* among other features by the presence of unipinnate leaves as against bipinnate in the latter. Similarly, the genus *Sorbus* has been separated from *Pyrus*, and genus *Sorbaria* separated from *Spiraea* based on pinnate leaves. Stipules are an important source for identification in *Viola* and *Salix*. Leaf venation is important for the identification of the species in *Ulmus* and *Tilia*. Interpetiolar stipules are useful for identification within family Rubiaceae.

Flowers

Floral characters are extensively used in delimitation of taxa. These may include the calyx (Lamiaceae), corolla (Fabaceae, *Corydalis*), stamens (Lamiaceae, Fabaceae-Mimosoideae), or carpels (Caryophyllaceae). A gynobasic style is characteristic of Lamiaceae. Similarly, the gynostegium characterizes Asclepiadaceae (now recognized as subfamily Asclepiadoideae of family Apocynaceae). Different species of *Euphorbia* have a distinctive cyathium inflorescence with clusters of male flowers each represented by a single stamen.

Fruits

Fruit characteristics are very widely used in identification. Coode (1967) used only fruit characteristics in delimitation of species of the genus *Valerianella*. Singh et al. (1972) used fruit morphology in identification of Indian genera of Compositae (Liguliflorae). In Asteraceae—the shape cypsela (usually called achene), presence or absence of pappus and whether the pappus is represented by hairs, scales or bristles, the presence or absence of beak, and its length, the number of ribs on the cypsela—constitute valuable identifying features. The number of capsule valves is used in segregating genera in family Caryophyllaceae (*Melandrium*, *Silene*, *Cerastium*). Seed characters are valuable identification features in the genus *Veronica*.

ANATOMY

Anatomical features have played an increasingly important role in elucidation of phylogenetic relationships. Anatomical characteristics are investigated with the help of a light microscope; whereas **ultrastructure** (finer details of contents) and **micromorphology** (finer details of surface features) are brought out brought out using an electron microscope. Anatomical work of taxonomic significance was largely undertaken by Bailey and his students. Carlquist (1996) has discussed the trends of xylem evolution, especially in the context of primitive angiosperms.

Wood Anatomy

Wood represents secondary xylem constituting the bulk of trees and shrubs, formed through the activity of vascular cambium. It primarily consists of **tracheids** and **vessels**. Tracheids are long narrow elements with tapering ends, imperforate at ends, and transfer of water and minerals occurring through pit-pairs (two adjacent pits of two tracheids, separated by primary cell walls). The vessels, on the other hand, are composed of vessel elements, much broader than tracheids and with **perforation plates** at ends (with opening not having primary walls unlike pit-pair). Vessel elements are joined end to end to form long tubes, the vessels. Perforation plate may be **simple** with a single opening, or **compound** with several openings. Latter with elongated openings in a row like a ladder is known as **scalariform**, a common type in primitive angiosperms.

Vessels are absent in Gymnosperms but present in Angiosperms. It is commonly believed that there has been a progressive evolution in angiosperms from tracheids to long, narrow vessel elements with slanted, scalariform perforation plates, to short, broad vessel elements with simple perforation plates. Studies on wood anatomy have contributed largely in arriving at the conclusion that Amentiferae constitute a relatively advanced group, and that Gnetales are not ancestral to angiosperms. Bailey (1944) concluded that vessels in angiosperms arose from tracheids with scalariform pitting, whereas in Gnetales they arose from tracheids with circular pitting, thus suggesting an independent origin of vessels in these two groups. Demonstration of vessel-less angiosperms (Winteraceae, Trochodendraceae), also having other primitive features, has led to the conclusion that angiosperm ancestors were vessel-less. The separation of *Paeonia* into a distinct family Paeoniaceae and *Austrobaileya* into a separate family Austrobaileyaceae has been supported by studies of wood anatomy.

Nodal anatomy has considerable significance in angiosperm systematics. The number of vascular traces entering leaf base and associated gaps (lacunae) left in the vascular cylinder of stem at each node are distinctive for several groups. The node may have single gap (**unilacunar**) from single leaf trace or three leaf traces (two additional commonly entering stipules) or three gaps (**trilacunar**) associated with three leaf traces (Figure 4.1) The genus *Illicium* has been separated from Winteraceae because of unilacunar nodes, continuous pseudosiphonostele and the absence of granular material in stomatal depressions.

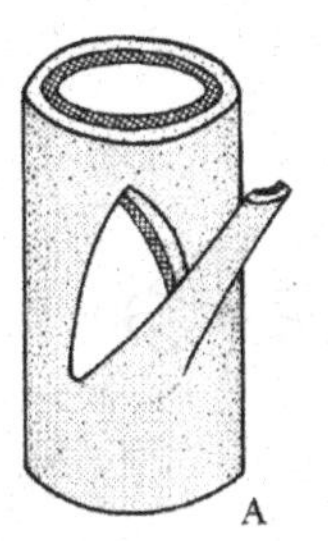

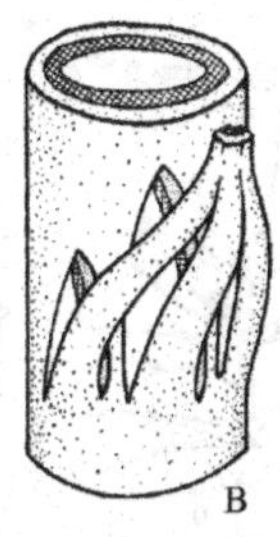

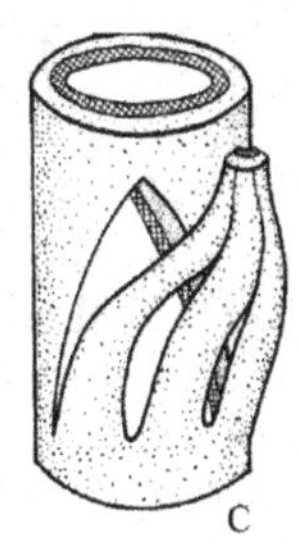

Figure 4.1: Nodal anatomy. A: Unilacunar node with one leaf trace; **B:** Trilacunar node with three leaf traces; **C:** Unilacunar node with three leaf traces.

Trichomes

Trichomes constitute appendages of epidermis which may be non-glandular or glandular. Non-glandular trichomes may be in the form of simple unicellular or multicellular hairs (common in Brassicaceae, Lauraceae and Moraceae), in the form of vesicles, peltate hairs (*Olea*) or flattened scales. Branched hairs may be dendroid, stellate (*Styrax*) or candelabrum-like (*Verbascum*). Glandular trichomes may be sessile or stalked and present a variety of forms.

Unicellular **glandular hairs** of *Atriplex* are bladder-like (Figure 4.2) with few-celled stalk and basal cell and they secrete salt. Others may secrete nectar (calyx of *Abutilon*), mucilage (leaf base of *Rheum* and *Rumex*). The **stinging hairs** of *Urtica* are highly specialized with silica tip which readily breaks when hair is touched. The broken tip is sharp like a syringe and easily penetrates the skin injecting irritating cell contents.

Trichomes hold considerable promise in systematics of angiosperms. Trichomes have been of considerable help in Cruciferae (Schulz, 1936), especially in the genera *Arabis* and *Arabidopsis*. Trichome characters are very useful in the large genus *Astragalus* (with more than 2000 species). The Himalayan species *Hedera nepalensis* is distinguished from its European relative *H. helix* in having scaly trichomes as against stellate in the latter. In family Combretaceae the trichomes are of immense significance in classification of genera, species or even varieties (Stace, 1973). Trichomes are also diagnostic characters for many species of *Vernonia* (Faust and Jones, 1973).

Epidermal Features

Epidermal features are also of considerable taxonomic interest (SEM epidermal features are discussed under ultrastructure and micromorphology). Prat (1960) demonstrated that one can distinguish a **Festucoid type** (simple silica cells, no bicellular hairs) and **Panicoid type** (complicated silica cells, bicellular hairs) of epidermis in grasses.

Stomatal types (Figure 4.3) are distinctive of certain families such as Ranunculaceae (**anomocytic**), Brassicaceae (**anisocytic**), Caryophyllaceae (**diacytic**), Rubiaceae (**paracytic**), and Poaceae (**graminaceous**). Anomocytic type has ordinary epidermal surrounding the stomata. In others the epidermal cells surrounding the stomata are differentiated as subsidiary cells. There may be two subsidiary cells at right angles to the guard cells (diacytic), two are more parallel to the guard cells (paracytic), or three subsidiary cells of unequal size (anisocytic). Other types include **actinocytic** type with stomata surrounded by a ring of radiating cells, **cyclocytic** with more than one concentric ring of subsidiary cells and **tetracytic** with four subsidiary cells. The stomatal complex of Poaceae is distinctive in having two dumb-bell shaped guard cells with two small subsidiary cells parallel to the guard cells.

Stace (1989) lists 35 types of stomata in vascular plants. Closely related families Acanthaceae and Scrophulariaceae are distinguished by the presence of diacytic stomata in the former as against anomocytic in the latter. The stomatal features, however, are not always reliable. In *Streptocarpus* (Sahasrabudhe and Stace, 1979) cotyledons have anomocytic while mature organs have anisocytic stomata. In *Phyla nodiflora* (syn = *Lippia nodiflora*) the same leaf may show anomocytic, anisocytic, diacytic and paracytic stomata (Pant and Kidwai, 1964).

Leaf Anatomy

The florets of Poaceae are reduced and do not offer much structural variability. Leaf anatomy has been of special taxonomic help in this family. The occurrence of the C-4 pathway and its association with **Kranz anatomy** (dense thick-walled chlorenchymatous bundle sheath, mesophyll simple), has resulted in revised classification of several genera of grasses. Melville (1962, 1983) developed his gonophyll theory largely on the basis of the study of venation pattern of leaves and floral parts. The rejection of *Sanmiguelia* and *Furcula* as angiosperm fossils from the Triassic has largely been on the basis of detailed study of the venation pattern of leaves (Hickey and Doyle, 1977). The more recent rediscovery of *Sanmiguelia* from the Upper Triassic of Texas (Cornet, 1986, 1989) points to presumed angiosperm incorporating features of both

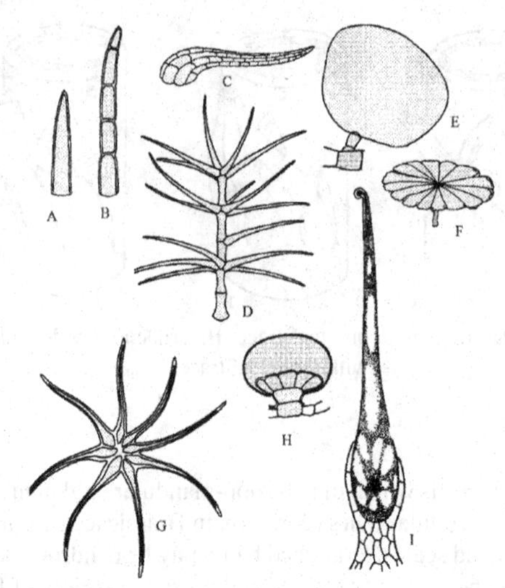

Figure 4.2: Trichomes. A: Simple unicellular hair; **B:** Multicellular hair; **C:** Scale; **D:** Candelabra trichome of *Verbascum*; **E:** Vesicular hair of *Atriplex*; **F:** Peltate hair; **G:** Stellate hair *Styrax*; **H:** Secretary gland of *Thymus*; **I:** Stinging hair of *Urtica*.

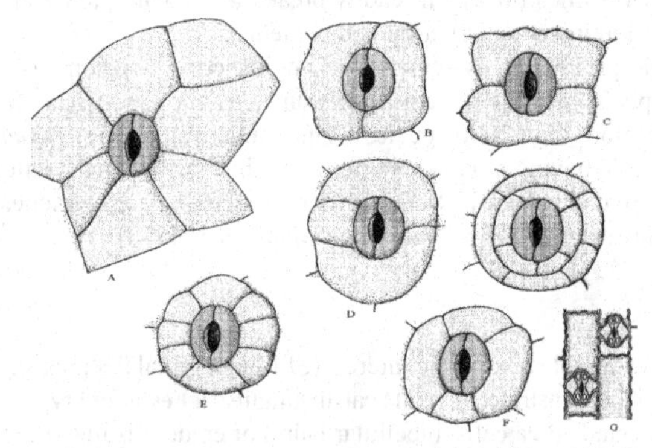

Figure 4.3: Stomatal apparatus in Angiosperms. A: Anomocytic type with epidermal cells around stomata not differentiated; **B:** Paracytic type with two or more cells parallel to the guard cells differentiated as subsidiary cells; **C:** Diacytic type with two subsidiary cells at right angles to the guards cells; **D:** Anisocytic type with three subsidiary cells of unequal size; **E:** Actinocytic type with stomata surrounded by a circle of radiating cells; **F:** Tetracytic type with four subsidiary cells; **G:** Cyclocytic type with concentric rings of subsidiary cells; **H:** Graminaceous type with dumb-bell shaped guard cells with two small subsidiary cells parallel to the guard cells.

monocots and dicots. Discovery of the Late Triassic *Pannaulika* (Cornet) from the Virginia-North Carolina border has reopened the possibilities of Triassic origin of angiosperms.

Floral Anatomy

Floral anatomy has been one of the thoroughly explored areas, with significant contributions to the understanding of the phylogeny of angiosperms. Vascular traces in the carpels of various genera of the family Ranunculaceae have confirmed the origin of achene (*Ranunculus*, *Thalictrum*, etc.) from follicle (*Delphinium*, *Aquilegia*, etc.) through successive reduction in the number of ovules ultimately to one. The additional traces which would have gone to other ovules, now aborted, can be observed in many genera. There, thus, is no justification for Hutchinson's separation of achene-bearing genera and follicle-bearing genera into separate families Ranunculaceae and Helleboraceae, respectively.

Melville (1962, 1983) developed his **gonophyll theory** after studying the vasculature of carpel and other floral parts through the clearing technique. He believed the angiosperm carpel to be a modified dichotomous fertile branch adnate to the petiole of a leaf. Sporne (1971) cautioned against such a drastic conclusion citing the example of bathroom loofah.

The genus *Melandrium* was segregated from *Silene* on the basis of the ovary being unilocular as against partly septate in *Silene*. Detailed floral anatomy revealed that in all the species of both genera, the ovary is multilocular, at least in the early stages of development. The septa break down to various degrees in different species as the ovary develops. Thus structurally, the ovaries are similar. The two genera were consequently merged into the single genus *Silene*.

The **inferior ovary** in angiosperms has been formed in two ways: **appendicular origin** (formed by fusion of calyx, corolla and their traces to the ovary wall; in this case, all vascular traces have normal orientation, i.e., phloem towards the outside) or by **axial invagination** (formed by depression of the thalamus; the inner vascular traces have reverse orientation, i.e., phloem towards the inside). Studies on floral anatomy have confirmed that in a large majority of families, the inferior ovary is of appendicular origin. Only in a few cases (*Rosa*, Cactaceae, etc.) is the origin by axial invagination of the thalamus.

Floral anatomy has also supported the inclusion of *Acer negundo* under *Acer* and does not support its separation into a distinct genus *Negundo*. Although this species is specialized in having a dioecious habit and anemophily, the anatomy of the flower shows unspecialized features of other species. Floral anatomy also supports the separation of *Menyanthes* from Gentianaceae into a distinct family Menyanthaceae. The genus *Centella* is separated from *Hydrocotyle* on the basis of inflorescence being a cyme, and ovules receiving vascular supply from alternate bundles. In *Hydrocotyle,* the inflorescence is an umbel and the ovules receive vascular supply from fusion of two adjacent bundles. *Paeonia* is a classic example of a genus, which was removed from family Ranunculaceae into a distinct family Paeoniaceae. The separation has been supported by evidence from morphology, embryology and chromosomes. Floral anatomy also supports this separation, as both sepals and petals have many traces, carpels have five traces and the stamens are centrifugal. Developmental studies have indicated that some flowers, such as Apiaceae and Ericaceae, that appear to have free petals, are gamopetalous early in development. They are, therefore, considered to have evolved from gamopetalous ancestors.

EMBRYOLOGY

Embryology has made a relatively lesser contribution in understanding taxonomic affinities. This is primarily because of long preparatory work needed for embryological studies. More often, the study of hundreds of preparations may reveal just a single embryological characteristic of any significance. It may take many years of laborious and painstaking research to study even a few representatives of a family. The embryological features of major significance include microsporogenesis, development and structure of ovule, embryo sac development, endosperm and embryo development.

Families Marked Out by Distinct Embryological Features

Several families of angiosperms are characterized by unique embryological features found in all members. These include:

Podostemaceae

Family Podostemaceae includes perennial aquatic herbs, which have a unique embryological feature in the formation of a **pseudoembryo sac** due to the disintegration of the nucellar tissue. The family is also characterized by the occurrence of pollen grains in pairs, bitegmic tenuinucellate ovules, bisporic embryo sac, solanad type of embryogeny, prominent suspensor haustoria, and absence of triple fusion and, consequently, endosperm.

Cyperaceae

Family Cyperaceae is characterized by the formation of only one microspore per microspore mother cell. Following meiosis, of the four microspore nuclei formed, only one gives rise to pollen grain. Besides Cyperaceae, only Epacridaceae in a few members shows the degeneration of three microspore nuclei. Cyperaceae is distinct from these taxa in pollen shedding at the 3-celled stage, as against the 2-celled stage shedding in Epacridaceae.

Onagraceae

Family Onagraceae is characterized by *Oenothera* type of embryo sac, not found in any other family except as an abnormality. This type of embryo sac is 4-nucleate and is derived from the micropylar megaspore of the tetrad formed.

Specific Examples of the Role of Embryological Data

There are a few examples of the embryological data having been very useful in the interpretation of taxonomic affinities:

Trapa

The genus *Trapa* was earlier (Bentham and Hooker, 1883) included under the family Onagraceae. It was subsequently removed to the family Trapaceae (Engler and Diels, 1936; Hutchinson, 1959, 1973) based on distinct aquatic habit, two types of leaves, swollen petiole, semiepigynous disc and spiny fruit. The following embryological features support this separation: (i) pyramidal pollen grains with 3 folded crests (bluntly triangular and basin shaped in Onagraceae); (ii) ovary semi-inferior, bilocular with single ovule in each loculus (not inferior, trilocular, with many ovules); (iii) *Polygonum* type of embryo sac (not *Oenothera* type); (iv) endosperm absent (not present and nuclear); (v) embryo Solanad type (not Onagrad type); (vi) one cotyledon extremely reduced (both not equal); and (vii) fruit large one-seeded drupe (not loculicidal capsule).

Paeonia

The genus *Paeonia* was earlier included under the family Ranunculaceae (Bentham and Hooker; Engler and Prantl). Worsdell (1908) suggested its removal to a distinct family, Paeoniaceae. This was supported on the basis of centrifugal stamens (Corner, 1946), floral anatomy (Eames, 1961) and chromosomal information (Gregory, 1941). The genus as such has been placed in a distinct monogeneric family, Paeoniaceae, in all modern systems of classification. The separation is supported by the following embryological features: (i) centrifugal stamens (not centripetal); (ii) pollen with reticulately-pitted exine with a large generative cell (not granular, papillate and smooth, small generative cell); (iii) unique embryogeny in which early divisions are free nuclear forming a coenocytic stage, later only the peripheral part becomes cellular (not onagrad or solanad type); and (iv) seed arillate.

Exocarpos

The genus *Exocarpos* (sometimes mis-spelled *Exocarpus*) is traditionally placed under the family Santalaceae. Gagnepain and Boureau (1947) suggested its removal to a distinct family Exocarpaceae near Taxaceae under Gymnosperms on the basis of articulate pedicel, 'naked ovule' and presence of a pollen chamber. Ram (1959) studied the embryology of this genus and concluded that the flower shows the usual angiospermous character, the anther has a distinct endothecium and glandular tapetum, pollen grains shed at the 2-celled stage, embryo sac of the *Polygonum* type, endosperm cellular, and the division of zygote transverse. This confirms that the genus *Exocarpos* is undoubtedly an angiosperm, and a member of the family Santalaceae, with no justification for its removal to a distinct family. The genus is as such placed in Santalaceae in all the major systems of classification.

Loranthaceae

The family Loranthaceae is traditionally divided into two subfamilies—Loranthoideae and Viscoideae—largely based on presence of a calyculus below the perianth in the former and its absence in the latter. Maheshwari (1964) noted that the Loranthoideae has triradiate pollen grains, *Polygonum* type of embryo sac, early embryogeny is biseriate, embryo suspensor present, and viscid layer outside the vascular supply in fruit. As against this, Viscoideae have spherical pollen grains, *Allium* type of embryo sac, early embryogeny many tiered, embryo suspensor absent, and viscid layer inside the vascular supply of fruit. He thus advocated separation of the two as distinct families Loranthaceae and Viscaceae. The separation was accepted by Takhtajan (1980, 1987, 1997), Dahlgren (1980), Cronquist (1981, 1988) and Thorne (1981, 2007), APG IV (2016) shifting Viscaceae under Santalaceae.

PALYNOLOGY

The pollen wall has been a subject of considerable attention, especially to establish the evolutionary history of angiosperms. Some families, such as Asteraceae, show different types of pollen grains (**eurypalynous**), whereas several others have a single morphological pollen type (**stenopalynous**). Such stenopalynous groups are of considerable significance in systematic palynology. Pollen grains present several features of taxonomist interest. The number of nuclei present at the time of shedding is also significant. Most primitive angiosperms are shed at 2-nucleate stage, whereas in more advanced groups pollen is shed at 3-nucleate stage.

Angiosperms mostly have pollen grains of radial symmetry; bilateral symmetry being found in several gymnosperms. Most pollen grains are globose in shape, although boat-shaped, ellipsoidal and fusiform are also met in different angiosperms. Since most pollen grains at least in early stages form tetrads, the outer end of grain is termed **distal pole**, whereas the inner end where grains meet as **proximal pole**, and the line joining the two poles as **polar axis**. The line running around the pollen at right angles to the polar axis is termed as **equator**.

Pollen Aggregation

Microsporogenesis yields four microspores which mature into pollen grains. In large majority of angiosperms, the pollen grains separate prior to release. Such single pollen grains are known as **monads**. In rare cases pollen grains are released fused in pairs, when they are known as **dyads**. In many angiosperms the four microspores do not separate, and the pollen grains form a **tetrad**. Five different types of tetrads are differentiated:

1. **Tetrahedral tetrad**—four pollen grains form a tetrahedron: four grains compacted in a sphere. Such pollen grains are found in family Ericaceae.
2. **Linear tetrad**—four pollen grains arranged in a straight line as in genus *Typha*.
3. **Rhomboidal tetrad**—four pollen grains in one plane, with two separated from one another by close contact of the other two.
4. **Tetragonal tetrad**—four grains are in one plane and equally spaced as in *Philydrum*.
5. **Decussate tetrad**—four grains in two pairs, arranged at right angles to one another, as in genus *Lachnanthes*.

In some genera, such as *Calliandra* of Mimosoideae, the pollen grains are connate in a group of more than four. Such pollen grains constitute a **polyad**. A polyad generally consists of eight pollen grains, and rarely of more than ten. In some members of family Orchidaceae, as for example genus *Piperia*, large number of pollen grains form irregular groups, of which there are more than one groups in a theca. These are known as **massulae**. In subfamily Asclepiadoideae of family Apocynaceae, and several members of Orchidaceae, all pollen grains of a theca are fused into a single mass known as **pollinium**.

Pollen Wall

The pollen grain wall is made of two principal layers, outer **exine** and inner **intine**. The exine is hard and impregnated with **sporopollenin**, a substance that makes it resistant to decay, and enables preservation in fossil record. Exine is further differentiated into two layers: outer **ektexine** and inner **endexine**. The ektexine is further distinguished into basal **foot layer,** radially elongate **columella** and roof like **tectum** (Figure 4.4). In some taxa the columella may be replaced by granular middle layer. Similarly, in some primitive angiosperms tectum is lacking (**atectate** pollen grain), and the exine appears granular. Above layers of exine are clearly visible under an electron microscope, but when observed under a light microscope, the inner layer known as **nexine**, includes endexine plus foot layer of ektexine. The upper layers consisting of columella, tectum and the supratectal sculpturing constitute **sexine**.

Pollen Wall Sculpturing

Present on the outer surface of tectum are often supratectal projections, which provide a variety of sculpturing to exine wall. In some cases, lacking tectum, the sculpturing is formed by columellae.

The common types of sculpturing include: **baculate** (rod-shaped elements, each known as **baculum**), **clavate** (club-shaped elements), **echinate** (spine-like elements longer than 1 micron), **spinulose** (spine-like elements shorter than 1 micron; **scabrate**), **foveolate** (pitted surface with pores), **reticulate** (forming network, each element known as murus and space in between as lumen), **fossulate** (longitudinal grooves), **verrucate** (short wart-like elements), **gemmate** (globose or ellipsoid elements), **psilate** (smooth surface), and **striate** (having thin striations on surface).

Pollen Aperture

Pollen aperture is a specialized region of pollen wall through which the pollen tube comes out. The exine may be **inaperturate** (without an aperture) or aperturate. An aperturate pollen may have a single pore (**monoporate**), a single

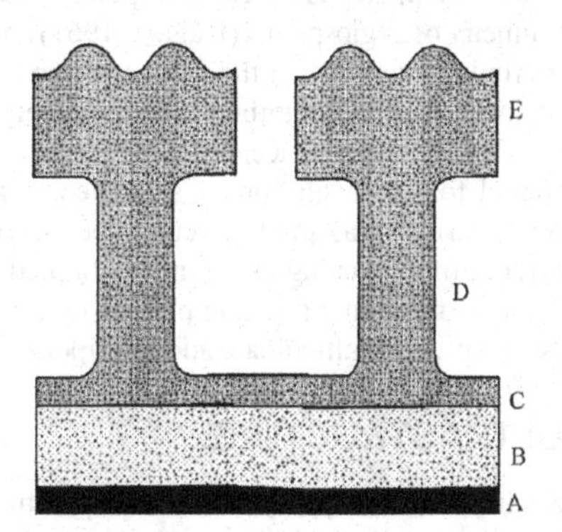

Figure 4.4: Fine structure of Pollen wall. A: Intine; **B:** Endexine; **C:** Foot layer; **D:** Baculum; and **E:** Tectum. Note the aperture formed due to break in the tectum and baculum layers.

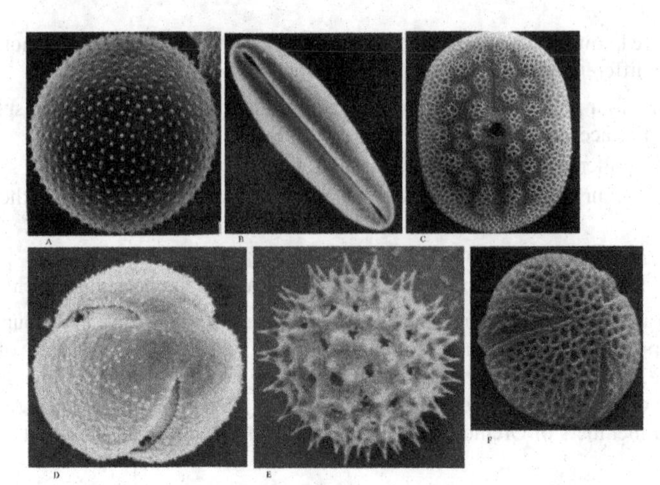

Figure 4.5: SEM of pollen grains. A: Nonaperturate pollen grain of *Persea americana*; **B:** Monosulcate pollen grain of *Magnolia grandiflora*; **C:** Monoporate pollen grain of *Siphonoglossa*; **D:** Tricolporate pollen grain of *Scaevola glabra*; **E:** Polyporate spinose pollen grain of *Ipomoea wolcottiana*; **F:** Tricolpate pollen grain of *Disanthus cercidifolius* (A, after Fahn, 1982; C, after Mauseth, 1998 courtesy R. A. Hilsenbebeck, Sul Ross State University; F, after Endress, 1977; rest, after Gifford and Foster, 1988).

slit running at right angles to the equator (**monocolpate**), three slits (**tricolpate**), three pores (**triporate**) three slits each with a geminate pore in middle (**tricolporate**), with many pores (**multiporate**) accompanied by a variety of surface ornamentations (Figure 4.5). Pollen with one or more slits located at the polar end is accordingly termed, **monosulcate**, **disulcate** and **trisulcate**, depending on the number of slits. Pollen grain with slits joined at poles is termed **syncolpate**. Aperture having three branches is termed **trichotomosulcate**.

Monocolpate condition is widely spread in primitive dicots and most monocots. The pollen of anemophilous plants is usually small, rounded, smooth, rather thin-walled and dry with shallow furrows. Anemophilous pollen is found in *Populus*, Poaceae, Cyperaceae, Betulaceae and several other families. Insect- and bird-pollinated pollen, on the other hand, is large, sculptured and often coated with adhesive waxy or oily substance. The pollen of Asteraceae is generally highly elaborate but simplification towards loss of sculpturing has occurred in several genera with wind pollination. The vestigial scattered patches of adhesive layer on wind pollinated pollen have been considered as evidence of the derivation of anemophily from entomophily.

Fossil studies over the last three decades have confirmed monosulcate pollen of *Clavitopollenites* described (Couper, 1958) from Barremian and Aptian strata of the Early Cretaceous of southern England (132 to 112 Mya) to be the oldest recorded angiosperm fossil with distinct sculptured exine, resembling the pollen of extant genus *Ascarina*.

Brenner and Bickoff (1992) recorded similar but inaperturate pollen grains from the Valanginian (ca 135 Mya) from the Helez formation of Israel, now considered being the oldest record of angiosperm fossils (Taylor and Hickey, 1996). This last discovery has led to the belief that the earliest angiosperm pollen were without an opening, the monosulcate types developing later.

Many claims of angiosperm records from the strata, earlier than the Cretaceous were made, but largely rejected. Erdtman (1948) described *Eucommiidites* as a tricolpate dicotyledonous pollen grain from the Jurassic. This, however, had bilateral symmetry instead of the radial symmetry of angiosperms (Hughes, 1961) and a granular exine with gymnospermous laminated endexine (Doyle et al., 1975). Among examples of the role of pollen grains in systematics is *Nelumbo* whose separation from Nymphaeaceae into a distinct family Nelumbonaceae is largely supported by the tricolpate pollen of *Nelumbo* as against the monosulcate condition in Nymphaeaceae.

Brenner (1996) proposed a new model for the evolutionary sequence of angiosperm pollen types. The earliest angiosperm pollen (from the Valanginian or earlier) was small, circular, tectate-columellate and without an aperture. In the Hauterivian, there was possible occurrence of thickening of the intine coupled with thickened endexine and evolution of the sulcus. A considerable diversification of these monosulcate pollens occurred in the Barremian. Tricolpate pollen evolved in northern Gondwana in the lower Aptian. Multicolpate and multiporate pollen arose at a later stage.

MICROMORPHOLOGY AND ULTRASTRUCTURE

Although widely used in lower plants, electron microscopy has been a comparatively new approach for flowering plants. The finer details of external features (**micromorphology**) have been explored in the recent years by Scanning Electron Microscopy (**SEM**), whereas the minute details of cell contents (**ultrastructure**) have been discerned through Transmission

Electron Microscopy (**TEM**). On an average basis, the resolution power of SEM is 250A (20 times as good as optical microscope, but 20 times lesser than TEM). Behnke and Barthlott (1983) have made extensive studies of SEM and TEM characters. In most of the examples studied, Electron Microscopy (EM) characters proved to be stable and unaffected by environmental conditions.

Micromorphology

SEM studies have been made primarily on pollen grains, small seeds, trichomes and surface features of various organs. In most of these organs (except pollen grains), the studies involved the epidermis. The value of epidermal studies lies in the fact that an epidermis covers almost all the organs and is always present, even in herbarium specimens. The epidermis is thick and stable in SEM preparations and is little affected by environment. However, it is important to note that only comparable epidermis should be studied (e.g., petals of all plants, leaves of all plants, not petals of some and leaves of others). Most of SEM studies have been concentrated on seed-coats which are usually thick-walled and stable in vacuum, thus facilitating quick preparation for SEM examination without the need for complicated dehydration techniques. The micromorphology of the epidermis includes the following aspects:

Primary Sculpture

This refers to the arrangement and shape of cells. The **arrangement of cells** is specific for several taxa. In Papaveraceae, seed-coat cells by a particular arrangement form a reticulate supercellular pattern (Figure 4.6D), which is a family character. The members of Caryophyllaceae, Portulacaceae and Aizoaceae exhibit a specific arrangement and orientation of smaller and larger cells known as "**centrospermoid**" pattern. There is specific distribution of long and short cells over the veins in the family Poaceae. The **shape of cells** is mainly determined by the **outline** of the cells, **boundaries** of the walls, **relief**, and cell wall **curvature** (flat, convex or concave). Outline of cells may be isodiametric (usually tetragonal or hexagonal: Figure 4.6B and C), elongated in one direction (Figure 4.6F). Cell boundaries of superficially visible anticlinal walls may be straight (Figure 4.6B and C), irregularly curved or undulated (S-, U-, omega-, V-types) and are of high taxonomic significance in family Cactaceae and Orchidaceae. Relief of the anticlinal boundary may be channeled or raised. In primitive members of Cactaceae, cell junctions are depressed, whereas in derived Cactinae they are raised. The curvature of outer periclinal walls may be flat, concave (Figure 4.6B) or convex.

Secondary Sculpture

The secondary sculpture (Microrelief) is formed by the deposition of cuticle over the outer wall or due to secondary wall thickenings, often shrinking and collapsing in desiccated cells. It may be smooth, striate (Figure 4.6C), reticulate

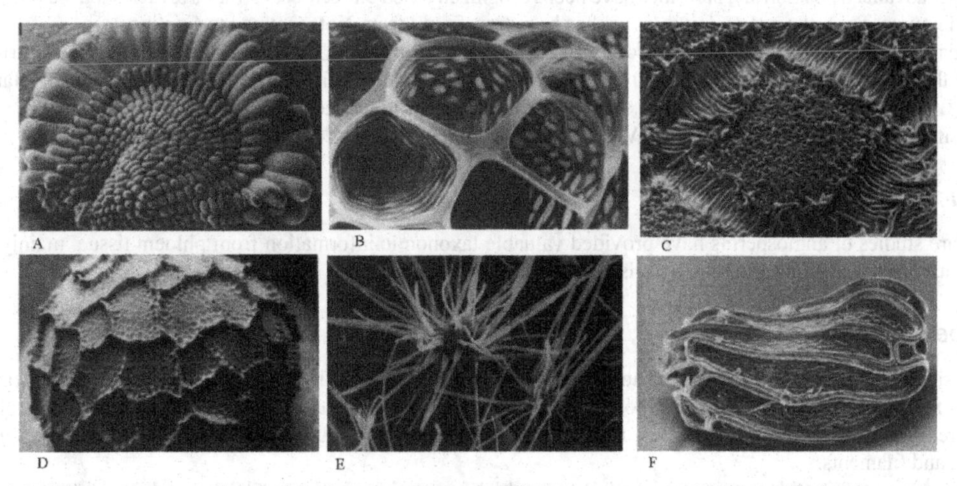

Figure 4.6: SEM seed characteristics of angiosperms. **A:** Seed of *Sceletium campactum* (Aizoaceae) showing centrospermoid cell arrangement; **B:** Seed-coat of *Aeginatia indica* (Orobanchaceae) with isodiametric deeply concave cells and reticulate secondary structure; **C:** Single isodiametric tetragonal cell of seed-coat of *Matucana weberbaueri* (Cactaceae) with heavy secondary sculpturing; **D:** Seed of *Eschscholzia californica* (Papaveraceae) with cells arranged to form a supercellular net-like pattern; **E:** Seed-coat of *Jacaranda macarantha* (Bignoniaceae) with stellate epicuticular sculpture; **F:** Seed of *Dichaea* sp. (Orchidaceae) almost one cell long with heavy marginal thickenings and irregular secondary sculpture (From Barthlott, 1984).

(Figure 4.6B) or micro-papillate (verrucose). All members of Urticales have curved trichomes with silicified cuticular striations at the base, and micro-papillations on the trichome body. This single character of trichomes allows for precise circumscription of the order Urticales (Barthlott, 1981). Loasiflorae is circumscribed by unicellular irregularly hooked trichomes. Secondary wall thickenings are always of a high taxonomic significance. In Orchidaceae, for example longitudinal striations caused by underlying secondary thickenings are restricted to all members of Catasetinae.

Tertiary Sculpture

Tertiary sculpture is formed by epicuticular secretions such as waxes and other mucilaginous adhesive lipophilic substances and shows a variety of patterns. Secondary and tertiary sculpturing are mutually exclusive as the presence of waxes would invariably mask the cuticle; the cuticle would be visible only if there are no wax deposits. Winteraceae have a particular type and distribution of wax-like secretions (alveolar material not soluble in lipid solvents) on their stomata, similar to gymnosperms, and absent in all other angiosperms.

In monocots orientation and pattern of epicuticular waxes seem to provide a new taxonomic character of high systematic significance. Four types of wax patterns and crystalloids have been distinguished (Barthlott and Froelich, 1983):

1. **Smooth wax layers** in the form of thin films, common in angiosperms.
2. **Non-oriented wax crystalloids** in the form of rodlets or platelets with no regular pattern. These are common in dicots and Lilianae groups of monocots.
3. **Strelitzia wax-type** with massive compound wax projections composed of rodlet-like subunits that form massive compound plates around the stomata. This wax type is found in Zingiberanae, Commelinanae, and Arecanae. It is also found in Velloziales, Bromeliales, and Typhales, which further differ from other Lilianae in a starchy endosperm.
4. **Convallaria wax-type** with small wax platelets arranged in parallel rows, which cross the stomata at right angle and form a close circle around each polar end of the stoma, like the lines of an electromagnetic field. This type is restricted to Lilianae only.

Tertiary sculpture is generally lacking from seeds. In orchidaceae, however, certain tribes possess epicuticular waxes on their seed-coats. Seed-coats of *Jacaranda* (Figure 4.6E) have stellate epicuticular sculpture known as 'star scales', a feature characteristic of this genus. Many members of Aizoaceae are characterized by seed-coat with epicuticular secretions forming long upright rodlets and small rodlets lying on the cell surface.

Cactaceae is a huge family commonly divided into three subfamilies, of which Cactoideae includes 90 percent species but its classification is difficult because of uniform floral characters, pollen morphology and plasticity. Barthlott and Voit (1979) analyzed 1050 species and 230 genera by SEM for seed coat structure in the family Cactaceae. The simple unspecialized testa of Pereskoideae supports its ancestral position. Opuntioideae has a unique seed with a hard aril, thus confirming its isolated position, also indicated by pollen morphology. Cactoideae shows complex diversity, confirming its advanced position and subtribes have been recognized based on seed-coat structure, each subtribe possessing distinctive features. Thus, the genus *Astrophytum* has been transferred from Notocactinae to Cactinae.

Orchidaceae is another large family with complicated phylogenetic affinities. Minute '**dust seeds**' show microstructural diversity of the seed-coat. Studies of over 1000 species (Barthlott, 1981) have helped in better subdivision into subfamilies and tribes. Barthlott also supports the merger of Cypripediaceae with Orchidaceae, a suggestion incorporated in several recent classification (Judd et al., 2002; APG IV, 2016; Thorne, 2007; Stevens, 2018).

Ultrastructure

Ultrastructure studies of angiosperms have provided valuable taxonomic information from phloem tissue, mainly sieve tube elements. Besides this, information has also come from studies of seeds.

Sieve-tube Plastids

Studies on sieve-tube plastids were first initiated by Behnke (1965) in the family Dioscoreaceae. Since then, nearly all angiosperm families have been investigated for the taxonomic significance of these plastids. All sieve-element plastids contain starch grains differing in number, size and shape. The protein accumulates in specific plastids in the form of crystalloids and filaments.

Thus, two types of plastids are distinguished: **P-type** which accumulate proteins and **S-type** which do not accumulate proteins. Starch accumulation is of no primary importance in classification, since it may be present or absent in both types of plastids. P-type plastids are further divided into six subtypes (Behnke and Barthlott, 1983) (Figure 4.7):

(i) **PI-subtype**. The plastids contain single crystalloids of different sizes and shapes and/or irregularly arranged filaments. This subtype is thought to be the most primitive in flowering plants, mainly Magnoliales, Laurales and Aristolochiales.

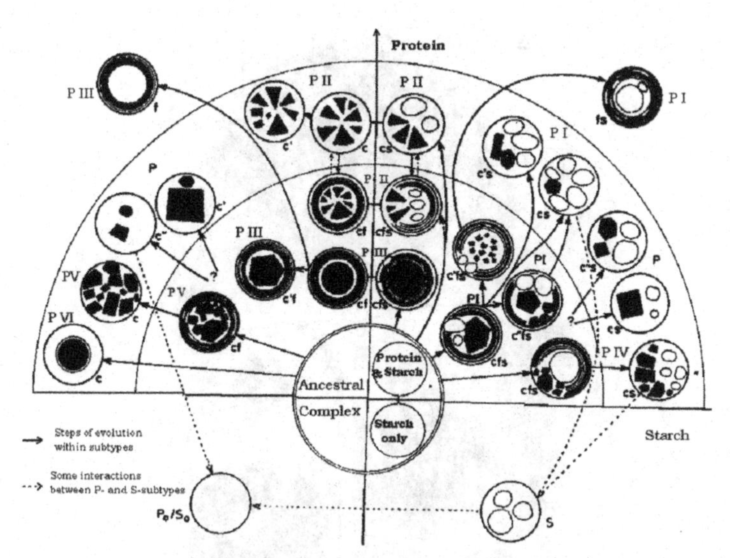

Figure 4.7: Various forms of **sieve-tube plastids** and their possible evolution (After Behnke and Barthlott, 1983).

(ii) **PII-subtype.** This subtype contains several cuneate crystalloids oriented towards the centre of the plastid. All investigated monocots contain this subtype. It is significant to note that only members of dicots with this subtype, *Asarum*, and *Saruma* of Aristolochiaceae are widely regarded among the most primitive members of dicots, a possible link between monocots and dicots.

(iii) **PIII-subtype.** This subtype contains a ring-shaped bundle of filaments. PIII-subtype is confined to Centrospermae (Caryophyllales) and the removal of Bataceae and Gyrostemonaceae has been supported by the absence of this subtype in these families. Further, forms are recognized based on the presence or absence of crystalloids (Figure 4.8) into **PIIIa** (globular crystalloid), **PIIIb** (hexagonal crystalloid) and **PIIIc** (without crystalloid). Based on the distribution of these forms, Behnke (1976) proposed division of the order into three family-groups which exactly correspond to the three suborders Caryophyllineae, Chenopodineae and Phytolaccineae, earlier established by Friedrich (1956). Whereas Takhtajan had recognized these three suborders in his 1983 revision, in his final revision of his classification, he merged Phytolaccineae with Caryophyllineae, thus recognizing only two suborders Caryophyllineae and Chenopodineae. Of the three orders recognized in Caryophyllidae of Takhtajan, only Caryophyllales contains PIII-subtype plastids while the other two orders, Polygonales and Plumbaginales, contain S-type plastids. Behnke (1977) as such, advocated retention of only Caryophyllales under Caryophyllidae and removal of the other two orders to subclass Rosidae whose members also contain S-type plastids. The suggestion was not accepted by Takhtajan (1987) and Cronquist (1988), who retained all the three orders under Caryophyllidae. Takhtajan, however, placed the three orders under separate superorders (Cronquist does not recognize superorders). On further intensive studies of plastids within the group, Behnke (1997) advocated the removal of genus *Sarcobatus* from the family Chenopodiaceae on the basis of the presence of **PIIIcf** plastids and absence of **PIIIf**, which are characteristic of family Chenopodiaceae. He places the genus in an independent family, Sarcobataceae, incorporated in APG IV (2016).

(iv) **PIV-subtype.** The plastid contains a few polygonal crystalloids of variable size. This subtype is restricted to the order Fabales.

(v) **PV-subtype.** The plastid contains many crystalloids of different sizes and shapes. This subtype is found in the order Ericales and family Rhizophoraceae.

(vi) **PVI-subtype.** The plastid contains a single circular crystalloid. This subtype is found in family Buxaceae.

Dilated Cisternae

Dilated Cisternae (**DC**) were first described by Bonnett and Newcomb (1965) as dilated sections of endoplasmic reticulum in the root cells of *Raphanus sativus*. Originally found in Brassicaceae and Capparaceae, DC have now been found in several other families of angiosperms but are concentrated in the order Capparales (Brassicaceae and Capparaceae) and form a part of the character syndrome of this order. The DC may be utricular, irregular or vacuole-like in form with filamentous, tubular or granular contents. They have been proposed to be functionally associated with glucosinolates and myrosin cells found in this order.

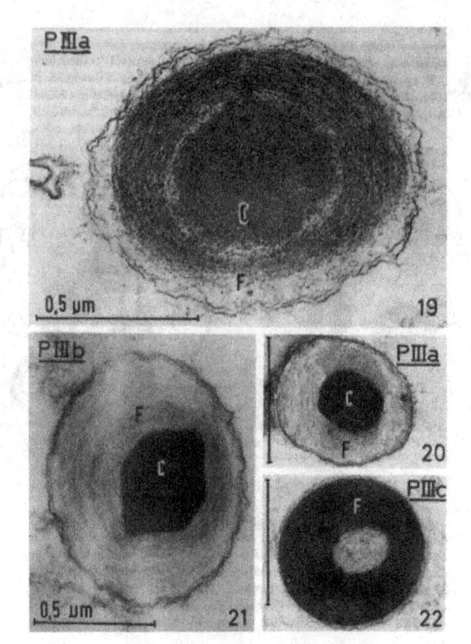

Figure 4.8: Different forms of PIII–subtype sieve-element plastids all with ring–like bundle of filaments (F). **19** and **20:** PIIIa with globular crystalloid (C); **21:** PIIIb with polygonal crystalloid; **22:** PIIIc without crystalloid. (From Behnke, 1977).

Phloem (p-) Proteins

P-proteins are found only in sieve elements of angiosperms and occur in the form of filaments or tubules. These assemble into large discrete bodies and are not dissolved during maturation of the sieve-element, unlike single membrane organelles. The composition and three-dimensional arrangement of these proteins exhibit taxonomic specificity. They are dispersed over the entire cell as the cell matures but, in some dicots, a single non-dispersive (crystalline) body of various shapes may be found in addition to dispersive ones. Crystalline bodies are absent in monocots. Their shape is often specific and thus of taxonomic importance. Globular crystalline bodies are found in Malvales and Urticales. Fabanae, which is characterized by PIV-subtype plastids, has spindle-shaped crystalline bodies in the family Fabaceae. The feature has supported the transfer of *Swartzia* to Fabaceae.

Nuclear Inclusions

Nuclear inclusions in the form of protein crystals occur in phloem- and Ray parenchyma, primarily in the families of Asteridae. Five types of crystals have been differentiated. Structural differences are significant for classification within Scrophulariaceae and Lamiaceae (Speta, 1979). Protein crystals in sieve-tube elements have also been reported in Boraginaceae and may prove useful.

Non-phloematic TEM Characters

Protein bodies in seeds through TEM, SEM and dispersive X-ray techniques have demonstrated their significance if qualitative and quantitative aspects are both considered. Similarly, SEM studies of starch grains are also potential sources of information of taxonomic significance.

CHROMOSOMES

Chromosomes are the carriers of genetic information and as such have a considerable significance in evolutionary studies. Increased knowledge about chromosomes and their behavior has largely been responsible for extensive biosystematic studies and development of the biological species concept. During the first quarter of the twentieth century, chromosomal data were relatively sparse. Such information has markedly increased over the last few decades, however, with ample useful information coming from studies of the banding pattern. Three types of chromosomal information have been of significance in Systematics.

Chromosomal Number

Extensive records of chromosome numbers are available in the works of Darlington and Janaki-Amal (1945), Darlington and Wylie (1955), Federov (1969) and Löve et al. (1977). The International Association of Plant Taxonomy (IAPT) has also been publishing an **Index to Plant Chromosome Numbers** in its series *Regnum Vegetabile*. Between 1967 and 1977, the series published 9 volumes mostly forming annual lists of chromosome numbers. An updated server of the Missouri Botanical Garden maintains the records of chromosome numbers and can be queried for online information about plant species. The chromosome counts are usually reported as diploid number ($2n$) from mitosis of **sporophytic** tissue but when based on mitosis in **gametophytic** tissue or on meiosis studies, counts are reported as haploid (n). The gametophytic chromosome number of diploid species is designated as **base-number** (x). In diploid species as such $n = x$, whereas in polyploid species n is in multiples of x. A hexaploid species with $2n = 42$ will thus have $n = 21$, $n = 3x$ and $2n = 6x$.

The chromosome number in angiosperms exhibits considerable variation. The lowest number ($n = 2$) is recorded in *Haplopappus gracilis* (Asteraceae) and the highest ($n = 132$) in *Poa littoroa* (Poaceae). The alga *Spirogyra cylindrica* also contains $n = 2$, whereas the record of the highest chromosome number ($n = 630$) is found in *Ophioglossum reticulatum* (Pteridophytes). Such a range of variation ($n = 2$ to $n = 132$), however, within nearly a quarter a million species of angiosperms, may not be very significant in taxonomic delimitation, but there have been instances of the isolated role of studies on chromosomes. Raven (1975) provided a review of chromosome numbers at the family level in angiosperms. He concluded that the original base-number for angiosperms is $x = 7$ and that comparisons at the family level are valid only when the base-number (and not n or $2n$) is used. The family Ranunculaceae is dominated by genera with large chromosomes (and $x = 8$). The two genera *Thalictrum* and *Aquilegia*—originally placed in two separate subfamilies or tribes (and even two separate families Ranunculaceae and Helleboraceae by Hutchinson, 1959, 1973 along with other achene bearing and follicle bearing genera, respectively)—are distinct in having small chromosomes (and $x = 7$) and as such have been segregated into a distinct tribe. The genus *Paeonia* with very large chromosomes (and $x = 5$) has been separated into a distantly related family Paeoniaceae, a placement which has been supported by morphological, anatomical and embryological data. Significant records in other families include Rosaceae with $x = 17$ in subfamily Pomoideae, whereas other subfamilies have $x = 7$, 8 or 9. In Poaceae similarly subfamily Bambusoideae has $x = 12$, whereas Pooideae has $x = 7$. *Spartina* was for long placed in the tribe Chlorideae ($x = 10$) although its chromosomes ($x = 7$) were at variance. Marchant (1968) showed the genus to have, in fact $x = 10$, thus securing placement within Chlorideae.

The classical study of the genus *Crepis* (Babcock, 1947) based separation from the closely related genera on chromosomal number and morphology. This led to the separation of the genus *Youngia* and merger of *Pterotheca* with *Crepis*. Similarly, in the genus *Mentha* which has small, structurally uniform chromosomes, the chromosome numbers provide strong support for subdivision into sections *Audibertia* ($x = 9$), *Pulegium* ($x = 10$), *Preslia* ($x = 18$) and *Mentha* ($x = 12$).

The duplication of chromosome numbers leading to **polyploidy** may prove to be of taxonomic significance. The grass genus *Vulpia* contains diploid ($2n = 14$), tetraploid ($2n = 28$) and hexaploid ($2n = 42$) species. The genus is divided into five sections, of which three contain only diploids, one diploids and tetraploids and one all three levels of ploidy. It is presumed that tetraploid and hexaploid species of *Vulpia* arose from diploid progenitors. The duplication of chromosome number of a diploid species may form a tetraploid (**autopolyploid**). Such a polyploid, however, does not show any or at most may show minor differences from the diploid species, and is rarely recognized as an independent taxonomic entity. The hybrid between two diploid species contains one genome from either parent and thus, generally doesn't survive due to failure of chromosomal pairing during meiosis. Hybridization followed by duplication of chromosomes establishes a tetraploid (**allopolyploid; amphiploid**) with normal pairing as both genomes are in pairs. Such a tetraploid hybrid with distinct characteristics may be recognized as an independent species. A triploid hybrid between a diploid species and a tetraploid species may, similarly, not survive as genome from the diploid parent would exhibit the problem of pairing at meiosis but the hybridization followed by duplication leading to hexaploidy can form a perfectly normal independent species. Such facts have led to the detection of hybrids or confirmation of suspected hybrids. *Senecio* (Asteraceae) includes the diploid *S. squalidus* ($2n = 20$), the tetraploid *S. vulgaris* ($2n = 40$) and the hexaploid *S. cambrensis* ($2n = 60$). The last is intermediate in morphology between the first two and is found in the area where these two grow. Additionally, sterile triploid hybrids between two species have been reported. It seems clear that *S. cambrensis* is an allohexaploid between the other two species (Stace, 1989). Similarly, based on chromosome number and karyotype, Owenby (1950) concluded that *Tragopogon mirus* ($2n = 24$), a tetraploid species arose as an amphiploid between two diploid species, *T. dubius* and *T. porrifolius* ($2n = 12$).

Whereas a species generally shows a single chromosome number, certain populations or infraspecific taxa (subspecies, variety, forma) may sometimes show a different chromosome number (or even different chromosomal morphology). Such populations or infraspecific taxa constitute **cytotypes**.

Chromosomal Structure

Chromosomes show considerable variation in size, position of centromere (Figure 4.9) and presence of secondary constriction. The chromosomes are commonly differentiated as **metacentric** (with centromere in middle), **submetacentric** (away from middle), **acrocentric** (near the end) or **telocentric** (at the end). The chromosomes are also characterized by their size. In addition, the occurrence and position of **secondary constriction,** which demarcates a **satellite** is important in chromosomal identification and characterization. The identification of satellites is often difficult, and especially when the secondary constriction is very long, a satellite may be counted as a distinct chromosome. This situation has often led to erroneous chromosome counts. The structure of the chromosome set (genome) in a species is termed **karyotype** and is commonly diagrammatically represented in the form of an **ideogram** (Figure 4.10) or **karyogram**. An analysis of a large number of studies has led to the conclusion that a **symmetrical karyotype** (chromosomes essentially similar and mainly metacentric) is primitive and an **asymmetric karyotype** (different types of chromosomes in a genome) advanced, the latter commonly found in plants with specialized morphological features, such as *Delphinium* and *Aconitum*.

An interesting example of utilization of chromosomal information is family Agavaceae. The family contains about 16 genera such as *Agave* (and others formerly placed in Amaryllidaceae due to inferior ovary) and *Yucca* (and others formerly placed in Liliaceae due to superior ovary). These genera were shifted and brought into Agavaceae based on great overall similarity. This was supported by the distinctive **bimodal karyotype** of Agavaceae consisting of 5 large chromosomes and 25 small ones. Rudall et al. (1997) advocated the transfer of *Hosta* (placed in Hostaceae; Hesperocallidaceae by Thorne, 1999), *Camassia* and *Chlorogalum* (both placed under Liliaceae by Hutchinson, 1973; Hyacinthaceae by Thorne, 1999) to family Agavaceae based on possession of bimodal karyotype, a suggestion incorporated by Judd et al. (2002) and Thorne (2003). Rousi (1973), from his studies on the genus *Leontodon,* showed that data on the basic number, chromosome length, centromeric position and the occurrence of satellites provide evidence for the relegation of the former genus *Thrincia* ($x = 4$) as a section of subgenus *Apargia* along with section *Asterothrix* ($x = 4, 7$). The subgenus *Leontodon* is distinct with $x = 6$ or 7, and a different chromosome morphology.

Cyperaceae and Juncaceae were earlier placed far apart due to distinct floral structure. Both families have small chromosomes without distinctive centromeres, the latter may be diffuse or non-localized. These families as such are now considered to be closely related. Such chromosomes (**holocentric chromosomes**) do not depend on a discrete centromere for meiosis and mitosis and may undergo fragmentation with no deleterious effect. This may result in variable chromosomal counts. In the *Luzula spicata* group, chromosomal counts are reported to be $2n = 12$, 14, and 24. Interestingly, the total

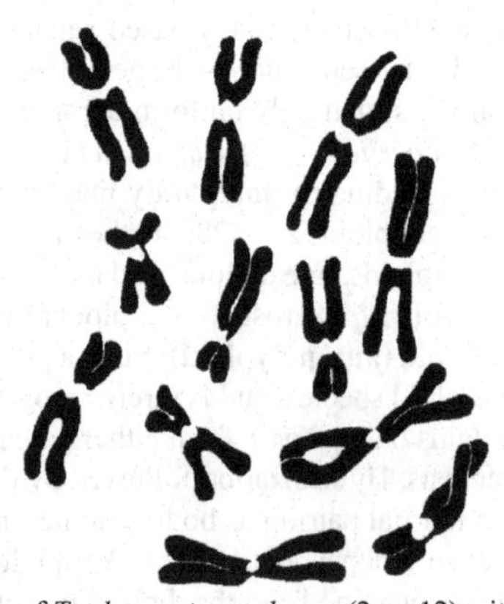

Figure 4.9: Mitotic chromosomes of *Tradescantia spathacea* (2n = 12) with sister chromatids and centromere.

Figure 4.10: Ideogram of the somatic complement of *Allium ampeloprasm.* Of the 32 somatic chromosomes, 8 show secondary constriction (courtesy Prof. R. N. Gohil).

chromosomal volume is the same and the higher chromosome number is the result of fragmentation (**agmatoploidy**) of these holocentric chromosomes. Different chromosome numbers may often occur in different cells of the same root-tip (**mixoploidy**). The occurrence of accessory chromosomes (known as **B-chromosomes**) in higher plants generally does not have a significant effect on morphology and, thus, is of little taxonomic importance. B-chromosomes in bryophytes, contrarily, are very small (termed **m-chromosomes**) and often highly diagnostic.

In recent years, considerable breakthrough has been achieved in the study of **banding patterns** of chromosomes using Giemsa and fluorochrome stains. Already different techniques such as **C-banding**, **G-banding**, **Q-banding** and **Hy-banding** are in use and help in clearly distinguishing the heterochromatic and euchromatic regions. C-banding is very useful in indicating the position of centromeres in cases where they cannot be identified by conventional staining.

The technique of **silver-staining** has been developed to highlight **NOR** (nucleolar organizing region). An interesting study of the chromosomes of top onion (variously recognized as *Allium cepa* var. *viviparum* or *A. fistulosum* var. *proliferum*) as also those of *A. Cepa* and *A. fistulosum* was done by Schubert, Ohle and Hanelt (1983). By Giemsa banding pattern and silver-staining studies, they concluded that some chromosomes of top onion resemble *A. cepa* and others resemble *A. fistulosum*. Of the two satellites, one resembles either species. Top onion is as such a **pseudodiploid** with no homologous pair. The study confirmed that top onion is a hybrid between the two aforesaid parents, and thus would be better known as *A.* × *proliferum* (Moench) Schrad. (based on *Cepa proliferum* Moench), and not as a variety of either species. Interestingly, the top onion owes its existence to the bulbils, which are produced in place of an inflorescence and ensure the multiplication of the hybrid, which is otherwise sterile.

Chromosomal Behavior

The fertility of a plant is highly dependent on the ability of meiotic chromosomes to pair (**synapsis**) and their subsequent separation. The meiotic behavior of chromosomes enables comparison between genomes to detect the degree of homology, especially when they are a result of hybridization. A greater degree of genomic non-homology results in either failure of pairing (**asynapsis**) or a loose pairing of chromosomes without chiasmata so that chromosomes fall apart before metaphase (**desynapsis**). In extreme cases, the entire genome may fail to pair. The genome analysis of suspected hybrids has helped in establishing the parentage of several polyploid species.

A diploid hybrid between two species generally exhibits failure of meiotic pairing due to non-homology of genomes resulting in hybrid sterility, but when hybridization is followed by duplication of chromosomes to form a tetraploid hybrid, the latter shows normal pairing between the two genomes derived from the same parent and is generally fertile. A triploid hybrid may, similarly, be sterile but a hexaploid one fertile. Genome analysis has confirmed that the hexaploid *Senecio cambrensis* is allohexaploid between tetraploid *S. vulgaris* and diploid *S. squalidus*. Similarly the tetraploid *Tragopogon mirus* is the result of hybridization between the two diploid species *T. dubius* and *T. porrifolius*. The most significant case, however, is the common bread wheat *Triticum aestivum*, a hexaploid with AABBDD genome. Genome analyses have confirmed that genome A is derived from the diploid *T. monococcum*, B from *Aegilops speltoides*, both genomes being represented in the tetraploid *T. dicoccum*. Genome D is derived from the diploid *Aegilops tauschii*.

CHEMOTAXONOMY

Chemotaxonomy of plants is an expanding field of study and seeks to utilize chemical information to improve upon the classification of plants. Chemical evidence has, in fact, been used ever since man first began to classify plants as edible and inedible, obviously based on their chemical differences. Chemical information about medicinal plants in herbals published nearly five centuries back was concerned with localization and application of physiologically-active secondary metabolites such as saponins and alkaloids. Knowledge about chemistry of plants greatly increased during the eighteenth and nineteenth centuries. The greatest interest has been generated over the last 40 years, however, with the development of improved techniques for studying biological molecules, especially proteins and nucleic acids. In recent years, interest has focused on the study of **allelochemy** and realization of the concept that the animal kingdom and the plant kingdom have experienced a chemical **coevolution**. Plants continuously evolve new defensive chemical mechanisms to save themselves from predators, and animals evolve methods to overcome these defenses. In the process, some plant species have developed animal hormones, thus disturbing the hormonal levels of animals if ingested.

A large variety of chemical compounds are found in plants and quite often the biosynthetic pathways producing these compounds differ in various plant groups. In many instances the biosynthetic pathways correspond well with existing schemes of classification based on morphology. In other cases, the results are at variance, thus calling for revision of such schemes. The natural chemical constituents are conveniently divided as under:

***Micromolecules*:** Compounds with low molecular weight (less than 1000).

Primary metabolites: Compounds involved in vital metabolic pathways—citric acid, aconitic acid, protein amino acids, etc.

Secondary metabolites: Compounds which are the by-products of metabolism and often perform non-vital functions—non-protein amino acids, phenolic compounds, alkaloids, glucosinolates, terpenes, etc.

***Macromolecules*:** Compounds with high molecular weight (1000 or more).

Non-semantide macromolecules: Compounds not involved in information transfer—starches, celluloses, etc.

Semantides: Information carrying molecules—DNA, RNA and proteins.

The utilization of studies on DNA and RNA for understanding of phylogenetic relations has received a great boost over the last decade, meriting the establishment of a new field referred to as **Molecular Systematics**, and would be dealt separately after chemotaxonomy. Only proteins would be described in this section.

Primary Metabolites

Primary metabolites include compounds, which are involved in vital metabolic pathways. Most of them are universal in plants and of little taxonomic importance. Aconitic acid and citric acid, first discovered from *Aconitum* and *Citrus* respectively, participate in Krebs cycle of respiration and are found in all aerobic organisms. The same is true of the 22 or so amino acids forming proteins, and the sugar molecules, which are involved in the Kalvin cycle of photosynthesis. The quantitative variations of these primary metabolites may, however, be of taxonomic significance sometimes. In *Gilgiochloa indurata* (Poaceae), **alanine** is the main amino acid in leaf extracts, **proline** in seed extracts and **asparagine** in flower extracts. Rosaceae is similarly rich in arginine.

Secondary Metabolites

Secondary metabolites perform non-vital functions and are less widespread in plants as compared to primary metabolites. These are generally the by-products of metabolism. They were earlier considered to be waste products, having no important role. Recently, however, it was realized that they are important in chemical defense against predators, pathogens, allelopathic agents and also help in pollination and dispersal (Swain, 1977). Gershenzon and Mabry (1983) have provided a comprehensive review of the significance of secondary metabolites in higher classification of angiosperms. The following major categories of secondary metabolites are of taxonomic significance:

Non-protein Amino Acids

A large number of amino acids not associated with proteins are known (more than 300 or so). Their distribution is not universal but specific to certain groups and, as such, holds promise for taxonomic significance. **Lathyrine** is, thus, known only from *Lathyrus*. **Canavanine** occurs only in Fabaceae and is shown (Bell, 1971) to be a protection against insect larvae. These amino acids are usually concentrated in storage roots and, as such, root extracts are generally used for their study.

Phenolics

Phenolic compounds form a loose class of compounds, based upon a phenol (C_6H_5OH). **Simple phenolics** are made of a single ring and differ in position and number of OH groups. These are water soluble and are localized in vacuoles in cell and are found in combination with sugars as glycoside. (**Simple phenolics** can be tested by extraction with ethanol. Take 5–6 gm of chopped leaf tissue in a beaker and add 30 ml of 70% ethanol; heat over water bath at 60–70 degree centigrade for 20 minutes; filter and concentrate filtrate over water bath till about 0.5 ml is left; load the sample on Whatman paper (No. 1) using BAW: Butanol, Acetic acid and water in ratio of 4:1:5; run chromatogram, dry and observe under UV light; spray with mixture (1:1) of 1% Ferric chloride and 1% Potassium ferricyanide and calculate Rf value). These are widely distributed in the plant kingdom; common examples being catechol, hydroquinone, phloroglucinol and pyragallol. Coumarins, a group of natural phenolics, have a characteristic smell. The crushed leaves of *Anthoxanthum odoratum* can thus be identified by this characteristic odor. More than 300 coumarins have been reported from nearly 80 families of plants. They are a group of lactones formed by ring closure of hydroxycinnamic acid. **Lignin** is a highly branched polymer of three simple phenolic alcohols. Whereas gymnosperm lignin is composed of coniferyl alcohol subunits, the angiosperm lignin is a mixture of coniferyl and sinapyl alcohol subunits. The alcohols are oxidized to free radicals by peroxidase enzyme, and the freed radicals react to form lignin.

Flavonoids, the more extensively studied compounds, are based on a flavonoid nucleus consisting of two benzene rings joined by a C3 open or closed structure (Figure 4.11) (Presence of **flavonoids** can be detected as follows: Finely chop 5 gm of flower petals or tepals in beaker; add 20 ml of 2N HCl, cover with aluminium foil, and heat at 80–90 degree centigrade for 30–40 minutes in water bath; filter and extract filtrate with 15–20 ml ethyl acetate in separating funnel, shake and allow solvent to evaporate; two layers are formed, upper organic and lower inorganic aqueous layer (mostly

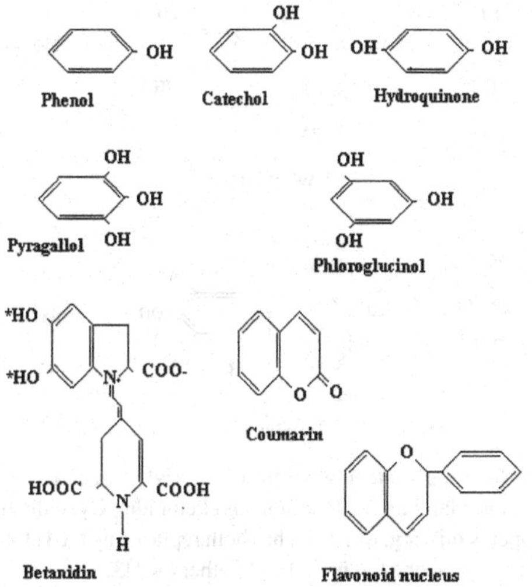

Phenol Catechol Hydroquinone

Pyragallol Phloroglucinol

Betanidin Coumarin Flavonoid nucleus

Figure 4.11: Structure of important phenolic molecules and a betalain (Betanidin). * indicates the position of sugar.

anthocyanins); collect them in two separate beakers; label beaker with upper organic layer as **B**; if aqueous layer is colored heat for 5–10 minutes to expel ethyl acetate; put back in separating funnel, add 2–4 ml of amyl alcohol, shake, transfer upper organic layer to beaker and mark it as **A**; heat both beakers to dryness on water bath uncovered; to each add 1% methanolic HCl; load each sample on two circular whatman (No. 1) filter paper, load spot in center using Forestall solvent: HCl, Acetic acid and water (3:10:30); mark spots visually; observe one set of chromatograms of A and B under UV; expose second set to ammonia vapours by rotating discs over open mouth of NH_3 bottle and observe under UV. **Phenyl propanoids** can be detected similarly except that extraction is done using diethyl ether instead of ethyl estate, only organic layer is retained and the solvent use for chromatography is BAW (Butanol:Acetic acid:Distilled water–4:1:5)). Common examples are flavonols (mainly colorless and commonly occurring as co-pigments, yield bright yellow spot in chromatogram after acid hydrolysis), flavones (similar, yield dull brown spots), glycoflavones, biflavonyls (similar, yield dull absorbing spots on BAW), isoflavones (colorless, often found in roots of legumes), flavanones (colorless, occur in leaves, citrus fruits, yield red color with HCl), chalcones or aurones (usually occur in yellow flowers, yield red color with NH_3), Anthocyanins (red, blue colored water soluble) and leucoanthocyanins (mainly colorless, mainly in heartwood and leaves of trees, yield anthocyanins).

Anthocyanins and **Anthoxanthins** are important pigments in the cell sap of petals providing red, blue (anthocyanins), and yellow (anthoxanthins) colors in a large number of families of angiosperms. They are formed by anthocyanadins combining with different sugars at different places. Six main categories (Figure 4.12) of anthocyanin forming molecules are recognized providing different colors: Cyanidin-magenta; Pelargonidin-orange-red; Delphinidin-purple, blue, mauve; Petunidin-purplish; Paeonidin-magenta and Malvidin-purple.

These pigments are absent in some families and replaced by highly different compounds, **betacyanins** and **betaxanthins** (together known as **betalains**), which consist of heterocyclic nitrogen-containing rings and having quite distinct metabolic pathways of synthesis. However, they carry the same functions as anthocyanins. Betalains are mutually exclusive with anthocyanins and concentrated in the traditional group Centrospermae of Engler and Prantl, now recognized as order Caryophyllales. Of the nine families which contain betalains, seven were included in Centrospermae, Cactaceae placed in Cactales or Opuntiales and the ninth was placed in Sapindales. Traditional Centrospermae also included Gyrostemonaceae, Caryophyllaceae and Molluginaceae which lack betalains and contain anthocyanins instead. Mabry et al. (1963) on the basis of separate structure and metabolic pathways, suggested the placement of only betalain-containing families in Centrospermae, thus advocating the inclusion of Cactaceae and Didiereaceae and exclusion of Gyrostemonaceae, Caryophyllaceae and Molluginaceae (Table 4.1). Whereas the inclusion of Cactaceae and Didiereaceae was readily accepted (thus bringing all betalain-containing families in the same order Centrospermae), the exclusion of Caryophyllaceae and Molluginaceae was strongly opposed based on structural data. This clash between orthodox and chemical taxonomy initiated renewed interest in the group.

Behnke and Turner (1971), based on ultrastructure studies, reported P-III plastids in all members of Centrospermae and thus suggested a compromise by including all families within subclass Caryophyllidae with betalain-containing families placed under the order Chenopodiales and the other two (Caryophyllaceae and Molluginaceae) placed under Caryophyllales. Interestingly, Mabry (1976), based on DNA/RNA hybridization studies, found closer affinities between these families

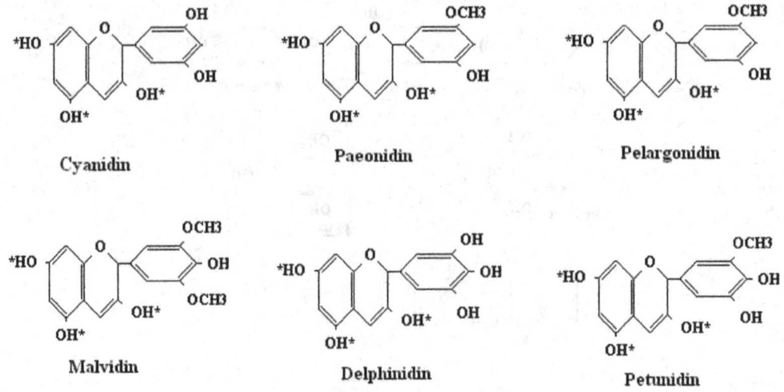

Figure 4.12: Structure of anthocyanin forming molecules, differing in right three positions, middle position absent in Cyanidin, Paeonidin and Pelargonidin, but having OH in Malvidin, Delphinidin and Petunidin. Cyanidin has OH at both other positions, Paeonidin has one replaced by OCH3, Pelargonin upper is missing, Malvidin has both replaced by OCH3, Delphinidin has OH at both and Petunidin one having OH and other OCH3.

Table 4.1: Classification of Betalain containing families and potential relatives.

Structural classification (Engler and Prantl)	Chemical classification (Mabry, 1963)	Compromise classification (Alston and Turner, 1971)	Phylogenetic classification (Thorne, 2007)	Phylogenetic classification (Takhtajan, 2009)	Phylogenetic classification (APG IV, 2016)
Centrospermae	***Chenopodiales***	***Caryophyllidae***	***Caryophyllales***	***Caryophyllales***	***Caryophyllales***
Chenopodiaceae	**Chenopodiaceae**	***Chenopodiales***	***Cactineae***	**Phytolaccaceae**	Caryophyllaceae
Amaranthaceae	**Amaranthaceae**	**Chenopodiaceae**	**Portulacaceae**	**Nyctaginaceae**	**Amaranthaceae***
Nyctaginaceae	**Nyctaginaceae**	**Amaranthaceae**	**Cactaceae**	**Aizoaceae**	**Aizoaceae**
Phytolaccaceae	**Phytolaccaceae**	**Nyctaginaceae**	**Basellaceae**	**Portulacaceae**	**Phytolaccaceae**
Gyrostemonaceae	**Aizoaceae**	**Phytolaccaceae**	**Didiereaceae**	**Basellaceae**	**Nyctaginaceae**
Aizoaceae	**Portulacaceae**	**Aizoaceae**	***Phytolaccineae***	**Cactaceae**	Molluginaceae
Portulacaceae	**Basellaceae**	**Portulacaceae**	**Nyctaginaceae**	**Didiereaceae**	**Didiereaceae**
Basellaceae	**Cactaceae**	**Basellaceae**	**Phytolaccaceae**	Molluginaceae	**Basellaceae**
Caryophyllaceae	**Didiereaceae**	**Cactaceae**	**Aizoaceae**	Caryophyllaceae	**Portulacaceae**
Molluginaceae		**Didiereaceae**	Molluginaceae	**Amaranthaceae**	**Cactaceae**
			Chenopodiineae	**Chenopodiaceae**	
Cactales	***Caryophyllales***	***Caryophyllales***	**Amaranthaceae**		
Cactaceae	Caryophyllaceae	Caryophyllaceae	**Chenopodiaceae**		
	Molluginaceae	Molluginaceae	***Caryophyllineae***		
Sapindales			Caryophyllaceae		**(*including**
Didiereaceae					**Chenopodiaceae)**

and suggested the placement of all these families under Caryophyllales with the betalain-containing families under the suborder Chenopodiineae and the two non-betalain families under Caryophyllineae. This final compromise has met with mixed response in recent years with the morphological, anatomical and DNA/RNA hybridization evidence overriding the betalain evidence. Takhtajan (1997) places only Chenopodiaceae and Amaranthaceae in Chenopodiineae, suborders abolished in 2009 revision. Dahlgren (1989) and Cronquist (1988) and APG II (2003) did not recognize suborders, and Thorne (2007) recognizing four suborders, Caryophyllaceae and Molluginaceae separated under different suborders.

It is interesting to note that the betalains have also been reported in Basidiomycetes (Fungi), in some cases the same substance found in both fungi and angiosperms. The above studies on the significance of distribution of betalains in Centrospermae bring home the fact that chemical data are useful in taxonomic realignments when such accord with data from other fields. The significance is reduced when larger evidence from elsewhere contradicts the chemical evidence. Thus, whereas no questions were ever asked about the removal of Gyrostemonaceae and the inclusion of Cactaceae and Didiereaceae, there has been no agreement about the removal of Caryophyllaceae and Molluginaceae as it goes against the evidence from morphology, anatomy, ultrastructure and DNA/RNA hybridization. This also highlights the danger of relying too much on one type of evidence.

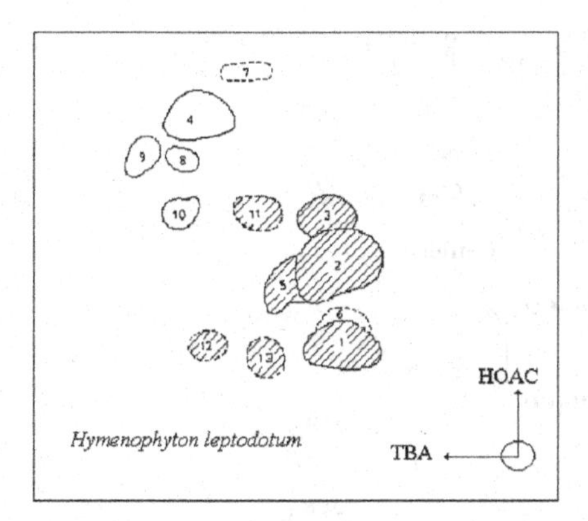

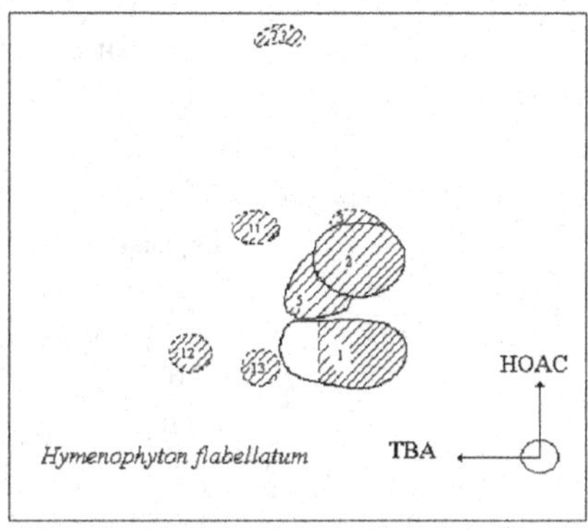

Figure 4.13: Two-dimensional paper chromatograms of the flavonoids in two species of *Hymenophyton* (after Markham et al., 1976).

Studies on phenolic compounds have helped in solving some specific problems. Bate-Smith (1958) studied five phenolic characters of different sections in the genus *Iris*. The chemical evidence supported the division into various sections, but *I. flavissima*, originally placed in the section *Pogoniris* resembled species of the section *Regelia* on the basis of phenolic characteristics. Chromosomal evidence also supported this transfer.

The technique of two-directional paper chromatography, which brings about a more pronounced separation of flavonoids, has proved very useful in taxonomic studies. *Hymenophyton* (Bryophytes) was considered by some researchers to be a monotypic genus, but by others to include two species. Markham et al. (1976) based on rapid flavonoid extraction, two-dimensional chromatographic analysis and identification (Figure 4.13) concluded that the genus contains two distinct species, *H. leptodotum* and *H. flabellatum*, and that there is no justification for their merger.

Similar studies in the genus *Baptisia* (Fabaceae) by Alston and Turner (1963) have been very useful in the detection of hybridization. Each species of the genus has a distinctive spectrum of flavonoids, and the hybrid can be easily identified by the combination of flavonoid pattern of both parental species in the suspected hybrid.

It is interesting to note that the ten taxa recognized (four parental and six hybrid), could not be differentiated based on morphological or biochemical characters alone, but a combination of both enabled a complete separation. The flavonoids in these studies were extracted from either flower or leaf.

Alkaloids

Alkaloids are organic nitrogen-containing bases, usually with a heterocyclic ring of some kind. They form one of the largest class of secondary metabolites, with nearly 10,000 different types reported. They are insoluble in water but soluble in organic solvents, but their salts are soluble in water and insoluble in organic solvents. Their distribution is restricted to some 20% of angiosperms. They are mostly present in storage tissues, seeds, fruits and roots. They act as chemical defense of plants against herbivory, and allelopathic reactions between plants.

Alkaloids are generally classified based on predominant ring system present in the molecule. They are synthesized from a few common amino acids like tyrosine, tryptophan, ornithine, argenine and lysine. Tobacco alkaloid **Nicotine** (*Nicotiana*) is synthesized from nicotinic acid and **caffeine** (coffee beans and tea leaves) from purine. Isoquinolene alkaloids **morphine**, **codeine** and **papaverine** are found in opium poppy (*Papaver somniferum*). Their distribution is often specific and thus taxonomically significant (Figure 4.14). **Conalium** is the simplest known alkaloid found in *Conium maculatum* (Apiaceae). Alkaloids are present in specialized parts of plant. Higher nicotine content is found in only older leaves. In *Datura* alkaloids occur only in seeds. Alkaloids are more widely distributed in dicots as compared to monocots. Some of them are of medicinal importance at low concentration, but toxic at high concentration. Some such as **lycotonine** (*Delphinium*), **scopolamine** (*Datura*), and **atropine** (*Atropa*) cause poisoning of livestock. Alkaloids are generally extracted from plants in weak acid alcoholic solution and precipitated by NH_3. Their presence is tested through chromatic method and quantified by solvent extraction method.

Mears and Mabry (1971), in studies conducted on the family Fabaceae, observed that the alkaloid **hystrine** occurs only in three genera *Genista*, *Adenocarpus* (both belonging to Genistae) and *Ammodendron* (originally placed in Sophorae). The latter, however, lacks **matrine,** characteristic of Sophorae. This indicates that the transfer of the last genus also to Genistae is warranted. Families Papaveraceae and Fumariaceae are closely related. This affinity is supported by the occurrence of the alkaloid **protopine** in both.

Gershenzon and Mabry (1983) reported that tropane alkaloids of Solanaceae and Convolvulaceae are similar, suggesting a close relationship. The families are placed in the same order in recent systems. Papaveraceae, earlier grouped

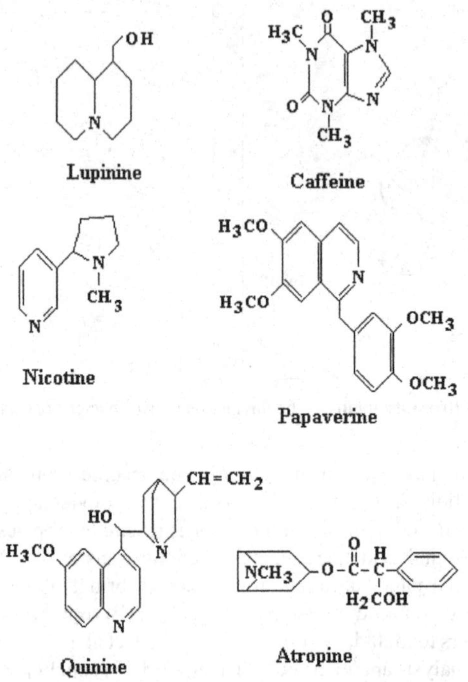

Lupinine   Caffeine

Nicotine   Papaverine

Quinine   Atropine

Figure 4.14: Main Examples of alkaloids found in plant kingdom, distribution of some is highly specific.

with Cruciferae and Capparaceae, is now removed to nearer Ranunculales on the basis of the absence of glucosinolates and presence of benzylisoquinolene. Nymphaeaceae and Nelumbonaceae differ in the sense that the former lacks benzylisoquinolene alkaloids. Benzylisoquinolene, and the alkaloids that can be derived from it, are characteristic of Magnoliidae, as also family Rutaceae, some Rhamnaceae and genus *Croton*.

Glucosinolates

Glucosinolates are sulphur containing compounds found in 15 families of angiosperms, mainly concentrated in the order Capparales (Figure 4.15). Mustard oils or isothiocyanates are hydrolytic products of glucosinolates. Originally Cruciferae, Capparaceae, Papaveraceae and Fumariaceae were placed in the same order, Rhoeadales. Chemical and other evidence, however, supported the placement of Cruciferae and Capparaceae in the order Capparales (on the basis of the presence of glucosinolates) and Papaveraceae and Fumariaceae in the order Papaverales—or suborder Papaverineae of Ranunculales (Thorne, 2003)—(on the basis of the absence of glucosinolates and the presence of the alkaloid benzylisoquinolene). Bataceae and Gyrostemonaceae were once placed in Centrospermae (Caryophyllales) but subsequently removed due to the absence of betalains. This removal was supported by the presence of glucosinolates, which are absent in Caryophyllales.

Cyanogenic Glycosides

Cyanogenic glycosides are phytotoxins which occur in at least 2000 plant species, of which several species are used as food in some areas of the world. They are hydrolysed by various enzymes to release **hydrogen cyanide**, the process known as cyanogenesis, and the plants as cyanogenic plants. Cassava and sorghum are especially important staple foods containing cyanogenic glycosides. There are approximately 25 cyanogenic glycosides known. The major cyanogenic glycosides found in the edible parts of plants used for human or animal consumption include **Amygdalin** (Figure 4.16; almonds, *Prunus dulcis*), **Dhurrin** (*Sorghum album, S. bicolor*). **Linamarin** (cassava, *Manihot esculenta*; lima beans, *Phaseolus lunatus*), **Lotaustralin** (cassava, *Manihot carthaginensis;* lima beans, *Phaseolus lunatus*). **Prunasin** (stone fruits, *Prunus avium, P. padus, P. persica, P. macrophylla*), and **Taxiphyllin** (bamboo shoots, *Bambusa vulgaris*). The potential toxicity of a cyanogenic plant depends primarily on the potential that its consumption will produce a concentration of HCN that is toxic to exposed animals or humans. Hydrogen cyanide is released from the cyanogenic glycosides when fresh plant material is macerated as in chewing, which allows enzymes and cyanogenic glycosides to come together, releasing hydrogen cyanide. Cyanides inhibit the oxidative processes of cells causing them to die very quickly. Because the body rapidly detoxifies cyanide, an adult human can withstand 50–60 ppm for an hour without serious consequences.

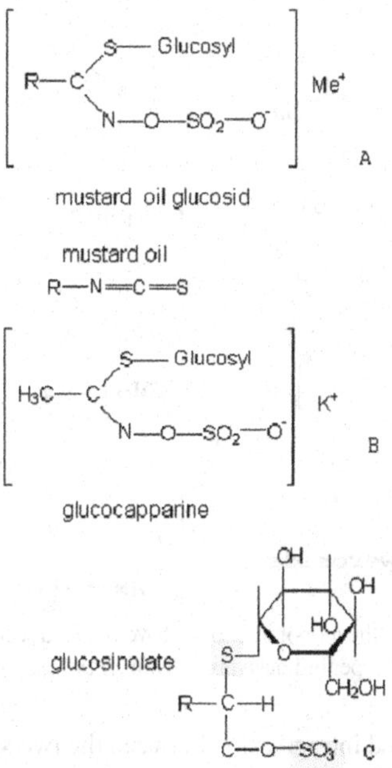

Figure 4.15: Glucosinolates. A: Mustard oil glucosid; **B:** Glucocapparine; **C:** General structure of Glucosinolates.

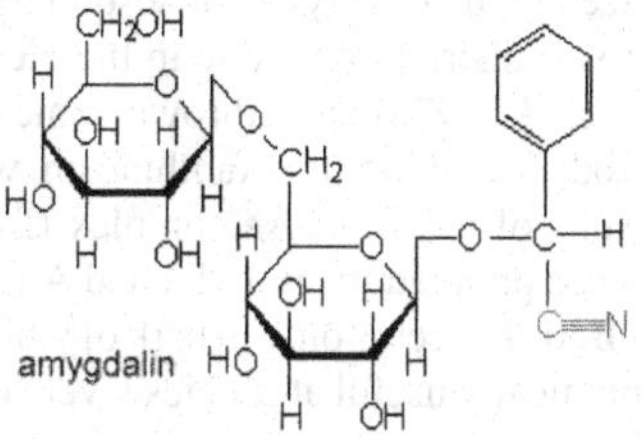

Figure 4.16: Structure of Amygdalin, cyanogenic glycosides found in seeds of almond.

However, exposure to concentrations of 200–500 ppm for 30 minutes is usually fatal. Whereas most of the cyanogenic glycosides are widely spread, others such as cyclopentenoid cyanogenic glycosides are restricted in distribution mainly to Flacourtiaceae, Passifloraceae, Turneraceae, and Malesherbiaceae. Leucine derived cyanogenic glycosides are found in Rosaceae, Fabaceae, and Sapindaceae. Several families belonging to Magnoliales and Laurales contain Cyanogenic glycosides derived from tyrosine.

Terpenes

Terpenes include a large group of compounds derived from the mevalonic acid precursor and are mostly polymerized isoprene derivatives. Common examples are **camphor** (*Cinnamomum*), **menthol** (*Mentha*), and **carotenoids** (Figure 4.17). They seem to have a definite role in the allelopathic effects of plants. They are lipid soluble found in single membrane bound liposomes, in glandular cells as essential oils, and can be easily extracted with petroleum ether and chloroform and can be separated by GLC (Gas Liquid Chromatography), enabling qualitative as well as quantitative measure of chemical differences.

Terpenes are isomeric unsaturated hydrocarbons of the basic 5-carbon **isoprene** ($CH_2=C(CH_3)-CH=CH_2$) present in *Hamamelis japonica.* **Terpenoids**, the common group of terpenes are distinguished as 10-Carbon **Monoterpenoids** (Fennel, Menthol-*Mentha*, Gymnosperms), 15-C **Sesquiterpenoids** (Asteraceae as sesquiterpenol, ABA, some essential oils), 20-C **Diterpenoids** (*Taxus*-Taxol, gibbrellins), 30-C **Triterpenoids** (sterols, steroids, saponins, betulin in *Betula papyrifera*), 40-C **Tetraterpenoids** (Carotenoids) and poly-C Polyterpenoids (Rubber). They have been largely used in distinguishing specific and subspecific entities, geographic races and detection of hybrids. Studies in *Citrus* have focused on determination of the origin of certain cultivars. Studies on *Juniperus viginiana* and *J. ashei* have refuted previous

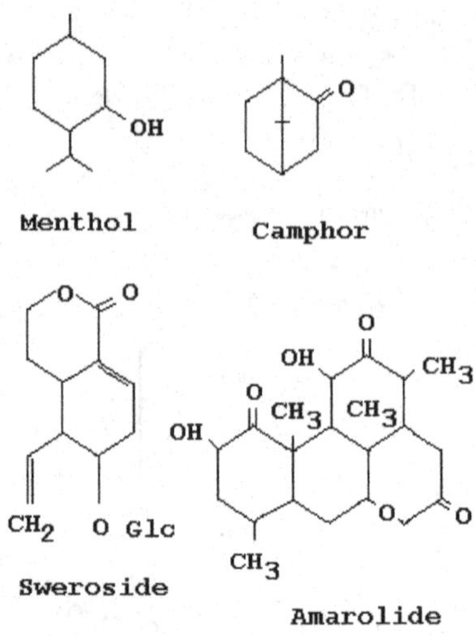

Menthol Camphor

Sweroside Amarolide

Figure 4.17: Terpenoids: Menthol and Camphor, the monoterpenoids; Sweroside, a seco-iridoid from *Swertia*, Gentianaceae; Amarolide, a quassinoid triterpenoid derivative from *Ailanthus*, Simaroubaceae.

hypotheses about extensive hybridization and introgression between the two species. Their distribution in *Pinus* has been used (Mirov, 1961) to understand relationships. *P. jeffreyi* has been considered a variety of *P. ponderosa*, but turpentine distribution showed that it strongly resembles the group Macrocarpae and not Australes to which *P. ponderosa* belongs. A major contribution of terpenoid chemistry has been the use of **sesquiterpene lactones** in the family Compositae. Many tribes within the family are characterized by distinct types of sesquiterpene lactones they produce. This helped to establish that genus *Vernonia* has two centres of distribution—one in the Neotropics and the other in Africa. Similarly, studies on *Xanthium strumarium* (McMillan et al., 1976) have thrown some light on the origin of Old World and New World populations. Old World populations produce xanthinin/or xanthinosin, whereas the New World populations contain xanthinin or its stereoisomer xanthumin. Plants of the *chinense* complex (from Louisiana) contain xanthumin and are believed to be the source of introduced *chinense* populations in India and Australia.

Triterpenoids occur in several families. **Betulin** occurs only in bark of white birch (*Betula papyrifera*) and its relatives, is waterproof highly flammable, and is taxonomically useful at species level. **Triterpene saponins** occur in Apiaceae and Pittosporaceae and support their close relationship.

Iridoids constitute another important group of terpenes (mostly monoterpene lactones). They are present in over 50 families and their presence is correlated with sympetaly, unitegmic tenuinucellate ovules, cellular endosperm and endosperm haustoria. Assuming that 'independent origin of several groups with this combination of independent attributes is unlikely', Dahlgren brought together all iridoid-producing families. The occurrence of iridoids in several unrelated families, e.g., Hamamelidaceae and Meliaceae, however, suggests that iridoids could have arisen independently several times in the evolution of angiosperms. The occurrence of a distinctive iridoid **aucubin** in *Budleja* has been taken to support its transfer from Loganiaceae to Budlejaceae. Aucubin and geniposide, have shown antitumoral activities.

Cronquist (1977) proposed that chemical repellents had an important role in the evolution of major groups of dicots. The alkaloid Isoquinolene of Magnoliidae gave way to tannins of Hamamelidae, Rosidae and Dilleniidae, which in turn gave way to the most effective iridoids in Asteridae, the family Compositae developing the most effective sesquiterpene lactones.

Non-semantide Macromolecules

In addition to DNA and RNA, which will be dealt under Molecular systematics, the macromolecules include proteins, and complex polysaccharides such as starches and celluloses. Starches are commonly found in the form of grains which may be concentric (*Triticum*, *Zea*) or eccentric (*Solanum tuberosum*) and present anatomical characteristics which can be seen under a microscope. Detailed studies of starch grains under SEM also hold promise for taxonomic significance.

Proteins

Proteins, together with nucleic acids, are often called Semantides, which are primary constituents of living organisms and are involved in information transfer. Based on their position in the information transfer DNA is a **primary semantide**, RNA

secondary semantide and proteins the **tertiary semantides**. Semantides are popular sources of taxonomic information, and most of this information has come from proteins. The information about DNA and RNA will be discussed under Molecular Systematics in the next section; only proteins are being discussed here.

Proteins are complex macromolecules made up of amino acids linked into a chain by peptide bond, thus forming a **polypeptide chain**, organized into a three-dimensional structure. Because of their complex structure, special techniques are necessary for the isolation, study and comparison of proteins. These methods include **serology**, **electrophoresis** and **amino acid sequencing**.

Serology

The field of **systematic serology** or **serotaxonomy** had its origin towards the turn of the twentieth century with the discovery of serological reactions and development of the discipline of immunology. Precipitin reactions were first reported by Kraus (1897). The technique was originally applied by Bordet (1899) in his work on birds, when he reported that immune reactions are relatively specific, and the degree of cross reactivity was essentially proportional to the degree of relationship among organisms. The present technique of serology is based on immunological reactions shown by mammals when invaded by foreign proteins. In the study of estimating relationships between plants, the plant extract of species A containing proteins (**antigens**) is injected into a mammal (usually a rabbit, mouse or goat). The latter will develop **antibodies**, each specific to an antigen with which it forms a precipitin reaction, coagulating and thus making it non-functional. These antibodies are extracted from the body of the animal as **antiserum**. This antiserum can coagulate all proteins in species A, but when mixed with the protein extract of species B, the degree of precipitin reaction would depend on the similarity between the proteins of the two species.

The antiserum obtained from the mammal normally contains several immunoglobulins that can bind to the same antigen, is said to be **polyclonal.** This is because an antigen activates several different lymphocytes within the animal, each producing a different antibody for the same antigen. Techniques have now been developed which can generate **monoclonal antibodies**. In a method developed by Milstein and Köhler (1975), antibody-producing lymphocyte of mammal (which cannot grow and divide in cultures) was fused with malignant myoloma cell (cancer cell which can grow rapidly in cultures) to produce hybrid cells called **hybridoma**. These hybridomas can grow, proliferate and produce large amount of monoclonal antibody.

Antigens are mostly extracted from seeds and pollen. In early works, crude total comparison of precipitin reactions was done but now more refined methods have been developed which can bring about individual antigen-antibody reactions. Major methods include:

Double-diffusion serology

In this technique the antigen mixture and antiserum are allowed to diffuse towards one another in a gel (Figure 4.18). The different proteins travel at different rates and thus the reactions occur at different places on the gel. This method allows comparison of precipitin reactions of several antigen mixtures from different taxa simultaneously on the same gel. In a modification of this method, the antiserum is placed in a circular well surrounded by a ring of several wells containing the samples of antigens.

Immuno-electrophoresis

In this technique the antigens are first separated unidirectionally in a gel by electrophoresis and then allowed to travel towards the antiserum (Figure 4.19). This method enables a better separation of constituent reactions but has the limitation that only one antigen mixture can be handled on a single gel.

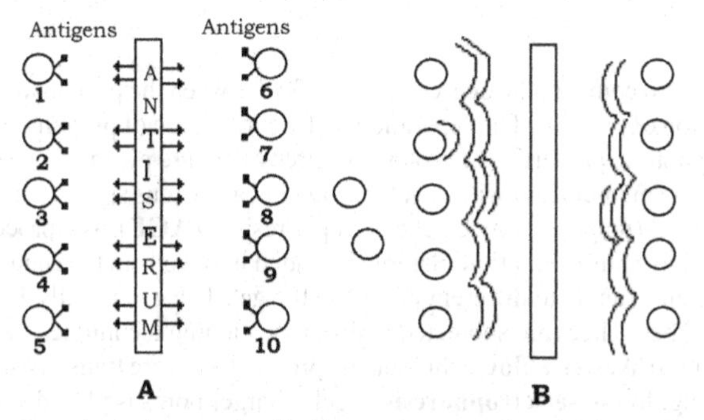

Figure 4.18: Double-diffusion serology. **A:** Antigens and antibodies moving towards each other on the gel. 1–10 refer to antigen mixtures from ten different taxa; **B:** Resultant precipitant lines.

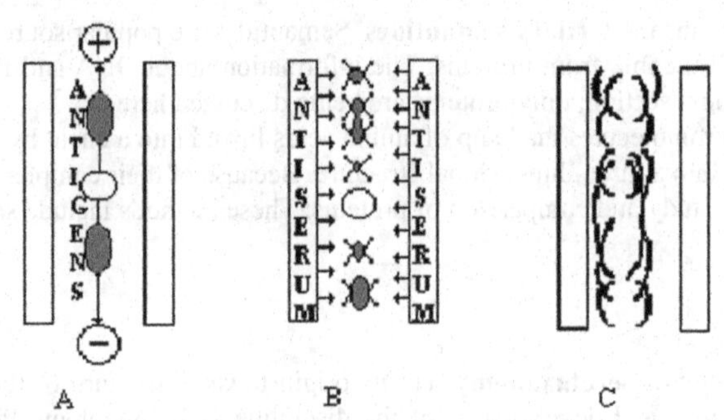

Figure 4.19: Immuno-electrophoresis. A: Antigen separation by electrophoresis; **B:** Antibodies and separated antigens diffusing towards each other; **C:** Resultant precipitant lines.

Absorption

Protein mixtures from different species often contain many common proteins, especially those involved in common metabolic processes. The antibodies for these common proteins (antigens) are first removed from the antiserum so that there is a more logical comparison of precipitin reactions.

Radio-immunoassay (RIA)

In this technique the antibodies or antigens are labelled with radioactive molecules enabling their detection even when present in minute quantities.

Enzyme-linked immunosorbent assay (ELISA)

In this technique either the antibodies or antigens are labelled linked with enzymes, thus enabling detection even in very small quantities.

It must be noted that there are specific sites on proteins (**determinants**), which are capable of initiating production of immunoglobulins in specific cells of mammals. Determinants are regions consisting of 10–20 amino acids and one protein may comprise several different determinants and thus several **antigens**.

Extensive studies of the immunoelectrophoretic patterns of the genus *Bromus* were done by Smith (1972, 1983). Results showed that North American diploids of the genus are reasonably diverse. The study also highlighted that antisera raised from different species could provide different results. Based on serological studies, Smith established the distinct identity of *B. Pseudosecalinus*, previously recognized as a variety of *B. secalinus*. This separation was supported by cytological evidence also. Serological studies have also supported the removal of *Nelumbo* from Nymphaeaceae into a separate family Nelumbonaceae, placement of *Hydrastis* in Ranunculaceae (and not Berberidaceae), and merger of *Mahonia* with *Berberis* (Fairbrothers, 1983).

Serology may be done through comparison of protein mixtures or the comparison of single isolated and purified proteins. Schneider and Liedgens (1981) developed a complex but excellent procedure of monoclonal culture of antibodies, but unfortunately used this for construction of a '**phylogenetic tree**' not parallel with accepted evolutionary schemes. Fairbrothers (1983) cautioned that an evolutionary tree should not be constructed on the reactions of a single enzyme or a single species. Lee (1981) using purified protein for antigen and using different techniques concluded that *Franseria* (Asteraceae) should be merged with *Ambrosia*.

Electrophoresis

The technique of serology serves to compare the degree of similarity between the protein mixtures of different species and does not involve the identification of proteins. The separation and identification of proteins can be done by electrophoresis. Separation is based on the amphoteric properties of proteins whereby they are positively or negatively charged to various extents according to the pH of the medium and will travel through gel at various speeds across a voltage gradient, usually carried out in a polyacrylamide gel (polyacrylamide gel electrophoresis—**PAGE**). The procedure involves homogenizing the tissues (containing proteins) in a buffer solution. Sample is loaded into wells in the center of the gel. The current is run for a specific time, and the proteins run up to different points on the gel. The gel, usually 1 cm thick, is cut into three thin slices, each about 3 mm thick. These slices are subjected to different staining techniques and proteins are identified using various criteria. In commonly used **Western blot** technique the protein bands are transferred from the gel to nitrocellulose membrane for further processing. In **disc-electrophoresis,** a gel of larger pores is placed over a gel of smaller pores. The former is used for crude separation and the latter for a complete separation.

In the technique of **isoelectric focusing,** a gel of a single pore size, is set up with a pH gradient (usually 3–10), so that proteins come to lie on the gradient corresponding to their isoelectric point. These can be subsequently separated more completely by disc-electrophoresis. Isoelectric focusing of **Rubisco** (Ribulose 1, 5 diphosphate carboxylase) has been very useful in determining relationship between species of *Avena, Brassica, Triticum*, and several other genera. It is an excellent protein for helping to evaluate hybridization.

Electrophoretic studies have supported the origin of hexaploid wheat (*Triticum aestivum*) from *A. tauschii* and *T. dicoccum*. Johnson (1972), working on storage proteins showed that *T. aestivum* (AABBDD) and *T. dicoccum* (AABB) possess all proteins of the A genome of the diploid *T. monococcum* (AA). They also share proteins of the B genome of uncertain origin. The D genome is believed to have come from *Aegilops tauschii* as evidenced by morphological and cytological data. By mixing proteins of *A. tauschii* and *T. dicoccum* it was seen that the electrophoretic properties of the mixture closely resemble those of *T. aestivum*, thus proving the origin of the latter from the two previous species. Electrophoretic studies have also helped to assess species relationships in *Chenopodium* (Crawford and Julian, 1976), by combining data from flavonoids with proteins. A flavonoid survey of seven species showed that in some taxa, the flavonoid data were fully compatible with interspecific protein differences, but in some cases, did not agree. Thus, *Chenopodium atrovirens* and *C. leptophyllum* had identical flavonoid patterns but could be distinguished by their different seed protein spectra. *C. desiccatum* and *C. atrovirens,* on the other hand, were closely similar in seed proteins but differed in flavonoids. Both flavonoid and protein evidence, however, distinguished *C. hians* from *C. leptophyllum*, thus providing support to their recognition as separate species. Vaughan et al. (1966) through the study of serology and electrophoresis have shown that *Brassica campestris* and *B. oleracea* are closer to each other than to *B. nigra.*

Electrophoresis has also made possible the separation of **allozymes** (different forms of the same enzyme with different alleles at one locus) and **isozymes** (or **isoenzymes** with different alleles at more than one locus). Barber (1970) showed that certain polyploids possess isozymes of all their progenitors plus some new ones. Backman (1964) crossed two strains of maize, each with three different isozymes. F1 possessed all six isozymes. The hybrids thus show **molecular complementation.**

Studies of the genus *Tragopogon* have confirmed that the tetraploid *T. mirus* is a hybrid between two diploid species, *T. dubius* and *T. porrifolius*. Whereas the parental diploids were found to be divergent at close to 40 per cent of the 20 enzyme loci examined, the tetraploid hybrid possessed completely additive enzyme patterns. The evidence thus supported the recognition of a hybrid based on morphological and chromosomal evidence.

Amino Acid Sequencing

Since only 22 amino acids are known to be the constituents of proteins, the primary differences between the proteins result from different sequences of amino acids in the polypeptide chain. It is now possible to break off the amino acids from the polypeptide chain one by one, identify each chromatographically and build up the sequence of amino acids step by step. **Cytochrome *c*** is the most commonly used molecule and out of 113 amino acids, 79 vary from species to species, but alteration of even one of the other 34 destroys the functioning of the molecule. Being present in all aerobic organisms, it is ideal for comparative studies. Boulter (1974) constructed a **cladogram** (Figure 4.20) of 25 species of spermatophytes using the '**ancestral sequence method**'. *Ginkgo biloba,* the only gymnosperm used occupied isolated position in the cladogram. *Ginkgo* with an isolated phylogenetic position is no new discovery, but rather a long-established fact. But the fact that amino acid sequencing also produces a similar cladogram establishes the significance of such studies in understanding phylogeny.

Recent data from various fields have pointed to the merger of *Aegilops* with *Triticum.* Autran et al. (1979) based on N-terminal amino acid sequencing supported this merger. In general, the number of amino acid differences is roughly parallel to the distance between the organisms in traditional classifications, suggesting that the method is broadly reliable. There are, however, certain contradictions. The number of differences between the cytochrome *c* of *Zea mays* and *Triticum aestivum* (both members of the same family Poaceae) is greater than between *Zea mays* and certain dicotyledons.

It has been found that cytochrome *c* and plastocyanin (another protein commonly used in amino acid sequencing studies) can exhibit a large number of **parallel substitutions** (identical changes from one amino acid to another at the same position in the protein in different organisms), thus rendering them unsuitable for constructing phylogenies. The practical solution is to use evidence from a wide range of proteins, preferably using different techniques.

MOLECULAR SYSTEMATICS

The closing years of the past century saw the concentration of macromolecular studies towards DNA and RNA, resulting in the establishment of an emerging field of **molecular systematics**. Although flavonoids and isozymes also constitute molecular data, molecular systematics commonly deals exclusively with the utilization of nucleic acid data. As molecular data reflects gene-level changes, it was believed to reflect true phylogeny better than morphological data. It has, however, been realized that molecular data may also pose similar problems, although there are more molecular characters available and comparison is generally easier.

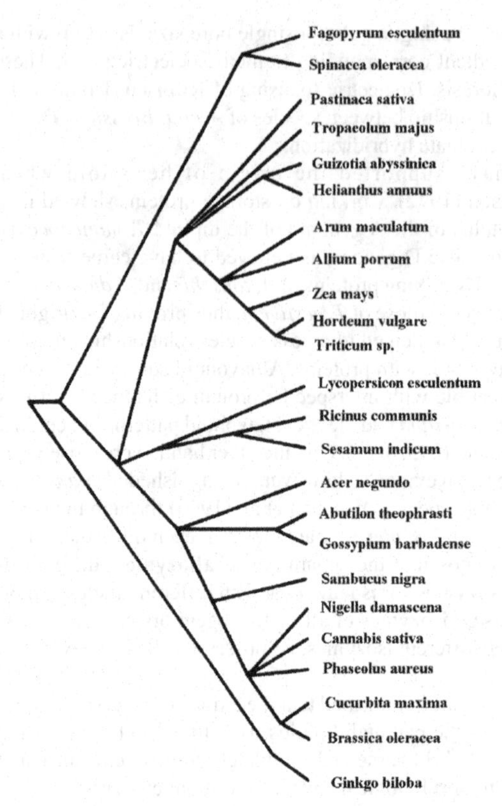

Figure 4.20: Cladogram of 25 species of seed plants based on the 'ancestral sequence method' used by Boulter (1974) (after Boulter).

Molecular Evolution

Traditionally, different taxa, especially the species have been characterized primarily on morphological differences (phenotypes). Additionally, differences in physiology, biochemistry, anatomy, palynology, embryology, gross chromosome structure and behavior, have been used in refining evolutionary trees. Although, it had been long recognized that evolution is based on genetic changes, only during the last two decades, there have been forceful drive to use genetic material for a better understanding of evolutionary relationships. Those species that are closely related, are expected to have greater similarities in their genetic material than the distantly related species. During the past decade, molecular genetics has taken a dominant role in enabling us to understand speciation and evolution clearly. Differences in the nucleotide sequences are quantitative and can be analyzed using mathematical principles, utilizing the help of computer programs. Evolutionary changes at the DNA level can be objectively compared among different species to establish evolutionary relationships.

Evolution of Nucleic Acids and Proteins

When Earth originated nearly 5000 million years ago the **primary atmosphere** consisted of only hydrogen and helium but being too small a planet to hold these light gases, they floated away into space. The earth accumulated its **secondary atmosphere** because of volcanic activity in early hot earth and the gases consisted of largely steam, variable amounts of CO_2, N_2, SO_2, H_2S, HCl, Sulphur and smaller quantities of H_2, CH_4, SO_3, NH_3. There was no free oxygen. Our **present atmosphere** is of biological origin, in which methane and ammonia have largely been consumed, inert components like nitrogen remained unchanged, and oxygen produced by photosynthesis. This happened nearly 2500 m years ago when Cyanobacteria, the first photosynthetic bacteria made their appearance. The ultraviolet radiations from the sun, together with lightening discharges caused the gases to react in the primeval atmosphere forming simple organic compounds such as amino acids, sugars and nucleic acid bases. This mostly happened because of gases dissolving in primeval oceans and continuing to react forming primitive soup, the precursor of life. Further reactions formed polymers, globules and eventually the **first primitive cell**.

The possibility of such reactions in the primitive atmosphere was demonstrated by a Russian biochemist Alexander Oparin in 1920s who proposed that life evolved before there was any free oxygen in the atmosphere. The oxygen if present at that stage would have reacted with precursor organic molecules formed in the atmosphere, oxidizing them back into carbon dioxide and water. These reactions were mimicked by biochemist Stanley Miller in 1950s, who subjected a mixture of methane, ammonia and water vapour to high voltage discharge or to ultraviolet light, and the products allowed to dissolve and react in water. As long as oxygen was excluded, the results were similar producing several organic molecules such as amino acids, formic acid, glycolic acid, lactic acid, acetic acid, propionic acid, succinic acid, urea, purines, pyrimidines and sugars. These energy sources can also destroy these organic molecules present in the atmosphere. The occurrence of primeval oceanic atmosphere helped shielding and preserving these organic molecules and prevented their destruction. Organic acids, particularly amino acids are soluble in water and non-volatile, and little chance of their returning to atmosphere.

The polymerization of amino acids and other monomers to form macromolecules requires energy for formation of bonds and removal of water. Such polymers known as **proteinoids** can be generated by simple heating of amino acids at around 1500 C for a few hours. Such heating could have occurred near volcanoes or when pools left behind by changing coastline evaporated. Inorganic polyphosphates present in the primeval times would have helped in condensation.

It is generally thought that RNA probably evolved first through polymerisation of nucleotides present in primeval environment. When RNA template is incubated with mixture of nucleotides and zinc as a catalyst, a complementary piece of RNA is synthesized. The complementary strand in turn will act as template to generate more of original RNA molecule.

It is assumed that RNA originated even before proteins. It is also believed that earliest organisms had both genes and enzymes made of RNA and formed **RNA world**. The examples of enzymatically active RNA are found in Ribozymes and self-replicating introns. Later proteins infiltrated and took over the role of enzymes. This was followed by the evolution of DNA as genetic material, and RNA relegated to the role of intermediate between genes and the enzymes.

Changes in DNA sequences (**mutations**) lead to the changes in the codons, that in turn determine the sequence of various amino acids, deciding the final structure and function of a protein. These changes commonly result from changes of one or more base pairs in a DNA sequence. Two types of nucleotide changes occur in the genome. Some changes give rise **nonsynonymous codons**, coding for different amino acid, and thus resulting in a corresponding change in the amino acid sequence of a protein. Other nucleotide changes give rise to **synonymous codons**, that code for the same amino acid.

Evolutionary Rates within a Gene

It is now well established that different parts of genes evolve at widely different rates, reflecting the extent of natural selection on each part. Some nucleotides code for amino acid sequence of a protein (**Coding sequences**), whereas others do not code for amino acids in a protein (**noncoding sequences**). Latter include **introns, leader regions, trailer regions** (all these are transcribed but not translated), and **5' and 3' flanking sequences** that are not transcribed. **Pseudogenes**, which are nucleotide sequences that no longer produce functional gene products as they have accumulated inactivating mutations, also constitute noncoding sequences. Even within coding regions of functional gene, not all nucleotide substitutions produce a corresponding change in the amino acid sequence of a protein. Many substitutions occurring at the third position of triplet codons have no effect on the amino acid sequence of the protein because such changes often produce **synonymous codons.**

Although synonymous and nonsynonymous nucleotide changes are likely to arise in equal frequency (because enzymes responsible for DNA replication and repair cannot differentiate between the two), yet the rate of synonymous nucleotide changes (**conservative substitutions** of Kimura) is about five times greater than observed rate of nonsynonymous changes (**disruptive substitutions** of Kimura). This is because synonymous changes do not alter protein structure and function and are tolerated by natural selection, but the nonsynonymous changes are usually detrimental and are excluded by natural selection. Synonymous substitution rates and not nonsynonymous nucleotide changes are, as such, the fair reflection of actual mutation rate within a genome. Pseudogenes and 3' flanking regions also show high evolutionary rates, comparable to synonymous changes. 5' flanking regions show a little slower rate, whereas leader and trailer regions show very low evolutionary rates, slightly higher than nonsynonymous changes. It is as such obvious that nucleotide changes in noncoding regions or coding that do not alter amino acid sequences, have high rate of evolution, whereas changes in coding regions, especially those affecting amino acid sequences show very low rate of evolution, as most of them get filtered out by natural selection.

It is important note that whereas **mutations** are changes in nucleotide sequences that occur because of mistakes in DNA replication or repair processes, the **substitutions** are mutations that have passed through the filter of selection at least at some level.

Location of Molecular Data

Systematists use molecular data from three different locations within a plant cell: **chloroplast**, **mitochondrion** and the **nucleus**, yielding three different types of **genome** (**DNA**). **Chloroplast genome** is the smallest ranging from 120 to 160 kbp (kilo base pairs) in higher plants (up to 2000 kbp in alga *Acetabularia*), **mitochondrion genome** 200 to 2500 kbp, whereas the **nuclear genome** is much larger often ranging between million to more than billion kbp. Although the former two are inherited from the maternal parent, the latter is biparental. Mitochondrion genome undergoes a lot of rearrangements, so that many different forms may be found within the same cell, and hence is of little significance in interpreting phylogenetic relationships, whereas the other two are highly stable not only within the same cell, but also within a species, and present useful taxonomic tools.

Mitochondial DNA

Mitochondrial DNA has been studied from several species of plants. Each mitochondrion contains several copies of **mtDNA**, and as each cell contains several mitochondria, the number of mtDNA molecules per cell could be very large. Most mtRNA molecules are circular, but linear in *Chlamydomonas reinhardtii*. In vascular plants, mtDNA is considerably larger, circular, containing many noncoding sequences, including some that are duplicated. The physical mapping of genes in vascular plants has shown that these are located in different positions on mtDNA circles of different species, even in fairly closely-related species. This renders mtDNA less useful in phylogenetic studies.

Chloroplast DNA

Studies of DNA in plants have largely been undertaken from chloroplast compared to the other two cellular genomes. This is because **chloroplast DNA (cpDNA)** can be easily isolated and analyzed. It is also not altered by evolutionary processes such as gene duplication and concerted evolution (in rRNA, having thousands of copies of repeated segments so that mutation in one sequence gets corrected to match other copies, this homogenization process is termed as **concerted evolution**). It also has an added advantage in that it is highly conserved in organization, size and primary sequence. Chloroplast DNA is closed circular molecule (Figure 4.21) with two regions that encode the same genes but in the opposite direction and known as inverted repeats. Between the inverted repeats are single copy regions. All cpDNA molecules carry basically the same set of genes, arranged differently in different species of plants. These include genes for ribosomal RNA, transfer RNA, ribosomal proteins and about 100 different polypeptides and subunits of enzyme capturing CO_2.

Most studies of chloroplast DNA have focused on chloroplast gene ***rbcL***, which encodes large subunit of photosynthetic enzyme **RuBisCO** (ribulose-1,5-biphosphate carboxylase/oxygenase), carbon acceptor in all photosynthetic eukaryotes and cyanobacteria. The gene occurs in all plants (except parasites), is fairly long (1428 bp), presents no problems of alignment, and has many copies available in the cell. Ready availability of PCR primers has made it possible to generate over 2000 sequences, primarily of seed plants. Other commonly used chloroplast genes include ***atpB*** (beta subunit of ATP synthetase involved in the synthesis ATP), ***matK*** (maturase involved in splicing type II introns from RNA transcripts), and ***ndhF*** (subunit of chloroplast NADH dehydrogenase), which functions in converting NADH to NAD+H, involved in reactions of respiration. Of these four commonly used genes ***rbcL***, ***atpB***, and ***matK*** belong to **large single copy region**, where as ***ndhF*** is located on **small single copy region**.

Nuclear DNA

The **nuclear DNA**, although more difficult to analyze, and hence used less frequently has two great advantages. Certain nuclear sequences evolve more rapidly than cpDNA sequences, and thus allow finer level of discrimination at population level than cpDNA. Also, whereas the nuclear genome is inherited biparentally, the chloroplast genome is inherited maternally. Thus, the hybrid plant will possess the nuclear complement of both parents but only the cpDNA of the maternal plant.

The study of nuclear genes has traditionally involved **ribosomal RNA.** Ribosomal genes are arranged in tandem arrays of up to several thousand copies. Each set of genes has a small subunit (18S) and a large subunit (26S) separated by a smaller (5.8S) gene (Figure 4.22). It must be noted that 5S RNA although also a part of the unit, but of the unknown function is synthesized separately outside nucleolus. The three subunits are separated by internal transcribed spacers (ITS: ITS1 and ITS2). Each set of genes is separated from adjacent one by a larger spacer (variously known as IGS-intergenic spacer, EGS-extragenic spacer or NTS-nontranscribed spacer). Sequences of 18S and 26S genes have been used in phylogenetic studies, because they have some highly conservative regions which help in alignment, and other variable regions, which help to distinguish phylogenetic groups. Recently ITS region has been used to determine relationships among species. In general, the ITS region has supported relationships inferred from chloroplast studies and morphology.

Molecular Techniques

The techniques of handling molecular data saw great advancements over past few decades, starting with comparison of whole DNA molecules. It is now possible to break DNA at specific sites, generate maps of individual genes, determine sequence of genes, and make multicopies of a DNA through Polymerase chain reaction (PCR) technique. These help in generating enough molecular data for comparison.

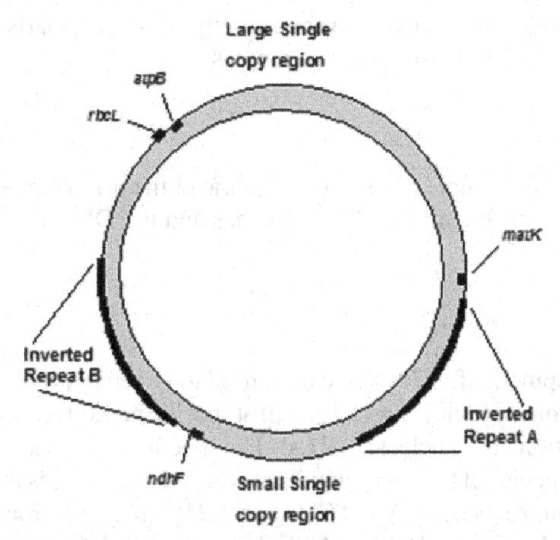

Figure 4.21: Chloroplast DNA with location of genes commonly used in molecular systematics.

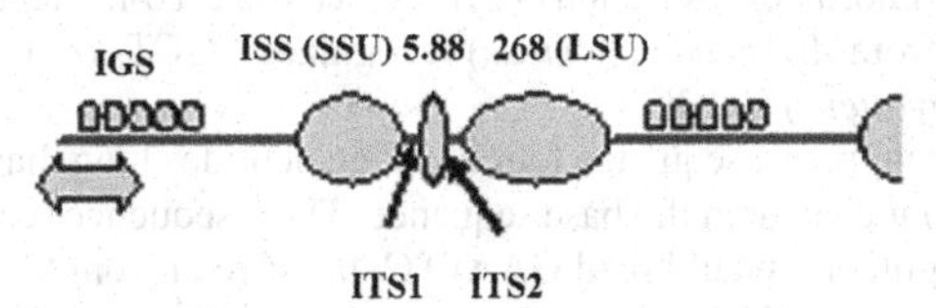

Figure 4.22: A portion an array of ribosomal genes. Each unit has three subunits separated by two ITS (internal transcribed spacer) regions. Adjacent units are separated by IGS (intergenic spacers).

Total DNA/DNA Hybridization

The early studies on utilization of nucleic acids in systematics involved **DNA/DNA hybridization** using the whole DNA for study. In a method developed by Bolton and Mecarthy (1962), the extracted DNA is treated to make it single stranded. The DNA of another organism is, similarly, made single stranded. The two are subsequently allowed to hybridize *in vitro*. The degree of reassociation (**annealing**) expresses the degree of similarity in sequences of nucleotides of the two organisms. Procedure involves heating DNA so that it becomes **denatured** into single strands (**ssDNA**). The temperature is lowered just enough to allow the multiple short sequences of **repetitive DNA** to rehybridize back into double-stranded DNA (**dsDNA**). The mixture of ssDNA (representing single genes) and dsDNA (representing repetitive DNA) is passed over a column packed with hydroxyapatite. The **dsDNA sticks** to the hydroxyapatite; **ssDNA does not** and flows right through. The purpose of this step is to be able to compare the information-encoding portions of the genome—mostly genes present in a single copy—without having to worry about varying amounts of noninformative repetitive DNA. The ssDNA of **species A** is made radioactive. The radioactive ssDNA is then allowed to rehybridize with nonradioactive ssDNA of the same species (**A**) as well as—in a separate tube—the ssDNA of species **B**. After hybridization is complete, the mixtures (**A/A**) and (**A/B**) are individually heated in small (2°–3°C) increments. At each higher temperature, an aliquot is passed over hydroxyapatite. Any radioactive strands (**A**) that have separated from the DNA duplexes pass through the column, and the amount is measured from their radioactivity. A graph showing the percentage of ssDNA at each temperature is drawn. The temperature at which 50% of the DNA duplexes (dsDNA) have been denatured ($T_{50}H$) is determined.

Bolton (1966) found that only half nucleotide sequences in the DNA of *Vicia villosa* are similar (homologous) with those of *Pisum*, while only 1/5th are homologous between *Phaseolus* and *Pisum*. In the technique of **DNA/RNA hybridization,** the RNA is hybridized with the complementary DNA of related plants. Mabry (1976) used this technique in Centrospermae (Caryophyllales) and concluded that the family Caryophyllaceae (although lacking betalains) is quite close to betalain-containing families, but not as close as the latter are to each other.

Chromosome Painting

The technique of chromosome painting provides another way to compare entire genomes. A fluorescent label is attached to the DNA of individual chromosomes of one species. These chromosomes are exposed to the chromosomes of another species. The regions of gene homology will hybridize taking up the fluorescent label and the 'painted' chromosomes can be examined under a microscope. The method is a modification of fluorescence *in situ* hybridization (FISH). Chromosome painting studies in humans have shown that human chromosome 6 has counterparts in chromosome 5 of chimpanzee, chromosome 7 of pig and chromosome 23 of cow as few examples.

Unravelling DNA Structure

Understanding DNA structure involves complex procedure to unravel the arrangement of genes in DNA, and sequence of arrangement of nucleotides which differentiates different genes and the DNA of different organisms. The procedure involves some distinct steps.

DNA Cleaving

This technique is a landmark development of 1970s that can be used to generate **physical maps** of individual genes or the entire genome. The DNA extracted from a species is cut (cleaved) at specific points (recognition-site; restriction site), yielding **restriction fragments** using restriction endonucleases (**REs**). The specific enzymes are named using the first letter of the genus and the first two letters of the species of the bacterium from which the enzyme is isolated. Thus, enzyme ***Eco*RI** which cleaves DNA at every site where it finds sequence GAATTC (Figure 4.24) is obtained from *Escherichia coli*. ***Hind*III** obtained from *Haemophilus influenzae* strain R_d cleaves DNA at AAGCTT, and ***Bam*HI** from *Bacillus amyloliquefaciens* cleaves GGATCC. More than 400 restriction enzymes have already been isolated. Their natural function is to inactivate invading viruses by cleaving the viral DNA. Majority of restriction enzymes recognize a 6-nucleotide sequence, but others recognize 4-nucleotide sequence. Thus *Alu*I (from *Arthrobacter luteus*) recognizes AGCT, *Taq*I (from *Thermus aquaticus*) TCGA, and *Hae*III (from *Haemophilus aegypticus*) GGCC.

Each restriction enzyme can recognize a sequence four to six nucleotides long, having twofold rotational symmetry, because it can be rotated 180° without change in the base sequence. Thus, sequence recognized by *EcoliRI*—if read from '5 to 3' in both strands, of DNA segment—would read GAATTC, but if read from '3 to 5' in both strands it would read CTTAAG. This symmetry is known as **palidrome** (as, for example, in nonsense phrase: AND MADAM DNA that is read similarly from both ends). This feature combined with the fact that most restriction enzymes give **staggered cuts** (and not straight cuts) wherein they cut two strands of DNA at different points, produces complementary single-stranded termini that can be rejoined later using enzyme DNA ligase. Such enzymes produce **sticky ends** or **cohesive ends**. Others like *Alu*I and *Hae*III, however, make simple double stranded cut in the middle of the recognition sequence, resulting in **blunt end** or **flush end**.

The use of restriction enzymes allows the DNA to be dissected into a precisely-defined set of specific segments. Using different enzymes, sites cleaved by different enzymes can be identified and ordered into a **restriction map** or **physical map**.

DNA Cloning

A detailed analysis of DNA requires availability in sufficient quantity of DNA or its restriction fragments. DNA **cloning** is a technique to produce large quantities of a specific DNA segment.

The technique of cloning has largely been made possible through **recombinant DNA** technology. The DNA molecules from two different sources are treated with restriction enzyme that makes staggered cuts in DNA, leaving single-stranded tails in either of the cleaved DNA. These tails act as sticky ends and complementary ends of two different DNA molecules join to form double stranded recombinant DNA, in the presence of DNA ligase. For successful cloning, one of the parental DNAs incorporated into recombinant DNA molecule is capable of self-replication and is known as **cloning vector**. In practice, the gene or DNA fragment of interest is inserted into a specially-chosen cloning vector, which is used as a vehicle for carrying foreign DNA into a suitable host cell, such as a bacterium.

Plasmid Vector

Plasmids are extra-chromosomal double-stranded circular DNA molecules present in microorganisms, especially bacteria. The plasmid chosen as vector contains a gene for antibiotic resistance. In the most commonly-employed technique (Figure 4.23-I), the recombinant plasmids (with foreign DNA inserted into plasmid) are added to an *E. coli* bacterial culture pretreated with calcium ions. When subjected to brief heat shock, such bacteria are stimulated to take up DNA from their surrounding medium. Once within the bacterial cell, the plasmid replicates autonomously and is passed on to the progeny during cell division. The bacteria containing recombinant plasmid can be separated by treatment with an antibiotic which removes bacterial cells without plasmid. Because many different recombinant plasmids are formed, incorporating different segments, the one of interest can be separated by combined procedure of **replica plating** and ***in situ* hybridization** (Figure 4.23-III). Through replica plating, numerous dishes with representatives of the same bacterial colony are prepared. In one of the replica plates, cells are lysed, and DNA fixed on to surface of nylon or nitrocellulose membrane. DNA is next denatured; membrane is incubated with labelled single stranded **DNA probe**, containing

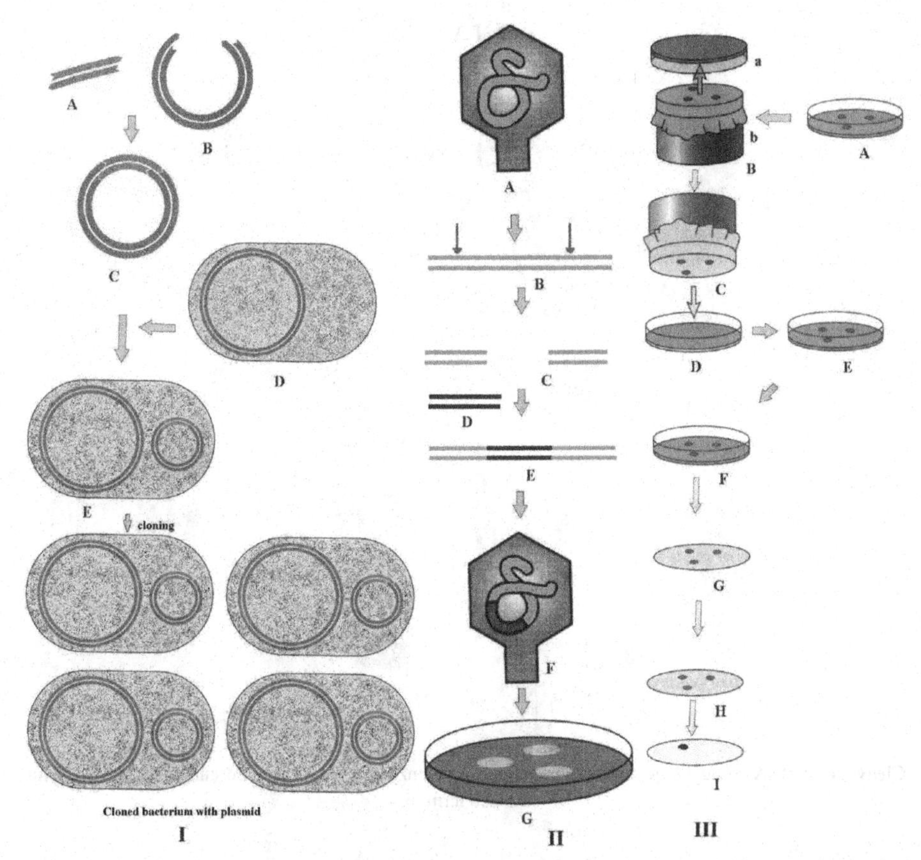

Figure 4.23: DNA Cloning. I. Cloning using plasmid vector. **A:** DNA fragment of an organism; **B:** Cleaved plasmid DNA; **C:** Recombinant DNA molecule (5–10 kb); **D:** Bacterium; **E:** Bacterium with recombinant DNA. **II.** Cloning eukaryotic DNA using lambda phage. **A:** Mutant strain of lambda phage with DNA having two *EcoRI* cleavage sites; **B:** Extracted DNA of phage treated with *EcoRI*; **C:** Two fragments of phage DNA, middle segment discarded; **D:** DNA segment from eukaryotic cell (about 25 kp); **E:** Recombinant DNA; **F:** Recombinant DNA packed into phage head; **G:** Culture dish with bacterial culture with clear plaques of phage infection. **III:** Combined procedure of replica plating and *in situ* hybridization. **A:** Dish with bacterial colonies; **B:** Transferring bacterial cells from dish (a) to a filter paper (b); **C:** Filter paper with bacterial colonies; **D:** inoculating empty culture dish by pressing filter paper; **E:** Dish with bacterial colonies; **F:** Culture dish with bacterial colonies for replica plating; **G:** Nitrocellulose membrane with replica of bacterial colonies; **H:** DNA separated by lysis of cells and denatured to become single stranded adhering to membrane; **I:** Radiograph of labelled hybrid.

complementary sequence being sought. The unhybridized probe is washed away, and the location of labelled hybrids determined by autoradiography. In refined technique of **fluorescence *in situ* hybridization (FISH)**, probe labelled with fluorescent dyes is used, and labelled hybrids localized with fluorescent microscope. The live representatives of the identified clones can be found on corresponding sites on the original plates, these cells are grown into large colonies, which serve to amplify recombinant DNA plasmid. After sufficient amplification, the DNA is extracted, and recombinant plasmid DNA is separated from bacterial DNA. The recombinant plasmid DNA is again treated with the same restriction enzyme that releases plasmid DNA from the cloned DNA segments. Latter can be separated from plasmid DNA by centrifugation.

Bacteriophage Vector

Bacteriophage l (lambda) is commonly used as a vector. The DNA of the phage is linear 50 kb in length. During treatment with restriction enzyme middle 15 kb segment of phage DNA which contains genes for lysis and can be dispensed with is replaced with foreign DNA. The resultant recombinant DNA is packed into phage heads *in vitro* (Figure 4.23-II).

Phage particles can inject the recombinant DNA molecules into *E. coli* cells, where they will replicate and produce clones of recombinant DNA molecules. As lambda heads can accommodate molecules of only 45 to 50 kb size, it can accommodate inserts (foreign DNA fragments) of only 10–15 kb.

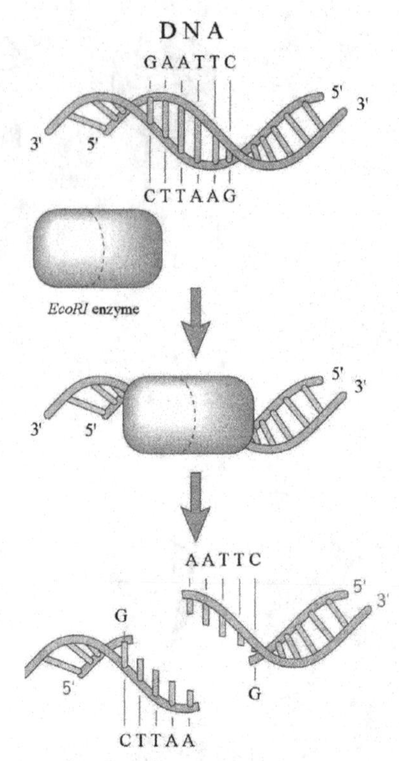

Figure 4.24: Cleavage of DNA using *EcoRI* restriction enzyme. The enzyme gives staggered cuts to ensure complimentary single stranded termini.

Cosmid Vector

For inserting larger DNA insertion, cosmid vectors are used. A cosmid is a hybrid between plasmid and lambda phage. Cosmids combine plasmid's ability to replicate autonomously with *in vitro* packaging capacity of lamda phage. A cosmid vector can carry out inserts of 35 to 45 kb.

Eukaryotic Shuttle Vectors

Some of the most useful cloning vectors are shuttle vectors that can replicate in both *E. coli* and another species. Such shuttle vectors are very useful for genetic dissections. A yeast gene can be cloned in shuttle vector, subjected to site-specific mutagenesis in *E. coli*, and then moved back to the yeast to examine the effects of induced modifications in native host cells.

Artificial Chromosome Vectors

Attempts have been made over the recent years to develop vectors which can accommodate DNA sequences larger than 45 kb. One of the most important of these vectors is **YAC (yeast artificial chromosome)**, which can accept DNA fragments as large as 1000 kb. More recently the use of **BAC (bacterial artificial chromosome)** has become more common. BACs are specialized bacterial plasmids (F factors) that contain bacterial origin of replication and can accommodate up to 300 kb of DNA segments.

Amplification through PCR

The earlier procedures for obtaining a large quantity of DNA were very cumbersome, involving the cloning of genes into bacteria, which replicate genes along with their own genome. The development of PCR (**polymerase chain reaction**) technique has now made it possible to obtain large number of copies of a gene using enzyme in place of bacteria. Small pieces of single-stranded DNA with known sequence are used as **primers** (Figure 4.25). These primers are built from templates of short regions of DNA that occur at either end (flanking) of DNA segment of interest, do not occur anywhere else in genome (unique), and are invariable (conserved) in all taxa to be investigated. The extracted DNA from a species

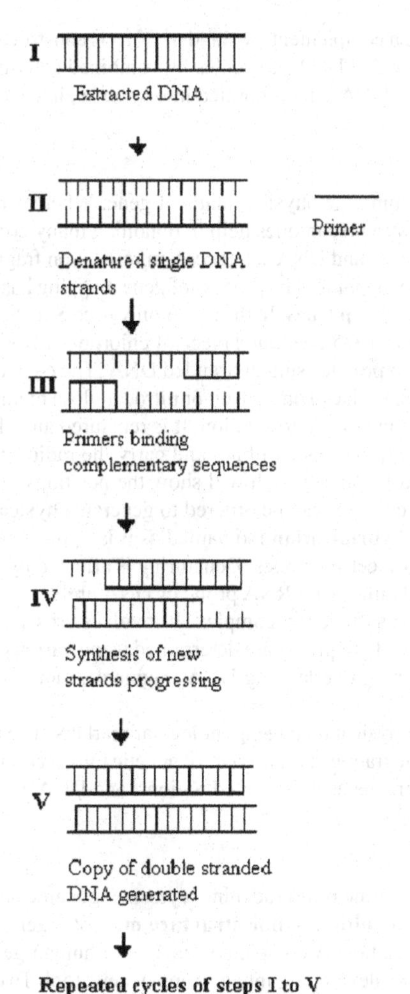

Figure 4.25: Polymerase chain reaction technique.

is mixed with the primer, DNA polymerase (usually **taq polymerase**, which can tolerate heat), buffers, salts and free nucleotides in a tube. The mixture is alternately heated and cooled. Heating denatures DNA making it single-stranded. The subsequent cooling allows primers to bind to the complementary DNA sequences. Polymers are designed so that they cannot bind with each other. The temperature is then raised to make polymerase active, bind to the already formed complex (DNA + polymerase), and begin synthesis of complementary strand (at DNA region not bound by primers) using free nucleotides. The temperature is raised further to denature DNA and the cycle repeated, thus making enough copies of DNA.

DNA Libraries

DNA libraries are collections of cloned DNA fragments. Two basic types of DNA libraries can be created. **Genomic libraries** are produced from the total DNA extracted from the nuclei and contain all the DNA sequences of the species. **cDNA libraries** (cDNA—complementary DNA) on the other hand, are derived from DNA copies of usually the messenger RNA, and thus represent DNA sequences which are expressed in the species. This is significant because many DNA sequences do not express themselves and are of little significance. Sometimes, individual chromosomes of an organism are isolated by a procedure that sorts chromosomes based on size and the DNA content. The DNAs from the isolated chromosomes are then used to construct **chromosome-specific DNA libraries**, which facilitates the search for a gene that is known to reside on a specific chromosome. This is particularly useful for organisms with large genomes, such as humans.

To construct a DNA library, the DNA from a species is randomly cleaved using enzymes which recognize short nucleotide sequences, the fragments are incorporated into lambda phage and multiple copies of each recombinant DNA obtained. These are stored and constitute a permanent collection of all DNA sequences present in the genome of a species. To construct a cDNA library using mRNA, a complementary stand of DNA is constructed by reverse transcriptase. RNA-DNA duplexes are converted into double-stranded DNA molecules by combined activity of ribonuclease H, DNA polymerase I, and DNA ligase. The double-stranded DNA is incorporated into lambda phage and further processed as detailed above.

Gene Mapping

Above techniques contribute in developing the physical maps of gene. Whereas restriction enzymes enable cleavage at specific sites, the cloning and amplification techniques help in obtaining many copies of fragments.

Identification of the location of genes and DNA sequences on restriction fragments separated by gel electrophoresis constitutes an important step of genome mapping. The process of gene mapping has been simplified with the availability of **cloned** organelle genomes which are used as **probes**. In the commonly used **Southern blot hybridization** method (named after E. M. Southern, who published it in 1975), a cloned piece of chloroplast DNA (to be used as probe) is labelled with radioactive phosphorus and denatured to produce single-stranded DNA. The cleaved DNA from the specific species, after electrophoretic separation of fragments, is placed on a nylon or nitrocellulose membrane, and denatured by using alkaline solution and finally immobilized by drying or UV irradiation. It is renatured and allowed to bind to the radioactive probe on a nylon membrane. Only matching sequences will bind and carry the radioactive tag. When transferred to an X-ray film the bound bands will appear as dark bands, which will show the positions of DNA sequences that have hybridized with the probe. The segments of different sizes can be ordered to generate physical maps.

The technique of **Northern blot hybridization** (so named as it is opposite of Southern blot technique), is used to hybridize RNA molecules separated by electrophoresis. Denaturing is affected by formaldehyde, and after transfer to the membrane, the RNA blot is hybridized either with RNA probe or DNA probe.

The procedure of gene mapping is sufficiently complex. It involves crossing two plants, selfing F1 and producing many F2 plants. Genotypes of parents and offsprings are determined using various markers. Although **physical maps** can be constructed by identifying and aligning overlapping DNA fragments, more elaborate **genetic maps** are constructed using **genetic markers.**

Genetic map can be unified with physical map using molecular markers. The physical map thus obtained will afford single framework for organizing and integrating diverse types of genetic information, including the position of chromosome bands, chromosome breakpoints, mutant genes, transcribed regions, and DNA sequences.

Gene Sequencing

Sequencing determines the exact order of the bases (adenine, cytosine, guanine and thymine) constituting nucleotides in a portion of a DNA and thus building an **ultimate fine structure map** of a gene or chromosome (Figure 4.26). Today, sequencing is a routine laboratory procedure. A complete sequence of human genome has been developed, as also the small annual weed *Arabidopsis thaliana*, developing into a strong genetic tool. Two main procedures of DNA sequencing are commonly used.

In the first procedure developed by Allan Maxam and Walter Gilbert, the DNA chain is cleaved using four different chemical reactions, each targeting A, G, C or C+T. In the second procedure developed by **Fred Sanger** (**chain termination method**) and colleagues, there is *in vitro* synthesis of DNA in presence of radioactive nucleotides and specific chain terminators to generate four populations of radioactively-labelled fragments that end with As, Gs, Cs and Ts, respectively.

The procedure begins with obtaining identical DNA fragments up to about 500 bp using a restriction enzyme. The preparation is divided into four samples. Each sample is denatured into single strands, incubated with a short radioactively-labelled oligonucleotide complementary to 3' end of single strands. To each sample is also added DNA polymerase and all the four deoxyribonuclease triphosphate precursors (**dNTPs**). To one sample is now added chain terminator ddATP (2', 3'-dideoxyadenosine triphosphate), to the second ddGTP (2', 3'-dideoxyguanosine triphosphate), to the third ddCTP (2', 3'-dideoxycytidine triphosphate), and to the fourth ddTTP (2', 3'-dideoxythymidine triphosphate). The first sample after reaction will have all the segments terminated at As, the second at Gs, the third at Cs and the fourth at Ts. The fragments are separated on gel electrophoresis, and their positions determined by autoradiography. Different bands, representing different segments will be arranged like a ladder. By reading the ladder, a complete nucleotide sequence of DNA chain can be determined. In conventional slab-gel procedure, four different samples are loaded in four different wells on a gel. Nowadays, automated DNA sequencing machines are used which make use fluorescent dyes instead of radioactive nucleotide. The products of all four samples are run through single well, and photocells are used to detect the fluorescence as they pass through the well (tube or gel). The output is directly analyzed by a computer, which analyses, records and prints out the results.

The PCR product can be sequenced directly using restriction enzymes. Since restriction sites are spread at several places on the DNA, the results are less sensitive to local vagaries of selection or differences in mutation rate. Sequencing of both the strands often minimizes errors.

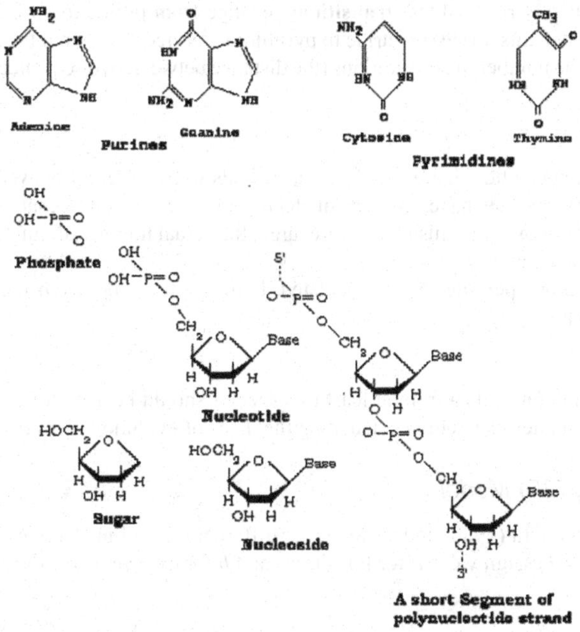

Figure 4.26: Structure of DNA. One of the four bases is joined to a deoxyribose sugar to form nucleoside, which links with a phosphate to yield a nucleotide. A long chain of nucleotides forms the DNA strand having OH (3' terminus) group at one end and phosphate (5' terminus) at the other end. Purines bases have double ring structure, whereas pyrimidines have single ring structure.

Analysis of Sequence Data

For the analysis of changes at the level of nucleotides and the amino acids, the alignment of DNA sequences derived from different taxa constitutes an important step. Alignment helps in detection of insertion, deletion or substitutions of one or more base pairs at different sites within a DNA. When comparing two sequences with L positions (nucleotides), of which D positions are different, the evolutionary distance counted Several different models have been proposed to explain evolutionary distance between two sequences on account of nucleotide changes.

Jukes-Cantor Model

Jukes and Cantor (1969) realized, even before the DNA sequences were available for analysis, that alignments between sequences with many differences might cause a significant underestimation of the actual number of substitutions that occurred since sequences last shared a common ancestor. They assumed that each nucleotide was as likely to change into any of the other three nucleotides. A can thus equally well change into T, C or G. Based on this assumption they created a mathematical model in which rate of change to any one of the three alternative nucleotides was assumed to be a, and the overall rate of substitution for any given nucleotide was $3a$. According to this model, if a site within a gene was occupied by a C ($t = 0$), then the probability (P) that this site would still be same nucleotide at time 1 ($t = 1$) would be $PC(1) = 1 - 3a$. On the other hand, if C changed to some other nucleotide, the probability that after time t, the site would contain C can be calculated as:

$PC(t) = (3/4)e–4at$

The probability rate matrix for the changes in four nucleotides can be represented as under:

	A	G	C	T
A	$1-3a$	a	a	a
G	a	$1-3a$	a	a
C	a	a	$1-3a$	a
T	a	a	a	$1-3a$

It was, however, subsequently realized that **transitions** (change from purine to purine; pyrimidine to pyrimidine) proceed at much faster rate than **transversions** (purine to pyrimidine or vice versa), but the Jukes-Cantor model can still be considered for calculating the number of substitutions (the distance between two sequences) per site (K) when multiple substitutions were possible:

K or $djc = -3/4 \ln(1 - (4/3)p)$

where p is the fraction of the nucleotide that a simple count reveals to be different between two sequences. It follows from the equation that if two sequences have fewer mismatches, p is small, and the chance of multiple substitutions is also small. On the other hand, if number of mismatches are large, the actual number of multiple substitutions per site will be considerably larger than that counted.

Once number of substitutions per site (K) is calculated, knowing the time taken for divergence (T), the **rate of substitution** (r) can be calculated as:

$r = K/(2T)$

For calculating substitution rates, data from at least two species should be available. If evolutionary rates between species are similar, substitution rates can help in calculating the dates of evolutionary events.

Kimura Two-parameter (K2P) Model

The model was proposed by Kimura (1980) and accounts for different rates of nucleotide changes involving transitions and transversions. Supposing we assign value a for transitions and b for transversions, the probability rate matrix would be represented as:

	A	G	C	T
A	$1-a-2b$	a	b	b
G	a	$1-a-2b$	b	b
C	b	b	$1-a-2b$	a
T	b	b	a	$1-a-2b$

The number of substitutions per site (distance between two sequences) could be calculated as:

K or $dK2P = 1/2 \ln (1-2P-Q) - 1/4 \ln (1-2Q)$

where P and Q are observed fractions of aligned sites whose two bases are related by a transition or a transversion, respectively.

Once the sequences are generated, they must be aligned. First the sequences of a given length are aligned by arranging homologous nucleotides in corresponding columns. Alignment is simpler for conserved genes, where all taxa will have same number of nucleotides per gene. Some other genes which have some deletions, additions, inversions or translocations in some taxa, are difficult to align. Similarly, DNA with multiple copies of a gene makes it difficult to assess homology. Several computer programs are available to produce alignment, but the assumptions used in each program should be carefully examined before the program can be used for a particular set of taxa. In phylogenetic analysis each nucleotide position is considered as one character, and each of the four nucleotides as one-character state. Many nucleotide positions, however don't show variation among taxa, and of others that are variable are often uninformative because of being autapomorphic for a given taxon. This leaves only a small proportion of nucleotide positions that can be used for phylogenetic analysis. Chromosomal mutations such as additions, deletions and translocations are identified as evolutionary novelty, and are generally given more weightage than individual nucleotides (Figure 4.27). Such chromosomal changes representing apomorphy are important and often used in establishing a lineage. Thus, all members of subfamily Faboideae lack one of the inverted repeats found in the chloroplast DNA of most angiosperms.

In our example illustrated in Figure 4.27, the four nucleotides are given coding from 0 to 3 for different nucleotides. Other strategies could also be used. **Transitions** (change from A to G or vice versa; or from C to T and vice versa) are more common than **transversions** (A to T, A to C, G to C, G to T; C to A, C to G, T to A, T to G). The latter are often given more weight depending upon the frequency of distribution in the taxa, more frequently the transitions are distributed, greater weight is consequently given to transversions. Thus, if transitions occur 4 times more than transversions, a transition may be given weight of 1 and transversion a weight of 4. Computer programs such as DNAPARS, DNADIST, etc., of PHYLIP are available, which can read and analyze the DNA sequence data directly. Details are described under chapter on Developing Classifications.

	DNA Alignment	Coding of characters
Species	2 0 2 5 3 1 3 1 4 0	1 2 3 4 5
A	GTCCAAGACTCTCAGTGGTTCAATCGTCTGTT	2 0 2 4 3
B	CTCCAAGTCTCTCACTG ------ TCGTCAGTT	1 3 1 5 0
C	CTCCAAGACTCTCAGTGGTTCAATCGTCTGTT	1 0 2 4 3
D	GTCCAAGTCTCTCACTGGTTCAATCGTCTGTT	2 3 1 4 3

Figure 4.27: Alignment of a DNA sequence for 32 nucleotide positions in four species. Nucleotides at a specific position showing variation can be coded as characters and are shown in bold. The codes assigned to nucleotide states are A = 0, C = 1, G = 2 and T = 3. Character number four here involves deletion of a specific sequence in species B. Presence of this deletion is coded as 5 and absence of deletion as 4. Computer programs (such as DNADIST and DNAPARS of PHYLIP) are available which can read DNA sequence data directly.

Whereas alignment of simple chloroplast genes such as *rbCL* is easier, others such as genes encoding RNAs, secondary structure (folding) of the molecule is also accounted for. The nucleotide differences that result in major changes in the structure of a product, such as ribosomal RNA or a protein, and may have greater effect in the plant function, often receive greater weight than those changes that do not affect the function. Several computer algorithms are available to evaluate and handle such analysis.

DNA Polymorphism

Utilization of sequence data in phylogenetic analysis involves the identification of unique sequences which show certain differences in different organisms or populations. These sequences, which could be used as **genetic markers** in identification of character-state differences between the target taxa, and ultimate construction of phylogenetic trees. The phenomenon is also known as **DNA Fingerprinting** or **DNA polymorphism.** The technique is now widely used in **forensic investigations**. A variety of methods have been developed to detect this polymorphism. Each method has its own advantages and limitations, and suitable for a specific situation. New methods are being continuously developed. Some of the commonly used procedures are discussed below.

Single-Nucleotide Polymorphisms (SNPs)

DNA differences in a population may often be the result of differences in single nucleotide pair at a locus, say from C-G to T-A. This may result in three genotypes in a population: homozygous with C-G at corresponding sites on both homologous chromosomes, homozygous with T-A at corresponding sites on both homologous chromosomes, and heterozygous with C-G in one chromosome and T-A in homologous chromosome. However, all SNPs are not located on coding sequences or genes. In human genome, for example, any two randomly chosen DNA molecules differ at one SNP site about every 1000–3000 bp in protein coding DNA, but 500–1000 bp in noncoding DNA segments. SNPs are most common types of genetic differences among human populations and are uniformly distributed over the chromosomes.

The SNPs can be easily detected if they are located in a cleavage site (Figure 4.28). Thus, a sequence GAATTC can be cleaved by *Eco*RI, but a corresponding GAACTC sequence can't be cleaved as T has been replaced with C (and on the complementary segment A replaced by G). This will result in larger DNA fragment in the latter case.

Restriction Fragment Length Polymorphisms (RFLPs)

RFLP results from the fact that a mutation that causes changes in base sequence may result in loss or gain of a cleavage site, thus alleles differing in the presence or absence of a cleavage site. This may also result from SNPs located at cleavage sites as indicated earlier. As a result, fragments of different lengths are yielded. The method is widely used for identification of individuals, species or populations. The DNA from a species is cleaved using a restriction enzyme (say *Eco*RI) yielding a certain number of fragments (Figure 4.29). These fragments can be separated using Southern blotting procedure, and a map of these constructed. These fragments are further fragmented using another enzyme (say *Hind*III), and the data incorporated into original map. Restriction site fragments obtained are coded as characters and character-states for phylogenetic analysis.

The absence or presence of a restriction site in closely related species and the presumed hybrids can also be detected by Southern blotting procedure. Species A, for example, lacks restriction site at 3000 bp position (allele a, genotype aa), where this site is present in species B (allele A, genotype AA). Southern blotting technique will yield longer first restriction

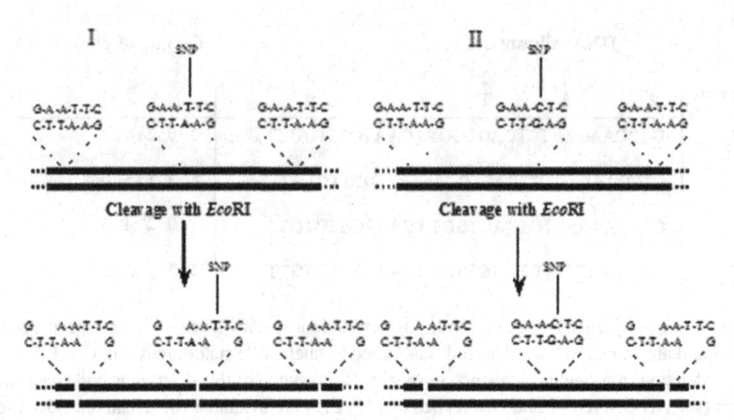

Figure 4.28: The effect of the location of SNPs in the restriction sites in two DNA molecules. DNA with TA nucleotide pair shows normal cleaving with *Eco*RI, resulting in two restriction fragments. In another DNA (II) substitution of CG pair prevents cleaving resulting in single large cleavage fragment. This results in unequal fragments in different DNA molecules.

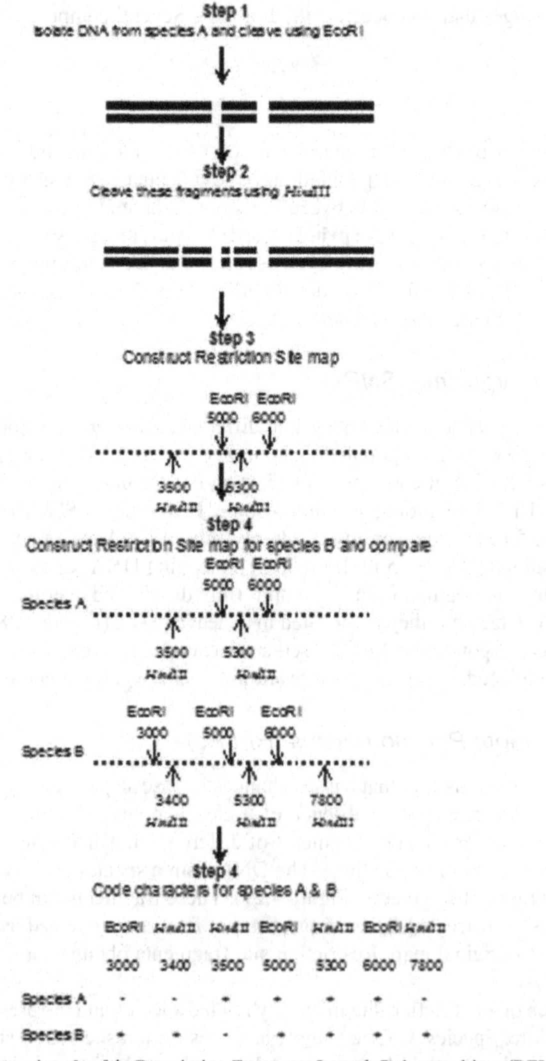

Figure 4.29: Major steps involved in Restriction Fragment Length Polymorphism (RFLP) procedure.

fragment of 5000 bp length for aa genotype, whereas it will yield fragment of 3000 bp length in AA genotypes. The heterozygous Aa genotype, presuming the alleles are codominant will yield two fragments from homologous chromosomes, one of 3000 bp length and another of 5000 bp length.

RFLP analysis, however, contains much lesser data than complete DNA sequencing, accounting only for presence or absence of sites 6–8 base pairs long, but the method affords advantage of surveying larger segments of DNA. The use of this method has, however, declined with the development of improved and less expensive sequencing techniques in the recent years.

Random Amplified Polymorphisms (RAPDs)

RAPD method is commonly used for population studies and involves short (10 bp) random PCR primers that will bind to the matching sequences on genome. The approach is useful for species where cloned DNA probes are not available (essential for Southern blotting method), or where DNA sequences are not known (necessary for PCR amplification where oligonucleotide primers have to be constructed). The method uses PCR primers of 8–10 nucleotides with **random** sequence. These primers are tried singly or in pairs in PCR reactions to amplify segments of DNA from a species. These short primers anneal at multiple sites on DNA, and those that anneal at suitable distance can amplify unknown region between them. The presence or absence of such amplified regions in different individuals can be suitably coded for analysis. The procedure helps in identifying different genotypes in the population. The morphologic characters of interest are mapped according to their linkage to markers. The results of one of several primers used are shown in Figure 4.30. Gel electrophoresis yields 13 bands, of which four show polymorphism, the rest nine are monomorphic. Each of the polymorphic allele can be represented similarly as + for the presence of band, – for its absence, and if + is dominant, both genotypes +/+ and +/– will show this band, whereas it will be lacking in –/– genotype. Thus, for the last band in the gel species B and C have –/– genotype, whereas A and D are either +/+ or +/–.

Amplified Fragment Length Polymorphisms (AFLPs)

AFLP (amplified fragment length polymorphism) technology is used for nucleic acid fingerprinting, exploiting molecular genetic variations existing between closely related genomes in the form of restriction fragment length polymorphisms.

AFLP procedure involves four basic steps (Figure 4.31). In first step DNAs from different sources are isolated and digested with appropriate **restriction endonucleases (REs)**. For most plant DNAs, two REs are used: one a rare cutter

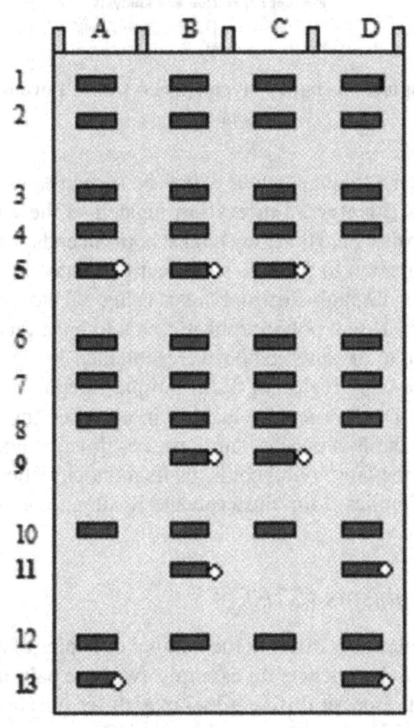

Figure 4.30: Results of one of several primers used for RAPD procedure on DNA from four species. A total of 13 bands appear on electrophoresis gel, 4 show polymorphism in species compared, whereas 9 are monomorphic.

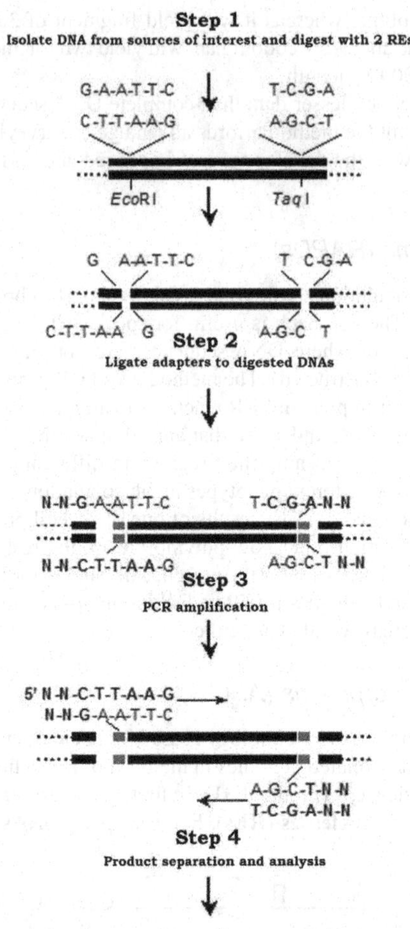

Figure 4.31: Basic steps in Amplified Fragment Length Polymorphism (AFLP) protocol. N-N represents a particular nucleotide sequence.

having 6-bp recognition site, and the other a frequent cutter with 4-bp recognition site. In the second step, specific double-stranded oligodeoxynucleotide adapters (**primer adapters**) are ligated to the ends of the digested DNAs to generate chimeric molecules. These primers are so designed that they bind at both cut ends of fragments. In the third step the chimeric fragments are subjected to PCR amplification to provide sufficient template DNA for fingerprinting PCRs. During the fourth step, PCR products are resolved on through electrophoresis using polyacrylamide sequencing gel, which separates the amplified DNA fragments that exhibit length polymorphisms, enabling the recognition of numerous genetic markers.

One of the earliest significant results of this method were obtained by Jansen and Palmer (1987), who found a unique order of genes in the large single-copy region of the chloroplast genome in Asteraceae. This unique order could be explained by single inversion of the DNA, a feature lacking in all other angiosperms, strongly confirming that the Asteraceae family is monophyletic. The family Poaceae, similarly, has three inversions in the chloroplast genome. Out of these three inversions, one is unique to the family and confirms its monophyletic status. Of the other two, one is shared with Joinvilleaceae and one with both families, Joinvilleaceae and Restionaceae, suggesting that these two are the sister groups of Poaceae.

Simple Tandem Repeat Polymorphisms (STRPs)

STRP results from the fact DNA molecules may differ in the number of copies of a sequence of few nucleotides repeated in tandem at a particular locus. In TGTGTG sequence, for example, two base pairs are repeated. Such repeated nucleotides are known as **tandem repeats.** STRPs present at different loci may differ in sequence and length of repeating unit, and in minimum and maximum number of tandem copies occurring in the DNA of a population. A repeating sequence of 2–9 bp is often known as **microsatellite** or **SSLP (Simple sequence length polymorphism)**, whereas one of 10–60 bp as

minisatellite. If these repeated sequences show variation within a population or a species, they are known as **variable number tandem repeats** (***VNTRs***). At a given locus in different individuals, the length of tandem repeats may vary, because of irregularities of crossing over and replication, and as such can be used as genetic marker. Identification of microsatellites involves constructing primers that flank tandem repeats, and then using PCR technology to generate multiple copies of tandem repeat DNA, whose length can be determined by gel electrophoresis (Figure 4.32). VNTR technology generates data quickly and efficiently and is often used for population studies, for examining relationships within a species, or between closely related species.

STRP is very useful in mapping, as many alleles present in the population often have high proportion of genotypes that are heterozygous for different alleles. *STRPs* are widely used in DNA typing (DNA fingerprinting) involving identification of human individuals in criminal investigation.

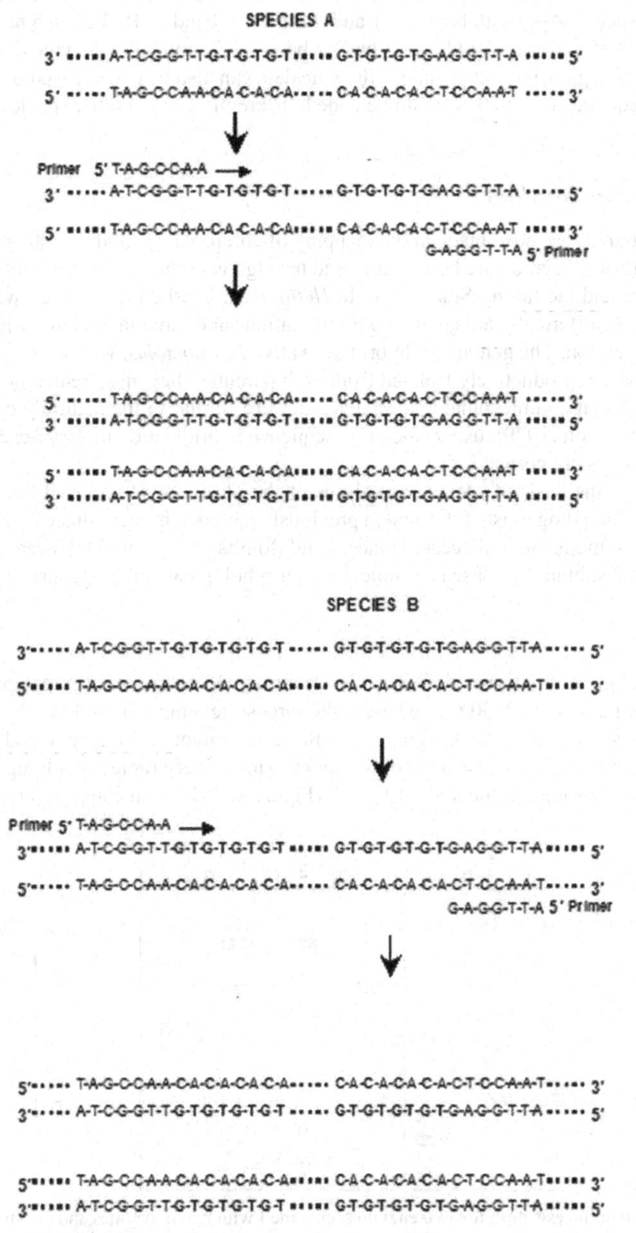

Figure 4.32: VNTR procedure. Specific primers to flank regions of tandem repeats are constructed and used for PCR amplification of DNA segments with tandem repeats, for comparison of different species.

Allozymes

Different forms of an enzyme differing in different alleles at the same locus constitute **allozymes**, as distinct from isozymes showing differences at different loci. Allozymes are separated and detected using starch electrophoresis as against gel electrophoresis for DNA sequencing. Allozymes differing slightly in amino acid composition will take different charges and migrate differently and can be identified using specific stains.

Allozymes have traditionally been used to assess genetic variation within a population or a species, but they can also be used for phylogenetic analysis of closely related species. Allozyme data can be coded in a variety of ways. Each allele may be coded as a character and its presence or absence as character states. Alternately a locus may be treated as character, and unique allele combinations as character states (Figure 4.33). A comparison electrophoresis bands of four species for enzyme I with two allozymes can be coded as 0 for allozyme separated at position 18 and 1 for allozyme separated at position 21. Similarly, Enzyme II with three allozymes can be coded as 0 for 27, 1 for 31 and 2 for 35. Enzyme I, as such would be coded as 1 for species A–C with band at 21 and 0 for D for band at 18. For enzyme II, similarly species A has band at 31 coded as 1, B at 35 coded as 2 and C as 3 having bands at 31 and 35. Allozyme data can also be coded as loss of each allele as one state and gain as another state. Allozyme data can also be coded based on allele frequency. A species with two alleles in the frequency of 90/10% would be coded differently from another species with same alleles but with frequency of 40/60%.

Examples of Molecular Studies

Whereas considerable progress has been made in the mapping of chloroplast genome, similar success in nuclear genome is at its infancy. Questions of speciation are being addressed through genome mapping in *Helianthus*. Some progress has also been made in grasses and the family Solanaceae. In *Helianthus*, Riesberg and his co-workers (1996) reported that *H. annuus* and *H. petiolaris* differed by at least seven translocations and three inversions, which affected recombination and possibilities of introgression. The genome of hybrid derivative *H. anomalus,* was rearranged relative to both parents, and the species was partially reproductively isolated from both parents. They also created new hybrids between the two parental species and found that chromosomal rearrangements were similar to the naturally occurring hybrid species, *H. anomalus*. Belford and Thomson (1979), using side-copy sequence hybridization in *Atriplex* concluded that division into two subgenera in this genus is not correct.

Bayer et al. (1999) on the basis of sequence analyses of the plastid atpB and *rbc*L DNA, found a support for an expanded order Malvales, including most of the genera previously included in Sterculiaceae, Tiliaceae, Bombacaceae and Malvaceae. They propose to merge Sterculiaceae, Tiliaceae and Bombacaceae with Malvaceae and subdivide this enlarged family Malvaceae into nine subfamilies based on molecular, morphological and biogeographical data.

Grass Genome

Genome analysis of cereal grasses has provided useful information. Of the common cereal grasses, rice has the smallest genome (400 mb). Maize genome is 2500 mb, whereas the largest genome is found in wheat (17,000 mb). In spite of large variations in chromosome number and genome size, there are a number of genetic and physical linkages between single-copy genes that are remarkably conserved amid a background of very rapidly evolving repetitive DNA sequences. By comparison of rice chromosomes numbered R1 to R12 (Figure 4.34-I), with conserved regions marked in lower case

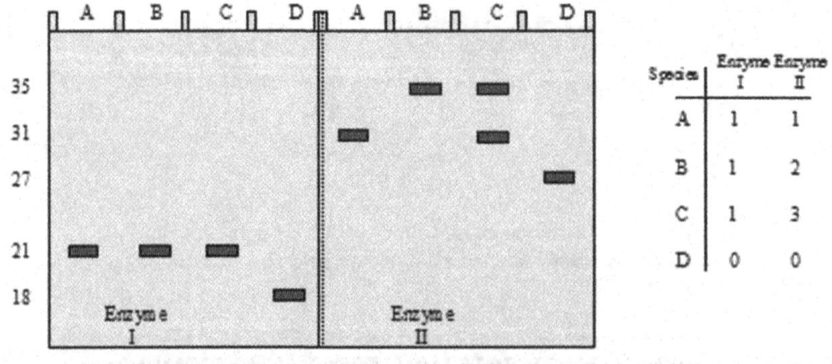

Figure 4.33: Allozyme electrophoresis data for two enzymes, enzyme I with two allozymes and enzyme II with three allozymes. The coding of data for four species, of which species D represents an outgroup is presented below. Each locus is treated as character and combination of character states as character states.

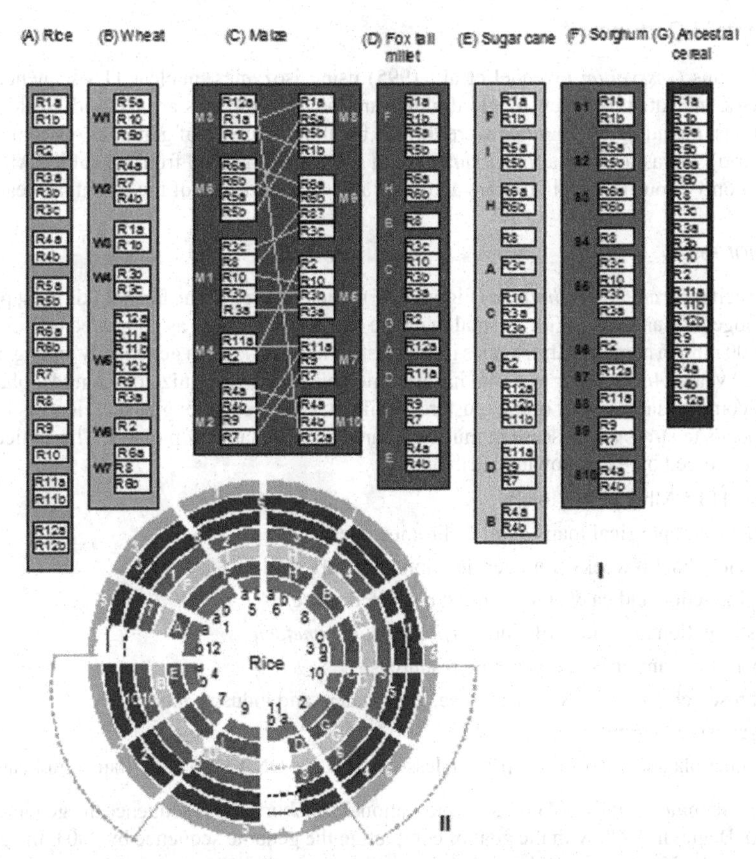

Figure 4.34: Grass genome evolution. I. Conserved linkages (synteny groups) between the rice genome and other grass species. **A:** Rice genome with chromosomes divided into blocks of linked genes; **B:** Wheat genome with chromosomes showing correspondence with rice segments; **C:** Maize genome with duplicated blocks indicating ancient tetraploidy; **D:** Foxtail millet genome; **E:** Sugar cane genome; **F:** Sorghum genome; **G:** Inferred or 'reconstructed' order of segments in a hypothetical ancestral cereal genome consisting of a single chromosome pair. **II.** Circular arrangement of synteny groups in above grasses. Thin dashed lines indicate connections between blocks of genes (After Moore et al., 1995).

(R1a, R1b, etc.), it is found that conserved regions homologous to rice are found in other cereals. The wheat monoploid chromosome set is designated W1 through W7. One region of W1 contains single-copy sequences that are homologous to those in rice segment R5a, another contains single-copy sequences that are homologous to those in rice segment R10, and still another contains single-copy sequences homologous to those in rice segment R5b. Each of such conserved physical and genetic linkages is called a **synteny group**. It is notable that maize genome has repetition of segments, confirming that maize is a complete, very ancient tetraploid with two duplicated genome blocks rearranged relative to each other.

Simultaneous comparison of above cereal grass genomes is better represented with the help of a circular diagram (Figure 4.34-II). The segments are arranged into a circle in the same order in which they were aligned in the hypothetical ancestral chromosome. Because of the synteny groups in the genomes, homologous genes can often be identified by location alone. It must, however, remembered that the circular diagram is only for convenient representation; there is no indication that the ancestral grass chromosome was actually circular. It was a normal linear chromosome.

Many workers have targeted the family Poaceae using different criteria and techniques. All molecular phylogenies point to the Stipeae to be an early-diverging lineage. The morphological characters of the Stipeae are thus a mixture of synapomorphies linking them with pooids and symplesiomorphies, which they share with many other grasses. The studies based on chloroplast gene: cpRFLP (Davis and Soreng, 1993), ndhF sequences (Catalan et al., 1997), and nuclear genes: through ITS (Hsiao et al., 1994), phytochrome b (Mathews and Sharrock, 1996), and granule bound starch synthase I (Mason-Gamer et al., 1998) all supported the same placement of Stipeae. Similar studies of comparison of results from chloroplast DNA and nuclear DNA in Triticeae, however, produced different results, although two chloroplast phylogenies constructed from RFLP (Mason-Gamer and Kellog, 1996) and rpoA sequences (Petersen and Seberg, 1997) produced similar results.

New World Tetraploid Cottons

Genomic studies in genus *Gossypium* (Wendel et al., 1995) using isozymes, nuclear ITS sequences, and chloroplast restriction site analysis, indicated that New World diploids are monophyletic, as are the Old-World diploids. The New World tetraploid cottons, including *G. hirsutum* were formed by allopolyploidy of genomes A (from the Old World) and D (from the New World). It was found that *H. hirsutum* has a chloroplast derived from one of the African species, and it must have acquired it only about 1–2 million years ago, well after the formation of the Atlantic Ocean.

Arabidopsis Genome

Insignificant small crucifer, *Arabidopsis thaliana* (Figure 4.35), often ignored in the field, holds great promise for opening new frontiers of phylogenetic analysis. With its small genome size of 114.5 mbp (as compared to 165 mbp in *Drosophila melanogaster* and 3000 mbp in humans), the species is the most completely known genetically among all flowering plants. During the last 8 to 10 years, *Arabidopsis thaliana* has become universally recognized as a model plant for such studies. Although it is a non-commercial member of the mustard family, it is favored among basic scientists because it develops, reproduces, and responds to stress and disease in much the same way as many crop plants. The choice of *Arabidopsis* as a genetic tool has been forced by the following attributes:

1. Small genome (114.5 Mb/125 Mb total).
2. Extensive genetic and physical maps of all 5 chromosomes.
3. A rapid life cycle (about 6 weeks from germination to mature seed).
4. Prolific seed production and easy cultivation in restricted space.
5. Efficient transformation methods utilizing *Agrobacterium tumefaciens*.
6. A large number of mutant lines and genomic resources.
7. Multinational research community of academic, government and industry laboratories.
8. Easy and inexpensive to grow.
9. Compared to other plants, it lacks the repeated, less-informative DNA sequences that complicate genome analysis.

The *Arabidopsis* Genome Initiative (AGI) is an international collaboration to sequence the genome of the model plant *Arabidopsis thaliana.* Begun in 1996 with the goal of completing the genome sequence by 2004, the genome sequencing was completed at the end of 2000. Comprehensive information on *Arabidopsis* genome is available on the internet via The **Arabidopsis Information Resource (TAIR)**, which provides a comprehensive resource for the scientific community working with *Arabidopsis thaliana.* TAIR is a collaboration between the Carnegie Institution of Washington Department of Plant Biology, Stanford, California, and the National Center for Genome Resources (NCGR), Santa Fe, New Mexico. Funding is provided by the National Science Foundation.

Important studies on *Arabidopsis thaliana* have been devoted to genetic control of development. Transgenic plants of this species have been created that either overexpress or underexpress cyclin B. Overexpression of cyclin B results in accelerated rate of cell division; underexpression results in decelerated rate. Plants with faster rate of cell division contain more cells and are somewhat larger than their wild type counterparts, but otherwise they look completely normal. Likewise, plants with the decreased rate of cell division have less than half the normal number of cells, but they grow at almost the same rate and reach almost the same size as wild-type plants, because as the number of cells decrease, the individual cells get larger. The plants thus have ability to adjust to abnormal growth conditions, as opposed to animals which frequently develop proliferative cancer cells.

The studies on genetic control of flower development in *Arabidopsis* have revealed interesting results. During floral development (as in other tetracyclic plants), each whorl of the floral parts (sepals, petals, stamens and carpels) arises from a separate whorl of initials. Three types of mutations result in three different phenotypes, one lacking sepals and petals, the second lacking petals and stamens and the third lacking stamens and carpels. Crosses between homozygous organisms have resulted in identification of four genetic groups (Table 4.2). Mutations in the gene *ap2* (*apetala-2*) result in phenotype without sepals and petals. The phenotype lacking petals and stamens is caused by mutation in either of two genes, *ap3* (*apetala-3*) or *pi* (*pistillata*). The genotype lacking stamens and carpels is caused by mutations in the gene *ag* (*agamous*). Each of these genes has been cloned and sequenced. They are all transcription factors, members of **MAD box** family of transcription factors, each containing a sequence of 58 amino acids.

An interesting finding from this study is that mutation in any of the genes eliminates two floral organs belonging to adjacent whorls. The pattern suggests that *ap2* is necessary for sepals and petals, *ap3* and *pi* are both necessary for stamens and *ag* necessary for stamens and carpels. As mutant phenotypes are caused by loss-of-function in alleles, it may be inferred that *ap2* is expressed in whorls 1 and 2, *ap3* and *pi* expressed in whorls 2 and 3, and *ag* is expressed in whorls 3 and 4. The floral development in this plant is thus controlled by combinational effect of these four genes. Sepals develop

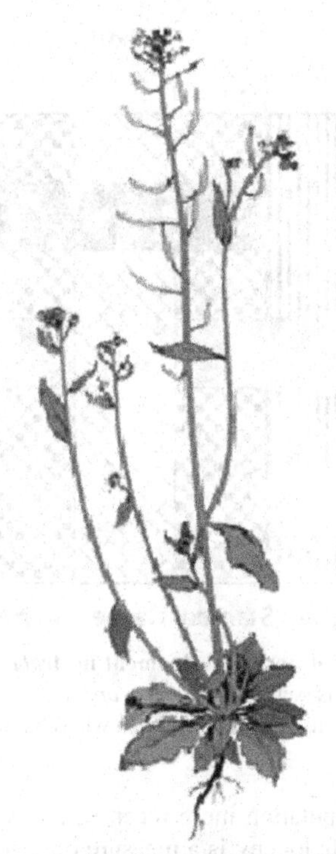

Figure 4.35: ***Arabidopsis thaliana,*** small annual herb from family Brassicaceae, whose genome is most completely known among the angiosperms, is aptly known as the guinea-pig of plant kingdom.

Table 4.2: Floral development in mutants of *Arabidopsis thaliana.*

Genotype	Whorl			
	1	2	3	4
wildtype	sepals	petals	stamens	carpels
ap2/ap2	carpels	stamens	stamens	carpels
ap3/ap3	sepals	sepals	carpels	carpels
pi/pi	sepals	sepals	carpels	carpels
ag/ag	sepals	petals	petals	sepals

from tissue in which *ap2* is active; petals by combination of *ap2, ap3* and *pi,* stamens by combination of *ap3, pi* and *ag;* and carpels where only gene *ag* is expressed. This is graphically represented in Figure 4.36.

It is pertinent to remember that *ap2* expression and *ag* expression are mutually exclusive. In presence of *ap2* transcription factor *ag* is repressed, and in the presence of *ag* transcription factor, *ap2* is repressed. Accordingly, in *ap2* mutants, *ag* expression spreads to whorls 1 and 2, and, in *ag* mutants, *ap2* expression spreads to whorls 3 and 4. This assumption enables us to explain the phenotypes of single and even double mutants. This pattern of gene expression has been assayed by *in situ* hybridization of RNA in floral cells with labelled probes for each of the genes. The results confirm the above assumption of repressive action of concerned genes. It is significant that triple mutation involves all the genes.

The phenotype of *ap2 pi ag* triple mutant does not have any normal floral organs. There are concentric whorls of leaves instead.

Species Trees vs Gene Trees

Whereas **Species trees** recover geneology of taxa including species and populations based on orthologous genes, the **Gene trees** represent evolutionary history of genes based on both homologous and orthologous genes. Gene trees can provide evidence for gene duplication events, as well as speciation events. A species tree shows us the overall pattern—

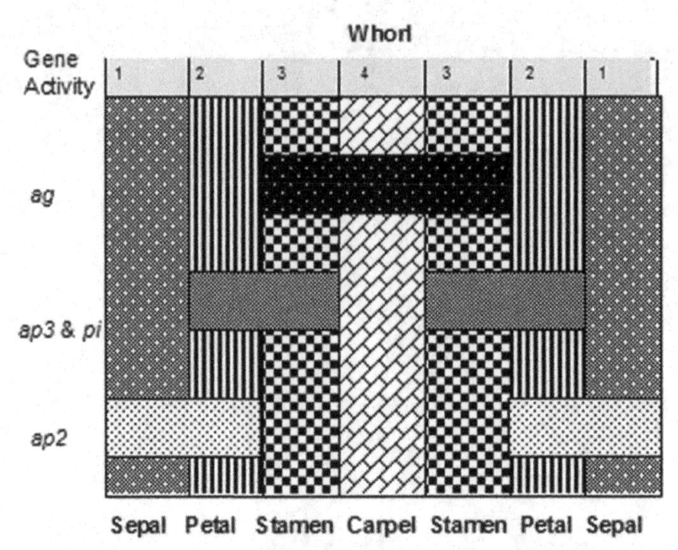

Figure 4.36: Graphic representation of control of floral development in *Arabidopsis thaliana* by the overlapping action of four genes. Gene *ap2* is expressed in the outer two whorls (sepals and petals), *ap3* and *pi* are expressed in the middle two whorls (petals and stamens) and *ag* in the inner two (stamens and carpels). Each whorl has a unique combination of active genes.

which species share a common ancestral population more recently, and which share a common ancestral population more distantly in the past. In other words, phylogeny is a measure of shared history and separate history for any two species. The longer two species have a common history, the more similar they are expected to be, on average. Gene trees concentrate on evolution of different alleles through mainly mutations resulting in variations. Gene trees and species trees can be incongruent for many reasons: Genes can have unequal rates of evolution, Gene loss and gene duplication are common, Gene flow can occur between lineages after their separation, and Recombination between neighboring regions can also lead to species phylogenies and gene histories that do not match. Molecular systematics presents powerful tools for constructing phylogenetic trees. Commonly used methods over the recent years include studies on chloroplast DNA using restriction site polymorphism (cpRFLP), analysis of chloroplast gene for subunit F of NADP dehydrogenase (ndhF, in the small copy region), for 'a' and 'b' subunits of RNA polymerase II (*rpoA* and *rpoC2*, in a large single copy region), for 'b' subunit of ATP synthase (*AtpB*), ITS region of ribosome, phytochrome B, and granule bound starch synthaseI. An encouraging congruence of results of these diverse studies was met in tribe Stipeae of grasses. In other cases, results from chloroplast phylogeny and nuclear phylogeny did not agree, suggesting caution in relying on any attribute singly for constructing molecular phylogenies. The gene trees constructed from *rbcL* have great utility in angiosperms. Chase et al. (1993) attempted to yield the phylogeny of all seed plants using 499 *rbcL* sequences. The analysis proved a few sequences to be pseudogenes, and entire families were represented by single sequences. The data set have been reanalyzed by other authors to yield parsimonious trees (Rice et al., 1997). *RbcL* data has supported that Caryophyllidae is monophyletic. It has also supported the union of family pairs Asclepiadaceae-Apocynaceae, Araliaceae-Apiaceae, and Brassicaceae-Capparaceae. The data also supported the polyphyletic nature of Saxifragaceae and Caprifoliaceae.

Swenson and El-Mabrouk (2012) inferred irreconcilable differences between Species trees and Genes trees. Under certain reasonable hypotheses based on the widely accepted link between function and sequence constraints, even a well-supported gene tree yields a reconciliation that does not correspond to the true history. Even the most recent speciation events must be old enough to allow for clear formation of isorthogroups. They assume that a single gene copy preserves the parental function after a duplication, an assumption widely used by the community and confirmed experimentally in many cases, and a single gene copy preserves the parental function after a duplication, an assumption widely used by the community and confirmed experimentally in many cases.

Chapter 5

Hierarchical Classification

With more than 3,90,000 species of vascular plants already known and thousands added every year, it would be total chaos to study and document information about them if there were no proper mechanism for grouping the same. Whatever may be the criterion for classification—artificial characters, overall morphology, phylogeny or phenetic relationship—the basic steps are the same. The organisms are first recognized and assembled into groups based on certain resemblances. These groups are in turn assembled into larger and more inclusive groups. The process is repeated until finally all the organisms have been assembled into a single, largest most inclusive group. These groups (**Taxonomic groups or Taxa**) are arranged in order of their successive inclusiveness, the least inclusive at the bottom, and the most inclusive at the top.

The groups thus formed and arranged are next assigned to various **categories**, having a fixed sequence of arrangement (**taxonomic hierarchy**), the most inclusive group assigned to the highest category (generally a **division**) and the least inclusive to the lowest category (usually a **species**). The names are assigned to the taxonomic groups in such a way that the name gives an indication of the category to which it is assigned. Rosales, Myrtales, and Malvales all belong to the **order** category and Rosaceae, Myrtaceae and Malvaceae to the **family** category. Once all the groups have been assigned categories and named, the process of classification is complete, or the **taxonomic structure** of the whole largest most inclusive group has been achieved. Because of the hierarchical arrangement of categories to which the groups are assigned, the classification achieved is known as **hierarchical classification**. This concept of categories, groups and taxonomic structure can be illustrated in the form of a **box-in-box** figure (Figure 5.1) or a **dendrogram** (resembling a pedigree chart, Figure 5.2).

TAXONOMIC GROUPS, CATEGORIES AND RANKS

Taxonomic groups, categories and ranks are inseparable once a hierarchical classification has been achieved. *Rosa alba* is thus nothing else, but a **species** and *Rosa* is nothing other than a **genus**. However, the differences do exist in concept and application. The **categories** are like shelves of an almirah, having no significance when empty, and importance and meaning only after something has been placed in them. Thereafter, the shelves will be known by their contents: books, toys, clothes, shoes, etc. Categories in that sense are artificial and subjective and have no basis in reality. They correspond to nothing in nature. However, they have a fixed position in the hierarchy in relation to other categories. But once a group has been assigned to a particular category the two are inseparable and the category gets a definite meaning because it now includes something actually occurring in nature. The word genus does not carry a specific meaning, but the genus *Rosa* says a lot. We are now talking about roses. There is practically no difference between **category** and **rank**, except in the grammatical sense. *Rosa* thus belongs to the **category genus** and has **generic rank**. If categories are like shelves, ranks are like partitions, each separating the given category from the category above. **Taxonomic groups**, on the other hand, are objective and non-arbitrary to the extent that they represent discrete sets of organisms in nature. Groups are biological entities or a collection of such entities. By assigning them to a category and providing an appropriate ending to the name (Rosaceae with ending—**aceae** signifies a family which among others also includes roses, belonging to the genus *Rosa*) we establish the position of taxonomic groups in the hierarchical system of classification. Some important characteristics, which enable a better understanding of the hierarchical system of classification, are enumerated below.

1. Different categories of the hierarchy are higher or lower according to whether they are occupied by more inclusive or less inclusive groups. Higher categories are occupied by more inclusive groups than those occupying lower categories.

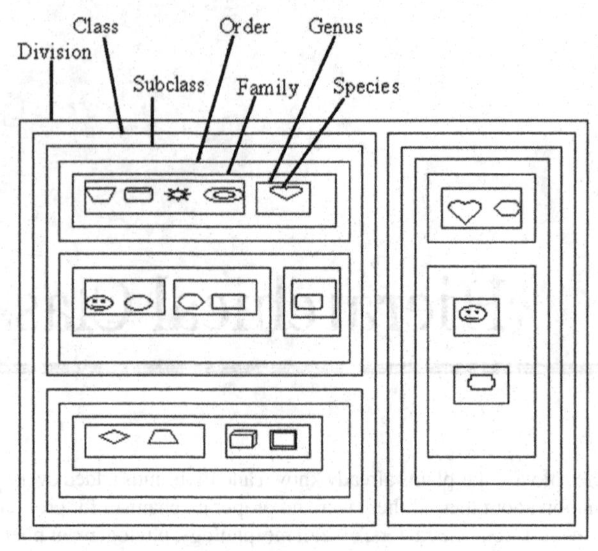

Figure 5.1: Processes of assembling taxonomic groups according to the hierarchical system, depicted by box-in-box method. In the above example, there are 18 species grouped into 10 genera, 6 families, 4 orders, 3 subclasses, 2 classes and 1 division.

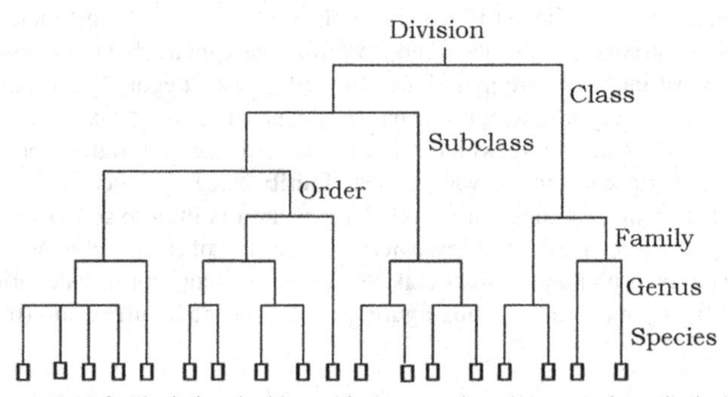

Figure 5.2: Dendrogram method for depicting the hierarchical system based on same hypothetical example as in Figure 5.1.

2. Plants are not classified into categories but into groups. It is important to note that a plant may be a member of several taxonomic groups, each of which is assigned to a taxonomic category, but is not itself a member of any taxonomic category. A plant collected from the field may be identified as *Poa annua* (assigned to species category). It is a member of *Poa* (assigned to genus category), Poaceae (assigned to family category) and so on, but the plant can't be said to be belonging to the species category.
3. A taxon may belong to other taxa, but it can be a member of only one category. *Urtica dioica,* thus, is a member of *Urtica*, Urticaceae, Urticales, and so on, but it belongs only to species category.
4. Categories are not made up of lower categories. The category family is not made up of the genus category, since there is only one genus category.
5. The characters shared by all members of a taxon placed in a lower category provide the characters for the taxon immediately above. Thus, the characters shared by all the species of *Brassica* make up the characters of the genus *Brassica*. The characters shared by *Brassica* and several other genera form distinguishing characters of the family Brassicaceae. It is important to note that the higher a group is placed in the hierarchy, the fewer will be the characters shared by the subordinate units. Many higher taxa, as such (e.g., Dicots: Magnoliopsida) can only be separated by a combination of characters; no single diagnostic character may distinguish the taxa. Dicots are thus conveniently separated from monocots by possession of two cotyledons, pentamerous flowers, reticulate venation and vascular bundles in a ring as against one cotyledon, trimerous flowers, parallel venation and scattered vascular bundles in monocots. But when taken individually, *Smilax* is a monocot with reticulate venation and *Plantago* is a dicot with parallel venation. Similarly, *Nymphaea,* is a dicot with scattered bundles, and the flowers are trimerous in *Phyllanthus*, which is a dicot.

UTILIZATION OF CATEGORIES

Taxonomic categories possess only relative value and an empty category has no foundation in reality and obviously can't be defined. An important step in the process of classification is to assign taxa to an appropriate category. It thus becomes imperative to decide what should be the properties of taxa to be included in a particular category? Only with a proper utilization of the concept of categories can their application in hierarchical systems be meaningful. The problem is far from resolved. An attempt will be made here to discuss the relevant aspects of the inclusion of type of entities or groups of entities under different categories.

Species Concept

Darwin aptly said: 'Every biologist knows approximately what is meant when we talk about species, yet no other taxon has been subjected to such violent controversies as to its definition'. A century and a half have passed, so much advancement in the taxonomic knowledge has been achieved, yet the statement of Darwin is as true today as it was then. Numerous definitions of species have been proposed, making it futile to recount all of them. Some significant aspects of the problem will be discussed here. Probably the best explanation of diversity of opinions can be explained as under.

'***The species is a concept. Concepts are constructed by the human mind, and as humans think differently, we have so many definitions of a species.***' Obviously, a concept can't have a single acceptable definition.

The word species has different meaning for different botanists. According to ICN (formerly ICBN), which has attempted to clarify the meaning of the word species, '***species are convenient classificatory units defined by trained biologists using all information available***'. The word species has a dual connotation in biological science. First, the species is a naturally-occurring group of individual organisms that comprises a basic unit of evolution. Second, the species is a category within a taxonomic hierarchy governed by various rules of nomenclature.

Species as Basic Unit of Taxonomy

The following information serves to substantiate the view that species constitutes the basic unit of classification or, for that matter, taxonomy (systematics):

1. Species is considered the basic unit of taxonomy, since in the greater majority of cases, we do not have infraspecific names. This is especially common in families such as Apiaceae (Umbelliferae) and Liliaceae.
2. Species, unlike other taxa, can be described and recognized without relating to the taxa at other ranks. Thus, we can sort herbarium sheets into different species without difficulty, without knowing or bothering to know how many genera are covered by these sheets. We cannot recognize genera or describe them without reference to the included species. Species is thus the only category dealing directly with the plants.
3. Whether defined in terms of morphological discontinuity or restriction of gene exchange, species is unique in being ***non-arbitrary to both inclusion and exclusion***. A group is non-arbitrary to inclusion if all its members are continuous by an appropriate criterion. It would be arbitrary to inclusion if it shows internal discontinuity. A group is non-arbitrary to exclusion if it is discontinuous from any other group by the same criterion. A group not showing discontinuity with other groups is arbitrary. All higher taxa although non-arbitrary to exclusion are arbitrary to inclusion, i.e., they exhibit internal discontinuity as now species with external discontinuity form part of these taxa.

Ideal Species

A perfect situation! Species that can be easily distinguished and have no problem of identity. Such species, however, are very few; common examples include Apiaceae, Asteraceae and the genera *Allium* and *Sedum*. The following characteristics are expected in an ideal species:

1. The species poses no taxonomic problems and is easily recognized as a distinct entity based on morphological characters.
2. It exhibits no discontinuity of variation within, i.e., it contains no subspecies, varieties or formas.
3. It is genetically isolated from other species.
4. It is sexually reproducing.
5. It is at least partially outbreeding.

Unfortunately, ideal species are rare among the plant kingdom and the greater majority of species pose situations contrary to one or more of the above criteria.

Idea of Transmutation

This is an ancient Greek idea which persisted as late as the seventeenth century. Greeks believed in the transmutation of wheat into barley, *Crocus* into *Gladiolus*, barley into oats, and many other plants, under certain conditions. The supporters of this notion often included professional botanists like Bobart (who swore that Crocus and Gladiolus, as likewise the Leucojum, and Hyacinths by a long standing without replanting have in his garden changed from one kind to the other) as reported by Robert Sharrock (1660) in his book *History of the propagation and improvement of vegetables by the concurrence of art and nature*. Sharrock fortunately, however, on investigation did not find any proof of this in the field. So called transmutation can be explained as nothing other than the result of unintentional mixing of seeds or other propagules of another plant with a particular crop before plantation.

The present author had a glimpse of this fallacy while studying the weeds in saffron (*Crocus sativus*) fields of Kashmir valley. With a few vegetative specimens of *Iris reticulata* (whose corms and leaves are closely similar to saffron; the flowers are quite distinct) in his hand, the author tried in vain to convince the saffron grower (who always thinks that he knows more about his crop) that the plant he was carrying was not saffron. The author managed to escape the assault but was more convinced that this *Iris* (which does not grow elsewhere in Kashmir valley) would have come unintentionally from Persia where it grows commonly, and from where the Kashmir saffron is supposed to have been introduced. The concept of transmutation is now firmly rejected.

Nominalistic Species Concept

This nominalistic species concept is also only of academic interest now. For the purpose of nomenclature, all organisms must be referable to species. Species, by this concept, ***can be defined by the language of formal relations and not by property of their organisms***. The concept considers species to be a category in taxonomic hierarchy and may correspond to a specific name in the binomial system of nomenclature. The concept is logically sound but scientifically irrelevant since the ultimate aim is to place a particular group of individuals in a species.

Typological Species Concept

This concept was first proposed by John Ray (1686) and further elaborated by C. Linnaeus in *Critica botanica* (1737). Linnaeus refuted the idea of transmutation of species. Linnaeus believed that although there is some variation within a species, the species by themselves are fixed (**fixity of species**) as created by the Almighty Creator. The species, according to the concept, is ***a group of plants which breed true within their limits of variation***. Towards the later part of his life, however, Linnaeus moved away from idea of fixity of species and was convinced that species can arise by hybridization. In his later publication (*Fundamenta fructificationis*, 1762), Linnaeus imagined that at the time of creation, there arose as many genera as were the individuals. These, in the course of time, were fertilized by others and thus arose species until so many were produced as now exist. These species were sometimes fertilized by other species of the same genus, giving rise to varieties. The typological concept, however, should not be confused with **typification,** which is a distinct methodology of nomenclature, providing names to taxonomic groups.

Taxonomic Species Concept

The doctrine of fixity was challenged by Lamarck (1809) and finally Darwin (1859), who recognized continuous and discontinuous variation and developed his taxonomic species concept based on morphology, more appropriately known as the **Morphological species concept**. According to this concept, the species is regarded as ***an assemblage of individuals with morphological features in common, and separable from other such assemblages by correlated morphological discontinuity in a number of features***. The supporters of this view believe in the concept of continuous and discontinuous variations. The individuals of a species show continuous variation, share certain characters and show a distinct discontinuity with individuals belonging to another species, with respect to all or some of these characters.

Du Rietz (1930) modified the taxonomic species concept by also incorporating the role of geographic distribution of populations and developed the **morpho-geographical species concept.** The species was defined as ***the smallest population that is permanently separated from other populations by distinct discontinuity in a series of biotypes***.

The populations recognized as distinct species and occurring in separate geographical areas are generally quite stable and remain so even when grown together. There are, however, examples of a few species pairs which are morphologically quite distinct, well adapted to respective climates, but when grown together, they readily interbreed and form intermediate fertile hybrids, bridging the discontinuity gap between the species. Examples are *Platanus orientalis* of the Mediterranean region and *P. occidentalis* of E. United States. Another well-known pair is *Catalpa ovata* of Japan and China and *C. bignonioides* of America. Such pairs of species are known as **vicarious species** or **vicariants** and the **phenomenon** as **vicariance** or **vicariism**.

Morphological and **morpho-geographical** types of taxonomic species have been widely accepted by taxonomists who even take into account the data from genetics, cytology, ecology, etc., but firmly believe that **species recognized must be delimited by morphological characters**.

The taxonomic species concept has several advantages:

1. It is useful for general taxonomic purposes especially the field and herbarium identification of plants.
2. The concept is very widely applied, and most species have been recognized using this concept.
3. The morphological and geographical features used in the application of this concept can be easily observed in populations.
4. Even experimental taxonomists who do not recognize this concept, apply this concept in cryptic form.
5. The greater majority of species recognized through this concept correspond to those established after experimental confirmation.

The concept, however, also has some inherent drawbacks:

1. It is highly subjective and different sets of characters are used in different groups of plants.
2. It requires much experience to practice this concept because only after considerable observation and experience can a taxonomist decide the characters which are reliable in a particular taxonomic group.
3. The concept does not take into account the genetic relationships between plants.

Biological Species Concept

The biological species concept was first developed by Mayr (1942) who defined species as ***groups of actually or potentially interbreeding natural populations, which are reproductively isolated from other such groups***. The words 'actually or potentially', being meaningless, were subsequently dropped by Mayr (1969). Based on the same criteria, Grant (1957) defined species as ***a community of cross-fertilizing individuals linked together by bonds of mating and reproductively isolated from other species by barriers to mating***. The recognition of biological species thus involve: (a) interbreeding among populations of the same species; and (b) reproductive isolation between populations of different species. Valentine and Love (1958) pointed out that species could be defined in terms of gene exchange. ***If two populations are capable of exchanging genes freely either under natural or artificial conditions, the two are said to be conspecific (belonging to the same species). On the other hand, if the two populations are not capable of exchanging genes freely and are reproductively isolated, they should be considered specifically distinct***. The concept has several advantages:

1. It is objective, and the same criterion is used for all the groups of plants.
2. It has a scientific basis as the populations showing reproductive isolation do not intermix and the morphological differences are maintained even if the species grow in the same area.
3. The concept is based on the analysis of features and does not need experience to put it into practice.

The concept, first developed for animals, holds true because animals as a rule are sexually differentiated and polyploidy is very rare. When applying this concept to plants, however, several problems are encountered:

1. A good majority of plants show only vegetative reproduction, and hence the concept of reproductive isolation as such cannot be applied.
2. Reproductive isolation is commonly verified under experimental conditions, usually under cultivation. It may have no relevance for wild populations.
3. Genetic changes causing morphological differentiation and those causing reproductive barriers do not always go hand in hand. *Salvia mellifera* and *S. apiana* are morphologically distinct (two separate species according to the taxonomic species concept) but not reproductively isolated (single species according to the biological species concept). Such species are known as **compilospecies**. Contrary to this, *Gilia inconspicua* and *G. transmontana* are reproductively isolated (two separate species according to the biological species concept) but morphologically similar (single species according to the taxonomic species concept). Such species are known as **sibling species**.
4. Fertility-sterility is only of theoretical value in allopatric populations.
5. It is difficult and time consuming to carry out fertility-sterility tests.
6. Occurrence of reproductive barriers has no meaning in apomicts.
7. Necessary genetic and experimental data are available for only very few species.

Stebbins (1950), it would appear, combined two concepts when he stated that ***species must consist of systems of populations that are separated from each other by complete or at least sharp discontinuities in the variation pattern, and that these discontinuities must have a genetic basis***. These populations with isolating mechanisms (different species) may occur either in the same region (**sympatric species**) or in different regions (**allopatric species**).

Fortunately, although the taxonomic and biological concepts are based upon different principles, the species recognized by one concept, in the majority of cases, stand the test of the other. Morphology provides the evidence for putting the genetic definition into practice.

Cladistic Species Concept

Unlike the Biological Species Concept, a cladistic species does not rely on reproductive isolation, so it is independent of processes that are integral in other concepts (Nixon and Wheeler, 1990). It works for asexual lineages, and can detect recent divergences, which the Morphological Species Concept cannot.

Evolutionary Species Concept

This concept was developed by Meglitsch (1954), Simpson (1961) and Wiley (1978). Although maintaining that interbreeding among sexually reproducing individuals is an important component in species cohesion, this concept is compatible with a broad range of reproductive modes. Wiley (1978) defines: ***an evolutionary species is a single lineage of ancestor-descendant populations which maintains its identity from other such lineages, and which has its own evolutionary tendencies and historical fate***. This concept avoids many of the problems of the biological concept. Lineage is a single series of demes (populations) that share a common history of descent, not shared by other demes. The identity of species is based on **recognition systems** that operate at various levels. In sexually reproducing species, such systems include recognition because of phenotypic, behavioral and biochemical differences. In asexual species phenotypic, genotypic differences maintain the identity of species. Identity in both sexual and asexual species may also be due to distinct ecological roles. Viewed from the standpoint of evolutionary species concept, however, the important question is not whether two species hybridize, but whether two species do or do not lose their distinct ecological and evolutionary roles. If, despite some hybridization, they do not merge, then they remain separate species in the evolutionary perspective.

Several other terms have been proposed to distinguish species based on specific criteria. Grant (1981) recognizes **microspecies** as 'populations of predominantly uniparental plant groups which are themselves uniform and are slightly differentiated morphologically from one another'; they are often restricted to a limited geographical area. Microspecies develop in inbreeding species but are usually not stable over longer periods. They may undergo cross-fertilization sooner or later forming recombinant types which themselves become new microspecies. Several microspecies have been found in *Erophila verna* mostly representing single biotypes or groups of similar biotypes some of which are marked by only one or two characters. These may be distinguished as **clonal microspecies** (reproducing by vegetative propagation, e.g., *Phragmites*), **agamospermous microspecies** (reproducing by agamospermy, e.g., *Rubus*), **heterogamic microspecies** (reproducing by genetic systems, e.g., *Oenothera biennis* or *Rosa canina*), and **autogamous microspecies** (predominantly autogamous and chromosomally homozygous, e.g., *Erophila*). The term microspecies was first suggested by Jordan (1873) and as such they are often termed as **Jardanons** to distinguish them from **Linnaeons**, the normal species, first established by Linnaeus. Microspecies are distinct from **cryptic species**, which are morphologically similar but cytologically or physiologically different. Stace (1989) uses the term **semi-cryptic species** for the latter.

Biosystematic Species Concept

The term **biosystematic species** has been used by Grant (1981) to refer to the categories based on fertility relationships as determined by artificial hybridization experiments. **Ecotype** refers to all members of a species that 'represent a product of genetic response of a species towards a particular habitat'. The ecotypes, which are able to exchange genes freely without loss of fertility or vigour in the offsprings, form an **ecospecies**. An ecospecies corresponds to a taxonomic species. A group of ecospecies capable of limited genetic exchange constitutes a **coenospecies**. A coenospecies is considered equivalent to a subgenus. A group of related coenospecies between which hybridization is possible—directly or through intermediates—constitutes a **comparium**, which is considered equal to a genus. Complete sterility barriers exist between genera.

Infraspecific Ranks

The species is regarded as the basic unit of classification and many works, including the *Flora of USSR*, do not recognize infraspecific taxa. Many European, American and Asian Floras, however, do recognize taxa below the rank of species. The **international Code of Nomenclature for Algae, Fungi and Plants** recognizes five infraspecific ranks: **subspecies**, **variety** (Latin, **varietas**), **subvariety**, **form** (Latin, **forma**) and **subform**. Of these, three (subspecies, variety and form) have been widely used in the literature.

Du Rietz (1930) defined **subspecies** as a ***population of several biotypes forming more or less a distinct regional facies of a species***. Facies stands for race. Morphologically distinct but interfertile populations of a species growing in different geographical regions are maintained as distinct subspecies due to the geographical isolation of the species.

Du Rietz defined variety as a ***population of several biotypes, forming more or less a local facies of a species***. The term variety is commonly used for morphologically distinct populations occupying a restricted geographical area. Emphasis is on a more localized range of the variety, compared with the large-scale regional basis of a subspecies. Several varieties are often recognized within a subspecies. The term variety is also used for variations whose precise nature is not understood, a treatment often necessary in the pioneer phase of taxonomy.

Forma is often regarded as sporadic variant distinguished by a single or a few linked characters. Little taxonomic significance is, however, attached to minor and random variations upon which the forms are normally based.

Genus

The concept of genus is as old as folk science itself as represented by names rose, oak, daffodils, pine and so on. A genus ***represents a group of closely-related species***. According to Rollins (1953), the function of the genus concept is to bring together species in a phylogenetic manner by placing the closest related species within the general classification. When attempting to place a species within a genus, the primary question would be, is it related to the undoubted species of that genus? Mayr (1957) defined genus as a ***taxonomic category which contains either one species or a monophyletic group of species and is separable from other genera by a decided discontinuity gap***. It was earlier believed that a genus should always be readily definable based on a few technical floral characters. A more rational recognition should take the following criteria into consideration:

1. The group, as far as possible, should be a natural one. The monophyletic nature of the group should be deduced by cytogenetic and geographic information in relation to morphology.
2. The genera should not be distinguished on a single character but a sum total of several characters. In several cases, genera are easily recognized based on adaptive characters (adaptations in response to ecological niches), as in the case of establishing aquatic species of *Ranunculus* under a separate genus *Batrachium*.
3. There is no size requirement for a genus. It may include a single species (**monotypic genus**) as *Leitneria, Ginkgo, Milula* or many (**Polytypic genus**): *Euphorbia* (2420 species), *Astragalus* (2000) *Carex* (1800), *Senecio* (1470) and *Acacia* (1300) being the examples of large genera. The genus *Senecio* was earlier included more than 2500 species, but it has now been split into several genera. The only important criterion is that there should be a decided gap between the species of two genera. If the two genera are not readily separable, then they can be merged into one and distinguished as subgenera or sections. Such an exercise should take into consideration the concept in other genera of the family, size of the genus (it is more convenient to have subgenera and sections in a larger genus) and traditional usage.
4. When generic limits are being drawn, it is necessary that the group of species should be studied throughout the range distribution of the group, because characters stable in one region may break down elsewhere.

Family

A **family,** similarly, represents a group of closely-related genera. Like genus, it is also a very ancient concept because the natural groups now known as families, such as legumes, crucifers, umbels, grasses have been recognized by laymen and taxonomists alike for centuries. Ideally, families should be monophyletic groups. Like the genus, the family may represent a single genus (Podophyllaceae, Hypecoaceae, etc.) or several genera (Asteraceae: nearly 1100). Most taxonomists favor broadly-conceived family concepts that lend stability to classification. Although there is no marked discontinuity between Lamiaceae (Labiatae) and Verbenaceae, the two are maintained as distinct families (although merged in recent APG IV classification). The same tradition prevents taxonomists from splitting Rosaceae, which exhibits considerable internal differences.

Chapter 6

Nomenclature of Plants

Nomenclature deals with the application of a correct name to a plant or a taxonomic group. In practice, nomenclature is often combined with identification, since while identifying an unknown plant specimen, the author chooses and applies the correct name. The favorite temperate plant is correctly identified whether you call it 'Seb' (vernacular Hindi name), Apple, *Pyrus malus* or *Malus malus*, but only by using the correct scientific name *Malus domestica* does one combine identification with nomenclature. The activity of botanical nomenclature is governed by the International Code of Nomenclature of algae, fungi and plants (**ICN**) (formerly ICBN) published by the International Association of Plant Taxonomy (**IAPT**). The Code is revised after changes at each International Botanical Congress. The naming of the animals is governed by the International Code of Zoological Nomenclature (**ICZN**) and that of bacteria by the International Code for the Nomenclature of Bacteria (**ICNB**; now known as Bacteriological Code-**BC**). Virus nomenclature is governed by International Code of Virus Classification and Nomenclature (**ICVCN**). Naming of cultivated plants is governed by the International Code of Nomenclature for Cultivated Plants (**ICNCP**), which is largely based on **ICBN** with a few additional provisions. Whereas within the provisions of a specific code no two taxa can bear the same correct scientific name, same names are allowed across the codes. The generic name *Cecropia* applies to showy moths as also to tropical trees. Genus *Pieris,* similarly, refers to some butterflies and shrubs.

During the last decade, there have been attempts at developing unified code for all living organisms, for convenient handling of combined database for all organisms. **Draft BioCode** and **PhyloCode**, have been concerted efforts in this direction, but it will take a long time before acceptability of these endeavors can be determined.

NEED FOR SCIENTIFIC NAMES

Scientific names formulated in Latin are preferred over vernacular or common names since the latter pose several problems:

1. Vernacular names are not available for all the species known to man.
2. Vernacular names are restricted in their usage and are applicable in a single or a few languages only. They are not universal in their application.
3. Common names usually do not provide information indicating family or generic relationship. Roses belong to the genus *Rosa;* woodrose is a member of the genus *Ipomoea* and primrose belongs to the genus *Primula.* The three genera, in turn, belong to three different families—Rosaceae, Convolvu-laceae and Primulaceae, respectively. Oak is similarly common name for the species of genus *Quercus*, but Tanbark oak is *Lithocarpus*, poison oak a *Rhus*, silver oak a *Grevillea* and Jerusalem oak a *Chenopodium*.
4. Frequently, especially in widely distributed plants, many common names may exist for the same species in the same language in the same or different localities. Cornflower, bluebottle, bachelor's button and ragged robin all refer to the same species *Centaurea cyanus*.
5. Often, two or more unrelated species are known by the same common name. Bachelor's button may thus be *Tanacetum vulgare, Knautia arvensis* or *Centaurea cyanus.* Cockscomb, is similarly, a common name for *Celosia cristata* but is also applied to a seaweed *Plocamium coccinium* or to *Rhinanthus minor.*

Why Latin?

Scientific names are treated as Latin regardless of their origin. It is also mandatory to have a Latin diagnosis for any new taxon published 1 January 1935 onwards (Latin or English diagnosis after 1 January, 2012). The custom of Latinized names and texts originates from medieval scholarship and custom continued in most botanical publications until the middle of nineteenth century. Descriptions of plants are not written in classical Latin of Cicero or of Horace, but in the '**lingua franca**' spoken and written by scholars during middle ages, based on popular Latin spoken by ordinary people in the classical times. The selection has several advantages over modern languages: (i) Latin is a dead language and as such meanings and interpretation are not subject to changes unlike, English and other languages; (ii) Latin is specific and exact in meaning; (iii) grammatical sense of the word is commonly obvious (white translated as **album**-neuter, **alba**-feminine or **albus**-masculine); and (iv) Latin language employs the Roman alphabet, which fits well in the text of most languages.

DEVELOPMENT OF BOTANICAL CODE

For several centuries, the names of plants appeared as polynomials—long descriptive phrases, often difficult to remember. A species of willow, for example, was named *Salix pumila angustifolia altera* by Clusius in his herbal (1583). Casper Bauhin (1623) introduced the concept of **Binomial nomenclature** under which the name of a species consists of two parts, the first the name of the genus to which it belongs and the second the **specific epithet**. Onion is thus appropriately named *Allium cepa, Allium* being the generic name and *cepa* the specific epithet. Bauhin, however, did not use binomial nomenclature for all the species and it was left to Carolus Linnaeus to firmly establish this system of naming in his *Species plantarum* (1753). The early rules of nomenclature were set forth by Linnaeus in his *Critica botanica* (1737) and further amplified in *Philosophica botanica* (1751). A. P. de Candolle, in his *Theorie elementaire de la botanique* (1813), gave explicit instructions on nomenclatural procedures, many taken from Linnaeus. Steudel, in *Nomenclator botanicus* (1821), provided Latin names for all flowering plants known to the author together with their synonyms.

The first organized effort towards the development of uniform botanical nomenclature was made by Alphonse de Candolle, who circulated a copy of his manuscript *Lois de la nomenclature botanique.* After deliberations of the First International Botanical Congress at Paris (1867), the **Paris Code,** also known as '**de Candolle rules**' was adopted. Linnaeus (1753) was made the starting point for plant nomenclature and the rule of priority was made fundamental. Not satisfied with the Paris Code, the American botanists adopted a separate **Rochester Code** (1892), which introduced the concept of **types**, strict application of rules of priority even if the name was a **tautonym** (specific epithet repeating the generic name, e.g., *Malus malus*).

The Paris Code was replaced by the **Vienna Code** (1905), which established *Species plantarum* (1753) of Linnaeus as the **starting point**; tautonym was not accepted, and **Latin diagnosis** was made essential for new species. In addition, a list of conserved generic names (**Nomina generic conservanda**) was approved. Not satisfied with the Vienna Code also, adherents of the Rochester Code adopted the **American Code** (1907), which did not accept the list of conserved names and the requirement for Latin diagnosis.

It was not until the 5th International Botanical Congress (IBC) at **Cambridge** (1930) that the differences were finally resolved and a truly International Code evolved, accepting the concept of type method, rejecting the tautonyms, making Latin diagnosis mandatory for new groups and approving conserved generic names. The Code has since been constantly amended at each International Botanical Congress. The 15th IBC was held at Tokyo in 1993, 16th at St. Louis in 1999 (published by Greuter et al., 2000), 17th at Vienna in 2005 (Published by McNeill et al., 2006—Code is generally published one year after the Congress). The 18th International Botanical Congress held in Melbourne, Australia in July, 2011 made some major changes including renaming the Code as International Code of Nomenclature of algae, fungi and plants, **ICN** (formerly International Code of Botanical Nomenclature, **ICBN**), and from January 1, 2012 allowing English diagnosis or description for making publication valid, and allowing Electronic publication for making it effective. The 19th International Botanical Congress was held in Shenzhen, China in July 2017. Shenzhen Code was formally published on June 26, 2018. It supercedes the Melborne Code. Important changes included mechanism for creating a framework for future registration of algal and plant names, provisions for improved clarity in the governance of the Code and the working of future Nomenclature Sections, and the sharing of governance of nomenclature by referring decisions on rules solely relating to fungi to International Mycological Congresses. The 20th International Botanical Congress to be held in Rio de Janeiro, Brazil in 2023.

CONTENTS OF BOTANICAL CODE

Publication of the Code is based on the realization that botany requires a precise and simple system of nomenclature used by botanists in all countries. The Code aims at provision of a stable method of naming taxonomic groups, avoiding and rejecting the use of names which may cause error or ambiguity or throw science into confusion. **Preamble** highlights the philosophy of the botanical Code. The Code is divided into 3 divisions:

I. Principles
II. Rules and recommendations
III. Provisions for the governance of the Code

In addition, the Code includes the following appendices:

I. Names of hybrids
IIA. Nomina familiarum algarum, fungorum, pteridophytorum et fossilium conservanda et rejicienda
IIB. Nomina familiarum bryophytorum et spermatophytorum conservanda
IIIA. Nomina generica conservanda et rejicienda
IIIB. Nomina specifica conservanda et rejicienda
IV. Nomina utique rejicienda (A. Algae, B. Fungi, C. Bryophyta, D. Pterido-phyta, E. Spermatophyta)
V. Opera utique oppressa

The last three useful appendices were included for the first time in the Tokyo Code. The first (IIIB) includes the names of conserved and rejected specific names; the second (IV) lists the names and all combinations based on these names, which are ruled as rejected under Art. 56, and none is to be used; and the last (V) the list of publications (and the category of taxa therein) which are not validly published according to the Code.

Principles form the basis of the system of botanical nomenclature. There are 62 main **rules** (set out as articles) and associated **recommendations.** The object of the rules is to put the nomenclature of the past into order and provide for that of the future; names contrary to the rules cannot be maintained. Recommendations deal with subsidiary points, and are meant for uniformity and clarity. Names contrary to the recommendations cannot, on that account, be rejected, but they are not examples to be followed. **Conserved names** include those that do not satisfy the principle of priority but are sanctioned for use. The various rules and recommendations are discussed here under relevant headings.

Preamble

1. Botany requires a precise and simple system of nomenclature used by botanists in all countries, dealing on the one hand with the terms which denote the ranks of taxonomic groups or units, and on the other hand with the scientific names which are applied to the individual taxonomic groups of plants. The purpose of giving a name to a taxonomic group is not to indicate its characters or history, but to supply a means of referring to it and to indicate its taxonomic rank. This **Code** aims at the provision of a stable method of naming taxonomic groups, avoiding and rejecting the use of names which may cause error or ambiguity or throw science into confusion. Next in importance is the avoidance of the useless creation of names. Other considerations, such as absolute grammatical correctness, regularity or euphony of names, more or less prevailing custom, regard for persons, etc., notwithstanding their undeniable importance, are relatively accessory.
2. Algae, Fungi and Plants are the organisms covered by this Code.
3. The Principles form the basis of the system of nomenclature governed by this Code.
4. The detailed Provisions are divided into Rules, which are set out in the Articles (Art.), and Recommendations. Examples (Ex.) are added to the rules and recommendations to illustrate them.
5. The object of the Rules is to put the nomenclature of the past into order and to provide for that of the future; names contrary to a rule cannot be maintained.
6. The Recommendations deal with subsidiary points, their object being to bring about greater uniformity and clarity, especially in future nomenclature; names contrary to a recommendation cannot, on that account, be rejected, but they are not examples to be followed.
7. The provisions regulating the governance of this Code form its last Division (Div. III).
8. The provisions of this Code apply to all organisms traditionally treated as plants, whether fossil or non-fossil, e.g., blue-green algae, Cyanobacteria, fungi, including chytrids, oomycetes, and slime moulds, photosynthetic protists and taxonomically related non-photosynthetic groups (but excluding Microsporidia). Provisions for Names of hybrids appear in Chapter H.
9. Names that have been conserved, protected or rejected, suppressed works, and binding decisions are given in Appendices I–VII.
10. Appendices form an integral part of this Code, whether published together with, or separately from, the main text.
11. The *International code of nomenclature for cultivated plants* is prepared under the authority of the International Commission for the Nomenclature of Cultivated Plants and deals with the use and formation of names for special categories of organisms in agriculture, forestry, and horticulture.
12. The only proper reasons for changing a name are either a more profound knowledge of the facts resulting from adequate taxonomic study or the necessity of giving up a nomenclature that is contrary to the rules.
13. In the absence of a relevant rule or where the consequences of rules are doubtful, established custom is followed.
14. This edition of the **Code** supersedes all previous editions.

Principles of ICN

The International Code of Nomenclature of algae, fungi and plants is based on the following set of six principles, which are the philosophical basis of the Code and provide guidelines for the taxonomists who propose amendments or deliberate on the suggestions for modification of the Code:

Principle I: Botanical nomenclature is independent of zoological and prokaryotic nomenclature. The Code applies equally to names of taxonomic groups treated as plants whether or not these groups were originally so treated.

Principle II: The application of names of taxonomic groups is determined by means of nomenclatural types.

Principle III: The nomenclature of a taxonomic group is based upon priority of publication.

Principle IV: Each taxonomic group with a particular circumscription, position, and rank can bear only one correct name, the earliest that is in accordance with the Rules, except in specified cases.

Principle V: Scientific names of taxonomic groups are treated as Latin regardless of their derivation.

Principle VI: The Rules of nomenclature are retroactive unless expressly limited.

Names of Taxa

Taxon (pl. taxa) refers to a taxonomic group belonging to any rank. The system of nomenclature provides a hierarchical arrangement of ranks. Every plant is treated as belonging to a number of taxa, each assigned a particular rank. Onion thus belongs to *Allium cepa* (species rank), *Allium* (genus rank), Alliaceae (now under Amaryllidaceae, family rank) and so on. The seven principal obligatory ranks of taxa in descending sequence are: **kingdom** (regnum), **division** or **phylum** (divisio, phylum), **class** (classis), **order** (ordo), **family** (familia), **genus** (genus), and **species** (species). The ending of the name indicates its rank: ending **-bionta** denotes a kingdom, **-phyta** a division, **-phytina** a subdivision, **-opsida** a class, **-opsidae** or **-idae** a subclass, **-ales** an order, **-ineae** a suborder and **-aceae** a family. The detailed hierarchy of ranks and endings with examples is given in Table 6.1. Stevens (2005) describes this system of naming where endings determine ranks of taxa and suggest relative positions of groups in local hierarchy as **flagged hierarchy**.

The names of the groups belonging to ranks above the level of genus are uninomials in the plural case. Thus, it is appropriate to say 'Winteraceae are primitive' and inappropriate when we say 'Winteraceae is primitive'. The focus changes when we are mentioning the rank with it. Thus, 'the family Winteraceae is primitive' is a logically correct statement.

The name of a taxon above the rank of family may be formed by replacing the termination ***-aceae*** in the name of an included family by the termination denoting their rank (order **Rosales** from family Rosaceae, class **Magnoliopsida** from family Magnoliaceae). The name of a family is a plural adjective used as a noun. It is formed from the name of the type genus by replacing the genitive singular (gender) ending with the termination ***-aceae*** in the genera of classical Latin or Greek origin (Family Rosaceae from genus *Rosa,* Potamogetonaceae from *Potamogeton*). For generic names of non-classical origin, when analogy with classical names is insufficient to determine the genitive singular, ***-aceae*** is added to the full word (**Ginkgoaceae** from *Ginkgo*). For generic names with alternative genitives the one implicitly used by the original author must be maintained (Nelumbonaceae from *Nelumbo—Nelumbonis* declined by analogy with *umbo* and *umbonis*).

The endings for ranks, subclass and above are recommendations, whereas for order and below these are mandatory rules. It is, thus, nothing strange that group names such as Gymnosperms, Angiosperms, Bryophytes, Pteridophytes, Lignosae, Herbaceae, Dicotyledoneae, Monocotyledoneae, etc., have been used as valid group names for supraordinal taxa. Recently developed versions of the APG classification recognize only informal group names such as Paleoherbs, Tricolpates (Eudicots), Asterids, Rosids, Euasterids, Eurosids above the order level as monophyletic clades. No formal taxonomic names are used above the level of the order. The name of a family ends in **-aceae**. The following eight families (nine if Papilionaceae is treated as distinct family) of angiosperms, however, whose original names are not in accordance with the rules but the use of these names has been sanctioned because of old traditional usage. The type genus of each family is listed:

Traditional name	***Alternate name***	***Type genus***
Cruciferae	Brassicaceae	*Brassica*
Guttiferae	Clusiaceae	*Clusia*
Leguminosae	Fabaceae	*Faba*
Umbelliferae	Apiaceae	*Apium*
Compositae	Asteraceae	*Aster*
Labiatae	Lamiaceae	*Lamium*
Palmae	Arecaceae	*Areca*
Graminae	Poaceae	*Poa*

Table 6.1: Ranks and endings provided by the ICN.

Rank	*Ending*	*Examples*
Kingdom	-bionta	Chlorobionta
Division	-phyta	Magnoliophyta
	-mycota (Fungi)	Eumycota
Subdivision	-phytina	Pterophytina
	-mycotina (Fungi)	Eumycotina
Class	-opsida	Magnoliopsida
	-phyceae (Algae)	Chlorophyceae
	-mycetes (Fungi)	Basidiomycetes
Subclass	-opsidae	Pteropsidae
	-idae (Seed plants)	Rosidae
	-physidae (Algae)	Cyanophysidae
	-mycetidae (Fungi)	Basidiomycetidae
Order	-ales	Rosales
Suborder	-ineae	Rosineae
Family	-aceae	Rosaceae
Subfamily	-oideae	Rosoideae
Tribe	-eae	Roseae
Subtribe	-inae	Rosinae
Genus	*-us, -um, -is, -a, -on*	*Pyrus, Allium, Arabis, Rosa, Polypogon*
Subgenus		*Cuscuta* subgenus *Eucuscuta*
Section		*Scrophularia* section *Anastomosanthes*
Subsection		*Scrophularia* subsection *Vernales*
Series		*Scrophularia* series *Lateriflorae*
Species		*Rosa canina*
Subspecies		*Crepis sancta* subsp. *bifida*
Varietas		*Lantana camara* var. *varia*
Forma		*Tectona grandis* f. *punctata*

The alternate names of these families which are in accordance with the ICN rules need to be encouraged. Under a unique exception to article 18 of the Code, the name Leguminosae is sanctioned as alternate name for Fabaceae only as long as it includes all the three subfamilies: Faboideae (Papilionoideae), Caesalpinioideae and Mimosoideae. In case these are upgraded as families, then the name Papilionaceae is conserved against Leguminosae for the first of these getting the name Fabaceae. The two alternate names allowed then are Papilionaceae and Fabaceae.

When a name of a family has been published with an improper Latin termination, the termination must be changed to conform with Art. 18.1, without change of authorship or date (Art. 32.2). However, if such a name is published with a non-Latin termination, it is not validly published. Coscinodisceae Kütz., 1844 would as such be cited as Coscinodiscaceae Kütz., 1844, and not attributed to De Toni who first made the correction in 1890. On the other hand Tricholomées Roze, 1876 is changed to Tricholomataceae Pouzar, 1983, as original name by Roze was in French.

Fossil taxa were hitherto treated as **morphotaxa,** nomenclatural purposes, comprising only the parts, life-history stages, or preservational states represented by the corresponding nomenclatural type. Melbourne Code of 2011 envisaged single name for each fossil and not separate morphotaxa.

Genus

The generic name is a uninomial **singular** word treated as a noun. The examples of the shortest generic name *Aa* as well as the longest name *Brassosophrolaeliocattleya* (26 characters), both belong to the family Orchidaceae. The genus may have a masculine, neuter or feminine form as indicated by the ending: **-us, -pogon** commonly stand for masculine genera, **-um** for neuter and **-a, -is** for feminine genera. The first letter of the generic name is always capitalised. The name may be based on any source, but the common sources for generic names are as under:

1. **Commemoration of a person** commonly an author such as *Bauhinia* for Bauhin, *Benthamia* and *Benthamida* for Bentham, *Darwinia* for Darwin, *Hutchinsonia* for Hutchinson, *Lamarckia* for Lamarck and *Linnaea* for Linnaeus. It may also be used for head of a state such as *Victoria* for Queen Victoria of England, *Washingtonia* for King George Washington, and *Zinobia* for Queen Zinobia of Palmyra. The names commemorating a person, man or woman always take the feminine form. The name of a genus is constructed by adding *-ia* if name of a person ends in a consonant (*Fuchsia* after Fuchs), *-a* if it ends in a vowel (*Ottoa* after Otto), but *-ea* is added if it ends in **-a** (*Collaea* after Colla). If the name ends in **-er** both are permitted (*Kernera* for Kerner; *Sesleria* for Seslar). For Latinized personal names ending with **-us**, this termination is dropped before adding appropriate ending (*Linnaea* after Linnaeus, *Dillenia* after Dillenius). The name may also be formed directly as in case of *Victoria* and *Zinobia,* as indicated above.
2. **Based on a place** such as *Araucaria* after Arauco a province of Chile, *Caucasia* for Caucasus in Russia, *Salvadora* for EL Salvadore, *Arabis* for Arabia and *Sibiraea* for Siberia. The name could also be based on names of two places such as *Austroamericium* (Australia and America) or place and author such as *Austrobaileya* (Australia and Bailey).
3. **Based on an important character** such as yellow wood in *Zanthoxylum*, liver-like leaves in *Hepatica*, marshy habit of *Hygrophila,* trifoliate leaves of *Trifolium*, and spiny fruit of *Acanthospermum*.
4. **Aboriginal names** taken directly from a language other than Latin without alteration of ending. *Narcissus* is the Greek name for daffodils named after the famous Greek god Narcissus, *Ginkgo* a Chinese, *Vanda* a Sanskrit and *Sasa* a Japanese aboriginal name.

The generic name of a tree, whatever be the ending, takes a feminine form, since trees are generally feminine in classical Latin. *Pinus*, *Quercus* and *Prunus* are, thus, all feminine genera. If two words are used to form a generic name, these must be joined by a hyphen (generic name *Uva-ursi*). In case, however, the two words were combined into one word by the original author, the use of hyphen is not needed (generic name *Quisqualis*). The name of a genus may not coincide with a technical term currently used in morphology unless it was published before 1 January 1912 and was accompanied by a specific name published in accordance with the binary system of Linnaeus. The generic name *Tuber* (published in 1780 was accompanied by a binary specific name *Tuber gulosorum* F. H. Wigg.) and is, therefore, validly published. On the other hand, the intended generic names '*Lanceolatus*' (Plumstead, 1952) is, therefore, not validly published. Words such as *'radix'*, *'caulis'*, *'folium'*, *'spina'*, etc., cannot now be validly published as generic names.

Species

The name of a species is a **binomial:** consisting of two words, a generic name followed by a specific epithet. The Code recommends that all specific epithets should begin with a lower-case initial letter. An upper-case initial letter is sometimes used, however, for specific epithets derived from a person's name, former generic name or a common name. The Code discourages such usage for specific epithets. A specific epithet may be derived from any source or composed arbitrarily. The following sources are commonly used:

1. **Name of a person.** The specific epithet named after a person may take genitive (possessive) or an adjectival form:
   (i) When used in the genitive form the epithet takes its form depending on the ending of the person's name. For names ending in a vowel or **-er** the letter *-i* is added for a male person (*roylei* after Royle, *hookeri* after Hooker), *-ae* for female person (*laceae* after Lace), and *-orum* for more than one person with the same surname (*hookerorum* after Hooker & Hooker). If the name, however, ends in **-a** then *-e* is added (*paulae* after Paula). If the name ends in a consonant *-ii* is added male person (*wallichii* after Wallich), *-iae* for a female person (*wilsoniae* after Wilson), and *-iorum* for more than one person with same surname and at least one male (*verlotiorum* after Verlot brothers), and *-iarum* if both are female (*brauniarum* for Braun sisters). For names of the persons already in Latin (e.g., Linnaeus), the Latin ending (-us in this case) must be dropped before adding the appropriate genitive ending. The specific epithets in genitive form are not related to the gender of the genus. Illustrative examples are listed in Table a.

   (ii) When used in adjectival form, the epithet takes its ending from the gender of the genus after adding *-ian* if name of the person ends in a consonant, adding *-an* if the name ends in a vowel except when it ends in **-a,** wherein *-n* is added. Illustrative examples are given in Table b.
2. **Place.** The specific epithet may, similarly, be formed by using the place name as an adjective, when again the genus determines the ending after the addition of *-ian* or *-ic* and then the relevant gender ending as determined by the genus. The specific epithet is also formed by adding *-ensis* (for masculine and feminine genera, e.g., *Hedera nepalensis*, *Rubus canadensis*) or *-ense* (for neuter genera, e.g., *Ligustrum nepalense*) to the place name. Different situations are illustrated in Table c.

Table a

Person	Sex	Specific epithet	Binomial
Royle	M	*roylei*	*Impatiens roylei*
Hooker	M	*hookeri*	*Iris hookeri*
Sengupta	M	*senguptae*	*Euphorbia senguptae*
Wallich	M	*wallichii*	*Euphorbia wallichii*
Todd	F	*toddiae*	*Rosa toddiae*
Gepp & Gepp	M	*geppiorum*	*Codiaeum geppiorum*
Linnaeus	M	*linnaei*	*Indigofera linnaei*

Table b

Author	Genus	Gender	Specific epithet	Binomial
Webb	*Rosa*	Feminine	*webbiana*	*Rosa webbiana*
Webb	*Delphinium*	Neuter	*webbianum*	*Rheum webbianum*
Webb	*Astragalus*	Masculine	*webbianus*	*Astragalus webbianus*
Kotschy	*Hieracium*	Neuter	*kotschyanum*	*Hieracium kotschyanum*
Lagasca	*Centaurea*	Feminine	*lagascana*	*Centaurea lagascana*

Table c

Place	Genus	Gender	Specific epithet	Binomial
Kashmir	*Iris*	Feminine	*kashmiriana*	*Iris kashmiriana*
	Delphinium	Neuter	*kashmirianum*	*Delphinium kashmirianum*
	Tragopogon	Masculine	*kashmirianus*	*Tragopogon kashmirianus*
India	*Rosa*	Feminine	*indica*	*Rosa indica*
	Solanum	Neuter	*indicum*	*Solanum indicum*
	Euonymus	Masculine	*indicus*	*Euonymus indicus*

3. **Character.** Specific epithets based on a character of the species are always in adjectival form and derive their gender from the genus. A name based on a white plant part may take the form *alba* (*Rosa alba*), *album* (*Chenopodium album*) or *albus* (*Mallotus albus*). A common epithet used for cultivated plants may similarly take the form *sativa* (*Oryza sativa*), *sativum* (*Allium sativum*) or *sativus* (*Lathyrus sativus*) depending on the gender of the genus to which the epithet is assigned. Some epithets, however, such as *bicolor* (two-colored) and *repens* (creeping) remain unchanged, e.g., *Ranunculus repens*, *Ludwigia repens* and *Trifolium repens*.
4. **Noun in apposition.** A specific epithet may sometimes be a noun in apposition carrying its own gender, and usually in the nominative case. Binomial *Pyrus malus* is based on the Greek name malus for common apple. In *Allium cepa*, similarly, cepa is the Latin name for onion.

Both the generic name and the specific epithet are underlined when written or typed. When printed, they are in Italics or boldface. After the generic name in a species has been spelled out at least once, if used for other species, it may be abbreviated using the initial capital, e.g., *Quercus dilatata, Q. suber*, *Q. Ilex,* etc. A specific epithet is usually one word but when consisting of two words, these must be hyphenated as in *Capsella bursa-pastoris* and *Rhamnus vitis-idaea*, or else the two words may be combined into one as in *Narcissus pseudonarcissus*.

Although not leading to rejection, the use of same name in genitive form as well as adjectival form in species of the same genus is to be avoided, e.g., *Iris hookeri* and *I. hookeriana; Lysimachia hemsleyana* Oliv. and *L. hemsleyi* Franch.

Infraspecific Taxa

The names of **subspecies** are **trinomials** and are formed by adding a subspecific epithet to the name of a species, e.g., *Angelica archangelica* ssp. *himalaica.* A variety (**varieta**) within a subspecies may accordingly be **quadrinomial** as in *Bupleurum falcatum* ssp. *eufalcatum* var. *hoffmeisteri,* or it may just be a **trinomial** when no subspecies is recognized within a species as in *Brassica oleracea* var. *capitata.* A **forma** may also be assigned a name in a similar manner, e.g., *Prunus cornuta* forma *villosa.* The formation of the infraspecific epithet follows the same rules as the specific epithet. Infraspecific name may sometimes be a polynomial as Saxifraga *aizoon* var. *aizoon* subvar. *brevifolia* f. *multicaulis* subf. *surculosa* Engl. & Irmsch.

The Type Method

The names of different taxonomic groups are based on the **type method** (Article 7), by which a certain representative of the group is the source of the name for the group. This representative is called the **nomenclatural type** or simply the **type**, and methodology as **typification**. The type need not be the most typical member of the group, it only fixes the name of a particular taxon and the two are permanently associated. Type may be correct name or even a synonym. Thus the tea family name (Theaceae) is derived from synonym *Thea* although the correct name for the genus is *Camellia. Mimosa* is the type for family Mimosaceae, but unlike most representatives of the family that have pentamerous flowers, the genus *Mimosa* has tetramerous flowers. The family Urticaceae, similarly, has *Urtica* as its type. When the originally large family was split into a number of smaller natural families, the name Urticaceae was retained for the group containing the genus *Urtica,* since the two cannot be separated. The other splitter groups with family rank got the names Moraceae, Ulmaceae and Cannabaceae with type genera *Morus, Ulmus* and *Cannabis,* respectively. The family Malvaceae has seen a lot of realignments, with Tiliaceae sometimes merged with Malvaceae. Thorne (2003) shifts *Tilia* to Malvaceae but retains rest of the genera. This necessitates name change for former Tiliaceae (excluding genus *Tilia*) to Grewiaceae, with *Grewia* as the type genus.

The type of a family and the higher groups is ultimately a genus (Article 10), as indicated above. A type of a particular genus is a species, e.g., *Poa pratensis* for *Poa.* The type of name of a species or infraspecific taxon, where it exists, is a single type specimen, preserved in a known herbarium and identified by the place of collection, name of the collector and his collection number. It may also be an illustration of the plant. The Code recognizes several kinds of type, depending upon the way in which a type specimen is selected. These include (Article 9):

1. **Holotype:** A particular specimen or illustration designated by the author of the species to represent type of a species. For the purpose of typification, a specimen is a gathering, or part of a gathering, of a single species or infraspecific taxon made at one time, disregarding admixtures. It may consist of a single plant, parts of one or several plants, or of multiple small plants. A specimen is usually mounted either on a single herbarium sheet or in an equivalent preparation, such as a box, packet, jar or microscope slide. Type specimens of names of taxa must be preserved permanently and may not be living plants or cultures. However, cultures of fungi and algae, if preserved in a metabolically inactive state (e.g., by lyophilization or deep-freezing), are acceptable as types. It is now essential to designate a holotype when publishing a new species. Any errors in data relating holotype such as locality, date, collector, collecting number, herbarium code, specimen identifier, or citation of an illustration if found later are to be corrected, provided intent of original author is not changed. Ommisions, however, are not correctable (Article 40.6–40.8).
2. **Isotype:** A specimen which is a duplicate of the holotype, collected from the same place, at the same time and by the same person. Often the collection number is also the same, differentiated as a, b, c, etc.
3. **Syntype:** Any one of the two or more specimens cited by the author when no holotype was designated, or any one of the two or more specimens simultaneously designated as types. Duplicate of a syntype is an isosyntype.
4. **Paratype:** A paratype is a specimen cited in the protologue that is neither the holotype nor an isotype, nor one of the syntypes if two or more specimens were simultaneously designated as types.
5. **Lectotype:** A specimen or illustration selected from the original material cited by the author when no holotype was originally selected or when it no longer exists. A lectotype is selected from isotypes or syntypes. In lectotype designation, an isotype must be chosen if such exists, or otherwise a syntype if such exists. If no isotype, syntype or isosyntype (duplicate of syntype) is extant, the lectotype must be chosen from among the paratypes if such exist. If no cited specimens exist, the lectotype must be chosen from among the uncited specimens and cited and uncited illustrations which comprise the remaining original material, if such exist.
6. **Neotype:** A specimen or illustration selected to serve as nomenclatural type as long as all of the material on which the name of the taxon was based is missing; a specimen or an illustration selected when no holotype, isotype, paratype or syntype exists.
7. **Epitype:** A specimen or illustration selected to serve as an interpretative type when the holotype, lectotype or previously designated neotype, or all original material associated with a validly published name, is demonstrably ambiguous and cannot be critically identified for purposes of the precise application of the name of a taxon. When an epitype is designated, the holotype, lectotype or neotype that the epitype supports must be explicitly cited. Designation of an epitype is not effected unless the herbarium, collection, or institution in which the epitype is conserved is specified or, if the epitype is a published illustration, a full and direct bibliographic reference.

In most cases in which no holotype was designated there will also be no paratypes, since all the cited specimens will be syntypes. However, when an author has designated two or more specimens as types, any remaining cited specimens are paratypes and not syntypes.

Topotype is often the name given to a specimen collected from the same locality from which the holotype was originally collected.

In cases where the type of a name is a culture permanently preserved in a metabolically inactive state, any living isolates obtained from that should be referred to as '**ex-type**' (ex typo), 'ex-holotype' (ex holotypo), 'ex-isotype' (ex isotypo), etc., in order to make it clear they are derived from the type but are not themselves the nomenclatural type.

When an infraspecific variant is recognized within a species for the first time, it automatically establishes two **infraspecific taxa**. The one, which includes the type specimen of the species, must have the same epithet as that of the species, e.g., *Acacia nilotica* ssp. *nilotica*. Such a name is called an **autonym**, and the specimen an **autotype**. The variant taxon would have its own holotype and is differentiated by an epithet different from the specific epithet, e.g., *Acacia nilotica* ssp. *indica*.

On or after 1 January 2001, lectotypification, neotypification, or epitypification of a name of a species or infraspecific taxon is not effected unless indicated by use of the term "lectotypus", "neotypus", or "epitypus", its abbreviation, or its equivalent in a modern language (Art. 7.11 and 9.10).

It must be borne in mind that the application of the type method or **typification** is a methodology different from **typology**, which is a concept based on the idea that does not recognize variation within the taxa and believes that an idealized specimen or pattern can represent a natural taxon. This concept of typology was very much in vogue before Darwin put forward his ideas about variations.

Author Citation

For a name to be complete, accurate and readily verifiable, it should be accompanied by the name of the author or authors who first published the name validly. The names of the authors are commonly abbreviated, e.g., **L.** or **Linn.** for Carolus Linnaeus, **Benth.** for G. Bentham, **Hook.** For William Hooker, **Hook.f.** for Sir J. D. Hooker (f. stands for filius, the son; J. D. Hooker was son of William Hooker), **R.Br.** for Robert Brown, **Lam.** for J. P. Lamarck, **DC.** for A. P. de Candolle, **Wall.** for Wallich, **A. DC.** for Alphonse de Candolle, **Scop.** for G. A. Scopoli and **Pers.** for C. H. Persoon.

Single Author

The name of a single author follows the name of a species (or any other taxon) when a single author proposed a new name, e.g., *Solanum nigrum* L.

Multiple Authors

The names of two or more authors may be associated with a name for a variety of reasons. These different situations are exhibited by citing the name of the authors differently:

1. **Use of *et*:** When two or more authors publish a new species or propose a new name, their names are linked by *et*, e.g., *Delphinium viscosum* Hook.f. *et* Thomson.
2. **Use of parentheses:** The rules of botanical nomenclature specify that whenever the name of a taxon is changed by the transfer from one genus to another, or by upgrading or downgrading the level of the taxon, the original epithet should be retained. The name of the taxon providing the epithet is termed a **basionym**. The name of the original author or authors whose epithet is being used in the changed name is placed within parentheses, and the author or authors who made the name change outside the parentheses, e.g., *Cynodon dactylon* (Linn.) Pers., based on the basionym *Panicum dactylon* Linn., the original name for the species.
3. **Use of *ex*:** The names of two authors are linked by *ex* when the first author had proposed a name but was validly published only by the second author, the first author failing to satisfy all or some of the requirements of the Code, e.g., *Cerasus cornuta* Wall. *ex* Royle.
4. **Use of *in*:** The names of authors are linked using *in* when the first author published a new species or a name in a publication of another author, e.g., *Carex kashmirensis* Clarke *in* Hook.f. Clarke published this new species in the *Flora of British India* whose author was Sir J. D. Hooker.
5. **Use of *emend*:** The names of two authors are linked using *emend* (***emendavit:*** person making the correction) when the second author makes some change in the **diagnosis** or in **circumscription** of a taxon without altering the type, e.g., *Phyllanthus* Linn. *emend.* Mull.
6. **Use of square brackets:** Square brackets are used to indicate prestarting point author. The generic name *Lupinus* was effectively published by Tournefort in 1719, but as it happens to be earlier than 1753, the starting date for botanical nomenclature based on *Species plantarum* of Linnaeus, the appropriate citation for the genus is *Lupinus* [Tourne.] L.

An alteration of the diagnostic characters or of the circumscription of a taxon without the exclusion of the type does not warrant a change of authorship of the name of the taxon.

When naming an **infraspecific taxon**, the authority is cited both for the specific epithet and the infraspecific epithet, e.g., *Acacia nilotica* (L.) Del. ssp. *indica* (Benth.) Brenan. In the case of an **autonym,** however, the infraspecific epithet does not bear the author's name since it is based on the same type as the species, e.g., *Acacia nilotica* (L.) Del. ssp. *nilotica.*

Publication of Names

The name of a taxon, when first published, should meet certain requirements so as to become a legitimate name for consideration when deciding on a correct name. A valid publication should satisfy the following requirements:

Formulation

A name should be properly formulated and its nature indicated by a proper abbreviation after the name of the author:

1. ***sp. nov.*** for ***species nova***, a species new to science; *Tragopogon kashmirianus* G. Singh, ***sp. nov***. (published in 1976).
2. ***comb. nov.*** for ***combinatio nova***, a name change involving the epithet of the basionym, name of the original author being kept within parentheses; *Vallisneria natans* (Lour.) Hara ***comb. nov***. (published in 1974 based on *Physkium natans* Lour., 1790).
3. ***comb. et stat. nov***. for ***combinatio et status nova,*** when a new combination also involves the change of status. Epithet of the basionym will accordingly be used in the combination intended; *Caragana opulens* Kom. var. *licentiana* (Hand.-Mazz.) Yakovl. ***comb. et stat. nov.*** (published in 1988 based on *C. licentiana* Hand.-Mazz., 1933; new combination also involved change of status from a species *C. licentiana* to a variety of *Caragana opulens* Kom.).
4. ***nom. nov.*** for ***nomen novum***, when the original name is replaced and its epithet cannot be used in the new name; *Myrcia lucida* McVaugh ***nom. nov***. (published in 1969 to replace *M. laevis* O. Berg, 1862, an illegitimate homonym of *M. laevis* G. Don, 1832).

These abbreviations are, however, used only when first published. In future references, these are replaced by the name of the publication, page number and the year of publication for full citation, or at least the year of publication. Thus when first published in 1976 as a new species, *Tragopogon kashmirianus* G. Singh sp. nov. appeared in a book titled *Forest Flora of Srinagar* on page 123, Figure 4, any successive reference to this species would appear as: *Tragopogon kashmirianus* G. Singh, *Forest Flora of Srinagar,* p 123, f. 4, 1976 or *Tragopogon kashmirianus* G. Singh, 1976. The other names would be cited as *Vallisneria natans* (Lour.) Hara, 1974, *Caragana opulens* Kom. var. *licentiana* (Hand.-Mazz.) Yakovl., 1988 and *Myrcia lucida* McVaugh, 1969, specifying the year of publication. A new combination, or an avowed substitute (replacement name, nomen novum), published on or after 1 January 1953 based on a previously and validly published name is not validly published unless its basionym (name-bringing or epithet-bringing synonym) or the replaced synonym (when a new name is proposed) is clearly indicated and a full and direct reference given to its author and place of valid publication, with page or plate reference and date. Authors should cite themselves by name after each new name they publish rather than refer to themselves by expressions such as 'nobis' (nob.) or 'mihi' (m.).

Description or Diagnosis

Names of all new species (or other taxa new to science) published 1 January 2012 onwards should have **description or diagnosis** (listing of essential features) in Latin or English language to make the publication valid. For publications before 1 January 1908, an **illustration with analysis** without any accompanying description is valid. Description in any language is essential from 1 January 1908 onwards and this accompanied by a **Latin diagnosis** from 1 January 1935, also allowing English description or diagnosis from 1 January 2012. If names are published in any other language, then a description/diagnosis must be provided in either English or Latin. For name changes or new names of already known species, a full reference to the original publication should be made.

Typification

A **holotype** should be designated. Publication on or after 1 January 1958 of the name of a new taxon of the rank of genus or below is valid only when the type of the name is indicated. For the name of a new taxon of the rank of genus or below published on or after 1 January 1990, an indication of the type must include one of the words 'typus' or 'holotypus', or its abbreviation, or even its equivalent in a modern language. For the name of a new species or infraspecific taxon published on or after 1 January 1990 whose type is a specimen or unpublished illustration, the herbarium or institution in which the type is conserved must be specified. Names published on or after 1 January 2007 would require a specimen (and not a mere illustration) as type, except only for microscopic algae or microfungi for which preservation of a type was technically difficult, and where illustration is accepted as type. On or after 1 January 2001, lectotypification or neotypification of a name of a species or infraspecific taxon is not affected unless indicated by use of the term 'lectotypus' or 'neotypus', its abbreviation, or its equivalent in a modern language. The specimen selected as type should belong to a single gathering. '*Echinocereus sanpedroensis*' (Raudonat and Rischer, 1995) was based on a 'holotype' consisting of a complete plant with roots, a detached branch, an entire flower, a flower cut in halves, and two fruits, which according to the label were taken from the same cultivated individual at different times and preserved, in alcohol, in a single jar. This material belongs to more than one gathering and cannot be accepted as a type. Raudonat & Rischer's name is thus not validly published.

Effective Publication

Effective from 1 January, 2012, Electronic material published online in Portable Document Format (PDF) with an International Standard Serial Number (ISSN) or an International Standard Book Number (ISBN) and accessible electronically through World Wide Web will also constitute effective publication as additional option to hitherto requirement of printed hard copy. Online material should be placed in multiple online digital ISO certified repositories in more than one area of the world, preferably different continents. Prior to this publication had to be in printed form, distributed through sale, exchange or gift to the general public or at least the botanical institutions with libraries accessible to botanists generally. It is not affected by communication of new names at a public meeting, by the placing of names in collections or gardens open to the public; by the issue of microfilm made from manuscripts, typescripts or other unpublished material, by publication on-line, or by dissemination of distributable electronic media. The publication in newspapers and catalogues (1 January 1953 onwards) and seed exchange lists (1 January 1977 onwards) is not an effective publication. A thesis submitted for a higher degree on or after 1 January, 1953 is considered effectively published, only if it carries a statement of its publication or an internal evidence (e.g., an ISBN, or a commercial publisher). Publication of handwritten material, reproduced by some mechanical or graphic process (**indelible autograph**) such as lithography, offset, or metallic etching before 1 January 1953 is effective. *Salvia oxyodon* Webb & Heldr. was effectively published in an indelible autograph catalogue placed on sale (Webb & Heldreich, *Catalogus plantarum hispanicarum ... ab A. Blanco lectarum,* Paris, July 1850, folio). The *Journal of the International Conifer Preservation Society,* Vol. 5[1]. 1997 ('1998'), consists of duplicated sheets of typewritten text with handwritten additions and corrections in several places. The handwritten portions, being indelible autograph published after 1 January 1953, are not effectively published. Intended new combinations (*'Abies koreana* var. *yuanbaoshanensis',* p. 53), for which the basionym reference is handwritten are not validly published. The entirely handwritten account of a new taxon (p. 61: name, Latin description, statement of type) is treated as unpublished.

The date of a name is that of its valid publication. When the various conditions for valid publication are not simultaneously fulfilled, the date is that on which the last condition was fulfilled. However, the name must always be explicitly accepted in the place of its validation. A name published on or after 1 January 1973 for which the various conditions for valid publication are not simultaneously fulfilled is not validly published unless a full and direct reference is given to the places where these requirements were previously fulfilled.

In order to be accepted, a name of a new taxon of fossil plants published on or after 1 January 1996 must be accompanied by a Latin or English description or diagnosis or by a reference to a previously and effectively published Latin or English description or diagnosis.

For groups originally not covered by ICN (or former ICBN), the Code accepts them as validly published if they meet the requirements of the pertinent non-botanical Code but are now recognized as organisms covered under botanical Code. This provision earlier covered organisms subsequently recognized as algae, but Vienna Code extended this provision also to organisms subsequently recognized as fungi. The provision has benefitted the recognition of Microsporidia, long considered protozoa and now recognized as fungi. Similarly, the species of *Pneumocystis*, not validly published because of lack of Latin diagnosis or description, are now treated as validly published, since Latin requirement is not mandatory under Zoological Code, which originally covered these mammalian pathogens, now treated as fungi.

Several fungal taxa used to have different names for sexual and asexual phases of same fungus. Starting from 1 January, 2012 only one correct name will apply to all fungi like other groups within this Code. This has been made possible due to molecular studies enabling easy linking the two phases. This shall require a major exercise to ensure that there is minimal nomenclatural disruption in the proper use of fungal names when the "Principle of Priority" is applied. From 1 January 2013, all new names of fungi shall have to cite an "identifier issued by a recognized repository" in order to be validly published. Since 2004, whenever a new fungal name is published, "the online database Mycobank (www.mycobank.org)" registers the description and illustrations. Upon registration, Mycobank issues a unique registration number which can be cited in the publication where the name appears. This requirement shall become mandatory for the valid publication of fungal names.

A correction of the original spelling of a name does not affect its date of valid publication.

Rejection of Names

The process of selection of correct name for a taxon involves the identification of **illegitimate names**, those which do not satisfy the rules of botanical nomenclature. A legitimate name must not be rejected merely because it, or its epithet, is inappropriate or disagreeable, or because another is preferable or better known or because it has lost its original meaning. The name *Scilla peruviana* L. (1753) is not to be rejected merely because the species does not grow in Peru. Any one or more of the following situations leads to the rejection of a name:

1. ***Nomen nudum*** (abbreviated ***nom. nud.***): A name with no accompanying description. Many names published by Wallich in his *Catalogue* (abbreviated *Wall. Cat.*) published in 1812 were ***nomen nudum***. These were either validated by another author at a later date by providing a description (e.g., *Cerasus cornuta* Wall. *ex* Royle) or if by that time the name has already been used for another species by some other author, the ***nomen nudum*** even if validated is rejected and a new name has to be found (e.g., *Quercus dilatata* Wall., a ***nom. nud.*** rejected and replaced by *Q. himalayana* Bahadur, 1972).
2. Name not effectively published, not properly formulated, lacking typification or without a Latin diagnosis.
3. ***Tautonym***: Whereas the Zoological Code allows binomials with identical generic name and specific epithet (e.g., *Bison bison*), such names in Botanical nomenclature constitute **tautonyms** (e.g., *Malus malus*) and are rejected. The words in the tautonym are exactly identical, and evidently names such as *Cajanus cajan* or *Sesbania sesban* are not tautonyms and thus legitimate. Repetition of a specific epithet in an infraspecific epithet does not constitute a tautonym but a legitimate **autonym** (e.g., *Acacia nilotica* ssp. *nilotica*).
4. ***Later homonym***: Just as a taxon should have one correct name, the Code similarly does not allow the same name to be used for two different species (or taxa). Such, if existing, constitute **homonyms**. The provision extends across Algae, Fungi and Plants, which are now covered under this Code. The one published at an earlier date is termed the **earlier homonym** and that at a later date as the **later homonym**. The Code rejects later homonyms even if the earlier homonym is illegitimate. *Ziziphus jujuba* Lam., 1789 had long been used as the correct name for the cultivated fruit jujube. This, however, was ascertained to be a later homonym of a related species *Z. jujuba* Mill., 1768. The binomial *Z. jujuba* Lam., 1789 is thus rejected and jujube correctly named as *Z. mauritiana* Lam., 1789. Similarly, although the earliest name for almonds is *Amygdalus communis* L., 1753 when transferred to the genus *Prunus* the name *Prunus communis* (L.) Archangeli 1882 for almond became a later homonym of *Prunus communis* Huds., 1762 which is a species of plums. *P. communis* (L.) Archangeli was as such replaced by *P. dulcis* (Mill.) Webb, 1967 as the name for almonds. When two or more generic or specific names based on different types are so similar that they are likely to be confused (because they are applied to related taxa or for any other reason) they are to be treated as homonyms. Names treated as homonyms include: *Asterostemma* Decne. (1838) and *Astrostemma* Benth. (1880); *Pleuropetalum* Hook. f. (1846) and *Pleuripetalum* T. Durand (1888); *Eschweilera* DC. (1828) and *Eschweileria* Boerl. (1887); *Skytanthus* Meyen (1834) and *Scytanthus* Hook. (1844). The three generic names *Bradlea* Adans. (1763), *Bradleja* Banks ex Gaertn. (1790), and *Braddleya* Vell. (1827), all commemorating Richard Bradley, are treated as homonyms because only one can be used without serious risk of confusion. The following specific epithets under the same genus would also form homonyms *chinensis* and *sinensis; ceylanica* and *zeylanica; napaulensis, nepalensis,* and *nipalensis.* Authors naming new taxa under this Code, should as far as practicable, avoid using such names as already exist for zoological and prokaryotic taxa.
5. ***Later isonym***: When the same name, based on the same type, has been published independently at different times by different authors, then only the earliest of these so-called 'isonyms' has nomenclatural status. The name is always to be cited from its original place of valid publication, and later 'isonyms' may be disregarded. Baker (1892) and Christensen (1905) independently published the name *Alsophila kalbreyi* as a substitute for *A. podophylla* Baker (1891) non Hook. (1857). As published by Christensen, *Alsophila kalbreyeri* is a later 'isonym' of *A. kalbreyeri Baker*, without nomenclatural status.
6. ***Nomen superfluum*** (abbreviated as ***nom. superfl.***): A name is illegitimate and must be rejected when it was nomenclaturally superfluous when published, i.e., if the taxon to which it was applied—as circumscribed by its author—included the type of a name or epithet which ought to have been adopted under the rules. *Physkium natans* Lour., 1790 thus when transferred to the genus *Vallisneria,* the epithet *natans* should have been retained but de Jussieu used the name *Vallisneria physkium* Juss., 1826 a name which becomes superfluous. The species has accordingly been named correctly as *Vallisneria natans* (Lour.) Hara, 1974. A combination based on a superfluous name is also illegitimate. *Picea excelsa* (Lam.) Link is illegitimate since it is based on a superfluous name *Pinus excelsa* Lam., 1778 for *Pinus abies* Linn., 1753. The legitimate combination under *Picea* is thus *Picea abies* (Linn.) Karst., 1880.
7. ***Nomen ambiguum*** (abbreviated as ***nom. ambig.***): A name is rejected if it is used in a different sense by different authors and has become a source of persistent error. The name *Rosa villosa* L. is rejected because it has been applied to several different species and has become a source of error.

8. ***Nomen confusum*** (abbreviated as ***nom. confus.***): A name is rejected if it is based on a type consisting of two or more entirely discordant elements, so that it is difficult to select a satisfactory lectotype. The characters of the genus *Actinotinus,* for example, were derived from two genera *Viburnum and Aesculus,* owing to the insertion of the inflorescence of *Viburnum* in the terminal bud of an *Aesculus* by a collector. The name *Actinotinus* must, therefore, be abandoned.
9. ***Nomen dubium*** (abbreviated as ***nom. dub.***): A name is rejected if it is dubious, i.e., it is of uncertain application because it is impossible to establish the taxon to which it should be referred. Linnaeus (1753) attributed the name *Rhinanthus crista-galli* to a group of several varieties, which he later described under separate names, rejecting the name *R. crista-galli* L. Several later authors, however, continued to use this name for diverse occasions until Schwarz (1939) finally listed this as ***Nomen dubium,*** and the name was finally rejected.
10. **Name based on monstrosity:** A name must be rejected if it is based on a monstrosity. The generic name *Uropedium* Lindl., 1846 was based on a monstrosity of the species now referred to as *Phragmidium caudatum* (Lindl.) Royle, 1896. The generic name *Uropedium* Lindl. must, therefore, be rejected. The name *Ornithogallum fragiferum* Vill., 1787, is likewise, based on a monstrosity and thus should be rejected.

Principle of Priority

The principle of priority is concerned with the selection of a single correct name for a taxonomic group. After identifying **legitimate** and **illegitimate** names, and rejecting the latter, a correct name has to be selected from among the legitimate ones. If more than one legitimate names are available for a taxon, the correct name is the earliest legitimate name in the same rank. For taxa at the species level and below the correct name is either the earliest legitimate name or a combination based on the earliest legitimate **basionym**, unless the combination becomes a tautonym or later homonym, rendering it illegitimate. The following examples illustrate the **principle of priority**:

1. The three commonly known binomials for the same species of *Nymphaea* are *N. nouchali* Burm.f., 1768, *N. acutiloba* DC., 1824, *N. stellata* Willd., 1799 and N. malabarica Poir., 1798. Using the priority criterion, *N. nouchali* Burm.f. is selected as the correct name as it bears the earliest date of publication. The other three names are regarded as **synonyms**. The citation is written as:

   Nymphaea nouchali Burm.f., 1768
   N. malabarica Poir., 1798
   N. stellata Willd., 1799
   N. acutiloba DC., 1824

   The following binomials for common maize plant exist: *Zea mays* Linn., 1753, *Z. curagua* Molina, 1782, *Z. indurata* Sturtev., 1885 and *Z. japonica* Von Houtte, 1867. *Zea mays* being the earliest validly published binomial is chosen as correct name, and others cited as its synonyms as under:

   Zea mays L., 1753
   Z. curagua Molina, 1782
   Z. japonica Von Houtte, 1867
   Z. indurata Sturtev., 1885

2. Loureiro described a species under the name *Physkium natans* in 1790. It was subsequently transferred to the genus *Vallisneria* by A. L. de Jussieu in 1826, but unfortunately, he ignored the epithet *natans* and instead used a binomial *Vallisneria physkium,* a **superfluous name**. Two Asiatic species with independent typification were described subsequently under the names *V. gigantea* Graebner, 1912 and V. asiatica Miki, 1934. Hara on making a detailed study of Asiatic specimens concluded that all these names are synonymous, and also that *V. spiralis* Linn. with which most of the Asiatic specimens were identified does not grow in Asia. As no legitimate combination based on *Physkium natans* Lour. existed, he made one—*V. natans* (Lour.) Hara—in 1974. The synonymy would be cited as under:

   Vallisneria natans (Lour.) Hara, 1974
   Physkium natans Lour.,1790—Basionym
   V. physkium Juss., 1826—nom. superfl.
   V. gigantea Graebner, 1912
   V. asiatica Miki, 1934
   V. spiralis auct. (non L., 1753)

   The correct name of the species in this case, is the most recent name, but it is based on the earliest **basionym.** It must be noted that *Physkium natans* and *Vallisneria physkium* are based on the same type as the correct name *V. natans* and are thus known as **nomenclatural synonyms** or **homotypic synonyms**. These three would remain together in all citations. The other two names *V. gigantea* and *V. asiatica* are based on separate types and may or may not be regarded as synonyms of *V. natans* depending on taxonomic judgement. Such a synonym, which is based on a type different from the correct name, is known as a **taxonomic synonym** or **heterotypic synonym**. *V. spiralis* auct. (auctorum-authors) is misplaced identification of Asian specimens with *V. Spiralis* L.

3. The common apple was first described by Linnaeus under the name *Pyrus malus* in 1753. The species was subsequently transferred to the genus *Malus* but the combination *Malus malus* (Linn.) Britt., 1888 cannot be taken as the correct name since it becomes a tautonym. The other binomial under *Malus* available for apple is *M. domestica* Borkh, 1803 which is accepted as correct name and citation written as:

   Malus domestica Borkh

   Pyrus malus Linn., 1753
   M. malus (Linn.) Britt., 1888—Tautonym
   M. pumila auct. (non Mill.)
   M. communis Desf., 1798—Nom. superfl.

   M. communis Desf. is based on same type as *Pyrus malus*, and is as such a nomen superfluum. Apple has been assigned by some authors to *M. pumila* Mill., 1768, which however is small fruited Paradise apple.

4. Almond was first described by Linnaeus under the name *Amygdalus communis* in 1753. Miller described another species under the name *A. dulcis* in 1768. The two are now regarded as synonymous. The genus *Amygdalus* was subsequently merged with the genus *Prunus* and the combination *Prunus communis* (L.) Archangeli made in 1882 based on the earlier name *Amygdalus communis* Linn. It was discovered by Webb that the binomial *Prunus communis* had already been used by Hudson in 1762 for a different species rendering *P. communis* (L.) Archangeli a **later homonym** which had to be consequently rejected. Webb accordingly used the next available **basionym** *Amygdalus dulcis* Mill., 1768 and made a combination *Prunus dulcis* (Mill.) Webb, 1967 as the correct name for almond. Another binomial, *Prunus amygdalus* Batsch, 1801, cannot be taken up as it ignores the earlier epithets. The citation for almond would thus be:

   Prunus dulcis (Mill.) Webb, 1967

   Amygdalus dulcis Mill., 1768—basionym
   A. communis L., 1753
   P. communis (L.) Arch., 1882 (non Huds., 1762)
   P. amygdalus Batsch, 1801

When two or more names simultaneously published are united, the first author to make such a merger has the right of choosing the correct name from these. Brown, 1818 was the first to unite *Waltheria americana* L., 1753 and *W. indica* L., 1753. He adopted the name *W. indica* for the combined species, and this name is accordingly treated as having priority over *W. americana*. The generic names *Thea* L. and *Camellia* L. are treated as having been published simultaneously on 1 May 1753. The combined genus bears the name *Camellia*, since Sweet, 1818, who was the first to unite the two genera, chose that name, and cited *Thea* as a synonym.

Limitations to the Principle of Priority

Application of the principle of priority has the following limitations:

Starting Dates

The principle of priority starts with the *Species plantarum* of Linnaeus published on 1-5-1753. The starting dates for different groups include:

Seed plants, Pteridophytes, Sphagnaceae and Hepaticae and some Algae and Fungi..1-5-1753 for ranks genus and below, 4-8-1789 for ranks above genus.

Mosses (excluding Sphagnaceae)........1-1-1801
Fungi 31-12-1801
Fossils 31-12-1820
Algae (Nostocaceae)................................1-1-1892
Algae (Oedogoniaceae).............................1-1-1900

The publications before these dates for respective groups are ignored while deciding the priority.

Starting date for suprageneric names was set at Vienna Congress as 4 August, 1789, the date of publication of de Jussieu's *Genera Plantarum.* Double author citation is not justified or permitted at suprageneric ranks.

Not Above Family Rank

The principle of priority is applicable only up to the family rank, and not above.

Not Outside the Rank

In choosing a correct name for a taxon, names or epithets available at that rank need to be considered. Only when a correct name at that rank is not available, can a combination be made using the epithet from another rank. Thus at the level of section the correct name is *Campanula* sect. *Campanopsis* R. Br., 1810 but when upgraded as a genus the correct name is *Wahlenbergia* Roth, 1821 and not *Campanopsis* (R. Br.) Kuntze, 1891. The following names are synonyms:

Lespedza eriocarpa DC. var. *falconeri* Prain, 1897
L. meeboldii Schindler, 1911
Campylotropis eriocarpa var. *falconeri* (Prain) Nair, 1977
C. meeboldii (Schindler) ......Schindler, 1912

The correct name at the species level under the genus *Campylotropis* would be *C. meeboldii*, ignoring the earlier epithet at the varietal level. If treated as a variety, the correct name would be *C. eriocarpa* var. *falconeri,* based on the earliest epithet at that rank. Under the genus *Lespedza*, at the species level the correct name would be *L. meeboldii,* whereas at the varietal level, it would be *L. eriocarpa* var. *falconeri.*

Magnolia virginiana var. *foetida* L., 1753 when raised to specific rank is called *M. grandiflora* L., 1759, not *M. foetida* (L.) Sarg., 1889.

Nomina Conservanda

Nomina conservanda (abbreviated as ***nom. cons.***)**:** Conservation aims at retention of those names that best serve stability of nomenclature. Strict application of the principle of priority has resulted in numerous name changes. To avoid name changes of well-known families or genera—especially those containing many species—a list of conserved generic and family names has been prepared and published in the Code with relevant changes. Such ***nomina conservanda*** are to be used as correct names replacing the earlier legitimate names, which are rejected and constitute ***nomina rejicienda*** (abbreviated ***nom. rejic.***). The family name Theaceae D. Don, 1825 is thus conserved against Ternstroemiaceae Mirbe, 1813. The genus *Sesbania* Scop., 1777 is conserved against *Sesban* Adans., 1763 and *Agati* Adans., 1763.

Conservation of Names of Species

Despite several protests from agricultural botanists and horticulturists, who were disgusted with frequent name changes due to the strict application of the principle of priority, taxonomists for a long period did not agree upon conserving names at the species level. The mounting pressure and the discovery that *Triticum aestivum* was not the correct name of common wheat, compelled taxonomists to agree at the **Sydney Congress** in 1981 upon the provision to conserve names of species of major economic importance. As a result, *Triticum aestivum* Linn. was the first species name conserved at **Berlin Congress** in 1987 and published in subsequent Code in 1988. Another species name also conserved along with was *Lycopersicon esculentum* Mill.

Linnaeus described two species, *Triticum aestivum* and *T. hybernum* in his *Species plantarum*, both bearing the same date of publication in 1753. According to the rules of nomenclature when two species with the same date of publication are united, the author who unites them first has the choice of selecting the correct binomial. For a long time, it was known that the first persons to unite the two species were Fiori and Paoletti in 1896 who selected *T. aestivum* L. as the correct name. It was pointed out by Kerguélen (1980), however, that the first person to unite these two species was actually Mérat (1821) and he had selected *T. hybernum* L. and not *T. aestivum.* This discovery led to the danger of *T. aestivum* L. being dropped in favour of *T. hybernum* L. A proposal for conserving the name *T. aestivum* L. was thus made by Hanelt and Schultze-Motel (1983), and being the number one economic plant, this was accepted at the Berlin Congress, removing any further danger to the name *Triticum aestivum* L.

In 1768 P. Miller proposed a new name, *Lycopersicon esculentum* for tomato, a species described earlier by Linnaeus (1753) as *Solanum lycopersicum.* Karsten (1882) made the name change *Lycopersicum lycopersicum* (L.) Karst., retaining the epithet used by Linnaeus, but since the name became a **tautonym** it was not considered the correct name for tomato. Nicolson (1974) suggested an orthographic correction *Lycopersicon lycopersicum* (L.) Karst., suggesting that *Lycopersicon* and *lycopersicum* are orthographic variants. Since the name *Lycopersicon lycopersicum* was no longer a tautonym, it was accepted as the correct name. But since *Lycopersicon esculentum* Mill., 1768 was a more widely known name, a proposal for its conservation was made by Terrel (1983) and accepted at the Berlin Congress along with that of *Triticum aestivum* L. A list from a mere 5 in Tokyo Code has grown to nearly 60 for Spermatophyta alone. Names listed in this Appendix fall under the special provisions of Art. 14.4. Neither a rejected name, nor any combination based on a rejected name may be used for a taxon that includes the type of the corresponding conserved name (Art. 14.7; see also Art. 14 Note 2). Combinations based on a conserved name are, therefore, in effect, similarly conserved. Given below are the major examples of species names which have been declared nomina conservanda (each name followed by the (=) sign, indicating taxonomic synonym or a (= =) sign, indicating nomenclatural synonym and then the binomial against which it has been conserved). Some names listed as conserved have no corresponding nomina rejicienda because they were conserved solely to maintain a particular type:

Allium ampeloprasum L., 1753 (=) *Allium porrum* L., 1753
Amaryllis belladonna L.
Bombax ceiba L.
Carex filicina Nees, 1834 (=) *Cyperus caricinus* D. Don, 1825
Hedysarum cornutum L., 1763 (= =) *Hedysarum spinosum* L., 1759
Lycopersicon esculentum Mill., 1768 (= =) *Lycopersicon lycopersicum* (L.) H. Karsen, 1882
Magnolia kobus DC., 1817
Silene gallica L., 1753
(=) *Silene anglica* L., 1753
(=) *Silene lusitanica* L., 1753
(=) *Silene quinquevulnera* L., 1753
Triticum aestivum L., 1753 (=) *Triticum hybernum* L., 1753

Names of Hybrids

Hybridity is indicated using the multiplication sign, or by the addition of the prefix 'notho-' to the term denoting the rank of the taxon, the principal ranks being nothogenus and nothospecies. A hybrid between named taxa may be indicated by placing the multiplication sign between the names of the taxa; the whole expression is then called a hybrid formula:

1. *Agrostis* × *Polypogon*
2. *Agrostis stolonifera* × *Polypogon monspeliensis*
3. *Salix aurita* × *S. caprea*

It is usually preferable to place the names or epithets in a formula in alphabetical order. The direction of a cross may be indicated by including the sexual symbols (♀ female; ♂ male) in the formula, or by placing the female parent first. If a non-alphabetical sequence is used, its basis should be clearly indicated.

A hybrid may either be **interspecific** (between two species belonging to the same genus) or **intergeneric** (between two species belonging to two different genera). A binary name may be given to the interspecific hybrid or nothospecies (if it is self-perpetuating and/or reproductively isolated) by placing the cross sign (if mathematical sign is available it should be placed immediately before the specific epithet, otherwise 'x' in lower case may be used with a gap) before the specific epithet as in the following cases (hybrid formula may be added within the parentheses if the parents are established):

1. *Salix* x *capreola* (*S. aurita* × *S. caprea*) or *Salix* ×*capreola* (*S. aurita* × *S. caprea*)
2. *Rosa* x *odorata* (*R. chinensis* × *R. gigantea*) or *Rosa* ×*odorata* (*R. chinensis* × *R. gigantea*)

The variants of interspecific hybrids are named **nothosubspecies** and **nothovarieties**, e.g., *Salix rubens* nothovar. *basfordiana.*

For an intergeneric hybrid, if given a distinct generic name, the name is formed as a **condensed formula** by using the first part (or whole) of one parental genus and last part (or whole) of another genus (but not the whole of both genera). A cross sign is placed before the generic name of the hybrid, e.g., ×*Triticosecale* (or x *Triticosecale*) from *Triticum* and *Secale,* ×*Pyronia* (or x *Pyronia*) from *Pyrus* and *Cydonia,* and *Agropogon* from *Agrostis* and *Polypogon.* The names may be written as under:

1. ×*Triticosecale* (*Triticum* × *Secale*)
2. ×*Pyronia* (*Pyrus* × *Cydonia*)

The nothogeneric name of an intergeneric hybrid derived from four or more genera is formed from the name of a person to which is added the termination **-ara**; no such name may exceed eight syllables. Such a name is regarded as a condensed formula:

×*Potinara* (*Brassavola* × *Cattleya* × *Laelia* × *Sophronitis*)

The nothogeneric name of a trigeneric hybrid is either: (a) a condensed formula in which the three names adopted for the parental genera are combined into a single word not exceeding eight syllables, using the whole or first part of one, followed by the whole or any part of another, followed by the whole or last part of the third (but not the whole of all three) and, optionally, one or two connecting vowels; or (b) a name formed like that of a nothogenus derived from four or more genera, i.e., from a personal name to which is added the termination **-ara**:

×*Sophrolaeliocattleya* (*Sophronitis* × *Laelia* × *Cattleya*)

When a nothogeneric name is formed from the name of a person by adding the termination **-ara**, that person should preferably be a collector, grower, or student of the group.

A binomial for the intergeneric hybrid may similarly be written as under:

×*Agropogon lutosus* (*Agrostis stolonifera* × *Polypogon monspeliensis*)

It is important to note that a binomial for an interspecific hybrid has a cross before the specific epithet, whereas in an intergeneric hybrid, it is before the generic name. Since the names of nothogenera and nothotaxa with the rank of a subdivision of a genus are condensed formulae or treated as such, they do not have types.

Since the name of a nothotaxon at the rank of species or below has a type, statements of parentage play a secondary part in determining the application of the name.

The grafts between two species are indicated by a plus sign between two grafted species as, for example, *Rosa webbiana* + *R. floribunda*.

Names of Cultivated Plants

The names of cultivated plants are governed by the International Code of Nomenclature for Cultivated Plants (**ICNCP**), last published in 1995 (Trehane et al.). Most of the rules are taken from ICBN with additional recognition of a rank **cultivar** (abbreviated **cv.**) for cultivated varieties. The name of a cultivar is not written in Italics, it starts with a capital letter, and is not a Latin but rather a common name. It is either preceded by **cv.** as in *Rosa floribunda* cv. Blessings or simply within single quotation marks, e.g., *Rosa floribunda* 'Blessings'. Cultivars may also be named directly under a genus (e.g., *Hosta* 'Decorata'), under a hybrid (e.g., *Rosa* ×*paulii* 'Rosea') or directly under a common name (e.g., Hybrid Tea Rose 'Red Lion'). The correct nothogeneric name for plants derived from the *Triticum* × *Secale* crosses is ×*Triticosecale* Wittmack ex A. Camus. As no correct name at the species level is available for the common crop triticales, it is recommended that crop triticales be named by appending the cultivar name to the nothogeneric name, e.g., ×*Triticosecale* 'Newton'. Since 1 January 1959 new cultivar names should have a description published in any language and these names must not be the same as the botanical or common name of a genus or a species. Thus, cultivar names 'Rose', 'Onion', etc., are not permitted as the name of a cultivar. It is recommended that cultivar names be registered with proper registering **authorities** to prevent duplication or misuse of cultivar names. Registering authorities exist separately for roses, orchids and several other groups or genera.

UNIFIED BIOLOGICAL NOMENCLATURE

Biology as a science is unusual in the sense that the objects of its study can be named according to five different Codes of nomenclature: International Code of Zoological Nomenclature (**ICZN**) for animals, International Code of Nomenclature for Algae, Fungi and Plants (**ICN** previously **ICBN**) for plants, International Code for the Nomenclature of Bacteria (**ICNB**) now called Bacteriological Code (**BC**) for bacteria, International Code of Nomenclature for Cultivated Plants (**ICNCP**) for plants under cultivation, and International Code of Virus Classification and Nomenclature (**ICVCN**) for viruses. For the general user of scientific names of organisms, there is thus inherent confusion in many aspects of this situation: different sets of rules have different conventions for citing names, provide for different forms for names at the same rank, and, although primarily each is based on priority of publication, differ somewhat in how they determine the choice of the correct name.

This diversity of Codes can also create more serious problems as, for example, in the determination of which Code to follow for those organisms that are not clearly plants, animals or bacteria, the so-called **ambiregnal organisms**, or those whose current genetic affinity may be well established but whose traditional treatment has been in a different group (e.g., the cyanobacteria). Moreover, the development of electronic information retrieval, by often using scientific names without clear taxonomic context, accentuates the problem of divergent methods of citation and makes homonymy between, for example, plants and animals a source of trouble and frequently confusion. **BioCode** and **PhyloCode** are two efforts towards a unified code, the former retaining the ranked hierarchy of Linnaean system, whereas the latter developing a rankless system based on the concepts of phylogenetic systematics.

Draft BioCode

The desirability of seeking some harmonization of all biological Codes has been appreciated for some time (see Hawksworth, 1995) and an exploratory meeting on the subject was held at Egham, UK in March 1994. Recognizing the crucial importance of scientific names of organisms in global communication, these decisions included not only agreement to take steps to harmonize the existing terminology and procedures, but also the desirability of working towards a unified system of biological nomenclature. The **Draft BioCode** is the first public expression of these objectives. The first draft was prepared in 1995. After successive reviews the fourth draft, named **Draft BioCode (1997)** prepared by the International Committee for Bionomenclature (ICB) and published by Greuter et al. (1998), it was further revised in 2010 and published as **Draft BioCode (2011)** is now available on the web: (http://www.bionomenclature.net/biocode2011.html). Draft BioCode has been substantially rewritten to take past experience into account. It is most appropriately viewed as framework over-arching the practices of current series of Codes, but one which also addresses ways by which some of the key issues of current concern in systematics could be handled by all Codes, for example the registration of new names and electronic publication. It no longer aims to replace the current (Special) Codes. It is therefore to be used alongside the special Codes, as their complement, and is ready for immediate implementation.

Salient Features

Largely on the pattern of the Botanical Code the salient features of this Draft BioCode include:

1. **General points:** No examples are listed, Notes omitted at the present stage, although some will no doubt be needed. A considerable number of articles and paragraphs have been dropped; the Draft BioCode 2011 has only **35 Articles (41 in 1997 version)** against 62 in recent Botanical Code.
2. **Preamble:** It lists six statements: Biology requires a precise, coherent, international system for naming clades and species of organisms. It is applicable to names of all Clades extant or extinct may be used concurrently with the rank-based codes. Although it relies on rank-based determining the acceptability of preexisting names, it governs the application of those names independently from the rank-based codes. Rules are mandatory, Recommendations should be followed but not mandatory. Its application is not retroactive.
3. **Taxa and ranks:** The present ranks of the Botanical Code are maintained in the Draft BioCode, and a few tentatively added: **domain** (above kingdom), in use for the pro-/eukaryotes, **tribe** between family and genus, **section** and **series** between genus and species, **superfamily** (in widespread use in zoology), and the option of adding the prefix **super**—to rank designations that are not already prefixed. The phrase '**family group**' refers to the ranks of superfamily, family and subfamily; '**subdivision of a family**' only to taxa of a rank between family group and genus group; '**genus group**' refers to the ranks of genus and subgenus; '**subdivision of a genus**' only to taxa of a rank between genus group and species group; '**species group**' to the ranks of species and subspecies; and the term '**infra-subspecific**' refers to ranks below the species group.
4. **Status:** For the purposes of this Code **Established names** are those that are published in accordance with relevant articles of this Code or that, prior to 1 January 200*n*, were validly published or became available under the relevant Special Code. **Acceptable names** are those that are in accordance with the rules and are not unacceptable under homonymy rule, and, for names published before 1 January 200*n*, are neither illegitimate nor junior homonyms under the relevant Special Code. In the family group, genus group, or species group, the accepted name of a taxon with a particular circumscription, position, and rank is the acceptable name which must be adopted for it under the rules. In ranks not belonging to the family group, genus group, or species group, any established name of a taxon adopted by a particular author is an accepted name. In this Code, unless otherwise indicated, the word 'name' means an established name, whether it be acceptable or unacceptable. The name of a taxon consisting of the name of a genus combined with one epithet is termed a binomen; the name of a species combined with an infraspecific epithet is termed a trinomen; binomina or trinomina are also termed combinations. Scientific names of all ranks should appear in the same distinctive, and preferably italic, type.
5. **Establishment of names:** In order to be established on or after 1 January 200*n*, a name of a taxon must be published as provided for by the rules for **publication**, which are essentially similar to the **Effective publication** in botany. The rules for establishment (valid publication of Botanical Code) are generally similar to the Botanical Code with certain changes. The new taxon may have a Latin or English description or diagnosis (thus Latin diagnosis is not mandatory). Change of rank within the family group or genus group, or elevation of rank within the species group do not require the formal establishment of a new name or combination. In order to be established, a name of a new fossil botanical taxon of specific or lower rank must be accompanied by an illustration or figure showing diagnostic characters, in addition to the description or diagnosis, or by a bibliographic reference to a previously published illustration or figure. This requirement also applies to the names of new non-fossil algal taxa at these ranks. Only if

the corresponding genus or species name is established can the name of a subordinate taxon. Establishment (valid publication) under the BioCode includes registration of names in the family group, genus group and species group as a last step after fulfillment of the present requirements for valid publication.

6. **Registration:** Registration is affected by submitting the published matter that includes the protologue(s) or nomenclatural acts to a registering office designated by the relevant international body. It is pertinent to mention that this requirement was based on the Botanical Code (Tokyo Code, 1994) where it has already been abandoned (St. Louis Code, 2000), removing all references to registration in the Botanical Code. The date of a name is that of its registration, which is the date of receipt of the relevant matter at the registering office. When alternative (homotypic) names are proposed simultaneously for registration for one and the same taxon (same rank and same position) neither is considered to be submitted. When one or more of the other conditions for establishment have not been met prior to registration, the name must be resubmitted for registration after these conditions have been fulfilled. The registering centres are empowered to register non-submitted items placed in the public domain that meet the requirements for establishment.
7. **Typification:** The type of a nominal taxon in the rank of genus or subdivision of a genus is a nominal species. The type of a nominal taxon of the family group, or of a nominal taxon of a higher rank whose name is ultimately based on a generic name, is the nominal genus. For the names of superspecies, species or infraspecific taxon is a specimen in a museum jar, herbarium sheet, slide preparation, or mounted set of freeze-dried ampoules. It should be in metabolically inactive state. Type designations must be published and registered. The typeless ('**descriptive**') names do not have a representative type and are formed based on some character/s, apply to taxa defined by circumscription, and may be used unchanged at different ranks above the rank of a family.
8. **Precedence (priority):** For purposes of precedence, the date of a name is either the date attributed to it in an adopted List of Protected Names or, for unlisted names, the date on which it was validly published under the botanical or bacteriological Code, or became available under the zoological Code, or was established under the present Code. Limitations of priority that under previous Codes affected names in certain groups or of certain categories—even if not provided for in the present Code—still apply to such names if they were published before 1 January 200*n*. The limitations to precedence are largely similar to botany. Conservation and rejection procedures would remain largely the same as at present. The botanical process of sanctioning concerns old names only and need be provided for in a future BioCode.
9. **Homonymy:** The major change with respect to the homonymy rule would be that in future, it would operate across the kingdoms. In order that this provision be applicable, it is necessary that lists of established generic names of all organisms be publicly available, ideally in electronic format; most such, apparently, already exist, but are not yet generally accessible. A list of across-kingdom generic homonyms in current use is being prepared, and, as a next step, a list of binomina in the corresponding genera is planned, so that future workers may avoid the creation of new (illegal) homonymous binomina. Existing across-kingdom homonyms would not lose their status of acceptable names but would be flagged for the benefit of biological indexers and users of indexes. Existing names are not affected by the proposed rules. The practice of '**Secondary Homonymy**' in ICZN is not followed in BioCode.
10. **Hybrids:** The Appendix for Hybrids in the Botanical Code is replaced by a single Article in the Draft BioCode. This extreme simplification should in no way disrupt the present and future usage of hybrid designations but has some philosophical changes as its basis. Most importantly, taxonomy and nomenclature are disentangled, in conformity with Principle I. Cultivated plants are not covered under the BioCode.

PhyloCode

The PhyloCode is being developed by International Committee on Phylogenetic Nomenclature on the philosophy of Phylogenetic taxonomy replacing the multirank Linnaean system with a rankless system recognizing only species and 'clades'. It is intended to cover all biological entities, living as well as fossil. Underlying principle of the PhyloCode is that the primary purpose of a taxon name is to provide a means of referring unambiguously to a taxon, not to indicate its relationships. The PhyloCode grew out of recognition that the current Linnaean system of nomenclature—as embodied in the pre-existing botanical, zoological, and bacteriological Codes—is not well suited to govern the naming of clades and species, the entities that compose the tree of life and are the most significant entities above the organism level. Rank assignment is subjective and biologically meaningless. The PhyloCode will provide rules for the express purpose of naming the parts of the tree of life—both species and clades—by explicit reference to phylogeny. In doing so, the PhyloCode extends 'tree-thinking' to nomenclature. The PhyloCode is designed so that it can be used concurrently with the pre-existing Codes or (after rules governing species names are added) as the sole code governing the names of taxa, if the scientific community ultimately decides that it should.

Table 6.2: Equivalence table of nomenclatural terms used in the Draft PhyloCode, the Draft BioCode and the current biological codes (excluding Code for Viruses).

PhyloCode	BioCode	Bacteriological Code	Botanical Code	Zoological Code
Publication and precedence of names				
published	published	effectively published	effectively published	published
precedence	precedence	priority	priority	precedence
earlier	earlier	senior	earlier	senior
later	later	junior	later	junior
Nomenclatural status				
established	established	validly published	validly published	available
converted	——	——	——	——
acceptable	acceptable	legitimate	legitimate	potentially valid
registration	registration	validation	——	——
Taxonomic status				
accepted	accepted	correct	correct	valid
Synonymy and homonymy				
homodefinitional	homotypic	objective	nomenclatural	objective
heterodefinitional	heterotypic	subjective	taxonomic	subjective
replacement name	replacement name	——	avowed substitute	new replacement name
Conservation and suppression				
conserved	conserved	conserved	conserved	conserved
suppressed	suppressed/ rejected	rejected	rejected	suppressed

The starting date of the PhyloCode has not yet been determined and is cited as 1 January 200*n* in the draft Code. Rules are provided for naming clades and will eventually be provided also for naming species. In this system, the categories 'species' and 'clade' are not ranks but different kinds of biological entities. A species is a segment of a population lineage, while a clade is a monophyletic group of species. Fundamental differences between the phylogenetic and traditional systems in how supraspecific names are defined lead to operational differences in the determination of synonymy and homonymy. For example, under the PhyloCode, synonyms are names whose phylogenetic definitions specify the same clade, regardless of prior associations with particular ranks; in contrast, under the pre-existing Codes, synonyms are names of the same rank based on types within the group of concern, regardless of prior associations with particular clades. The requirement that all established names be registered will reduce the frequency of accidental homonyms.

Phylogenetic nomenclature was presumed to have several advantages over the traditional system. In the case of clade names, it eliminates a major source of instability under the pre-existing Codes—name changes due solely to shifts in rank. It also facilitates the naming of new clades as they are discovered and not waiting till a full classification is developed as in the case of existing Codes. This is a particularly significant when new advances in molecular biology and computer technology have led to a burst of new information about phylogeny, much of which is not being translated into taxonomy at present. The availability of the PhyloCode will permit researchers to name newly discovered clades much more easily than they can under the pre-existing Codes. At present PhyloCode has rules only for clades but when extended to species, it will improve nomenclatural stability here as well, by removing the linkage to a genus name. Under the PhyloCode, phylogenetic position can easily be indicated by associating the species name with the names of one or more clades to which it belongs. Another benefit of phylogenetic nomenclature is that abandonment of ranks eliminates the most subjective aspect of taxonomy. The arbitrary nature of ranking is not widely appreciated by non-taxonomists.

The PhyloCode is designed so that it can be used concurrently with the rank-based codes or (after rules governing species names are added) as the sole code governing the names of taxa, if the scientific community ultimately decides that it should. The intent is not to replace existing names but to provide an alternative system for governing the application of both existing and newly proposed names. In developing the PhyloCode, much thought has been given to minimizing the disruption of the existing nomenclature. Thus, rules and recommendations have been included to ensure that most names will be applied in ways that approximate their current and/or historical use. However, names that apply to clades will be redefined in terms of phylogenetic relationships rather than taxonomic rank and therefore will not be subject to the subsequent changes that occur under the rank-based systems due to changes in rank. Because the taxon membership associated with particular names will sometimes differ between rank-based and phylogenetic systems, suggestions are provided for indicating which code governs a name when there is a possibility of confusion.

The concept of PhyloCode was first introduced by de Queiroz and Gauthier (1992). The theoretical development of PhyloCode resulted from a series of papers from 1990 onwards and three symposia first in 1995, the second in 1996 at the Rancho Santa Ana Botanic Garden in Claremont, California, U.S.A., organized by J. Mark Porter and entitled "The Linnean Hierarch: Past Present and Future," and the third at the XVI International Botanical Congress in St. Louis, Missouri, U.S.A. (1999), entitled 'Overview and Practical Implications of Phylogenetic Nomenclature'.

Practical shape to the PhyloCode was given at the first workshop held in 1998 at the Harvard University Herbaria, Cambridge, Massachusetts, U.S.A. The initial philosophy of unification of biological world was based on draft BioCode. The first public draft of the PhyloCode was posted on the internet in April 2000. A second workshop was held at Yale University in July 2002 wherein it was decided to publish separate documents governing clade names and species names. Modified versions of PhyloCode were posted in October 2003 (PhyloCode2), December 2003 (Phylocode2a). The efforts crystallized into the establishment of the **International Society for Phylogenetic Nomenclature (ISPN)** at the First International Phylogenetic Nomenclature Meeting, which took place in July 2004 in Paris, attended by about 70 systematic and evolutionary biologists from 11 nations, and resulted in publication of PhyloCode**2b**. **The Second International Phylogenetic Nomenclature Meeting** was held between June 28–July 2, 2006 at Yale University (New Haven, Connecticut, U.S.A.) **PhyloCode3** published in June, followed by **PhyloCode4a** in July 2007 and **PhyloCode4b** in September 2007. The **Third meeting** was held in July 21–23, 2008 at Dalhousie University, Halifax and took decisions on preparation of the "companion volume" and development of *RegNum*, the name registration database. The result was the publication of version **PhyloCode4c** in January 2010.

The latest version **PhyloCode5** was approved at **Fourth Meeting** of the Committee on Phylogenetic Nomenclature in January 2014. It includes substantial Changes from version 4c.

Preamble

1. Biology requires a precise, coherent, international system for naming clades. Scientific names have long been governed by the traditional codes (listed in Preamble item 4), but those codes do not provide a means to give stable, unambiguous names to clades. This code satisfies that need by providing rules for naming clades and describing the nomenclatural principles that form the basis for those rules.
2. This code is applicable to the names of all clades of organisms, whether extant or extinct.
3. This code may be used concurrently with the rank-based codes.
4. Although this code relies on the rank-based codes (i.e., *International Code of Nomenclature for Algae, Fungi and Plants* (*ICN*), *International Code of Zoological Nomenclature* (*ICZN*), *International Code of Nomenclature of Bacteria: Bacteriological Code* (*BC*), *International Code of Virus Classification and Nomenclature* (*ICVCN*)) to determine the acceptability of preexisting names, it governs the application of those names independently from the rank-based codes.
5. This code includes rules, recommendations, notes and examples. Rules are mandatory in that names contrary to them have no official standing under this code. Recommendations are not mandatory in that names contrary to them cannot be rejected on that basis. Systematists are encouraged to follow them in the interest of promoting nomenclatural uniformity and clarity, but editors and reviewers should not require that they be followed. Notes and examples are intended solely for clarification.
6. This code will take effect on the publication of *Phylonyms: a Companion to the PhyloCode*, and it is not retroactive.

Principles

1. **Reference.** The primary purpose of taxon names is to provide a means of referring to taxa, as opposed to indicating their characters, relationships, or membership.
2. **Clarity.** Taxon names should be unambiguous in their designation of particular taxa. Nomenclatural clarity is achieved through explicit definitions, which describe the concept of the taxon designated by the defined name.
3. **Uniqueness.** To promote clarity, each taxon should have only one accepted name, and each accepted name should refer to only one taxon.
4. **Stability.** The names of taxa should not change over time. As a corollary, it must be possible to name newly discovered taxa without changing the names of previously discovered taxa.
5. **Phylogenetic context.** This code is concerned with the naming of taxa and the application of taxon names in the context of phylogenetically conceptualized of taxa.

6. **Taxonomic freedom.** This code does not restrict freedom of taxonomic opinion with regard to hypotheses about relationships; it only concerns how names are to be applied within the context of any relevant phylogenetic hypothesis.
7. **There is no "case law" under this code.** Nomenclatural problems are resolved by the Committee on Phylogenetic Nomenclature (CPN) by direct application of the code; previous decisions will be considered, but the CPN is not obligated by precedents set in those decisions.

Salient Features

At present the PhyloCode has rules only for clades. Rules for species will be added later.

1. **Taxa:** Groups of organisms or species considered potential recipients of scientific names are called taxa. Only clade names are governed by the PhyloCode. In this code, a clade is an ancestor (an organism, population, or species) and all its descendants. Every individual organism (on Earth) belongs to at least one clade (i.e., the clade comprising all extant and extinct organisms, assuming that they share a single origin). Each organism also belongs to several nested clades (though the ancestor of the clade comprising all life—again assuming a single origin—does not belong to any other clade). It is not necessary that all clades be named. Clades are often either nested or mutually exclusive; however, phenomena such as speciation via hybridization, species fusion, and symbiogenesis can result in clades that are partially overlapping. This code does not prohibit, discourage, encourage, or require the use of taxonomic ranks. In this code, the terms 'species' and 'clade' refer to different kinds of biological entities, not ranks. The concepts of synonymy, homonymy, and precedence adopted in this code are, in contrast to the pre-existing codes, independent of categorical rank.
2. **Publication:** The provisions of the Code apply not only to the publication of names, but also to the publication of any nomenclatural act (e.g., a proposal to conserve a name). Publication, under this code, is defined as distribution of text (but not sound), with or without images, in a peer-reviewed book or periodical. To qualify as published, works must consist of at least 50, simultaneously obtainable, identical, durable, and unalterable copies, some of which are distributed to major institutional libraries (in at least five countries on three continents) so that the work is generally accessible as a permanent public record to the scientific community, be it through sale or exchange or gift, and subject to the restrictions and qualifications in the present article.
3. **Names-status and establishment:** Established names are those that are published in accordance with rules of PhyloCode. In order to indicate which names are established under this Code and therefore have explicit phylogenetic definitions (and whose endings are not reflective of rank), it may be desirable to distinguish these names from the supraspecific names governed by pre-existing codes, particularly when both are used in the same publication. The letter 'P' (bracketed or in superscript) might be used to designate names governed by the PhyloCode, and the letter 'L' to designate names governed by the pre-existing Linnaean codes. Using this convention, the name '*Ajugoideae*[L]' would apply to a plant subfamily which may or may not be a clade, whereas '*Teucrioideae*[P]' would apply to a clade which may or may not be a subfamily. **Establishment** of a name can only occur on or after 1 January 200*n*, the starting date for this code. In order to be established, a name of a taxon must be properly published, be adopted by the author(s), be registered, and the registration number must be cited in the protologue. The **accepted name** of a taxon is the name that must be adopted for it under this code. It must; (1) be established; (2) have precedence over alternative uses of the same name (homonyms) and alternative names for the same taxon (synonyms); and (3) not be rendered inapplicable by a qualifying clause in the context of a particular phylogenetic hypothesis.
4. **Registration:** In order for a name to be established under the PhyloCode, the name and other required information must be submitted to the PhyloCode registration database. A name may be submitted to the database prior to acceptance for publication, but it is not registered (i.e., given a registration number) until the author notifies the database that the paper or book in which the name will appear has been accepted for publication.
5. **Clade names:** The names of clades may be established through conversion of preexisting names or introduction of new names. In order to be established, the name of a clade must consist of a single word and begin with a capital letter. In order to be established, converted clade names must be clearly identified as such in the protologue by the designation '**converted clade name**' or '**nomen cladi conversum**'. New clade names must be identified as such by the designation '**new clade name**' or '**nomen cladi novum**'. In order to be established, a clade name must be provided with a phylogenetic definition, written in English or Latin, linking it explicitly with a particular clade. The name applies to whatever clade fits the definition. Examples of phylogenetic definitions are node-based, stem-based, and apomorphy-based definitions. A node-based definition may take the form 'the clade stemming from the most recent common ancestor of A and B' (and C, D, etc., as needed) or 'the least inclusive clade containing A and B' (and C, D, etc.), where A–D are specifiers. A node-based definition may be abbreviated as Clade (A+B). A stem-based definition may take the form 'the clade consisting of Y and all organisms that share a more recent common ancestor with Y than with W' (or V or U, etc., as needed) or 'the most inclusive clade containing Y but not W' (or

V or U, etc.). A stem-based definition may be abbreviated as Clade (Y<—W). An apomorphy-based definition may take the form 'the clade stemming from the first species to possess character M synapomorphic with that in H'. An apomorphy-based definition may be abbreviated as Clade (M in H). When giving a new name for total clade, prefix *Pan*-must be used to the name of crown clade (separated by hyphen) and designated as panclade.

6. **Specifiers and qualifying clauses:** Specifiers are species, specimens, or synapomorphies cited in a phylogenetic definition of a name as reference points that serve to specify the clade to which the name applies. All specifiers used in node-based and stem-based definitions of clade names, and one of the specifiers used in apomorphy-based definitions of clade names, are species or specimens. The other specifier used in an apomorphy-based definition of a clade name is a synapomorphy. If subordinate clades are cited in a phylogenetic definition of a more inclusive clade, their specifiers must also be explicitly cited within the definition of the more inclusive clade. An internal specifier is one that is explicitly included in the clade whose name is being defined; an external specifier is one that is explicitly excluded from it. All specifiers in node-based and apomorphy-based definitions are internal, but stem-based definitions must always have at least one specifier of each type. When a species is used as a specifier, the author and publication year of the species name must be cited. When a type specimen is used as a specifier, the species name that it typifies, and the author and publication year of that species name must be cited.
7. **Precedence:** Although the entity to which precedence applies in this code is referred to as a name, it is really the combination of a name and its definition. In different cases, one or the other of these components is more important. Specifically, in the case of synonyms, precedence refers primarily to the name, whereas in the case of homonyms, precedence refers primarily to the definition. Precedence is based on the date of establishment, with earlier-established names having precedence over later ones, except that later-established names may be conserved over earlier ones. In the case of homonymy involving names governed by two or more preexisting codes (e.g., the application of the same name to a group of animals and a group of plants), precedence is based on the date of establishment under the PhyloCode. However, the International Committee on Phylogenetic Nomenclature has the power to conserve a later-established homonym over an earlier-established homonym. This might be done if the later homonym is much more widely known than the earlier one. For the determination of precedence, the date of establishment is considered to be the date of publication, and not the date of registration.
8. **Synonymy:** Synonyms are names that are spelled differently but refer to the same taxon. In this code, synonyms must be established and may be homodefinitional (based on the same definition) or heterodefinitional (based on different definitions). Homodefinitional synonyms are synonyms regardless of the phylogenetic context in which the names are applied. However, in the case of names with different definitions, the phylogenetic context determines whether the names are heterodefinitional synonyms or not synonymous. When two or more synonyms have the same publication date, the one that was registered first (and therefore has the lowest registration number) takes precedence.
9. **Conservation, supression and emendation:** Conservation of names is possible only under extraordinary circumstances and requires approval of the CPN. Once a name has been conserved, the entry for the affected name in the registration database is to be annotated to indicate its conserved status relative to other names that are simultaneously suppressed. An emendation is a formal change in a phylogenetic definition. A restricted emendation (changes in definitional type, clade category, specifiers, and/or qualifying clauses) requires approval by the CPN, while an unrestricted emendation (changes in specifiers or qualifying clauses) may be published without CPN approval.
10. **Provisions for hybrids:** Hybrid origin of a clade may be indicated by placing the multiplication sign (×) in front of the name. The names of clades of hybrid origin otherwise follow the same rules as for other clades. An organism that is a hybrid between named clades may be indicated by placing the multiplication sign between the names of the clades; the whole expression is then called a hybrid formula.
11. **Authorship of names:** A taxon name is to be attributed to the author(s) of the protologue, even though authorship of the publication as a whole may be different. In some cases, it may be desirable to cite the author(s) who established a name. If the author of a converted name is cited, the author of the pre-existing name on which it is based must also be cited, but in square brackets []. If the author of a replacement name is cited, the author of the definition of the replaced name must also be cited, but in braces {}. If the author of a homonym that has been conserved for the purpose of emending a definition is cited, the author of the original definition must also be cited, but by using '<' and '>' symbols (e.g., *Hypotheticus* <Stein> Maki). Phylocode follows the use of *in* but not *ex*.
12. **Species names:** This code does not govern the establishment or precedence of species names. To be considered available (*ICZN*) or validly published (*ICBN*, *BC*), a species name must satisfy the provisions of the appropriate rank-based code. Because this code is independent of categorical ranks, the first part of a species binomen is not interpreted as a genus name but instead as simply a **prenomen**, first part of the species name, and the second part of a species binomen is associated with the species as a kind of biological entity, not as a rank. A prenomen has no

necessary tie to any categorical rank under this code. This code also does not govern the establishment of names associated with ranks below that of species under the rank-based codes ("infraspecific names"); however, such names may be used in conjunction with phylogenetic nomenclature.

13. **Governance:** The PhyloCode will be managed by The Society for Phylogenetic Nomenclature (SPN) through its two committees: International Committee on Phylogenetic Nomenclature (ICPN) and the Registration Committee.

The desirability of PhyloCode has been reviewed in several papers published over last few years. Nixon and Carpenter (2000) showed that Phylogenetic nomenclature would be less stable than existing systems. A critique of draft PhyloCode is presented by Carpenter (2003), pointing out that its stated goals can't be met by proposals in current draft, which also fails to uphold its stated Principles. The internal contradictions include a cumbersome reinvention of the very aspect of the current Linnaean System that the advocates of PhyloCode most often decry. The incompleteness is due to the fact that the drafters cannot agree on what form the species names should take. Keller et al. (2003) pointed out inherent instabilities, fundamental flaws in its very foundation by exposing unsubstantiated philosophical assumptions preceding and subtending it.

A strong opposition to the PhyloCode was offered by Nixon et al. (2003) who concluded that "The PhyloCode is fatally flawed, and the Linnaean System can be easily fixed". They argued that the proponents of the PhyloCode have offered nothing real to back up claims of greater stability for their new system. A rank free system of naming would be confusing at the best and would cripple our ability to teach, learn and use taxonomic names in the field or publications. They assured that the separate issue of stability in reference to rules of priority and rank can be easily addressed within the current Codes, by implementation of some simple changes. Thus, there is no need to 'scrap' the current Linnaean Codes for a poorly reasoned, logically inconsistent and fatally flawed new Code that will only bring chaos.

Rieppel (2006) argues the issue, then, is not so much a replacement of the Linnaean system by the PhyloCode, but the naturalization of the Linnaean System. The issue, accordingly, is the distinction between nominal and natural kinds. Given that the diagnoses of the Linnaean system specify the descriptive properties of a stereotype only, there always remains the possibility that groupings so diagnosed are nominal (artificial) kinds (i.e., paraphyletic groupings) instead of natural kinds (monophyletic groups). The naturalization of the Linnaean system is complete when nominal kinds have been replaced by natural kinds, when paraphyletic groups have been replaced by monophyletic groups.

Chapter 7

Variation, Biosystematics, Population Genetics and Evolution

Species of organisms as understood now are not fixed entities but systems of populations which exhibit variation and wherein no two individuals are identical. This concept of variations was first proposed by Lamarck and further developed by Darwin, culminating in his famous book *Origin of Species* (1859). Systematics is a unique natural science concerned with the study of individual, population and taxon relationships for purposes of classification. The study of plant systematics is based on the premise that in the tremendous variation in the plant world, there exist conceptual discrete units (usually named as species) that can be recognized, classified, described, and named, on the further premise that logical relationships developed through evolution exist among these units.

The studies on variations, experimental studies and hybridization studies in light of genetic information are commonly covered under the term **biosystematics**. The term was first proposed by Camp and Gilly (1943 as Biosystematy) to delimit natural biotic units and to apply to these units a system of nomenclature adequate to the task of conveying precise information regarding their defined limits, relationships, variability and dynamic structure. Clausen et al. (1945) regard genetics, cytology, comparative morphology and ecology as furnishing the critical data which together, when applied to the study of organic evolution, make up biosystematics. These two different approaches aim at the same problem, the study of variations.

The study of biosystematics, mainly the **experimental systematics** and **population genetics** approach the common aim, although the methodology is different. The experimental systematist usually begins with classical interpretation of species and works backwards so as to understand the genetic mechanisms involved. The population geneticist, on the other hand, begins with raw population, discarding any classical concept in mind. He works into a series of group concepts which may or may not be comparable to the taxonomists concept of species.

TYPES OF VARIATION

The recognition of taxonomic units is based on the identification of the occurrence and the degree of discontinuity in variation in the populations. The variation may be **continuous** when the individuals of a population are separable by infinitely small differences in any of the attributes. In a discontinuous **variation**, however, there is a distinct gap between two populations, each showing its own continuous variation for a particular attribute. The discontinuity between the populations primarily results from **isolation** in nature. Isolation plays a major role in establishing and widening the gap between the populations, allowing evolution to take its destined course with no disturbance. Variation in plants includes three fundamental types: developmental, environmental and genetic.

Developmental Variation

A distinct change in attributes is often found during different stages of development. Juvenile leaves of *Eucalyptus*, *Salix* and *Populus* are often different from the mature leaves, and may often cause much confusion, but may prove equally useful when both types of leaves are available from a plant. The first leaves of *Phaseolus* are opposite and simple, the later ones alternate and pinnately compound. As the seedling stage is most critical in a plant's life, the characters present during this period surely have survival value. Takhtajan proposed a **neotenous** origin for angiosperms on the assumption of juvenile simple leaves of seed ferns having persisted in the adult forms, which were the direct progenitors of angiosperms.

Environmental Variation

Environmental factors often play major role in shaping the appearance of a plant. Heterophylly is the common manifestation of environmental variation. The submerged leaves of *Ranunculus aquatilis* are finely dissected, whereas the emergent leaves of the same plant are broadly lobed. The first submerged leaves of *Sium suave* are pinnately dissected and flaccid; the older emerged leaves are pinnately compound and stiff. The individuals of a species often exhibit **phenotypic plasticity**, expressing different phenotypes under different environmental conditions. Such populations are named **ecophenes**. In *Epilobium*; the sun-plants have small, thick leaves, many hairs and a short stature, whereas the shade-plants have larger thinner leaves with fewer hairs and a taller stature.

Genetic Variation

Genetic variation may result from **mutation** or **recombination**. Mutation is the occurrence of heritable change in the genotype of an organism that was not inherited from its ancestors. It is the ultimate source of variation in a species and replenishes the supply of genetic variability. A mutation may be as minute as the substitution of a single nucleotide pair in the DNA (**point mutation**), change in a sequence nucleotide controlling gene action (**Gene mutation**) or as great as a major change in the chromosome structure (**chromosomal mutation**). Chromosomal mutation may be due to deletion, inversion, aneuploidy or polyploidy. Recombination is a reassortment of chromosomes, bringing together via meiosis and fertilization the genetic material from different parents and producing a new genotype.

VARIANCE ANALYSIS

Since no two individuals in a population are similar, there is need for some objective analysis for useful comparison. It is, however, often impossible to collect information about all the individuals of a population, and as such is reasonable to analyze a representative **sample**. It is essential that this sample should represent random subset of the population. The simplest tool is to calculate the **mean** or **average** by adding the series of values and dividing the total by the number of values. The formula for calculating the mean is:

$$\bar{X} = \frac{\sum X_i}{n}$$

where $\bar{X}$ represents the mean, $\sum$ summation of all values of X, X_i represents the individual values of an attribute under study and n represents the number of values. Thus, five plants of a species with height 15 cm, 12 cm, 10 cm, 22 cm and 16 cm would have a mean of 15 cm ((15 + 12 + 10 + 22 + 16)/5). The extent of variation within a population of a species is best represented by determining the **variance**. It is a measure of the spread of individual observations around the mean, i.e., how variable the individuals and their measurements are. It is defined as the ***average squared deviation from the mean***. If various individuals were not far from this mean the variance would be minimum. On the other hand, if many individuals were far removed from mean, the variance would be higher. The variance may either be calculated for a population, or a sample from the population. The variance for a population may be calculated as:

$$\sigma^2 = \frac{\sum_{i=1}^{n}(X_i - \bar{X})^2}{n}$$

To obtain the variance, the difference between each value of the attribute (X) and the mean is squared, and a sum of these squares is divided by the number of observation (n). For calculating sample variance (s^2) the sum of squares is divided by $n - 1$ instead of n. The formula for sample variance may be written as:

$$s^2 = \frac{\sum_{i=1}^{n}(X_i - \bar{X})^2}{n-1}$$

For the calculation of sample variance, the reason for dividing by $n - 1$ and not by n, is related to the **degrees of freedom**. If we have a single value, we can't compare it, if we have two, we have one comparison (2 – 1), if we have three values, we have two comparisons (3 – 1), and if there are n values, $n - 1$ comparisons are possible. While calculating population variance, with large number of values, the difference of one would be irrelevant, and as such the sum is divided directly by n. Two samples may have the same mean, but different variance (Table 7.1). The square root of variance is represented by **standard deviation**. Latter is often preferred over the variance because it shares the same units as the original measurements, whereas the variance is in the units squared. The reason for first squaring the values and then determining the square root, is to obtain the real picture of variation. If simple difference of each value and mean is summed, the negative values (measurements lower than the mean) may get cancelled by positive values (measurements

Table 7.1: Mean, variance and standard deviation of two samples based to plant height. The two samples have the same mean but different variance and standard deviation, highlighting the significance of these calculations.

Height (cm) (X_i) Sample A	Sample B	(X_i - $\bar{X}$) Sample A	Sample B	(X_i - $\bar{X}$)^2 Sample A	Sample B
18	22	18 – 15 = 3	22 – 15 = 7	9	49
14	10	14 – 15 = -1	10 – 15 = -5	1	25
16	06	16 – 15 = 1	06 – 15 = -9	1	81
15	16	15 – 15 = 0	16 – 15 = 1	0	1
17	12	17 – 15 = 2	12 – 15 = -3	4	9
19	19	19 – 15 = 4	19 – 15 = 4	16	16
12	11	12 – 15 = -3	11 – 15 = -4	9	16
14	27	14 – 15 = -1	27 – 15 = 12	1	144
13	14	13 – 15 = -2	14 – 15 = -1	4	1
16	21	16 – 15 = 1	21 – 15 = 6	1	36
14	09	14 – 15 = -1	09 – 15 = -6	1	36
12	13	12 – 15 = -3	13 – 15 = -2	9	4
ΣX_i = 180	180				
Mean $\bar{X}$ = 15	15				
Variance $S^2 = \frac{\Sigma (X_i - \bar{X})^2}{n-1}$ =				5.09	38
Standard deviation $S = \sqrt{S^2}$				2.256	6.17

higher than the mean), and the result would be zero, and thus meaningless. The squaring converts all values to plus and thus a real diversion from the mean on either side is taken into account. We may thus determine the standard deviation of a population as:

$$\sigma = \sqrt{\sigma^2}$$

and that for a sample as:

$$s = \sqrt{s^2}$$

For our sample data, the sample variance would be [(15–15)^2 + (12–15)^2 + (10–15)^2 + (22–15)^2 + (16–15)^2]/4 = 21 and the sample standard deviation:

$$\sqrt{21} = 4.5825$$

The determination of sample size from a population is crucial for the calculation of variance and standard deviation. Sample size n can be computed from magnitude of standard deviation (this can be estimated from the smallest and the largest value of an attribute (say smallest 5, largest 45, mean 25, deviation 20), level of confidence desired (z, say 0.95%) and maximum width of units from true value (d, say 5)):

$$n = \frac{z^2\sigma^2}{d^2}$$

For the above parameters the adequate sample size is (0.95 × 0.95) = 0.902 × (20 × 20) = 360.8/(5 × 5) = 14.4. Thus 15 would be ideal sample size with these parameters.

The analysis of variance data is often complicated, especially where more than one factors are responsible for variation. The technique of **Analysis of variance (ANOVA)** developed by Sir Ronald Fisher (1930) is commonly used for dividing the variance into components. It is a powerful statistical procedure for determining whether the differences from the mean are significant, i.e., larger than expected by chance. Thus, for example, if probability value of less than 0.002 (There is less than one per cent chance than variation obtained is due to chance) is obtained through variance analysis, the results are due to factors others than chance. The analysis involves **partitioning the variance** and comparing the role of various factors (environmental, genetical, etc.). This is significant in economical plants where it is important to determine whether the attributes are related to environmental variations or genotype. If former is true, it is advisable to improve cultural practices, if it is related to genotype, selective breeding would be the answer.

REPRODUCTIVE SYSTEMS

The diverse mechanisms of reproduction in seed plants can be classified under four major categories. **Allogamy** involves cross-fertilization between closely related individuals growing at a suitable distance from each other, and results in the formation of hybrids. Cross-fertilization promotes heterozygosity, resulting in considerable variation and diversity in individuals. **Autogamy** involves self-fertilization, resulting in inbred offsprings. It promotes homozygosity, yielding uniform populations. **Agamospermy** involves production of seeds resulting from the development of embryos from maternal tissue without fertilization. Finally, the reproduction may result from **vegetative propagation** of somatic regions such as shoot segments, bulbs, rhizomes, corms and other vegetative structures. Both allogamy and autogamy are examples of **sexual reproduction**, involving meiosis and fertilization. The last two, circumvent sexual reproduction and multiplication occurs through **asexual reproduction,** and are often termed **apomixis**. The products of asexual reproduction are known as **ramets**, whereas products of sexual reproduction, which show genetic variation as **genets**.

Outbreeding

Outbreeding, as mentioned earlier is largely responsible for genetic and phenetic diversity in populations. It is also known as **outcrossing**, **allogamy** or **xenogamy**. It enables plants to adapt to wide range of environmental conditions and increases likelihood of survival and evolutionary change. A variety of mechanisms promote outbreeding. These are briefly described below:

1. **Dioecy:** The phenomenon involves the occurrence of unisexual flowers, with male and female flowers in different individuals. Some variations of this are also encountered as for example, some individuals having male flowers others bisexual flowers (**androdioecy**), some individuals having female flowers others bisexual flowers (**gynodioecy**), or some individuals having male flowers others female and still others bisexual flowers (**trioecy**).
2. **Dichogamy:** The situation reflects the maturation of male and female floral parts at different times. In some members of Apiaceae and Asteraceae, pollen grains mature and are released before gynoecium is mature and receptive, the phenomenon known as **protandry**. In others, like members of Chenopodiaceae, the gynoecium is mature and receptive before the pollen maturation and release, a feature known as **protogyny**.
3. **Herkogamy:** It results from physical separation and stamens and carpels. This could be achieved by **heterostyly**, differences in the length of stamens and carpels. In the phenomenon known as **distyly** some flowers have short style and longer stamens (**thrum flowers**), whereas others have long style and short stamens (**Pin flowers**). In a rarer situation known as **trisyly**, three types of stamen and carpel lengths occur. In other cases the style is curved away from stamens, either towards right (right-handed flowers) or left (left-handed flowers), the situation known as **enanciostyly**. In some genera like *Mimulus*, the stigmas close after being touched by a pollinator, thus preventing pollination from same flower (*movement herkogamy*). In others like *Kalmia*, the pollinator triggers the movement of one or more stamens, dusting insect with pollen (**trigger herkogamy**).
4. **Self-incompatibility:** The phenomenon refers to the prevention of fertilization between the gametes derived from the same flower. **Gametophytic self-incompatibility** results from genetic composition of male gametophyte, and **sporophytic self-incompatibility** by genetic composition of sporophytic tissue such as style and stigma.

Hybridization

Although occurrence of breeding barriers is dominant criterion for distinction between the species, several cases of interspecific hybridization have been reported. Based on the studies of the Flora British Isles, Stace (1989) concluded that there are approximately 70,000 different naturally occurring interspecific hybrids, accounting for more than one fourth of the total number of species of seed plants on this planet. Natural hybridization is common in *Salix*, *Helianthus*, *Quercus*, *Senecio* and *Tragopogon*. It is more common in perennials as compared to annuals.

Hybridization between different species usually results in sterile offsprings, due to failure of pairing at meiosis, but in several genera like *Senecio* and *Tragopogon*, interspecific hybridization is often followed by chromosomal duplication, the resulting **polyploid (Allopolyploid; Amphiploid)** generation is sexually stable due to normal meiosis, of paired genomes. Many such polyploid species with distinct characters have been reported in these genera.

Occurrence of **intergeneric hybrids** is much rarer, and there may be less than 300 naturally intergeneric hybrids world-wide. Such hybrids are reported mostly in Poaceae and Orchidaceae, although in the latter family there are many artificially synthesized intergeneric hybrids, often involving five different genera.

Introgressive Hybridization

The process of introgressive hybridization, also known as **introgression** involves the gradual infiltration of one species into that of another, and commonly involves species with some degree of reproductive isolation. The phenomenon involves three steps: the formation of F1 hybrids, their backcrossing with one or another parental species, and natural selection of certain favorable recombinant types. The hybrids generally produce a lot of variability through backcrossing and F2 segregation, and may produce hybrid swarms, occupying a variety of habitats. Backcrossing with parental species frequently results in reversion of hybrid offsprings towards parental types. Backcrossing may also result in movement of genes from one species to another via the hybrids and backcrosses. Introgression may lead to three diverse consequences. In some cases, as *Gilia capitata*, it may lead to merging of species. The species has eight geographical races, of which three are believed to have been distinct in Pliocene. Subsequent gene flow led to intergrading races, and as such they are included under single species. Introgression may also transfer genetic material from one species to another without merging them. Introgressants, which get stabilized, may lead to the formation of new species.

Two types of introgression are commonly recognized. **Sympatric introgression** commonly occurs between species occurring in the same general geographical area but occupying different habitats. In California the introduced *Helianthus annuus* has introgressed with native serpentine species *H. bolanderi,* and the vigorous weedy variant of latter has spread into irrigated areas. Such introgression usually results in the wider spread of one species as compared to another. In England, for example *Silene alba* is spreading in weedy areas, whereas *S. dioica*, a woodland species is contracting. In Scotland, on the other hand, the more humid climate allows *S. dioica* to flourish outside woodlands, on hedgebanks and cliff ledges.

Allopatric introgression occurs between species which are now fully allopatric but had contact in the past. Such species are centered in different areas but share intermediate area largely occupied by products of hybridization. *Juniperus virginiana*, a mesophytic tree of eastern North America has shown introgression of bushy xerophyte *J. ashei* from dolomitic outcrops in Texas and Okhlahoma. Throughout the intermediate area are seen plants with partial recombination of characters between the two well differentiated species.

It must be mentioned that the process of introgression may lead to the development of variants with no taxonomic status, their recognition as ecotypes, subspecies, or if the intermediate species is sufficiently distinct, recognition as distinct species.

Recognition of Hybrids

The identification of hybrid nature of an offspring is possible through the use of some important criteria.

Phenetic intermediacy: Hybrids tend to have **phenetic intermediacy** between the putative parents. It is easier to recognize morphological characters, which can be plotted on a scatter diagram (Figure 7.1). Hybrids can also be detected by calculating **hybrid index**. A list of characters by which the two species differ is prepared. Each character-state of one species is assigned zero score, whereas each contrasting character-state of another species given a score of 2. The hybrid index of each species is calculated by summing up the score. Thus, one species will have hybrid index of 0, and another species 2n (n refers to the number of characters by which two species differ).

Reduced fertility: Hybrids between different species commonly tend to have reduced fertility, some being totally sterile. The degree of sterility is reflected upon the degree of heterozygosity between genomes of parental species. A hybrid which perishes at zygote stage would represent maximum heterozygosity, whereas a hybrid which manages to produce viable seeds, although less vigorous than either parents, would depict least heterozygosity between parents.

F2 segregation: Although F1 hybrids may tend to be normal, the next (F2) generation might show a lot of variability, exhibiting the segregation of parental characters. There may thus be reappearance of parental forms, as also many new recombinations of parental characters.

Distributional area: The hybrids between two species can also be verified by studying their distribution. In case of parental species occupying the same area, the hybrid populations would commonly be located in the same area. In case of species occupying different but adjacent areas, hybrids would commonly be located in the contact area, or transitional area between the parental species.

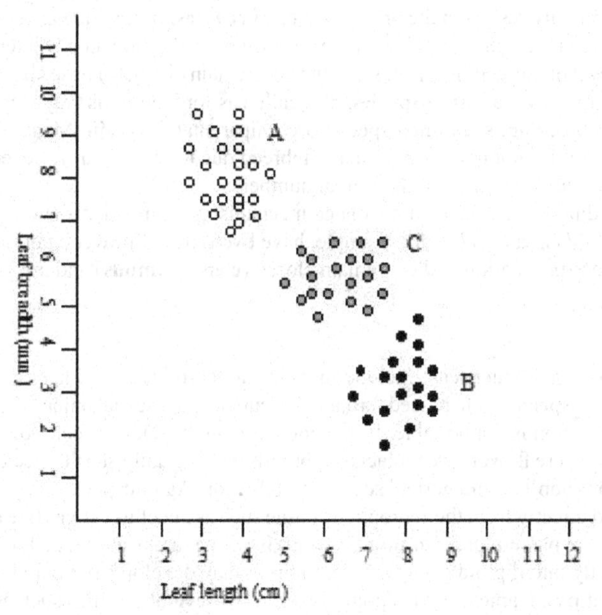

Figure 7.1: Scatter diagram of populations of presumed parental species (A and B) and hybrid population (C). Latter shows structural intermediacy. More characters can be added and depicted by appropriate symbols.

Artificial synthesis: Hybrids can often be created artificially through breeding techniques. The comparison of these artificial hybrids with suspected natural hybrids can help in confirming their identity.

Stabilization of Hybrids

Hybrids generally tend to obscure distinction between parental species, due to intermediacy, segregation and consequent character combinations. The hybrids, however, often establish themselves as distinct taxa through several methods. Commonest of these is bypass sexual processes and perpetuate by asexually means such as **vegetative propagation** and **agamospermy**. The hybrid may similarly become established sexually by hybridization followed by duplication of chromosomes (**Amphiploidy**), a phenomenon common in several genera, such as *Senecio* and *Tragopogon*. The hybrids may also establish through **translocation heterozygosity,** wherein multivalent rings of chromosomes are formed at meiosis, as seen in *Oenothera*. In other cases, **unbalanced polyploidy**, wherein female parent contributes greater number of chromosome sets, as compared to male parent. In genus *Rosa*, for example hybrids are often established because female parent contributes four sets of chromosomes, and male parent only one set.

In some genera such as *Quercus*, there is frequent hybridization between closely related species, resulting in the production hybrid swarms. Such sets of hybridizing species constitute a **syngameon**, or **semispecies**.

Outbreeders with Internal Barriers

Several genera are reported to include species or species complexes, which include races which are not morphologically very distinct, but are unable to interbreed, owing to differences in chromosome number or structure (**structural hybridity**), in others the chromosomal differences are not clear (**cryptic structural hybridity**). Such intersterile races are often known as **semi-cryptic species**. Intersterile populations with no apparent morphological distinction are known as **cryptic** or **sibling species**.

Inbreeding

Also known as selfing, involves union of gametes from the same plant. It may either occur within the same flower (**autogamy**), or between different flowers of the same plant (**geitonogamy**). Although it ensures reproduction, even when there are fewer individuals, or pollinators are not available. It, however, reduced variation in populations and may even result in accumulation of **deleterious alleles**, a phenomenon known as **inbreeding depression**. Inbreeding species tend to exist as relatively uniform populations, often differing considerably from one another, because of absence of gene flow between them. This commonly results in the production of **pure lines**. In several genera of flowering plants, distinct inbreeding populations have been recognized as distinct species. Although they are mostly interfertile, but very low level of outbreeding, and very high level of inbreeding ensures that the taxa remain distinct. Those taxa with very minor differences, but reproductively isolated, are known as **microspecies**, also called as Jordanons, as they were first recognized by Jordon (1873). Being the result of inbreeding, such microspecies are uniparental in origin. Many microspecies are recognized with *Andropogon virginicus* species complex, where many inbreed due to cleistogamy. Several authors, however, avoid recognizing them as distinct species, because of their great numbers.

Outbreeding and inbreeding are, however, not isolated mechanisms. Some plants show both, a phenomenon known as **allautogamy**. Species of *Viola* and *Clarkia*, for example, have two types of flowers: normally open (**chasmogamous**) flowers which experience cross-pollination. Others remain closed (**cleistogamous**) and are self-pollinated.

Apomixis

The phenomenon of apomixis in a broader sense includes non-sexual reproduction, either through vegetative propagation (**vegetative apomixis**) or **agamospermy**, where seed formation occurs without sexual union. Vegetative apomixis is common in plants where sexual reproduction is not possible. It is encountered in dioecious species of *Elodea* with flowers of only one sex, some species of *Poa* where flowers are replaced by bulbils, and sexually sterile species of *Potentilla*, *Mentha* and *Circaea*, where genetic reasons don't permit normal sexual reproduction. Agamospermy may be manifested in a variety of ways. Embryo may be formed directly from the sporophytic tissue such as nucellus (**adventive embryony**), or from diploid gametophyte where meiosis is bypassed (**gametophytic apomixis**), either archesporial cells (**diplospory**) or somatic cells (**apospory**) developing directly into diploid gametophyte. Embryo may develop from unreduced egg (**parthenogenesis**) or from a somatic cell (**apogamy**). Gametophytic apomixis occurs in several families such as Rosaceae and Asteraceae. Although male parent does not contribute towards embryonic tissue, nevertheless pollination is necessary because one of the male nucleus has to fuse with female nucleus to produce endosperm, the phenomenon known as **pseudogamy**. Apogamy is prevalent in ferns.

The populations produced by agamospermy may often show smaller differences, because any genetic mutation is preserved in population, and as these are stable through generations, they are usually recognized as distinct taxa, often as **apomictic microspecies (agamospecies)**. Such microspecies are more stable than those produced through inbreeding (Jardanons). Agamospecies with better dispersal mechanisms, as in *Taraxacum* and *Hieracium*, are widely spread, whereas those with poorly developed dispersal mechanisms as in *Ranunculus* are narrowly distributed.

POPULATION GENETICS

Population genetics deals with the application of genetic principles to populations of a particular species. A population constitutes a group of individuals growing in a specific area and freely interbreeding. A group of interbreeding individuals who share a common set of genes constitute a **Mendelian population**. The widely distributed species often have separate populations in different geographical regions, known as **subpopulations**. Each subpopulation growing in a particular geographical area constitutes a **local population**. The entire set of genetic information covering all alleles in a population, forms its **gene pool**. The evolutionary process is best understood by studying the gene pool of Mendelian population and not the individual members.

Allele Frequencies

While analyzing genetic data, it is more logical to talk in terms of **genotype frequency** and **allele frequency**, instead of absolute numbers. Thus, in a population with alleles *A* and *a*, and 100 *aa* individuals, 300 heterozygous *Aa* individuals and 600 homozygous *AA* individuals, the genotypic frequencies are calculated as:

$AA = 600/1000 = 0.6$
$Aa = 300/1000 = 0.3$
$aa = 100/1000 = 0.1$

It must be noted *AA* genotype has 600 + 600 = 1200 *A* alleles, *Aa* genotype 300 *A* and 300 *a* alleles, and *aa* genotype 100 + 100 = 200 *a* alleles. This totals 1500 *A* alleles and 500 *a* alleles. Allele frequency as such would be calculated as:

$A = 1500/2000 = 0.75$
$a = 500/2000 = 0.25$

Please note, lower the allele frequency, rarer it is. Once the value touches 0, the allele is **lost**, and the other allele with value of 1 would get **fixed**.

It is more appropriate to analyze populations in terms of alleles, and not genotypes, because genotypes are disrupted during the process of segregation in subsequent generations.

Mating Systems

Three types of mating patterns are recognized, which determine the genotype frequencies of populations. In **random mating**, the two genotypes mate in proportion to their relative frequencies in the population. A population may undergo random mating with respect to some traits, but nonrandom with respect to others at the same time. **Assortative mating** is a type of nonrandom mating in which the mates are formed based on their degree of similarity in phenotype. In human population **positive assortative mating** is more common, as mating partners are more similar in phenotype, for example skin color. In several species of plants **negative assortative mating** is prevalent, the mating partners differing in phenotype. In *Primula officinalis*, for example **pin type** flowers (long style and short stamens) produce pollen lower down in flower but receive pollen higher up, whereas the thrum type flowers (long stamens short style) receive pollen lower down and produce higher up. Consequently, the insect pollinators that work deeper into flower collect pollen from pin types flowers and deposit on stigma of thrum type flowers. Pollinators working higher up (due to shorter mouth parts) do the reverse, collecting pollen from thrum type flowers and depositing on pin type flowers.

The third type of mating involves inbreeding, mating between relatives, and in bisexual flowers, generally between gametes of the same flower.

Hardy-Weinberg Law

Most species of animals and plants, except inbreeding plants, random mating is prevalent form of reproduction. Thus, each type of mating pair is formed as often as would be expected by chance encounter between the genotypes. In a randomly mating population with genotype consisting of alleles *A* and *a,* with allele frequency of *p* and *q*, respectively (note $p + q = 1$), genotypes formed from fusion of *A* and *a* gametes of either parent would be AA, Aa, Aa and aa. In terms of allele frequencies, the genotypes could be written as:

AA: p2 AA: 2pq aa: q2

The foundations for these calculations—a landmark contribution in population genetics—were laid by Godfrey Hardy and Wilhelm Weinberg, independently in 1908. The law is based on assumptions that in a infinitely large, randomly mating population, free from mutation, migration and natural selection (five assumptions) the frequencies of the alleles don't change over time. The law also concluded that as long as mating is random, the genotype frequencies remain in the proportion of *p2*, *2pq* and *q2*. The sum of genotype frequencies equals 1, i.e., $p2 + 2pq + q2 = 1$. The ***allelic frequencies remain constant from generation to generation***, in such randomly mating populations.

Although it is difficult for a population to be infinitely large in size, but a fairy large population satisfies the requirement. If the size of the population is limited, chance deviations from the expected rates can result in changes in allelic frequency, a phenomenon known as **genetic drift**. It must also be remembered, however, that ***random mating does not always mean that populations must be interbreeding randomly for all the traits*** for the law to hold true. In human populations, for example, where marriages are chosen on the basis of religion, cast and color, the mating partners do not select blood groups or other such traits, which may thus satisfy the Hardy Weinberg law. The law thus applies to any locus for which random mating occurs, even if mating is nonrandom for other loci (traits).

For Hardy-Weinberg law to apply, the populations must be free from mutation, migration and natural selection. It is important to note that the condition of no evolutionary change applies only to the trait (locus) in question. A population may be subject to evolutionary processes acting on some genes, while still meeting the Hardy-Weinberg assumptions for other loci. The populations which satisfy the requirements of the law are said to be in genetic equilibrium or **Hardy-Weinberg equilibrium**. If the observed genotype proportions are different from the expected, one or more assumptions of the law have been violated.

Null Hypothesis and Chi-Square Test

A population with random mating should represent the progeny with numbers of various phenotypes closer to the expected numbers. For any population analysis it is important to determine, whether the observed values match (fit) the expected values or not. **Null hypothesis** states that there is no real difference between the observed and the predicted data. If the statistical analysis shows that the difference between predicted and observed values is due to chance, the null hypothesis is proved, if not, it is rejected. Although we could never expect a perfect match, the calculation of **goodness of fit**, gives us an indication of the departure from the expected values. It helps us to conclude whether the departure is within the probability limits, or due to some other phenomenon operating on the population. Goodness of fit is conventionally measured in terms of **chi-square**. It is calculated as follows:

$$c^2 = \frac{\sum (\text{Observed value} - \text{Expected value})^2}{\text{Expected value}}$$

Supposing phenotype X represents a population heterozygous for one pair of alleles, and Y homozygous recessive. Mating between the two is expected to produce progeny with the two phenotypes in the ration of 1:1 (X = A + a, Y = a + a; phenotypes Aa, Aa, aa, aa; X = 2, Y = 2). Supposing there is a progeny of 30 individuals with 18 of X phenotype and 12 of Y phenotype, whereas the expected number for each phenotype is 15.

$$c^2 = \frac{(18-15)^2}{15} + \frac{(12-15)^2}{15} = \frac{9}{15} + \frac{9}{15} = 12$$

Closer the observed values are to the expected, lower the value of c^2. If the two match perfectly the values would be zero, although it never happens in nature. Once the c^2 value has been calculated, the goodness of fit of this value to the expected numbers is determined. Two parameters are essential for this interpretation. First the number of **degrees of freedom** for a particular c^2 test is calculated, as number of classes of data minus one. In our case there are two classes of data (phenotypes X and Y), hence one degree of freedom. The second parameter is the **probability** *p*, which can be determined from graph of the Chi-square test. The graph shows range of c^2 values along the X axis, probability along the Y axis and curves of different degrees of freedom running from base towards the right top. The vertical line starting from the specific c^2 value is located at the point it touches the relevant curve, and a horizontal line from this point towards the left touching the Y axis determines the probability values. The probability can also be read from the statistical table for Chi-square (Table 7.2). A value of *p* between 0.01 (1 per cent; worse fit; one in 100 studies, this value would appear by chance) and 0.05 (1 percent; bad fit; one in 20 studies, this value would appear by chance) is considered significant. A value lower than 0.01 is considered highly significant and the hypothesis is rejected outright. It is, however, safer to reject any value lower than 0.05. A value of *p* higher than 0.05 shows that the departure of observed values from expected values is not significant, and the hypothesis is supported.

It is important to note that Chi-square value is calculated on the basis of actual numbers and not on the basis of percentages or ratios of various phenotypes.

Table 7.2: Chi-square table showing the relationship between Chi-square values, degrees of freedom and the probability. For a particular degree of freedom, the nearest Chi-square value is located from the row, the the appropriate probability value read from the top row. Probability values lower than 0.05 are highly significant and do not support the hypothesis being tested. The values higher than 0.05 support the hypothesis.

Degrees of freedom	Probability									
	0.95	0.90	0.70	0.50	0.30	0.20	0.10	0.05	0.01	0.001
	Chi-square values									
1	0.004	0.016	0.15	0.46	1.07	1.64	2.71	3.84	6.64	10.83
2	0.10	0.21	0.71	1.39	2.41	3.22	4.61	5.99	9.21	13.82
3	0.35	0.58	1.42	2.37	3.67	4.64	6.25	7.82	11.35	16.27
4	0.71	1.06	2.20	3.36	4.88	5.99	7.78	9.49	13.28	18.47
5	1.15	1.61	3.00	4.35	6.06	7.29	9.24	11.07	15.09	20.52
6	1.64	2.20	3.83	5.35	7.23	8.56	10.65	12.59	16.81	22.46
7	2.17	2.83	4.67	6.35	8.38	9.80	12.02	14.07	18.48	24.32
8	2.73	3.49	5.53	7.34	9.52	11.03	13.36	15.51	20.09	26.13
9	3.33	4.17	6.39	8.34	10.66	12.24	14.68	16.92	21.67	27.88
10	3.94	4.87	7.27	9.34	11.78	13.44	15.99	18.31	23.21	29.59
15	7.26	8.55.	11.72	14.34	17.32	19.31	22.31	25.00	30.58	37.70
20	10.85	12.44	16.27	19.34	22.78	25.04	28.41	31.41	37.57	45.32
30	18.49	20.60	25.51	29.34	33.53	36.25	40.26	43.77	50.89	59.70
50	34.76	37.69	44.31	49.34	54.72	58.16	63.17	67.51	76.15	86.66

Extension of Hardy-Weinberg Law

Hardy-Weinberg Law may also be extended to situations such as **multiple alleles** and **sex-linked alleles**. Consider a population with three alleles A, B and C with allele frequencies of p, q and r respectively. The frequencies of various genotypes would be represented as:

$$(p + q + r)^2 = p^2\ (AA) + 2pq\ (AB) + q^2\ (BB) + 2pr\ (AC) + 2qr\ (BC) + r^2\ (CC)$$

In a population with allele frequencies of p = 0.52, q = 0.31 and r = 0.17, the following genotypes are frequencies are expected if the population is in Hardy-Weinberg equilibrium:

$AA = p^2 = (0.52)^2 = 0.27$

$AB = 2pq = 2(0.52 \times 0.31) = 0.32$

$BB = q^2 = (0.31)^2 = 0.10$

$AC = 2pr = 2(0.52 \times 0.17) = 0.18$

$BC = 2qr = 2(0.31 \times 0.17) = 0.11$

$CC = r^2 = (0.17)^2 = 0.03$

The law can similarly be applied to sex-linked alleles. In human populations, for example males are XY and females XX. For X-linked alleles the female genotypes show normal Hardy-Weinberg distribution, whereas the male genotypes are distributed in the same frequencies as respective alleles

Female = XX = $p^2(X^AX^A)$; $2pq\ (X^AX^a)$; $q^2(X^aX^a)$

Male = XY = $p(X^AY)$; $q(X^aY)$

This is the obvious reason for the prevalence of X-linked recessive traits such as color blindness and haemophilia among human males.

Inbreeding and the Hardy-Weinberg Law

Inbreeding constitutes another important departure from random mating. It is often measured in terms of **coefficient of inbreeding** (F), which can be calculated as:

$$F = \frac{\text{(Observed heterozygosity} - \text{Expected heterozygosity)}}{\text{Expected heterozygosity}}$$

Greater the value of F, the greater the reduction in heterozygosity relative to that expected from the Hardy-Weinberg expectation. If genotypes are in Hardy-Weinberg proportions, $F = 0$, because observed heterozygosity equals expected one. In self-fertilization, common in plants, however, the decreases with every generation and homozygosity increases consequently. Supposing we start with a completely heterozygous population *Aa* reproducing by self-fertilization. After one generation the progeny will consist of 1/4 *AA*, 1/2 *Aa* and 1/4 *aa.* In next generation homozygotes *AA* will produce only *AA* progeny, aa will produce only *aa* progeny, whereas only heterozygotes will again segregate into half heterozygotes and half homozygotes (1/4 *AA*, 1/4 *aa*). This will reduce heterozygotes to 1/4 of the total population. After large number of generations, there will be no heterozygotes, and the population will be divided into half *AA* and half *aa*.

It should, however, be noted that although genotype frequencies change from one generation to another, the allele frequencies remain constant.

EVOLUTION

Evolution consists of progressive changes in the gene pool, associated with progressive adaptation of a population to its environment. The evolution is the end result of four distinct processes, which result in changes in allelic frequencies, ultimately resulting in cumulative changes in the genetic characteristics of populations. It is believed that these processes over a geological time lead to the evolution of species.

Mutation

It is now agreed that variation in heritable traits results from mutations. Mutation is an important process in evolution. It was earlier believed that variations result mostly from adaptive inheritable change induced by environment. This **adaptation theory**, based on **Lamarckism,** believed in the inheritance of acquired characteristics. The middle of twentieth Century saw the emergence of **mutation theory**, to explain changes in several bacterial populations. Although some mutations occur only in the somatic cells, and not passed to the next generation (**somatic mutation**), others occurring in germ cells are transmitted from one generation to another (**genetic mutation**; **germ-line mutation**). A mutation may involve changes within the same base type (purine to purine; pyrimidine to pyrimidine), when its is known as **transition mutation**; in others it may involve changes from a purine to a pyrimidine or vice versa, when it is termed **transversion mutation**. A mutation in which base pair change in DNA causes a change in mRNA codon so that a different amino acid is inserted into the polypeptide chain constitutes **missense mutation**. On the other hand, a change that results in mRNA codon for an amino acid to stop is known as **nonsense mutation**.

Most of the newly arising mutations are harmful to the organism and are eliminated from the population in successive generation. Some mutations may result in amino acid changes that cause detectable change in structure or function of the organism. Such **neutral mutations** also do not participate in evolution. Similarly, a base pair may change mRNA codon that inserts the same amino acid, when it is known as **silent mutation**. Mutations may be **irreversible** or **reversible**. Change from *A* to *a* constitutes **forward mutation**, whereas from *a* to *A* as **reverse mutation**. In reversible mutations, forward mutations are offset by reverse mutations and as such an equilibrium in allele frequencies is reached.

Migration

Migration is similar to mutation in that new alleles are introduced into the local population, although the new alleles are derived from another population, and not from mutation within the same population. In populations with no migrations, the genetic changes cause considerable differentiation in subpopulation of a population. This genetic differentiation among subpopulations gets minimized when exchange of individuals through migration occurs. Only a small amount of migration is sufficient to prevent the accumulation of high level of genetic differentiation. However, despite migration, the genetic differentiation may continue if other evolutionary forces, such as natural selection for adaptation to local environment, are operating.

Random Genetic Drift

Although population is supposed to be infinite in size as per Hardy-Weinberg Law, in practice they are finite or limited in size, although large enough so that chance factors have little effects on allelic frequencies. Some populations, however, may be small and the chance factors may produce large changes in allelic frequencies. Random change in allelic frequency due to chance is called **random genetic drift** or simply **genetic drift**. Ronald Fisher and Sewall Wright, who laid the foundations of population genetics, were the first to describe how genetic drift affects the evolution of populations.

Imagine a small population with equal number of individuals with either of the two biallelic traits. In a population of twenty individuals, say 15 carry homozygous dominant genotype, 5 heterozygous genotype and 10 homozygous recessive genotype the frequency of a allele would be 5 + 20 = 25, i.e., 0.62. Supposing all the individuals carrying the dominant allele (*AA* or *Aa* genotypes) perish, the population will be left with only *a* allele, and hence allelic frequency of *a* equals 1. The genetic drift, and consequently evolution has occurred in the population due to chance.

In addition to random factors such as floods, fires or cyclones which cause genetic drift due to mortality, the genetic drift may also result from other causes. In small populations, although large number of gametes are produced, only a few will participate in fertilization and represented in the zygotes of next generation. This process of **random sampling (sampling error)** of gametes from generation to generation, may result in changes in allele frequencies by chance, and consequent genetic drift.

Genetic drift may cause several evolutionary changes in addition to changes in allelic frequencies. As different populations may experience genetic drift in different directions, and as a result, the populations diverge in allelic frequencies. Also, we expect more genetic drift in smaller populations as compared to larger populations.

Mutation versus Random Genetic Drift

Mutation continues to introduce new genetic changes in populations, whereas random genetic drift removes certain traits from the population. How does the balance between two opposite forces is achieved? **Infinite alleles model** attempts to explain this balance. According to this model, each mutation in a gene is assumed to generate a novel allele that has never been seen before. The model also assumes that random genetic drift occurs by the repeated random sampling, as described earlier. In this situation the mutation and genetic drift balance each other, and an equilibrium is reached. In this state of balance, some alleles may bounce a bit, but do not stray too much from steady state.

Natural Selection

Natural selection is the driving force of adaptive evolution. The theory of natural selection was first developed independently by Charles Darwin and Alfred Russel Wallace and presented at the Linnean Society of London in 1858. Darwin further pursued this theory and published in his famous book, *The Origin of Species* (1859). The theory has undergone considerable refinement through the incorporation of genetic concepts in the succeeding years. The theory is based on two premises. First—In all organisms, more offspring are produced than survive and reproduce (originally proposed by Thomas Malthus); Second—Organisms differ in their ability to survive and reproduce, and some of these differences are due to genotype. It is deduced that in every generation, the genotypes that promote survival in the prevailing environment (**favored genotypes**) must be present in excess among individuals of reproductive age, and hence the favored genotypes will contribute disproportionally to the offsprings of the next generation. As a result of this process, the alleles that enhance survival and reproduction increase in frequency from generation to generation, and the population becomes progressively better equipped to survive and reproduce in the prevailing environment (**Survival of the fittest**). This progressive genetic improvement constitutes the process of **evolutionary adaptation**.

Darwinian Fitness and Fitness Coefficient

For natural selection to operate, survival of the fittest is not the only criteria, the population should be able to reproduce and pass on the fit genes to the next generation. The relative reproductive ability of a genotypes is represented as **Darwinian fitness**. It is represented by W, and when comparing populations, the one able to produce most offspring is given value of 1. Other populations are assigned value relative to this. Supposing out of the four populations A produces 15 offspring, B 10, C 8 and D 5 offspring. The population A will be assigned W value of 1, B 10/15 = 0.66, C 8/15 = 0.53, and D 5/15 = 0.33.

It is, however, important to consider the number of offspring surviving, rather than number produced. Darwinian fitness can also be used to calculate **Selection coefficient**, which is a measure of relative intensity of selection against a genotype, or selective disadvantage of a disfavored genotype. It is symbolized as s and calculated as:

$$s = 1 - W$$

Thus, for above populations the selection coefficient would be calculated as: Population A $s = 1 - 1 = 0$, B $1 - 0.66 = 0.44$, C $1 - 0.53 = 0.47$, D $1 - 0.33 = 0.67$. Thus, Population A has zero selective disadvantage, B 44%, C 47%, and D 67% selective disadvantage.

Whereas the estimation of fitness is relatively easier with microorganisms with shorter life span, the same in higher organisms may take several years of study. In such organisms it is convenient to divide fitness into its component parts and estimate these parts separately. The commonly used components include: **viability**, the probability that the zygote of a genotypes survives up to reproductive stage, and **fertility**, the average number of offsprings produced by a genotype during the reproductive period.

Selection-Mutation Balance

The process of natural selection reduces the frequency of harmful alleles in a population. These harmful alleles, however, are never eliminated, since mutations from wild types continuously produce harmful alleles. These two opposite forces maintain an equilibrium within a population. In selection-mutation balance new mutations exactly offset selective eliminations. In populations (with A a alleles) with harmful (lethal) allele being recessive, s equals 1, and the number of a alleles eliminated by selection will be proportional to $q2s$ according to Hardy-Weinberg proportions, whereas the number of new a alleles introduced by mutation will be proportional to m. At equilibrium the two should balance and $\hat{q} = \sqrt{\frac{\mu}{s}}$ as such $q2s$ = m where q stands for **equilibrium value**. In those cases where the harmful allele shows partial dominance in having small detrimental effect on fitness of the heterozygous carriers then the fitness of genotypes AA, Aa and aa can be written as $1 : 1 - hs : 1 - s$, where hs stands for selection coefficient against heterozygous carriers, h refers to the **degree of dominance**. When $h = 1$, the harmful allele is completely dominant, but when $h = 0$ it is completely recessive, and when $h = 1/2$ the fitness of heterozygous genotype is average of homozygous genotypes. For most harmful alleles that show partial dominance, the value of h is smaller than 1/2.

Heterozygote Superiority

Whereas in most cases of selection, the fitness of heterozygote is intermediate between dominant and recessive genotypes, or equaling one of them, some cases of selection lead to superior heterozygote. The phenomenon is known as

overdominance, heterozygote superiority or **heterosis**. An equilibrium of allelic frequencies is maintained, because both alleles are favored in heterozygous state. Selection will lead to changes in allelic frequencies, but once the equilibrium is reached, the frequencies will stabilize. With overdominance the fitness of genotypes *AA*, *Aa* and *aa* may be written as $1 - s : 1 : 1 - t$, where s and t are the selection coefficients against *AA* and *aa*, respectively. The proportion of A alleles eliminated by selection in event of random mating is $p2s/p = ps$, and the proportion of a alleles eliminated by selection is $q2t/q = qt$. When the equilibrium is reached and the selective elimination of the two alleles is balanced.

$$\hat{p}s = \hat{q}t$$

$$\hat{p} = \frac{t}{s+t}$$

With overdominance of one allele, the allele frequencies ultimately reach equilibrium, but the rate of approach depends on magnitudes of s and t. The equilibrium is reached much faster, when there is a strong selection against homozygotes.

Molecular Evolution

Molecular evolution involves changes in populations at the level of DNA and protein sequences. It attempts to correlate these changes with evolution of new genes and organisms. Whereas the population genetics is concerned with changes in gene frequencies from generation to generation, the molecular evolution considers much longer time frames, associated with speciation.

The field of molecular evolution is multidisciplinary, involving data from genetics, ecology, evolutionary biology, statistics and even computer science. The analysis of molecular data can help in unravelling the historical records preserved in genomes and identifying the dynamics behind evolutionary processes to understand and reconstruct the chronology of change. Molecular evolution mostly operates by substitutions that lead to the change of codon for one amino acid to another. Leucine codon CUU, for example can be changed to isoleucine codon AUU by change in single base pair of DNA, where the change to codon for chemically dissimilar asparagine (AAU), two base pair changes must occur. The methods of analyzing DNA sequence data, and its utilization in phylogenetic analysis is discussed in detail in Chapter 4 under **Molecular Systematics**.

Neutral Theory of Evolution

Molecular studies over the last few decades have also seen emergence of alternate views concerning the evolution of species. Kimura (1968) proposed **neutral theory of evolution**, according to which most genetic variation observed in natural populations due to accumulation of **neutral mutations**. Whereas the **nonneutral mutation** affects the phenotype of the organism and can be acted upon by **natural selection**, the neutral mutation does not affect the phenotype of the organism and not acted on by natural selection. Since neutral mutations do not affect phenotype, they spread throughout a population according to their frequency of appearance and to genetic drift. This **non-Darwinian evolution** has been called as **survival of the luckiest** as opposed to **survival of the fittest** as advocated by Darwinian theory. Although Kimura agreed with Darwin that natural selection is responsible for adaptive changes in species during evolution, but he stressed that modern variation in gene sequences is explained by neutral variation rather than adaptive variation.

In further elaboration of their theory, Kimura et al. (1974) developed five principles that govern the evolution of genes at molecular level:

1. For each protein, the rate of evolution, in terms of amino acid substitutions, is approximately constant with regard to neutral substitutions that do not affect protein structure or function.
2. Proteins that are functionally less important for the survival of an organism, or parts of a protein that are less important for its function, tend to evolve faster than more important proteins or regions of a protein. In other words, during evolution, less important proteins will accumulate amino acid substitutions more rapidly than important proteins.
3. Those amino acid substitutions that do not disrupt the existing structure and function of a protein (**conservative substitutions**) occur more frequently in evolution than those which disrupt (**disruptive substitutions**) existing structure and function.
4. Gene duplication must always precede the emergence of a gene having a new function.
5. Selective elimination of definitively deleterious mutations and random fixing of selectively neutral or very slightly deleterious alleles occur far more frequently in evolution than Darwinian selection of definitely advantageous mutants.

Although, the DNA sequencing of hundreds of thousands of different genes from thousands of species has provided compelling support for the five principles, there are, however, some geneticists called **selectionists**, who oppose neutralist theory. It is, however, agreed by all that genetic drift and natural selection both play key roles in evolution.

Speciation

Speciation is a general term for a number of different processes which involve the production of new species during course of evolution. It involves splitting of single evolutionary lineage into two or more genetically independent lineages with an array of observable physical characteristics (Phenotypic differentiation). The speciation commonly results from the development of barriers to gene flow. Different types are isolating mechanisms are responsible for the development of barriers.

Isolating Mechanisms

Isolation is the key factor preventing intermixing of distinct species through prevention of hybridization. Based on whether isolating mechanisms operate before or after sexual fusion, two main types of mechanisms are distinguished: **prezygotic mechanisms** and **postzygotic mechanisms**. A detailed classification of isolating mechanisms is presented below.

A. Prezygotic mechanisms

(operating before sexual fusion)

I. Pre-pollination mechanisms

1. ***Geographical isolation***: Two species are separated geographically by a gap larger than their pollen and seed dispersal. *Platanus orientalis* (Mediterranean region) and *P. occidentalis* (North America) are well established species but readily interbreed when brought into the same area (vicarious species).
2. ***Ecological isolation***: Two species occupy the same general area but occupy different habitats. *Silene alba* grows on light soils in open places while *S. dioica* on heavy soils in shade. Their habitats rarely overlap, but when they do, hybrids are encountered.
3. ***Seasonal isolation***: Two species occur in the same region but flower at different seasons. *Sambucus racemosa* and *S. nigra* flower nearly 7 weeks apart.
4. ***Temporal isolation***: Two species flower during the same period but during different times of the day. *Agrostis tenuis* flowers in the afternoon, whereas *A. stolonifera* flowers in the morning.
5. ***Ethological isolation***: Two species are interfertile but have different pollinators. Humming-birds for example, are attracted to red flowers and hawk-moths to white ones.
6. ***Mechanical isolation***: Pollination between two related species is prevented by structural differences between flowers, as for example between *Ophrys insectifera* and *O. Apifera*.

II. Post-pollination mechanisms

1. ***Gametophytic isolation***: This is the commonest isolating mechanism wherein cross-pollination occurs, but the pollen tube fails to germinate or if germinated, it can't reach and penetrate the embryo sac.
2. ***Gametic isolation***: In such cases, reported in several crop plants, the pollen tube releases the male gametes into the embryo sac, but gametic and/or endospermic fusion does not occur.

B. Postzygotic mechanisms

(operating after sexual fusion)

1. ***Seed incompatibility***: The zygote or even immature embryo is formed but fails to develop and as such a mature seed is not formed. The phenomenon is commonly encountered in cross between *Primula elatior* and *P. veris*.
2. ***Hybrid inviability***: Mature seed is formed and manages to germinate but the F1 hybrid dies before the flowering stage is reached. The phenomenon is commonly encountered in crosses between *Papaver dubium* and *P. rhoeas*.
3. ***F1 hybrid sterility***: F1 hybrids are fully viable and reach flowering stage but flowers may abort or abortion may occur as late as F2 embryo formation, with the result that the F1 hybrid fails to produce viable seeds.
4. ***F2 hybrid inviability or sterility***: F2 hybrid dies much before reaching the flowering stage or fails to produce seeds.

Once prezygotic isolation is partially achieved, there is a snowball effect in which the rate of accelerates. Individuals who engage in interspecific mating suffer an increasing disadvantage until at last the barrier of gene flow is complete. New species may develop through the mechanism of **abrupt speciation** or **gradual speciation**.

Abrupt Speciation

The phenomenon of abrupt speciation is commonly involves sympatric populations of two different species and as such is also known as **sympatric speciation**. It is commonly the result of a sudden change in chromosome number or structure, producing instantly an almost irreversible barrier between populations and thus effectively isolating them. Phenomenon is met in genera such as *Tragopogon, Bromus* and *Senecio* (see examples under Chromosomes in chapter on Taxonomic evidence). The species are often well isolated, and any chance hybridization fails to culminate into successful hybrids because of genomic differences. In some cases, however, hybridization may be accompanied by chromosome duplication resulting in the formation of allopolyploids. Such allopolyploids depict normal pairing at meiosis and thus represent well-isolated, phenotypically as well as genotypically distinct species. Classical example hybridization followed by polyploidy is provided by evolution of bread wheat *Triticum aestivum*, although it has taken a period of several thousand years starting from Neolithic times. The two diploid species *Triticum monococcum* (AA) and *Aegilops speltoides* (BB) were involved in the evolution of tetraploid *T. dicoccum* (AABB). Subsequent hybridization between the latter and *Aegilops tauschii* (Syn: *A. squarrosa*) (DD) and subsequent duplication resulted in the evolution of hexaploid *T. aestivum* (AABBDD). Abrupt speciation also results from the phenomenon of apomixes, and an increasing number of species are being formed in genera such as *Taraxacum, Euphrasia* and *Alchemilla*, where the normal reproductive process is bypassed, and any expressive mutations are retained within the population.

Gradual Speciation

This is a more common phenomenon in nature. It may involve **phyletic evolution** when one species might evolve into something different from its ancestor over a period of time (**phyletic speciation**). Alternatively, a population belonging to a single species might differentiate into two evolutionary lines through **divergent evolution** (**additive speciation**).

Phyletic Speciation

The concept of phyletic speciation has been the subject of considerable debate. It is the sequential production of species within a single evolutionary lineage. Species A might, over a period of time, change through species B and C into species D without ever splitting. The new species produced in this manner are variously called **successional species**, **palaeospecies**, and **allochronic species**. The species which have become extinct in the process are termed **taxonomic extinctions**. Wiley (1981), while agreeing with the concept of phyletic character transformation, rejects the concept of phyletic speciation on the grounds that:

1. Recognition of phyletic species is an arbitrary practice. Mayr (1942) argues that delimitation of species, which do not belong to the same time-scale, is difficult.
2. Arbitrary species result in arbitrary speciation mechanisms.
3. Phyletic speciation has never been satisfactorily demonstrated.

Additive Speciation

Additive speciation is the commonest mode of speciation, which adds to the diversity of living organisms. Mayr (1963) suggested the occurrence of **reductive speciation**, whereby two previously independent species fuse into a third, new species, themselves becoming extinct. Hybridization likewise produces new species, but this always leads to an addition in the number of species.

It is impossible to imagine that two evolutionary species can fuse to produce a third species and themselves become extinct. This may happen in a specific region, but not over the entire range of these species. The various modes of additive speciation are described below:

1. **Allopatric speciation:** Lineage independence and consequent speciation result from geographical separation of lineages, i.e., the actual physical separation of two relatively large populations of a single species by barriers such as mountain range, desert and river. Over a period of time, such separation would enable these geographical races to develop and maintain gene combinations controlling their morphological and physiological characters. The development of reproductive isolation would sooner or later result in the establishment of distinct species (Figure 7.2B). Allopatric speciation may also result from the development of new species along the boundaries of a large central population. These marginal populations (races) get separated from the main population during environmental differentiation. They then undergo adaptive radiations to develop physical and physiological differences, which sooner or later get genetically fixed (ecotypes). With further morphological and physiological differentiation, they form distinct varieties (or subspecies). Development of reproductive isolation establishes these as distinct species that will retain their identity even if a future chance should draw them together (Figure 7.2A).
2. **Allopatric introgressive speciation:** Although origin of species through hybridization is commonly results from sympatric species, examples of speciation involving two allopatric species, which had contacts in the past, are also reported. *Quercus brandegei*, now confined to Cape Region of Baja California, extended to west in Tertiary times and had a narrow zone of contact in Edwards Plateau escarpment area, with *Q. virginiana* of S. E. Coastal plain of U.S.A. Allopatric introgression between the two species occurred at the contact zone, but the genes spread

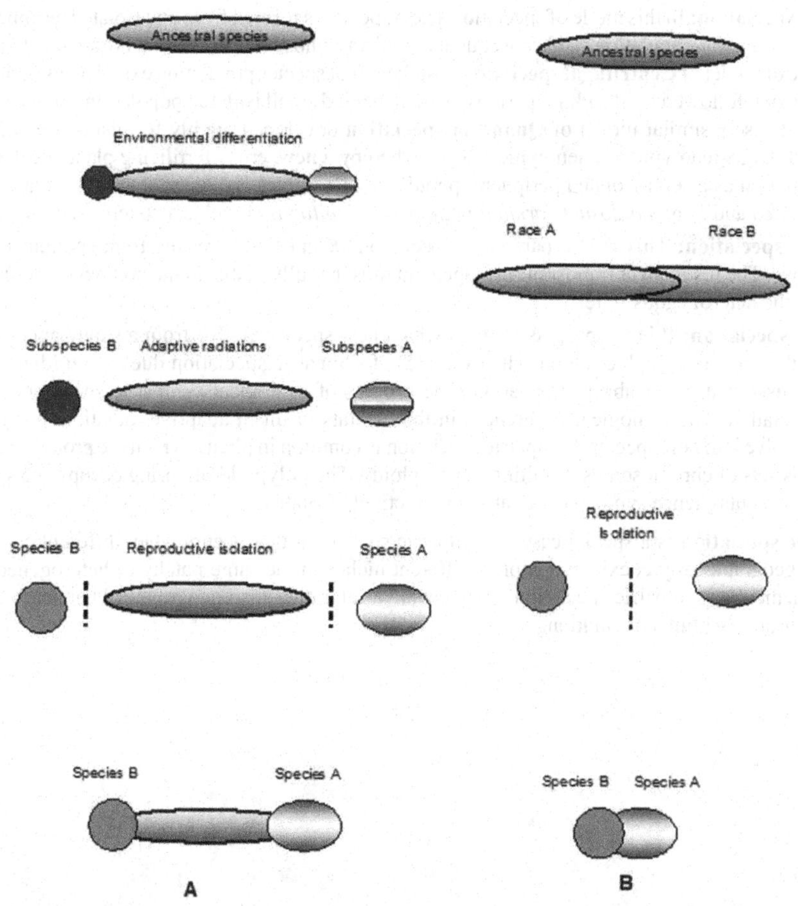

Figure 7.2: Allopatric speciation. A: Allopatric speciation through environmental differentiation, successive adaptive radiations and development of reproductive isolation. **B:** Allopatric speciation resulting from geographical separation of populations of an ancestral species.

slowly because of limited contact of parental species and predominantly rhizomatous propagation. The introgressed population, now in contact with only *Q. virginiana*, is sufficiently stabilized to be classified as a distinct species *Q. fusiformis*.

3. **Allo-parapatric speciation:** Such speciation occurs when two populations of an ancestral species are separated, differentiate to a degree that is not sufficient for lineage independence, and then develop lineage independence during a period of parapatry (limited sympatry). It differs from allopatric speciation in the sense that speciation is completed after a period of sympatry and the process of attaining lineage independence is potentially reversible because it is possible that two partly differentiated populations could form a single evolutionary lineage showing clinal variation after they meet rather than the period of sympatry reinforcing differences between them.
4. **Parapatric speciation:** This occurs when two populations of an ancestral species differentiate despite the fact that no complete disjunction has occurred. The daughter species may share a small fraction of their respective ranges and interbreed within this narrow contact zone and yet still differentiate. This mode of speciation has three distinguishing characteristics: mating occurs non-randomly, gene flow occurs unequally, and populations exist in either continuous or discontinuous geographic ranges. Parapatry is often used to describe the relationship between organisms whose ranges do not significantly overlap but are immediately adjacent to each other; they do not occur together except in a narrow contact zone. Parapatric speciation is extremely rare. It occurs when populations are separated not by a geographical barrier, such as a body of water, but by an extreme change in habitat. While populations in these areas may interbreed, they often develop distinct characteristics and lifestyles. Reproductive isolation in these cases is not geographic but rather temporal or behavioral. For example, plants that live on boundaries between very distinct climates may flower at different times in response to their different environments, making them unable to interbreed.

5. **Peripatric speciation:** In this mode of speciation a new species is formed from and isolated peripheral much smaller population. Peripatric populations and consequent organisms whose ranges are closely adjacent but do not overlap. An alternate of model of **Centrifugal speciation** postulates that species population experiences periods of geographic range expansion followed by shrinking periods, leaving behind small isolated populations on the periphery of main population. Closely similar model of **Quantum speciation** developed mainly for plants (Grant, 1971) is a rapid process with large genotypic or phenotypic effects, whereby a new, cross-fertilizing plant species buds off from a larger population as a semi-isolated peripheral population. Examples are found in closely related pairs of species *Layia discoidea* and *L. glandulosa*, *Clarkia lingulata* and *C. biloba*. Model is also termed as **Budding speciation**.
6. **Stasipatric speciation:** This is like parapatric speciation except that it results from spontaneous chromosomal modifications. The resultant chromosome arrangement must be fully viable in the homozygous state but of reduced viability in the heterozygous state.
7. **Sympatric speciation:** It is the process through which new species evolve from a single ancestral species while inhabiting the same geographic region. The examples of sympatric speciation due to hybridization and apomixis have been discussed under abrupt speciation. The process of ecological sympatric speciation is a slow one of gradual speciation. The ecological differences in the habitats result in adaptive radiations in populations which gradually evolve into new species. Sympatric speciation is common in plants, which are prone to acquiring multiple homologous sets of chromosomes, resulting in polyploidy. The polyploid offspring occupy the same environment as the parent plants (hence sympatry) but are reproductively isolated.

Heteropatric speciation is a special case of sympatric speciation that occurs when different ecotypes or races of the same species geographically coexist but exploit different niches in the same patchy or heterogeneous environment. It is thus is a refinement of sympatric speciation, with a behavioral, rather than geographical barrier to the flow of genes among diverging groups within a population.

Chapter 8

Numerical and Cladistic Methods

Systematics aims at developing classifications based on different criteria and, often a distinct methodology is employed for the analysis of data. Data handling to establish relationships between the organisms often makes use of one of the two methods: **numerical** or **phenetic methods** and **phylogenetic methods**, often providing different types of classification. Distinction is sometimes also made between phylogenetic and evolutionary classification schemes. Phylogenetic methods aim at developing a classification based on an analysis of phylogenetic data, and developing a diagram termed **cladogram** or **phylogenetic tree**, or more recently, simply **tree,** which depicts the genealogical descent of taxa. Biologists practicing this methodology are known as **cladists**, and the field of study as **cladistics**. The term, however, is slowly being replaced by **phylogenetic systematics**. The phylogenetic concepts present a huge diversity of variation, unfortunately often contradictory, leading to different interpretations of similar results. A brief understanding of these is, therefore, necessary before attempting to explore this complex field. Before the development of modern methods of cladistics, the numerical methods were largely used for drawing phylogenetic inferences from the data analysis. The modern Phylogenetic methods, however, integrate the concepts and practices of numerical taxonomy with cladistic methods. It is, however, essential to understand the concepts of each, and the final integration in phylogeny reconstruction.

NUMERICAL (PHENETIC) METHODS

Numerical taxonomy received a great impetus with the development and advancement of computers. This field of study is also known as **mathematical taxonomy** (Jardine and Sibson, 1971), **taxometrics** (Mayr, 1966), **taximetrics** (Rogers, 1963), **multivariate morphometrics** (Blackith and Reyment, 1971) and **phenetics**. The modern methods of numerical taxonomy had their beginning from the contributions of Sneath (1957), Michener and Sokal (1957), and Sokal and Michener (1958) which culminated in the publication of Principles *of Numerical Taxonomy* (Sokal and Sneath, 1963), with an expanded and updated version *Numerical Taxonomy* (Sneath and Sokal, 1973). The latter authors define Numerical taxonomy as ***grouping by numerical methods of taxonomic units into taxa on the basis of their character states***. Before the development of modern methods of cladistics, the numerical methods were also used for drawing phylogenetic inferences from the data analysis.

The last few decades have witnessed an intense debate on the suitability of the **empirical approach** or **operational approach** in systematic studies. Empirical taxonomy forms the classification on the basis of taxonomic judgment based on observation of data and not assumptions. Operational taxonomy, on the other hand, is based on operational methods, experimentation to evaluate the observed data, before a final classification. Numerical taxonomy finds a balance between the two as it is both empirical and operational (Figure 8.1).

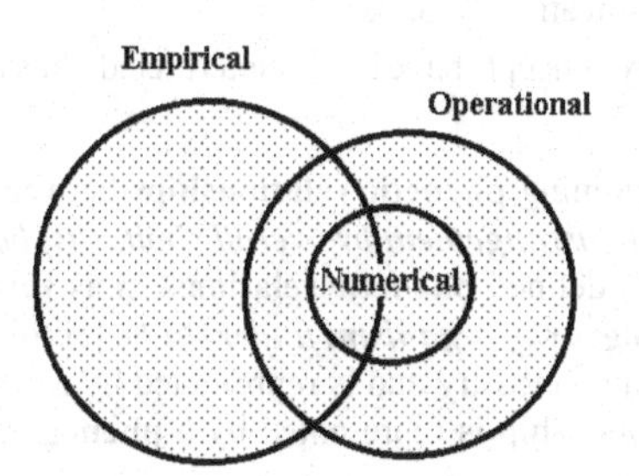

Figure 8.1: Relationship between empirical, operational and numerical taxonomy (after Sneath and Sokal, 1973).

It must be remembered that numerical taxonomy does not produce new data or a new system of classification but is rather a new method of organizing data that could help in better understanding of relationships. Special classifications are based either on one or a few characters or on one set of data. Numerical taxonomy seeks to base classifications on a greater number of characters from many sets of data in an effort to produce an entirely phenetic classification of maximum predictivity.

Principles of Taxometrics

The philosophy of modern methods of numerical taxonomy is based on ideas that were first proposed by the French naturalist Michel Adanson (1763). He rejected the idea of giving more importance to certain characters and believed that natural taxa are based on the concept of similarity, which is measured by taking all the characters into consideration. The principles of modern numerical taxonomy developed by Sneath and Sokal (1973) are based on the modern interpretation of the Adansonian principles and as such are termed **neo-Adansonian principles**. It would, however, be wrong to visualize Adanson as the founder of numerical taxonomy, because he worked in a different academic environment from that of today, when tools of investigation were much different. These **principles** of numerical taxonomy are enumerated below.

1. The greater the content of information in the taxa of a classification and the more characters it is based upon, the better a given classification will be.
2. *A priori*, every character is of equal weight in creating natural taxa.
3. Overall similarity between any two entities is a function of their individual similarities in each of the many characters in which they are being compared.
4. Distinct taxa can be recognized because correlations of characters differ in the groups of organisms under study.
5. Phylogenetic inferences can be made from the taxonomic structures of a group and also from character correlations, given certain assumptions about evolutionary pathways and mechanisms.
6. Taxonomy is viewed and practiced as an empirical science.
7. Classifications are based on phenetic similarity.

The methodology of numerical taxonomy involves the selection of operational units (populations, species, genera, etc., from which the information is collected) and characters. The information from these is recorded, and similarity (and/or distance) between units is determined using various statistical formulae. The ultimate analysis involves comparison of similarity data and constructing diagrams or models, which provide a summary of the data analysis. These diagrams or models are used for final synthesis and better understanding of the relationships. The **major advantages** of numerical taxonomy over conventional taxonomy include:

1. Numerical taxonomy has the power to integrate data from a variety of sources such as morphology, physiology, phytochemistry, embryology, anatomy, palynology, chromosomes, ultrastructure and micromorphology. This is very difficult to do by conventional taxonomy.
2. Considerable automation of the data processing promotes efficiency and the work can be handled by even less skilled workers.
3. Data coded in numerical form can be integrated with existing data-processing systems in various institutions and used for the creation of descriptions, keys, catalogues, maps and other documents.
4. The methods, being quantitative, provide greater discrimination along the spectrum of taxonomic differences, and can provide better classifications and keys.
5. The creation of explicit data tables for numerical taxonomy necessitates the use of more and better described characters, which will necessarily improve conventional taxonomy as well.
6. The application of numerical taxonomy has posed some fresh questions concerning classification and initiated efforts for re-examination of classification systems.
7. Several biological and evolutionary concepts have been reinterpreted, thus introducing renewed interest in biological research.

Numerical taxonomy aims at determining **phenetic relationships** between organisms or taxa. Cain and Harrison (1960) defined phenetic relationship as ***an arrangement by overall similarity, based on all available characters without any weighting***. Sneath and Sokal (1973) defined **phenetic relationship** as ***similarity (resemblance) based on a set of phenotypic characteristics and not phylogeny of organisms under study***. It is distinct from a cladistic relationship, which is an expression of the recency of common ancestry and is represented by a branching network of ancestor-descendant relationships. Whereas the phenetic relationship is represented by a **phenogram**, the cladistic relationship is depicted through a **cladogram**.

CLADISTIC METHODS

Although phylogenetic diagrams (now appropriately known as **phylograms**) have been used by Bessey (1915), Hutchinson (1959, 1973), and contemporary authors of classification systems to depict the relationships between taxa, the cladograms are distinct in the sense that they are developed using a distinct methodology. This method was first proposed by Hennig (1950, 1957), a German zoologist who founded the subject of **phylogenetic systematics**. The term **cladistics** for this methodology was coined by Mayr (1969). An American Botanist, W. H. Wagner, working independently, developed a method of constructing phylogenetic trees, called the groundplan-divergence method, in 1948. Over the years, cladistics has developed into a forceful methodology of developing phylogenetic classifications.

Cladistics is a methodology that attempts to analyze phylogenetic data objectively, in a manner parallel to taxometrics, which analyses phenetic data. Cladistic methods are largely based on the **principle of parsimony** according to which, the most likely evolutionary route is the shortest hypothetical pathway of changes that explains the pattern under observation. Taxa in a truly phylogenetic system should be monophyletic. It has been found that **symplesiomorphy** (possession of primitive or plesiomorphic character-state in common by two or more taxa) does not necessarily indicate monophyly. **Synapomorphy** (possession of derived or apomorphic character-state in common by two or more taxa), on the other hand, is a more reliable indicative of monophyly. It is thus common to use homologous shared and derived character-states for cladistic studies. Before analyzing the methodology of handling data for phylogenetic analysis, it is important to understand the major terms and concepts used in Phylogenetic Systematics.

Phylogenetic Terms

Many important terms have been repeatedly used in discussions on the phylogeny of angiosperms, with diverse interpretation, which has often resulted in different sets of conclusions. A prominent case in point is Melville (1983), who regards the angiosperms as a monophyletic group. His justification—several ancestral forms of the single fossil group Glossopteridae gave rise to angiosperms—renders his view as polyphyletic in the eyes of most authors who believe in the strict application of the concept of monophyly. The involvement of more than one ancestor makes angiosperms a polyphyletic group, a view that has been firmly rejected. A uniform thorough evaluation of these concepts is necessary for proper understanding of angiosperm phylogeny.

Plesiomorphic and Apomorphic Characters

A central point to the determination of the phylogenetic position of a particular group is the number of primitive (**plesiomorphic**) or advanced (**apomorphic**) characters (although the term character is often used broadly in literature, more appropriately primitive or advanced and similarly plesiomorphic and apomorphic refer to different character-states of a character, and not different characters) that the group contains. In the past, most conclusions on primitiveness were based on circular reasoning: 'These families are primitive because they possess primitive characters (or character-states) states, and primitive characters (or character-states) are those which are possessed by these primitive families'. Over the recent years, a better understanding of these concepts has become possible. It is generally accepted that evolution has proceeded at different rates in different groups of plants so that among the present-day organisms, some are more advanced than others. The first step in the determination of relative advancement of characters, is to ascertain which characters are plesiomorphic and which are apomorphic. Stebbins (1950) argued that it is wrong to consider the characters as separate entities, since it is through the summation of characters peculiar to an individual, that natural selection operates. Sporne (1974) while agreeing with this, believed that it is scarcely possible initially to avoid thinking in terms of separate characters, which can be treated better statistically. Given insufficient fossil records of the earliest angiosperms, comparative morphology has been largely used to decide the relative advancement of characters. Many doctrines have been proposed but unfortunately most rely on circular reasoning. Some of the important doctrines are described below:

The **Doctrine of conservative regions** holds that certain regions of plants have been less susceptible to environmental influence than others and, therefore, exhibit primitive features. Unfortunately, however, over the years, every part of the plant has been claimed as conservative region. Also, the assumption that a flower is more conservative than the vegetative parts is derived from classifications which are based on this assumption.

The **doctrine of recapitulation** holds that early phases in development are supposed to exhibit primitive features, i.e., '***ontogeny repeats phylogeny***'. Gunderson (1939) used this theory to establish the following evolutionary trends: polypetaly to gamopetaly (since the petal primordia are initially separate, the tubular portion of the corolla arises later); polysepaly to gamosepaly; actinomorphy to zygomorphy and apocarpy to syncarpy. The concept originally applied to animals does not always hold well in plants where ontogeny does not end with embryogeny but continues throughout the adult life. **Neoteny** (persistence of juvenile features in mature organism) is an example wherein a persistent embryonic form represents an advanced condition.

The **doctrine of teratology** was advocated by Sahni (1925), who argued that when a normal equilibrium is upset, an adjustment is often effected by falling back upon the surer basis of past experience. Thus, teratology (abnormality) is seen as reminiscent of some remote ancestor. According to Heslop-Harrison (1952), some teratological phenomena are just likely to be progressive or retrogressive, and each case must be judged on its own merit.

The **doctrine of sequences** advocates that if organisms are arranged in a series in such a way as to show the gradation of a particular organ or structure, then the two ends of the series represent apomorphy and plesiomorphy. The most crucial decision, however, is from which end should the series be read.

The **doctrine of association** advocates that if one structure has evolved from another, then the primitive condition of the derived one will be similar to the general condition of the ancestral structure. Thus, if vessels have evolved from tracheids, then the vessels similar to tracheids (vessels with longer elements, smaller diameter, greater angularity, thinner walls and oblique end walls) represent a more primitive condition than vessels with broader, shorter, more circular elements with horizontal end walls.

The **doctrine of common ground plan** advocates that characters common to all members of a group must have been possessed by the original ancestor and must, therefore, be primitive. The doctrine, however, cannot be applied to angiosperms in which there is an exception for almost every character.

The **doctrine of character correlation** was acknowledged during the second decade of the previous century when it was realized that certain morphological characters are statistically correlated and the fact can be used in the study of evolution. Sinnot and Bailey (1914) demonstrated a positive correlation between trilacunar node and stipules. Frost (1930) believed that correlation between characters arises because rates of their evolution have been correlated. Sporne (1974) has, however, argued that correlation can be shown to occur even though the rates of evolution of characters are not the same. Within any taxonomic group, primitive characters may be expected to show positive correlation merely because their distribution is not random. By definition, primitive members of that group have retained a relatively high proportion of ancestral (plesiomorphic) characters, while advanced members have dispensed with a relatively high proportion of these same characters—either by loss or replacement with different (apomorphic) characters. It follows, therefore, that the distribution of plesiomorphic characters is displaced towards primitive members, which have a higher proportion of plesiomorphic characters, than the average for the group as a whole. Departure from the random can be statistically calculated in order to establish correlation among characters. Based on these calculations, Sporne (1974) prepared a list of 24 characters in Dicotyledons and 14 in Monocotyledons, which exhibit positive correlation. These characters, because of their distribution, have been categorized as **magnoloid** and **amarylloid**, respectively. Based on the distribution of these characters, Sporne calculated an **advancement index** for each family and projected the placement of different families of angiosperms in the form of a **circular diagram**, with the most primitive families near the center, and the most advanced along the periphery. That the earliest members of angiosperms are extinct is clear from the fact that none of the present-day families has the advancement index of zero. All living families have advanced in some respects.

The concept of apomorphic and plesiomorphic characters in understanding the phylogeny of angiosperms has been considerably advanced with the recent development of **cladistic methods**. These employ a distinct methodology, somewhat similar to taxometric methods in certain steps involved, leading to the construction of **cladograms** depicting evolutionary relationships within a group. Certain groups of angiosperms are reported to have a combination of both plesiomorphic and apomorphic characters, a situation known as **heterobathmy**. *Tetracentron* has primitive vesselless wood but the pollen grains are advanced, being tricolpate.

Homology and Analogy

Different organisms resemble one another in certain characters. Taxonomic groups or taxa are constructed based on overall resemblances. The resemblances due to homology are real, whereas those due to analogy are generally superficial. A real understanding of these terms is, thus, necessary in order to keep organisms with superficial resemblance in separate groups. The two terms as such play a very important role in understanding evolutionary biology.

These terms were first used and defined by Owen (1848). He defined **Homology** as ***the occurrence of the same organ in different animals under every variety of forms and functions.*** He defined **Analogy** as ***the occurrence of a part or an organ in one animal which has the same function as another part or organ in a different animal.*** If applied to plants, the rhizome of ginger, the corm of colocasia, tuber of potato, and runner of lawn grass are all homologous, as they all represent a stem. The tuber of potato and the tuber of sweet potato, on the other hand, are analogous as the latter represents a root.

Darwin (1959) was the first to apply these terms to both animals and plants. He defined homology as ***that relationship between parts which results from their development from corresponding embryonic parts.*** The parts of a flower in different plants are thus homologous and these, in turn, are homologous with leaves because their development is identical.

During the latter half of the present century, phylogenetic interpretation has been applied to these terms. Simpson (1961) defined homology as ***the resemblance due to inheritance from a common ancestry.*** Analogy, similarly, represents ***functional similarity and not due to inheritance from a common ancestry.*** Mayr (1969) similarly defined homology as

the occurrence of similar features in two or more organisms, which can be traced to the same feature in the common ancestor of these organisms. It is, as such, imperative that homology between two organisms can result only from their having evolved from a common ancestor, and the ancestor must also contain the same feature or features for which the two organisms are homologous.

Wiley (1981) has provided a detailed interpretation of these terms. Homology may either be between two characters, two character states, or between two organisms for a particular character or character state. ***Two characters (or character-states) are homologous if one is directly derived from the other.*** Such a series of characters is called an **evolutionary transformation series** (also called **morphoclines** or **phenoclines**). The original, pre-existing character (or character-state) is termed **plesiomorphic** and the derived one as **apomorphic** or **evolutionary novelty**.

Three or more character-states may be homologous if they belong to the same evolutionary transformation series (ovary superior—> half-inferior—> inferior). The terms plesiomorphic and apomorphic are, however, relative. In an evolutionary transformation series representing characters A, B and C (Figure 8.2), B is apomorphic in relation to A but it is plesiomorphic in relation to C.

Two or more organisms may be homologous for a particular character (or character-state) if their immediate common ancestor also had this character. Such a character is called **shared homologue**. If the character-state is present in the immediate common ancestor, but not in the earlier ancestor (Figure 8.3), i.e., the character-state is a derived one, the situation is known as **synapomorphy**. If the character-state is present in the immediate common ancestor, as well as in the earlier ancestor, i.e., it is an original character-state, the situation is known as **symplesiomorphy** (note sym).

The homology between different organisms is termed **special homology**, as represented by different types of leaves in different species of plants. Different leaves in the same plant such as foliage leaves, bracts, floral leaves would also be homologous, representing **serial homology**. The following criteria may be helpful in identifying homology in practice:

1. Morphological similarity with respect to topographic position, geometric position, or position in relation to other parts. A branch, for example, occurs in the axil of a leaf, although it may be modified in different ways.
2. Similar ontogeny.
3. Continuation through intermediates, as for example, the evolution of mammalian ear from gills of fishes, evolution of achene fruit from follicle in Ranunculaceae. Similarly, vessels having evolved from tracheids, the primitive forms of vessels are more like tracheids, with elongated narrower elements with oblique end walls.
4. When the same relatively simple character is found in large number of species, it is probably homologous in all the species. Sets of characters may similarly be homologous.
5. If two organisms share the characters of enough complexity and judged homologous, other characters shared by the organisms are also likely to be homologous.

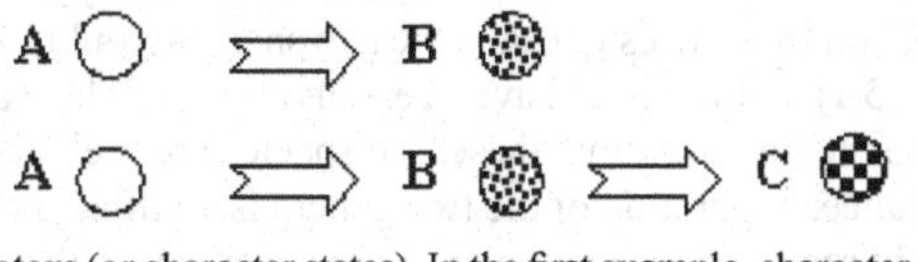

A ⟹ B

A ⟹ B ⟹ C

Figure 8.2: Homology between characters (or character states). In the first example, character A is plesiomorphic and B is apomorphic. In the second example, B is apomorphic in relation to A but plesiomorphic in relation to C as all three belong to an evolutionary transformation series.

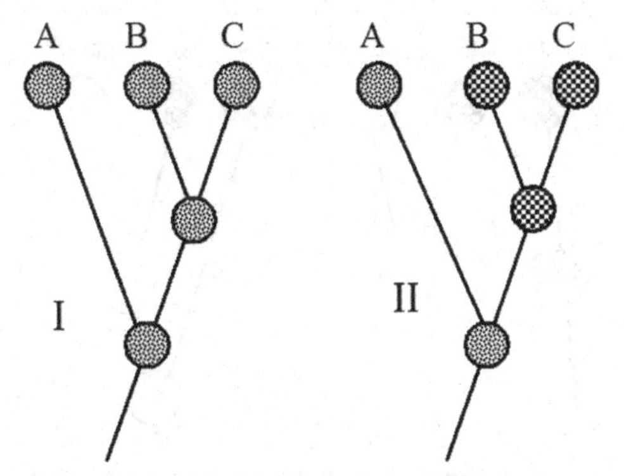

Figure 8.3: Homology between two organisms B and C. In diagram I, similarity is due to symplesiomorphy as the character was unchanged in the previous ancestor. In II, it is due to synapomorphy as the previous ancestor had a plesiomorphic character and the two now share a derived character.

Parallelism and Convergence

Unlike homology, if the character shared by two organisms is not traced to a common ancestor, the similarity may be the result of **homoplasy** (sometimes considered synonym of analogy). It can result in three different ways. One, the organisms have a common ancestor but the character-state was not present in their common ancestor (parallelism). It could also result from two different characters in different ancestors evolving into identical character-states (convergence). Similarity could also arise from loss of a particular character (**reversal**), thus reverting to ancestral condition (loss of perianth in some families). All the three situations represent **false synapomorphy** because the similar character-state is derived and not traced to a common ancestor.

Simpson (1961) defined **parallelism** as the ***independent occurrence of similar changes in groups with a common ancestry, and because they had a common ancestry***. The two species *Ranunculus tripartitus* and *R. hederacea* have a similar aquatic habit and dissected leaves and have acquired these characters by parallel evolution. The development of vessels in Gnetales and dicotyledons also represents a case of parallelism.

Convergence implies ***increasing similarity between two distinct phyletic lines, either with regard to individual organ or to the whole organism.*** The similar features in convergence arise separately in two or more genetically diverse and not closely related taxa or lineages. The similarities have arisen in spite of lack of affinity and have probably been derived from different systems of genes. Examples may be found in the occurrence of pollinia in Asclepiadaceae and Orchidaceae, and the '**switch habit**' (circular sheath at nodes) in *Equisetum, Ephedra* and *Polygonum.* The concepts of parallelism and convergence are illustrated in Figure 8.4.

Convergence is generally brought about by similar climates and habitats, similar methods of pollination or dispersal. Once the convergence has been identified between two taxa, which have been grouped together, they are separated to make the groups natural and monophyletic. The following criteria may help in the identification of convergence:

1. Convergence commonly results from **adaptation to similar habitats**. Water plants thus usually lack root hairs and root cap but contain air lacunae. Annuals are predominant in deserts, which also have a good number of succulent plants. The gross similarity between certain succulent species of Euphorbiaceae and Cactaceae is a very striking example of convergence.
2. Convergence may also result from **similar modes of pollination** such as wind pollination in such unrelated families as Poaceae, Salicaceae and Urticaceae, pollinia in Asclepiadaceae and Orchidaceae.
3. Convergence may also be due to **similar modes of dispersal**, as seen in hairy seeds of Asteraceae, Asclepiadaceae and some Malvaceae.
4. Convergence commonly occurs between relatively advanced members of respective groups. *Arenaria* and *Minuartia* form natural groups of species which were earlier placed within the same genus *Arenaria.* The two species *Arenaria leptocladus* and *Minuartia hybrida* show more similarity than between any two species of these two genera. If the similarity is **patristic** (result of common ancestry), then the two species would represent the most primitive members of respective groups (Figure 8.5-I) and it would have been advisable to place all of the species in the same genus *Arenaria.* The studies have shown, however, that these two species are the most specialized in each group (Figure 8.5-II) and thus show convergence. Separation of the two genera is justified, because placing all the species within the same genus *Arenaria* would render the group polyphyletic, a situation that evolutionary biologists avoid.

It is pertinent to mention that although the concepts parallelism and convergence seem to be distinct and theoretically sound, and often easy to apply when discussing **homoplasious** (non-homologous) similarity in the case of closely related

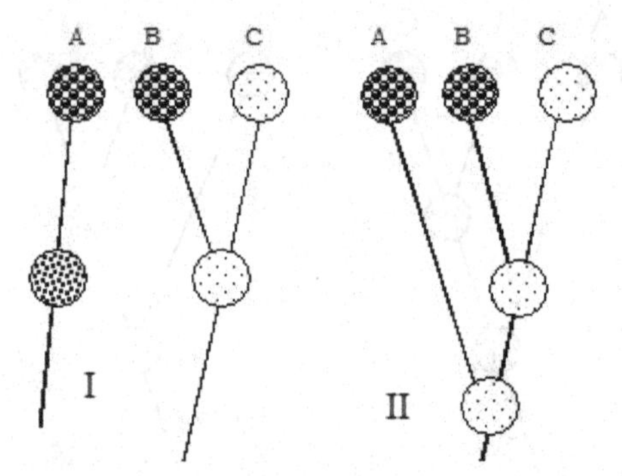

Figure 8.4: Examples of **convergence** (I) and **parallelism** (II) between organisms A and B. In convergence, similarity is between organisms derived from different lineages. In parallelism, the ancestor is common but both A and B have evolved an apomorphic character independently. In both cases, similarity represents false synapomorphy. Dissimilarity between B and C in both diagrams is due to divergence.

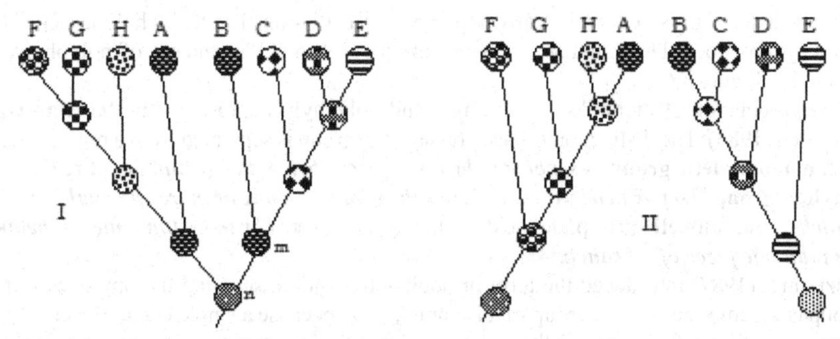

Figure 8.5: Two possible reasons for **similarity between species** A and B. In (I), A (cf. *Arenaria leptocladus*) and B (cf. *Minuartia hybrida*) are the most primitive members of respective lineages FGHA (cf. *Arenaria*) and BCDE (cf. *Minuartia*). The two lineages have common ancestry and thus constitute a single monophyletic group (cf. *Arenaria* s. l.). In (II), A and B happen to be the most advanced members of the respective groups, the two lineages are distinct and as such similarity between A and B is superficial due to convergence, justifying the independent recognition of two lineages (cf. distinct genera *Arenaria* and *Minuartia*).

organisms (parallelism), or distantly related organisms (convergence), the distinction is not always clear. In Figure 8.5-I, for example if we did not know the evolutionary history of the group before level **m**, there was no way of telling whether all the eight species had a common ancestor or not. For practical reasons, it is always safer to refer homoplasious situations together. Some recent authors like Judd et al. (2002) treat parallelism and convergence as same.

Reversal is a common evolutionary process, wherein loss of a particular character may lead to apparent similarity with ancestral condition. The occurrence of reduced unisexual flowers without perianth or with reduced perianth in Amentiferae was once considered to be primitive situation, but the evidence from wood anatomy, floral anatomy and palynology have shown that apparent simplicity of these flowers is due to evolutionary reduction (reversal), and as such the assumed similarity to angiosperm ancestral condition is representation of homoplasy, a false similarity between an evolutionary advancement (secondary reduction) and ancestral simple condition.

Monophyly, Paraphyly and Polyphyly

These terms have been commonly used in taxonomy and evolutionary literature with such varied interpretation that much confusion has arisen in their application. Defined broadly, the terms monophyly (derivation from a single ancestor) and polyphyly (derivation from more than one ancestor) would have different meanings depending upon how far back we are prepared to go in evolutionary history. If life arose only once on Earth, all organisms (even if you place an animal species and a plant species in the same group) are ultimately monophyletic in origin. There is thus a need for a precise definition of these terms, to make them meaningful in taxonomy.

Simpson (1961) defined **monophyly** as ***the derivation of a taxon through one or more lineages from one immediately ancestral taxon of the same or lower rank***. Such a definition would be true if, say, genus B evolved from genus A through one species of the latter, since in that case, the genus B would monophyletic at the same rank (genus) as well as at the lower (species) rank. On the other hand, if genus B evolved from two species of genus A, it would be monophyletic at the genus level but polyphyletic at the lower rank.

Most authors, however, including Heslop-Harrison (1958) and Hennig (1966), adhere to a stricter interpretation of monophyly, namely the group should have evolved from a single immediately ancestral species which, may be considered as belonging to the group in question. There are thus two different levels of monophyly: a **minimum monophyly** wherein one supraspecific taxon is derived from another of equal rank (Simpson's definition), and a **strict monophyly** wherein one higher taxon is derived from a single evolutionary species.

Mayr (1969) and Melville (1983) follow the concept of minimum monophyly. Most authors, including Heslop-Harrison (1958), Hennig (1966), Ashlock (1971) and Wiley (1981), reject the idea of minimum monophyly. All supraspecific taxa are composed of individual lineages that evolve independent of each other and cannot be ancestral to one another. Only a species can be an ancestor of a taxon. The supraspecific ancestors and, for that matter, supraspecific taxa are not biologically meaningful entities and are only evolutionary artifacts.

Hennig (1966) defined a monophyletic group as ***a group of species descended from a single ('stem') species, and which includes all the descendants from this species***. Briefly, a monophyletic group comprises all the descendants that at one time belonged to a single species. A useful analysis of Hennig's concept of monophyly was made by Ashlock (1971). He distinguished between two types of monophyletic groups: those that are **holophyletic** when ***all descendants of the most recent common ancestor are contained in the group*** (monophyletic sensu Hennig) and those that are **paraphyletic** and ***do not contain all descendants of the most recent common ancestor of the group***. A **polyphyletic** group, according to him, is one whose most recent ancestor is not cladistically a member of that group. The terms holophyletic and monophyletic are now considered synonymous. Diagrammatic representations of Ashlock's concept of polyphyly, monophyly and paraphyly is presented in Figure 8.6.

An excellent representation of monophyly, paraphyly and polyphyly is presented by **'cutting rules'**, devised by Dahlgren and Rasmusen (1983). The distinction is based on how the group is separated from a representative evolutionary tree (Figure 8.7). A **monophyletic group** is ***separated by a single cut below the group***, i.e., it represents ***one complete branch***. A **paraphyletic group** is ***separated by one cut below the group and one or more cuts higher up***, i.e., it represents ***one piece of a branch***. A **polyphyletic group,** on the other hand, is ***separated by more than one cut below the group***, i.e., it represents ***more than one piece of a branch.***

Gerhard Haszprunar (1987) introduced the term orthophyletic while discussing the phylogeny of Gastropods. An **orthophyletic** group is a stem group, i.e., a group that is paraphyletic because a single clade (the crown group), has been excluded. The term has not been followed in other groups, especially in botanical systematics. Sosef (1997) compares the existent hierarchical models of classification. He argues that a phylogenetic tree can be subdivided according to a monophyletic hierarchical model, in which only monophyletic units figure or, according to a 'Linnaean' hierarchical model, in which both mono- and paraphyletic units occur. Most present-day phylogeneticists try to fit the monophyletic model within the set of nomenclatural conventions that fit the Linnaean model. However, the two models are intrinsically incongruent. The monophyletic model requires a system of classification of its own, at variance with currently accepted conventions. Since, however, the monophyletic model is unable to cope with reticulate evolutionary relationships; it is unsuited for the classification of nature. The Linnaean model is to be preferred. This renders the acceptance of paraphyletic supraspecific taxa inevitable.

As is true for the distinction between parallelism and convergence, similarly, the concepts of paraphyly and polyphyly (both of which are rejected by modern phylogenetic systematics while constructing classification), hold good, when the former is applied to a group of closely related organisms and latter to distantly related organisms. The concepts become ambiguous when a small group of organisms is considered. In Figure 8.6-III, taxa B and C—if brought together—would form a polyphyletic group, because they are derived from two separate ancestors at level **m**. If, however, A, B, and C are

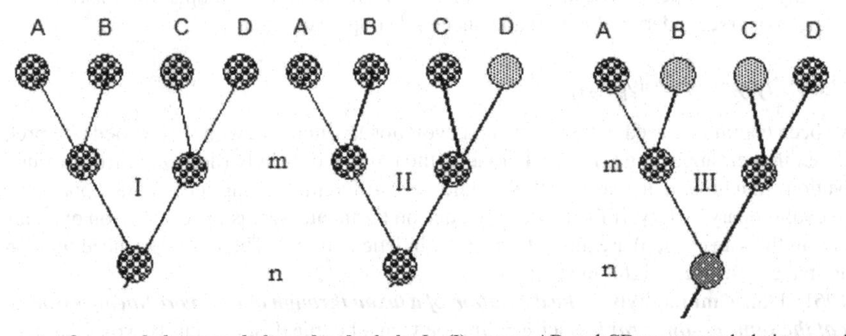

Figure 8.6: Concepts of monophyly, paraphyly and polyphyly. In (I) groups AB and CD are monophyletic as each has a common ancestor at level m. Similarly. group ABCD is monophyletic as it has a common ancestor at level n. In (II) group ABC is paraphyletic as we are leaving out descendant D of the common ancestor at level n. In (III) group BC is polyphyletic as their respective ancestors at level m do not belong to this group.

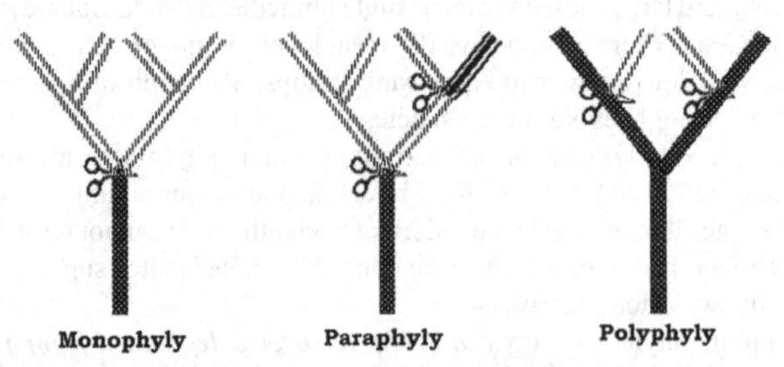

Figure 8.7: The application of cutting rules to distinguish between monophyly, paraphyly and polyphyly. The group is represented by lighter portion of the tree. Monophyletic group can be separated by a single cut below the group, a paraphyletic group by one cut below the group and one or more higher up. A polyphyletic is separated by more than one cut below the group. A monophyletic group represents one complete branch, a paraphyletic group one larger portion of the branch; whereas the polyphyletic group represents more than one pieces of a branch (based on Dahlgren et al., 1985).

under one group, B and C would still now be components of a paraphyletic group, because one descendant of the common ancestor at level **n** is kept out of the group. A natural group would be one, which includes all descendants of the common ancestor, or the group is monophyletic.

Phylogenetic Diagrams

The affinities between the various groups of plants are commonly depicted with the help of diagrams, with several innovations. These diagrams also help in understanding the classification of included taxa. An understanding of these terms is necessary for a correct interpretation of putative relationships. These branching diagrams are broadly known as **dendrograms**. Any diagram showing the evolutionary history of a group in the form of branches arising from one or more points has often been referred as a **phylogenetic tree**, but the use of terms is now becoming more precise, and more innovative diagrams are being developed often providing useful information about different taxa mapped in the diagram.

The most common form of diagram is one where the length of branch indicates the degree of apomorphy. Such diagrams were sometimes classified as **cladograms** (Stace, 1980), but the term has now been restricted to diagrams constructed through the distinct methodology of cladistics (Stace, 1989). Diagrams with vertical axis representing the degree of apomorphy are now more appropriately known as **phylograms**. The earliest well-known example of such a phylogram is **'Bessey's cactus'** (see Figure 10.11, Chapter 10). In such diagrams the most primitive groups end near the base and the most advanced reach the farthest distance.

Hutchinson (1959, 1973) presented his phylogram in the form of a line diagram (Figure 10.13, Chapter 10). The recent classifications of Takhtajan (1966, 1980, 1987) and Cronquist (1981, 1988) have more innovative phylograms in which the groups are depicted in the form of balloons or bubbles whose size corresponds to the number of species in the group (an approach also found in Besseyan cactus). Such phylograms thus not only depict phylogenetic relationships between the groups, they also show the degree of advancement as also the relative number of species in different groups. Such diagrams have been popularly known as **bubble diagrams**. The bubble diagram of Takhtajan (Figure 10.16, Chapter 10) is more detailed and shows the relationship of the orders within the 'bubble'; as mentioned earlier, Woodland (1991) aptly described it as **'Takhtajan's flower garden'**.

The **phylogenetic tree** is a commonly used diagram in relating the phylogenetic history. The vertical axis in such a diagram represents the geological time scale. In such a diagram, the origin of a group is depicted by the branch diverging from the main stock and its disappearance by the branch termination. Branches representing the fossil groups end in the geological time when the group became extinct, whereas the extant plant groups extend up to the top of the tree. As already mentioned, the relative advancement of the living groups is indicated by their distance from the centre, primitive groups being near the center, and advanced groups towards the periphery. A phylogenetic tree representing possible relationships and the evolutionary history of seed plants is presented in Figure 8.8.

Dahlgren (1975) presented the phylogenetic tree (preferred to call it phylogenetic 'shrub' in 1977) of flowering plants with all extant groups reaching the top, and the cross-section of the top of the phylogenetic tree was shown as top plane of this diagram (Figure 8.9). In subsequent schemes of Dahlgren (1977, 1983, 1989), the branching portion of the diagram was dropped and only the top plane (cross-section of the top) presented as a two-dimensional diagram (Figure 8.10), and this has been very useful in mapping the distribution of various characters in different groups of angiosperms, and the comparison of these provides a good measure of correspondence of various characters in phylogeny. This diagram has been popularly known as **'Dahlgrenogram'**.

Thorne's diagram (2000) is similarly the top view of a **phylogenetic shrub** (Figure 10.23, Chapter 10), in which the center representing the extinct primitive angiosperms, now absent, is empty.

A **cladogram** represents an evolutionary diagram utilizing cladistic methodology, which attempts to find the shortest hypothetical pathway of changes within a group that explains the present phenetic pattern, using the **principle of parsimony**. A cladogram is a representation of the inferred historical connections between the entities as evidenced by synapomorphies. The vertical axis of the cladogram is always an implied, but usually non-absolute time scale. Cladograms are ancestor-descendant sequences of populations. Each bifurcation of the cladogram represents a past speciation that resulted in two separate lineages.

It must be pointed out, however, that considerable confusion still exists between application of terms cladogram and phylogenetic tree. Wiley (1981) defines a cladogram as ***a branching diagram of entities where the branching is based on inferred historical connections between the entities as evidenced by synapomorphies. It is, thus, a phylogenetic or historical dendrogram.*** He defines a phylogenetic tree as ***a branching diagram portraying hypothesized genealogical ties and sequences of historical events linking individual organisms, populations, or taxa.*** At the species and population level, the number of possible phylogenetic trees could be more than cladograms for particular character changes, depending on which species is ancestral and being relegated lower down on the vertical axis. In the case of higher taxa, the number of cladograms and phylogenetic trees could possibly be equal, because higher taxa cannot be ancestral to other higher taxa since they are not units of evolution, but historical units composed of separately evolving species.

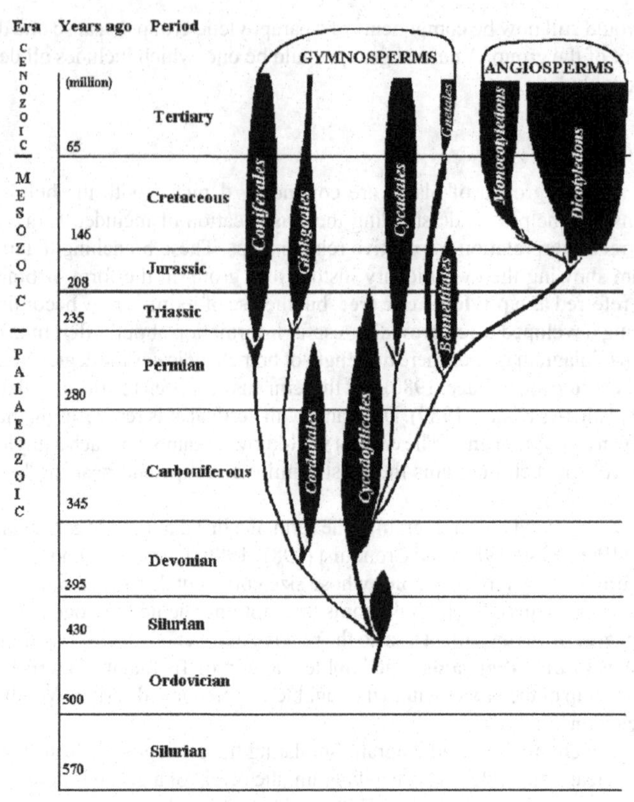

Figure 8.8: A phylogenetic tree representing the evolutionary history of plants including angiosperms. The vertical axis represents the geological time scale. Only extant (living) plants are shown reaching the top.

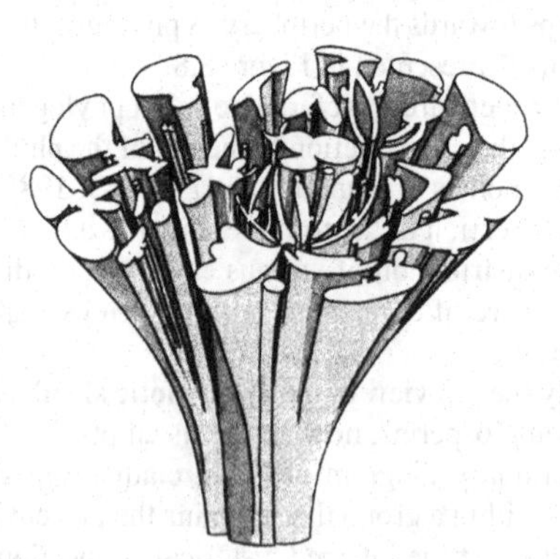

Figure 8.9: Phylogenetic tree of angiosperms presented by Dahlgren (1975) with a section of the top (subsequently named phylogenetic 'shrub' by Dahlgren, 1977).

Over the recent years, it has been thus becoming increasingly common to construct evolutionary diagrams using cladistic methodology, assuming that these character-state changes (represented as evolutionary scale or tree length) correspond to the geological time scale, and call these evolutionary diagrams as **evolutionary tree** (Judd et al., 2008), **phylogenetic tree**, or simply **tree** (Stevens, 2008), synonymous with a **cladogram**.

A **Phenogram** is a diagram constructed on the basis of numerical analysis of phenetic data. Such a diagram is the result of utilization of many characters, usually from all available fields, and involves calculating the similarity between taxa and constructing a diagram through **cluster analysis**. Such a diagram (Figure 8.20) is very useful, firstly because it is based on many characters, and secondly because a hierarchical classification can be achieved by deciding upon the threshold levels of similarity between taxa assigned to various ranks.

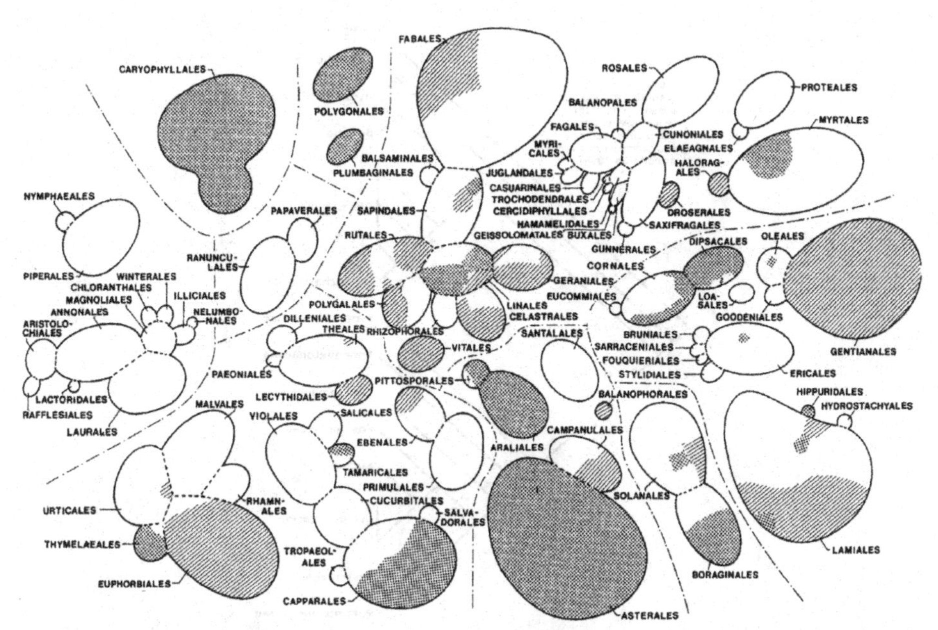

Figure 8.10: Mapping of pollen grain dispersal stage in different dicotyledons on a two-dimensional diagram (Dahlgrenogram) of Dahlgren, representing transverse section through the top of a phylogenetic shrub. Pollen grain dispersal in 2-celled stage (unshaded), 3-celled stage (dotted), or mixed (hatched) (Courtesy Gertrud Dahlgren).

It must be pointed out that the modern phylogenetic methods, which aim at constructing phylogenetic trees, also sometimes use large number of characters for comparison, especially when dealing with morphological data, and there seem to be a lot of similarities in data handling and computation but are unique in the utilization of evolutionary markers and, consequently, produce slightly different results. With the incorporation of distance methods in the construction of trees, the classical difference between the terms is largely disappearing. Modern cladistic programs develop trees in which branch lengths are indicated, and plotting programs offer the choice to indicate branch lengths (and often called phylograms) or not. In latter case branches may be square (line running vertically and horizontally- and often called phenogram (Figure 8.21) or V-shaped cladogram (Figure 8.11). These may be presented as upright or as horizontal trees (**prostrate trees**). Modern trees contain information about evolutionary markers such as bootstrap support, branch length, and Bremer support, as discussed in subsequent pages.

Phylogeny and Classification

The construction of phylogenetic classification involves two distinct steps: determining the **phylogeny** or evolutionary history of a group, and construction classification based on this history. Imagine a **lineage** (or **clade**—a group of individuals producing successively, similar and genetically related individuals, generally represented by lines in a cladogram) with woody habit, alternate leaves, cymose inflorescence, 5 red petals, 5 stamens, 2 free carpels, and dry fruit with many seeds. Over a period of time, some population acquires herbaceous habit and the original lineage splits into two, one with woody habit and the second with herbaceous habit (Figure 8.12). In the lineage with woody habit, one lineage emerges with fused carpels, while the other loses one of the two carpels. The one with fused carpels loses 3 of the five stamens in one or more populations, and that with a single carpel doubles the number of stamens to ten in one or more populations. The herbaceous lineage, similarly, splits into one with yellow petals and one with white petals, the former developing fleshy fruits in one or more populations, and the latter having the number of seeds reduced to one in some populations. The present descendants of the original ancestor are thus represented by eight lineages, which have developed a few apomorphic character-states, but also share plesiomorphic character-states such as alternate leaves, 5 petals, and cymose inflorescence. There must be hundreds of more plesiomorphic states, but of little significance in classification, as the above three. Note that the ancestral species at level I, II (woody habit, 2 free carpels), IA (herbaceous habit, red petals), III (herbaceous habit, yellow petals and free carpels and dry fruit), and IV (herbaceous habit, white petals, 5 stamens and many seeds) and have disappeared, whereas those at level V and VI are still represented (although with minor changes) in the form of E and G, respectively. Also note that united carpels have arisen twice independently. The same is also true for the loss of three stamens.

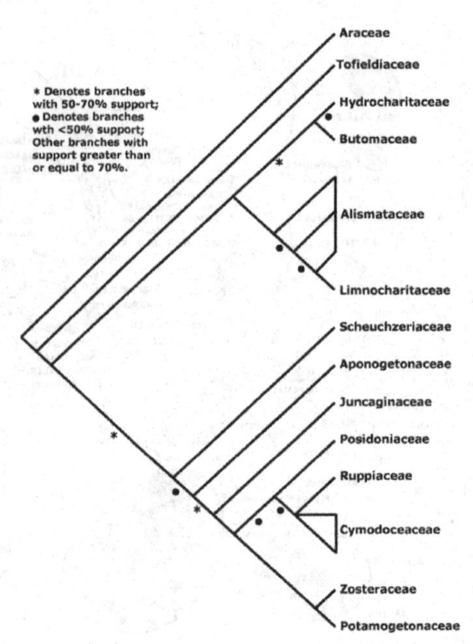

Figure 8.11: Tree (cladogram) for different families of the order Alismatales. Support indicated for branches refers to bootstrap support, discussed in subsequent pages. (Reproduced from APweb vesion 7 (June, 2008), with permission from Dr. P. F. Stevens).

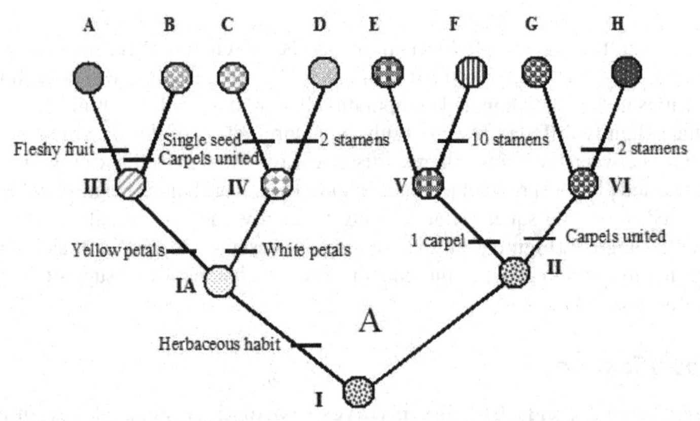

Figure 8.12: Evolutionary history of a hypothetical group of organisms which started with the ancestral species with woody habit, alternate leaves, cymose inflorescence, 5 red petals, 5 stamens, 2 free carpels and dry fruit with many seeds. Eleven character-state transformations at different stages have resulted in 8 present day species. Note that two of the changes (carpel union and loss of three stamens) have occurred twice, and as such only nine **genetic switches** are involved. The ancestral species at levels I, IA, II, III and IV have disappeared.

Having known the evolutionary history of the group, we could use synapomorphy and the concept of monophyly. Assuming that all eight lineages (groups of populations) are sufficiently distinct to be recognized as distinct species, we would have eight species. The simplest way would be to group these eight species into four genera, each having a common ancestor. Two of these common ancestors have disappeared, but two are still living and would also be included in the respective genera (it will be more appropriate to regard E as ancestral to F and G ancestral to H). These could be further assembled into two families of four species each (two genera each) having a common ancestor at level IA and II, and these two families into one order with a common ancestor at level I. Please note that common ancestor at level I, IA, II, III and IV are also no longer living.

The second option would be to include ABCD in one genus, and EFGH in another genus, and include all 8 species (2 genera) in one family (and, of course, depending upon the degree of diversity from related families, this could still be a constituent of a monotypic order). The third option would be to have a single genus of eight species.

Note the importance of synapomorphy in determining monophyly. Character-states alternate leaves, cymose inflorescence and 5 petals (character-states of different characters not same) have been passed unchanged in all the eight descendants (species), which as such are **symplesiomorphic** for each of these, and this symplesiomorphy will be valid between any two (or more) species that you choose to combine into a genus, say, D and E, or C and F, or say ABCDE. On the other hand, if we consider only synapomorphy, the monophyly is easily deciphered. A and B are accordingly synapomorphic for yellow petals, C and D for white petals, E and F for one carpel and G and H for united carpels. Similarly, A, B, C and D are synapomorphic for herbaceous habit. The development of fleshy fruit in A, single seed in C, 10 stamens in F represent the occurrence of a derived character state in a single taxon and termed as **autapomorphy**. Such character states are not helpful in cladogram construction, although indicative of divergence. Development of 2 stamens in D and H independently represents **homoplasy**, and may lead to artificial grouping of these two, if history of the group was not properly known. It must, however, be noted that symplesiomorphy may sometimes be helpful in detecting monophyly, especially where in some other taxa, it has evolved into another character-state. As such, out of the four plesiomorphic character-states listed here, only the woody habit has changed to herbaceous habit, and as such in the remaining taxa (E, F, G and H) symplesiomorphy of woody character-state identifies monophyly of the group ABCD. It must be remembered that synapomorphy and symplesiomorphy reflect **homology** for a particular character-state (or more than one character-states, each belonging to a different character).

All above options would render (genus, family or order) monophyletic groups, the ultimate goal of phylogenetic systematics. Any other options won't work. Keeping D and E in one genus (or CDE or DEF) would make it **polyphyletic**, promptly rejected once detected, because the group is derived from more than one ancestor. Keeping ABC under one genus, or FGH under one genus would make **paraphyletic** groups because we are not including all the descendants of the common ancestor (we are leaving out D in first genus and E in second). In the same way, putting more than four species (but less than eight) under the same genus would make it paraphyletic. Paraphyletic taxa are strongly opposed by phylogenetic systematists; the classical case is the demise of traditional division of angiosperms into monocots and dicots, over the last decade.

It must be noted that all eight species—whatever way you classify—share alternate leaves, cymose inflorescence and five petals. The species at level IA and above additionally share woody habit, VI and above additionally united carpels, V and above one carpel (and not fused carpels). Similarly, species at level II and above share herbaceous habit (instead of woody habit) in addition to three common, at level III and above additionally yellow petals, and at level IV and above additionally (in place of yellow petals) white petals.

The situation depicted above can be more easily represented through the concept of nested groups, more conveniently represented as a set of ovals, a **Venn diagram** (Figure 8.13A). The diagram drawn here is based on the assumption that we have information from a large group in which there are 21 woody species of which 13 are with united carpels and 8 with free carpels. Of the 13 with united carpels 7 have 2 stamens whereas 6 have more stamens.

The information is presented in the form of an **unrooted network (unrooted tree)** (Figure 8.13B). The herbaceous species are shown towards the left of left double arrow, and the woody species towards the right. Similarly, the species towards the left of the middle double arrow are with free carpels while those towards the right with fused carpels. The species towards the left of the right double arrow have more than 2 stamens and those towards the right just 2 stamens.

It must be noted that in constructing the Venn diagram and the unrooted network, only three character-state transformations are accounted for. We have completely left out grouping of herbaceous plants and the woody plants with a single carpel. Inclusion of these would make the diagrams much more complicated and present several alternatives. Also, the more meaningful trees have to be **rooted** (the most primitive end at the base), to reflect the phylogeny. Even with the phylogenetic history of the group known, there could be several variations of the rooted tree, two simple ones being shown in the Figure 8.13C and 8.13D. If we did not know the evolutionary history of the group, a number of variations would be possible, depending upon which character-state is plesiomorphic, and which character (habit, carpel fusion or stamen number) forms the root, and what would be the sequence of the character changes on the tree.

The character-states chosen for analysis should necessarily be homologous (one derived from another) and non-overlapping. The analysis becomes more meaningful when it is established that the evolution of a particular character-state has been the result of a corresponding **genetic change**, and not a mere plastic environmental influence. This fact underlies the importance of the emerging field of molecular analysis in **phylogenetic systematics**. It is believed that the recognition of molecular character-states (nucleotide sequences) is often easier and more precise, although there are always accompanying problems.

The problem with vascular plants, especially the angiosperms, is that we know very little about their evolutionary history. The fossil records, which generally give fair information about evolution, are very scarcely represented. What we have available with us is a mixture of primitive, moderately advanced and advanced groups. Almost each group has some plesiomorphic and some apomorphic character-states, and relative proportion of one or the another delimit the relative advancement of various groups. Attempt to reconstruct the evolutionary history of the group involves comparative study its living members, sorting out plesiomorphic and apomorphic character states, and distribution of these in various members.

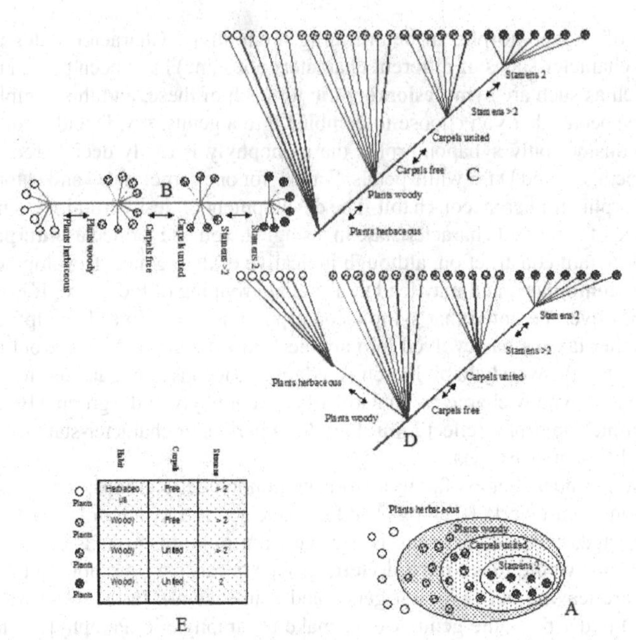

Figure 8.13: Diagrams based on evolutionary pattern depicted in Figure 8.12. A: Venn diagram based on the assumption that there are 21 woody species of which, 13 are with united carpels, and 8 with free carpels. Of the 13 with united carpels, 7 have 2 stamens whereas 6 have more stamens. **B:** An unrooted tree based on Venn diagram. **C:** A possible rooted tree, if evolutionary history of the group was not known, 15 possible rooted trees could be drawn. **D:** Rooted tree based on knowledge that herbaceous habit arose from woody habit, and we know further evolution of woody species. Other possibilities are discussed subsequently. **E:** Data matrix of above, where the number of species are not indicated.

Once the evolutionary history of the group has been constructed, monophyletic groups at various levels of inclusiveness are identified, assigned ranks, and given appropriate names, to arrive at a working system of classification.

PHYLOGENETIC DATA ANALYSIS

The methodology of cladistics with incorporation of numerical methods involves several steps.

Taxa-Operational Units

The first step in data analysis involves the selection of Taxa for data collection, often called Operational Taxonomic Units (OTUs) in Taxometrics, Operational Evolutionary Units (OEUs) in cladistics, referring to the sample from which the data is collected. Although it would be ideal to select different individuals of a population, practical considerations make it necessary to select the members of the next lower rank. Thus, for the analysis of a species would need selection of various populations, for the study of a genus they would be different species, and for a family they would be different genera. It is not advisable, however, to use genera and higher ranks, as most characters would show variation from one species to another and thus would not be suitable for comparison. The practical solution would be to use one representative of each taxon. Thus, if a family is to be analyzed and its genera to be compared, the data from one representative species of each genus can be used for analysis. Once the taxa are selected, a list of such taxa is prepared. A unique feature of cladistic studies, however, is that the list of taxa generally includes a **hypothetical ancestor**, the comparison with which reveals crucial phylogenetic information, and is used for rooting of the tree. It is, increasingly being realized that only a species is the valid evolutionary entity, and all taxa at higher ranks are artifacts, constructed for the sake of convenience. A meaningful analysis would always be one derived from data from various species (taken from populations) and not any higher rank directly.

Characters

A conventional definition of a taxonomic character is ***a characteristic that distinguishes one taxon from another***. Thus, white flowers may distinguish one species from another with red flowers. Hence, the white flower is one character and red flower another. A more practical definition espoused by numerical taxonomists defines character (Michener and Sokal, 1957) as ***a feature, which varies from one organism to another***. By this second definition, flower color (and not white flower or red flower) is a **character**, and the white flower and red flower are its two **character-states**. Some authors (Colless, 1967) use the term **attribute** for character-state but the two are not always synonymous. When selecting a character for numerical analysis, it is important to select a **unit character**, which may be defined as ***a taxonomic character of two or more states, which within the study at hand cannot be subdivided logically, except for the subdivision brought about by the method of coding***. Thus, trichome type may be glandular or eglandular. A glandular trichome may be sessile or stalked. An eglandular trichome may, similarly, be unbranched or branched. In such a case, a glandular trichome may be recognized as a unit character and an eglandular trichome as another unit character. On the other hand, if all glandular trichomes in OTUs are of the same type and all eglandular trichomes are of the same type, the trichome type may be selected as a unit character.

The first step in the handling of characters is to make a list of unit characters. A preliminary step involves **character compatibility** study in which each character is examined to determine the proper sequence of character-state changes that take place as the evolution progresses (**morphoclines** or **transformation series**). The list should include all such characters concerning which information is available. ***A priori***, all characters should be weighted equally (no weighting to be given to characters). Although some authors advocate that some characters should subsequently be assigned more weightage than others (***a posteriori weighting***), such considerations generally get nullified when many characters are used. It is generally opined that numerical studies should involve not less than 60 characters, but more than 80 are desirable. For practical consideration, there may be some characters concerning which information is not available (many plants in a population are not in fruit) or the information is irrelevant (trichome type if many plants are without trichomes), or the characters which show a much greater variation within the same taxon. Such characters are omitted from the list. This constitutes ***residual weighting*** of characters. The characters (leaves, bracts, carpels) or character states (simple leaf, palmate compound leaf, pinnate compound leaf) chosen should also be homologous, in terms of sharing common ancestry or belonging to same evolutionary transformation series. The 'petals' of *Anemone* are modified sepals and thus not homologous with the petals of *Ranunculus* and hence not comparable. Similarly, the tuber of sweet potato (a modified root) cannot be compared with the tuber of potato (a modified stem).

Binary and Multistate Characters

The characters most suitable for computer handling are **two-state** (**binary** or **presence-absence**) characters (habit woody or herbaceous). However, all characters may not be two-state. They may be qualitative multistate (flowers white, red, blue) or quantitative multistate (leaves two, three, four, five at each node). Such multistate characters can be converted into two-state (flowers white or colored; leaves four or more vs leaves less than four). Or else the characters may be split (flowers white vs not white, red vs not red, blue vs not blue; leaves two vs not two, three vs not three and so on). Such a splitting may, however, give more weightage to one original character (flower color or number of leaves). It is essential that different character states identified are **discrete** or discontinuous from one another. Discreteness of character states can be evaluated by comparing the means, ranges, and standard deviations of each character for all taxa in analysis. Additionally, t-tests and multivariate analysis may also be used for evaluating character state discreteness.

Ordering of Character-states

A binary character will have single step or **switch** (Figure 8.14-I) necessary for change. The minimum number of switches possible (**Wagner parsimony**) in a multistate character will depend whether the character states are **ordered** or left **unordered**. In an unordered transformation series each character state can evolve into every other character state and represents a single switch (Figure 8.14-II, III). A three-state character will have two switches or steps, and three possible morphoclines (Figure 8.14-IV), four-state character three switches and several morphoclines. Whereas ordering of two-state characters is relatively easy, multi-state characters are often difficult to order, and changes may often be reversible, and it is advisable to leave them **unordered**, and identify only one switch (**Fitch parsimony**). The molecular characters are different DNA sequences, that may differ in having one of the four bases (adenine, thymine, guanine and cytosine) at a particular locus, and as such present four character-states. As reversals are common in these, are always left unordered.

Assigning Polarity

It is, however, necessary to determine the relative ancestry of the character-states, or the assignment of **polarity**. The designation of polarity is often one of the more difficult and uncertain aspects of phylogenetic analysis. For this, the comparison may be made within the concerned group (**in-group comparison**) or relatives outside the group (**out-group comparison**). The latter may often provide useful information, especially when the out-group used is the **sister-group** of the concerned group. If two character-states of a character are found in a single monophyletic group, the state that is

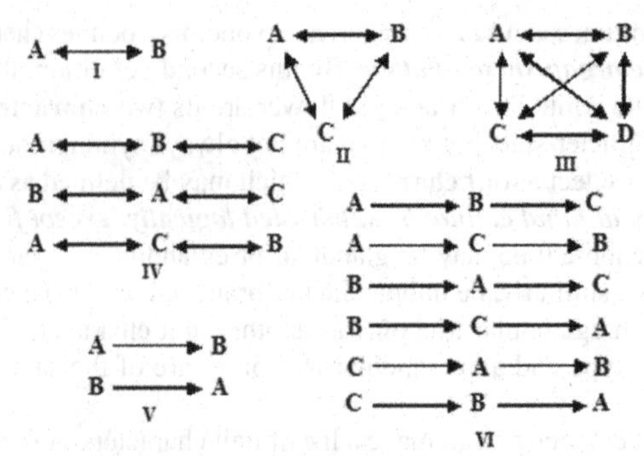

Figure 8.14: Ordering and polarity of character states. I: Binary character with single possible switch. **II:** Unordered three-state character with single possible switch. **III:** Unordered four-state character with single possible switch. **IV:** Ordered three-state character with two possible switches and three possible morphoclines. **V:** Polarized binary character with two possible morphoclines. **VI:** Ordered and polarized three-state character with 6 possible morphoclines.

also found in a sister-group is likely to be plesiomorphic and that found only within the concerned monophyletic group is likely to be apomorphic. **Ingroup comparison** (also known as **common ground plan** or **commonality principle**) is based on the presumption that in a given group (presumably monophyletic), the primitive structure would tend to be more common. Thus all 8 species of cladogram in Figure 8.12 share plesiomorphic character states: alternate leaves, cymose inflorescence and five petals. Five species have plesiomorphic 5 stamens, two derived 2 stamens and one with 10 derived assumed that the evolution of a derived condition will occur in only one of potentially numerous lineages of the group; thus, the ancestral condition will tend to be in the majority. As is evident from Figure 8.14, the number of possible morphoclines increases after the polarity criterion is included and the selection of single appropriate morphocline representing the true sequence even more challenging.

Character Weighting and Coding

The **coding** of character states is done by assigning non-negative integer values. Binary characters are conveniently assigned 0 and 1 for two states. If possible to distinguish, plesiomorphic state is assigned 0 and apomorphic state 1 code (Figure 8.15-IV). It is often assumed that whereas the same character-state may arise more than once within a group between closely related species (**parallelism**), or between remotely related species (**convergence;** the distinction between parallelism and convergence is sometimes omitted) for a simple character, it is highly unlikely for more complex characters. It is also assumed that whereas many genes must change in order to create a morphological structure, one gene change is enough for its loss (**reversal**). This **Dollo's law** is considered when choosing trees, gains of structures counted more than losses, a process known as **Dollo parsimony**. Such **weighting** of characters is often common in phylogenetic analysis. In transformation series leaf simple —> pinnately lobed —> pinnately compound, the development of pinnate compound leaf from simple leaf occurred in two steps and needs to be given more weightage. The coding may accordingly be done as 0 for most primitive character-state (simple leaf), 1 for intermediate character-state (pinnately lobed leaf) and 2 more most advanced state (pinnately compound leaf) (Figure 8.15-I). In molecular data, transversions (Purine to pyrimidine or pyrimidine to purine changes) are given more weightage (Figure 8.15-V) over **transitions** (purine to purine or pyrimidine to pyrimidine), because the latter occur more frequently and are easy to reverse, whereas the former is a less likely biochemical change. Restriction site gains may similarly be weighted over site losses. A complex character, presumably controlled by many genes, may change less easily than a simple character controlled by fewer genes. The former is often given more weighting over a simple character. It may be assumed that leaf anatomy may not change easily but hairiness may change readily. The number of steps between two character-states is conveniently represented through **character step matrix** (Figure 8.15). One may, however, be tempted to count leaf anatomy character as equivalent to two changes in hairiness. This may often be the result of bias to obtain desired results. It is reasonable, however, to adopt the approach of numerical taxonomy to give **equal weighting** to all the characters in the preliminary analysis, identify those characters which show the least homoplasy and give them more weightage in the subsequent analysis, a process known as **successive weighting**. This avoids a bias towards a particular character, and as such enables rational treatment of available data.

Residual weighting involves excluding a character from the list when information for many taxa is not available or is irrelevant. But in certain cases, information may be available for a specific character for large number of OTUs but not for a few. Alternately, the information may be irrelevant for a few taxa (say, the number of spurs in a taxon, which lacks spurs). Such characters are used in analysis but for the taxa for which information is not available or is irrelevant, an NC code (Not Comparable) is entered in the matrix. Whenever the NC code is encountered, the program bypasses that particular character for comparing the concerned taxon. For data handling by computers, the NC code is assigned a specific (not 0 or 1) numeric value. Such residual weighting should, however, be avoided when appreciable number of taxa are not comparable for a particular character. The coded data may be entered in the form of a matrix with ***t*** number of rows (OTUs) and ***n*** number of columns (character-states) with the dimension of the matrix (and the number of attributes) being $t \times n$ (Table 8.1).

	A	B	C
A	0	1	2
B	1	0	1
C	2	1	0

I

	A	B	C	D
A	0	1	2	3
B	1	0	1	2
C	2	1	0	1
D	3	2	1	0

II

	A	B	C	D
A	0	1	1	1
B	1	0	1	1
C	1	1	0	1
D	1	1	1	0

III

	A	B
A	0	1
B	1	0

IV

	A	B	C	D
A	0	1	5	5
B	1	0	5	5
C	5	5	0	1
D	5	5	1	0

V

Figure 8.15: Data matrix of coded character states. I: Ordered three-state character. **II:** Ordered four-state character. **III:** Unordered character. **IV:** Binary character. **V:** Differential weighting to character state changes; imagine A and B represent Purines (Adenine and Guanine), C and D Pyrimidines (Cytosine and Thymine), purine to purine or pyrimidine to pyrimidine change (transition) is given 1 step weight, but purine to pyrimidine change or reverse (transversion) given 5 steps weight.

Table 8.1: A portion of the data matrix with hypothetical **t** OTUs and **n** characters. Binary coding involves 0 for state *a* and 1 for state *b*. The NC code stands for characters not comparable for that OTU. In this analysis a total of 100 characters were used but only nine are pictured here.

Characters → OTUs (*t*)	Habit 0-woody 1-herbaceous	Fruit 0-follicle 1-achene	Ovary 0-superior 1-inferior	Leaves 0-simple 1-compound	Habitat 0-terrestrial 1-aquatic	Pollen 1-triporate 0-monosulcate	Ovule 1-unitegmic 0-bitegmic	Carpels 0-free 1-united	Plastids 1-PI-type 0-PII-type
1.	1	0	1	1	0	1	1	1	1
2.	1	0	1	1	1	1	0	0	1
3.	0	NC	0	1	0	0	1	1	1
4.	1	1	1	0	1	0	0	0	0
5.	1	0	1	1	0	1	1	0	1
6.	1	1	0	1	1	NC	0	1	0
7.	0	1	1	0	1	1	0	0	1
8.	0	0	0	1	0	0	1	1	1
9.	1	1	1	0	0	1	1	0	0
10.	0	0	0	1	1	0	1	1	1
11.	1	0	1	1	0	1	0	0	1
12.	1	1	0	0	1	1	0	0	1
13.	0	0	1	0	1	0	1	0	1
14.	1	0	1	0	1	0	1	0	1
15.	0	1	1	0	0	1	1	0	0

Certain characters in plants evolve together. Occurrence of stipules and trilacunar nodes is usually **correlated.** Similarly, sympetalous members tend to have epipetaly and tenuinucellate ovules. Such **correlated characters** receive lesser weighting. If two characters are correlated, each gets 1/2 weighting, if three 1/3 weighting and so on.

It is always advisable to identify and include the most ancestral taxon (outgroup) as last taxon (or first taxon, as certain programs choose first taxon for rooting) in the list of taxa. If it is possible to identify plesiomorphic and apomorphic character-states, 0 represents plesiomorphic character-state and 1 the apomorphic character-state of a particular character. Outgroup taxon in the matrix gets 0 code for all character states (Table 8.4). For multistep changes, or unlikely events appropriate codes as indicated in Figure 8.15 are transferred to the matrix. Outgroup taxon in the matrix is essential in final rooting of the most parsimonious cladogram (tree).

Measure of Similarity

Once the data have been codified and entered in the form of a matrix, the next step is to calculate the degree of resemblance between every pair of OTUs. A number of formulae have been proposed by various authors to calculate **similarity** or **dissimilarity (taxonomic distance)** between the OTUs. If we are calculating the similarity (or dissimilarity) based on binary data coded as 1 and 0, the following combinations are possible.

		OTUk 1	0
OTUj	1	*a*	*b*
	0	*c*	*d*

Number of matches $m = a + d$

Number of positive matches a

Number of mismatches $u = b + c$

Sample size $n = a + b + c + d = m + u$

j and **k** are two OTUs under comparison

Some of the common formulae are discussed below:

Simple Matching Coefficient

This measure of similarity is convenient and highly suitable for data wherein 0 and 1 represent two states of a character, and 0 does not merely represent the absence of a character-state. The coefficient was introduced by Sokal and Michener (1958). The coefficient is represented as:

$$S_{SM} = \frac{\text{Matches}}{\text{Matches} + \text{Mismatches}}$$

or

$$\frac{m}{m+u}$$

It is more convenient to record similarity in percentage (Table 8.2). In that case, the formula would read:

$$S_{SM} = \frac{m}{m+u} \times 100$$

When comparing a pair of OTUs, a match is scored when both OTUs show 1 or 0 for a particular character. On the other hand, if one OTU sows 0 and another 1 for a particular character, a mismatch is scored.

Jaccard Coefficient of Association

The coefficient was first developed by Jaccard (1908) and gives weightage to scores of 1 only. This formula is thus suitable for data where absence-presence is coded and 1 represents the presence of a particular character-state, and 0 its absence. The formula is presented as:

$$S_J = \frac{a}{a+u}$$

where a stands for number of characters that are resent (scored 1) in both OTUs. This can similarity be represented as percentage similarity.

Table 8.2: Similarity matrix of the representative hypothetical taxa presented as percentage simple matching coefficient.

OTUs	1	2	3	4	5	6	7	8	9	10	11	12	13	14	15
1	100														
2	47.0	100													
3	54.0	47.0	100												
4	49.0	54.0	52.0	100											
5	50.0	51.0	44.0	48.5	100										
6	46.0	59.0	46.0	47.0	48.0	100									
7	47.0	48.0	48.0	46.0	65.0	47.0	100								
8	56.0	51.0	56.0	51.5	46.0	58.0	25.0	100							
9	50.0	45.0	49.0	50.0	60.0	40.0	79.0	30.0	100						
10	50.0	45.0	54.0	50.5	58.0	41.0	77.0	36.0	92.0	100					
11	53.0	54.0	49.0	45.5	65.0	51.0	92.0	31.0	75.0	73.0	100				
12	48.0	47.0	49.0	50.0	58.0	42.0	81.0	30.0	96.0	94.0	75.0	100			
13	47.0	44.0	49.0	49.5	59.0	44.0	68.0	41.0	81.0	83.0	62.0	81.0	100		
14	55.0	46.0	55.0	51.5	57.0	44.0	72.0	39.0	81.0	81.0	72.0	81.0	74.0	100	
15	56.0	45.0	57.0	53.0	54.0	44.0	67.0	40.0	78.0	72.0	67.0	74.0	67.0	87.0	100

Yule Coefficient

This coefficient has been less commonly used in numerical taxonomy. It is calculated as:

$$S_Y = \frac{ac - bc}{ad - bc}$$

Taxonomic Distance

Taxonomic distance between the OTUs can be easily calculated as a value 1 minus similarity, or 100 minus percentage similarity. It can also be directly calculated as Euclidean distance using formula proposed by Sokal (1961):

$$\Delta jk = [\sum_{i=1}^{n} (X_{ij} - X_{ik})^2]^{1/2}$$

The average distance would be represented as:

$$d_{jk} = \sqrt{\Delta^2_{jk}/n}$$

Other commonly used distance measures include **Mean character difference** (M.C.D.) proposed by Cain and Harrison (1958), **Manhattan metric distance coefficient** (Lance and Williams, 1967) and **Coefficient of divergence** (Clark, 1952).

Once the similarity or distance between every pair of taxa has been calculated, the data are presented in a second matrix with **t** × **t** dimensions where both rows and columns represent taxa (Table 8.2, Table 8.3). It must be noted that diagonal **t** value in the matrix represents self-comparison of taxa and thus 100% similarity. These values are redundant as such. The values in the triangle above this diagonal line would be like the triangle below. The effective number of similarity values as such would be **t** × (**t** – 1)/2. Thus if 15 OTUs are compared the number of values calculated would be 15 × (15 – 1)/2 = 105.

A data matrix with coded character-states for each taxon can be used for calculating the distance (and, consequently, the similarity) between every pair of taxa, including the hypothetical ancestor. The distance is calculated as the total number of character-state differences between two concerned taxa, the data presented as **t** × **t** matrix (Table 8.5).

This method is closer to taxometric methods, because both plesiomorphic and apomorphic character-states are given equal weightage, but the inclusion of hypothetical ancestor is always crucial for the study.

Another method of calculating distance involves calculation of the number of apomorphic character-states common between the pairs of concerned taxa, ignoring the possession of plesiomorphic character-states in common (Table 8.6). Since only synapomorphy is likely to define monophyletic groups, this method is closer to the original cladistic concept.

Table 8.3: Dissimilarity matrix of the representative hypothetical taxa based on the similarity matrix in Table 8.2.

OTUs	1	2	3	4	5	6	7	8	9	10	11	12	13	14	15
1	0.0														
2	53.0	0.0													
3	46.0	53.0	0.0												
4	51.0	46.0	48.0	0.0											
5	50.0	49.0	56.0	51.5	0.0										
6	54.0	41.0	54.0	53.0	52.0	0.0									
7	53.0	52.0	52.0	54.0	35.0	53.0	0.0								
8	44.0	49.0	44.0	48.5	54.0	42.0	75.0	0.0							
9	50.0	55.0	51.0	50.0	40.0	60.0	21.0	70.0	0.0						
10	50.0	55.0	46.0	49.5	42.0	59.0	23.0	64.0	8.0	0.0					
11	47.0	46.0	51.0	54.5	35.0	49.0	8.0	69.0	25.0	27.0	0.0				
12	52.0	53.0	51.0	50.0	42.0	58.0	19.0	70.0	4.0	6.0	25.0	0.0			
13	53.0	56.0	51.0	50.5	41.0	56.0	32.0	59.0	19.0	17.0	38.0	19.0	0.0		
14	45.0	54.0	45.0	48.5	43.0	56.0	28.0	61.0	19.0	19.0	28.0	19.0	26.0	0.0	
15	44.0	55.0	43.0	47.0	46.0	56.0	33.0	60.0	22.0	28.0	33.0	26.0	33.0	13.0	0.0

Table 8.4: Data matrix of t taxa and n characters scored as 0 (plesiomorphic) and 1 (apomorphic) character-states. Multistate character is assigned 0 for ancestral state, 1 for intermediate and 2 for most advanced state.The matrix is similar to Table 8.1 but only 9 characters pictured are used for calculations. Also, the last taxon included is the hypothetical ancestor in which all character-states are scored as 0 (plesiomorphic), as it is presumed that the ancestor would possess all characters in a plesiomorphic state.

Characters (*n*)→ Taxa (*t*)↓	Habit 0-woody 1-herbaceous	Fruit 0-follicle 1-achene	Ovary 0-superior 1-inferior	Leaves 0-simple 1-lobed 2-compound	Habitat 0-terrestrial 1-aquatic	Pollen 1-triporate 0-monosulcate	Ovule 1-unitegmic 0-bitegmic	Campels 0-free 1-united	Plastids 1-PI-type 0-PII-type
1.	1	0	1	1	0	1	1	1	1
2.	1	0	1	1	1	1	0	0	1
3.	0	1	0	1	0	0	1	1	1
4.	1	1	1	0	1	0	0	0	0
5.	1	0	1	2	0	1	1	0	1
6.	1	1	0	1	1	1	0	1	0
7.	0	1	1	0	1	1	0	0	1
8.	0	0	0	1	0	0	1	1	1
9.	1	1	1	0	0	1	1	0	0
10.	0	0	0	1	1	0	1	1	1
11.	1	0	1	2	0	1	0	0	1
12.	1	1	0	0	1	1	0	0	1
13.	0	0	1	0	1	0	1	0	1
14.	1	0	1	0	1	0	1	0	1
15.	0	0	0	0	0	0	0	0	0

Table 8.5: t × t matrix presenting distance between taxa expressed as the number of character-state differences between pairs of taxa.

Eus	1	2	3	4	5	6	7	8	9	10	11	12	13	14	15
1	0														
2	3	0													
3	4	7	0												
4	7	4	8	0											
5	2	3	6	7	0										
6	5	5	6	4	7	0									
7	6	3	7	3	6	6	0								
8	3	6	1	8	5	6	7	0							
9	4	5	7	3	4	6	4	7	0						
10	4	5	2	7	6	5	6	1	8	0					
11	3	2	7	6	1	6	5	6	5	7	0				
12	6	3	7	3	6	4	2	7	4	6	5	0			
13	5	4	5	4	5	8	3	4	5	3	6	5	0		
14	4	3	6	3	4	7	4	5	4	4	5	4	1	0	
15	7	6	5	4	7	6	5	4	5	5	6	5	4	5	0

Table 8.6: t × t matrix presenting distance between taxa expressed as number of derived (apomorphic) character-states common between pairs of taxa.

EUs	1	2	3	4	5	6	7	8	9	10	11	12	13	14	15
1	X														
2	5	X													
3	4	2	X												
4	2	3	0	X											
5	5	4	2	2	X										
6	3	3	2	2	2	X									
7	3	4	1	3	3	2	X								
8	4	2	4	0	2	2	1	X							
9	4	3	1	3	4	2	3	1	X						
10	4	3	4	1	2	3	2	4	1	X					
11	4	4	1	2	5	2	3	1	3	1	X				
12	3	4	1	3	3	3	4	1	3	2	3	X			
13	3	3	2	2	3	1	3	2	2	3	2	2	X		
14	4	4	2	3	4	2	3	2	3	3	3	3	4	X	
15	0	0	0	0	0	0	0	0	0	0	0	0	0	0	X

Construction of Trees

Different methods are available for the final analysis of cladistic information. Three of these commonly used in phylogenetic analysis include Parsimony-based methods, Distance methods and Maximum likelihood method.

Parsimony-based Methods

The methods are largely based on the biological principle that mutations are rare events. The methods attempt to minimize the number of mutations that a phylogenetic tree must invoke for all taxa under consideration. A tree that invokes minimum

number of mutations (changes) is considered to be the tree of **maximum parsimony**. The evolutionary polarity of taxa is decided for construction of such trees. The **Wagner groundplan divergence method**, an example of this, was first developed by H. W. Wagner in 1948 as a technique for determining the phylogenetic relationships among organisms that he hoped would replace intuition with analysis. The method was based on determining the apomorphic character-states present within a taxon and then linking the subtaxa based on relative degree of apomorphy. Interestingly, whereas the method found little favour with zoologists, it has been used in many botanical studies. Kluge and Farris (1969) and Farris (1970) developed a comprehensive methodology for the development of Wagner trees, based on the principle of parsimony. The method is the basis of many phylogeny computer algorithms currently in use. A given dataset may, however, yield many possible equally parsimonious trees due to homoplasy, as more than one character-state change may occur during the evolutionary process of a particular group of organisms.
The following steps are involved in the analysis:

1. Determine which of the various characters (or character-states) in a series of character transformations are apomorphic.
2. Assign the score of 0 to the plesiomorphic character and 1 to the apomorphic character in each transformation series. If the transformation series contains more than two homologues, then these 'intermediate apomorphies' may be scaled between 0 and 1. Thus, a transformation series of three characters may be scored as 0, 0.5 and 1 (or 0, 1 and 2 depending on the weighting assigned).
3. Construct a table of taxa (EUs) and coded characters (or character-states: see Table 8.3).
4. Determine the **divergence index** for each taxon by totaling up the values. Since apomorphic character-states are coded 1, the divergence index in effect represents the number of apomorphies (character-states) in a taxon, except in cases of weighted coding. For the data matrix in Table 8.4, the divergence index for 15 taxa would be calculated as:

Taxon	Divergence index
1	7
2	6
3	5
4	4
5	7
6	6
7	5
8	4
9	5
10	5
11	6
12	5
13	4
14	5
15	0

Note that the hypothetical ancestral taxon 15 has an index of 0.

5. Plot the taxa on a graph, placing each taxon on a concentric semicircle that equals its divergence index. The lines connecting the taxa are determined by shared synapomorphies (see Table 8.6). The cladogram (**Wagner tree**) is presented in Figure 8.16.

Not all cladistic methods apply the principle of parsimony. The methods of **compatibility analysis** or **clique analysis** utilize the concept of character compatibility. Such methods can detect and thus omit homoplasy. They can be carried out manually or using a computer program and can generate both rooted as well as unrooted trees. Groups of mutually compatible characters are termed **cliques**. Let us consider two characters, A and B, with two character-states each. Four character-state combinations are possible:

Assuming the evolution has proceeded from A1 to A2 and from B1 to B2. If all the four combinations are met in nature then obviously there must have been at least one reversal (A2 to A1) or parallelism (A1 to A2 occurring twice), and as such A and B are incompatible. On the other hand, if only two or three of the combinations occur, then A and B are compatible. Cliques are formed by comparing all pairs of characters and finding mutually compatible sets. The largest clique is selected from the data to produce a cladogram. Finally, a rooted tree or network is obtained according to whether or not a hypothetical ancestor was included in the analysis.

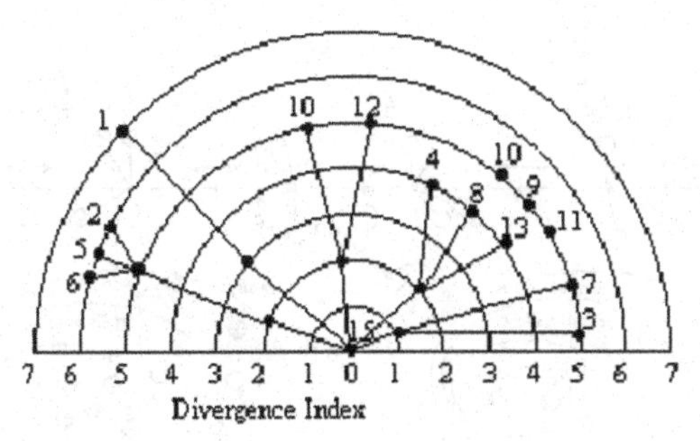

Figure 8.16: General representation of a **Wagner tree**.

Multiple Trees

The **unrooted tree** constructed in Figure 8.13B represented only a small portion of the evolutionary sequence. Extension of this tree would make it more complicated and present a lot of possibilities. Let us add a small portion of the herbaceous lineage with yellow petals, again assuming that there are a total of 15 herbaceous species of which six are with red petals and 9 with yellow petals. Of these nine 4 are with united carpels and 5 with free carpels. The additional Venn diagram for the herbaceous species and the extended unrooted tree is presented in the Figure 8.17C. It must be noted that here we know the evolutionary history of the group—which normally is never known—and the aim of phylogenetic analysis is to reconstruct and depict this evolutionary history through trees. The unrooted tree here has five character-state changes (actual changes, tree length) involved. The change from free to united carpels has occurred twice, and as such there are only four genetic switches involved. If we did not have the knowledge about the evolutionary history of the group, we would try a number of variations. One possible variation of the unrooted tree would be to link 4 herbaceous species with united carpels to the woody species with united carpels, thus presenting a single change of free to united carpels. But this brings in further changes. Now, change from woody habit has occurred twice, change from red to yellow petals has occurred twice, and more significantly the number of actual changes (tree length) has increased to six (Figure 8.18), with same four genetic switches involved. With more descendents being included in the tree, the number of options would increase. Also, we have to convert each unrooted tree into a **rooted tree** so that the most primitive basal end of the tree is known, and different lineages presented as the more advanced branches. This brings in many more options, as indicated earlier. In our example, where we know the history of the tree, the tree can be rooted at R, as indicated by an arrow (Figure 8.17C), but in a large majority of cases, it is a complicated process, and a lot of hypotheses, strategies and algorithms come into play.

A number of sophisticated computer algorithms are available which compare trees and calculate their lengths. The widely used ones include NONA, PAUP, and PHYLIP. These programs determine the number of possible trees, and then sort out the shortest of all these. If we are dealing with three species, three rooted trees are possible [A (B, C)], [B (A, C)] and [C (A, B)] (Figure 8.19), if 4 taxa are mapped the 15 trees are possible, 5 then 105, for 10 taxa 34,459,425 trees and so on. The number of possible rooted trees for ***n*** number of taxa can be calculated as:

$$\boldsymbol{Nr} = (2n-3)^{*}/[(2^{n-2}) \text{ X } (n-2)!]$$

It can also be calculated as:

$$\boldsymbol{Nr} = \text{P}\,(2\,i - 1)$$

where P represents the product of all factors $(2\,i - 1)$ from $i = 1$ to $i = \text{n} - 1$. A simpler way to calculate the possible number of rooted trees is as follows:

$$\boldsymbol{Nr} = (2(n + 1) - 5) \text{ X number of trees for } (n - 1) \text{ taxa}$$

As noted above, the number of possible rooted trees is much more than number of possible unrooted trees. Latter can be calculated as:

$$\boldsymbol{Nu} = (2n - 5)! / [(2^{n-3}) \text{ X } (n-3)!]$$

or more simply as:

$$\boldsymbol{Nu} = \text{number of rooted trees for } (n - 1) \text{ taxa}$$

Thus, for 3 taxa, 3 rooted trees and 1 unrooted trees are possible (Figure 8.19), for 4 taxa 15 rooted trees and 3 unrooted trees and for 5 taxa, 105 rooted trees are possible but only 15 unrooted trees. For our 8 species in Figure 8.12, if evolutionary history was not known we should expect 135135 rooted trees and 10395 unrooted trees. The figures also highlight the enormous challenges in reconstructing the evolutionary history of any group.

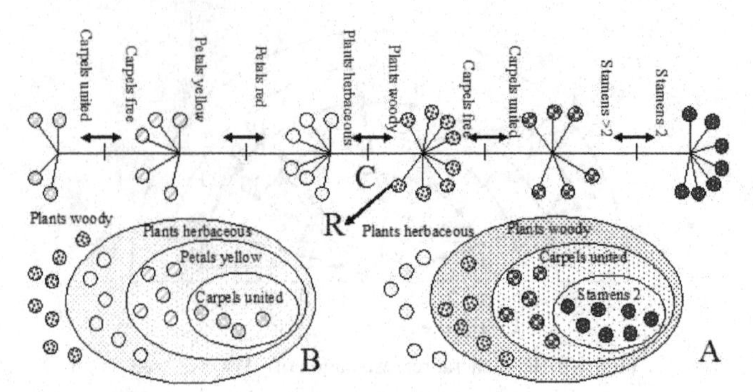

Figure 8.17: A: The Venn diagram for woody species, the same as Figure 8.12A. **B:** The Venn diagram for a small portion of the herbaceous lineage of assumed 15 species of which 6 are with red petals and 9 with yellow petals, latter with 4 species having united carpels and 5 free carpels. **C:** Extension of the unrooted tree of Figure 10.12B to include the species depicted in the Venn diagram B here. There are 5 actual character state changes but with 4 switches as united carpels have arisen twice.

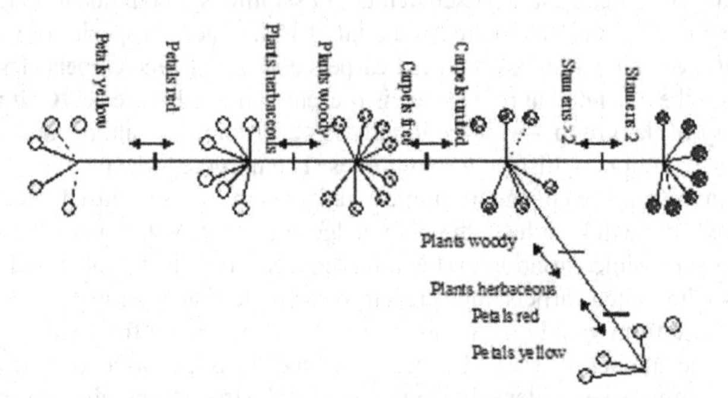

Figure 8.18: Possible variation of the unrooted tree presented in Figure 8.17C, if we did not have any idea about the evolutionary history of the group. Note that tree length has increased to six, and habit has changed twice from woody to herbaceous and from red to yellow petals. Such homoplasious situations are uncommon.

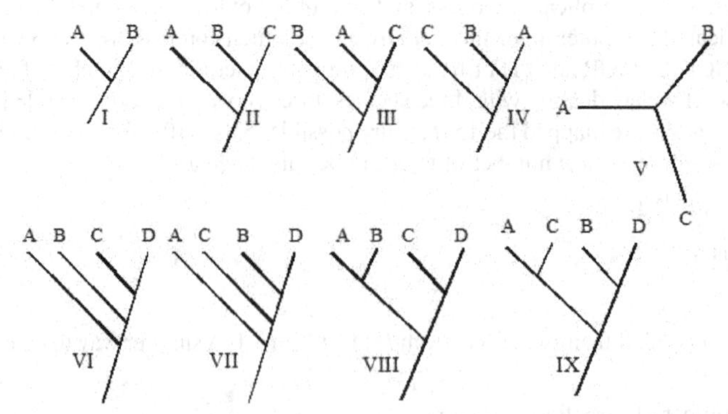

Figure 8.19: Possible number of rooted and unrooted trees. I: Single rooted tree for two taxa. **II–IV:** Three possible rooted trees for three taxa. **V:** One possible unrooted tree for three taxa. **VI–IX:** Some of the possible 15 rooted trees for four taxa.

A large number of trees generated are sorted and, ones presenting the shortest evolutionary path, in agreement with the **principle of parsimony**, are shortlisted.

Distance Methods

Distance methods were originally developed for handling phenetic information and construction of phenograms, some of these have now been incorporated in cladistic methodology. Cluster analysis the most commonly used method of constructing trees.

Cluster Analysis

Data presented in OTUs × OTUs (**t** × **t**) matrix are too exhaustive to provide any meaningful picture and need to be further condensed to enable a comparison of units. Cluster analysis is one such method in which OTUs are arranged in the order of decreasing similarity. The earlier methods of cluster analysis were cumbersome and involved shifting of cells with similar values in the matrix so that OTUs with closely similar similarity values were brought together as clusters. Today, with the advancement of computer technology, programs are available which can perform an efficient cluster analysis and help in the construction of cluster diagrams or phenograms. The various clustering procedures are classified under two categories.

Agglomerative Methods

Agglomerative methods start with **t** clusters equal to the number of OTUs. These are successively merged until a single cluster has finally been formed. The most commonly used clustering method in biology is the **Sequential Agglomerative Hierarchic Non-overlapping clustering method** (SAHN). The method is useful for achieving hierarchical classifications. The procedure starts with the assumption that only those OTUs would be merged which show 100% similarity. As no two OTUs would show 100% similarity, we start with **t** number of clusters. Let us now lower the criterion for merger as 99% similarity; still no OTUs would be merged as in our example the highest similarity recorded is 96.0%. The best logical solution would be to pick up the highest similarity value (here 96.0) and merge the two concerned OTUs (here 9 and 12). By inference, if our criterion for merger is 96.0 we will have **t**–1 clusters. Subsequently the next lower similarity value is picked up and the number of clusters reduced to **t**–2. The procedure is continued until we are left with a single cluster at the lowest significant similarity value. Since at various steps of clustering a candidate OTU for merger would cluster with a group of OTUs, it is important to decide the value that would link the clusters horizontally in a cluster diagram. A number of strategies are used for the purpose. In the commonly used **single linkage clustering** method (**nearest neighbour** technique or **minimum method**), the candidate OTU for admission to a cluster has similarity to that cluster equal to the similarity to the closest member within the cluster. The connections between OTUs and clusters and between two clusters are established by single links between pairs of OTUs. This procedure frequently leads to long straggly clusters in comparison with other SAHN cluster methods. The phenogram for our data using this strategy is shown in Figure 8.20.

The highest similarity value in our matrix (see Table 8.2) is 96.0 between OTUs 9 and 12, and as such they are linked at that level. The next similarity value of 94.0 is between OTUs 10 and 12, but since 12 has already been clustered with 9, 10 will join this cluster linked at 94.0. The process is repeated till all OTUs have been agglomerated into single cluster at similarity value of 53.0.

In the **complete linkage clustering** method (**farthest neighbour** or **maximum method**) the candidate OTU for admission to a cluster has similarity to that cluster equal to its similarity to the farthest member within the cluster. This method will generally lead to tight discrete clusters that join others only with difficulty and at relatively low overall similarity values.

In the **average linkage clustering** method, an average of similarity is calculated between a candidate OTU and a cluster or between two clusters. Several variations of this average method are used. The unweighted pair-group method using arithmetic averages (**UPGMA**) computes the average similarity or dissimilarity of a candidate OTU to a cluster,

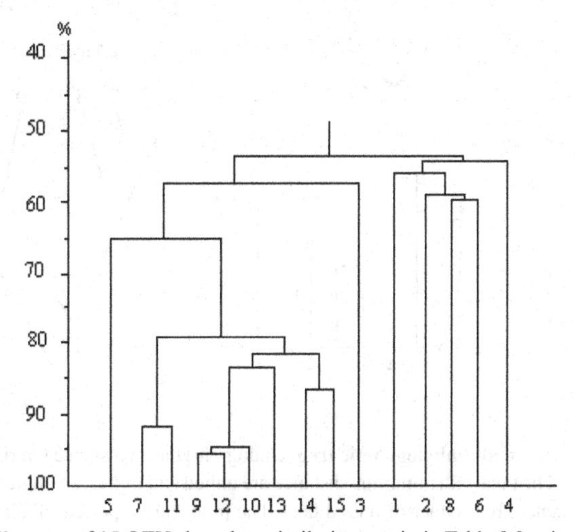

Figure 8.20: Cluster diagram of 15 OTUs based on similarity matrix in Table 8.2 using single linkage strategy.

weighting each OTU in the cluster equally, regardless of its structural subdivision. The method originally developed for the procedures of numerical taxonomy has been applied in phylogenetic analysis with relevant modifications, and used for the construction of trees. **UPGMA method** procedure begins with as many clusters as the number of taxa. The two taxa with **minimum distance** merged to reduce the number of clusters by one. In the next step average distance between new cluster and remaining taxa are determined by taking the average distance between these two members and all other remaining taxa, weighting each taxon in the cluster equally regardless of its structural subdivision, and merging the taxon with smallest distance to the first cluster. The process is repeated with this new cluster of three taxa, and the procedure continues till all the taxa are merged, the most distant taxon joining last of all. From measure of similarity or dissimilarity of taxa (OEUs) as presented in Tables 8.5 and 8.6, a network presenting minimum dissimilarity is constructed. Analysis of data from first six taxa of Table 8.4 is presented in Figure 8.21. The procedure begins by uniting nearest taxa A and E (with minimum distance of 2). Next matrix in now constructed in which distance between (AE) and rest of the taxa is recalculated. The lowest value in this matrix (step 1 matrix) is between (AE) and E, which are next united at distance level 3 into (AE)B. The distance between this cluster and rest of the taxa is now recalculated as presented in step 2 matrix. The lowest distance in this matrix is 4 between D and F which are united into one cluster. With this merger the distance between taxa/clusters is recalculated and presented in step 3 matrix. The lowest distance 5.5 is now between clusters (AE) B and DF, which are next united. Finally, the distance between this enlarged cluster and C is recalculated as presented in step 4 matrix. Finally the two clusters (((AE)B)(DF)) and C are united at distance of 6.5 to form final cluster ((((AE)B)(DF)) C). The resulting phenogram and reconstructed phylogenetic tree constructed from the analysis are presented in Figure 8.21.

The distance matrix can similarly be generated from single nucleotide differences between homologous DNA sequences derived from different species. Six such hypothetical sequences from different species are presented in Figure 8.22. The distance matrix is based on number nucleotide differences between different sequences. A and F with lowest distance are merged, followed by recalculation of new matrix in which A and F form one cluster and value of each taxon is calculated as average distance from A and F. Now lowest value is shared by B and D which form second cluster. The values are recalculated similarly, and successively C joins AF cluster, and then E joins (AF)C cluster. The two clusters are finally merged to enable construction of phylogenetic tree either as phenogram or as cladogram. Some types of genetic polymorphism data such as *RAPD* are best handled when sharing of 0 code by two taxa in the matrix is ignored (when both taxa lack a given polymorphic band in gel electrophoresis. Jaccard coefficient is best suited for handing such data. Figure 8.23 presents results of *RAPD* analysis of 8 taxa, where only polymorphic bands are shown, monomorphic bands being omitted. The distance matrix based on similarity matrix was processed using PHYLIP, as shown in Figure 8.24. Distance methods are suitable for handling both morphological and molecular data, or a combination of both. These methods use all data with usually equal importance, whereas the parsimony methods use only informative molecular data. In general, for a site to be informative, when handling sequence data, irrespective of how many sequences are aligned,

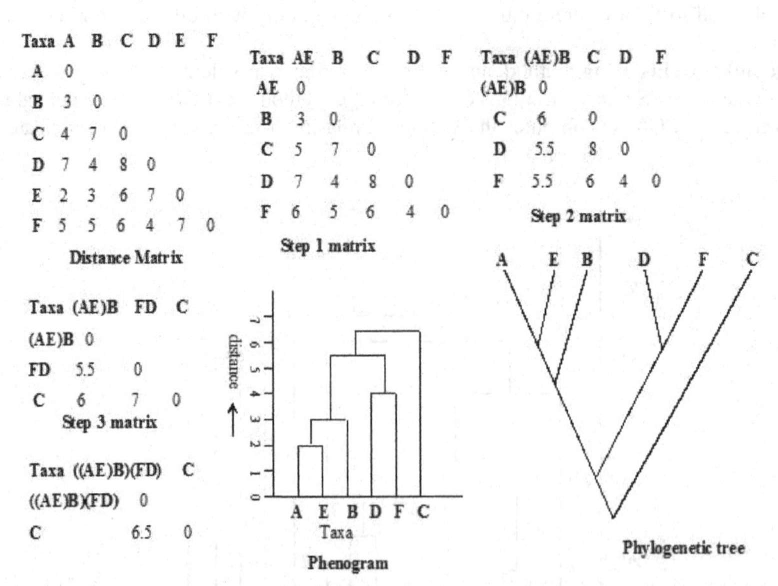

Figure 8.21: Construction of phenogram and phylogenetic tree (cladogram) based distance matrix concerning first six taxa in Table 8.4 using UPGMA clustering method. The taxa with minimum distance are united and treated as single cluster in next matrix, and distance values recalculated as average of distance from either of united taxa. The procedure repeated till all taxa are united. The phylogenetic tree is constructed based on sequence of clustering of taxa.

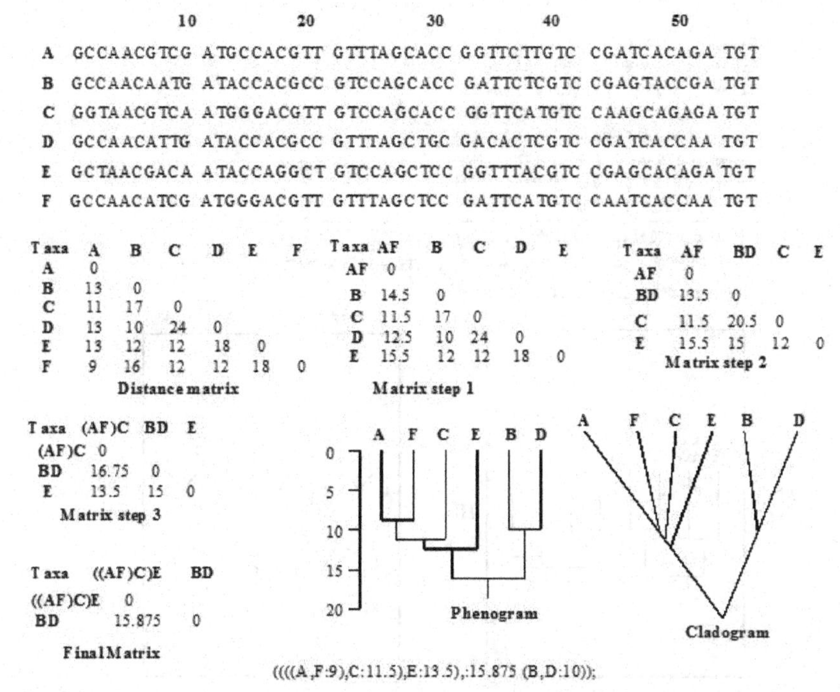

Figure 8.22: Construction of phylogenetic tree based on single nucleotide differences in hypothetical DNA sequences of six species. Distance matrix is constructed based on the number of nucleotide differences between each pair of DNA sequences and presented in distance matrix. Further analysis proceeds as detailed in Figure 8.21, and also in the text on these pages.

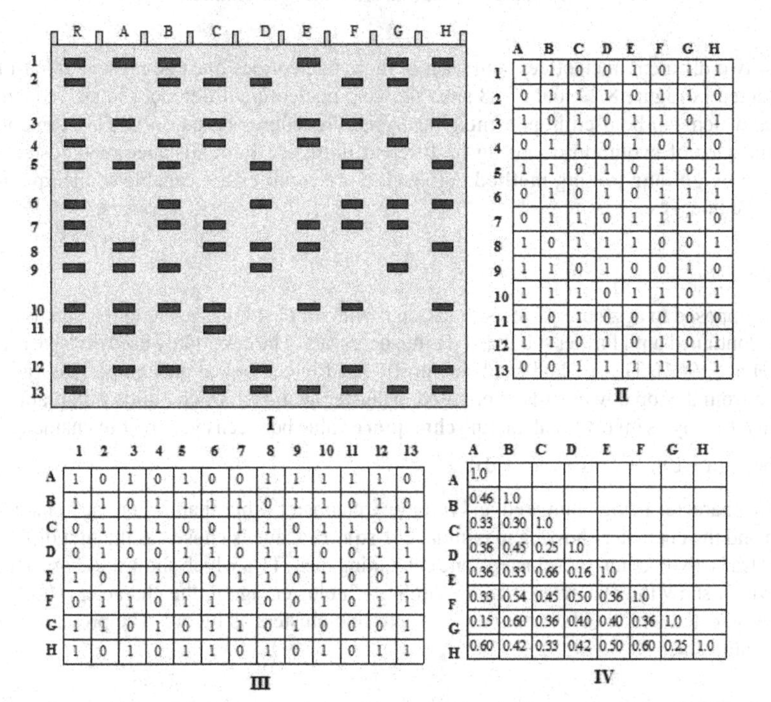

Figure 8.23: Phylogenetic analysis of data concerning polymorphic bands from gel electrophoresis from DNA of 8 taxa. **I:** Polymorphic bands of DNA from 8 taxa (A–H), R representing reference bands; **II:** Binary coded matrix of the polymorphic bands; **III:** The same matrix presented in conventional format; **IV:** Lower triangular matrix of similarity matrix using Jaccard coefficient, wherein sharing of 0 state (absence of bands) is ignored. Further handling of data using UPGMA program of PHYLIP is presented in Figure 8.24.

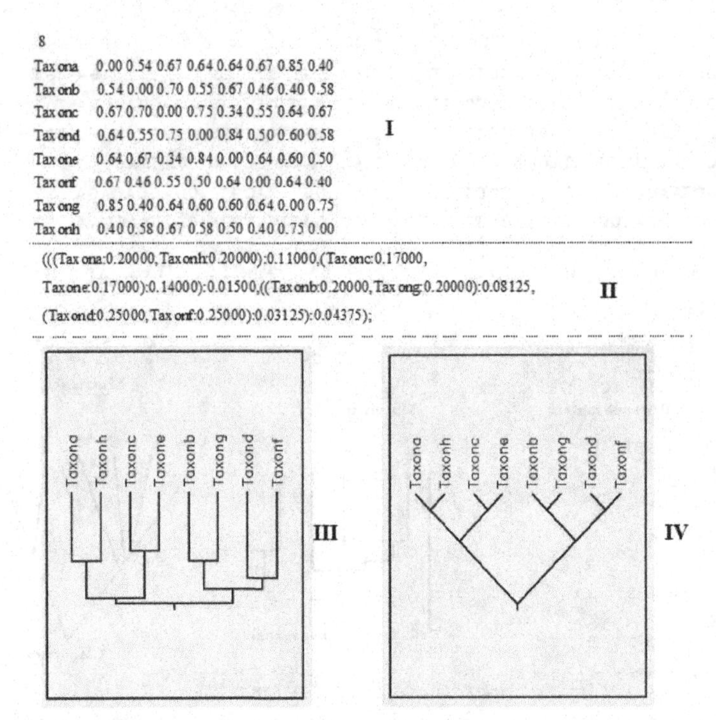

Figure 8.24: Construction of phylogenetic tree based on polymorphic bands from gel electrophoresis from DNA of 8 taxa using UPGMA program of PHYLIP; **I:** Square distance generated from Figure 8.23IV, each value calculated 1-similarity value. **II:** Outtree file generated by UPGMA option of NEIGHBOUR program; **III:** Upright square tree (Phenogram) plotted through DRAWGRAM program; **IV:** Cladogram, but with branch lengths omitted.

it must have at least two different nucleotides, and each of these nucleotides must be present at least twice. Thus, in the sequence data presented in Figure 8.22, out of 28 sites showing nucleotide differences in six sequences, there are only 12 informative sites which can be used in parsimony analysis. Procedures based on UPGMA method, however, don't account for different rates of evolution occurring in different lineages. Some distance methods such as **transformed distance method** and **neighbour-joining method**, although more complex are capable of incorporating different rates of evolution within the lineages.

Divisive Methods

Divisive methods as opposed to agglomerative methods, start with all **t** OTUs as a single set, subdividing this into one or more subsets; this is continued until further subdivision is not necessary. The commonly used divisive method is **association analysis** (William et al., 1966). The method has been mostly used in ecological data employing two state characters. It builds a dendrogram from the top downwards as opposed to cluster analysis, which builds a diagram from the bottom up.

The first step in the analysis involves calculating **chi-square** value between every pair of characters using the formula:

$$X^2_{hi} = n(ad - bc)^2/[(a + b)(a + c)(b + d)(c + d)]$$

where *i* stand for the character being compared and *h* for any character other than *i*. For each character the sum of chi-square is computed and the character showing maximum **chi-square** value is chosen as the first differentiating character. The whole set of OTUs is divided into two clusters, one containing the OTUs which show the character-state *a* and another containing OTUs which show the character-state *b*. Within each cluster, again, the character with the next value of the sum of chi-square is selected and the cluster subdivided into two clusters as before. The process is repeated till further subdivision is not significant.

Hierarchical Classifications

The phenogram constructed using any technique or strategy can be used for attempting hierarchical classification, by deciding about certain threshold levels for different ranks. One may tentatively decide 85 per cent similarity as the threshold

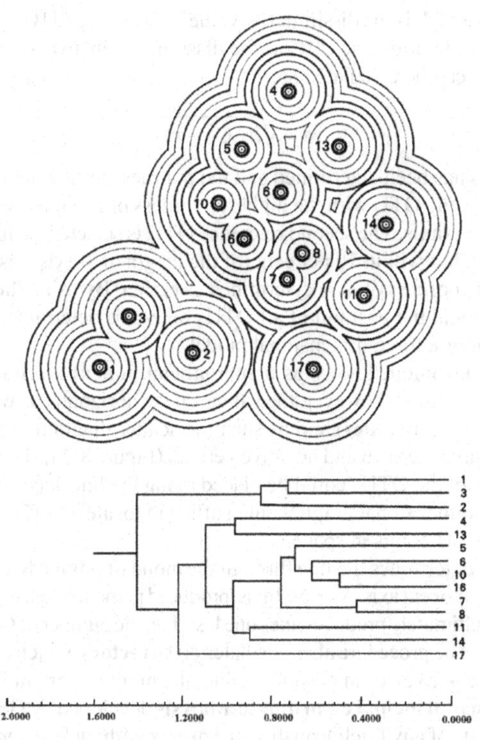

Figure 8.25: Contour diagram based on the phenogram shown alongside.

for the species, 65 for genera and 45 for families and recognize these ranks based on number of clusters established at that threshold. Whereas such an assumption can help in hierarchical classification, the point of conflict would always be the threshold level for a particular rank. Some may argue—and are justified in doing so—to suggest 80 per cent (or any other value) as the threshold for species. It is more common, therefore, to use terms **85 per cent phenon line, 65 per cent phenon line, and 45 per cent phenon lines.** These terms may conveniently be used till such time that sufficient data are available to assign them formal taxonomic ranks to the various phenon lines.

The results of cluster analysis are commonly presented as dendrograms known as **phenograms**. They can also be presented as **contour diagrams** (Figure 8.25), originally developed under the name **Wroclaw diagram** by Polish phytosociologists. The contour diagram may also incorporate the levels at which clustering has taken place.

Ordination

Ordination is a technique which determines the placement of OTUs in two-dimensional or three-dimensional space. The results of two-dimensional ordination are conveniently represented with the help of a scatter diagram and those of three-dimensional ordination with the help of a three-dimensional model. The procedure works on distance values calculated directly from the coded data or indirectly from the already calculated similarity values as 100 minus similarity (if similarity values are in percentage) or 1 minus similarity (if similarity values range between 0 to 1). A dissimilarity matrix based on Table 8.2 is presented in Table 8.3.

The first step in the ordination starts with construction of the x-axis (horizontal axis). In the commonly used method of **polar ordination,** the two most distant OTUs are selected as the end points (A and B) on x-axis. In our example, these are OTU 8 and 7 with a distance (dissimilarity value) of 75. The position of all other OTUs on this axis can be plotted one by one. OTU 10 has a distance of 64 from A (OTU 8) and a distance of 23 from B (OTU 7). A compass with a radius of 64 units is swung from A and a compass with a radius of 23 units is swung from B, forming two arcs. A line joining the intersection of two arcs forms a perpendicular on the x-axis, and the point at which the line crosses the x-axis is the position of the OTU. The distance between the x-axis and the point of intersection of arcs is the **poorness of fit** of the concerned OTU. The location of OTU on the axis from the left (point A) can also be calculated directly instead of plotting:

$$x = \frac{L^2 + dAC^2 - dBC^2}{2L}$$

where x is the distance from the left end, L is the dissimilarity value between A and B (length of x-axis), *d*AC is dissimilarity between A and the OTU under consideration and *d*BC as the dissimilarity between B and the OTU under consideration. The poorness of fit (e) of this OTU can be calculated as:

$$e = \sqrt{d\mathrm{AC}^2 - x^2}$$

After the position of all OTUs has been determined and the poorness of fit calculated, a second axis (vertical axis or y-axis) has to be calculated. For this, the OTU with the highest poorness of fit (most poorly fitted to x-axis) is selected and this forms the first reference OTU of y-axis. The second reference OTU is selected as that one with the highest dissimilarity to the first reference OTU of y-axis, but within 10 per cent (of the length of x-axis) distance on x-axis. The position of all other OTUs on the y-axis and their poorness of fit is determined as earlier. By using the values of poorness of fit to y-axis, a z-axis can be similarly generated and the position of all OTUs on z-axis determined similarly. The values can be used for constructing a **scatter diagram** or a three-dimensional model.

A commonly used ordination technique known as **principal component analysis** also calculates values for a two-dimensional scatter diagram. In this method, however, the values on the horizontal as well as the vertical axis are non-zero, ranging from –1 to 1 (calculated as **eigenvalues**) and as such the scatter diagram is presented along four axes: positive horizontal, negative horizontal, positive vertical and negative vertical (Figure 8.26). The technique assumes that if a straight line represented a single character, all the OTUs could be placed along the line according to their value for that character. If two characters were used, a two-dimensional graph would suffice to locate all OTUs. With n characters, n-dimensional space is required to locate all OTUs as points in space.

Principal component analysis determines the line through the cloud of points that accounts for the greatest amount of variation. This is the first principal component axis. A second axis, produced perpendicular to the first, accounts for the next greatest amount of variation. The procedure ultimately produces axes one less than the number of OTUs. The first two axes are generally plotted to produce a scatter diagram. The procedure also calculates **eigenvectors**, which indicate the importance of a character to a particular axis. The larger the eigenvector in absolute value, the more important is that particular character.

A related method of ordination is **principal co-ordinate analysis** developed by Gower (1966). This technique enables computation of principal components of any Euclidean distance matrix without being in possession of original data matrix. The method is also applicable to non-Euclidean distance and association coefficients as long as the matrix has no large negative eigenvalues. Principal co-ordinate analysis also seems to be less disturbed by NC entries than principal components.

Maximum Likelihood Method

The method is similar to distant method in that all data is taken into consideration. In this method, similarly character-state transformations are compared, and the probability of changes determined. These probabilities are used to calculate the likelihood that a given tree would lead to the particular data set observed, and the tree with maximum likelihood is selected. The method is especially suited to molecular data, where the probability of genetic changes can be modeled more

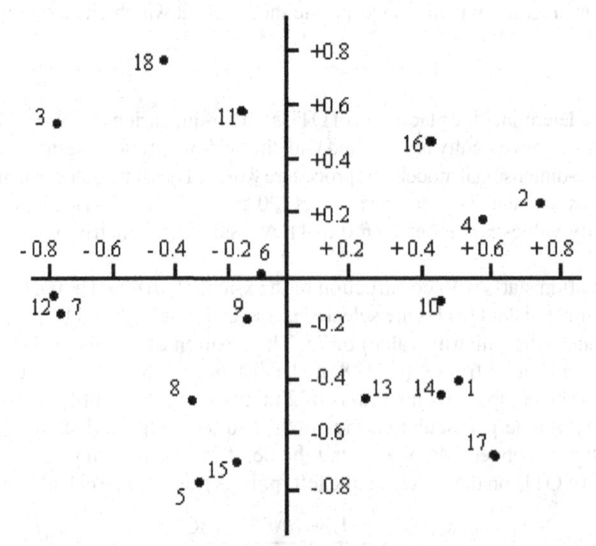

Figure 8.26: Plot of the results of the **principal component analysis** of 18 hypothetical taxa.

easily. With this approach, the probabilities are considered for every individual nucleotide substitution in a set of sequence alignments. It is commonly understood that transitions occur three times more frequently as compared to transversions. Thus, if C, T, and A occur in one column (representing one site), the sequences with C and T (pyrimidines) are more likely to be closely related than sequence with A (Purine). Using objective criteria probability for each site and every possible tree that describes the relationship of sequences. The tree with highest aggregate probability is selected as representation of a true phylogenetic tree.

Using any one of the methods, a large dataset commonly used, and which includes many homoplasies, large number of shortest trees may be generated by these automated algorithms. These short-listed trees have to be further compared.

The Consensus Tree

The use of automated methods based on parsimony, even after applying relevant strategies yield several trees, all presenting shortest pathways, based on parsimony but with different linkages among the taxa (OEUs), and often presenting different evolutionary history. Molecular studies of Clusiaceae by Gustafsson et al. (2002) for example, including 1409 nucleotide of chloroplast gene *rbcL* positions using PAUP*4.0b8a parsimony analysis method, yielded 8473 most parsimonious trees for the 26 species compared. Interestingly, the number of trees generated was so large that search for trees 3 steps longer than most parsimonious trees was aborted. More significantly different data sets (molecular, morphology) may yield different trees. While selecting the consensus tree, the commonest approach is to identify the groups, which are found in all the shortlisted trees, and build a **consensus tree**. This could be achieved in different ways.

Strict Consensus Tree

A more conservative approach in building a consensus tree involves including only monophyletic groups that are common to all the trees. The tree developed this way is known as **strict consensus tree**. Consider the two most parsimonious trees (although there could often be numerous trees of same shortest length available for comparison) as shown in Figure 8.27-I and 8.27-II.

Imagine that all groups A to J are monophyletic. Tree I shows that A and B are very closely related, and so are H and I. C, D, E, and F are shown arising successively and are related in that sequence. Tree II shows a similar relationship between H and I, and between A and B (but group J is shown related to these two). The tree also shows that C and D are closely related. As relationships between E, F, and G are ambiguous, they are shown arising from the same point in evolutionary history. The consensus tree III would thus omit taxon J (which is absent from tree I), show A and B, as also H and I as in the two trees I and II. The other taxa C, D, E, F, and G are shown arising from the same point.

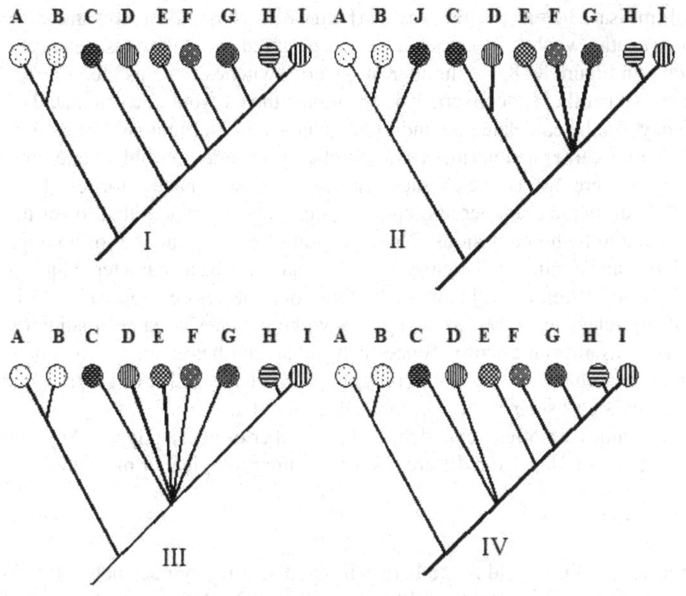

Figure 8.27: Two most parsimonious trees for a particular group of organisms, with monophyletic taxa A to J. **I:** showing C, D, E and F arising successively. **II:** E, F, G are shown arising at the same time from a common point, C and D being closely related. **III:** Strict consensus tree of trees I and II. **IV:** Semi-strict consensus tree of trees I and II.

Majority-rule Consensus Tree

Majority-rule consensus tree shows all the groups which appear in a majority of trees, say, more than 50 per cent of the trees. It is useful to indicate for each group on the consensus tree the percentage number of the most parsimonious trees in which the group appeared. Such a consensus tree, however, provides a partial summary of the phylogenetic analyses, and may be inconsistent with the trees from which it is derived.

Semi-strict Consensus Tree

A semi-strict consensus tree is useful when comparing trees from different data sets, or with different terminal taxa. The consensus tree developed indicates all the relationships supported by both type of trees or any one of these, but not contradicted by any. Thus, in Figure 8.27, tree II does not give us any information about the time of origin of E, F and G, the tree I indicates that they originated successively. Similarly, tree I does not indicate any close relationship between C and D, whereas the tree II does. The semi-strict consensus tree IV as such presents such information, not contradicted by either tree.

Evaluating Consensus Tree

Developing a consensus tree involves the use of intuition, making guesses and developing hypothesis. A number of evaluation strategies are used to test the soundness of the tree and measuring support for either the tree as a whole or for its individual branches. These values are generally published along with the tree, to allow the fair assessment of the final results for comparison of trees based on different datasets.

Consistency Index

The principle of Parsimony is based on a basic rule of science known as **Ockham's razor**, which says '***do not generate a hypothesis any more complex than is demanded by the data***'. Some information in the data may be representing homoplasy (reversals, parallelisms). Dollo parsimony (as indicated above) minimizes the use of homoplasious characters. The commonest measure of homoplasy is the **Consistency Index (CI)**, which is calculated by dividing the number of genetic switches by actual genetic changes on the tree.

Consistency Index $CI = Min/L$

Min stands for the minimum possible tree length or genetic switches, and *L* for the actual tree length or actual number of genetic changes. In the tree shown in Figure 8.13B, there are three character-state changes, each involving one switch, and, as such, the consistency index would measure 3/3 = 1. The tree shown in Figure 8.17C has five actual character-state changes (tree length is 5), but it involves only four genetic switches. As carpel fusion has occurred twice, the consistency index would accordingly be 4/5 = 0.8. In the tree shown in Figure 8.18, the number of genetic switches remains the same as four but the tree length has increased to six due to two parallel (or convergent) evolutions; the CI would be calculated as 4/6 = 0.66.

Consistency Index may also be calculated for individual characters. In Figure 8.13B as such CI for all characters is one, while in 8.17C, it is 0.5 for carpel fusion (minimum number of changes possible—one for binary character divided by actual number of changes—here 2 since the character has changed twice) and 1 for rest. In Figure 8.18, CI is 0.5 for habit and petal color, and 1 for stamen number and carpel fusion. The characters that lower the CI of a tree (or which have lower CI) are considered to be homoplasious. The inclusion of a larger number of homoplasious characters in the analysis lowers CI for the tree and contradicts phylogeny. There may also be a character, which changes only in one (or a very few) species, and may be of no relevance in others. Suppose one species develops spiny fruits. The length or number of spines would not be of any relevance in rest of the species without spines. Such a situation (a single species having a particular character) is known as **autapomorphy**. Since such a character has changed only once, it gives CI of 1, and as such the inclusion of many such characters would increase the consistency index of the tree, and provide false support. Such uninformative characters are as such omitted before calculating CI.

The Consistency index values are often dependent on the number of taxa analyzed. Any increase in number of taxa lowers CI values, and this is true for data from different sources, morphological or molecular.

Retention Index

Although theoretically the value of CI could range between 0 and 1, it rarely goes below 0.5. For a character that, has changed five times on a tree (this is a remote possibility), CI will be 0.2. More so, the value of CI for a tree, very rarely may go below 0.5, and the values thus range between 0.5 and 1. The **Retention Index (RI)** corrects this narrow range of CI by comparing maximum (and not minimum as in CI) possible number of changes in the character with actual number of changes in the character. RI is computed by first calculating the maximum possible tree length, if the apomorphic character-state originated independently in every taxon that it appears in, or say, the taxa are unrelated for the said character-state. The value of RI is calculated as:

Retention Index $RI = (Max - L)/(Max - Min)$

Max stands for the maximum tree length possible, *L* the actual tree length and *Min* the minimum tree length possible. The tree in Figure 8.17C thus has a maximum possible tree length of 9 (minimum length of 4 and actual length of 5 as we already know) and the RI would be (9–5)/(9–4) = 0.8. Higher the RI, sounder is the tree.

Bremer Support (Decay Index)

The principle of parsimony, followed in phylogenetic analyses, aims at selecting the shortest tree. Some parts of the tree may be more reliable than others. This is commonly evaluated by comparing the shortest tree with those one or more steps longer. **Decay index** or **Bremer Support** is the measure of how many extra steps are needed before the original clade (group) is not retained. Thus if an internode has decay index of 3, then the clade (monophyletic group) arising from it is maintained even in the cladogram 3 steps longer than the shortest tree (see Figure 8.29). Certain branches of the tree which appear in the shortest tree, but disappear, or 'collapse' in the tree one step longer, are not drawn in the strict consensus tree. Greater the decay index value, more robust is that internode of the cladogram.

Branches of the tree may also be tested by comparing the number of genetic changes leading up to a particular group, and the CI of individual characters involved. Doyle et al. (1994) on the basis of morphological data, developed a tree having 18 character changes leading to angiosperms. Of these 18 characters 11 had CI of 1, thus supporting the view that angiosperms form a unique group of plants.

Bootstrap Analysis

Any realistic analysis requires that the data used is randomized. Many techniques are available for randomizing the data. **Bootstrap analysis** is the commonly used method developed by Bradley Efron (1979). Its use in phylogeny estimation was introduced by Felsenstein (1985). Matrix in the Figure 8.28A contains information on the basis of which the unrooted tree in Figure 8.17C is constructed. Without touching the rows, any column is chosen at random to become the first column; similarly, any other as second and the process is repeated till the number of columns in the new matrix is the same as in the original matrix. As the columns are picked up from the original matrix, the new matrix may contain some characters represented several times (the same column may have been picked up at random more than once), while others may have been omitted (the columns were not picked up at all). The method is known as random sampling with replacement. The resultant matrix B shows that character carpel fusion was picked up twice, whereas the random selection process missed the stamen number.

Repeating the method of random selection, multiple such matrices (usually more than 100) are constructed, and for each matrix the most parsimonious tree/trees found. The consensus tree is developed from these most parsimonious trees. In this consensus tree, the percentage number of trees (generated by bootstrap analysis) that contain that clade is indicated as **bootstrap support** value of that clade. Bootstrap analysis based on the assumption that differential weighting

A

	Habit	Carpels	Stamens	Petals
Plants	Herbaceous	United	>2	Yellow
Plants	Herbaceous	Free	>2	Yellow
Plants	Herbaceous	Free	>2	Red
Plants	Woody	Free	>2	Red
Plants	Woody	United	>2	Red
Plants	Woody	United	2	Red

B

	Petals	Carpels	Habit	Carpels
Plants	Yellow	United	Herbaceous	United
Plants	Yellow	Free	Herbaceous	Free
Plants	Red	Free	Herbaceous	Free
Plants	Red	Free	Woody	Free
Plants	Red	United	Woody	United
Plants	Red	United	Woody	United

Figure 8.28: A: Matrix based on the tree 8.13. **B:** One possible matrix after procedure of random sampling with replacement.

by resampling of the original data will tend to produce same clades if the data are good and reflect actual phylogeny and very little of homoplasy. A bootstrap value of 70 per cent or more is generally considered as good support to the clade.

Several variations of bootstrap analysis are available. **The partial bootstrapping** involves sampling fewer than the full number of characters. The user is asked for the fraction of characters to be sampled. **Block-bootstrapping is** useful for handling correlated characters. When this is thought to have occurred, we can correct for it by sampling, not individual characters, but blocks of adjacent characters. Block bootstrap and was introduced by Künsch (1989). If the correlations are believed to extend over some number of characters, you choose a block size, *B*, that is larger than this, and choose *N/B* blocks of size *B*. In its implementation here, the block bootstrap "wraps around" at the end of the characters (so that if a block starts in the last *B-1* characters, it continues by wrapping around to the first character after it reaches the last character). Note also that if you have a DNA sequence data set of an exon of a coding region, you can ensure that equal numbers of first, second, and third coding positions are sampled by using the block bootstrap with $B = 3$. **Partial block-bootstrapping is** similar to partial bootstrapping except sampling blocks rather than single characters.

Jackknife analysis (Jackknifing) is similar to bootstrap analysis but differs in that each randomly selected character may be resampled only once, and not multiple times, and the resultant resampled data matrix is smaller than the original. **Delete-half-jackknifing** involves sampling a random half of the characters, and including them in the data but dropping the others. The resulting data sets are half the size of the original, and no characters are duplicated. The random variation from doing this should be very similar to that obtained from the bootstrap. The method is advocated by Wu (1986). **Delete-fraction jackknifing** was advocated by Farris et al. (1996) and involves deleting a fraction 1/e (1/2.71828). This retains too many characters and will lead to overconfidence in the resulting groups when there are conflicting characters. This and the preceding options form a part of the **SEQBOOT** program of **Phylip** software, and the user is asked to supply the fraction of characters that are to be retained. The program also offers **permuting** method, with following alternatives. **Permuting species within characters** involves permuting the columns of the data matrix separately. This produces data matrices that have the same number and kinds of characters but no taxonomic structure. It is used for different purposes than the bootstrap, as it tests not the variation around an estimated tree but the hypothesis that there is no taxonomic structure in the data: if a statistic such as number of steps is significantly smaller in the actual data than it is in replicates that are permuted, then we can argue that there is some taxonomic structure in the data (though perhaps it might be just the presence of a pair of sibling species). **Permuting characters** simply permutes the order of the characters, the same reordering being applied to all species. It is included as a possible step in carrying out a permutation test of homogeneity of characters (such as the Incongruence Length Difference test). **Permuting characters separately for each species** permute data so as to destroy all phylogenetic structure, while keeping the base composition of each species the same as before. It shuffles the character order separately for each species.

It is a common practice, and consequently more informative, to indicate the branch length (number of steps needed to reach that clade), bootstrap or jackknife support and Bremer support (decay index) for each clade in the consensus tree (Figure 8.29).

Effect of Different Outgroups

An important component of procedures generating rooted trees is the incorporation of an outgroup in the analysis. In morphological data, the outgroup choice can influence phylogenetic inference. In molecular data, one specific concern is the levels of sequence divergence between outgroups and ingroups and the subsequent possibility of spurious long-branch attraction (Albert et al., 1994). The robustness of tree can be tested by using randomly-generated outgroup sequences, excluding all outgroups, and using outgroups selectively.

Sytsma and Baum (1996), investigating the molecular phylogenies of angiosperms, found that removal of all outgroups generated 27 shortest unrooted trees. Using *Ginkgo* only as outgroup yielded lineages identical with baseline study (which included all outgroups); when only conifers were used as outgroup, the consensus tree was less resolved and many nodes collapsed. Use of Gnetales us outgroup increased the number of steps needed to yield baseline topologies, and interestingly, *Ceratophyllum* is shown as sister to all angiosperms except eudicots.

Effect of Lineage Removal

Lineage removal strategy highlights the problems of lineage extinction, which often leads to a particular group (especially critical in angiosperms where fossil record is meager) not being sampled in analysis, thus giving distorted phylogenies. The same may also be true for extant taxa, for which very little data is available. The removal of all major lineages, one at a time (Sytsma and Baum, 1994), provided useful information. The removal of *Ceratophyllum*, paleoherbs IIb (Chloranthaceae, and Magnoliales) had no effect on the remaining angiosperm topology, whereas the removal of paleoherbs I (Aristolochiales and Illiciales), Laurales and eudicots showed substantial changes.

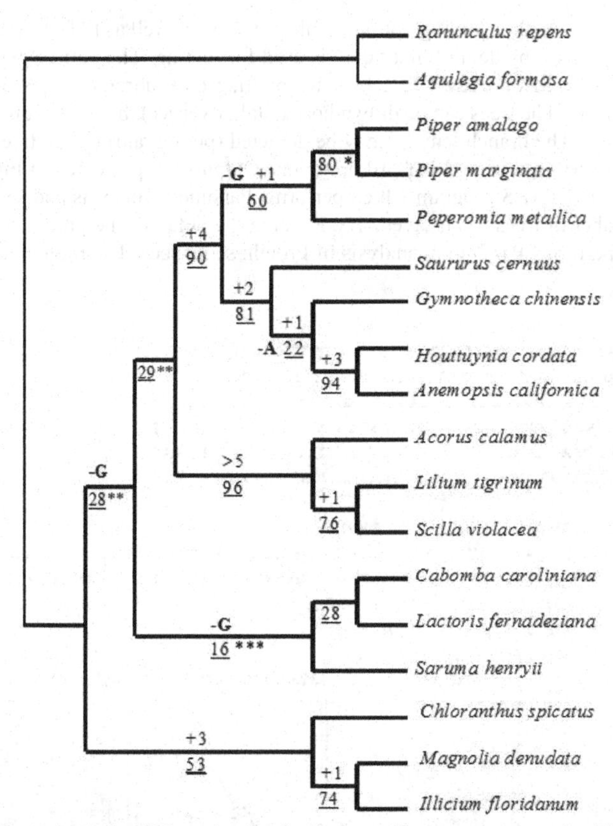

Figure 8.29: Tree developed from the study of 16 species of paleoherbs and 2 outgroup taxa, using 58 morphological and ontogenetic characters. The cladogram requires 214 steps and has CI = 0.51 and RI = 0.65. Bootstrap values are underlined and indicated below a branch. Decay Index is indicated above the branch. *Ranunculus repens* and *Aquilegia formosa* were chosen as outgroup taxa (Drawn from Tucker and Douglas, 1996).

Effect of Exemplars

The large computational load in handling a large data is often reduced by using **placeholders** or **exemplars**. These are often used to represent large lineages. The use of exemplars can warn about the possible artifacts when sparsely-sampled lineages appear in basal positions. In such cases, more taxa can be added to the data set for further analyses. But in the case of basal clade where a large number of taxa are extinct, the results could be ambiguous. The results from angiosperms have shown that clades shift around with ease when the number of taxa sampled for each lineage is reduced, and the use of exemplars at times could give misleading results.

Automated Trees

Number sophisticated computer programs are available to construct phylogenetic trees. These programs are basically similar to those designed for development of phenograms but differing essentially in the requirement to select one taxon for rooting in most programs. **PHYLIP** (Phylogeny Inference package), is a commonly used set of programs for inferring phylogenies (evolutionary trees) by parsimony, compatibility, distance matrix methods, and likelihood. It can also compute consensus trees, compute distances between trees, draw trees, resample data sets by bootstrapping or jacknifing, edit trees, and compute distance matrices. It can handle data that are nucleotide sequences, protein sequences, gene frequencies, restriction sites, restriction fragments, distances, discrete characters, and continuous characters.

Distance matrix can be generated using programs such as **DNADIST** (which handles nucleotide sequence data; it gives you choice to set weightage for transversions/transitions), **PRODIST** (which works with protein sequences) and **RESTDIST** (which works with restriction site data). The most commonly used programs of PHYLIP for handling distance matrix data include **FITCH**, **KITSCH**, and **NEIGHBOR**. These deal with data which comes in the form of a matrix of pairwise distances between all pairs of taxa **NEIGHBOR** offers **UPGMA** option in which no taxon needs to be selected for rooting, whereas neighbor-joining option of this program, as well as FITCH and KITSCH need one taxon to be selected for rooting, otherwise by default first taxon is used for rooting. The outtree generated by these programs can be plotted using DRAWGRAM or **DRAWTREE**, latter plotting only unrooted trees. DRAWGRAM provides a variety of options to choose from. The trees can be drawn horizontal or vertical, branches square (phenogram), v-shaped (cladogram), curved or circular. The branch lengths may be depicted (phylogram) on the tree. The DNA sequence data presented in Figure 8.22 was analysed using PHYLIP programs. Outputs are presented in Figure 8.30. DNA sequence data can also be handled by DNAPARS program which performs Parsimony analysis and selects the best tree. It gives you choice to select the number of trees to be saved, 10000 being the default. The program directly yields the outtree file. PROTPARS, similarly performs Parsimony analysis of Protein sequences. The protein sequences are given by the

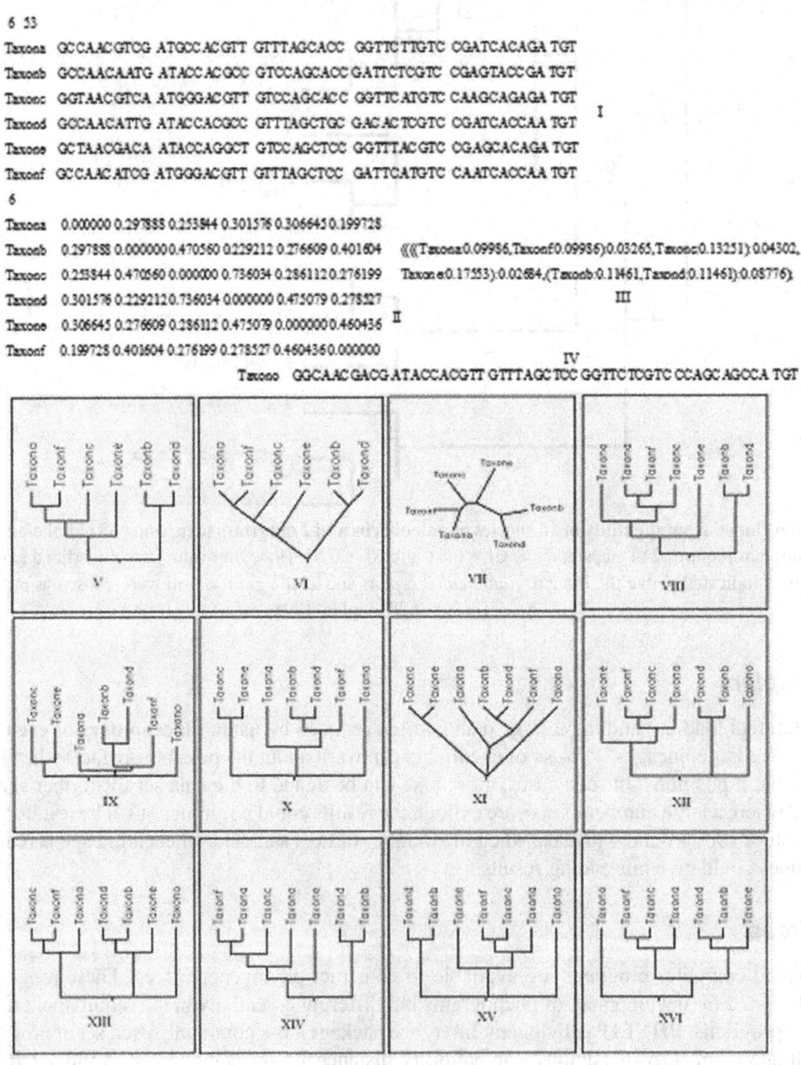

Figure 8.30: Analysis of the DNA sequence data presented in Figure 8.22 using PHYLIP. **I:** Infile, first line indicating number of taxa and number of nucleotides in each sequence; **II:** Square distance matrix (outfile) generated by DNADIST program; **III:** Outtree file generated by NEIGHBOR program using UPGMA option; **IV:** DNA sequence of the 7th hypothetical taxon (taxono) used for rooting; **V–VI:** Square tree (Phenogram) and V-shaped tree (cladogram); **VII:** Unrooted tree of same; **VIII–XVI:** Diagrams based on 7-taxa sequences; **VIII:** Phenogram, UPGMA option; **IX–XI:** Phylogram, Phenogram and Cladogram based on neighbour-joining option of NEIGHBOR; **XII:** Phenogram based on DNAML program; **XIII:** Phenogram based on FITCH program; **XIV:** Phenogram based on KITSCH program; **XV:** Tree (Phenogram) generated based on DNAPARS program; **XVI:** Majority-rule consensus tree based on CONSENSE program, using outtree files of above six programs (All trees except VII (plotted using DRAWTREE) plotted using different options of DRAWGRAM program).

one-letter code used by the late Margaret Dayhoff's group in the Atlas of Protein Sequences, and consistent with the IUB standard abbreviations. DNAMOVE which handles data similar to DNAPARS, allows the user to choose an initial tree, and displays this tree on the screen. The user can look at different sites and the way the nucleotide states are distributed on that tree, given the most parsimonious reconstruction of state changes for that particular tree. The user then can specify how the tree is to be rearranged, rerooted or written out to a file. By looking at different rearrangements of the tree the user can manually search for the most parsimonious tree and can get a feel for how different sites are affected by changes in the tree topology. DNAML program carries out analysis of DNA sequences using Maximum Likelihood Method. The program uses both informative and non-informative sites and yields the outtree file directly. RESTML similarly handles restriction site data using maximum likelihood method. Binary data coded as 0 (ancestral state) and 1 (advanced state) is handled by MIX, which performs parsimony analysis and generates outtree which can be plotted using DRAWGRAM. Input data from Table 8.4 and most parsimonious tree generated using MIX program is presented in Figure 8.31. For this analysis fourth multistate character was converted into binary character (simple and compound leaves). Using Wagner parsimony, the program was able to generate 34 trees. Taxon 15 was used for rooting. CONSENSE program was used to select the majority rule consensus tree. MOVE handles binary data and is an interactive program which allows the user to choose an initial tree, and displays this tree on the screen. The user can look at different characters and the way their states are distributed on that tree, given the most parsimonious reconstruction of state changes for that particular tree. The user then can specify how the tree is to be rearraranged, rerooted or written out to a file. By looking at different rearrangements of the tree the user can manually search for the most parsimonious tree and can get a feel for how different characters are affected by changes in the tree topology.

Multistate data can similarly be handled by PARS and can be converted into binary data by FACTOR program. Data from Gene frequencies and continuous characters is handled by CONTML (constructs maximum likelihood estimates of the phylogeny; handles both types of data), GENDIST (computes genetic distances for use in the distance matrix programs; handles data from gene frequencies) and CONTRAST (examines correlation of traits as they evolve along a given phylogeny; handles continuous characters data). The data matrix for gene frequencies contains number of species (or populations) and number of loci, whereas the second line contains number of alleles for each locus the default number of data for each species (A-all) contains one allele less for each locus. Thus, for three loci with 2, 3 and 2 alleles respectively there would be four values. Without A option, there should be 7 values. The values in dataset are preceded and followed by blanks. The data from continuous characters does not contain the second line, the data would include number of species and the number of characters in the first line (only line above species data).

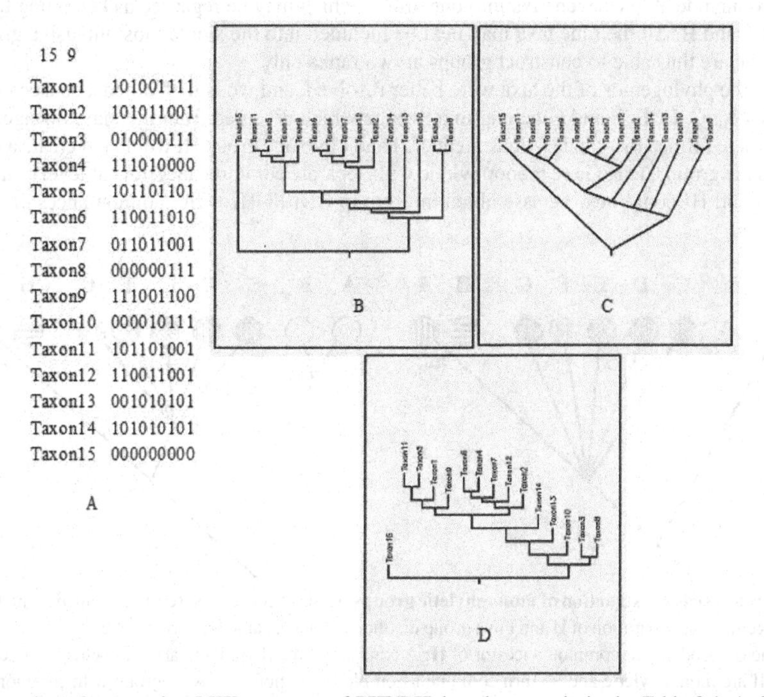

Figure 8.31: Construction of trees using MIX program of PHYLIP based on matrix in the Table 8.4. **A:** Input file with fourth character converted into binary (simple and compound leaves) character. Out of the 34 parsimonious trees generated by Mix, Consensus tree generated by CONSENSE program presented as Phenogram **B:** Cladogram **C:** and Phylogram **D.**

PHYLIP also offers programs to yield consensus tree (CONSENSE), Bootstrapping (SEQBOOT) and a host of related programs.

Binary data can similarly be input in infile with just replacing nucleotide alphabets with binary 0 and 1 data as presented in Figure 8.31 and handled by various programs mentioned earlier.

Developing Classification

Once the phylogeny of a group has been developed, the evolutionary process within the group can be reconstructed, the morphological, physiological and genetic changes can be described, and the resultant information used in the classification of the group. Phylogenetic classifications are based on the recognition of monophyletic groups and avoid including paraphyletic and often completely reject paraphyletic groups. Such classifications are superior over classifications based on overall similarity in several respects:

1. Such a classification reflects the genealogical history of the group much more accurately.
2. The classification based on monophyletic groups is more predictive and of greater value than classification based on some characteristics.
3. Phylogenetic classification is of major help in understanding distribution patterns, plant interactions, pollen biology, dispersal of seeds and fruits.
4. The classification can direct the search for genes, biocontrol agents and potential crop species.
5. The classification can be of considerable help in conservation strategies.

The evolutionary history of the group of 8 living species shown in Figure 8.12 was known with precise point of character transformations, and the construction of monophyletic groups, assembled into successively more inclusive groups, did not pose much problem. But it is often not the case. Even, most resolved consensus trees are often ambiguous in several respects.

Consider the strict consensus tree represented in the Figure 8.27-III. This tree is reproduced in the Figure 8.32-I. As noted earlier, the phylogenetic relationships between taxa (these could be different species, genera, etc.) C, D, E, F, and G are poorly resolved, and as such they are shown arising from the common point, and consequently common ancestor as level **o**. Although H and I form a distinct group with a common ancestor as **m** but leaving these two out of the group including CDEFG would render latter as paraphyletic (cf. traditional separation of dicots and monocots). The safest situation would be to include all the seven taxa into one group, which may be regarded as belonging to the same rank as the group including A and B. All the nine taxa may next be included into the single most inclusive group with common ancestry at level **p**. We are thus able to construct groups at two ranks only.

Now supposing the phylogenies of the taxa were better resolved, and we had obtained a consensus tree as shown in Figure 8.32-II. Now taxa C, D, E, F and G belong to a lineage which diverged from the main lineage, successive to the divergence of the lineage formed by A and B. Placement of H and I into one group HI would not create any problem as both this group as well as the group CDEFG are monophyletic with separate common ancestors at level **m** and **q**, respectively. The groups CEDFG and HI could next be assembled into group CDEFGHI with common ancestor at **o**. Note that the

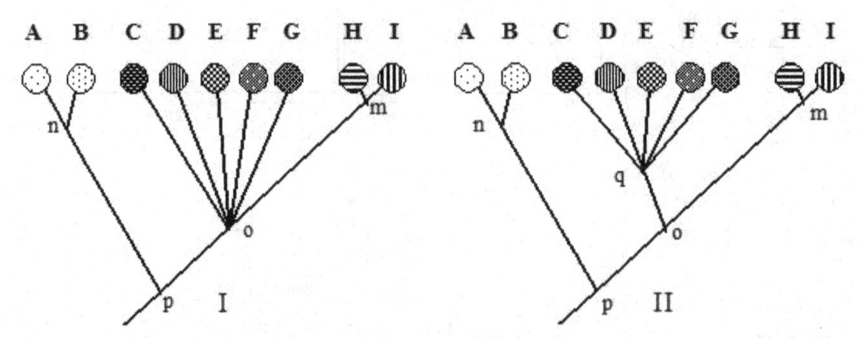

Figure 8.32: Attempts towards construction of monophyletic groups. I: Strict consensus tree as presented in the Figure 8.27-III. With poorly resolved phylogenies, the separation of H and I in a group distinct and of the same rank as group CDEFG would create a paraphyly, as HI are left out of the descendants of common ancestor o. **II:** A consensus tree (hypothetical) with better resolved phylogenies. Both groups CDEFG and HI are monophyletic and, in turn could be assembled into more inclusive group with common ancestor at level o, now containing all descendants of the common ancestor. This group (CDEFG, HI) and AB (also monophyletic) could be assembled into one most inclusive monophyletic group, containing all descendants of the common ancestor at p.

group AB can next be merged with CDEFGHI to form single most inclusive group ABCDEFGHI. Now we have been able to construct taxa at three ranks instead of two from tree I.

Supposing the taxa A to I included in the tree, are different species. From tree II, thus we are able to recognize three genera AB, CDEFG and HI. The last two are next assembled into family CDEFGHI and the former a monotypic family AB. The other alternative was to place A and B in two separate monotypic genera (depending on the degree of morphological and genetic divergence obtained) which are then assembled into family AB. The two families may next be assembled into order ABCDEFGHI. There could be other possibilities also. The second rank could be a subfamily and the third a family. Similarly, a third rank could be a suborder instead of an order. These final decisions are often made, based on the size of the group, degree of divergence, and the reliability of characters. All the groups recognized above would be monophyletic at the respective ranks.

Next, let us look at the tree shown in the Figure 8.29, a study on paleoherbs. *Ranunculus* and *Aquilegia* were used as outgroup representing family Ranunculaceae; their isolated position from paleoherbs is clearly depicted in the tree. Paleoherbs constitute a group of taxa of uncertain affinities, which have been placed differently in various classification schemes, but a few points seem to have been resolved. *Piper*, and *Peperomia* (both belong to Piperaceae) form a distinct group, and so do *Saururus*, *Gymnotheca*, *Houttuynia* and *Anemopsis* (all four belonging to Saururaceae), and the two families a well-supported (bootstrap support of 90 per cent). This was confirmed by comparison of seven published trees of paleoherbs. *Cabomba*, *Lactoris* and *Saruma* have least resolved affinities with very poor support, with highly unstable position.

The final decisions on the recognition of groups are, however, often based on personal interpretation of phylogenies. *Chloranthus*, in this tree as well in several others, is closer to *Magnolia* (Magnoliales) and *Laurus* (Laurales), but often finds different treatment. APG II places Chloranthaceae after Amborellaceae at the start of Angiosperms, Chloranthaceae unplaced independent lineage in APG IV (2016). Judd et al., had earlier (1999) placed Chloranthaceae under order Laurales of Magnoliid complex, but later (2008) shifted the family among basal ANITA Grade with uncertain position. APweb of Stevens (2017) places the family under order Chloranthales. Thorne had earlier (1999, 2000) placed Chloranthaceae in Magnoliidae—> Magnolianae—> Magnoliales—> Chloranthineae (other suborders within the order being Magnoliineae, and Laurineae), but subsequently (2003) included the family after Amborellaceae under order Chloranthales, the first order of Magnoliidae, finally (2006, 2007) separated under subclass Chloranthidae, a placement somewhat similar to APG II. Further discussion on angiosperm affinities will be resumed in Chapters 9 and 13.

Chapter 9

Phylogeny of Angiosperms

Angiosperms form the most dominant group of plants with at least 295,383 (Christenhusz and Byng, 2016), a number much greater than all other groups of plants combined. Not only in numbers, angiosperms are also found in a far greater range of habitats than any other group of land plants. The phylogeny of angiosperms has, however, been a much-debated subject, largely because of very poor records of the earliest angiosperms. These earliest angiosperms probably lived in habitats that were not best suited for fossilization. Before trying to evaluate the phylogeny, it would be useful to have an understanding of the major terms and concepts concerning phylogeny in general, and with respect to angiosperms in particular.

ORIGIN OF ANGIOSPERMS

The origin and early evolution of angiosperms are enigmas that have intrigued botanists for well over a century. They constituted an '**abominable mystery**' to Darwin. The mystery is slowly being 'sleuthed' and at the present pace of 'Sherlock Holms research', it may be no more mysterious within the next two decades than for any other major group. With the exception of conifer forest and moss-lichen tundra, angiosperms dominate all major terrestrial vegetation zones, account for the majority of primary production on land, and exhibit bewildering morphological diversity. Unfortunately, much less is known about the origin and early evolution of angiosperms, resulting in several different views regarding their ancestors, the earliest forms and course of evolution. The origin of angiosperms may be conveniently discussed under the following considerations.

What are Angiosperms?

Angiosperms form a distinct group of seed plants sharing a unique combination of characters. These important characters include carpels enclosing the ovules, pollen grains germinating on the stigma, sieve tubes with companion cells, double fertilization resulting in triploid endosperm, and highly reduced male and female gametophytes. The angiosperms also have vessels. The pollen grains of angiosperms are also unique in having non-laminate endexine and ectexine differentiated into a foot-layer, columellar layer and tectum (tectum absent in Amborellaceae). The angiosperm flower typically is a hermaphrodite structure with carpels surrounded by stamens and the latter by petals and sepals, since insect pollination prevails. Arbuscular mycorrhizae are also unique to angiosperms (except Amborellaceae, Nymphaeales and Austrobaileyales). The vessel elements of angiosperms typically possess scalariform perforations.

There may be individual exceptions to most of these characters. Vessels are absent in some angiosperms (Winteraceae) while some gymnosperms have vessels (Gnetales). The flowers are unisexual without perianth in several Amentiferae, which also exhibit anemophily. In spite of these and other exceptions, this combination of characters is unique to angiosperms and not found in any other group of seed plants.

What is the Age of Angiosperms?

The time of origin of angiosperms is a matter of considerable debate. For many years, the earliest well-documented angiosperm fossil was considered to be the form-genus *Clavitopollenites* described (Couper, 1958) from Barremian and Aptian strata of Early Cretaceous (Table 9.1) of southern England (132 to 112 mya-million years), a monosulcate pollen with distinctly sculptured exine, resembling the pollen of the extant genus *Ascarina*. Brenner and Bickoff (1992) recorded similar but inaperturate pollen grains from the Valanginian (ca 135 mya) of the Helez formation of Israel, now considered to be the oldest record of angiosperm fossils (Taylor and Hickey, 1996). Also found in Late Hauterivian (Brenner, 1996) of Israel

Table 9.1: Geological time scale.

Time	Era	Period	Epoch	Stage
m years (mya)				
_0.01_		Quaternary	Holocene	
_2.5 _			Pleistocene	
_7 _	Cenozoic		Pliocene	
_26 _			Miocene	
_38 _		Tertiary	Oligocene	
_54 _			Eocene	
_65 _			Palaeocene	
_74 _				Maestrichtian
_83 _				Campanian
_87 _				Santonian
_89 _			Upper	Coniacian
_90 _				Turonian
_97 _				Cenomanian
_112_		Cretaceous		Albian
_125_				Aptian
_132_	Mesozoic			Barremian
_135_			Lower	Hauterivian
_141_				Valanginian
_146_				Berriasian
			Upper	
		Jurassic	Middle	
_208_			Lower	
			Upper	
		Triassic	Middle	
_235_			Lower	
_280_		Permian		
_345_		Carboniferous		
_395_		Devonian		
_430_	Palaeozoic	Silurian		
_500_		Ordovician		
_570_		Cambrian		
_2400_	Precambrian	Algonkian		
_4500_		Archaean		

(ca 132 mya) were Pre-*Afropollis* (mostly inaperturate, few weakly monosulcate), *Clavitopollenites* (weakly monosulcate to inaperturate), and *Liliacidites* (monosulcate, sexine similar to monocots). From Late Barremian have been recorded *Afropollis* and *Brenneripollis* (both lacking columellae) and *Tricolpites* (the first appearance of tricolpate pollen grains).

The number and diversity of angiosperm fossils increased suddenly and by the end of the Early Cretaceous (ca 100 mya) period major groups of angiosperms, including herbaceous Magnoliidae, Magnoliales, Laurales, Winteroids and Liliopsida were well represented. In Late Cretaceous, at least 50 per cent of the species in the fossil flora were angiosperms. By the end of the Cretaceous, many extant angiosperm families had appeared. They subsequently increased exponentially and constituted the most dominant land flora, continuing up to the present.

The trail in the reverse direction is incomplete and confusing. Many claims of angiosperm records before the Cretaceous were made but largely rejected. Erdtman (1948) described *Eucommiidites* as a tricolpate dicotyledonous pollen grain from the Jurassic. This, however, had bilateral symmetry instead of the radial symmetry of angiosperms (Hughes, 1961) and granular exine with gymnospermous laminated endexine (Doyle et al., 1975). This pollen grain was also discovered in the micropyle of seeds of the female cone of uncertain but clearly gymnospermous affinities (Brenner, 1963). Several other fossil pollens from the Jurassic age attributed to Nymphaeaceae ultimately turned out to be gymnosperms.

In the last few years Sun et al. (1998, 2002) have described fossils of *Archaefructus* from Upper Cretaceous (nearly 124 mya) of China, with clearly defined spirally arranged conduplicate carpels enclosing ovules, a feature not reported in earlier angiosperms. The fruit is a follicle. This is considered to be the oldest record of angiosperm flower.

Several vegetative structures from the Triassic were also attributed to angiosperms. Brown (1956) described *Sanmiguilea* leaves from the Late Triassic of Colorado and suggested affinity with Palmae. A better understanding of the plant was made by Cornet (1986, 1989), who regarded it as a presumed primitive angiosperm with features of monocots and dicots. Although its angiosperm venation was refuted by Hickey and Doyle (1977), Cornet (1989) established its angiosperm venation and associated reproductive structures. Our knowledge of this controversial taxon, however, is far from clear.

Marcouia leaves (earlier described as *Ctenis neuropteroides* by Daugherty, 1941) are recorded from the Upper Triassic of Arizona and New Mexico. Its angiosperm affinities are not clear.

Harris (1932) described *Furcula* from the Upper Triassic of Greenland as bifurcate leaf with dichotomous venation. Although it seems to approach dicots in venation and cuticular structure, it has several non-angiospermous characters including bifurcating midrib and blade, higher vein orders with relatively acute angles of origin (Hickey and Doyle, 1977).

Cornet (1993) has described *Pannaulika,* a dicot-like leaf form from Late Triassic from the Virginia-North Carolina border. It was considered to be a three-lobed palmately veined leaf. The associated reproductive structures were attributed to angiosperms, but it is not certain that any of the reproductive structures were produced by the plant that bore *Pannaulika.* Taylor and Hickey (1996), however, do not accept its angiosperm affinities, largely based on the venation pattern, which resembles more that of ferns. Much more information is needed before the Triassic record of angiosperms can be established.

Cornet (1996) described *Welwitschia* like fossil as *Archaestrobilus cupulanthus* from the Late Triassic of Texas. The plant had similarly constructed male and female spikes, each possessing hundreds of spirally arranged macrocupules. The fossil has revived renewed interest in gnetopsids.

Given the inconclusive pre-Cretaceous record of angiosperms, it is largely believed that angiosperms arose in the Late Jurassic or very Early Cretaceous (Taylor, 1981) nearly 130 to 135 mya ago (Jones and Luchsinger, 1986).

Melville (1983), who strongly advocated his gonophyll theory, believed that angiosperms arose nearly 240 mya ago in the Permian and took nearly 140 mya before they spread widely in Cretaceous. The Glossopteridae which gave rise to angiosperms met with a disaster in the Triassic and disappeared, this disaster slowing down the progress of angiosperms slow until the Cretaceous when their curve entered an exponential phase. This idea has, however, found little favor.

There has been increasing realization in recent years (Troitsky et al., 1991; Doyle and Donoghue, 1993; Crane et al., 1995) to distinguish two dates—one in the Triassic when the **stem angiosperms** ('angiophytes' sensu Doyle and Donoghue, 1993 or 'proangiosperms' sensu Troitsky et al., 1991) separated from sister groups (Gnetales, Bennettitales and Pentoxylales) and the second in the Late Jurassic when the crown group of angiosperms (crown angiophytes) split into extant subgroups (Figure 9.1).

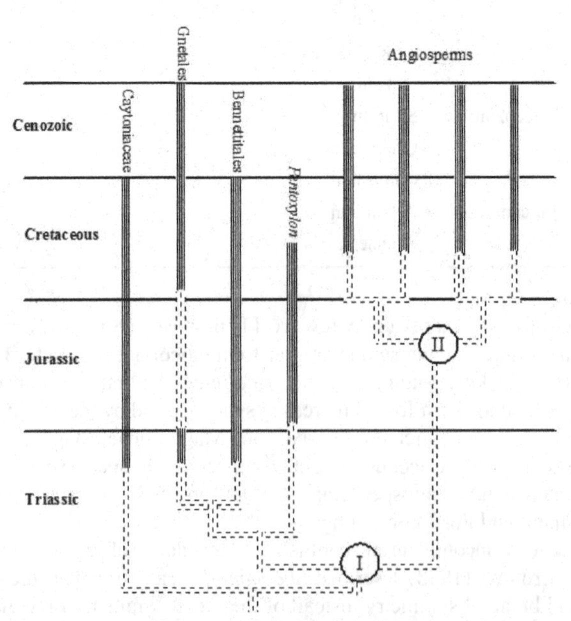

Figure 9.1: Phylogenetic tree of anthophytes (angiosperm lineage and sister groups). Point (I) marks when angiosperm lineage separated from sister groups in the Late Triassic, and (II) marks the splitting of crown angiosperms into extant subgroups in the Late Jurassic. Dotted line represents conclusions for which fossil record is not available (diagram based on Doyle and Donoghue, 1993).

Molecular Dating

There have been several attempts to estimate the time of divergence of angiosperms (node B in Figure 9.1) by applying a molecular clock to nucleotide sequence data. The results mostly pointing to much earlier origin of angiosperms have, however, been contradictory. The first detailed attempt was made by Martin et al. (1989) using nine angiosperm sequences from *gapC,* the nuclear gene encoding GADPH (cytosolic glyceraldehydes-3-phosphate dehydrogenase). The observed number of nonsynonymous substitutions between each pair of species was compared to estimated rates of substitutions per site per year inferred from known divergence times (e.g., plants-animals, plants-yeast, mammal-chicken, human-rat). The results implied separation of monocots and dicots at 319 + 35 mya, a dicot radiation at 276 + 33 mya, and cereal grass divergence at 103 + 22 mya. The results were questioned by several authors, since the study used a single gene. Wolfe et al. (1989) attempted to date the monocot-dicot split using a large number of genes in chloroplast genome and using a three-tiered approach. They suggested Late Triassic (200 mya) as the likely estimate of monocot-dicot split. Martin et al. (1993) provided new data to support Carboniferous origin (~ 300 Mya) of angiosperms. They used both *rbcL* and *gapC* sequences for this study. Sytsma and Baum (1996) conclude that the results strongly caution using the molecular clock for dating unless extensive sampling of taxa and genes with quite different molecular evolution is completed. Thus, the resolution of angiosperm phylogeny may have to wait for a more complete molecular data and its proper appraisal.

What is the Place of their Origin?

It was earlier believed that angiosperms arose in the Arctic region (Seward, 1931), with subsequent southwards migration. Axelrod (1970) suggested that flowering plants evolved in mild uplands (**upland theory**) at low latitudes. Smith (1970) located the general area of South-East Asia, adjacent to Malaysia as the site where angiosperms evolved when Gondwana and Laurasia were undergoing initial fragmentation. Stebbins (1974) suggested that their origin occurred in exposed habitats in areas of seasonal drought. Takhtajan (1966, 1980), who believed in the neotenous origin of angiosperms, suggested that angiosperms arose under environmental stress, probably as a result of adaptation to moderate seasonal draught on rocky mountain slopes in areas with monsoon climate.

Retallack and Dilcher (1981) believed that the earliest angiosperms were probably woody, small-leaved plants occurring in the Rift valley system adjoining Africa and South America. Some of these angiosperms adapted to the coastal environments and became widespread following changing sea levels during the Early Cretaceous.

Although agreeing with the role of environmental stress, many authors in recent years (Hickey and Doyle, 1977; Upchurch and Wolfe, 1987; Hickey and Taylor, 1992) have suggested that early angiosperms lived along stream- and lake-margins (**lowland theory**). Later, they appeared in more stable backswamp and channel sites, and lastly, on river terraces. Taylor and Hickey (1996) suggested that ancestral angiosperms were perennial rhizomatous herbs and evolved along rivers and streams on sites of relatively high disturbance with moderate amounts of alluviation. These sites would have been characterized by high nutrient levels and frequent loss of plant cover due to periodic disturbances.

Are Angiosperms Monophyletic or Polyphyletic?

Engler (1892) considered angiosperms to be polyphyletic, monocotyledons and dicotyledons having evolved separately. Considerable diversity of angiosperms in the Early Cretaceous and the extant angiosperms led several authors, including Meeuse (1963) and Krassilov (1977) to develop models for polyphyletic origin of angiosperms. This view is largely supported by considerable diversity in the early angiosperm fossils.

Most recent authors, including Hutchinson (1959, 1973), Cronquist (1981, 1988), Thorne (1983, 1992, 2000, 2007), Dahlgren (1980, 1989), Takhtajan (1987, 1997), Judd et al. (2002, 2008), Bremer et al. (APG II, 2003), and Stevens (APweb, 2008) believe in the monophyletic origin of angiosperms, monocotyledons having evolved from primitive dicotyledons. This view is supported by a unique combination of characters such as closed carpels, sieve tubes, companion cells, four microsporangia, triploid endosperm, 8-nucleate embryo sac and reduced gametophytes. Sporne (1974), on the basis of statistical studies, also concluded that it is highly improbable that such a unique combination of characters could have arisen more than once, independently from gymnosperm ancestors.

It is interesting to note that Melville (1983) considered angiosperms to be monophyletic but the explanation that he offers clubs him with the proponents of polyphyletic origin. He believes that angiosperms arose from several different genera of Glossopteridae. According to him, the species is not always to be considered as the ancestor for determining a monophyletic nature. A species from another species is monophyletic, as is a genus from a genus, a family from a family. The principle, according to him, is that to be monophyletic, a taxon of any rank must be derived solely from another taxon of the same rank. Glossopteridae and Angiospermidae belong to the same rank subclass. Both taxa consist of minor lineages that may be likened to a rope with many strands, a situation called **pachyphyletic**. This explanation, however, conforms to the concept of minimum monophyly and does not satisfy the rule of strict monophyly, which is now, the accepted criterion for monophyly.

What are the Possible Ancestors?

Ancestry of angiosperms is perhaps one of the most controversial and vigorously debated topics. In the absence of direct fossil evidence, almost all groups of fossil and living gymnosperms have been considered as possible ancestors by one authority or the other. Some authors even suggested the **Isoetes origin** of monocotyledons because the plant has a superficial resemblance with onion, albeit with no trace of seed habit. The various theories have revolved around two basic theories, viz., the **Euanthial theory** and the **Pseudanthial theory** of angiosperm origin. Some other theories projecting herbaceous ancestry for the angiosperms have also recently received attention, making the question of ancestry of angiosperms rather more ambiguous:

Euanthial Theory

Also known as **Anthostrobilus theory,** Euanthial theory was first proposed by Arber and Parkins (1907). According to this theory, the angiosperm flower is interpreted as being derived from an unbranched bisexual strobilus bearing spirally arranged ovulate and pollen organs, similar to the hermaphrodite reproductive structures of some extinct bennettitalean gymnosperms. The carpel is thus regarded as a modified megasporophyll (**phyllosporous** origin of carpel). The bisexual flower of Magnoliales has been considered to have evolved from such a structure. Also agreeing with this general principle, various authors have tried to identify different gymnosperm groups as possible angiosperm ancestors:

Cycadeoidales (Bennettitales)

The group, now better known as Cycadeoidales, appeared in the Triassic and disappeared in the Cretaceous. Their potential as angiosperm ancestors was largely built upon the studies of Wieland (1906, 1916). Lemesle (1946) considered the group to be ancestral to angiosperms, primarily because of the hermaphrodite nature of *Cycadeoidea*, which had an elongated receptacle with perianth-like bracts, a whorl of pollen-bearing microsporophylls surrounding the ovuliferous region having numerous ovules and interseminal scales packed together. There were, however, signs of abscission at the base of the male structure, which would have shed, exposing the ovular region.

The plant was believed to look like cycads with a short compact trunk and a crown of pinnate compound leaves (Figure 9.2A). It was earlier suggested that the microsporophylls opened at maturity but the subsequent studies of Crepet (1974) showed that microsporophylls were pinnate, and distal tips of pinnae were fused, the opening of the region was not structurally possible, and they later disintegrated internally (Figure 9.2B). The ovules were terminal in contrast to their position in carpels of angiosperms.

Caytoniaceae

Opinion has strongly inclined in the recent years towards the probability that angiosperms arose from **Pteridosperms** or **seed ferns,** often placed in the order Lyginopteridales but more commonly under Caytoniales. Caytoniaceae was described from the Jurassic of Cayton Bay in Yorkshire by Thomas, and subsequently from Greenland, England and Canada. The group appeared in the Late Triassic and disappeared towards the end of the Cretaceous.

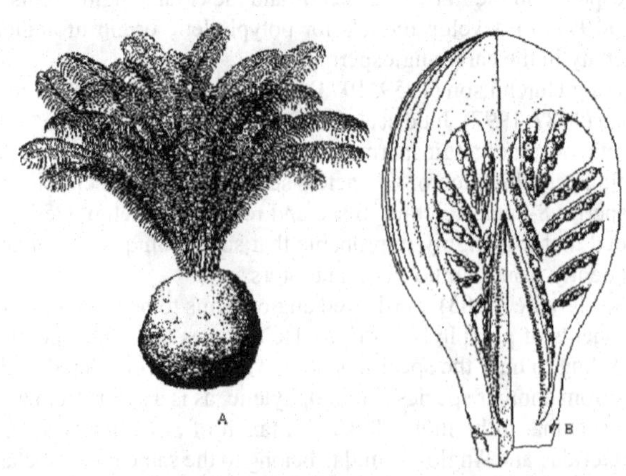

Figure 9.2: ***Cycadeoidea.*** **A:** Suggested reconstruction of plant with a compact trunk and numerous pinnate leaves. **B:** Suggested reconstruction of the cone cut open to show the arrangement of microsporangia. Ovulate receptacle is in the centre (A, after Delevoryas, 1971; B, after Crepet, 1974).

The leaves (*Sagenopteris*) were borne on twigs and not the trunk. These had two pairs of leaflets (rarely 3 to six leaflets) and were net veined. Male structures (*Caytonanthus*) had rachis with branching pinnae, each with a synangium of four microsporangia. The seed-bearing structure (*Caytonia*) had rachis with two rows of stalked cupules (Figure 9.3B). Each cupule contained several ovules borne in such a way that the cupule is recurved, with a lip like projection (often called stigmatic surface) near the point of attachment (Figure 9.3C).

The discovery of pollen grains within the ovules was thought to suggest their true gymnosperm position, however, rather than being angiosperm ancestors. Krassilov (1977) and Doyle (1978) regarded the cupule as homologous to the carpel, whereas Gaussen (1946) and Stebbins (1974) considered it the outer integument of the ovule. Cladistic studies of Doyle and Donoghue (1987) support the caytoniales-angiosperm lineage. Thorne (1996) agreed that angiosperms probably evolved during the Late Jurassic from some group of seed ferns.

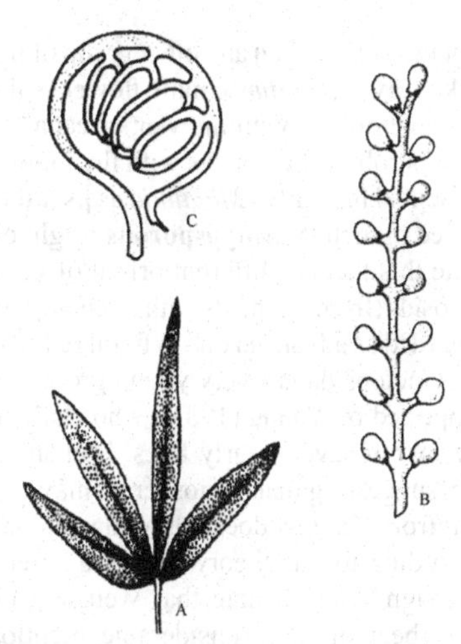

Figure 9.3: Caytoniaceae. A: Palmately compound leaf of *Sagenopteris phillipsi*. **B:** *Caytonia nathorstii* with two rows of cupules. **C:** Reconstruction of cupule of *Caytonia sewardii* (B and C from Dilcher, 1979; C, from Stewart and Rothwell, 1993).

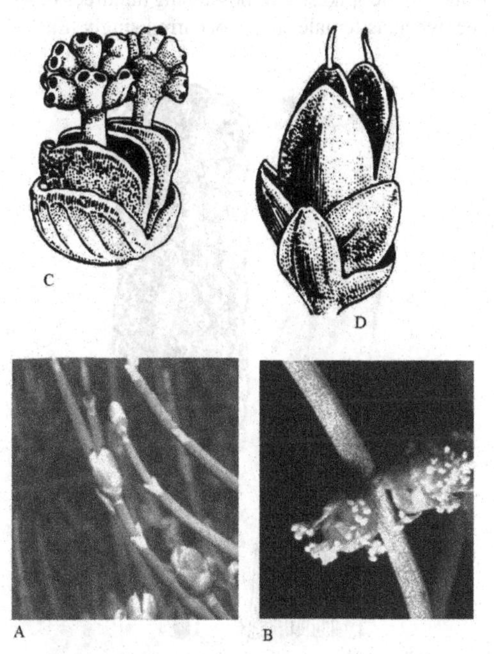

Figure 9.4: ***Ephedra*. A:** A small portion of plant with opposite scale-like leaves. **B:** Male strobili on a branch. **C:** A male strobilus with series of opposite bracts, apical bract subtending male stalk with several microsporangia. **D:** Female strobilus with series of whorled bracts, uppermost closely clasping ovules.

Cycadales

Sporne (1971) suggested possible links between Cycadales and angiosperms in the palm-like habit of Cycadales, the ovules being borne on leaf-like microsporophylls, trends in the reduction of sporophyll blade as seen in various species of *Cycas*. Although it may be difficult to assume Cycadales as ancestral to angiosperms, the fact that they have been derived from pteridosperms, and yet resemble angiosperms further supports the origin of angiosperms from pteridosperms.

Pseudanthium Theory

Commonly associated with the Englerian School, the theory was first proposed by Wettstein (1907), who postulated that angiosperms were derived from the Gnetopsida, represented by *Ephedra*, *Gnetum* and *Welwitschia* (formerly all placed in the same order Gnetales).

The group shows more angiosperm characteristics than any other group of living or fossil gymnosperms. These include the presence of vessels, reticulate dicot-like leaves (*Gnetum*), male flower with perianth and bracts, strong gametophyte reduction, and fusion of the second male gametophyte with the ventral canal nucleus. *Ephedra* resembles *Casuarina* in habit. Wettstein homologized the compound strobili of Gnetales with the inflorescences of wind-pollinated Amentiferae, and regarded the showy insect pollinated bisexual flowers of *Magnolia* as **pseudanthia** derived by aggregation of unisexual units, the carpel thus representing a modified branch (**Stachyosporous** origin of carpel).

A number of features, however, refute this theory: different origin of vessels (Bailey, 1944) in angiosperms (from tracheids with scalariform pitting) and Gnetosids (from tracheids with circular pitting), several vesselless living angiosperms (cf. Winteraceae). Amentiferae are now regarded as advanced due to floral reduction. Tricolpate pollen grains also represent an advanced condition. More importantly, Gnetopsida is a very young group.

But this theory has been strongly supported by Young (1981), who challenged the view that first angiosperms were vesselless and assumed that vessels were lost in several early lines. Muhammad and Sattler (1982) found scalariform perforations in vessel-elements of *Gnetum*, suggesting that angiosperms may be derived from Gnetales after all. Carlquist (1996), however, concludes that this claim from *Gnetum* does not hold when large samples are examined.

The basal group of angiosperms according to this theory included amentiferous-hamamelid orders Casuarinales, Fagales, Myricales and Juglandales. It is significant to note that Wettstein (1907) also included in this basal group, Chloranthaceae and Piperaceae, which have been inviting considerable attention in recent years.

The importance of Gnetopsids in angiosperm phylogeny has been further strengthened by the discovery of *Welwitschia* like fossil described by Cornet (1996) as *Archaestrobilus cupulanthus* from the Late Triassic of Texas (Figure 9.5). The plant had similarly constructed male and female spikes, each possessing hundreds of spirally arranged macrocupules. Male spikes were borne in clusters of three, whereas female spikes occurred singly. Each female macrocupule had an axially

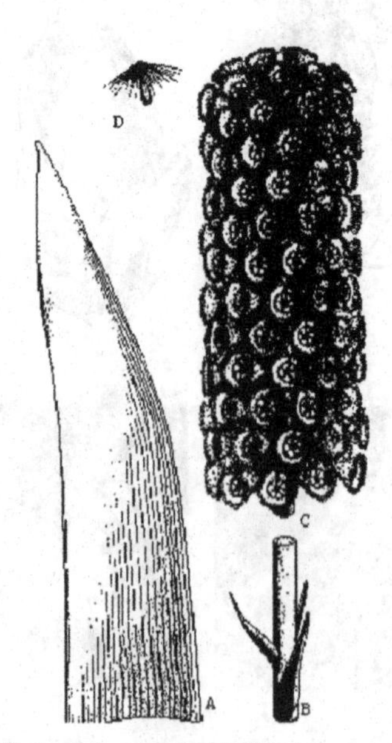

Figure 9.5: Resconstruction of *Archaestrobilus cupulanthus* and isolated organs. **A:** Associated leaf of *Pelourdea poleoensis*. **B:** Associated sterile lower part of strobilus. **C:** Female strobilus with numerous spirally arranged macrocupules. **D:** Dispersed seed (After Cornet, 1996).

curled (tubular) bract-like organ with a narrow shaft and expanded funnel shaped apex. The macrocupules contained an ovule (or seed) surrounded by sterile scales. Three to four very small bracts were present attached near the base and surrounding the macrocupule.

Each male macrocupule contained filament like appendages instead of sterile scales within. Outside, the macrocupule was crowded with numerous bivalved microsporophylls, each with four pollen sacs attached to an inflated stalk. On the outside of the female macrocupule were present gland-like structures resembling the stalks bearing pollen sacs on the male macrocupule. This suggests an origin from a bisexual macrocupule. The pollen grains are radially symmetrical and monosulcate. The plant is regarded as a gnetophyte more primitive than extant Gnetales.

Ephedra is generally considered to be the most primitive of the three living genera of gnetopsids. Cornet believes that *Archaestrobilus* possessed characters that may be plesiomorphic even for *Ephedra*, such as radial symmetry of floral parts which are spirally arranged and not opposite. He believes that Bennettitales, Gnetales, Pentoxylales and angiosperms had a common ancestry sometimes before Late Triassic. Gnetales are relatives of angiosperms and Bennettitales that underwent drastic floral reduction and aggregation in response to wind pollination.

Taylor and Hickey (1996) have presented a hypothesis for the derivation of the flower of Chloranthaceae from the inflorescence unit (anthion) of gnetopsids, with considerable reduction in the reproductive parts.

Anthocorm Theory

This theory is a modified version of the pseudanthial theory and was proposed by Neumayer (1924) and strongly advocated by Meeuse (1963, 1972). According to this theory, the angiosperm flower ('functional reproductive unit') has several separate origins (i.e., angiosperms are polyphyletic). In most Magnoliidae and their dicotyledonous derivatives, they are modified pluriaxial systems (**holanthocorms**) that originated from the gnetopsids via the Piperales, whereas the modification of an originally uniaxial system (**gonoclad** or **anthoid**) gave rise to flowers of Chloranthaceae. Meeuse (1963) postulated a separate origin of monocotyledons from the fossil order Pentoxylales through the monocot order Pandanales.

Pentoxylales (Figure 9.6) were described from the Jurassic of India and New Zealand. The stem (*Pentoxylon*) had five conducting strands. The pollen-bearing organ (*Sahnia*) was pinnate: free above and fused into a cup below. The seed-bearing structure was similar to a mulberry with about 20 sessile seeds, each having an outer fleshy sarcotesta and the inner hard sclerotesta. The sarcotesta was considered homologous to the cupule of seed ferns. The carpel of angiosperms was regarded as a composite structure being an ovule-bearing branch fused with a supporting bract. It is interesting, however, to note that Taylor and Hickey (1996) no longer include *Pentoxylon* as a member of **Anthophytes**, which include angiosperm lineage and its sister groups Bennettitales and gnetopsids (see Figure 9.12). According to them, *Pentoxylon* lacks key anthophyte characteristics such as distal, medial and proximal positioning of female, male and sterile organs on the reproductive axis, as well as the enclosure of ovules by bract-derived organs.

Gonophyll Theory

The **Gonophyll theory** was developed by Melville (1962, 1963, 1983) largely on the basis of a study of the venation pattern. He derived angiosperms from Glossopteridae, which formed important elements in the flora of Gondwanaland. He further derived angiosperm flower from **gonophyll**, a fertile branching axis adnate to a leaf. In simple Glossopterids *Scutum* and *Ottokaria*, the fertile branch consisted of a bivalved scale (having two wings) called the **scutella** with terminal ovules on dichotomous groups of branches. Folding of the scutella along the cluster of its ovules forms the angiosperm condition, an indication of this closure being found in the Permian fossil *Breytenia*. In *Lidgettonia*, the fertile branch consists of four

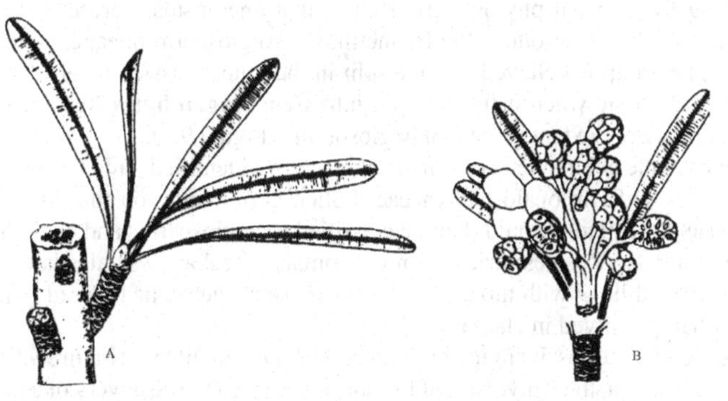

Figure 9.6: Pentoxylales. A: Suggested reconstruction of *Pentoxylon sahnii* with strap-shaped leaves. **B:** Suggested reconstruction of seed cones (From Sahni, 1948).

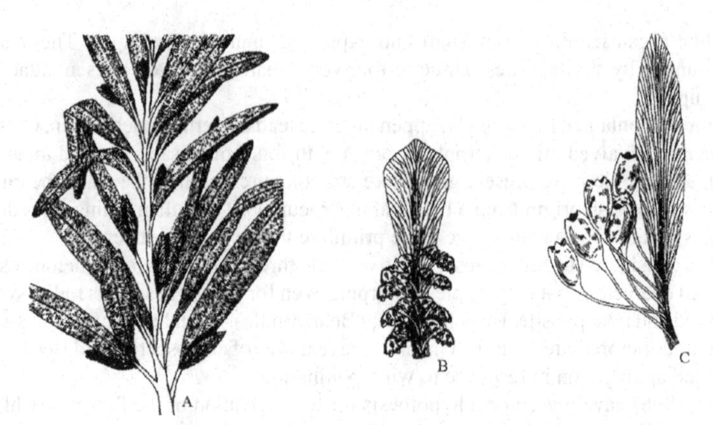

Figure 9.7: Glossopteridae. A: *Dictypteridium feistmantelii* (*Glossopteris tenuinervis*) vegetative branch. **B:** Fertile branch (Gonophyll) of *Lidgettonia mucronate.* **C:** Fertile branch of *Denkania indica* with cupules (A: from Chandra and Surange, 1976; B and C from Surange and Chandra, 1975).

to eight disc-like bearing several seeds. In *Denkania*, described from Raniganj, India, about six seed-bearing **cupules** are attached to long stalks borne from the midrib of fertile scale.

The leaves of *Glossopteris* (Figure 9.7) are lanceolate, with distinct reticulate venation. In *Glossopteris,* the fertile region is cone-like with a transition from leaves to fertile scales, spirally arranged and conforming to the **anthostrobilus**. In *Mudgea,* there is a suggestion of **anthofasciculi**, i.e., leafy structures with two fertile branches, one male and the other female, forming the angiosperm flowers as found in *Ranunculus* and *Acacia*.

Melville believed that angiosperms arose 240 mya ago in the Permian and took around 140 mya before they spread widely in the Cretaceous. It is pertinent to point out, as explained earlier, that although he considered angiosperms to be monophyletic, his justification puts him among the proponents of the polyphyletic origin of angiosperms.

Herbaceous Origin Hypothesis

The herbaceous origin hypothesis resembles the Pseudanthial theory, but the ancestral plant is considered to be a perennial rhizomatous herb instead of a tree. The term **paleoherb** was first used by Donoghue and Doyle (1989) for a group of derivative (not ancestral) forms of Magnoliidae having anomocytic stomata, two whorls of perianth and trimerous flowers, including Lactoridaceae Aristolochiaceae, Cabombaceae, Piperales, Nymphaeaceae and monocots.

According to this hypothesis, ancestral angiosperms were small herbaceous plants with a rhizomatous to scrambling perennial habit. They had simple leaves that were reticulately veined and had a primary venation pattern that would have been indifferently pinnate to palmate, whereas the secondary veins branched dichotomously. The vegetative anatomy included sieve-tube elements and elongate tracheary elements with both circular-bordered and scalariform pitting and oblique end walls. The flowers occurred in cymose to racemose inflorescences. The small monosulcate pollen had perforate to reticulate sculpturing. Carpels were free, ascidiate (ovules attached proximally to the closure) with one or two orthotropous, bitegmic, crassinucellate ovule and dicotyledonous embryo. The aforesaid authors cite extreme rarity of fossil angiosperm wood and abundance of leaf impressions in early fossils.

Consensus is emerging from recent phylogenetic studies that gnetopsids represent the closest living relatives of angiosperms, whereas the closest fossil group is the Bennettitales. Angiosperm lineage, together with these two groups, constitutes **Anthophytes**. The group is believed to have split in the Late Triassic, the angiosperm lineage continuing as **Angiophytes** up to the Late Jurassic when it further split into **stem Angiophytes** (the early extinct angiosperms) and **crown Angiophytes** constituting the extant groups of angiosperms (Figure 9.8).

Krassilov, who believed in the polyphyletic origin of angiosperms, identified three Jurassic groups as proangiosperms: Caytoniales, Zcekanowskiales and Dirhopalostachyaceae. Pollen germinating on the lip, according to him, would be rather disappointing because these plants would then be classified as angiosperms and excluded from discussion of their ancestors. He evolved the Laurales-Rosales series from Caytoniales. Zcekanowskiales had bivalved capsules provided with stigmatic bands and showed links with monocots. Dirhopalostachyaceae had paired ovules exposed on shield-like lateral appendages and probably evolved in Hamamelidales.

Using the oldest, most complete fossil angiosperm on record, David Dilcher (Figure 9.9), a palaeobotanist with the Florida Museum of Natural History at the University of Florida, announced the discovery of a new basal angiosperm family of aquatic plant, Archaefructaceae. It was published in the journal *Science* with coauthors Ge Sun of the Research Center

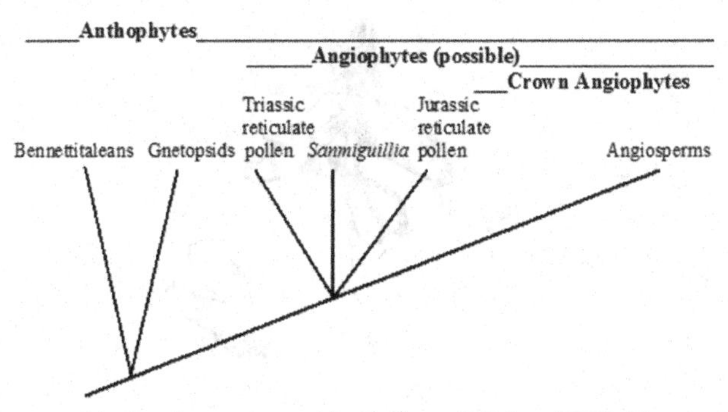

Figure 9.8: A consensus phylogeny of Anthophytes proposed by Taylor and Hickey (1996). Note that *Pentoxylon* has been excluded from sister groups (now only Bennettitaleans and Gnetopsids) of angiophytes.

Figure 9.9: Dr. D. L. Dilcher, Professor Emeritus with Indiana University, formerly palaeobotanist with the Florida Museum of Natural History at the University of Florida, who has pioneered research on Angiosperm fossils with specimen (above left) and reconstruction (above right) of fossil described (Sun, Dilcher et al., 2002) *Archaefructus sinensis*, believed to be the oldest angiosperm fossil nearly 124 mya old.

of Palaeontology at Jilin University, Qiang Ji of the Geological Institute of the Chinese Academy of Geosciences at Beijing and three others (Sun et al., 2002). The family is based on a single genus *Archaefructus* with two species, *Archaefructus sinensis* and *Archaefructus lianogensis*. These were probably aquatics herbs and living at least 124 mya. *Archaefructus* has perfect flowers rather unlike those of extant angiosperms—there is no perianth, the receptacle is very elongated, and the stamens are paired. The fruits are small follicles formed from conduplicate carpels helically arranged. Adaxial elongate stigmatic crests are conspicuous on each carpel. Earlier to this, Dilcher and Crane (1984) had described *Archaeanthus linnenbergeri* (Figure 9.10) from uppermost Albian/mid-cenomanian (approx 110 mya) of middle Cretaceous as a primitive flowering plant with simple bilobed leaves, terminal flower with numerous free carpels producing follicle fruit.

Archaefructus was about 50 cm high, rooted in the lake bottom and was partially supported by the water. Thin stems reached to the water's surface. Pollen and seed organs extended above the water. The leaves were possibly submerged. Seeds probably dispersed on the water and floated towards the shore where they germinated in shallower areas, he added. This is considered to be the oldest record of an angiosperm flower. It is placed in a distinct family Archaefructaceae, probably sister to all extant angiosperms. According to Stevens (2005), It is unclear as to how it relates to extant angiosperms and its flowers are perhaps better interpreted as inflorescences (Zhou et al., 2003; Ji et al., 2004; Crepet et al., 2004).

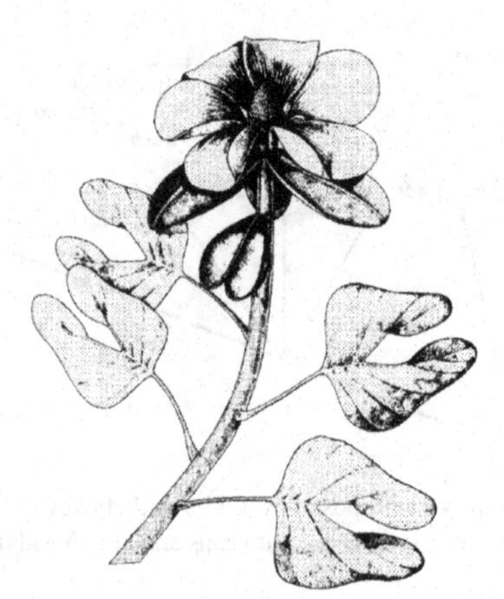

Figure 9.10: Reconstruction of leafy branch with flower of *Archaeanthus linnenbergeri* from middle Cretaceous (After Dilcher and Crane, 1984).

Transitional-Combinational Theory

Stuessy (2004) published a **transitional-combinational theory** for the origin of angiosperms, initiating renewed interest in angiosperms and explaining several recent divergent viewpoints and findings. The theory suggests that the angiosperms evolved slowly from the seed ferns in the Jurassic. Carpel was the first to develop, followed by the double fertilization and then the development of flower. These three fundamental transitions may have taken more than 100 million years to complete. The theory is proposed in view of the difficulty in finding ancestors for angiosperms, yet also considering their sudden appearance and explosive evolutionary success. The extant angiosperms did not appear until Early Cretaceous when the final combination of all three important angiosperm features occurred, as presented by fossil record. This combination provided the opportunity for explosive evolutionary diversification, especially in response to selection from insect pollinators, as also the accompanying modifications in compatibility and breeding systems. The theory attempts to explain discrepancy between fossil and molecular phylogenetic data, latter suggesting pre-Cretaceous origin of angiosperms when DNA (and protein) sequences showed first changes accompanying carpel evolution, much earlier than final combination of all the three angiosperm features. The theory also suggests that barring extinct seed ferns, from which the carpel arose, other gymnosperms had no direct phylogenetic connections to modern angiosperms.

Stuessy suggests that meaningful morphological cladistic analyses should focus on ties between pteridosperms and angiosperms directly, and not include rest of the gymnosperms. He believes that new biology of pollination and breeding systems that favoured outcrossing and developed angiosperm pollen, took place only after the flower had developed, explaining the absence of angiosperm pollen record prior to 130 mya.

Origin of Monocotyledons

It was originally believed (Engler, 1892) that monocotyledons arose before dicotyledons and are polyphyletic (Meeuse, 1963). It is now largely believed that monocotyledons evolved from dicots monophyletically. According to Bailey (1944) and Cheadle (1953), vessels had independent origin and specialization in monocots and dicots, and thus monocots arose from vesselless dicots. Cronquist did not agree with the independent origin of vessels in two groups. He considered monocots to have an aquatic origin from ancestors resembling modern Nymphaeales. This was strongly refuted, however, by studies of vessels done by Kosakai et al. (1970). They considered it difficult to believe that putatively primitive Alismataceae evolved advanced vessels in an aquatic environment yet gave rise to terrestrial monocots with more primitive vessel elements in the metaxylem of roots. They thus favoured the origin of Alismataceae from terrestrial forms.

According to Hutchinson (1973), monocots arose from Ranales along two lines, one (Ranunculoideae) giving rise to Alismatales and other (Helleboroideae) giving rise to Butomales. Takhtajan (1980, 1987) proposed a common origin for Nymphaeales and Alismatales from a hypothetical terrestrial herbaceous group of Magnoliidae. Dahlgren et al. (1985) believed that monocots appeared in the Early Cretaceous some 110 mya ago when the ancestors of Magnoliiflorae must have already acquired some of the present attributes of that group but were less differentiated; some other dicotyledonous groups had already branched off from the ancestral stock. Thorne (1996) believes that monocotyledons appear to be very early offshoot of the most primitive dicotyledons. In their *rbcL* sequence studies, Chase et al. (1993) and Qiu et al. (1993) found the monocots to be monophyletic and derived from within monosulcate Magnoliidae. *Acorus*, Melanthiaceae, and *Butomus* are regarded to be the least specialized Monocotyledons.

Fossil remains of monocotyledons are common in Cenozoic floras where they are represented by many different organs, including stems, leaves, rhizomes, flowers, fruits, seeds, pollen and phytoliths. There are also reliable monocot fossils from the Late Cretaceous, including stems, fruits, seeds and pollen. However, the record of monocotyledons is much less extensive in the Cretaceous than in the Cenozoic. Based on presence of monoaperturate pollen with a graded reticulum, lumina decreasing in size either towards the polar areas (e.g., some palms) or towards restricted areas of the equatorial regions, Walker and Walker (1984) suggested that many of the dispersed pollen grains from the Early Cretaceous that are assigned to extinct pollen genera, such as *Retimonocolpites* and *Liliacidites*, were produced by monocots. Support for a monocot affinity of some *Liliacidites* grains was also inferred by Doyle et al. (2008) from a recent phylogenetic analysis.

BASAL LIVING ANGIOSPERMS

Angiosperms are now increasingly believed to have evolved in very Late Jurassic or very Early Cretaceous. The course of evolution within the group is being thoroughly examined with newer tools of research.

There was general agreement for nearly a century that the early angiosperms were woody shrubs or small trees (herbaceous habit being derived), with simple evergreen entire and pinnately veined leaves with stipules. Concerning the most primitive living angiosperms, there have been two opposing points of view: **Englerian school** (Considering Amentiferae, particularly Casuarinaceae to be the most primitive dicots) and the **Besseyan school** (Bisexual flowers of Magnoliales to be the most primitive). During the last few years the **paleoherbs** are emerging as the strong contenders. The candidate basal groups are discussed below:

Casuarinaceae

According to the **Englerian School**—the view now largely rejected—Amentiferae with reduced unisexual flowers in catkins (or aments) constitute the most primitive living angiosperms. Engler, as well as Rendle (1892) and Wettstein (1935) considered Casuarinaceae (Figure 9.11) to be the most primitive dicot family, and the one derived from Ephedraceae. It is now agreed that Casuarinaceae and the other members of Amentiferae have advanced tricolpate pollen grains, wood anatomy is relatively advanced, and the simplicity of flowers is due to reduction rather than primitiveness. They have also secondarily achieved wind pollination. Other advanced features include trilacunar nodes, cyclic stamens, syncarpous pistil with axile placentation.

Magnoliids

The alternative **Besseyan School (Ranalian School)** considers the Ranalian complex (including Magnoliales), having bisexual flowers with free, equal, spirally arranged floral parts, representative of the most primitive angiosperms.

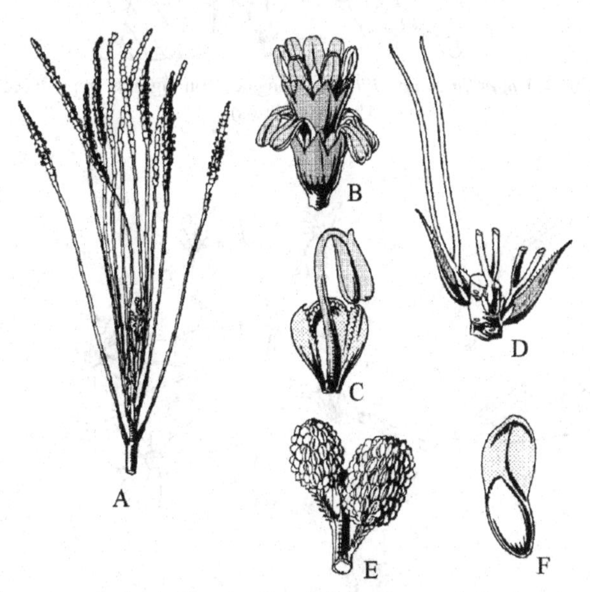

Figure 9.11: Casuarinaceae. *Casuarina suberba*. **A:** Branch with male inflorescences; **B:** Portion of male inflorescence; **C:** Male flower with single stamen; **D:** Part of female inflorescence showing 3 flowers; **E:** Fruits; **F:** Seed with broad wing.

Magnoliaceae

Bessey (1915), Hutchinson (1959, 1973), Takhtajan (1966, 1969) and Cronquist (1968) all believed that large solitary flower of *Magnolia* (Figure 9.12) ('**Magnolia the primitive theory**') with an elongated floral axis bearing numerous spirally arranged stamens and carpels, is the most primitive living representative. The stamens of *Magnolia* and other closely related genera are laminar, perianth undifferentiated, and pollen grains monosulcate and boat-shaped. In the subsequent works, however, Takhtajan agreed that the flower of Magnolia is more advanced than those found in Winteraceae and Degeneriaceae. *Archaeanthus linnenbergeri* from latest Albian–earliest Cenomanian Dakota Formation is very similar in floral structure to extant Magnoliaceae, but the fruits differ from all extant taxa in producing many small seeds (Dilcher and Crane, 1984).

Winteraceae

After several decades of *Magnolia* being considered as the most primitive living angiosperm, the view was challenged by Gottsberger (1974) and Thorne (1976), who considered the most primitive flowers to have been middle sized, with fewer stamens and carpels, and grouped in lateral clusters as in the family Winteraceae, to which such primitive genera as *Drimys* (Figure 9.13) have been assigned. This view is supported by the occurrence of similar stamens and carpels, absence of vessels, morphology similar to pteridosperms, high chromosome number suggesting a long evolutionary history, and less specialized beetle pollination of *Drimys* compared to *Magnolia.*

Takhtajan (1980, 1987) later acknowledged that moderate sized flowers of *Degeneria* and Winteraceae are primitive, and the large flowers of *Magnolia* and Nymphaeaceae are of secondary origin. However, he considered Degeneriaceae to be the most primitive family of living angiosperms. Cronquist (1981, 1988) also discarded *Magnolia* while considering Winteraceae to be the most primitive. Dispersed permanent tetrads of pollen described as *Walkeripollis* (Doyle et al., 1990) probably represent stem-group Winteraceae.

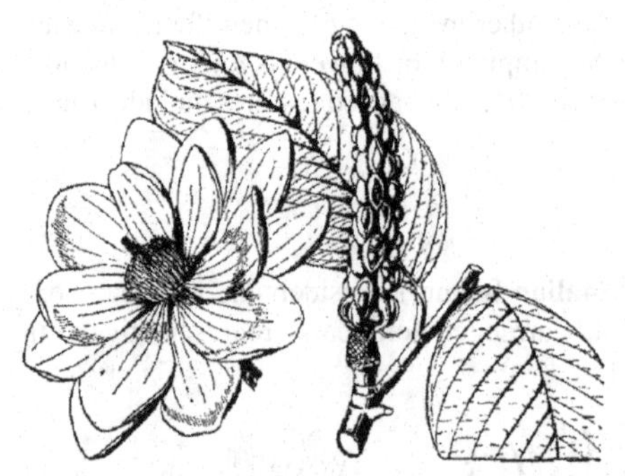

Figure 9.12: Flower and a twig of *Magnolia campbellii* with elongated fruiting axis (reproduced with permission from Oxford University Press).

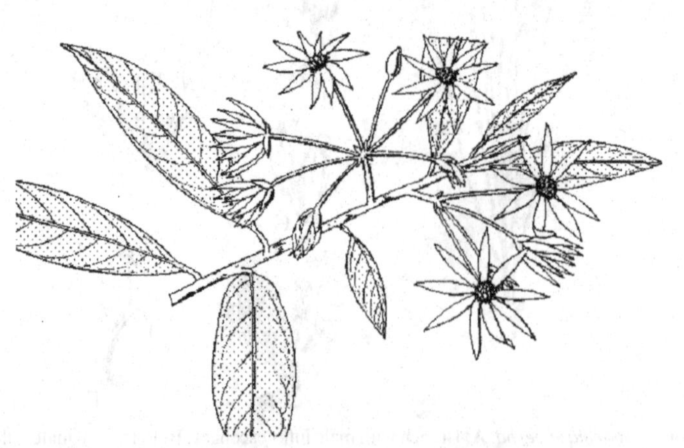

Figure 9.13: Winteraceae. Flowering twig of *Drimys winteri.*

Degeneriaceae

Takhtajan, who was earlier a strong supporter of *Magnolia* as the most primitive living angiosperm, has abandoned this view in favour of Degeneriaceae and Winteraceae to be the basal angiosperm families, but has maintained since 1980 to regard Degeneriaceae as the most primitive.

Degeneriaceae (Figure 9.14) may be recognized by their spiral, entire, exstipulate leaves and large, axillary flowers with many tepals and a single carpel. Vessel elements have scalariform perforations. Leaves are spirally arranged and pollen boat-shaped. The most significant plesiomorphic features include stigma running the entire length of the carpel, laminar stamens with three veins, the fruit a follicle and embryo with 3 to 4 cotyledons.

Calycanthaceae

Suggestions have also come projecting Calycanthaceae (Loconte and Stevenson, 1991) as basic angiosperms with a series of vegetative and reproductive angiosperm plesiomorphies such as shrub habit, unilacunar two-trace nodes, vessels with scalariform perforations, sieve-tube elements with starch inclusions, opposite leaves, strobilar flowers, leaf-like bracteopetals, poorly differentiated numerous spirally arranged tepals, and few ovulate carpels. Food bodies terminating the stamen connectives indicate beetle pollination.

The family (Figure 9.15) is regarded as the most basal family of Laurales. It is interesting to note that genus *Idiospermum* (which was recognized as new genus based on *Calycanthus australiensis* by S. T. Blake in 1972) was considered as the

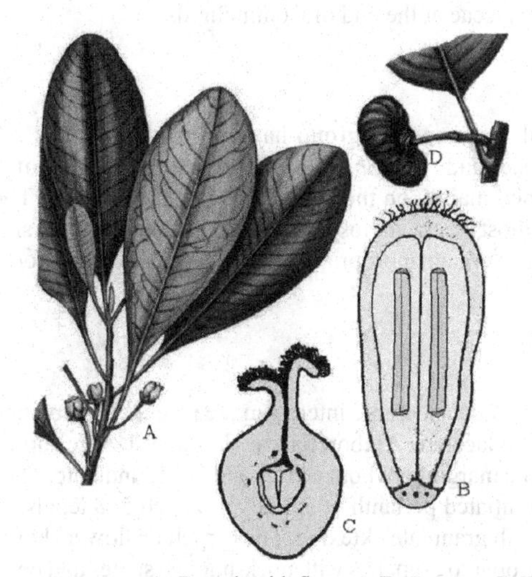

Figure 9.14: Degeneriaceae. *Degeneria vitiensis.* **A:** Branch with flowers; **B:** Stamen; **C:** Transverse section of carpel; **D:** Fruit.

Figure 9.15: Calycanthaceae. *Calycanthus occidentalis.* **A:** Flowering twig with solitary terminal flower. **B:** L. S. of flower showing free carpels. **C:** Flower with some tepals and stamens removed.

most primitive flowering plant by these authors. Endress (1983) had described 'In all respects, *Ideospermum* gives the impression of a strange living fossil'.

Paleoherbs

The last decade of the twentieth century has seen the strong development of an alternative **herbaceous origin hypothesis** for angiosperms (Taylor and Hickey, 1996) originally developed as paleoherb hypothesis. The most primitive angiosperms are rhizomatous or scrambling perennial herbs with simple net-veined leaves, flowers in racemose or cymose inflorescences, with free carpels containing one or two ovules. A number of families are included in the group. Thorne (2000) had placed all of them under Magnoliales, along with Magnoliaceae and Winteraceae. In his later revision (2003), however, placed Amborellaceae and Chloranthaceae (together with Trimeniaceae and Austrobaileyaceae) under Chloranthales, the first order of Magnoliidae (and accordingly angiosperms), the families arranged in that order. Subsequently (2006, 2007) he separated them under distinct subclass Chloranthidae, at the begining of angiosperms. The family Ceratophyllaceae is placed after the monocot families, towards the beginning of Ranunculidae. The placement of Amborellaceae at the beginning of angiosperms is found in the classification schemes of Judd et al. (2003), APG IV (2016) and APweb (Stevens, 2018). The position of the other two families is, however, not settled. Judd et al. considers both Chloranthaceae (towards the end of basal families before Magnoliid complex) and Ceratophyllaceae (towards the end of Magnoliid complex) as having uncertain position. APG IV places Chloranthaceae in unplaced independant lineage after Magnoliids, whereas family Ceratophyllaceae is placed after monocot, before and prabable sister group of Eudicots. APweb places Chloranthaceae among basal families, but Ceratophyllaceae at the end of Commelinids.

ANITA Lineage

More recently five groups of basal Angiosperms group have been identified as **ANITA** (consisting of *Amborella*, Nymphaeales, Illiciales, Trimeniaceae, and *Austrobaileya*), representing a series of the earliest-diverging lineages of the angiosperm phylogeny, determined mainly on the bases of Multigene analysis. These analyses demonstrate that the identification of **ANITA** as the basal most extant angiosperms was based on historical signals preserved in the gymnosperm sequences and that the gymnosperms were an appropriate outgroup with which to root the angiosperm phylogeny in the multigene sequence analysis.

Amborellaceae

The family Amborellaceae has attracted considerable interest in the recent years, being unique in angiosperms in lacking pollen tectum and being inaperturate to lacerate. Amborellaceae (Figure 9.18) are shrubs without vessels, with unilacunar nodes, 2-ranked, exstipulate leaves; the margins are both serrate and rather undulate. The plant is dioecious, and the flowers are small in cymes, with an undifferentiated perianth of spirally arranged 5–8 tepals. The staminate flowers have 10–25 stamens, sessile anthers and pollen with granulate ektexine. The carpellate flowers have 1–2 staminodes and 5–6 whorled incompletely closed carpels that develop into drupelets with pock-marked stones and pockets of almost resinous substances.

Relationships at the base of the angiosperm lineage are being clarified. Amborellaceae are most likely to be sister to other angiosperms, Nymphaeaceae sister to the rest, then Austrobaileyales.

Nymphaeales

Nymphaeales consist of three families of aquatic plants, the Hydatellaceae, the Cabombaceae, and the Nymphaeaceae, the family Nymphaeaceae, which was earlier unplaced in APG and APG II, has been included under distinct order along with other two families. The family Hydatellaceae was placed among the monocots in previous systems, but a 2007 study found that the family belongs to the Nymphaeales. Genus Nelumbo formerly placed under Nymphaeaceae has been removed to distinct family Nelumbonaceae. The Cretaceous fossil record of Nymphaeales is much more meagre. Dispersed Nymphaea-like pollen, described as *Zonosulcites scollardensis* and *Zonosulcites parvus*, is known from the Maastrichtian of Canada. Despite their sparse Late Cretaceous record, and even though the extant genera may not have diversified until the Cenozoic, Nymphaeales were clearly present at an early stage in angiosperm evolution.

Chloranthaceae

Taylor and Hickey (1996) consider Chloranthaceae (Figure 9.16) the basic angiosperm family. The family shows several plesiomorphic characters such as flowers in an inflorescence, plants dioecious, carpels solitary, placentation apical, and fruit drupaceous with small seeds. The family is the oldest in the fossil record, the fossil genus *Clavitopollenites* being assigned to Chloranthaceae and closer to the genus *Ascarina*. The stems of *Sarcandra* are primitively vesselless, but Carlquist (1996) has reported vessels in this genus. The family is considered to be earliest to record wind pollination in angiosperms.

The plants are mostly herbaceous, some species being shrubs. The flowers are highly reduced, subtended by a bract and without any perianth, and arranged in decussate pairs. The flowers are unisexual in *Ascarina*, *Hedyosmum* and *Ascarinopsis* but bisexual in *Chloranthus* and *Sarcandra*. Stamens vary in number from 1 to 5. The carpel lacks style, and single orthotropous ovule is bitegmic.

Taylor and Hickey believe in the origin of Chloranthaceae from gnetopsids, hypothesizing that the ovule and the bract subtending the floral unit in Chloranthaceae are homologous with one of the terminal ovules and proanthophyll subtending the anthion (inflorescence unit) of gnetopsids. Chloranthaceae has undergone considerable reduction in its number of parts as well as general level of elaborateness.

They also believed that the outer integument of the angiosperm bitegmic ovule has ring-like origin and is homologous with the ovular bracts that form the second integument in the gnetopsids.

Ceratophyllaceae

Chase et al. (1993) based on *rbcL* had expressed the view that Ceratophyllaceae represents the basal angiosperm family. The family has fossil record extending back to the Early Cretaceous. Cladistic studies by Sytsma and Baum (1996) based on molecular data support the placement of *Ceratophyllum* (Figure 9.17) at the base of angiosperms, but the authors cautioned that resolution of basal angiosperm relationships may have to await both the collection of additional molecular and morphological data as well as further theoretical advances in phylogenetic systematics. Hickey and Taylor (1996)

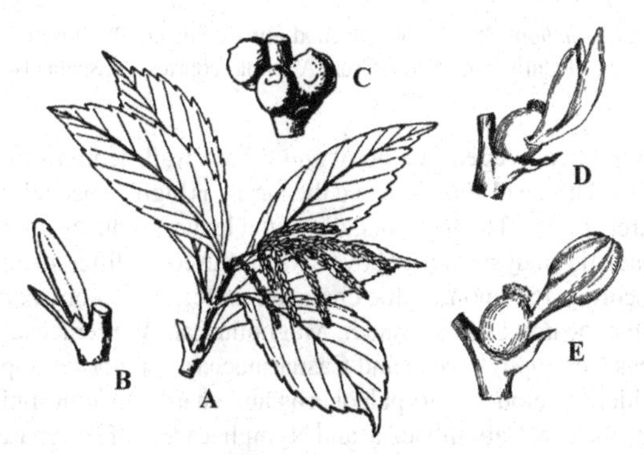

Figure 9.16: Chloranthaceae. A: *Ascarina lanceolata*, flowering branch. **B:** A male flower. **C:** Fruit. **D:** Bisexual flower of *Chloranthus henryi* with bract, three stamens and pistil with tufted stigma. **E:** Bisexual flower of *Sarcandra glabra*.

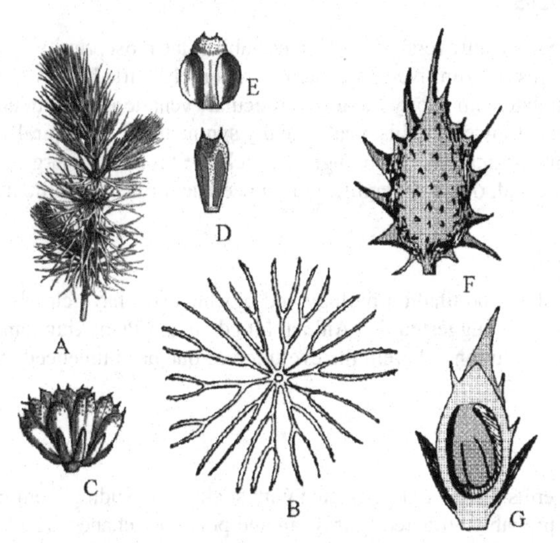

Figure 9.17: Ceratophyllaceae. *Ceratophyllum submersum*. **A:** A portion of plant; **B:** Whorl of leaves at node; **C:** Male flower; **D:** Young stamen; **E:** Dehiscing stamen; **F:** Fruit; **G:** Longitudinal section of fruit with pendulous seed.

Figure 9.18: Amborellaceae. *Amborella trichopoda*. **A:** Fully opened female flower; **B:** Close up of a branch; **C:** Male flower (photo B, courtesy University of California, Santa Cruz; A, photo courtesy Missouri Botanical Garden).

felt that aquatic plant with highly reduced vegetative body and pollen wall, tenuinucellate, unitegmic ovules is a poor candidate for the basal-most position. Thorne (1996) believed the family is highly specialized and its relationships are highly obscured. Other specialized features include lack of roots, dissected leaves, reduced vasculature and the lack of stomata.

Loconte (1996) carried out cladistic analysis of the above taxa proposed by different authors as most basal angiosperms. He included 69 taxa in the study scoring 151 apomorphic character-states. Parsimony analysis using PAUP resulted in 10 trees at 590 steps. Calycanthaceae appeared as first branch. Magnoliaceae, Winteraceae and Chloranthaceae hypotheses appeared two steps longer, whereas Ceratophyllaceae and Casuarinaceae hypotheses appeared six steps longer.

Other families that are considered belonging to paleoherbs and share plesiomorphic features include Saururaceae, Piperaceae, Aristolochiaceae, Barclyaceae, Cabombaceae and Nymphaeaceae. They share characters of herbaceous habit, tectate-columellate monosulcate pollen, apocarpous gynoecia, and simple floral units.

EVOLUTIONARY TRENDS

Although there has been some recent controversy regarding the habit of the most primitive living angiosperms being woody or herbaceous, the general features of primitive angiosperms are largely settled. They have simple alternate exstipulate leaves, which are entire and petiolate with poorly organized reticulate venation and with unilacunar, two-trace nodes. The vessels are absent or tracheid-like. Flowers are bisexual, radially symmetrical with spirally arranged floral parts. Stamens are broad, undifferentiated with marginal microsporangia. Carpels are broad with large number of ovules, stigma along the margin and not completely sealed, ovules bitegmic, crassinucellate. Fruits are follicular.

Coevolution with Animals

Studies on comparative morphology, pollination biology and biochemistry have clearly elucidated the role of animals in the evolution of angiosperms. It is suggested that Animal kingdom and Plant kingdom, particularly the Angiosperms have undergone a process of co-evolution, wherein the evolution of one has influenced the other. This has proceeded in various ways.

Pollination

Early seed plants, the gymnosperms were wind pollinated with sticky sap exuding from micropyles trapping the pollen. Early insects, the beetles were probably attracted to this sap and pollen by chance. The better pollination and increased seed set encouraged the selection towards showy flowers more attractive to insects, edible flower parts, protein rich pollen, nectaries and bisexual flowers so that same insect visit can both deposit the pollen and pick up for visit to another flower. Increased visits by insects posed danger to the exposed seeds, resulting in selection towards protection of seeds in closed carpel, a major step towards the evolution of angiosperms. Increased protection of seeds encouraged smaller seeds in increased numbers and shorter life cycle to overcome drought conditions. Complete closure of carpel was accompanied by the differentiation of stigmatic region for receiving pollen, and the distinct style to keep the stigma within the reach of insects. To suite to the floral mechanisms the early beetles were slowly replaced by higher insects such as moths, butterflies, bees, wasps and flies, coinciding with the floral diversification of angiosperms.

Beetle pollinated flowers are typically dull or white with fruity odors, edible petals and heavily protected seeds. Bee pollinated flowers are brightly colored (blue or yellow but not red) with honey guides and with lot of pollen and nectar. Butterfly pollinated flowers are red, blue or yellow. Moth pollinated flowers mostly open at night and have heavy fragrance to attract moths. Moth and Butterfly pollinated flowers generally have long corolla tubes with nectaries at the base. Bird pollinated flowers are bright red or yellow, produce large amount of nectar, with little or no fragrance. Bat pollinated flowers are dull colored, open at night and have fruity odor.

Biochemical Coevolution

Plants and their insect predators are believed to have undergone adaptive radiation in stepwise manner, with the plant groups evolving new and highly effective chemical defenses against herbivores and the latter continually evolving means of overcoming these defenses. Mustard oils of Brassicaceae are toxic for many animals, yet they attract other herbivores such as cabbage worm which uses the mustard oils to locate the cabbage plant for laying its eggs. The chemical hypericin in genus *Hypericum* repels almost all herbivores but the beetle genus *Chrysolina* can detoxify hypericin and use it to locate the plant.

The evolution of new chemical defense of plant has resulted in plants often acquiring the growth hormones found in insect larvae. Proper levels of juvenile hormone in insect larvae are essential for the hatching of insect larvae into normal sexual adults. Several species of plants such as *Ageratum* contain hormone **juvabione**, similar to the juvenile hormone of insects. Such plants if ingested by the insect larvae elevate the level of hormone, resulting in their development into abnormal asexual adults. The larvae as such, learn to avoid such plants.

Some plant products help insects against predators. Monarch butterfly, for example, ingests cardiac glucoside from milkweed *Asclepias*. Such butterflies if ingested by blue jays make latter violently sick. Blue jays learn to recognize the toxic brightly colored monarch butterflies. The milkweed thus helps to protect monarch butterfly from blue jay.

Basic Evolutionary Trends

Evolution within Angiosperms has proceeded along different lines in different groups. Numerous trends in the evolution of angiosperms have been recognized from comparative studies of extant and fossil plants. The general processes involved in attaining diversity of angiosperms are underlined below.

Fusion

During the course of evolution in angiosperms, fusion of different parts has led to floral complexity. Fusion of like parts has led to the development of gamosepaly, gamopetaly, synandry and syncarpy in various families of angiosperms. Stamens have shown fusion to different degrees: fusion of filaments only (monadelphous condition in Malvaceae), fusion of anthers only (syngenesious condition found in Asteraceae) or complete fusion (synandry as in *Cucurbita*). Carpels may similarly be fused only by ovaries (Synovarious: Caryophyllaceae), only by styles (synstylous: Apocynaceae) or complete fusion of both ovaries and styles (Synstylovarious: Solanaceae, Primulaceae). Fusion of unlike parts has resulted in an epipetalous condition (fusion of petals and stamens), formation of gynostegium (the fusion of stamens and gynoecium: Asclepiadaceae) and formation of an inferior ovary (fusion of calyx with ovary: Apiaceae, Myrtaceae, etc.).

Reduction

Relatively simple flowers of many families have primarily been the result of reduction. The loss of either stamens or carpels has resulted in unisexual flowers. The loss of one perianth whorl has resulted in monochlamydeous forms, and their total absence in achlamydeous forms. There has also been individual reduction in the number of perianth parts, number of stamens and carpels. Within the ovary different genera have shown reduction in the number of ovules to ultimately one, as seen in the transformation of follicle into achene within the family Ranunculaceae. There has also been reduction in the size of flowers, manifested in diverse families such as Asteraceae and Poaceae. Reduction in the size of seeds has been extreme in Orchidaceae. Male flower of *Euphorbia* presents a single stamen, there being no perianth or any trace of a pistillode, only a joint indicates the position of thalamus and the demarcation between the pedicel and the filament.

Change in Symmetry

From simple radially symmetrical actinomorphic flowers in primitive flowers developed zygomorphic flowers in various families to suit insect pollination. The size of corolla tube and orientation of corolla lobes changed according to the mouthparts of the pollinating insects, with striking specialization achieved in the turn-pipe mechanism of *Salvia* flowers, and female wasp like flowers of orchid *Ophrys*.

Elaboration

This compensating mechanism has been found in several families. In Asteraceae and Poaceae, the reduction in the size of flowers has been compensated by an increase in the number of flowers in the inflorescence. Similarly, reduction in the number of ovules has been accompanied by an increase in the size of ovule and ultimately seed, as seen in *Juglans* and *Aesculus*.

Remoration

The term was suggested by Melville (1983) to refer to evolutionary retrogression found in angiosperms and their fossil relatives. The fertile shoots of angiosperms, according to him, show venation pattern changes progressively from vegetative leaves through successive older evolutionary stages in bracts and sepals, and the most ancient in petals. The innermost parts in a bud as such represent the most primitive evolutionary condition, and the outermost the most recent condition.

There has been some shift in the understanding of angiosperm phylogeny to support the stachyosporous origin of angiosperm carpel (Taylor and Kirchner, 1996). With the acceptance of such a viewpoint, the reproductive axis with many flowers, few carpels per flower and few ovules per carpel are ancestral. Evolution proceeded along two directions from this: one with few flowers, each of which had many carpels and few ovules and the other with few flowers, each containing few carpels and many ovules. The evolutionary trends in angiosperms are thus often complicated and frequent reversal of trends may be encountered, as for example the secondary loss of vessels in some members.

Xylem Evolution

Xylem tissue of angiosperms largely consists of dead tracheids and vessels, supporting fibers and living ray cells. Tracheids are elongate, imperforate water conducting cells found in almost all lower vascular plants, gymnosperms and angiosperms. Vessels are perforate elements largely restricted to angiosperms, although also found in extant gnetopsids, some species of *Equisetum*, *Selaginella*, *Marsilea* and *Pteridium*. The presence of vessels in gnetopsids, the closest relatives of angiosperms, had given rise to the speculation that the latter arose from former. The studies of Bailey and associates, however, showed that the vessels in the two groups arose independently. In gnetopsids, they developed from tracheids with circular pitting and in angiosperms from tracheids with scalariform pitting. It is also pointed out by Carlquist (1996) that circular pitting in the vessels of gnetopsids, as also the gymnosperm tracheids in general, is different from angiosperms in having pits with torus and pit margo with pores much larger than those of angiosperms. Although some angiosperms do have pit membrane with torus, the pit margo is always absent.

Because all fossil and almost all extant gymnosperms possess tracheids with circular-bordered pitting, it has led to the conclusion that the tracheid is the most primitive type of tracheary element in the angiosperms. As tracheids have given rise to vessel elements, the most primitive type of vessels have long narrow vessel elements with tapering ends. The tracheids in angiosperms have scalariform pitting, and as such it is assumed that these tracheids arose from tracheids of gymnsperms with circular pitting. In the transformation of tracheids with scalariform pitting to vessel elements with scalariform pitting, the earliest elements had perforation plates with numerous scalariform bars. During further evolution of vessels, elements became smaller and broader, and perforation plates more horizontal. There has been an accompanied reduction in the number of scalariform bars, resulting in shortest broadest vessel elements with simple perforation plate and transverse end wall (Figure 9.19) in most advanced forms. Hamamelididae were once regarded to be primitive due to their simple floral structure but have advanced vessel elements and thus considered advanced over Magnoliidae with primitive elongated narrow vessel elements. This is supported by studies on floral anatomy and palynology.

Carlquist (1996), based on a survey of wood anatomy, has identified a number of distinct evolutionary trends in angiosperms. The cambial initials have shortened, the ratio of length accompanying fibers to vessel elements (F/V ratio) has shown an increase from 1.00 in primitive dicotyledons to about 4.00 in the most specialized woods, and angular outline of vessels changed to circular outline together with widening of their diameter. There was also a progressive reduction in the number of scalariform bars, ultimately resulting in simple perforation, facilitating an easier flow of water. The lateral walls showed a shift from scalariform to the opposite circular pits and finally to alternate circular pits, so as to provide better mechanical strength. Imperforate tracheids have shown a shift to fibre-tracheids to finally libriform fibres. Shortening of fusiform initials is correlated with storeying of woods. It is interesting to note that cladistic studies have shown that present-day vessel-less angiosperms do not form a single clade and are distributed in diverse groups such as

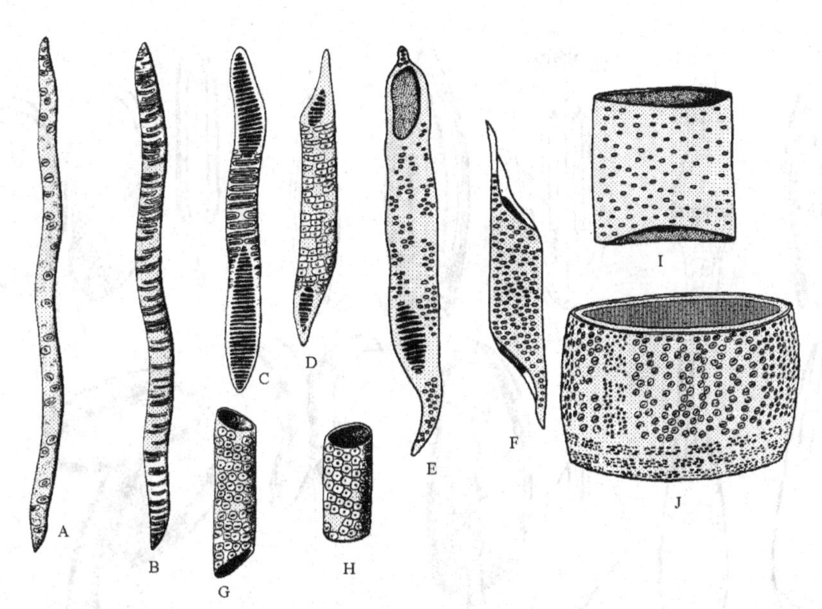

Figure 9.19: Presumed evolutionary transformation of gymnosperm tracheid with circular pitting (**A**) to angiosperm tracheid with scalariform pitting (**B**), further to vessel-element with oblique perforation plate with numerous scalariform bars (**C**). Further shortening and broadening of vessel-element, perforation plate becoming more and more horizontal and reduction in the number of bars in the perforation plate ultimately led to shortest, broad vessel element with transverse simple perforation plate. Vessel-element **E** shows one scalariform and one simple perforation plate. Note conspicuous tails, a reminder of tracheids in vessel-elements **D, E** and **F** with still oblique perforation plates. plates, **G, H, I** stages in shortening and broadening of vessel elements. **J** represents the vessel-element of *Quercus alba*, being broader than long and with a simple large perforation.

Hamamelidales (*Trochodendron* and *Tetracentron*), Magnoliales (Amborellaceae, Winteraceae), and Laurales (*Sarcandra*: Vessels have been, however, reported in root secondary xylem by Carlquist, 1987 and in stem metaxylem by Takahashi, 1988). This led Carlquist to conclude that vessels have originated numerous times in dicotyledons.

It had often been held that vessels arose first in the secondary xylem and later in the metaxylem, and that specialization has gradually advanced from the secondary to primary xylem. Carlquist pointed out that scalariform pitting is widespread in the metaxylem of vascular plants and if primitive angiosperms were herbs, in accord with paleoherb hypothesis, metaxylem would be expected to have scalariform pitting of tracheary elements. The development of woody habit—if featured paedomorphosis—would extend scalariform patterns in secondary xylem.

Stamen Evolution

The most primitive type of Stamen in angiosperms represented in genera like *Degeneria*, *Austrobaileya*, *Himantandra* and *Magnolia* (Figure 9.20) and other primitive genera is laminar, 3-veined leaf-like organ without any clear-cut distinction of fertile and sterile parts. The pollen sacs (sporangia) are borne near the centre either on the abaxial (dorsal) side (*Degeneria*, Annonaceae and Himatandraceae) or on adaxial (ventral) side (*Austrobaileya* and *Magnolia*). Semilaminar stamens occur in some other primitive families like Nymphaeaceae, Ceratophyllaceae and Eupomatiaceae. In further specialization of stamen, there has been reduction of sterile tissues and retraction of marginal areas. The proximal part became filament and distal part the anther. The midvein region formed the connective, and distal part the appendage, as seen in several genera.

In primitive families, connective forms a major part of anther. In more advanced families, the connective is highly reduced (Acanthaceae, Plantaginaceae) or may be almost absent. In some families such as Betulaceae, the connective as well as the upper part of filament may become divided and two anther lobes get separated. In more primitive families the connective is produced above into an appendage which disappears progressively in more advanced families.

Broad laminar filament, merging with the rest of stamen represents the most primitive state. It becomes narrower and finally terete in advanced families. The stamens with well marked narrow filament may have basal, dorsal or versatile attachment with anthers. Basifixed condition is the most primitive, versatile the most advanced often commonly seen in grasses and Amaryllidaceae.

It is generally agreed that primitive stamen was laminar, with two pairs of sporangia, borne on adaxial or abaxial side, because both situations are met in primitive families. During the course of evolution, the stamen became slenderer,

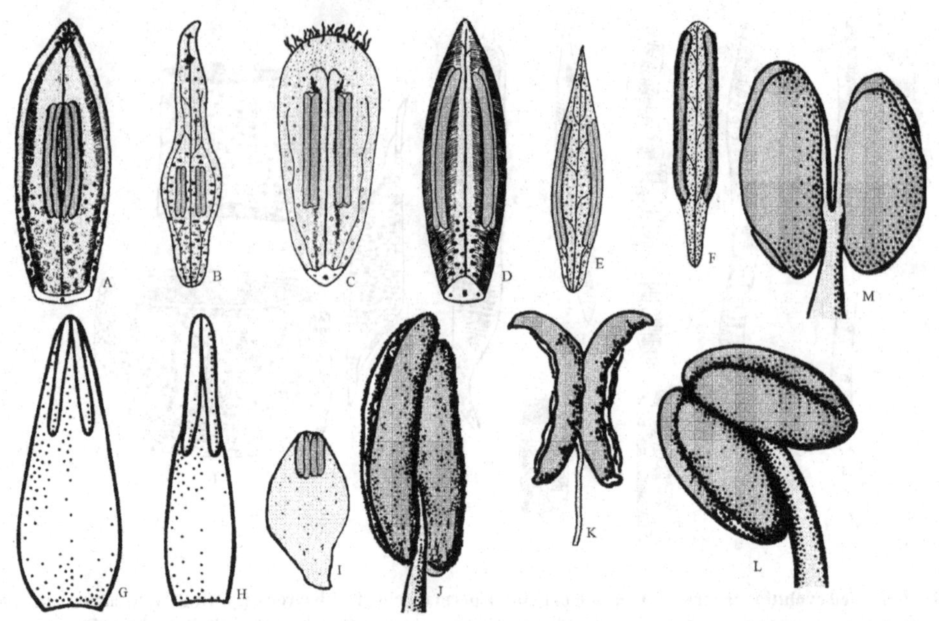

Figure 9.20: Evolution of stamen in angiosperms. A–D: Primitive laminar stamens without clear distinction of anthers and filaments; **A:** *Austrobaileya scandens* with adaxial pollen sacs (microsporangia); **B:** *Himantandra baccata* with abaxial pollen sacs; **C:** *Degeneria vitiensis* with abaxial pollen sacs; **D:** *Magnolia maingayi* with adaxial pollen sacs; **E:** Laminar stamen of *Magnolia nitida* with marginal pollen sacs and prolonged sterile appendage; **F:** Semilaminar stamen of *Michelia fuscata* with marginal pollen sacs and narrowed filament; **G:** Outer semilaminar stamen of *Nymphaea odorata* with petaloid filament and narrow anthers; **H:** Inner stamen with narrower filament and differentiated anther region; **I:** Stamen of *Illicium parviflorum* with reduced anther region and broad filaments; **J:** Stamen of *Opuntia pusilla* with well differentiated anthers and filament; **K:** Stamen of *Poa pratensis* with reduced connective and thread-like filament; **L:** Stamen of *Penstemon canescens* with well-defined filament and large anthers with distinct anther lobes but reduced connective; **M:** Stamen of *Betula nigra* with divided connective and filament.

its laminar form slowly disappearing, and sporangia occupying marginal position. The transformation of broad laminar to narrow stamens is clearly depicted in *Nymphaea* from outer to inner stamens and in different species of the genus.

A typical anther of angiosperms is bithecous with two anther lobes with the two anther sacs in each, finally merging into one. The anther with a single anther lobe, as in Malvaceae, has a single final sac (theca) and as such monothecous. The monothecous anthers of Malvaceae and some other families result from splitting of stamens, thus separating the two anther lobes. In others, like *Salix,* there may be a partial connation of two stamens resulting in apparent dichotomy. Anatomical evidence shows two independent vascular supplies derived from opposite sides of the receptacle, as opposed to families where splitting of stamens occurs.

Pollen Grain Evolution

Many families in monocots and several primitive dicots of the magnoliid complex bear monosulcate pollen grains, a condition generally considered to be the primitive one in angiosperms. Walker and Doyle (1975) and Walker and Walker (1984) suggested that the primitive angiosperm pollen grain is large- to medium-sized, boat-shaped, smooth-walled, with homogenous or granular infratectal layer, the tectum being absent (pollen atectate) and endexine (a layer unique to angiosperms, being absent in gymnosperms) either missing or poorly developed under the apertural area. This type of pollen is found among extant angiosperms in Annonaceae, Degeneriaceae and Magnoliaceae. The prototype of this was gymnosperm pollen which was monosulcate, large, boat-shaped with laminated nexine, homogenous sexine or with granular infratectal layer (Figure 9.21), a pollen type common in Bennettitales, Gnetopsids (excluding *Gnetum*) and *Pentoxylon.* In evolution of angiosperm pollen from this, the laminated nexine disappeared, nexine became granular to tectate with columella and reticulate surface to those with intectate collumellae.

Brenner and Bickoff (1992) recorded globose inaperturate pollen grains from the Valanginian (ca 135 mya) of the Helez formation of Israel, now considered to be the oldest record of angiosperm fossils. These pollen grains resemble those of *Gnetum* in general shape and lack of aperture, and also found in Chloranthaceae, Piperaceae, and Saururaceae, which are gaining increased attention as basal angiosperm families. This led Brenner (1996) to postulate a new model for

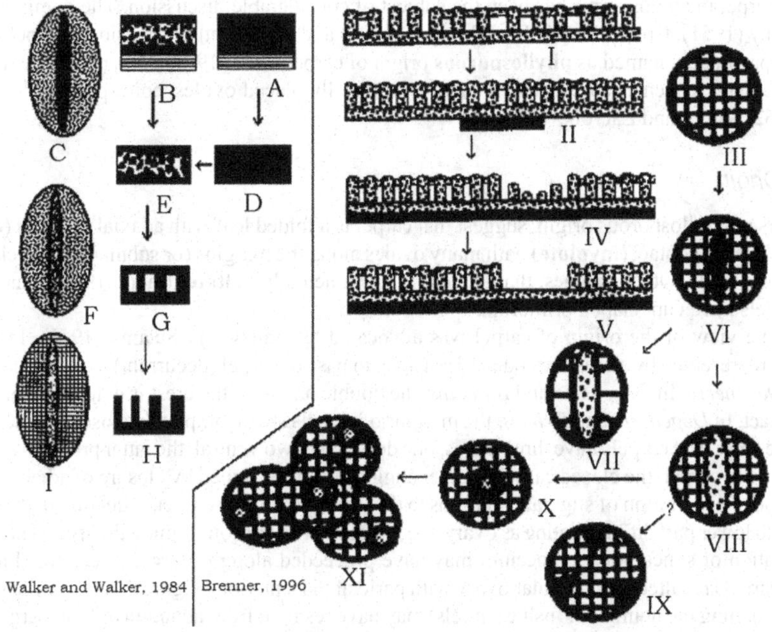

Figure 9.21: Two different models for evolution of exine and pollen grains in Angiosperms. Walker and Walker, 1984 (on left). **A:** Exine of ancestral gymnosperm with homogenous sexine and laminated nexine; **B:** Same but sexine with granular infratectal layer; **C:** Pollen grain of same, boat shaped and monosulcate; **D:** Exine of most primitive angiosperm with smooth sexine and disappearance of laminated nexine; **E:** Same but with homogenous sexine; **F:** Monosulcate pollen of same; **G:** Exine with development of tectum, infratectal layer and homogenous nexine; **H:** Same but with loss of tectum; **I:** Monosulcate pollen grain with intectate exine. Brenner, 1996 (on right). **I:** Exine of early angiosperm, tectate-columellate and without aperture; **II:** Exine with initiation of endexine (shaded solid black); **III:** Nonaperturate early pollen with circular outline; **IV:** Exine with complete endexine layer and initiation of sulcus; **V:** Exine with developed sulcus; **VI:** Monosulcate pollen of basal angiosperms; **VII:** Monosulcate boat-shaped pollen of Magnoliids and monocots; **VIII:** Circular monosulcate pollen of dicots; **IX:** Tricolpate pollen of Eudicots which might have developed from monosulcate or inaperturate forms; **X:** Uniporate pollen; **XI:** Uniporate pollen of Winteraceae in tetrads (Modified from Brenner, 1996).

the evolution of angiosperm pollen. The earliest pollen in angiosperms, developed in Valanginian or earlier from stock that also gave rise to *Gnetum*, had small pollen, circular in outline, tectate columellate, and without aperture. A possible intine thickening was accompanied by developments of endexine layer above intine in Hauterivian. The next step involved evolution of sulcus and divergence of monocot and dicot pollen types from basic dicot stock. In Barremian diversification of monosulcate pollen grains occurred with migration to different geographical regions. In Lower Aptian tricolpate pollen evolved from either monosulcate or inaperturate forms in northern Gondwana, resulting in evolution of eudicots.

It is suggested by Brenner that the formation of sulcus during Early Cretaceous may have been an adaptation that was a more effective way of releasing recognition proteins involved in pollen-tube development while the later development of tricolpate condition in the Aptian would be a further extension of this process.

The formation of endexine before aperture development may reflect the development of intine thickening, which is related to sulcus development. In extant angiosperms, the intine beneath the aperture stores recognition proteins.

Carpel Evolution

Carpel is a structure unique to angiosperms enclosing and protecting ovules. The evolution of carpel probably played a major role in diversity and success of angiosperms as it not only protected seeds from predators, but also carried associated benefits. These included seed dissemination via the evolution of numerous dispersal mechanisms, effective fertilization by transport of pollen grains to the stigma and growth of a pollen tube, promotion of outbreeding by insect pollinators through the evolution of special structural mechanisms and through the developments of intraspecific and interspecific incompatibilty.

It is more common in recent years to differentiate three types of carpels, a terminology developed by Taylor (1991). **Ascidiate** carpels have ovules attached proximally to the closure, **plicate** ones have ovules attached along margins of the closure and **ascoplicate** carpels are intermediate between the two.

The nature of carpel in angiosperms has been the subject of considerable discussion. The dominant view supported by Bailey and Swamy (1951), Cronquist (1988), Takhtajan (1997) and several others considers carpel as homologous to megasporophyll, appropriately named as **phyllosporous** origin of carpel (Lam, 1961). Others believe that carpel consists of a subtending bract with placenta representing a shoot with distally placed ovules, concept named as **stachyosporous** origin (Pankow, 1962; Sattler and Lacroix, 1988).

Phyllosporous Origin

Among the believers of phyllosporous origin, suggest that carpel is a folded leaf with adaxial surfaces (**conduplicate**), or involute abaxial surfaces in contact (**involute**) with many ovules along the margins (or submargins) of closure (Bailey and Swamy, 1951; Eames, 1961). Others suggest that the leaf is fundamentally **peltate** (Baum, 1949; Baum and Leinfellner, 1953), as many carpels have cup-shaped primordia.

The conduplicate view of the origin of carpel was advocated by Bailey and Swamy (1951). In primitive type of carpel, the stigma is represented by a crest extending from apex to base of carpel (decurrent) as in some species of *Drimys*, *Himantandra* and *Degeneria*. In *Degeneria* and *Butomus,* the double nature of the crest is evident in margins flaring back from the line of contact. In *Degeneria* and *Drimys,* the margins of carpels are incompletely closed by interlocking papillose cells of the stigmatic crest. The carpels have three traces, one dorsal and two ventral, the latter providing vascular supply to ovules. From this type of carpel, the closed carpel of other angiosperms developed by closure of adjacent adaxial surfaces (Figure 9.22D–F) and concentration of stigmatic margins to the upper part of the carpel, reduction in the number of ovules and their restriction to lower part differentiating as ovary, the middle sterile portion forming the style. The fusion of adjacent carpels in the formation of syncarpous gynoecium may have proceeded along different directions. Lateral cohesion of open conduplicate carpels resulted in unilocular ovary with parietal placentation (Figure 9.22G). Axile placentation (with number of locules equalling the number of fusing carpels) may have resulted from adnation of free margins to the thalamus (Figure 9.22H) or cohesion of ventral sutures of carpels (Figure 9.22I). Different families of angiosperms exhibit different degrees of fusion, some like Caryophyllaceae with free styles and stigmas, others like Solanaceae with complete fusion of ovaries, styles and stigmas. Free central placentation may result from dissolution of septa from ovary with axile placentation (Caryophyllaceae) the placental column being attached to the base and top of the ovary. It may also result from protruding

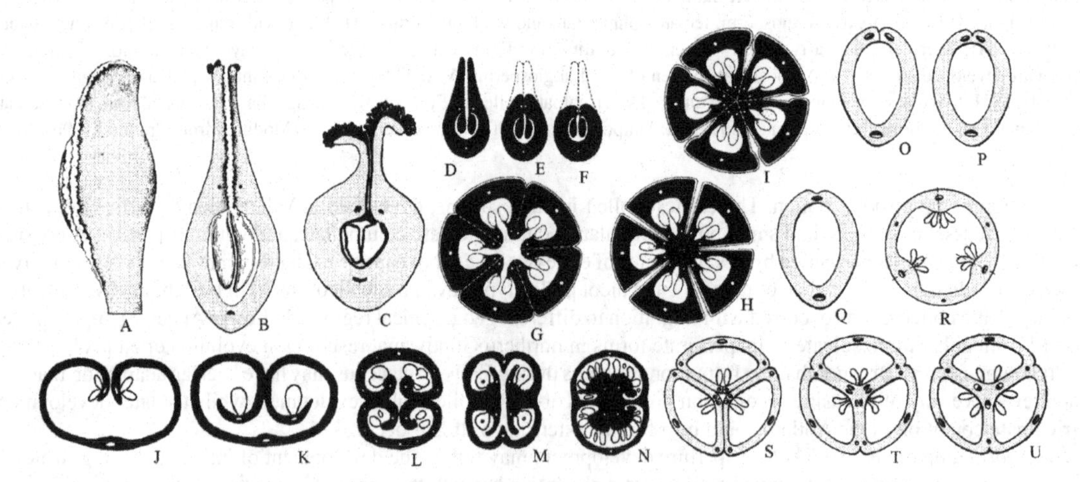

Figure 9.22: Phyllosporous concept of carpel evolution. A–I, conduplicate closure; **J–N**, involute closure; **O–U**, closer along margins. **A:** Carpel of *Drimys piperita* with long stigmatic crest; **B:** Transverse section of same showing partially closed margins; **C:** Transverse section of carpel of *Degeneria vitiensis* with flared up margins and conspicuous papillose growth; **D–F:** Stages in conduplicate closing of carpel with disappearance of stigmatic region (broken lines) from the body of carpel and its localization towards the tip, resulting in marginal placentation; **G:** Fusion of adjacent carpels by lateral cohesion of open conduplicate carpels forming parietal placentation; **H:** Fusion by adnation of free margins of conduplicate carpels to the thalamus forming axile placentation; **I:** Fusion by cohesion of ventral surfaces of the carpels formong axile placentation; **J–K:** Involute closing of carpels by meeting of dorsal surfaces of carpels. **M–N:** Examples of fused carpels with involute closure. **L:** *Erythraea centaurium*; **M:** *Isanthus brachiatus*; **N:** *Limnophila heterophylla*. **O–Q:** Closure by fusing margins of carpel with ultimate merging of ventral bundles; **R:** Fusion of margins of adjacent opens carpels with merging of ventral bundles and formation of parietal placentation. **S–U:** Fusion of sides of closed carpels with merging of adjacent lateral bundles and ventral bundles resulting in axile placentation (**A–H** based on Bailey and Swamy, 1951; **J–N** based on Eames, 1961; **O–U** after Eames and MacDaniels, 1947).

thalamus carrying the placenta from the base of the ovary (Primulaceae). The basal placentation with the number of ovules reduced to basal one may be derived from one (Alismataceae) or more than one carpel (Asteraceae).

Laminar placentation with ovules scattered over the entire inner surface of the ovary wall is considered to be the most primitive type, present in Nymphaeaceae, Cabombaceae, Butomaceae and other relatively primitive families. The ovules in these carpels derive their vascular supply chiefly from a smaller meshwork of the bundles, and rarely from dorsal bundle or ventral bundles. There was consequent reduction in the number of ovules and their restriction to submarginal position with vascular supply coming from ventral bundles. The evidence of this transition is seen in Winteraceae and Degeneriaceae.

Closing of carpels may also result from fusing of incurved margins of carpels. Progressive fusion results in final fusion of adjacent ventral bundles (Figure 9.22O–P) in follicular carpel. Fusion of margins of adjacent open carpels results in parietal placentation with only ventral bundles ultimately merging, whereas the fusion of sides of closed carpels results in axile placentation with both adjacent lateral bundles as well as ventral bundles merging.

Involute closing of carpels was advocated by Joshi (1947), Puri (1960) and several other workers. In such carpels, the margins of the carpel are involuted and abaxial (and not adaxial surfaces) or margins are in contact. The example of such carpels involved in the formation of syncarpous gynoecium is seen in several genera (Figure 9.22L–N). Although there have been suggestions that involute types may have evolved from conduplicate types, Eames (1961) considered it highly unlikely, as such a derivation would involve change from contact by adaxial sides to contact by abaxial sides, a major change, far more complicated and circuitous than usually found in evolutionary derivation. He suggested that several independent closure of carpels occurred in different phylogenetic lines.

The theory that the carpel is a **peltate** leaf was strongly advocated by Baum and Leinfellner (1953). The peltate form of carpel is assumed to have arisen by turning upward (ventrally of the basal lobes of the lamina and their fusion, margin to margin, as in the formation of peltate leaves). A transverse meristem, known as **cross zone,** develops where the two marginal meristems meet. As the carpel primordium elongates, the cross zone, continuous with the marginal meristems, is said to build up a ventral strip of carpel wall, which, united with the lateral walls, forms a tubular organ. Under this theory, the ovules are borne on the wall formed by the cross zone. Peltate carpels may be **manifest peltate** carpels with well-defined stalks and tubular lamina with well-marked cross zone (*Thalictrum*) or **latent peltate** with short tubular base and cross zone present only in early ontogeny (*Calycanthus*). According to this theory, the achene of *Ranunculus* having a single ovule on latent cross zone is considered to be the first stage in the building of tubular follicle. This theory is reverse of the most commonly held view that the follicle represents a more primitive state, and that the achene of *Ranunculus* is derived.

Stachyosporous Origin

The idea was first developed by Hagerup (1934, 1936, 1938) who suggested that conduplicate carpels have two growth areas. The theory was further developed by Lam (1961), Melville (1962, 1983—who proposed a variation of this as **gonophyll theory**) and more recently Taylor (1991) and Taylor and Kirchner (1996). The theory is gaining increased interest with the renewed interest in gnetopsids as close relatives of angiosperms. The theory holds that the carpel envelope represents a bract and placenta homologous with a shoot bearing distally placed ovules. This bract-terminal ovule system is directly homologous to the one found in outgroups including Gnetales, the closest living sister-group. According to this theory, ascidiate carpel with few ovules represents an ancestral stage. It is believed that the origin of plicate (conduplicate) and ascoplicate carpel types would be due to integration of the gynoecial primordia and ovular (placental) growth areas. Taylor and Kirchner found further evidence for stachyosporous origin from:

1. Ingroup phylogeny based on structural and DNA sequences, *rbcL cpDNA* datasets which place either woody or herbaceous magnoliids as basal clades suggest that ascidiate carpels with 1 or 2 ovules represent ancestral state.
2. Outgroup analysis involving Bennettitales, Gnetales and Cordaitales which suggest that female reproductive structures are compound organs.
3. Morphogenic analysis involving understanding of floral development in *Antirrhinum* and *Arabidopsis* through mutagenic analysis, study of the development of carpel in *Datura* by examining chromosomal chimeras (carpel wall is similar to petals and leaves in development; carpel with two types of primordia, one forming the wall and one with distinct central ridge which develops into septum, placenta and false septum and functions like floral apex), study of carpel development in *Nicotiana* using *Ac*-GUS reporter system using GUS bacterial gene as marker (showing that whereas carpel wall is composed only L1 and L2 layers, the placental region has an additional L3 layer).

Based on new evidence, Taylor and Kirchner concluded that ancestral carpel is ascidiate with marginal stigma and basal to slightly lateral placentation of one or two orthotropous ovules. The evolution of curved ovules and placement of the ovules in other positions was to direct the micropyle away from the stigma or pollen-tube transmission-tissue. They suggested that reproductive axes with many flowers, few carpels per flower, and few ovules per carpel were ancestral (Figure 9.23).

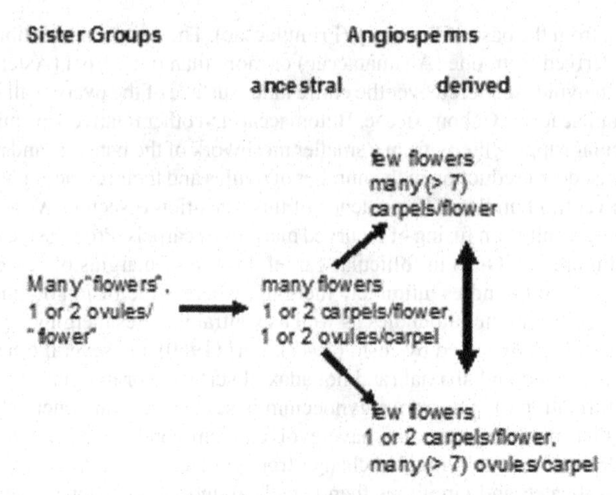

Figure 9.23: Evolution of carpel and floral types in angiosperms based on stachyosporous model proposed by Taylor and Kirchner (1996). Ancestral type is based on suggested homologies between female structures of sister groups.

From such an ancestral type developed two types of inflorescences: one with few flowers, each of which had many carpels and few ovules and the other with few flowers each containing few carpels and many ovules.

Gonophyll Theory

Melville (1962, 1983) developed his gonophyll theory largely on the basis of studies of vasculature in leaves and floral whorls. This theory is a variation of stachyosporous origin. According to him, the ovary consists of sterile leaves and ovule-bearing branches attached to the petiole of the leaf. Each leaf, together with the fertile branch, is considered a unit and termed as **gonophyll** instead of carpel. This theory has already been discussed under probable ancestors of angiosperms.

Evolution of Inferior Ovary

It has been universally agreed that the inferior ovary in angiosperms is a derived state. The nature of origin of this type had two opposing views. Linnaeus, de Candolle and many early botanists believed in the origin of an inferior ovary through adnation of bases of outer floral whorls to the gynoecium, a view known as **appendicular theory** (Candollean theory, concrescence theory). Others believed in the **receptacular theory (axial theory)** developed by German school of botanists and supported by Schleide, Eichler, Sachs and others, according to which inferior ovary resulted from invagination of the floral receptacle which surrounds the ovary.

The accumulating anatomical evidence has shown that inferior ovary has evolved a number of times in different groups of angiosperms, in some due to adnation of floral parts and in others due to axial invagination. In certain plants like *Hedera*, separate traces related to different floral organs are found, in others like *Juglans*, different stages of bundle fusion can be found in the inferior ovary. Such plants also have normal orientation of vascular bundles (phloem outside, xylem inside) and evidently, the inferior ovary is appendicular in origin.

An axial invagination of floral receptacle will eventually result in inverted vascular bundles (xylem on outside, phloem inside) in the inner part of inferior ovary, with normal orientation in the outer part. This has been observed in the inferior ovaries of Cactaceae and Santalaceae. In others like *Rosa*, the lower part of the fleshy receptacle has invaginated receptacular tissue whereas the upper portion consists of fused floral parts. The adnation of floral parts above or surrounding the ovary in a large number of plants forms hypanthium, a structure distinct from but often confused with the calyx tube, the latter involving the cohesion of sepals only. The development of inferior ovary has occurred within several families, as genera with superior ovary and those with inferior ovary may be encountered in the same family as seen in Rosaceae, Gesneriaceae, Nymphaeaceae and several others. In Nymphaeaceae, *Nuphar* has superior ovary, *Nymphaea* semi-inferior and *Euryale* superior ovary.

Chapter 10

Systems of Classification

The urge to classify plants has been with man since he first set his foot on this planet, borne of a need to know what he should eat, avoid, use as cures for ailments and utilize for his shelter. Initially, this information was accumulated and stored in the human brain and passed on to generations through word of mouth in dialects restricted to small communities. Slowly, man learned to put his knowledge in black and white for others to share and improve upon. We have now reached a stage whereby a vast amount of information can be conveniently stored and utilized for far-reaching conclusions aimed at developing ideal systems of classification, which depict the putative relationships between organisms. Historical development of classification has passed through four distinct approaches, beginning with simple classifications based on gross morphology to the latest phylogenetic systems incorporating all types of phenetic information.

CLASSIFICATIONS BASED ON GROSS MORPHOLOGY

Classifications based on features studied without microscopic aids continued until the seventeenth century, when the naked eye was the sole tool of observation. The trail backwards leads us to preliterate man.

Preliterate Mankind

Although no written records of the activities of our preliterate ancestors are available, it is safe to assume that they were practical taxonomists having acquired knowledge as to which plants were edible and which cured their ailments. Primitive tribes in remote areas of the world still carry the tradition of preserving knowledge of the names and uses of plants by word of mouth from one generation to another. Such classifications of plants developed by isolated communities through the need of the society and without the influence of science are termed **Folk taxonomies**; often parallel modern taxonomy. The common English names grass and sedge are equivalent to the modern families Poaceae and Cyperaceae and illustrate this parallel development between folk taxonomy and modern taxonomy.

Early Literate Civilizations

Early civilizations flourished in Babylonia, Egypt, China and India. Though the written records of Indian botany appeared several centuries before those of the Greeks, they remained in obscurity, not reaching the outside world. Moreover, they were written in Sanskrit, a language not easily understood in the West. Crops such as wheat, barley, dates, melons and cotton were grown during the Vedic Period (2000 BC to 800 BC). Indians obviously knew about descriptive botany and cultural practices. The first world symposium on medicinal plants was held in the Himalayan region in the seventh century BC. The *Atharva Veda*, written around 2000 BC describes the medicinal uses of various plants.

Theophrastus—Father of Botany

Theophrastus (372 BC to 287 BC), the successor of Aristotle in the Peripatelic School (those following the philosophy propagated by Aristotle), was a native of Eresus in Lesbos. His original name was Tyrtamus, but he later became known by the nickname 'Theophrastus' given to him—it is said—by Aristotle, to indicate the grace of his conversation. After receiving his first introduction to philosophy in Lesbos from one Leucippus or Alcippus, Theophrastus then proceeded to Athens, and became a member of the Platonic circle. After Plato's death, he attached himself to Aristotle, and after the latter's death, he inherited his library and the garden. He rose to become the head of the Lyceum at Athens.

Figure 10.1: Theophrastus (372 BC to 287 BC) the Greek philosopher, credited to be the father of botany, wrote more than 200 manuscripts.

Theophrastus (Figure 10.1) is credited with having authored more than 200 works most of which survive as fragments or as quotations in the works of other authors. Two of his botanical works have survived intact, however, and are available in English translations: *Enquiry into plants* (1916) and *The Causes of plants* (1927). Theophrastus described about 500 kinds of plants, classified into four major groups: the trees, shrubs, subshrubs and herbs. He also recognized the differences between flowering plants and non-flowering plants, superior ovary and inferior ovary, free and fused petals and fruit types. He was aware of the fact that many cultivated plants do not breed true. Several names used by Theophrastus in his *De Historia plantarum*, e.g., *Daucus*, *Crataegus* and *Narcissus,* to name a few, are in use even today.

Theophrastus was fortunate to have the patronage of Alexander the Great. During his conquests, Alexander made arrangements to send back materials to Athens, enabling Theophrastus to write about exotic plants such as cotton, cinnamon and bananas. Botanical knowledge at the Lyceum in Athens thus flourished during this truly **golden age of learning,** whose botanical advance Theophrastus was privileged to steer.

Parasara—Indian Scholar

Parasara (250 BC to 120 BC) was an Indian scholar who compiled *Vrikshayurveda* (Science of plant life), one of the earliest works dealing with plant life from a scientific standpoint, a manuscript discovered a few decades ago. The book has separate chapters on morphology, properties of soil, forest types of India and details of internal structure, which suggest that the author possessed a magnifying apparatus of some kind. He also described the existence of cells (rasakosa) in the leaf, transportation of the soil solution from the root to the leaves where it is digested by means of chlorophyll [ranjakena pacyamanat] into nutritive substance and the by-products. Plants were classified into numerous families [ganas] based on morphological features not known to the European classification until the eighteenth century. Samiganyan [Leguminosae] were distinguished by hypogynous flowers, five petals of different sizes, gamosepalous calyx and fruit being a legume. Svastikaganyan [Cruciferae] similarly, were differentiated as having calyx resembling a swastika, ovary superior, 4 free sepals and petals each, six stamens of which 2 are shorter and 2 carpels forming a bilocular fruit. Unfortunately, this great scientific advance did not reach Europe at that time, where scientific knowledge was just making its debut.

Among the other Indian scholars, **Caraka** (**Charaka**—Ist century AD) wrote *Caraka samhita* (*Charaka samhita*) in which he recognized trees without flowers, trees with flowers, herbs which wither after fructification and other herbs with spreading stems as separate groups. This huge treatise on Indian medicine, containing eight divisions, is largely based on a much earlier treatise published by Agnivesh. A. C. Kaviratna translated it into English in 1897.

Caius Plinius Secundus—Pliny the Elder

The decline of the Greek Empire witnessed the emergence of the Romans. Pliny (23 AD to 79 AD), a naturalist who served under the Roman army, attempted a compilation of everything known about the world in an extensive 37-volume work *Historia naturalis*, 9 volumes of which were devoted to medicinal plants. In spite of a few errors and fanciful tales from travellers, the Europeans held this work in reverential awe for many centuries. Pliny died during the eruption of Vesuvius.

Pedanios Dioscorides

Dioscorides (first Century AD), of Greek parentage, was a native of the Roman province Cicilia. Being a physician in the Roman army, he travelled extensively and gained firsthand knowledge about plants used for treating various ailments. He wrote a truly outstanding work, *Materia medica*, presenting an account of nearly 600 medicinal plants, nearly 100 more than Theophrastus. Excellent illustrations were added later. Written in a straightforward style, the book was an asset for any literate man for the next 15 centuries. No drug was recognized as genuine unless mentioned in *Materia medica.* It is no less a tribute to Dioscorides that a beautiful illustrated copy of the book was prepared for Emperor Flavius Olybrius Onycius around 500 AD, who presented it as a gift to his beautiful daughter Princess Juliana Anicia. The manuscript, better known as *Codex Juliana*, is a prize manuscript preserved in Vienna. *Materia medica* was not a deliberate attempt at classification but legumes, mints and umbels were described as separate groups.

Medieval Botany

During the Middle Ages (fifth to fifteenth century AD), little or no progress was made in botanical investigation. During this dark period in history, Europe and Asia witnessed wars, famine and epidemics, and the only worthwhile contribution was copying and recopying of earlier manuscripts, unfortunately often with errors added. The strawberry plant was thus shown to have five leaflets instead of three in several manuscripts. The manuscripts were lost at a faster rate than they could be copied.

Islamic Botany

The ascent of the Muslim Empire between 610–1100 AD saw the revival of literacy. Greek manuscripts were translated and preserved. Being practical people, they concentrated on agriculture and medicine and produced lists of drug plants. Ibn-Sina, better known as Avicenna authored *Canon of medicine*, a scientific classic along the lines of *Materia medica.* Another Muslim scholar, Ibu-al-Awwan, in the twelfth century described nearly 600 plants, and interpreted their sexuality as well as the role of insects in fig pollination. Although Muslim scholars produced several practical lists of drug plants, but did not develop any significant scheme of classification.

Albertus Magnus—Doctor Universalis

Albertus Magnus (1193–1280 AD), called **Doctor Universalis** by his contemporaries and **Aristotle of the Middle Ages** by historians, is the best remembered naturalist of that period. He wrote on many subjects. The botanical work *De vegetabilis* dealt with medicinal plants and provided descriptions of plants based on firsthand information. Magnus is believed to be the first to recognize monocots and dicots based on stem structure. He also separated vascular and non-vascular plants.

Renaissance

The fifteenth century saw the onset of the Renaissance in Europe, with technical innovations, mainly the printing machine and the science of navigation. Invention of the printing machine with movable type around 1440 ensured wide circulation of manuscripts. Navigation led to the successful exploitation of botanical wealth from distant places.

Herbalists

Printing made books cheap. The first to become popular were medically-oriented books on plants. Specialists started producing their own botanical-medical books, which were easily understood as compared to ancient manuscripts. These came to be known as **herbals** and the authors who wrote these were known as **herbalists.** The first herbals were published under the name *Gart der Gesundheit* or *Hortus sanitatus*. These were cheaply done and of poor quality. The outstanding herbals came from German herbalists Otto Brunfels, Jerome Bock, Valerius Cordus and Leonard Fuchs, constituting the **German Fathers of Botany**. Otto Brunfels (1464–1534) wrote *Herbarium vivae eicones* in three volumes (1530–1536), a herbal that marked the beginning of modern taxonomy, and contained excellent illustrations prepared from living plants. The text, however, was of little value, comprising extracts from earlier writers. Jerome Bock (Hieronymus Tragus), who lived between 1498 and 1554, wrote *New kreuterbuch* in 1539, which contained no illustrations but did include accurate descriptions based on firsthand knowledge, also mentioning the localities and habitats. He described 567 species classified as herbs, shrubs and trees. The herbal, written in German, was widely understood as compared to the manuscripts of earlier scholars, which were in Greek and Latin, languages which had by then become obsolete. Leonard Fuchs (1501–1566), regarded as more meritorious than his contemporaries, wrote *De Historia stirpium* in 1542, containing descriptions as well as illustrations of 487 species of medicinal plants (Figure 10.2).

Valerius Cordus (1515–1544), whose tragic early death prevented him from becoming the greatest of all herbalists, undertook to study plants from living material. He travelled in the forests of Germany and Italy, where unfortunately he fell ill and died at the young age of 29. His work *Historia plantarum,* published in 1561, many years after his death, contained accurate descriptions of 502 species, 66 apparently new. He was perhaps the first to show how to describe plants from nature accurately. Unfortunately, Konrad Gesner, the editor of his work chose to add illustrations, which were not only of poor quality, but also wrongly identified, and the work suffered for no fault of Valerius Cordus.

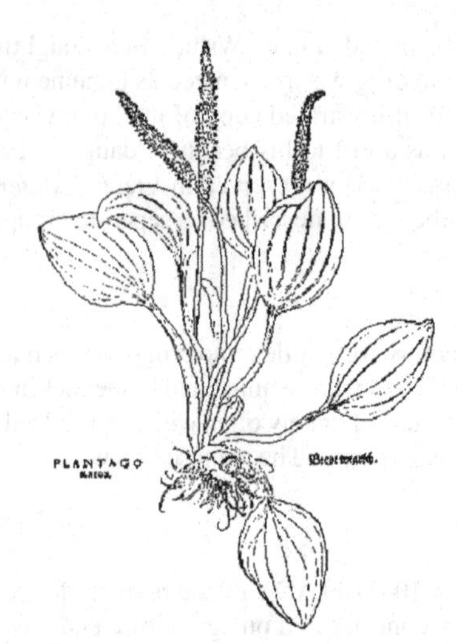

Figure 10.2: Illustration of *Plantago major* from Fuch's *De Historia Stirpium* (1542) (reproduced from Arber: *Herbals, their origin and Evolution*, 1938: used with permission from Cambridge University Press).

At the times when herbals flourished in Germany, Pierandrea Mathiola was active in Italy, producing *Commentarii in sex libros Pedacii Dioscorides* in 1544, adding many illustrations, though it was a commentary on Dioscorides. The **Dutch Big Three**—Rembert Dodoens, Carolus Clusius and Mathias de L'obel—spread the botanical knowledge to Holland and France through their herbals. William Turner, in his *Herball* (1551–1568), swept out many old superstitions concerning plants. *Herball* won for Turner the title of Father **of British Botany**.

Herbalism also saw the emergence of the **doctrine of signatures,** a result of the urge to search for clues from plants. Many medicinal plants, the doctrine held, are stamped with a clear indication of their medicinal use. This was based upon the belief that plants and plant parts cured that particular portion of the human body, which they resembled. Thus, herbs with yellow sap would cure jaundice, the walnut kernel would comfort the brain and maidenhair fern would prevent baldness. Paracelsus and Robert Turner were the main proponents of this doctrine, later ridiculed when more knowledge concerning medicinal plants was acquired from the seventeenth century onwards.

Early Taxonomists

With renewed interest in plants and extensive explorations of Europe, Asia, Africa and the New World, the list of plant names increased enormously, signifying the need for a formalized scheme of classification, naming and description of plants. Botany, hitherto dependent on medicine, started to spread its wings as a science per se.

Andrea Cesalpino (1519–1603)—***The first plant taxonomist***

Andrea Cesalpino was an Italian botanist who studied botany under Luca Ghini and became the Director of the Botanical Garden and later professor of botany and medicine at Bologna. He went to Rome in 1592 as the personal physician to Pope Clement VIII. He prepared a herbarium of 768 well-mounted plants in 1563, which is still preserved in the Museum of Natural History at Florence. His work *De Plantis libri* in 16 volumes appeared in 1583 and contained descriptions of 1520 species of plants grouped as herbs and trees and further differentiated on fruit and seed characters. Cesalpino subscribed to the Aristotelian logic, taking decisions based on reasoning and not the study of features. It was not surprising, therefore, that he considered pith akin to the spinal cord of animals and leaves having the sole role of protecting the apical bud. However, he highlighted the significance of reproductive characters, an attitude not liked by his contemporaries, but having much bearing on the subsequent classifications of Ray, Tournefort and Linnaeus.

Joachin Jung (1587–1657)—***The first terminologist***

A brilliant teacher in Germany, Jung succeeded in defining several terms such as nodes, internodes, simple and compound leaves, stamens, styles, capitulum composed of ray and disc florets. Though he left no publications of his own, two of his pupils preserved records of his teaching.

Gaspard (Caspar) Bauhin (1560–1624)—***Legislateur en botanique***

A Swiss botanist, Bauhin travelled extensively and formed a herbarium of 4000 specimens. He published *Phytopinax* (1596), *Prodromus theatri botanici* (1620) and, lastly, *Pinax theatri botanici* (1623), containing a list of 6000 species of plants giving **synonyms** (other names used for a species by earlier authors) and introducing the **binomial nomenclature** for several species which he named. He sought to clarify in a single publication the confusion regarding multiplicity of names for all species known at that time. Although he did not describe genera, he did recognize the difference between species and genera and several species were included under the same generic name. His elder brother Jean Bauhin (1541–1613) had earlier compiled a description of 5000 plants with more than 3500 figures, a work published under the name *Historia plantarum universalis* in 1650–51, several years after his death. It is tragic that the two brothers never collaborated and rather worked on identical lines independently.

John Ray (1627–1705)

Ray was an English botanist who travelled extensively in Europe and published numerous works, the most significant being *Methodus plantarum nova* (1682) and *Historia plantarum* (1686–1704), a three-volume work. The last edition of *Methodus,* published in 1703, included 18000 species. Ray divided the plant kingdom as shown in the outline of his classification presented in Table 10.1.

Table 10.1: Outline of classification of plants published by John Ray in *Historia plantarum* (1686–1704).

1. Herbae (Herbs)
    - A. Imperfectae (Cryptogams)
    - B. Perfectae (Seed plants)
        - i. Monocotyledons
        - ii. Dicotyledons
2. Arborae (Trees)
    - A. Monocotyledons
    - B. Dicotyledons

John Ray was the first to group together plants that resembled one another and separated those that differed more. His classification was a great advancement in plant sciences. It was evidently ahead of his time, groping at what later developed as natural systems, which were perfected by de Jussieu, de Candolle and Bentham and Hooker.

J. P. de Tournefort (1656–1708)—***Father of genus concept***

A French botanist, de Tournefort studied under Pierre Magnol in the University of Montpellier and later became the professor of botany at Jardin du Roy in Paris and later, Director of Jardin des Plantes in Paris. He published *Elements de botanique* in 1694, including 698 genera and 10,146 species. A Latin translation of this work with additions was published as *Institutions rei herbariae* in 1700. Tournefort travelled extensively in Greece and Asia Minor and brought back 1356 plants, which were fitted into his system by his admirers. He was perhaps the first to give names and description of genera, merely listing the species. Casper Bauhin, who did recognize genera and species, provided no such description. Tournefort was, thus, the first to establish genera. His system of classification, though inferior to that of Ray, was useful for identification, recognizing petaliferous and apetalous flowers, free and fused petals, and regular and irregular flowers. No doubt the system became very popular in Europe during the eighteenth century.

SEXUAL SYSTEM

A turning point in the classification approach was establishing the fact of sexuality in flowering plants by Camerarius in 1694. He concluded that stamens were male sex organs and pollen was necessary for seed set. He showed that the style and ovary form female sex organs of a flower. The thought regarding sexuality in plants, ridiculed by the church hitherto, once established saw renewal in botanical interest, amply exploited by Linnaeus for classifying flowering plants.

Carolus Linnaeus—Father of Taxonomy

Carolus Linnaeus (1707–1778), was also known as Carl Linnaeus, Carl Linne, or Carl Von Linne. Whereas Darwin dominated botanical thinking during the nineteenth century, Linnaeus did so during the eighteenth. Carl Linne, Latinized as Carl Linnaeus or Carolus Linnaeus (Figure 10.3), born in Rashult, Sweden on 23 May 1707, had botany attached to him at birth, since Linnaeus is the Latin for Linn or Linden tree (*Tilia* spp.). His father, a country Parson, wanted his son to become a priest, but Linnaeus chose to enter the University of Lund in 1727 to learn medicine. Although he had no money to buy books, his dedication impressed Professor Kilian Stobaeus, who not only allowed him full use of his library but also gave him free boarding at his house. Lund not being a suitable place for Medicine, Linnaeus shifted to the University of Uppsala in 1729. In recognition of his enthusiasm for plants, Dean Olaf Celsius introduced him to botanist

Figure 10.3: Carolus Linnaeus (1707–1778), the Father of taxonomy (reproduced with permission from Royal Botanic Gardens Kew).

Professor Rudbeck. Under the able guidance of Prof. Rudbeck, Linnaeus published his first paper on the sexuality of plants in 1729. Following favourable publicity of his paper, he was appointed as Demonstrator and subsequently promoted as Docent. In 1730, he published *Hortus upplandicus*, enumerating the plants in the Uppsala Botanical Garden according to the Tournefort's system. Faced with problem of increasing numbers of plants which he found hard to fit in Tournefort's system, he published a revised edition of *Hortus upplandicus* with plants classified according to his own sexual system.

Linnaeus was sent on an expedition to Lapland in 1732, a trip that widened his knowledge. He brought back 537 specimens. The results of the expedition were later published as *Flora Lapponica* (1737). Linnaeus went to the Netherlands in 1735 and obtained an M. D. degree from the University of Haderwijk. While in the Netherlands, he met several prominent naturalists including John Frederick Gronovius and Hermann Boerhaave, the former financing the publication of *Systema naturae* (1735), presenting an outline of the **sexual system** of Linnaeus. He became the personal physician of a wealthy person named George Clifford who was the Director of the Dutch East India Company, and this gave Linnaeus an opportunity to study numerous tropical and temperate plants grown by Clifford in his garden. It was at Clifford's expense that Linnaeus published several manuscripts, including *Hortus cliffortianus* and *Genera plantarum* (1737). Linnaeus then went to England, where he met Professor John Jacob Dillen, who initially thought of Linnaeus as 'this is he who is bringing all botany into confusion', but he soon became the advocate of the Linnaean system in England. He also met the de Jussieu brothers in France.

Following the death of Professor Rudbeck, Linnaeus was appointed Professor of medicine and botany at the University of Uppsala, a position he held until his death in 1778. He published his best-known *Species plantarum* in 1753. His growing fame and publications attracted large number of students, their number increasing every year and the botanical garden at Uppsala enriched considerably.

Linnaeus' botanical excursions every summer also included an annotator to take notes, a Fiscal to maintain discipline and marksmen to shoot birds. At the end of each trip, they marched back to the town with Linnaeus at the head, aided with French horns, kettledrums and banners.

In recognition of his contributions, Linnaeus was made **Knight of the Polar Star** in 1753, the first Swedish scientist to get this honor. In 1761, he was granted the **patent of nobility** and from this date came to be known as **Carl von Linne.**

Among his enthusiastic students were Peter Kalm and Peter Thunberg. Kalm collected plants extensively in Finland, Russia and America and when he returned with bundles of collection from America, Linnaeus was bedridden, but forgot his ailment and transferred his concern to plants. Thunberg collected extensively in Japan and South Africa.

Linnaeus first outlined his system in *Systema naturae*, which classified all known plants, animals and minerals. In his *Genera plantarum,* he listed and described all the plant genera known to him. In *Species plantarum,* he listed and described all the known species of plants. For each species there was (Figure 10.4):

(i) a generic name;
(ii) a polynomial descriptive phrase or phrase-name commencing with generic name and of up to twelve words, intended to serve as description of the species;
(iii) a trivial name or specific epithet on the margin;
(iv) synonyms with reference to important earlier literature; and
(v) habitats and countries.

The generic name followed by the trivial name formed the name for each species. Linnaeus thus established the **binomial nomenclature**, first started by Caspar Bauhin and the generic concept, started by Tournefort.

The system of Linnaeus, very simple in its application, recognized 24 classes (Table 10.2), mostly based on stamens. These classes were further subdivided on the basis of carpel characteristics into orders such as Monogynia, Digynia, etc.

DIADELPHIA HEXANDRIA. 699
Classis XVII

DIADELPHIA.

HEXANDRIA.

FUMARIA.

Corollis bicalcaratis,
I. FUMARIA Scapo nudo. *Hort. Cliff. 251:* * *Gron.* *cucullaria,*
virg. 171. Rov.lugdb,. 393·
Fumaria tuberosa insipida. *Corn. canad. 127.*
Fumaria siliquosa, radice grumosa, flore bicorporeo ad labia conJucto, virginiana. *Plak, alam. 162. t. 90. f* 3· *Raj. suppl.* 475·
Cucullaria. *Juss. act. paris*. 1743·
Habitat in Virginia, Canada 2:
Radix *tuberosa;*Folium *radicale tricompositum.* Scapus *nudus, Racemo simplici; bracteae vix ullae;* Nectarium *duplex corollam basi bicornem efficiens.*
2. FUMARIA floribus *postice* bilobis, caule folioso. *spectabilis.*
Habitat in Sibiria. *D. Demidoff:*
Planta *eximia floribus speciofissimis, maximis.* Habitus *Fumariae bulbosae, sed majora omnia.* Rami *ex alis rarioris.* Caulis *erectus.* Racemus *absque bracteis.* Corollae *magnitudine extimi articuli pollicis, pone in duos lobos aequales, rotundatos divisae.*

* *Corollis unicalcaratis.*
3. FUMARIA caule simplici, bracteis longitudine florum *bulbosa*

Figure 10.4: A portion of a page from ***Species plantarum*** of Linnaeus (1753). Specific epithet (trivial name) is indicated towards the margin.

Table 10.2: Outline of the 24 classes recognized by Linnaeus in his *Species plantarum* (1753) on the basis of stamens.

Classes
1. Monandria-stamen one
2. Diandria-stamens two
3. Triandria-stamens three
4. Tetrandria-stamens four
5. Pentandria-stamens five
6. Hexandria-stamens six
7. Heptandria-stamens seven
8. Octandria-stamens eight
9. Ennandria-stamens nine
10. Decandria-stamens ten
11. Dodecandria-stamens 11–19
12. Icosandria-stamens 20 or more, on the calyx
13. Polyandria-stamens 20 or more, on the receptacle
14. Didynamia-stamens didynamous; 2 short, 2 long
15. Tetradynamia-stamens tetradynamous; 4 long, 2 short
16. Monadelphia-stamens monadelphous; united in 1 group
17. Diadephia-stamens diadelphous; united in 2 groups
18. Polyadelphia-stamens polyadelphous; united in 3 or more groups
19. Syngenesia-stamens syngenesious; united by anthers only
20. Gynandria-stamens united with the gynoecium
21. Monoecia-plants monoecious
22. Dioecia-plants dioecious
23. Polygamia-plants polygamous
24. Cryptogamia-flowerless plants

Such a classification based on stamens and carpels resulted in the artificial grouping of unrelated taxa and separation of relatives.

Linnaeus knew that his system was more convenient than natural, but it was the need of the day when there was a tremendous increase in the number of plants known to man, which necessitated quick identification and placement. This is exactly what the **sexual system** of Linnaeus achieved with merit. His *Species plantarum* (1753) marks the starting point of botanical nomenclature today. Linnaeus did aim at natural classification and in the 6th edition of his *Genera plantarum* (1764), he appended a list of 58 natural orders. It was, however, left to others to carry forward. Linnaeus had done his job according to the demands of the day.

Following Linnaeus' death in 1778, his son Carl received the post of Professor as well as the collections at Uppsala. When the latter died in 1783, the collections went to the widow of Linnaeus, whose sole aim was to sell it the highest bidder. Fortunately, this highest bidder of 1000 guineas was J. E. Smith, an English botanist. Smith founded the Linnaean Society of London in 1788 and handed over the herbarium to this society. Herbarium specimens have since been photographed and are available in microfiche and online photographs.

The Linnaean classification remained dominant for a long time. The 5th edition of *Species plantarum* appeared as late as in 1797–1805, greatly enlarged and edited by C. L. Wildenow in four large volumes.

NATURAL SYSTEMS

Linnaeus had provided a readily referable cataloguing scheme for a large number of plants, but it soon became evident that unrelated plants came together in such groupings. A need was realized for a more objective classification. France, which was undergoing an intellectual ferment and where the Linnaean system never became popular, took the lead in developing natural systems of classification.

Michel Adanson (1727–1806)

A French botanist, unimpressed with artificial choice of characters, Michel Adanson devised a classification of both animals and plants, on the equal use of as many features as possible. In his two-volume work *Familles des plantes* (1763), he recognized 58 **natural orders** according to their natural affinities. Present-day **Numerical taxonomy** is based on the idea conceived by Adanson and now developed into **Neo-Adansonian principles.**

Jean B. P. Lamarck (1744–1829)

A French naturalist, Jean B. P. Lamarck authored *Flore Francaise* (1778), which in addition to a key for identification of plants, contained principles concerning the natural grouping of species, orders and families. He is better known for his evolutionary theory, **Lamarckism**.

de Jussieu Family

Four well-known botanists belonged to this prominent French family. Of the three brothers—Antoine (1686–1758), Bernard (1699–1776) and Joseph (1704–1779), the youngest spent many years in South America, where after losing his collections of five years, he became insane. The elder two studied at the University of Montpellier under Pierre Magnol. Antoine succeeded Tournefort as Director **de Jardin des Plantes, Paris** and later added Bernard to the staff. Bernard started arranging plants in the garden at La Trianon, Versailles, according to the classification that was initially similar to *Fragmenta methodi naturalis* of Linnaeus with some similarities to Ray's *Methodus plantarum*, introducing changes, so that when finally set, it had no resemblance with the Linnaean system. Bernard based his classification on the number of cotyledons, presence or absence of petals and their fusion. He never published his system and it was left to his nephew Antoine Laurent de Jussieu (1748–1836; Figure 10.5) to publish this classification, along with his own changes in *Genera plantarum* (1789).

In this classification, the plants were divided into three groups, further divided on corolla characteristics and ovary position to form 15 classes and 100 orders (till the beginning of present century class and order were mostly used as names of categories now understood as order and family, respectively). An outline of the classification is presented below:

1. Acotyledones
2. Monocotyledones
3. Dicotyledones
   i. Apetalae
   ii. Monopetalae
   iii. Polypetalae
   iv. Diclines irregulares

Acotyledones, in addition to cryptogams, contained some hydrophytes whose reproduction was not known then. **Diclines irregulares** contained Amentiferae, Nettles, Euphorbias as also the Gymnosperms.

Figure 10.5: Antoine Laurent de Jussieu (1748–1836) the author of *Genera plantarum* (1789) largely based on the work of his Uncle Bernard de Jussieu (reproduced with permission from the Royal Botanic Gardens, Kew).

de Candolle Family

The de Candolles were a Swiss family of botanists. Augustin Pyramus de Candolle (1778–1841) was born in Geneva, Switzerland but took his education in Paris, where he became the Professor of Botany at Montpellier (Figure 10.6). He published several books, the most important one being *Theorie elementaire de la botanique* (1813), wherein he proposed a new classification scheme, outlined the important principles and introduced the term **taxonomy**. From 1816, until his death **Augustin de Candolle** worked in Geneva and undertook a monumental work, intended to describe every known species of vascular plants under the title *Prodromus systematis naturalis regni vegetabilis*, the first volume appearing in 1824. He published seven volumes himself. His son Alphonse de Candolle and grandson Casimir de Candolle continued the work. Alphonse published ten more volumes, the last one in 1873, resulting in revision of several families by specialists.

The classification by A. P. de Candolle delimited 161 natural orders (the number was increased to 213 in the last revision of *Theorie elementaire….*, edited by Alphonse in 1844), grouped primarily based on the presence or absence of vascular structures (Table 10.3).

Ferns were provided a place co-ordinate with monocots and in contrary to de Jussieu, Gymnosperms were given a place, although among dicots. The importance of anatomical features was highlighted and successfully employed in the classification.

Robert Brown (1773–1858)

Robert Brown was an English botanist, who did not propose a classification of his own but demonstrated that Gymnosperms were a group discrete from dicotyledons and had naked ovules. He also clarified the floral morphology and pollination of Asclepiadaceae and Orchidaceae, morphology of grass flower structure of cyathium in Euphorbiaceae and established several families.

Figure 10.6: Augustin Pyramus de Candolle (1778–1841) who first introduced the term 'taxonomy' in his *Theorie elementaire de la botanique* (1813) (reproduced with permission from the Royal Botanic Gardens, Kew).

Table 10.3: Outline of classification proposed by A. P. de Candolle in his *Theorie elementaire de la botanique* (1813).

I. Vasculares (vascular bundles present)
  Class 1. Exogenae (dicots)
    A. Diplochlamydeae
      Thalamiflorae
      Calyciflorae
      Corolliflorae
    B. Monochlamydeae (also including gymnosperms)
  Class 2. Endogenae
    A. Phanerogamae (monocots)
    B. Cryptogamae
II. Cellulares (no vascular bundles)
  Class 1. Foliaceae (Mosses, Liverworts)
  Class 2. Aphyllae (Algae, Fungi, Lichens)

George Bentham & Sir J. D. Hooker

The system of classification of seed plants presented by Bentham and Hooker, two English botanists, represented the most well-developed natural system. The classification was published in a three-volume work *Genera plantarum* (1862–83). George Bentham (1800–1884) was a self-trained botanist (Figure 10.7). He was extremely accomplished and wrote many important monographs on families such as Labiatae, Ericaceae, Scrophulariaceae and Polygonaceae. He also published *Handbook of British Flora* (1858) and *Flora Australiensis* in 7 volumes (1863–78). Sir J. D. Hooker (1817–1911), who succeeded his father William Hooker as Director, Royal Botanic Gardens in Kew, England was a very well known botanist, having explored many parts of the world (Figure 10.8). He published *Flora of British India* in 7 volumes (1872–97), *Student's Flora of the British Islands* (1870) and also revised later editions of *Handbook of British Flora*, which remained a major British Flora until 1952. He also supervised the publication of *Index Kewensis* (2 volumes, 1893), listing the names of all known species and their synonyms.

The *Genera plantarum* of Bentham and Hooker provided the classification of seed plants describing 202 families and 7569 genera. They estimated the seed plants to include 97,205 species. The classification was a refinement of the systems proposed by A. P. de Candolle and Lindley, which in turn were based on that of de Jussieu. The delimitation of families and genera was based on natural affinities and was pre-Darwinian in concept. The descriptions were based on personal studies from specimens and not a mere compilation of known facts, an asset which made the classification so popular and authentic. Many important herbaria of the world have specimens arranged according to this system.

The system divided Phanerogams or seed plants into three classes: **Dicotyledons, Gymnospermae** and **Monocotyledons.** Dicotyledons were further subdivided into three subclasses: **Polypetalae, Gamopetalae** and **Monochlamydeae** based on the presence or absence of petals and their fusion. These subclasses, in turn, were subdivided into series, orders (called cohorts by the two authors) and families (called natural orders). No orders (cohorts) were recognized within Monochlamydeae and Monocotyledons, the series being directly divided into families (natural orders). A broad outline of the classification is presented in Table 10.4.

Merits

The fact notwithstanding that the system does not incorporate phylogeny and is more than 100 years old, it still enjoys a reputation of being a very sound system of classification, owing to the following merits:

1. The system has great practical value for identification of plants. It is very easy to follow for routine identification.
2. The system is widely followed for the arrangement of specimens in the herbaria of many countries, including Britain and India.
3. The system is based on a careful comparative examination of actual specimens of all living genera of seed plants and is not a mere compilation of known facts.
4. Unlike de Candolle, the **Gymnosperms** are not placed among dicots but rather in an independent group.
5. Although the system is not a phylogenetic one, **Ranales** are placed in the beginning of Dicotyledons. The group Ranales (in the broader sense including families now separated under order Magnoliales) is generally regarded as primitive by most of the leading authors.
6. Dicotyledons are placed before the Monocotyledons, a position approved by all present-day authors.

Figure 10.7: George Bentham (1800–1884), co-author of *Genera plantarum* (with J. D. Hooker, 1862–1883), and the author of the 7-volume *Flora Australiensis* and several monographs on major families (reproduced with permission from Royal Botanic Gardens, Kew).

Figure 10.8: Sir Joseph Dalton Hooker (1817–1911), the famous British botanist who co-authored *Genera Plantarum* with George Bentham, besides authoring the 7-volume *Flora of British India* and several other publications. He was the Director of the Royal Botanic Gardens, Kew (reproduced with permission from Royal Botanic Gardens, Kew).

7. The description of families and genera are precise. Keys to the identification are very useful. Larger genera have been divided into subgenera in order to facilitate identification.
8. The arrangement of taxa is based on overall **natural affinities** decided on the basis of morphological features, which can be easily studied with the naked eye or with a hand lens.
9. Although a few important characters have been chosen to name a few groups, the grouping itself is based on a **combination of characters**, rather than any single character in the majority of cases. Thus, although *Delphinium* has fused petals, it has been kept in **Ranunculaceae** along with the related genera and placed in Polypetalae. Similarly, some gamopetalous genera of **Cucurbitaceae** are retained along with the polypetalous ones and placed in Polypetalae.
10. Heteromerae is rightly placed before Bicarpellatae.

Demerits

The system being pre-Darwinian in approach, suffers from the following drawbacks:

1. The system does not incorporate phylogeny, although it was published after Darwin published his evolutionary theory.
2. **Gymnosperms** are placed between Dicotyledons and Monocotyledons, whereas their proper position is before the former, as they form a group independent from angiosperms.
3. **Monochlamydeae** is an unnatural assemblage of taxa, which belong elsewhere. The creation of this group has resulted in the separation of many closely-related families. **Caryophyllaceae**, **Illecebraceae** and **Chenopodiaceae** are

Table 10.4: Outline of the system of classification presented by Bentham and Hooker in *Genera plantarum* (1862–1883).

Phanerogams or seed plants	
Class 1. Dicotyledons	(Seed with 2 cotyledons, flowers pentamerous or tetramerous, leaves netveined)
	14 series, 25 orders and 165 families
Subclass 1. Polypetalae	(sepals and petals distinct, petals free)
Series 1. Thalamiflorae	(flowers hypogynous, stamens many, disc absent)
	6 orders: Ranales, Parietales, Polygalineae, Caryophyllineae, Guttiferales and Malvales
2. Disciflorae	(Flowers hypogynous, disc present below the ovary)
	4 orders: Geraniales, Olacales, Celastrales and Sapindales
3. Calyciflorae	(flowers perigynous or epigynous)
	5 orders: Rosales, Myrtales, Passiflorales, Ficoidales and Umbellales
Subclass 2. Gamopetalae	(sepals and petals distinct, petals united)
Series 1. Inferae	(ovary inferior)
	3 orders: Rubiales, Asterales and Campanales
2. Heteromerae	(ovary superior, stamens in one or two whorls, carpels more than 2)
	3 orders: Ericales, Primulales and Ebenales
3. Bicarpellatae	(ovary superior, stamens in one whorl, carpels 2)
	4 orders: Gentianales, Polemoniales, Personales and Lamiales
Subclass 3. Monochlamydeae	(flowers apetalous; perianth lacking or if present not differentiated into sepals and petals)
Series 1. Curvembryeae	(embryo coiled, ovule usually 1)
2. Multiovulatae aquaticae	(aquatic plants, ovules many)
3. Multiovulatae terrestres	(terrestrial plants, ovules many)
4. Microembryeae	(embryo minute)
5. Daphnales	(carpel 1, ovule 1)
6. Achlamydosporae	(ovary inferior, unilocular, ovules 1–3)
7. Unisexuales	(flowers unisexual)
8. Ordines anomali	(relationship uncertain)
Class 2. Gymnospermae	(ovules naked)
	3 families
Class 3. Monocotyledons	(flowers trimerous, venation parallel)
	7 series, 34 families
Series 1. Microspermae	(ovary inferior, seeds minute)
2. Epigynae	(ovary inferior, seeds large)
3. Coronarieae	(ovary superior, carpels united, perianth colored)
4. Calycinae	(ovary superior, carpels united, perianth green)
5. Nudiflorae	(ovary superior, perianth absent)
6. Apocarpae	(ovary superior, carpels more than 1, free)
7. Glumaceae	(ovary superior, perianth reduced, flowers enclosed in glumes)

closely related families to the extent that they are placed in the same order in all major contemporary classifications. Several authors including Takhtajan (1987) merge Illecebraceae with Caryophyllaceae. In Bentham and Hooker's system, however, Caryophyllaceae are placed in **Polypetalae**, and the other two in **Monochlamydeae**. Similarly, **Podostemaceae** which are placed in a separate series Multiovulatae aquaticae, better belong to the order Rosales (Cronquist, 1988). **Chloranthaceae** placed by Bentham and Hooker under Microembryeae and Laurineae placed under Daphanales are closely allied to the order **Magnoliales** (Ranales ***s. l.***) and are thus placed in the same subclass Magnoliidae by Cronquist (1988).

4. In Monocotyledons, **Liliaceae** and **Amaryllidaceae** are generally regarded as closely related and often included in the same order, some authors, including Cronquist merging Amaryllidaceae with Liliaceae. In this system they are placed under different series, Amaryllidaceae under **Epigynae** and Liliaceae under **Coronarieae**.
5. **Unisexuales** is a loose assemblage of diverse families, which share only one major character, i.e., unisexual flowers. Cronquist (1988) separates these families under two distinct subclasses **Hamamelidae** and **Rosidae** and Takhtajan (1987) under **Hamamelididae** and **Dilleniidae**.

6. Bentham and Hooker did not know the affinities of the families placed under Ordines anomali, and the families were tentatively grouped together. Cronquist (1988) and Takhtajan (1987) place **Ceratophyllaceae** under subclass **Magnoliidae** and the other three under **Dilleniidae.**
7. Many large families, e.g., **Urticaceae, Euphorbiaceae, Liliaceae** and **Saxifragaceae**, are unnatural assemblages and represent **polyphyletic** groups. These have rightly been split by subsequent authors into smaller, natural and **monophyletic** families.
8. **Orchidaceae** is an advanced family with inferior ovary and zygomorphic flowers, but the family is placed towards the beginning of Monocotyledons.
9. In Gamopetalae, Inferae with an inferior ovary is placed before the other two series having a superior ovary. The inferior ovary is now considered to have been derived from a superior ovary.
10. The system divides angiosperms into dicotyledons and monocotyledons, whereas the modern phylogenetic systems place paleoherb families and Magnoliids before monocotyledons and Eudicots.

PHYLOGENETIC SYSTEMS

The publication of *The Origin of species* by Charles Darwin, with every copy of the first edition sold on the first day, 24 November 1859 revolutionized biological thinking. The species was no longer regarded as a fixed entity having remained unchanged since its creation. Species were now looked upon as systems of populations, which are dynamic and change with time to give rise to lineages of closely-related organisms. Once the existence of this evolutionary process was acknowledged, the systems of de Candolle as also of Bentham and Hooker were found to be inadequate and classifications, which made an attempt to reconstruct evolutionary sequence, found immediate takers.

Transitional Systems

The early systems were not intended to be phylogenetic. Rather, they were attempts to rearrange earlier natural systems in the light of the prevalent phylogenetic theories.

A. W. Eichler

Eichler (1839–1887) was a German botanist who proposed the rudiments of a system in 1875. This was elaborated into a unified system covering the entire plant kingdom and finally published in the third edition of *Syllabus der vorlesungen...* (1883). The plant kingdom was divided into two subgroups: Cryptogamae and Phanerogamae, the latter further subdivided into Gymnospermae and Angiospermae. Angiospermae was divided into two classes: Monocotyledons and Dicotyledons. Only two groups **Choripetalae** and **Sympetalae** were recognized in Dicotyledons. Gymnosperms thus found their separate identity before angiosperms, Monochlamydeae found itself abolished and dispersed among the two groups. Monocotyledons, strangely, found a place before Dicotyledons.

Adolph Engler and Karl A. Prantl

This is a system of classification of the entire plant kingdom, proposed jointly by two German botanists: Adolph Engler (1844–1930) (Figure 10.9) and Karl A. E. Prantl (1849–1893). The classification was published in a monumental work *Die Natürlichen pflanzenfamilien* in 23 volumes (1887–1915). Engler was Professor of Botany at the University of Berlin and later Director, Berlin Botanic Garden. The system provided classification and description down to the **genus level**, incorporating information on morphology, anatomy and geography.

The system is commonly known under Engler's name, who first published classification up to the family level under the title *Syllabus der pflanzenfamilien* in 1892. This scheme was constantly revised by Engler and continued by his followers after his death, the latest 12th edition appearing in 2 volumes, 1954 (ed. H. Melchior and E. Werdermann) and 1964 (ed. M. Melchior). In this last edition, however, dicots were placed before monocots.

Engler also initiated an ambitious plan of providing **taxonomic monographs** of various **families** up to species level under the title *Das pflanzenreich.* Between 1900 and 1953, 107 volumes were published covering 78 families of seed plants and one family (Sphagnaceae) of mosses.

This system, often considered the beginning in **phylogenetic** schemes, was not strictly phylogenetic in the modern sense. It was an arrangement of linear sequence starting with the simplest groups and arranged in the order of **progressing complexity.** In doing so, unfortunately, Engler misread angiosperms, where in many groups, the **simplicity** is a result of **evolutionary reduction**.

The system, however, had significant improvements over Bentham and Hooker: **Gymnosperms** were placed before angiosperms, group **Monochlamydeae** was abolished and its members distributed along with their polypetalous

Figure 10.9: Adolph Engler (1844–1930), the famous German botanist who produced the most comprehensive classification of the plant kingdom along with K. Prantl in a 20-volume work *Die Natürlichen pflanzenfamilien* (1887–1915).

Table 10.5: An outline of the system of classification presented by Engler and Prantl.

Plant Kingdom

- **Division 1.** }
  - } ....Thallophytes
- **Division 11.** }
- **Division 12.** .... Embryophyta Asiphonogama
  - Subdivision 1. Bryophyta
  - Subdivision 2. Pteridophyta
- **Division 13.** .... Embryophyta Siphonogama
  - Subdivision 1. Gymnospermae
  - Subdivision 2. Angiospermae
    - Class 1. Monocotyledoneae—***11 orders, 45 families***
      - Order 1. Pandanales (first family Pandanaceae) ....
      - Order 11. Microspermae (last family Orchidaceae)
    - Class 2. Dicotyledoneae—***44 orders, 258 families***
      - Subclass 1. Archichlamydeae (petals absent or free)—***33 orders, 201 families***
        - Order 1. Verticillatae (family Casuarinaceae only) ....
        - Order 33. Umbelliflorae (Last family Cornaceae)
      - Subclass 2. Metachlamydeae (petals united)—***11 orders, 57 families***
        - Order 34. Diapensiales (family Diapensiaceae only) ....
        - Order 44. Campanulatae (Last family Compositae)

relatives, and many large unnatural families were split into smaller natural families. The placement of monocots before dicots, another change made by this system did not, however, get subsequent support. The placement of the so-called group **Amentiferae** comprising families Betulaceae, Fagaceae, Juglandaceae, etc., in the beginning of dicots, also did not find much subsequent support. The system (Table 10.5) became very popular, like that of Bentham of Hooker, due to its comprehensive treatment and is still being followed in many herbaria of the world. Some recent floras including *Flora Europaea* (1964–1980) follow this system.

In this scheme of classification, the plant kingdom was divided into 13 divisions (in the 11th edition of *Syllabus der pflanzenfamilien* published in 1936, 14 divisions and in the 12th edition edited by Melchior 17 divisions were recognized), of which the first 11 dealt with Thallophytes, the 12th **Embryophyta Asiphonogama** (embryo formed, no pollen tube) included Bryophytes and Pteridophytes.

The 13th and last division **Embryophyta Siphonogama** (embryo formed, pollen tube developing) included seed plants.

Merits

The classification of Engler and Prantl has the following improvements over that of Bentham and Hooker:

1. This was the first major system to incorporate the ideas of **organic evolution**, and the first major step towards phylogenetic systems of classification.
2. The classification covers the entire plant kingdom and provides description and identification keys down to the level of **family** (in *Syllabus der pflanzenfamilien*), **genus** (in *Die Natürlichen pflanzenfamilien*) and even **species** for large number of families (in *Das pflanzenreich*). Valuable illustrations and information on anatomy and geography are also provided.
3. **Gymnosperms** are separated and placed before angiosperms.
4. Many large unnatural families of Bentham and Hooker have been split into smaller and natural families. The family **Urticaceae** is thus split into **Urticaceae, Ulmaceae** and **Moraceae.**
5. Abolition of **Monochlamydeae** has resulted in bringing together several closely related families. Family **Illecebraceae** is merged with **Caryophyllaceae. Chenopodiaceae** and Caryophyllaceae are placed in the same order, **Centrospermae.**
6. **Compositae** in dicots and **Orchidaceae** in monocots are advanced families with inferior ovary, zygomorphic and complex flowers. These are rightly placed towards the end of dicots and monocots, respectively.
7. Several recent systems of classification place monocots before true dicots (eudicots).
8. Consideration of gamopetalous condition as advanced over polypetalous condition is in line with current phyletic views.
9. The classification, being very thorough has been widely used in textbooks, Floras and herbaria around the world.
10. The terms **cohort** and **natural order** have been replaced by the appropriate terms order and family, respectively.
11. Closely related families Liliaceae and Amaryllidaceae have been brought under the same order Liliiflorae.

Demerits

With better understanding of the phylogenetic concepts in recent years, many drawbacks of the system of Engler and Prantl have come to light. These primarily result from the fact that they applied the concept of 'simplicity representing primitiveness' even to the angiosperms, where evolutionary reduction is a major phenomenon, not commonly seen in the lower groups. The major drawbacks of the system include:

1. The system is not a phylogenetic one in the modern sense. Many ideas of Engler are now outdated.
2. Monocotyledons are placed before Dicotyledons. In the recent systems, paleoherbs and sometimes Magnoliids are placed before monocots.
3. The so called **Amentiferae** including such families as Betulaceae, Juglandaceae and Fagaceae with reduced unisexual flowers, having few floral members and borne in catkins, were considered primitive. It has been established from studies on wood anatomy, palynology and floral anatomy that Amentiferae is an advanced group. The simplicity of flowers is due to **evolutionary reduction** and not primitiveness. Cladistic studies of Loconte (1996) have shown that tree based on this hypothesis is six steps longer than the shortest tree.
4. **Dichlamydeous** forms (distinct calyx and corolla) were considered to have evolved from the **monochlamydeous** forms (single whorl of perianth). This view is not tenable.
5. Angiosperms were considered a **polyphyletic** group. Most of the recent evidence points towards monophyletic origin.
6. **Araceae** in Monocotyledons are now believed to have evolved from **Liliaceae**. In this classification, Araceae are included in the order **Spathiflorae** which is placed before Liliiflorae, including family Liliaceae.
7. **Helobiae** (including families Alismaceae, Butomaceae and Potamogetonaceae) is a primitive group, but in this classification it is placed after **Pandanales,** which is a relatively advanced group.
8. Derivation of **free central** placentation from the **parietal** placentation, and of the latter from **axile** placentation is contrary to the evidence from floral anatomy. Free central placentation is now believed to have evolved from axile placentation through the disappearance of septa.
9. **Ranales** (in the broader sense_***s. l.***) are now considered as a primitive group with bisexual flowers, spirally arranged floral parts and numerous floral members. In this classification, they are placed much lower down, after Amentiferae.
10. Family Liliaceae of Engler and Prantl is a large unnatural assemblage, which has been split into several smaller monophyletic families like Liliaceae, Alliaceae, Asparagaceae, Asphodelaceae in the recent classification of Judd et al. (2002), APG II (2003) and Thorne (2006).

The above two systems of classification have been widely followed in different herbaria around the world, as also in various regional and local Floras. Although based on basically different criteria, the two are similar in being exhaustive in treatment, allowing the placement and identification of various genera with the help of valuable keys and detailed descriptions. Such a treatment is very necessary for distribution of specimens in a herbarium. It is also valuable in preliminary identification of a specimen up to the generic level. Most of the contemporary systems of classification lack treatment beyond the family level. Such systems of classification may be very sound in the placement of higher groups but have little practical value for the purpose of actual identification.

A comparison of the classification of Bentham and Hooker with that of Engler and Prantl is presented in Table 10.6.

Table 10.6: Comparison of classification system of Bentham and Hooker with that of Engler and Prantl.

Bentham and Hooker	**Engler and Prantl**
1. Published in *Genera Plantarum* in 3 volumes (1862–83).	1. Published in *Die Naturlichen Pflanzen-familien* in 23 volumes (1887–1915).
2. Includes only seed plants.	2. Includes the entire plant kingdom.
3. Gymnosperms placed in between Dicotyledons and Monocotyledons.	3. Gymnosperms separated and placed before angiosperms.
4. Dicotyledons placed before Monocotyledons.	4. Dicotyledons placed after Monocotyledons.
5. Dicotyledons divided into 3 subclasses: Polypetalae, Gamopetalae and Monocotyledoneae.	5. Dicotyledons divided into 2 subclasses: Archichlamydeae and Metachlamydeae.
6. Subclasses are further subdivided into series, cohorts (representing orders), and natural orders (representing families).	6. Subclasses are further subdivided into orders and families, series not recognized.
7. Monocotyledons include 7 series and 34 natural orders.	7. Monocotyledons include 11 orders and 45 families.
8. Pre-Darwinian in concept.	8. Post-Darwinian in concept.
9. Dicotyledons start with Ranales having bisexual flowers.	9. Dicotyledons start with Verticillatae having unisexual flowers.
10. Monocotyledons start with Microspermae, including Orchidaceae.	10. Monocotyledons start with Pandanales, Microspermae placed towards the end of Monocotyledons.
11. Closely related families Caryophyllaceae, Illecebraceae and Chenopodiaceae are kept apart, the first under Polypetalae and the other two under Monochla-mydeae.	11. Family Illecebraceae merged with Caryophyllaceae. Chenopodiaceae and Caryophyllaceae are placed in the same order Centrospermae.
12. Closely related families Amaryllidaceae and Liliaceae are placed in separate series Epigynae and Coronariae, respectively.	12. Liliaceae and Amaryllidaceae placed in the same order Liliiflorae.
13. Many large families, e.g., Urticaceae, Saxifragaceae and Euphorbiaceae are unnatural heterogenous groups.	13. Several larger families of Bentham and Hooker split into smaller homogeneous families. Urticaceae split into Urticaceae, Ulmaceae and Moraceae.

Intentional Phylogenetic Systems

The natural systems rearranged in the light of phylogenetic information soon gave way to systems that reflect evolutionary development. A beginning in this direction was made by an American botanist, Charles Bessey.

Charles Bessey

C. A. Bessey (1845–1915) was an American botanist, who laid the foundations of modern phylogenetic classifications (Figure 10.10). He was a student of Asa Gray and later became Professor of botany at the University of Nebraska. He was the first American to make a major contribution to plant classification, and also the first botanist to develop intentional phylogenetic classification. He based his classification on Bentham and Hooker, modified in the light of his 28 **dicta** and published in *Ann. Mo. Bot. Gard.* under the title 'The phylogenetic taxonomy of flowering plants' (1915).

Bessey considered angiosperms to have evolved monophyletically from Cycadophyta belonging to implied **bennettitalean ancestry**. He was the pioneer to consider that the large-sized bisexual flowers Magnoliaceae with spirally arranged floral parts represent the most primitive condition in angiosperms, a theory followed by many subsequent authors.

Bessey believed in the **strobiloid theory** of the origin of the flower, the latter having originated from a vegetative shoot with spiral phyllomes, of which some modified to form sterile perianth, fertile stamens and carpels. Two evolutionary lines from such a flower formed **strobiloideae** (Ranalian line) with connation of like parts and **cotyloideae** (Rosalian line) with connation of unlike parts (Table 10.7).

Ranales in dicots and Alismatales in monocots were considered to be the most primitive in each group, a fact recognized by most subsequent authors. Ranalian plants were considered to be primitive angiosperms having given rise to monocots, but unfortunately monocots were placed before dicots.

Figure 10.10: Charles Bessey (1845–1915) who initiated the modern phylogenetic systems of classification. He proposed his ideas in *Ann. Mo. Bot. Gdn.* (1915).

Table 10.7: Outline of the classification of angiosperms proposed by Charles Bessey (1915).

- Class 1. Alternifoliae (Monocotyledoneae)
  - Subclass 1. Strobiloideae (5 orders)
  - Subclass 2. Cotyloideae (3 orders)
- Class 2. Oppositifoliae (Dicotyledoneae)
  - Subclass 1. Strobiloideae
    - Superorder 1. Apopetalae-polycarpellatae (7 orders)
    - Superorder 2. Sympetalae-polycarpellatae (3 orders)
    - Superorder 3. Sympetalae-dicarpellatae (4 orders)
  - Subclass 2. Cotyloideae
    - Superorder 1. Apopetalae (7 orders)
    - Superorder 2. Sympetalae (3 orders)

Bessey also initiated the representation of evolutionary relationships through an evolutionary diagram, a phylogram with primitive groups at the base and the most advanced at the tips of branches (Figure 10.11). His diagram, resembling a cactus plant is better known as **Besseyan cactus**.

Hans Hallier

Hallier (1868–1932) was a German botanist who developed a classification resembling Bessey's and starting with Ranales. Dicots were, however, placed before monocots. **Magnoliaceae** were separated from Ranales and placed in a separate order **Annonales.**

Wettstein

Wettstein (1862–1931) was an Austrian systematist who published his classification in *Handbuch der systematischen botanik* (1930, 1935). The classification resembled that of Engler in considering unisexual flowers primitive but treated monocots advanced over dicots; and considered **Helobiae** to be primitive and Pandanales advanced. Many of his conclusions on phylogeny have been adopted in subsequent classifications.

Alfred Rendle

Rendle (1865–1938), an English botanist associated with the British Museum of Natural History, published *Classification of Flowering Plants* (1904, 1925), resembling that of Engler in considering monocots more primitive than dicots and Amentiferae a primitive group under dicots. He recognized three **grades** in dicots: **Monochlamydeae**, **Dialapetalae** (petals free) and **Sympetalae**. In monocots Palmae were separated as a distinct order and Lemnaceae considered to be advanced over Araceae.

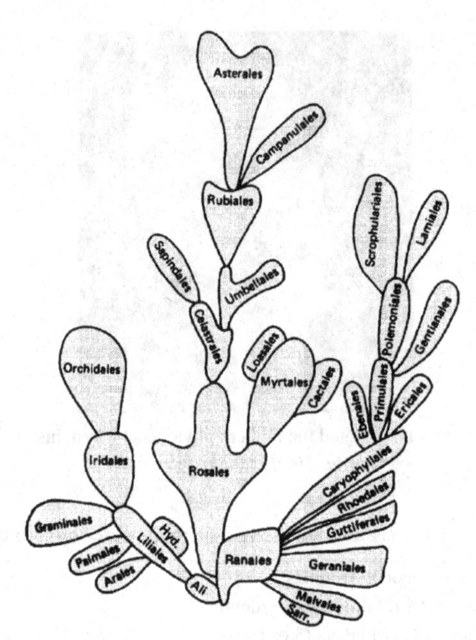

Figure 10.11: **Besseyan cactus** or **Opuntia Besseyi** showing the relationship of orders recognized by Bessey.

John Hutchinson

John Hutchinson (1884–1972) was a British botanist associated with the Royal Botanic Gardens, Kew, England who also served as keeper of Kew herbarium for many years (Figure 10.12). Hutchinson first proposed his classification of angiosperms in his book *The Families of Flowering Plants*, the first volume on Dicotyledons appearing in 1926 and the second on Monocotyledons in 1934. The classification was revised periodically, second edition in 1959 and the 3rd in 1973, one year after his demise.

In addition to presenting his system of classification for angiosperms, Hutchinson also published valuable works such as *Flora of West Tropical Africa* (1927–29), *Common Wild Flowers* (1945), *A Botanist in South Africa* (1946), *Evolution and Classification of Rhododendrons* (1946), *British Flowering Plants* (1948), *More Common Wild Flowers* (1948), *Uncommon Wild Flowers* (1950), *British Wild Flowers* (1955), *Evolution and Phylogeny of Flowering Plants* (1969) and *Key to the Families of Flowering Plants of the World* (1968).

Hutchinson also embarked upon an ambitious plan of revising *Genera plantarum* of Bentham and Hooker under the title *The Genera of Flowering Plants*. Unfortunately, he could complete only 2 volumes of this work, published in 1964 and 1967, the project cut short by his demise.

The classification system of Hutchinson dealt only with the flowering plants, included under **Phylum Angiospermae** as distinct from **Phylum Gymnospermae**. The classification was based on **24 principles** including General principles, Relating to General Habit, Relating to General Structure of Flowering plants and those Relating to Flowers and Fruits. These **principles** are outlined below:

Following Bessey, Hutchinson considered flowering plants to be **monophyletic**, having evolved from the hypothetical cycadeoid ancestral group which he gave the name of **Proangiosperms.** He recognized a number of smaller groups, bound together by a combination of characters. He established **Magnoliales** as an order distinct from **Ranales**, as he considered them to have evolved on parallel lines. Hutchinson regarded **Magnoliaceae** as the most primitive family of the living angiosperms. He considered Dicotyledones to be more primitive and placed them (Table 10.8) before Monocotyledones, giving them a rank of Subphylum.

The groups Polypetalae, Gamopetalae and Monochlamydeae were totally abolished; instead Hutchinson recognized two evolutionary lines: division **Lignosae** (fundamentally woody group) and **division Herbaceae** (fundamentally herbaceous group) within Dicotyledones, the former starting with Magnoliaceae and ending with Verbenaceae. The Herbaceae started with Paeoniaceae and ended with Lamiaceae. Within Monocotyledones he recognized three evolutionary lines: division Calyciferae (calyx bearers), division Corolliferae (corolla-bearers) and division Glumiflorae (glume-bearers). A total of 411 families are recognized, 342 in Dicotyledones and 69 in Monocotyledones. Lignosae includes 54 orders, Herbaceae 29, Calyciferae 12, Corolliferae 14 and Glumiflorae 3. A diagram (appropriately **phylogram**) showing phylogeny and evolution within Dicotyledones is presented in Figure 10.13.

Figure 10.12: John Hutchinson (1884–1972), the British botanist who worked as keeper of Kew Herbarium and published classification of angiosperms as *Families of Flowering Plants* (1973), as also the *Genera of Flowering Plants* (reproduced with permission from Royal Botanic Gardens Kew).

Table 10.8: Outline of the system of classification of flowering plants presented by Hutchinson in 3rd edition of *The Families of Flowering Plants* (1973).

Phylum I. Gymnospermae

Phylum II. Angiospermae

  Subphylum I. Dicotyledones

    Division I. Lignosae_ *54 orders*

      Order 1. Magnoliales (first family Magnoliaceae)

      ....

      Order 54. Verbenales (last family Verbenaceae)

    Division II. Herbaceae_ *28 orders*

      Order 55. Ranales (first family Paeoniaceae)

      ....

      Order 82. Lamiales (last family Lamiaceae)

  Subphylum II. Monocotyledones

    Division I. Calyciferae_ *12 orders*

      Order 83. Butomales (first family Butomaceae)

      ....

      Order 94. Zingiberales (last family Marantaceae)

    Division II. Corolliferae_ *14 orders*

      Order 95. Liliales (first family Liliaceae)

      ....

      Order 108. Orchidales (family Orchidaceae only)

    Division III. Glumiflorae_ *3 orders*

      Order 109. Juncales (first family Juncaceae)

      ....

      Order 111. Graminales (family Poaceae only)

Whereas Hutchinson considered the woody habit to be primitive in dicots, in monocots he considered the herbaceous habit to be primitive, and the woody forms derived from the herbaceous forms. He also considered Monocotyledones also to be a monophyletic group derived from Ranales, Butomales having a link with Helleboraceae and Alismatales with Ranunculaceae. The presence of endosperm in seeds of Ranunculaceae and its absence from Butomaceae and Alismataceae, otherwise considered closer, is explained by Hutchinson to be the result of aquatic habit in the last two. A diagram (**phylogram**) showing the probable phylogeny of various orders in Monocotyledones is presented in Figure 10.14.

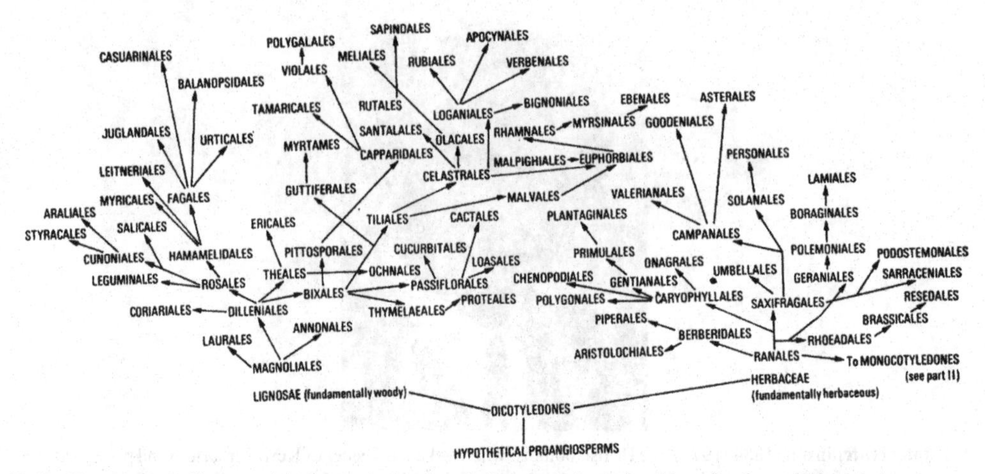

Figure 10.13: Hutchinson's diagram (phylogram) showing phylogeny and relationships of orders of Dicotyledones as presented in his 1973 classification.

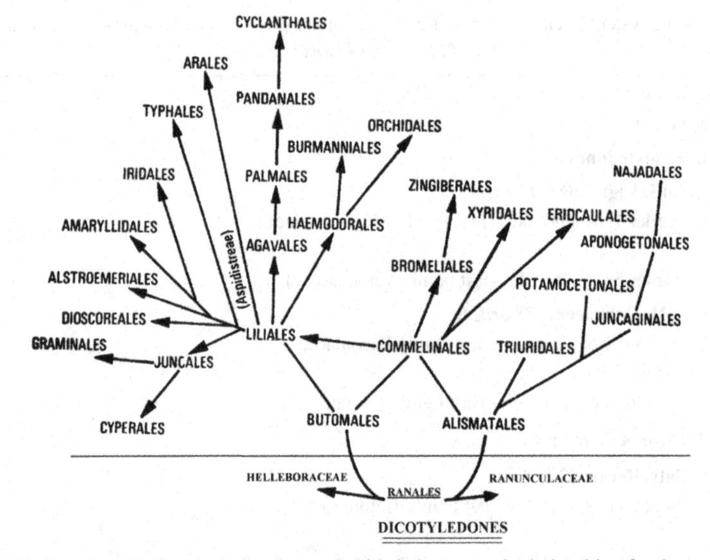

Figure 10.14: Hutchinson's diagram (phylogram) showing probable phylogeny and relationship of orders within Monocotyledones.

Merits

The system of Hutchinson, being based on a number of sound phylogenetic principles, and studies of a large number of plants at his disposal at Kew, shows the following improvements over earlier systems:

1. The system is more **phylogenetic** than that of Engler and Prantl, as it is based on phylogenetic principles, generally recognized by most authors.
2. The treatment of **Magnoliales** as the starting point in the evolutionary series of Dicotyledones is in agreement with prevalent views.
3. The abolition of Polypetalae, Gamopetalae, Monochlamydeae, Archichlamydeae and Metachlamydeae and rearrangement of taxa on the **combination of characters** and not one or a few characters as in earlier systems is more logical.
4. Many large unnatural families have been split into smaller natural ones. **Euphorbiaceae** of Bentham and Hooker is split into Euphorbiaceae, Ricinaceae and Buxaceae. The family **Urticaceae** is similarly split into Urticaceae, Moraceae, Ulmaceae and Cannabinaceae.
5. Standards of description are very high. Useful keys are provided for the identification of families.

6. **Phylograms** for dicots and monocots are more superior to the Besseyan cactus.
7. The classification of **Monocotyledones** is sounder and generally appreciated, even keys to the identification of **genera** have been provided.
8. The derivation of Monocotyledones from Dicotyledones is widely agreed.
9. The placement of **Alismatales** towards the beginning of Monocotyledones finds general acceptance.
10. Detailed classification up to the generic level, together with identification keys and description has been provided for some families in the two volumes of *The Genera of Flowering Plants.*

Demerits

The classification of Hutchinson has largely been ignored, as it mostly did not proceed beyond family level, and gave much importance to the habit. The major drawbacks of the system are listed below:

1. The system is not useful for practical identification, arrangement in Floras and herbaria, as it does not proceed beyond the family level in the greater majority of taxa.
2. The division of Dicotyledones into **Lignosae** and **Herbaceae** is most artificial and has resulted in separation of closely related families **Araliaceae** and **Apiaceae**, in Lignosae and Herbaceae respectively. Similarly, Lamiaceae and Verbenaceae are very closely related and often placed in the same order in contemporary systems of classification. Hutchinson, based on habit, separated them under distinct orders and even separate divisions—Herbaceae and Lignosae, respectively.
3. Hutchinson did not provide a full explanation for the majority of his evolutionary concepts.
4. He derives angiosperms from **proangiosperms** but does not provide information about the nature of this hypothetical ancestral group.
5. Although he has split several large unnatural families into natural units, in some cases he has even split some families which were already natural monophyletic groups. The family **Ranunculaceae** has been split into Ranunculaceae and **Helleboraceae** on the basis of achene and follicle fruit, respectively. Studies on the floral anatomy have shown that evolutionary stages in the reduction of ovule number can be seen in the genera of Helleboraceae, and many genera of Ranunculaceae show traces which would have gone to now aborted ovules. Thus, the Ranunculaceae of Bentham and Hooker represents a monophyletic group and need not be split.
6. The family **Calycanthaceae** is related to Laurales, but placed here in Rosales.
7. Hutchinson regards Magnoliaceae as the most primitive family of living Dicotyledones, but most contemporary authors consider vessel-less **Winteraceae,** or **paleoherbs** be the most primitive.
8. The monocotyledons are placed after dicotyledons, whereas the recent classifications place them between primitive angiosperms and the eudicots.
9. Family Liliaceae of Hutchinson is a large unnatural assemblage, which has been split into several smaller monophyletic families like Liliaceae, Alliaceae, Asparagaceae, Asphodelaceae in the recent classification of Judd et al. (2002), APG II (2003) and Thorne (2003).

Lyman Benson

Lyman Benson developed a classification designed for teaching botany and published in his book *Plant classification* (1957). Dicotyledons are divided into five groups on features derived from the classifications of Bentham and Hooker and Engler and Prantl. Monocotyledons are divided directly into 13 orders, starting with Alismales and ending with Pandanales. Although several realignments have been made by Benson, de Candolle as also Bentham and Hooker have been followed for grouping in dicots and Bessey's outline for classification of monocots:

1. Thalamiflorae (hypogynous, free or no petals)
2. Corolliflorae (hypogynous, petals fused)
3. Calyciflorae (perigynous or epigynous, petals free or none)
4. Ovariflorae (epigynous, petals fused)
5. Amentiferae (catkin bearing)

Modern Phylogenetic Systems

Several contemporary workers are involved in improving schemes of classification based on new information from various sources. Recent data from paleobotany, phytochemistry, ultrastructure and improved techniques of the numerical analysis of available data have helped in developing classifications that have several features in common, though differing in some basic concepts. It is now largely agreed upon that Angiosperms are a **monophyletic** group with dicots being more primitive than monocots. Vesselless **Winteraceae** and the paleoherb families are now generally regarded as among the basal living angiosperms. Authors of the four major contemporary systems of classification, A. Cronquist, R. Dahlgren, A. Takhtajan and R. Thorne have unfortunately left us during the past three decades. There is, however, a positive trend of frequent updating of classification schemes in electronic versions. During the last decade, the Angiosperm Phylogeny Group (APG) has been working towards realization of monophyletic groups.

Armen Takhtajan

Armen Takhtajan (1910–2009) was a leading Russian plant taxonomist (Figure 10.15) and chief of the Department of higher plants in V. L. Komarov Botanic Institute, USSR Academy of Sciences, Leningrad (now named St. Petersburg). He was an international authority on phytogeography, origin and phylogeny of flowering plants. He was the President of the 12th International Botanical Congress held in Leningrad in 1975.

His classification was first published in 1954 in Russian but came to be known outside the Soviet Union only after its English translation *Origin of Angiospermous Plants* was published in 1958. The system was elaborated in *Die Evolution der Angiospermen* (1959), and *Systema et Phylogenia Magnoliophytorum* (1966), both in Russian. The classification became popular with the English translation of the latter as *Flowering Plants—Origin and Dispersal* by C. Jeffrey in 1969. The classification was published in a revised form in *Botanical Review* in 1980. A more elaborate revision of this classification appeared in the Russian work *Sistema Magnoliophytov* (Latin facsimile *Systema Magnoliophytorum*) in 1987. Between 1980 and 1987, he proposed smaller revisions in 1983 (revision of dicots only in Metcalfe and Chalk: *Anatomy of Dicotyledons*, vol. 2) and 1986 (Takhtajan: *Floristic Regions of the World*). His comprehensive system of classification was published in 1997 (*Diversity and Classification of Flowering Plants*), considerably revised in final version published in second edition (2009) of *Flowering Plants*. Earlier Takhtajan, along with Cronquist and Zimmerman, had also provided a broad classification of **Embryobionta** (1966).

Takhtajan, who provided a classification of angiosperms up to the family level, belongs to the Besseyan School and was strongly influenced by Hutchinson, Hallier and the other more progressive German workers. He believes in the **monophyletic origin** of angiosperms; the group having evolved from seed ferns **Lyginopteridophyta.** According to Takhtajan, the angiosperms are of **neotenous origin** (retention of juvenile characters in the adult plant_also called **paedomorphosis**). The simple entire, pinnately veined leaves of primitive angiosperms representing the juvenile stage of frond-like leaves of the seed ferns.

Takhtajan was of the opinion that angiosperms arose under environmental stress, probably as a result of adaptation to moderate seasonal drought on rocky mountain slopes, in an area with monsoon climate.

For many years, Takhtajan considered Winteraceae along with Degeneriaceae to represent the most primitive angiosperms. Subsequently (1997), however, he chose **Degeneriaceae** as the most primitive family, placed under the order Magnoliales. In latest revision (2009), he like APG III and Thorne (2007), placed Amborellaceae at the beginning of angiosperms. He shifted Winteraceae along with Canellaceae to a distinct order Canellales, placed before Magnoliales (which includes Degeneriaceae). While deciding the placement of various groups, Takhtajan has used a number of criteria based on his understanding of the available information. His major conclusions are summarized below:

1. The most primitive angiosperms are regarded to be the small evergreen trees or shrubs, taller trees and deciduous habit being later developments.
2. Simple leaves with entire margin and pinnate venation are primitive. Pinnately and palmately lobed leaves arose subsequently followed finally by the pinnate and palmate compound leaves.
3. Trilacunar nodal structure is most primitive. Unilacunar node is a secondary development, having evolved from tri-pentalacunar nodal types. Vessels (rather vessel elements) evolved from tracheids with scalariform pitting.
4. Cymose inflorescence is most primitive and the racemose type is derived. Solitary flowers represent reduction from both other types.
5. Primitive flower is moderate in size, in few flowered cymes, as in *Degeneria*. The large flowers of *Magnolia* and Nymphaeaceae are of secondary origin.
6. Petals have dual origin, from the bracts in Magnoliales (**bracteopetals**), and from the stamens in Caryophyllales (**andropetals**). Early angiosperms have numerous spirally arranged perianth of modified bracts. Distinct sepals and petals are secondary developments.
7. Primitive stamens were broad, laminar, 3-veined, not differentiated into filament and connective. The common ancestral type had marginal sporangia and later on gave rise to the abaxial types (**extrorse** as in *Degeneria*) and the adaxial types (**introrse** as in *Magnolia*).

Figure 10.15: Armen Takhtajan (1910–2009), leading Soviet authority on phytogeography and classification of flowering plants. Published last version of his classification in 2010, incorporating several modifications in his system.

8. Monocolpate pollen grains are primitive, from which arose tricolpate and then the polycolpate types. Inaperturate type, according to him is more specialized.
9. Primitive carpels are free, unsealed, conduplicate, containing numerous ovules and with laminar placentation (as in *Tasmania and Degeneria*). Fusion is a later development. Fusion of closed carpels laterally resulted in **syncarpous** gynoecium with axile placentation, the dissolution of septa subsequently resulting in **lysicarpous** gynoecium with free-central placentation. Lateral fusion of open conduplicate carpels formed **paracarpous** gynoecium with parietal placentation.
10. Outer integument arose from the cupule of ancient gymnospermous ancestor. Unitegmic ovules arose by fusion of two integuments or abortion of one.
11. Takhtajan (and Cronquist) earlier regarded monocotyledons as being of aquatic origin from Nymphaeales via Alismatales. Later he regarded the latter only as a lateral side branch of monocotyledons, and proposed that Nymphaeales and Alismatales had a common origin from hypothetical extinct terrestrial group of Magnoliidae, the main monocotyledonous stock being terrestrial in origin.

Inspite of several refinements, Takhtajan's classification still sticks to traditional distinction between dicots and monocots, placing monocots after dicots. It approaches more closely that of Cronquist (1981, 1988) in naming angiosperms as division **Magnoliophyta.** Dicots and monocots are given the rank of a class and named **Magnoliopsida** and **Liliopsida,** respectively. These are further subdivided into subclasses (ending in **-idae**, e.g., **Rosidae**), superorders (ending in **-anae**, e.g., **Rosanae**), orders and families. Cronquist, however, does not recognize superorders. Also, as against 8 and 4 (11 and 6 in 1997 classification) subclasses of dicots and monocots respectively in Takhtajan's system, Cronquist recognizes 6 and 5, respectively. Both systems are developed based on **phylogenetic,** as well as **phenetic information** from every field of study. However, whereas Cronquist gives more importance to phenetic information, Takhtajan relies more heavily on phylogenetic data.

These two systems of classification show a general agreement with the other two major classifications of angiosperms, developed by Thorne (1981, 1983, 1992) and Dahlgren (1981, 1983, 1989), although the recent revisions by Thorne (2000, 2003, 2007) are in more agreement with APG classifications, in abandoning traditional division into monocots and dicots.

Both Takhtajan and Cronquist prefer the name Magnoliophyta for angiosperms and appropriate names **Magnoliopsida** and **Liliopsida** for dicots and monocots, respectively. An outline of the classification (2009 version) is presented in Table 10.9.

As against the classification proposed in 1987, in 1997 revision he added three new subclasses **Nymphaeidae, Nelumbonidae** (separated from Magnoliidae) and **Cornidae** (separated from Rosidae) to Magnoliopsida and two **Commelinidae** (separated from Liliidae) and **Aridae** (separated from Arecidae) to Liliopsida. In 2009 revision, he has dropped all three dicot subclasses and two in monocots (retaining Commelinidae but discarding Triurididae), thus recognising 8 subclasses in dicots and 4 in monocots, same number as in 1987, but with changed circumscription. Also, the sequence of monocot subclasses has been altered placing Alismatidae at the beginning. He has also reduced the number of superorders (from 71 to 44), orders (232 to 156) and families (589 to 562). The book incidently mentions 560 families.

An interesting aspect about the 1987 classification of Takhtajan was uncertainty about the placement of the family **Cynomoriaceae**. The single genus *Cynomorium* earlier placed in family Balanophoraceae (Hutchinson, 1973; Cronquist, 1988), was removed to the family Cynomoriaceae and placed next to Balanophoraceae under the order Balanophorales by

Table 10.9: Outline of the system of classification of Angiosperms proposed by Takhtajan in 2009.

Phylum Magnoliophyta* **(Flowering Plants)** *2 classes, 12 subclasses, 44 superorders, 156 orders, 562 families (2 classes, 17 subclasses, 71 superorders, 232 orders, 589 families in 1997 classification); estimated genera_13,000, species_ 2,50,000*	
Class 1. Magnoliopsida (Dicotyledons)	**6. Rosidae**
Family of uncertain position: Haptanthaceae	1. Rosanae (9, 24)
Subclass 1. Magnoliidae	2. Myrtanae (1, 14)
Superorder 1. Nyphaeanae (7 orders, 11 families)	3. Fabanae (2, 2)
2. Magnolianae (5, 7)	4. Rutanae (11, 45)
3. Lauranae (1, 8)	5. Celastranae (1, 9)
4. Piperanae (2, 7)	6. Santalanae (1, 10)
5. Rafflesianae (2, 5)	7. Balanophoranae (2, 2)
2. Ranunculidae	8. Rhamnanae (1, 4)
1. Proteanae (3, 3)	**7. Asteridae**
2. Ranunculanae (9, 15)	1. Cornanae (9, 46)
3. Hamamelididae	2. Asteranae (5, 16)
1. Trochodendranae (2, 3)	**8. Lamiidae**
2. Myrothamnanae (1, 1)	1. Lamianae (6, 55)
3. Hamamelidanae (7, 11)	**Class 2. Liliopsida (Monocotyledons)**
4. Juglandanae (2, 3)	**Subclass 1. Alismatidae**
4. Caryophyllidae	Superorder 1. Petrosavianae (1, 4)
1. Caryophyllanae (2, 23)	2. Alismatanae (3, 14)
2. Polygonanae (3, 4)	3. Aranae (1, 4)
3. Nepenthanae (1, 5)	**2. Liliidae**
5. Dilleniidae	1. Lilianae (10, 52)
1. Dillenianae (1, 1)	2. Pandananae (5, 7)
2. Ericanae (12, 34)	3. Dioscoreanae (1, 4)
3. Primulanae (3, 10)	**3. Arecidae**
4. Violanae (5, 19)	1. Arecanae (1, 1)
5. Capparanae (6, 18)	**4. Commelinidae**
6. Malvanae (2, 21)	1. Bromelianae (1, 1)
7. Euphorbianae (1, 5)	2. Zingiberanae (1, 8)
	3. Commelinanae (3, 14)
	4. Juncanae (1, 3)
	5. Poanae (3, 9)

*125 orders 441 families (including Haptanthaceae) in dicots, 31 orders 121 families in monocots

Takhtajan (1980), Thorne (1983, 1992, 2003) and Dahlgren (1983, 1989). In his 1987 classification, Takhtajan had placed this family under the order **Cynomoriales**, but not being certain about its affinities, had inserted this order tentatively towards the end of Rosidae. In his 1997 classification he had brought Cynomoriales under Magnoliidae within superorder **Balanophoranae**. In 2009 revision, he shifted Balanophoranae back to the end of Rosidae. It is interesting to note that APG III (2009) was uncertain about the placement of Cynomoriaceae, but like APweb (2009, 2017) APG IV (2016) places it under Saxifragales.

Like other phylogenetic systems of classification, the presumed relationship of various subclasses and superorders is indicated with the help of a **bubble diagram** (Figure 10.16 for dicots; Figure 10.17 for monocots)—more appropriately a **phylogram**—the size of each bubble or balloon representing the relative size of each group, the branching pattern the phylogenetic relationship, and the length of bubble its evolutionary advancement (**degree of apomorphy**).

Merits

The latest classification of Takhtajan (2009) shows several improvements in light of recent information on phylogeny and phenetics. Many merits achieved in the earlier versions have also been retained in the latest revision. The major achievements of this system include:

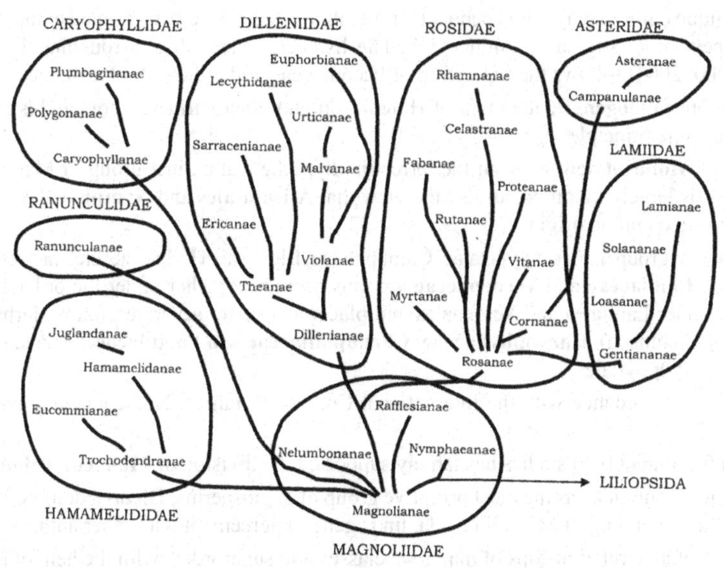

Figure 10.16: Bubble diagram of Takhtajan showing the probable relationship between different subclasses and superorders of **dicotyledons** (based on Takhtajan, 1987). 1997 and 2009 classifications do not include a bubble diagram.

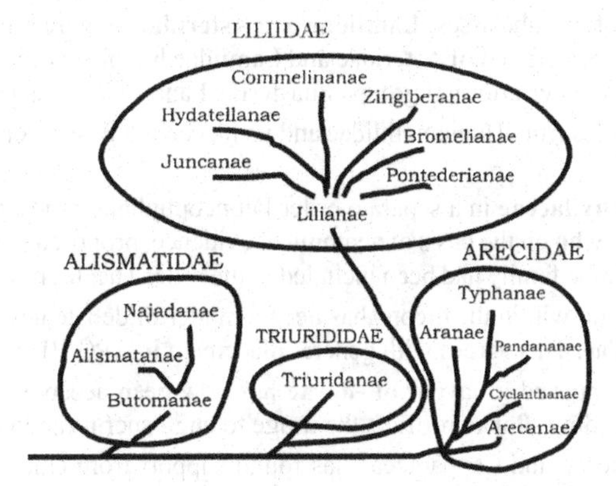

Figure 10.17: Bubble diagram of Takhtajan showing the probable relationship between different subclasses and superorders of **monocotyledons** (based on Takhtajan, 1987).

1. A general agreement with the major contemporary systems of Cronquist, Dahlgren and Thorne (earlier versions up to 1992) and incorporation of phylogenetic as well as phenetic information for the delimitation of orders and families. The genus *Nelumbo* was earlier placed in the family Nymphaeaceae under Nymphaeales. Takhtajan separated it to **Nelumbonaceae** under the order **Nelumbonales** on the basis of the occurrence of tricolpate pollen grains, embryo structure, absence of laticifers and chromosome morphology. He finally separated it to a separate superorder **Nelumbonanae** under the distinct subclass **Nelumbonidae.** Thorne (1983, 1992, 2000, 2003) also follows the separation into Nelumbonales (closer to Ranunculales) but under superorder Ranunculanae. APG IV (2016) also places Nelumbonaceae closer to Ranunculales (under Eudicots), but in order Proteales. Similarly, the genus *Eucommia* was earlier placed in the family Hamamelidaceae. Takhtajan removed it to the family **Eucommiaceae** under the order **Eucommiales** based on the presence of stipules, unilacunar nodes, unitegmic ovule and cellular endosperm, a separation followed by Cronquist (1988). Thorne (1983, 1992) gave it a rank of a suborder under the order Hamamelidales (Hamamelididae) but has now shifted it to Lamiidae under order Garryales, somewhat similar to APG II (2003; Garryales of Euasterids I). de Soo (1975) placed it under a separate subclass Eucommiidae. Similarly, the genus *Paeonia,* placed under the family Ranunculaceae in earlier classifications, was separated by Takhtajan to the family **Paeoniaceae** under the order **Paeoniales** on the basis of evidence from chromosomes (5 large chromosomes), floral anatomy (centrifugal stamens, many traces in sepals and petals, 5 in carpels), and

embryology (unique embryogeny with coenocytic proembryo stage, reticulately pitted exine, large generative cell, thick fleshy carpels, broad stigmas, prominent lobed fleshy nectariferous disc surrounding the gynoecium). Thorne (1983, 1992, 2000, 2003) follows the separation of Paeoniaceae under the order Paeoniales.

2. The system is more **phylogenetic** than that of Hutchinson and other earlier authors and is based on now widely accepted phylogenetic principles.
3. The derivation of **Monocotyledons** from the terrestrial hypothetical extinct group of Magnoliidae (often called proangiosperms), is largely favoured, as also the view that **Alismatales** and **Nymphaeales** represent ancient side branches and have a common origin.
4. Abolition of artificial group names Polypetalae, Gamopetalae, Lignosae, Herbaceae, etc., has resulted in more natural grouping of taxa. **Lamiaceae** and **Verbenaceae** are thus brought together under the order **Lamiales** (as against their separation under Lamiales and Verbenales and placement under separate groups Herbaceae and Lignosae, respectively, by Hutchinson). **Caryophyllaceae**, **Chenopodiaceae** and **Portulacaceae** have similarly been placed under the order Caryophyllales.
5. Nomenclature is in accordance with the International Code of Botanical Nomenclature, even up to the level of division.
6. Clifford (1977) from numerical studies has largely supported the division of **Monocotyledons** into subclasses.
7. The placement of Magnoliidae as the most primitive group of angiosperms, Dicotyledons before Monocotyledons, **Magnoliales** at the beginning of Magnoliopsida, finds general agreement with other authors.
8. Depiction of the putative relationships of major subclasses and superorders with the help of a **bubble diagram** is very useful. It gives some idea about the **relative size** of different groups, point of **cladistic divergence** and **degree of advancement** (apomorphy) reached. Larger groups are represented by larger bubbles, vertical length the degree of advancement, and the point of separation of a branch its cladistic divergence.
9. By splitting **Asteridae** into two subclasses: **Lamiidae** and **Asteridae,** a more rational distribution of sympetalous families has been achieved. Separation of Asteridae and Lamiidae has also been followed by Thorne (2000, 2003) and APG II (2003; although under informal groups Euasterids I and Euasterids II).
10. Removal of the order Urticales from Hamamelididae and its placement in superorder **Malvanae**, under Dilleniidae is more appropriate.
11. The placement of **Dioncophyllaceae** in a separate order Dioncophyllales is in line with the opinion presented by Metcalfe and Chalk (1983), who on the basis of anatomical evidence, proposed that the family occupied an isolated taxonomic position. Earlier, the family had been included in the order Theales next to the family Ancistrocladaceae.
12. Nymphaeales, whose position within the dicots, has been a matter of debate have finally been placed towards the beginning under Magnoliidae. This agrees with general placement in APG III and Thorne (2007).
13. The ending **-anae**, earlier opposed in favour of **-florae** has now been accepted by Dahlgren (1989) and Thorne (1992 onwards) since the ending **-florae** restricts the usage to angiosperms and is not universal in application.
14. The separation of Brassicaceae and Capparaceae has found support from chloroplast sequence data (Hall et al., 2002), consistent with morphological data.
15. The merger of Asclepiadaceae with Apocynaceae has been supported by molecular analyses by Judd et al. (1994) and Sennblad and Bremer (1998). Recognition of distinct Asclepiadaceae would render Apocynaceae as paraphyletic (Judd et al., 2002).
16. Amborellaceae are rightly placed towards the beginning of Angiosperms, in line with APG III and Thorne (2007).
17. Subclass Aridae has been abolished and Arales appropriately placed towards the end of Alismatidae under superorder Aranae, a placement similar to Thorne (2007). The recent studies have, shown the affinities of Araceae with Alismatales. As such, the family is included under Alismatales in APG III.
18. Subclass Triurididae which was raised earlier has rightly been abolished and order Triuridales placed under Pandananae, in line with recent hylogenetic schemes. The evidence from 18S rDNA sequencing (Chase et al., 2000) justifies its placement near Pandanaceae. Thorne (2007) shifts Triuridaceae under Pandananae, but distinct order Triuridales.
19. The Families Winteraceae and Canellaceae, which were placed in two separate orders in 1997 version have finally been brought under the same order. Multigene analyses (Soltis et al., 1999; Zanis et al., 2002, 2003) have provided 99–100 per cent **bootstrap support** in their relationship. The two are accordingly placed in the same order in APG III, APweb and Thorne (2007).
20. Alismatidae has rightly been brought to the beginning of monocots. In 1997 version, it was shifted lower down and Liliidae placed in the beginning.

Demerits

With the latest revision of his classification in 2009, Takhtajan attempted to remove deficiencies in the earlier versions of his system. The critical appraisal of his latest version, in future, may bring out some further drawbacks. The following limitations of the system can be recorded:

1. The system, although very sound and highly phylogenetic, is not helpful for identification and for adoption in herbaria, as it provides classification only up to the family level.
2. Dahlgren (1980, 1983) and Thorne (1983, 1992, 2003) consider that the **angiosperms** deserve a **class** rank equivalent to the main groups of gymnosperms such as Pinopsida, Cycadopsida, etc.
3. Although the system is based on data derived from all sources, in final judgment more weightage is given to **cladistic** information compared to phenetic information.
4. Ehrendorfer (1983) points out that **Hamamelididae** do not represent an ancient side branch of Magnoliidae but are remnants of a transition from Magnoliidae to Dilleniidae-Rosidae-Asteridae. Thorne (2009) admits there are some links between lower Hamamelididae and some lower Dilleniidae and Rosidae but points out that higher members are highly specialized that they show no evident link with either of the two.
5. Behnke (1977) and Behnke and Barthlott (1983) point out that Caryophyllales have PIII-type plastids whereas Polygonales and Plumbaginales have S-type plastids, and thus advocate their removal from Caryophyllidae to Rosidae, retaining only Caryophyllales in the subclass Caryophyllidae. Though not agreeing on their removal, Takhtajan (1987, 1997) partly incorporated Behnke's suggestion by placing the three orders under superorders Caryophyllanae and Polygonanae, but within the same subclass, Caryophyllidae. Thorne (2003, 2007) places Plumbaginaceae and Polygonaceae under the same order Polygonales.
6. Further splitting and increase in the number of families to 562 (589 in 1997, 533 in 1987) has resulted in a very narrow circumscription by the creation of numerous **monotypic families** such as Pottingeriaceae, Barclayaceae, Hydrastidaceae, Nandinaceae, Griseliniaceae, Hypecoaceae, etc., and numerous **oligotypic** ones such as Balanophoraceae, Sarraceniaceae, Peganaceae and Agrophyllaceae.
7. Takhtajan has made substantial changes in his scheme of classification in 1997 and 2009 over his earlier version of 1987. Unfortunately, however, he has failed to provide a bubble diagram, which was a positive feature of his earlier versions and was very useful in relating affinities between the groups as also to know the relative sizes of the various groups. This is especially significant, as circumscription of groups has greatly changed.
8. Takhtajan suggested that smaller families are more 'natural'. According to Stevens (2003), this is incorrect. Monophyletic groups that include fewer taxa—Takhtajan's smaller families—do not necessarily have more apomorphies, even if all members of such groups are certainly likely to have more features in general in common.
9. The monocotyledons are placed after dicotyledons, whereas the recent classifications place them between primitive angiosperms and the eudicots.
10. Family Hyptanthaceae has been laced in the start as family of uncertain position. APG III considers it as unplaced under eudicots, APweb assigned to Buxales and Thorne (2007) as unplaced genus Haptanthus.

Arthur Cronquist

Arthur Cronquist (1919–1992), a leading American taxonomist, associated with the New York Botanical Garden (Figure 10.18), produced a broad classification of **Embryobionta** along with Takhtajan and Zimmerman (1966). He produced a detailed classification of angiosperms in 1968 in his book *The Evolution and Classification of Flowering Plants.* The classification was further elaborated in 1981 in his book *An Integrated System of Classification of Flowering Plants.* The final revision was published in the second edition (1988) of *The Evolution and Classification of Flowering Plants.* Some realignments in Dicotyledons were published in *Nordic Journal of Botany* in 1983.

The classification is conceptually similar to that of Takhtajan's system but differs in details. The classification, like that of Takhtajan, is based on evidence derived from all sources, but in contrast to Takhtajan who gives more importance to cladistics, Cronquist gave more importance to morphology (Ehrendorfer, 1983).

Following Takhtajan, the angiosperms are given the name Magnoliophyta and divided into Magnoliopsida (dicots) and Liliopsida (monocots). Cronquist includes only six subclasses in dicots and recognizes five in monocots. In dicotyledons, the Ranunculidae of Takhtajan are merged with Magnoliidae and Lamiidae are not given a separate rank at subclass level, but retained in Asteridae.

In monocotyledons, Zingiberidae are treated separate from Liliidae and Triuridales kept under Alismatidae. As a major departure from the systems of Takhtajan, Dahlgren and Thorne, no superorders are recognized, the subclasses are divided into orders directly. Also, as against 233 orders and 592 families recognized by Takhtajan, Cronquist recognizes

Figure 10.18: Arthur Cronquist (1919–1992) leading American Plant taxonomist who published 2nd edition of his *Evolution and Classification of Flowering Plants* in 1988. His classification is similar to that of Takhtajan in general outline (photograph courtesy Allen Rokach, The New York Botanical Garden, Bronx, New York).

83 orders and 386 families. Cronquist agrees with Thorne (earlier versions up to 1992) in keeping the family **Winteraceae** (and not Degeneriaceae as done by Takhtajan) at the beginning of dicotyledons, and included along with Degeneriaceae, Magnoliaceae, Annonaceae, etc., in the same order **Magnoliales**. Paeoniaceae, unlike other contemporary authors, are not separated by Cronquist into a distinct order Paeoniales, but instead shifted to the order Dilleniales under Dilleniidae.

Another significant departure from Takhtajan's system is the merger of Amaryllidaceae with Liliaceae, under the order Liliales. Takhtajan places these two families in separate orders Amaryllidales and Liliales, respectively. Unlike most recent authors, Cronquist believed in the aquatic origin of monocotyledons, from a primitive vessel-less ancestor resembling present-day Nymphaeales.

In contrast to Takhtajan's system, Nelumbonaceae are placed in Nymphaeales (and not a separate order Nelumbonales), Typhales in Commelinidae (and not Arecidae) and sympetalous families of dicotyledons placed in a large subclass Asteridae (and not three subclasses Asteridae, Cornidae and Lamiidae). Urticales are included along with wind-pollinated families under Hamamelididae (and not with its related orders Malvales and Euphorbiales), and Malvales and Euphorbiales are kept in separate subclasses Dilleniidae and Rosidae respectively (and not the same subclass Dilleniidae). Cronquist has provided a synoptic arrangement of taxa, facilitating the process of identification up to the family level. An outline of Cronquist's system is presented in Table 10.10. The system is widely used in the USA.

The relationships of various subclasses and orders (Figure 10.19) are shown with the help of a phylogram which takes the form of a **bubble diagram**, like other contemporary systems of classification.

Merits

The classification of Cronquist is largely based on principles of phylogeny that find acceptance with major contemporary authors. The system is merited with the following achievements over the previous systems of classification:

1. It shows general agreement with major contemporary systems of Takhtajan, Dahlgren and Thorne (earlier versions), and incorporates evidence from all sources in arrangement of various groups. *Paeonia* and *Nelumbo* are thus placed under **Paeoniaceae** and **Nelumbonaceae**, although the orders Paeoniales and Nelumbonales are not recognized. *Eucommia* is also kept in a separate family Eucommiaceae under a distinct order **Eucommiales**.
2. The revision of the classification in 1981 and 1988 was presented a in comprehensive form, giving detailed information on phytochemistry, anatomy, ultrastructure and chromosomes besides morphology.
3. The text, being in English, has been readily adopted in books and floristic projects originating in the USA.
4. The system is highly **phylogenetic** and is based on now largely accepted phylogenetic principles.
5. The placement of **Winteraceae** at the beginning of dicotyledons is generally favoured by most authors including Ehrendorfer (1968), Gottsberger (1974) and Thorne (up to 1992). The family has vessel-less wood similar to gymnosperms, great similarity between micro- and megasporophylls, unifacial stamens and carpels, morphology similar to **pteridosperms**, high chromosome number suggesting long evolutionary history and less specialized beetle pollination as compared to the genus *Magnolia*.

Table 10.10: Broad outline of the classification of angiosperms presented by **Cronquist** (1988).

Division. Magnoliophyta_ ***2 classes, 11 subclasses, 83 orders and 386 families; 219,300 species***		
Class 1. Magnoliopsida (Dicotyledons)_ ***6 subclasses, 64 orders, 320 families; 169,400 species***		
Subclass	1. Magnoliidae	(12 orders: Magnoliales, Laurales, Piperales, Aristolochiales, Illiciales, Nymphaeales, Ranunculales and Papaverales)
	2. Hamamelidae	(11 orders: Trochodendrales, Hamamelidales, Daphniphyllales, Didymelales, Eucommiales, Urticales, Leitneriales, Juglandales, Myricales, Fagales and Casuarinales)
	3. Caryophyllidae	(3 orders: Caryophyllales, Polygonales and Plumbaginales)
	4. Dilleniidae	(13 orders: Dilleniales, Theales, Malvales, Lecythidales, Nepenthales, Violales, Salicales, Capparales, Batales, Ericales, Diapensiales, Ebenales and Primulales)
	5. Rosidae	(18 orders: Rosales, Fabales, Proteales, Podostemales, Haloragales, Myrtales, Rhizophorales, Cornales, Santalales, Rafflesiales, Celastrales, Euphorbiales, Rhamnales, Linales, Polygalales, Sapindales, Geraniales and Apiales)
	6. Asteridae	(11 orders: Gentianales, Solanales, Lamiales, Callitrichales, Plantaginales, Scrophulariales, Campanulales, Rubiales, Dipsacales Calycerales and Asterales)
Class 2. Liliopsida (Monocotyledons)_ ***5 subclasses, 19 orders, 66 families; 49,900 species***		
Subclass	1. Alismatidae	(4 orders: Alismatales, Hydrocharitales, Najadales, and Triuridales)
	2. Arecidae	(4 orders: Arecales, Cyclanthales, Pandanales and Arales)
	3. Commelinidae	(7 orders: Commelinales, Eriocaulales, Restionales, Juncales, Cyperales, Hydatellales and Typhales)
	4. Zingiberidae	(2 orders: Bromeliales and Zingiberales)
	5. Liliidae	(2 orders: Liliales and Orchidales)

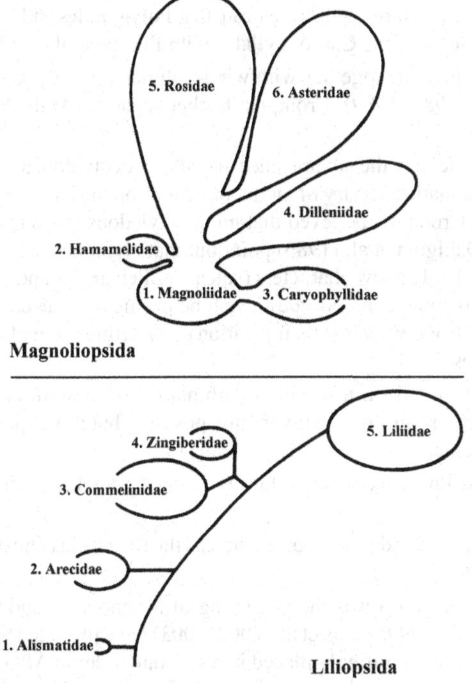

Figure 10.19: Phylogram showing the relationship between various subclasses and orders as **presented by Cronquist** (based on Cronquist, 1988).

6. Abolition of artificial group names such as Polypetalae, Gamopetalae, Lignosae, Herbaceae, etc., has resulted in more natural grouping of taxa. Verbenaceae and Lamiaceae are thus brought under the order **Lamiales**. Caryophyllaceae, Chenopodiaceae and Portulacaceae are similarly placed in the same order **Caryophyllales**.
7. **Nomenclature** is in accordance with the International Code of Botanical Nomenclature.
8. Placement of **Magnoliidae** as the most primitive group of angiosperms, dicotyledons before monocotyledons, Magnoliales at the beginning of Magnoliidae and Butomaceae at the beginning of Liliopsida, finds general agreement with other authors.
9. Compositae in dicotyledons and Orchidaceae in monocotyledons are generally regarded as advanced families, and are rightly placed towards the end of each group, respectively.
10. The relationship of various groups has been depicted with diagrams, which provide valuable information on relative advancement, cladistic relationship and size of various subclasses.
11. The separation of Brassicaceae and Capparaceae has found support from chloroplast sequence data (Hall et al., 2002), consistent with morphological data.

Demerits

The system is becoming increasingly popular, especially in the USA, where many books are following this system. The following drawbacks, however, may be pointed out:

1. In spite of being a highly phylogenetic and popular in the USA, the system is not very useful for **identification** and adoption in herbaria since identification keys for genera, their distribution and description are not provided.
2. Dahlgren (1983, 1989) and Thorne (1981, 2003) considered angiosperms to deserve a **class** rank, and not that of a division.
3. **Asteridae** represent a loose assemblage of several diverse sympetalous families.
4. Clifford (1977) on the basis of numerical studies has shown that Typhales are better placed in Arecidae. Cronquist places Typhales in Commelinidae.
5. **Superorder,** as a rank above the order, is not recognized, thus showing a significant departure from the contemporary systems of Takhtajan, Thorne and Dahlgren.
6. Ehrendorfer (1983) pointed out that **Hamamelididae** (Hamamelidae by Cronquist) do not represent an ancient side-branch of Magnoliidae but are remnants of a transition from Magnoliidae to Dilleniidae-Rosidae-Asteridae.
7. Behnke (1977) and Behnke and Barthlott (1983) advocate that **Polygonales** and **Plumbaginales,** with S-type plastids, should be removed to **Rosidae** and only Caryophyllales with PIII-type plastids retained in Caryophyllidae.
8. Urticales are placed in Hamamelidae together with wind-pollinated families, whereas they are close to Malvales and Euphorbiales (Dahlgren, 1983, 1989). Cronquist further separates Malvales in Dilleniidae and Euphorbiales in Rosidae.
9. Most recent authors do not believe in the aquatic ancestry of monocotyledons. Kosakai et al. (1970) have provided ample evidence to refute the aquatic ancestry of monocotyledons on the basis of study of primary xylem in the roots of *Nelumbo* (Nymphaeales). Cronquist believed that monocotyledons arose from vesselless ancestors resembling present-day Nymphaeales. Dahlgren et al. (1985) point out that Nymphaeales and Alismatales demonstrate a case of **multiple convergence**, and only a few characters (sulcate pollen grains and trimerous flowers) are due to shared ancestry. The presence of two cotyledons, S-type sieve tube plastids, occurrence of ellagic acid and perispermous seeds in Nymphaeales argue strongly against their position as a starting point of monocotyledons, and none of these attributes occur in Alismatales.
10. Metcalfe and Chalk (1983), based on a unique combination of anatomical features, suggested that family **Dioncophyllaceae** should occupy an isolated taxonomic position, but it was placed by Cronquist in order Violales before family Ancistrocladaceae.
11. Cronquist (1988) recognized Physenaceae as a family under Order Urticales but was not sure about its exact placement.
12. The monocotyledons are placed after dicotyledons, whereas the recent classifications place them between primitive angiosperms and the eudicots.
13. The family Winteraceae is placed towards the beginning of Magnoliales and Canellaceae towards the end. The multigene analyses (Soltis et al., 1999; Zanis et al., 2002, 2003) have provided 99–100 per cent **bootstrap support** in their relationship. The two are accordingly placed in a separate order in APG II and APweb, and under the same suborder in Thorne (2003). The affinities between these two families is also supported by morphological studies of Doyle and Endress (2000).

Rolf Dahlgren

Rolf Dahlgren (1932–87), a Danish botanist working in Botanical Museum of the University of Kopenhagen first proposed his system and a new method of illustrating phylogenetic relationships in a text book in Danish in 1974. The revised system in English and subsequent revisions were published in 1975, 1980, 1981, 1983. A useful detailed treatment of Monocotyledons was presented in a book *The families of Monocotyledons* (Dahlgren et al.) in 1985. His diagram, a cross-section through the top of an imaginary phylogenetic tree became very popular for mapping the distribution of character-states in various orders of angiosperms and is popularly known as **Dahlgrenogram**.

After Dahlgren's tragic death in a car accident in 1987, his wife **Gertrud Dahlgren** (Figure 10.20) continued his work and finally published the 'last Dahlgrenogram' for dicotyledons, followed by a classification of monocotyledons, both in 1989, incorporating the latest ideas of Dahlgren, and bringing up an updated classification of angiosperms. She also changed the endings for the superorders from **-florae** to **-anae**, since the use of former term restricted its application to only flowering plants, and the change to -anae was in the interest of nomenclatural uniformity. This practice of using -anae was initially started by Takhtajan and has now been incorporated by Thorne (since 1992), who earlier like Dahlgren, preferred the ending -florae. Gertrud followed this up (1991) with the mapping of various embryological character-states.

The classification is closely similar to the earlier versions of Thorne in using the name **Magnoliopsida** for angiosperms, **Magnoliidae** for dicots, and **Liliidae** for monocots. The realignments are based on a large number of phenetic characteristics, mainly phytochemistry, ultrastructure and embryology. The system includes 25 superorders in dicots and 10 in monocots. Several hundred such maps have been developed by Dahlgren and his associates. Dahlgren pointed out that recognition of Dicotyledons and Monocotyledons would not be allowed if one followed rigid **cladistic approaches**, but he nevertheless, considered Monocotyledons as a unique group worthy of subclass rank.

The Dahlgrenogram (Figure 10.21) is a bubble diagram in which different orders are represented as bubbles, whose size is relative to the number of species in the order, and their related positions reflect phylogenetic affinities. The orders are combined into superorders, thus forming bubble complexes. While presenting a revision of Dahlgren's system in 1989, his wife Gertrud Dahlgren (1931–2009) made significant changes in the superorders Theanae, Malvanae, Rutanae and Cornanae. Similarly, in monocotyledons, the minor changes included recognizing Acoraceae as a family in Arales, merging Sparganiaceae in Typhaceae, Thismiaceae in Burmanniaceae, and Geosiridaceae in Iridaceae, plus shifting of a few families. Gertrud also included the position of families in the bubble diagram. A broad outline of the classification, as presented by **Gertrud Dahlgren** is presented in Table 10.11.

Merits

The system of classification presented by Dahlgren has several advancements over the previous systems of classification. The salient advantages of the system include:

1. The system is a highly phylogenetic one incorporating evidence from morphology, phytochemistry and embryology.
2. The angiosperms are given a more agreeable rank of a **class** like Thorne and other recent systems.
3. Unlike recent phylogenetic systems, no family of angiosperms is left unplaced.
4. The Dahlgrenogram in the form of a bubble diagram is very useful in giving an idea about the relationships of superorders, orders and even families. It also gives an idea about the relative number of species in each group.

Figure 10.20: Rolf F. Dahlgren and his wife Gertrud Dahlgren who continued his work on the classification of Angiosperms since his death in 1987. Gertrud (1931–2009) concentrated on evolutionary botany and species differentiation after 1990 (Photographs courtesy Gertrud Dahlgren).

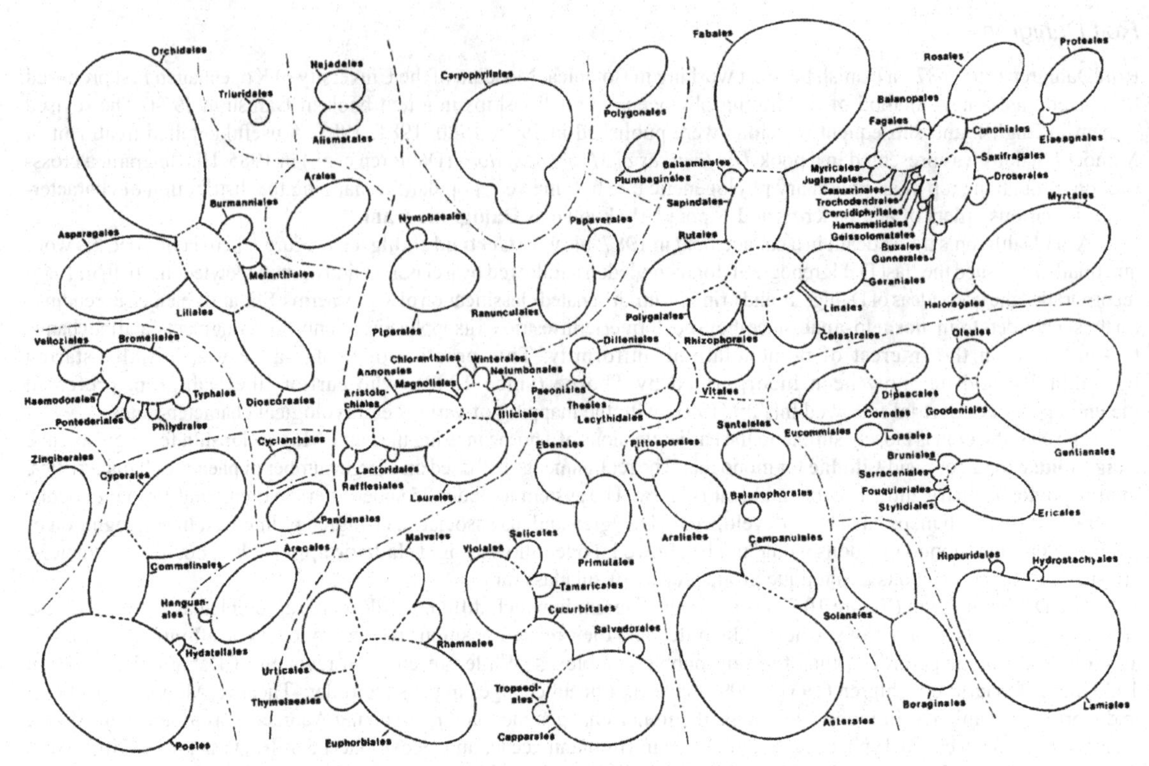

Figure 10.21: Two-dimensional diagram of angiosperm classification (both dicots and monocots included) showing orders, combined into superorders (Courtesy G. Dahlgren).

Table 10.11: Outline of the updated Dahlgren's classification of angiosperms as presented by his wife Gertrud Dahlgren (1989).

Superorder	Superorder
Dicotyledons *25 superorders, 87 orders and 343 families*	
1. Magnolianae (10 *orders*)	14. Rutanae (9 *orders*)
2. Nymphaeanae (2 *orders*)	15. Vitanae (1 *order*)
3. Ranunculanae (2 *orders*)	16. Santalanae (1 *order*)
4. Caryophyllanae (1 *order*)	17. Balanophoranae (1 *order*)
5. Polygonanae (1 *order*)	18. Aralianae (2 *orders*)
6. Plumbaginanae (1 *order*)	19. Asteranae (2 *orders*)
7. Malvanae (5 *orders*)	20. Solananae (2 *orders*)
8. Violanae (7 *orders*)	21. Ericanae (5 *orders*)
9. Theanae (4 *orders*)	22. Cornanae (3 *orders*)
10. Primulanae (2 *orders*)	23. Loasanae (1 *order*)
11. Rosanae (15 *orders*)	24. Gentiananae (3 *orders*)
12. Proteanae (2 *orders*)	25. Lamianae (3 *orders*)
13. Myrtanae (2 *orders*)	
Monocotyledons *10 superorders, 24 orders and 104 families*	
1. Alismatanae (2 *orders*)	6. Zingiberanae (1 *order*)
2. Triuridanae (1 *order*)	7. Commelinanae (3 *orders*)
3. Aranae (1 *order*)	8. Arecanae (2 *orders*)
4. Lilianae (6 *orders*)	9. Cyclanthanae (1 *order*)
5. Bromelianae (6 *orders*)	10. Pandananae (1 *order*)

5. The Dahlgrenogram has been widely used for plotting and comparing the distribution of various character-states in angiosperms.
6. The use of a superorder rank similar to Thorne and Takhtajan has resulted in a more realistic arrangement of families and orders. The use of ending **-anae** is in line with other two authors.
7. The separation of genus *Acorus* from Araceae into a distinct family **Acoraceae** has been followed by recent systems of Takhtajan (1997), Thorne (2000, 2003) and APG II, who have even separated the family under a distinct order Acorales. The genus is distinct from Araceae in ensiform leaves, glandular tapetum, and the type of endothecial cells.
8. The merger of Scrophulariales with Lamiales has been followed in the recent classifications of Cronquist and APG.
9. The placement of Cornales closer to Ericales is justified by Judd et al. (2002) and also followed up in Thorne (2003) and APG II (2003).
10. The separation of Brassicaceae and Capparaceae has found support from chloroplast sequence data (Hall et al., 2002), consistent with morphological data.

Demerits

Although the system of classification of Dahlgren shows several improvements over the earlier systems, it suffers from the following drawbacks:

1. The system covers only flowering plants and does not proceed below the family level, as such is not usefully for arranging specimens in a herbarium or for following in the Floras.
2. Dahlgren places **Asteranae, Cornanae and Aralianae** before Lamianae, whereas the data from molecular studies justifies placement of the group (with circumscription somewhat similar to Euasterids II) after Lamiidae (Comparable to Euasterids I of APG II).
3. Dahlgren divides angiosperms into dicotyledons and monocotyledons, whereas the recent classification of APG II (2003) and Thorne (2000, 2003), the primitive angiosperms are placed separately.
4. Monocotyledons are placed after dicotyledons, whereas the recent classifications place them between primitive angiosperms and the eudicots.
5. The family Ceratophyllaceae is placed under order Nymphaeales, but the studies of Zanis et al. (2002) and Whitlock et al. (2002) have shown that the family is a sister group of monocots as indicated by microsporogenesis and structure of leaf margin. It is accordingly placed just before monocots in APG II.
6. Family Acoraceae is placed under order Arales, but according to molecular studies of Chase et al. (2000) and Fuse and Tamura (2000), it deserves placement before the rest of the monocots.
7. Family Winteraceae is placed in a separate order much after Canellaceae, whereas the multigene analyses (Soltis et al., 1999; Zanis et al., 2002, 2003) have provided 99–100 per cent **bootstrap support** in their relationship. The two are accordingly placed in the same order in APG II and APweb, and under the same suborder in Thorne (2000, 2003). The affinities between these two families are also supported by morphological studies of Doyle and Endress (2000).
8. Dahlgren had regarded Budlejaceae and Myoporaceae as distinct from Scrophulariaceae but morphological studies of Bremer et al., (2001) and molecular (three gene analysis) by Olmstead et al. (2001) supported their merger, which was followed by APG II (2003) and Thorne (2003).

Robert F. Thorne

Robert F. Thorne (1920–2015), an American taxonomist, associated with the Rancho Santa Ana Botanic Garden, Claremont, California, developed and periodically revised a system of classification. Earlier versions of the classification closely approached the system proposed by Dahlgren in giving angiosperms a rank of a class, and dicots and monocots as subclasses. These were further subdivided into superorders, orders, suborders and families. In general approach of arrangement of orders and families, there was a considerable parallel development with other three contemporaries Cronquist, Dahlgren and Takhtajan.

Thorne (Figure 10.22) first put forward his classification in 1968 and proposed revisions in 1974, 1976, 1981, 1983, 1992, 1999, 2000, 2003 and 2007. He earlier preferred the ending **-florae** over **-anae** of Takhtajan for superorders but has now (1992 onwards) accepted the ending **-anae**. Dr. Thorne is the recipient of the prestigious Asa Gray Award (2001) from the American Society of Plant Taxonomists and the Merit (1996) and Centennial Award (2006) from the Botanical Society of America.

Figure 10.22: Robert Thorne of Rancho Santa Botanic Garden. His latest revision of the classification of Angiosperms was published in 2007.

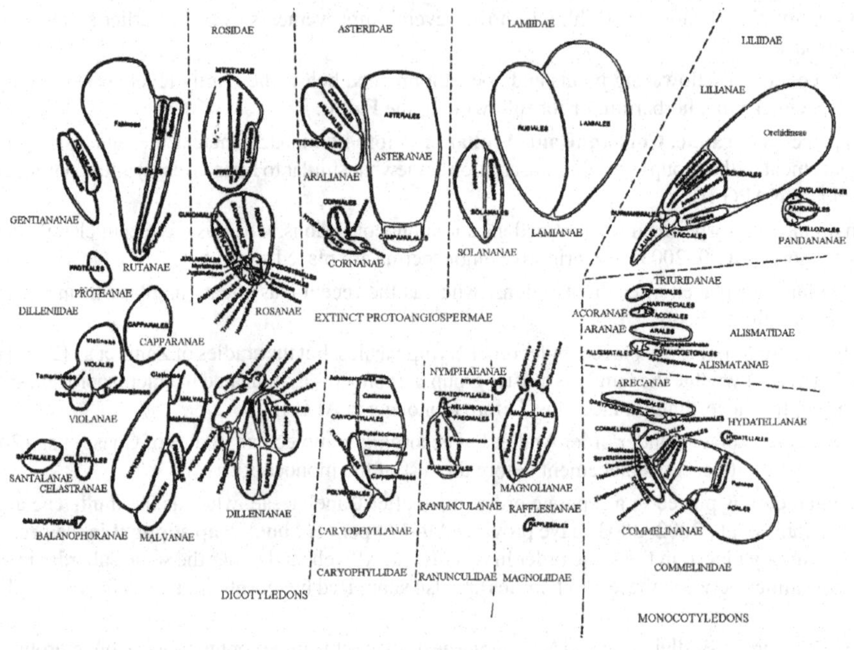

Figure 10.23: Thorne's Phylogenetic shrub of Angiospermae (2000 version of classification).

Thorne incorporated the role of phytochemistry in realignment of taxa, recognizing subfamilies more frequently and applied the principle of priority up to the class rank thus preferring name **Annonopsida** for angiosperms, **Annonidae** for dicots, replacing Magnoliflorae by **Anniniflorae** and Magnoliales by **Annonales.** Since 1992 he has, however, abandoned this departure from contemporary systems and adopted the generally accepted names **Magnoliopsida, Magnoliidae** and **Magnoliales**.

Thorne's diagram indicating the relationship between different groups, is a **phylogenetic shrub** viewed from above, with the centre of the diagram left empty to indicate extinct early angiosperms; those nearer the centre being the primitive groups and those nearer the periphery the advanced ones. The relative number of species in different groups is indicated by balloons of different sizes (Figure 10.23).

The classification has undergone a major revision after the publication of APG II, in 2003 (10 subclasses, 33 superorders, 90 orders and 489 families), 2006 (11 subclasses, 35 superorders, 89 orders and 486 families) classification displayed on the website of the Rancho Santa Ana Botanic Garden as www.rsabg.org/angiosperms/angiosperms.pdf. The latest revision published in 2007 (New York Botanical Garden Press) recognizes 12 subclasses, 36 superorders, 85 orders and 485 families. The significant changes include:

1. The number of subclasses has been increased to 12 (as against 10 in 2000, 2003, and 11 in 2006) by adding Chlorandidae (earlier placed under Magnoliidae] at the beginning of classification and Malvidae (containing five superorders segregated from Rosidae) after Rosidae. Certaophyllaceae placed in 2006 under Ranunculidae has been shifted under order Nymphaeales of Chloranthidae.
2. Magnoliidae now has only one superorder Magnolianae (Nymphaeanae removed to Chloranthidae and Rafflesianae to Malvidae) with four orders (as against 1 in 2000 and four in 2003). Winteraceae (like in 2003) loses its position as the first family of angiosperms.
3. Superorder Triuridanae in Alismatidae (Takhtajan 1987, 1997 took it to distinct subclass Triurididae) has been abolished, as also the order Triuridales. As in 2003, Family Petrosaviaceae is placed order Petrosaviales (under superorder Acoranae). Triuridaceae is shifted to Liliidae—> Pandananae—> Pandanales.
4. Liliidae contains the same number of three superorders as 2003 (there were two in 2000) but Taccanae has been replaced by Dioscoreanae. The number of orders in have, however been reduced from 9 in 2003, 2006 to 5 in 2007, all 5 in Pandananae merged into a single order Pandanales.
5. Subclass Commelinidae has 2 superorders (same as in 2003) instead of 3 (in 2000), Hydatellanae being abolished.
6. Ranunculidae finds one superorder added to 2000 classification. Proteanae is shifted from Dilleniidae and broadened to include Platanaceae (shifted from Rosidae—> Rosanae—> Hamamelidales), Buxaceae and Didymelaceae (shifted from Rosidae—> Rosanae—> Balanopales) and Sabiaceae (shifted from Dilleniidae—> Rutanae—> Rutales). These are recognized under four distinct orders Proteales, Platanales, Buxales and Sabiales. The placement broadly remains the same as in 2003.
7. The largest subclass Dilleniidae (with 10 superorders, 19 orders and 160 families in 1999, 2000 version), which was abolished in 2003 and its contents distributed mainly under Rosidae (see under Rosidae). Rest of the members are distributed under Caryophyllidae and Asteridae
8. The new subclass Hamamelididae, established in place of Dilleniidae in 2003, includes orders Hamamelidales, Saxifragales, Juglandales and Betulales, all shifted from Rosidae—> Rosanae (of 2000). The revisions of 2006 and 2007, however, shift Vitaceae of Gunnerales to Malvidae, Gunneraceae to Asteridae and Berberidopsidales to Caryophyllidae.
9. Subclass Caryophyllidae has seen major revision in 2006, recognising five superorders as against only one in 2003.
10. Malvidae has been added as new superorder after Rosidae in 2007, including four super Rosidae is the largest subclass with 11 superorders Malvanae, Rafflesianae, Capparanae, Huerteanae and Rutanae all segregated from Rosidae.
11. Rosidae with the removal of 5 superorders from 2006 revision includes the remaining 6 superorders.
12. Subclass Asteridae has same four superorders as 2003 (Ericanae was added in 2003 mostly containing the members of Dillenianae which has now been restricted to include only Dilleniaceae under Rosidae). The new version has two orders lesser. Hydrangeales has been abolished with its families distributed between Cornales and newly created order Desfontainiales.
13. Subclass Lamiidae has one order Garryales added (to 2000 version) including families Garryaceae, Aucubaceae, Eucommiaceae, Oncothecaceae and Icacinaceae. The arrangement is more or less same as in 2003 and 2006.
14. Genus *Guametela*, which was earlier listed as genus of uncertain position has been placed in 2007 revision under a distinct family Guemetelaceae under superorder Crossosomatales, Myrtanae, Rosidae; *Pottingeria* has been listed as genus of uncertain position.

Many of these changes are in line with APG II. Significant parallels include the (a) Placement of Amborellaceae, Chloranthaceae and Austrobaileyaceae towards the beginning of angiosperms; (b) Recognition of Canellales, Piperales, Laurales and independent orders within Magnoliidae; (c) Shifting of Triuridaceae and Stemonaceae closer to Pandanaceae; (d) Bringing closer families Proteaceae, Platanaceae, Buxaceae and Didymelaceae; (e) Shifting of Hamamelidaceae, Saxifragaceae, Vitaceae away from Rosidae to Hamamelididae; (f) Placement of Ericaceae and related families under Asteridae and away Rosidae (and abolished Dilleniidae) and (g) Recognition of Garryales as distinct order including Garryaceae, Aucubaceae and Eucommiaceae.

Thorne, in his 2003 version had also introduced the concept of assigning the degree of confidence in hierarchical level, circumscription and alignment of taxa, continued in 2006 and 2007 revisions. A represents limited confidence, B for probably correct assignment and C implies considerable confidence in assignment. Such an indication is very useful for future placements of the groups, and focussing those which need further investigation. Salient features of 2007 version are given in Table 10.12.

Table 10.12: Outline of the **system of classification** of Angiosperms proposed by **Thorne** in 2007.

Class Magnoliopsida

12 subclasses, 36 superorders, 85 orders, 485 families; estimated genera_ 13,372, species_2,53,300

Subclass	
1. **Chloranthidae**	*1 superorder, 2 order , 9 families, 19 genera, 250 species*
	Superorder 1. Chloranthanae
2. **Magnoliidae**	*1 superorder, 4 orders, 20 families, 276 genera, 8805 species*
	Superorder 1. Magnolianae
3. **Alismatidae**	*3 superorders, 6 orders, 18 families, 235 genera, 3660 species*
	Superorder 1. Acoranae
	2. Aranae
	3. Alismatanae
4. **Liliidae**	*3 superorders, 5 orders, 51 families, 1261 genera, 29085 species*
	Superorder 1. Pandananae
	2. Dioscoreanae
	3. Lilianae
5. **Commelinidae**	*2 superorders, 10 orders, 35 families, 1116 genera, 23270 species*
	Superorder 1. Arecanae
	2. Commelinanae
6. **Ranunculidae**	*2 superorders, 8 orders, 17 families, 298 genera, 6350 species*
	Superorder 1. Proteanae
	2. Ranunculanae
7. **Hamamelididae**	*1 superorder, 4 orders, 22 families, 145 genera, 3870 species*
	Superorder 1. Hamamelidanae
8. **Caryophyllidae**	*5 superorders, 9 orders, 46 families, 889 genera, 13875 species*
	Superorder 1. Berberidopsidanae
	2. Caryophyllanae
	3. Dillenianae
	4. Santalanae
	5. Balanophoranae
9. **Rosidae**	*7 superorders, 12 orders, 83 families, 2258 genera, 48127 species*
	Superorder 1. Celastranae
	2. Violanae
	3. Podostemanae
	4. Oxalidanae
	5. Geranianae
	6. Rosanae
	7. Myrtanae
10. **Malvidae**	*5 superorders, 8 orders, 61 families, 1430 genera, 20430 species*
	Superorder 1. Malvanae
	2. Rafflesianae
	3. Capparanae
	4. Huerteanae
	5. Rutanae
11. **Asteridae**	*4 superorders, 13 orders, 78 families, 2677 genera, 44970 species*
	Superorder 1. Cornanae
	2. Ericanae
	3. Aralianae
	4. Asteranae
12. **Lamiidae**	*2 superorders, 4 orders, 45 families, 2752 genera, 50310 species*
	Superorder 1. Solananae
	2. Lamianae

Four genera (*Haptanthus, Heteranthia, Pottingeria and Pteleocarpa*) of uncertain position

Merits

The classification of Thorne has kept pace with recent developments and is being regularly updated. The system is merited with the following achievements over the previous and contemporary systems of classification:

1. It is a highly phylogenetic system, incorporating the recent evidence from molecular systematics and chemotaxonomy, and balancing it with evidence from other sources.
2. The angiosperms are given a more agreeable rank of a **class** like Dahlgren and other recent systems.
3. The system is more exhaustive than the contemporary systems in that families, where necessary have been divided into **subfamilies**. Similarly, **suborders** are recognized under several orders.
4. The system, unlike the APG II has found a place for all unplaced families of APG.
5. The placement of **Amborellaceae, Chloranthaceae, Austrobaileyaceae, Nymphaeaceae** and **Cabombaceae** (a major shift from 2006 version) at the beginning of angiosperms is generally favoured in the recent cladistic schemes of APG (and supported by Qui et al., 2000; Soltis et al., 2000). These have been placed under an independent subclass **Chloranthidae**.
6. Abolition of traditional groups dicots and monocots, and dividing angiosperms directly into various subclasses (with circumscription largely paralleling informal groups of APG) is in line with the recent phylogenetic thinking.
7. The subclass Magnoliidae placed after paleoherb families is in line with APG classification.
8. The system is superior over APG II classification in that formal group names are given for all supraordinal ranks.
9. The recognition of superorders with ending **-anae** has resulted in more realistic arrangement of the orders within subclasses.
10. The monocots families are arranged in between primitive angiosperms and more advanced dicots, and not towards the end of angiosperms, as in previous systems of Takhtajan, Dahlgren and Cronquist. This treatment is in agreement with Angiosperm Phylogeny Group.
11. Creation of superorder Malvanae and shifting several families of Rosidae here has resulted in more realistic arrangement.
12. Family Winteraceae and Canellaceae are brought together under the same order. Their affinities are strongly supported by morphological studies and multigene analyses.
13. The separation of Brassicaceae and Capparaceae has found support from chloroplast sequence data (Hall et al., 2002), consistent with morphological data.
14. The merger of Budlejaceae in Scrophulariaceae is supported by morphological studies of Bremer et al. (2001) and molecular (three gene analysis) by Olmstead et al. (2001).
15. Shifting Triuridaceae and Stemonaceae closer to Pandanaceae is in line with recent APG schemes. The evidence from 18S rDNA sequencing (Chase et al., 2000) justifies placement under Pandanales. Triurididae as an independent subclass is not justified as indicated by recent evidence.
16. The placement of Cornales and Ericales together under Asteridae is in line with recent thinking of APG.
17. Family Liliaceae of Hutchinson and earlier authors has been split into a number of monophyletic families such as Liliaceae, Alliaceae, Asphodelaceae, Asparagaceae, etc., in line with the arrangement in APG classifications.
18. The concept of assigning the degree of confidence (A, B or C) in hierarchical level, circumscription and alignment of taxa is very useful for better understanding of phylogenetic affinities.
19. The merger of Asclepiadaceae with Apocynaceae has been supported by molecular analyses by Judd et al. (1994) and Sennblad and Bremer (1998). Recognition of distinct Asclepiadaceae would render Apocynaceae as paraphyletic (Judd et al., 2002).
20. Placement of Ceratophyllaceae under Chloranthidae before monocots is in line with recent data. Studies of Zanis et al. (2002) and Whitlock et al. (2002) have shown that the family is a sister group of monocots as indicated by microsporogenesis and structure of leaf margin.

Demerits

Despite several improvements, the following drawbacks, however, may be pointed out:

1. Although highly phylogenetic the system is not very useful for **identification** and adoption in herbaria since identification keys for genera, their distribution and description are not provided.
2. Thorne places **Asteridae** before Lamiidae, whereas the data from molecular studies justifies placement of the group (with circumscription somewhat similar to Euasterids II) after Lamiidae (Comparable to Euasterids I of APG II).
3. Thorne is not clear about the affinities of four genera of angiosperms.

4. Grewiaceae (former Tiliaceae with *Tilia* excluded) is recognized as an independent family, whereas recent APG classifications (Judd et al., APG and APweb) place all members of Tiliaceae, Bombacaceae and Sterculiaceae under Malvaceae.
5. Thorne separates *Cabomba* and *Brassenia* under Cabombaceae on the basis of trimerous flowers, with distinct sepals and petals, 2–3 free carpels, and fruit a follicle, whereas the cladistic analyses support their placement under Nymphaeaceae as done by APG, APweb and Judd et al. The separation of Cabombaceae renders Nymphaeaceae as paraphyletic.

C. R. de Soo

From Budapest, Hungary, C. R. de Soo proposed (1975) a classification essentially similar to Takhtajan's but preferring the name **Angiospermophyta** for angiosperms, **Dicotyledonopsida** for dicots and **Monocotyledonopsida** for monocots. Five subclasses are included in dicots and 3 in monocots; 54 orders in dicots and 14 in monocots are recognized.

Zheng-Yi Wu

During the last decades Zheng-Yi Wu (1916–2013) and his associates developed a system of classification of angiosperms, much different from contemporary systems in logic and treatment. Professor Wu, Director Emeritus, Academician of Chinese Academy of Sciences was a leading Chinese taxonomist. He was appointed as a deputy director of the Beijing Institute of Botany, Chinese Academy of Sciences in 1950, became a member of the Chinese Academy in 1955, and the director of the Kunming Institute of Botany in from 1974 through 1983. In addition to his numerous publications on many plant families and genera, Professor Wu is credited for 28 major works in taxonomy, vegetation, floristics, biogeography, Chinese herbals, and diversity. He led or joined several botanical expeditions, especially to Xizang. He described about 300 new species and proposed 11 genera. He devoted himself to the research of the flora of China and East Asia since 1930s. He was the Chairman of editorial board for the publication of "Flora of China", which describes all the diverse species of plants in China, a large-scale scientific work projected to cover 80 volumes and 125 issues in Chinese. An English version has been published under the co-editorship of Dr. Peter H. Raven, Director of the Missouri Botanical Garden. From a global perspective, Dr. Wu showed deep involvement with the attempts to protect natural flora, specially the human-induced extinction of plant species and their impact on the global environment. His efforts contributed to the establishment of national parks and natural reserves in China.

The ideas for this **polyphyletic-polychronic-polytypic** classification dividing angiosperms into **eight classes**, were presented in two papers of Wu et al. (1998a, 1998b). The synopsis of classification was published in 2002, and detailed description of families and genera represented in China in 2003.

The classification was developed on the basis of assumption that although angiosperms are monophyletic in their earliest origin yet owing to some intrinsic factors in plants themselves and different extrinsic factors appearing on the Earth after the Early Cretaceous explosion of angiosperms, some groups might have become isolated and continued to flourish for many generations. These groups might have given rise to many lineages, just like the situation that many branches and leaves may sprout from a single shoot. Thus, viewed from certain cross section of time, some lineages are **monophyletic-monochronic-monotopic**, whereas others are polyphyletic-polychronic-polytopic. By "polyphyletic" the authors meant that during Early Cretaceous explosion of angiosperms, there were many monophyletic groups due to extinction of many ancient species. By polytopic, the authors did not mean that the same group could have occurred on different continents at the same time, they rather believed that the modern inter-continental disjunctive distribution patterns of angiosperms can be explained only by using plate tectonics and vicariance biogeography. After their origin from Pangaea during Late Triassic to the Early Jurassic, the pro-angiosperms might have undergone a process of differentiation, extinction and re-differentiation of several dozen million years, and then undergone a great explosive radiation in the Early Cretaceous. By that time, the major groups of angiosperms might have appeared, forming 8 major lineages at early stage of differentiation of angiosperms. These eight lineages are circumscribed as 8 classes, thus proposing a **new 8-class system** of classification. The system is outlined in Table 10.13.

Angiosperm Phylogeny Group (APG)

First serious attempts towards developing a **cladistic classification** were made by Bremer (Figure 10.25) and Wanntorp (1978, 1981), who suggested that angiosperms should be treated as subclass **Magnoliidae** of class **Pinatae** (seed plants). They argued that Monocotyledons and Dicotyledons should not be recognized because it will make the group **paraphyletic**, suggesting that angiosperms should be directly divided into a number of superorders. The proposal was not taken seriously because monocots and dicots as separate groups were recognized in all major system of classifications up to the last decade of last century.

Table 10.13: Broad outline of 8-class **polyphyletic-polychronic-polytopic classification** of angiosperms proposed by **Zheng-Yi Wu et al.** (2002).

Phylum Magnoliophyta (Angiospermae) (8 classes, 40 subclasses, 202 orders, 572 families)			
Class	**Subclass**	**Class**	**Subclass**
1. Magnoliopsida			
	(5 subclasses, 11 orders, 17 families)		20. Commelinidae
	1. Magnoliidae		21. Juncidae
	2. Annonidae		22. Poaoidae
	3. Illiciidae		23. Arecidae
	4. Ceratophyllidae	**6. Ranunculopsida**	
	5. Nymphaeidae		*(4 subclasses, 9 orders, 17 families)*
2. Lauropsida			24. Nelumbonidae
	(3 subclasses, 4 orders, 9 families)		25. Ranunculidae
	6. Lauridae		26. Paeoniidae
	7. Calycanthidae		27. Papaveridae
	8. Chloranthidae	**7. Hamamelidopsida**	
3. Piperopsida			*(3 subclasses, 11 orders, 21 families)*
	(2 subclasses, 4 orders, 8 families)		28. Trochodendridae
	9. Aristolochidae		29. Hamamelididae
	10. Piperidae		30. Betulidae
4. Caryophyllopsida		**8. Rosopsida**	
	(3 subclasses, 8 orders, 20 families)		*(10 subclasses, 112 orders, 361 families)*
	11. Caryophyllidae		31. Dilleniidae
	12. Polygonidae		32. Malvidae
	13. Plumbaginidae		33. Ericidae
5. Liliopsida			34. Rosidae
	(10 subclasses, 43 orders, 119 families)		35. Myrtidae
	14. Alismatidae		36. Rutidae
	15. Triurididae		37. Geraniidae
	16. Aridae		38. Cornidae
	17. Liliidae		39. Asteridae
	18. Bromelidae		40. Lamiidae
	19. Zingiberidae		

Figure 10.24: Zheng-Yi Wu, a leading Chinese Taxonomist who spearheaded the publication of Flora of China and published his new eight-class classification of angiosperms.

There has been a considerable revival of the cladistic concepts with the utilization of molecular data and development of powerful tools of data handling. During the last decade, concept has developed into APG classification by collaborative efforts of a group of dedicated workers of '**Angiosperm Phylogeny Group**' (K. Bremer, A. Backlund, B. Briggs, B. Bremer, M. W. Chase, M. H. G. Gustafsson, S. B. Judd, F. A. Kellogg, P. F. Stevens, M. Thulin and several others), who published a classification of 462 families of Angiosperms in 1998. These families were grouped into 40 putative monophyletic orders under a small number of informal monophyletic higher groups: **monocots, commelinoids, eudicots, core eudicots, rosids, eurosids I, eurosids II, asterids, euasterids I** and **euasterids II**. Under these informal groups there were also listed a number of families without assignment to order. Eleven unclassified families were included in the beginning. Also, in the beginning under Angiosperms directly were 4 orders with no supraordinal grouping into informal groups. At the end of the system there was an additional list of 25 families of uncertain position for which no firm data existed regarding placement anywhere within the system.

Recent cladistic analyses are revealing the phylogeny of flowering plants in increasing detail, and there is support for the monophyly of many major groups above the family level.

With many elements of the major branching sequence of phylogeny established, a revised suprafamilial classification of flowering plants becomes both feasible and desirable. Cladistic information strongly points to the realization that simplistic division of angiosperms into monocots and dicots do not reflect phylogenetic history.

Some modifications of APG classification were presented Judd et al. (1999) recognizing a total of 51 orders and shifting some families from informal groups where they were placed directly in 1998 classification, to these orders. The book, however, lists only major families and as such nearly 200 families, have been left out. A revision presented in the 2nd edition (2002) has further improvements in line with thinking of APG, and is largely similar to the APG II classification, with minor differences.

APG classification has undergone considerable improvements with APG II (2003), APG III (2009) and APG IV (2016) with continuous upgradation on Angiosperm Phylogeny website (APweb) by P. F. Stevens (Figure 10.25)—http://www.mobot.org/MOBOT/research/APweb/, with more and more families (and some orders) coming out of the list of unplaced taxa. Latest version has 64 orders and 416 families as against 59 orders and 415 families in APG III. COM Clade with uncertain position is inserted among Rosids and Commelinids merged with Monocots. A broad outline of APG IV is presented in Table 10.14. 3 orders are placed at the beginning under Basal Angiosperms. Chloranthaceae is placed under independent unplaced Lineage. Genus *Peltanthera* finds unplaced position under Gesneriaceae. Three orders Celastrales, Oxalidales and Malpighiales are placed in COM Clade with uncertain position and inserted among Rosids. Five new orders added in APG IV include Boraginales, Dilleniales, Icacinales, Metteniusales and Vahliales. The concept of bracketed families (which made APG II unpopular) had been abolished in APG itself.

The short history of APG classification makes interesting reading. A few trends are also emerging fast. The monocots are better placed under two groups, the commelinids and the rest of the monocots. These two groups find their place after primitive angiosperms (and possibly the Magnoliids).

The 1998 edition of APG classification had 82 unplaced families, of which 12 were placed towards the beginning and 25 towards the end, and 45 unplaced in the informal ten groups. In addition, 18 families classified in four orders, placed in the beginning did not have any taxon at supraordinal rank. The number of these unplaced families was reduced to 48 (10 without any place and 39 under informal groups without ordinal placement) in APG II (in addition to 7 unplaced genera). APG III reduced unplaced families to mere 10 (two of uncertain position and eight under informal groups), and 3 genera are of uncertain position. APG IV has only one family and one genus unplaced. A Cladogram presented by APG IV, depicting the relationship of orders, informal higher groups, and some families is presented in Figure 10.26.

Figure 10.25: Kåre Bremer who first proposed a cladistic classification of angiosperms and has played a leading role in development of APG classification along with his wife **Birgitta Bremer**. These two, along with several colleagues of the Angiosperm Phylogeny Group, have been working at Phylogenetic classification (Published with permission from Kåre Bremer).

Table 10.14: Broad outline of APG IV (2016) classification of Angiosperm Phylogeny Group.

ANGIOSPERMS/FLOWERING PLANTS			
Group	**Order**	**Group**	**Order**
1. Basal Angiosperms			
	1. Amborellales		4. Rosales
	2. Numphaeales		5. Fagales
	3. Austrobaileyales		6. Cucurbitales
2. Magnoliids		**9. COM Clade (Placement Uncertain)**	
	1. Canellales		1. Celastrales
	2. Piperales		2. Oxalidales
	3. Magnoliales		3. Malpighiales
	4. Laurales		
3. Unplaced Independent Lineage		**8a. Rosids continued**	
	1. Chloranthales		7. Geraniales
4. Monocots			8. Myrtales
	1. Acorales		9. Crossosomatales
	2. Alismatales		10. Picramniales
	3. Petrosaviales		11. Huertales
	4. Dioscoreales		12. Sapindales
	5. Pandanales		13. Malvales
	6. Liliales		14. Brassicales
	7. Asparagales	**10. Superasterids**	
	8. Arecales		1. Berberidopsidales
	9. Commelinales		2. Santalales
	10. Zingiberales		3. Caryophyllales
	11. Poales	**11. Asterids**	
Probable sister to Eudicots			1. Cornales
	1. Ceratophyllales		2. Ericales
5. Eudicots			3. Icacinales
	1. Ranunculales		4. Metteniusales
	2. Proteales		5. Garryales
	3. Trochodendrales		6. Gentianales
	4. Buxales		7. Boraginales
6. Core Eudicots			8. Vahliales
	1. Gunnerales		9. Solanales
	2. Dilleniales		10. Lamiales
7. Superrosids			11. Aquifoliales
	1. Saxifragales		12. Asterales
8. Rosids			13. Escalloniales
	1. Vitales		14. Bruniales
	2. Zygophyllales		15. Paracryphiales
	3. Fabales		16. Dipsacales
			17. Apiales

Unplaced: Independent lineage Chloranthaceae, genus Peltanthera Benth. unplaced under Gesneriaceae. **Uncertain Position:** COM Clade with three orders Celastrales, Oxalidales and Malpighiales inserted among Rosids.

Stevens in the periodically updated APweb (current version 14, July 2017; Table 10.15) has number of unplaced families reduced to just 2. He has added four orders of gymnosperms Cycadales, Ginkgoales, Pinales and Gnetales in APweb. The main tree (Figure 10.28) shows relationships of orders (there are no unplaced families in the beginning). There are also useful tree links which lead to the trees for individual orders.

Table 10.16 presents a comparison of the treatment given to the unplaced families (of APG or any other system) in five recent systems of classification. The number of unplaced families has been drastically reduced in recent treatments of APG IV (2016) and Stevens, APweb, version 14 (2017). Thorne does not belong to the Angiosperm Phylogeny Group,

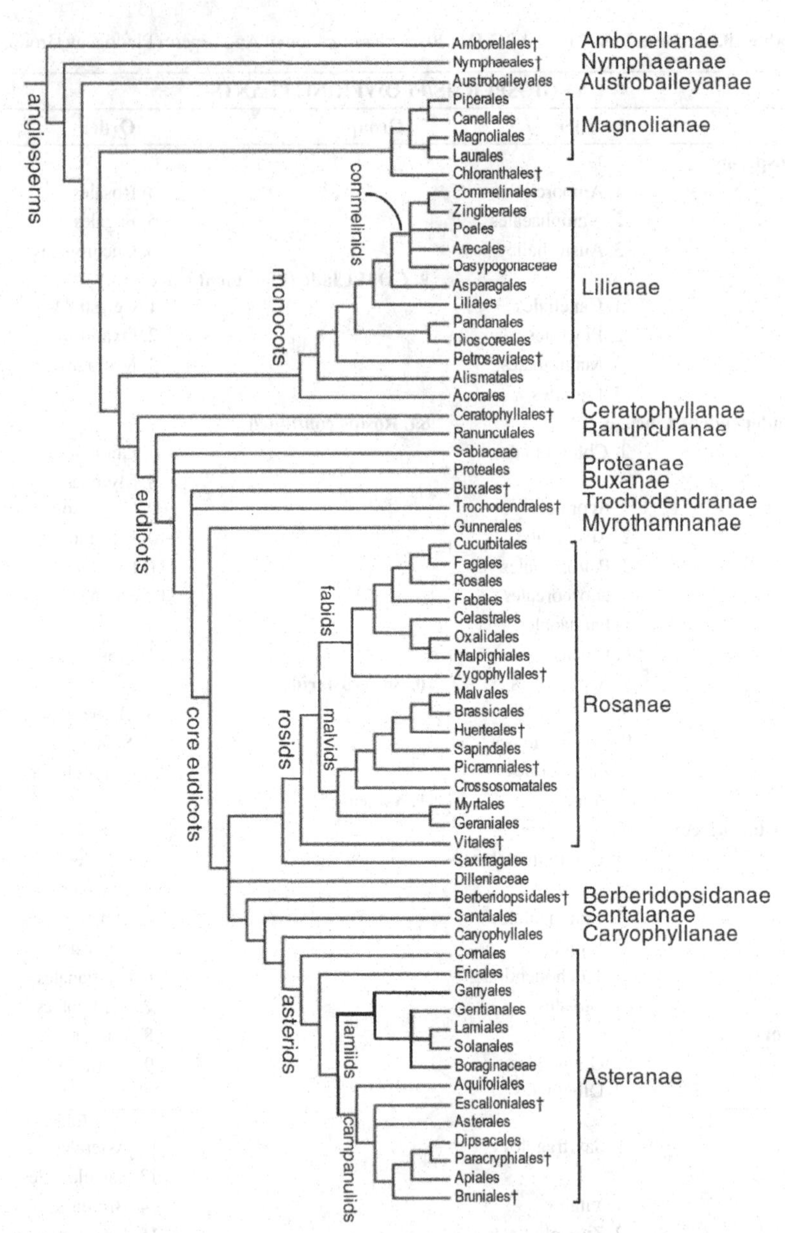

Figure 10.26: APG IV (2016) tree showing relationships of orders of Angiosperms. COM clade including Celastrales, Oxalidales and Malpighiales indicated by slashes. New orders added in APG IV indicated by †.

but has kept pace with the recent molecular developments, and continued to to balance hierarchical classification with the concept of monophyly, so sacred with the Angiosperm Phylogeny Group.

The true phylogenetic picture of angiosperms is still far from clear. There seems to be considerable unanimity in removing some of the primitive herbaceous families of Magnoloid complex such as Amborellaceae, Cabombaceae, Paeoniaceae, Austrobaileyaceae, Trimeniaceae, Illiciaceae, and Schizandraceae and place them towards the beginning of the angiosperms. This renders both monocots and eudicots as truly monophyletic groups. The position of Magnoliids (after the removal of herbaceous families) seems to be settled before monocots. Earlier versions of APweb (up to Version 6) and first edition of Judd et al. (1999) of their book, had placed Magnoliids after monocot, but from version 9 of APweb (2008) in the second (2002) and Third (2008) editions of Judd et al., like APG II and APG III, Magnoliid complex is placed before Monocots.

Table 10.15: Broad outline of APweb (version 14, July 2017) classification of Seed plants presented on Angiosperm Phylogeny website of P. F. Stevens. Update November, 2018.

Seed Plants			
Group	**Order**	**Group**	**Order**
Gymnosperms			
	1. Cycadales		3. Pinales
	2. Ginkgoale		4. Gnetales
Angiosperms/Flowering Plants			
	1. Amborellales	**7. Rosid/Fabid/Rosid I**	
	2. Nymphaeales		1. Zygophyllales
	3. Austrobaileyales		2. Celastrales
	4. Chloranthales		3. Oxalidales
1. Magnoliids			4. Malpighiales
	1. Magnoliales	**8. N-Fixing Clade**	
	2. Laurales		1. Fabales
	3. Canellales		2. Rosales
	4. Piperales		3. Cucurbitales
2. Monocots			4. Fagales
	1. Acorales	**9. Malvid/Rosid II**	
	2. Alismatales		1. Geraniales
	3. Petrosaviales		2. Myrtales
	4. Dioscoreales		3. Crossosomatales
	5. Pandanales		4. Picramniales
	6. Liliales		5. Sapindales
	7. Asparagales		6. Huerteales
3. Commelinids			7. Malvales
	1. Arecales		8. Brassicales
	2. Poales		9. Berberidopsidales
	3. Commelinales		10. Santalales
	4. Zingiberales		11. Caryophyllales
	5. Ceratophyllales	**10. Asterids**	
4. Eudicots			1. Cornales
	1. Ranunculales		2. Ericales
	2. Proteales	**11. Euasterids/Lamiid/Asterid I**	
	3. Trochodendrales	1. Icacinales	5. Solanales
	4. Buxales	2. Mettenuisales	6. Boraginales
5. Core Eudicots		3. Garryales	7. Lamiales
	1. Gunnerales	4. Gentianales	8. Vahliales
6. Pentapetalae			
	1. Dilleniales	**12. Campanulid/Asterid II**	
	2. Saxifragales	1. Aquifoliales	5. Apiales
	3. Vitales	2. Asterales	6. Paracryphiales
		3. Escalloniales	7. Dipsacales
		4. Bruniales	

Unplaced families: 1 in Saxifragales (Cynomoriaceae), 1 in Cucurbitales (Apodanthaceae), 1 in Santalales (Balanophoraceae), 1 in Asterid I (Vahliaceae).

Perhaps we have broken the jinx of dicot-monocot grouping of angiosperms, position of Magnoliid complex, which includes some of the most primitive representatives of angiosperms, also seems to be settled. The position of Piperales seems to be more or less stabilized towards the end of Magnoliids, but family Ceratophyllaceae has still to find a stable position. Judd et al., place the family under order Ceratophyllales after Piperales within Magnoliid complex with uncertain

Figure 10.27: Peter F. Stevens of Missouri Botanic Garden, who has been upgrading his APweb classification at the Angiosperm Phylogeny website.

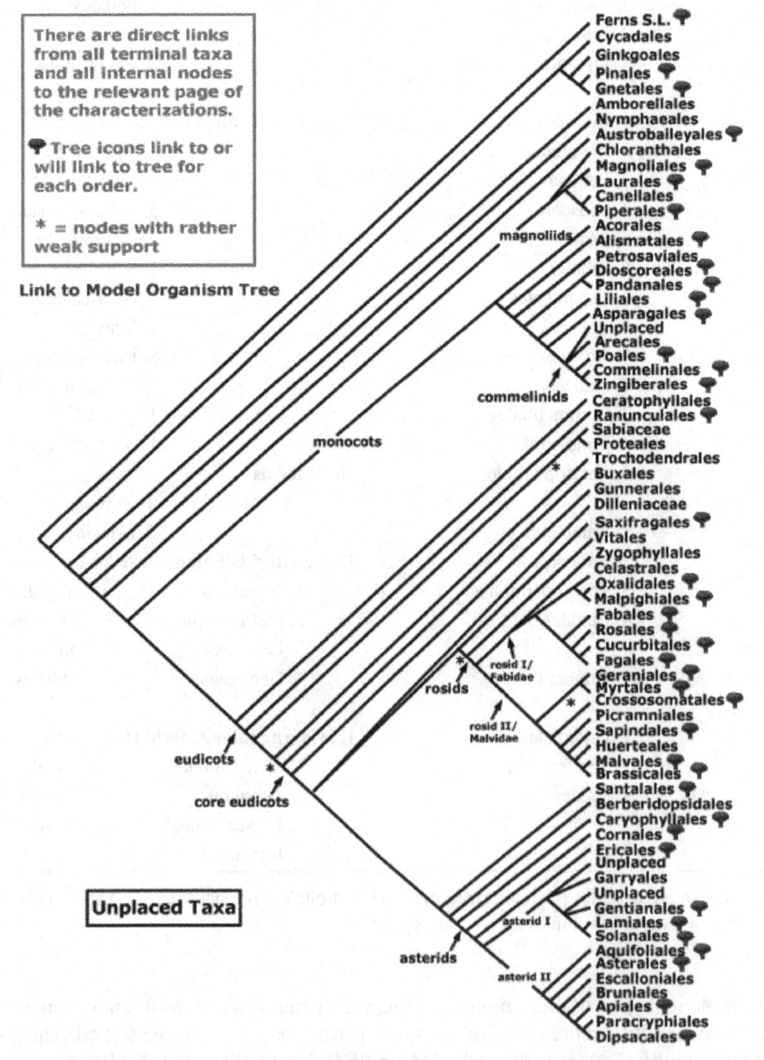

Figure 10.28: Main tree showing relationships between various orders of angiosperms and the informal higher clades presented by Stevens in APweb (version 14, 2017). Tree icons link to the trees of the respective orders on the web (Reproduced with permission from P. F. Stevens).

Table 10.16: Comparison of most recent phylogenetic systems of classification indicating major groups and the number of unplaced taxa compared to APG, 1998.

APG (1998)	APG IV (2016)	APweb (2017)	Takhtajan (2009)	Thorne (2007)
*(@ 12 families + 4 orders)	***(4 orders)	* (4 orders)	Magnoliidae Nymphaeanae	Chloranthidae
	Magnoliids	Magnoliids	Magnolianae	Magnoliidae
Monocots (@ 5)	Monocots	Monocots	!Alismatidae !Liliidae	!Alismatidae !Liliidae
Commelinoides (@6)	Commelinids	Commelinids	!Arecidae !Commelinidae	Commelinidae
Eudicots (@ 4)	Eudicots	Eudicots	Ranunculidae	Ranunculidae
Core Eudicots (@ 6)	Core Eudicots	Core Eudicots	Hamamelidae	Hamamelididae
	Superrosids	Pentapetalae	(@ 1)Caryophylidae	Caryophyllidae
Rosids (@ 7)	Rosids	Rosids	Rosidae	Rosidae
Eurosids I (@)	Fabids	Rosid I	Rosidae	Rosidae
	COM Clade**	N-Fixing Clade (@ 1)		
Eurosids II (@ 1)	Rosids	Rosid II (@ 1)	Dilleniidae	Malvidae
Asterids	Asterids	Asterids	Ericanae Asteridae Cornanae Asteranae	Asteridae Cornanae Ericanae !Aralianae !Asteranae
Euasterids I (@ 3)	Asterids	Asterid I (@ 1) (EuAsterid/Lamiid)	Lamiidae Lamianae	Lamiidae Solananae Lamianae
Euasterids II	Asterids	Asterid II (Campanuliid)	Asteridae partly	Asteridae partly
** (25 families)	** (three orders)	** (2 families, 1 genus)	** (1 family)	** (4 genera)

*** Unplaced at the start of Angiosperms, orders have no superordinal groups**
**** Families of uncertain position**
***** 3 Placed under Basal Angiosperms, Chloranthales unplaced independant lineage after Magnoliids**
@ Families unplaced in various groups
! Thorne and Takhtajan place Asteridae before Lamiidae which covers members of Euasterids I. The position of APG classification and these two authors is reversed to some degree. Takhtajan differs in including all monocots after dicots, sticking to traditional placement.

position, APG II place it before monocots, towards the beginning of angiosperms, and APweb towards end of Commelinids, a position followed by APG III, although regarded as sister to eudicots.

Merits

This newly emerging system of classification, which has undergone dramatic modification over the last fifteen years and is fast evolving, due to concerted efforts of a group of dedicated workers has several merits in APG IV (and APweb IV):

1. The system is based on the sound phylogenetic principle of constructing taxa based on established **monophyly**.
2. The system is based on a synthesis of information from mainly morphology, anatomy, embryology, phytochemistry and more strongly on molecular studies.
3. Formal group names have been given mostly only up to the level of the **order**, where monophyly of the group has been firmly established.
4. The traditional division of angiosperms has been abandoned and various monocot taxa placed in between primitive angiosperms and eudicots, thus overcoming the problem of paraphyly in the earlier recognized two groups monocot and dicots.

5. Although no formal names have been given for groups above the rank of order, there is constant endeavour to construct supraordinal monophyletic **clades**.
6. A number of cladograms are being presented for general affinities between various groups of angiosperms based on molecular as also on information from other fields.
7. The families with several primitive features are placed towards the beginning of angiosperms. The family Amborellaceae, which is unique in angiosperms in having granular and not tectate ectexine is placed at the start.
8. The number of **unplaced families** in various informal groups and uncertain families towards the end have been sufficiently reduced in APG IV (2016) and APweb (2017), finding ordinal places for many unplaced families of APG (1998).
9. The merger of Budlejaceae and Myoporaceae with Scrophulariaceae has the support of morphological studies of Bremer et al. (2001) and molecular (three gene analysis) by Olmstead et al. (2001).
10. Winteraceae and Canellaceae are brought together under the same order. Their affinities are strongly supported by morphological studies and multigene analyses.
11. Liliaceae of Hutchinson and earlier authors has been split to form several monophyletic families such as Liliaceae, Alliaceae, Asparagaceae, Asphodelaceae, etc.
12. Circumscription of Agavaceae has been further strengthened to include other genera like *Hosta, Camassia* and *Chlorogalum,* which also have bimodal karyotype. The placement has been adopted by Judd et al. (2002, 2008), Thorne and APweb.
13. Circumscription of Malvaceae has been broadened to also include Tiliaceae, Sterculiaceae and Bombacaceae, thus forming monophyletic Malvaceae, as supported by morphological and molecular evidence.
14. The merger of Asclepiadaceae with Apocynaceae has been strengthened by molecular evidence Judd et al. (1994) and Sennblad and Bremer (1998). Recognition of distinct Asclepiadaceae would render Apocynaceae as paraphyletic (Judd et al., 2002, 2008).
15. Family Capparaceae which was earlier merged with Brassicaceae, has now been recognized as distinct family. The Chloroplast sequence data points to the separation of these two families as also Cleomacae.
16. The concept of bracketed families in APG II, and which received a lot of criticism has finally been abandoned in APG III and APG IV.

Demerits

Although the system is still evolving and continuously improving, and will take a considerable time before it stabilizes and is tested by various parameters, a few shortcomings are obvious:

1. Classification having not proceeded below the family level, the system is not useful in practice and for adoption in herbaria and floras.
2. Although a large number of families have been assembled into more or less monophyletic orders, there still exist a number of unplaced families, and a few unplaced genera in both APG IV and APweb.
3. Although most of the orders have been assembled into informal groups, no proper names conforming to the Botanical Code have been given for these groups.
4. Angiosperms have been given the rank of a division, but there are no formal taxa between the rank of an order and division, a rather unusual phenomena for classification systems.
5. Placement of Celastrales, Oxalidales and Malpighiales is unceratain in APG IV. As such they have been placed under COM Clade among Rosids. APweb places them under Rosid 1.
6. Chloranthales including family Chloranthaceae is an independant unplaced lineage placed between Magnoliids and Monocots in APG IV.
7. A large number of Taxa are listed in APweb in being in urgent need of study.

The developments of the last few years have seen clear emergence of a few facts. The angiosperms are no longer to be divided into traditional dicots and monocots. Commelinids are distinct from other monocots, and these two, forming the traditional monocots are better placed between primitive angiosperms and the Eudicots. Primitive angiosperms include paleoherbs (Nymphaeaceae, Cabombaceae, Piperaceae, Amborellaceae, etc.) and true magnoliids (Magnoliales, Laurales, etc.) are better placed before monocots. Thorne (1999, 2000, 2003, 2006, 2007) has come up with a major revision of his classification, bringing it on lines of APG, but maintaining the hierarchical structure, and finding a place almost all families, with only 4 genera remaining unplaced. It is also interesting to note that his eleven subclasses are more or less complementary to the eleven informal groups of APG-II and APG III, the relationship somewhat reversed in Asterids.

Stevens (2018, APweb update) stresses the need for a more comprehensive understanding of plant morphology is needed to enable broad evolutionary questions to be answered with a greater degree of confidence. There is a particular need for targeted surveys of the morphological variation of particular groups. Given limited time and resources, one can also profitably focus on small clades an understanding of the morphology of which will have the greatest effect on our understanding of character evolution. Neither phylogeny without good comparative morphological knowledge nor comparative morphology without a good phylogeny is of much use.

Recent years have also seen major focus on genomic studies and utilization of information for better understanding of phylogeny of living organisms, which will go in long way in better classification systems. **Open Tree of Life** is one such collaborative effort funded by the National Science Foundation, a United States Government Agency. Open Tree of Life aims to construct a comprehensive, dynamic and digitally-available tree of life by synthesizing published phylogenetic trees along with taxonomic data. The project is a collaborative effort between 11 PIs across 10 institutions. Initial draft in 2015 included 2.5 million taxa, that has jumped to 3,453,840 in September 2016 count (OTT 2.10) and 3,594,550 in February 2017 count (OTT 3.0). The Viruses, which were included in earlier counts have now been excluded. Many names have been relegated to synonymy, and count 1842403, a big increase from earlier versions. Taxonomy module of the project hosted at https://tree.opentreeoflife.org/about/taxonomy-version/ott3.0. The Interactive graph allows the user to zoom in to taxonomic classifications, phylogenetic trees, and information about a node. Clicking on a species will return its source and reference taxonomy.

Chapter 11

Families of Pteridophytes

Pteridophytes, Gymnosperms and Angiosperms constitute Tracheophytes, a dominant group of green plants, characterized by the presence of a well-developed branched, independent and dominant sporophyte, with a vascular system consisting of xylem (tracheids-hence the name Tracheophytes, vessels in angiosperms) and phloem (sieve elements, sieve tubes in angiosperms). The group evolved nearly 420 million years ago and is regarded as monophyletic. Pteridophytes, the **seedless vascular plants** differ from higher Tracheophytes in lacking seed habit and absence of pollen tube, spores developing freely into gametophytes, although few members exhibit heterospory and the reduction of megaspore number to one, forerunner of seed habit.

CLASSIFICATION

Pteridophytes form a complex heterogenous group reflecting antiquity and divergent evolutionary clades and have been classified variously. Engler and Prantl recognized Bryophyta and Pteridophyta as two subdivisions of the division Embryophyta Asiphonogama. Cronquist et al. (1966) recognized four groups within Pteridophytes, each given the rank of a division: Psilophyta, Lycopodiophyta, Equisetophyta and Polypodiophyta

Bold, Alexopoulos and Delevoryas (1987) included the same four group but preferred name Microphyllophyta for Lycopodiophyta and Arthrophyta for Equisetophyta. Recent evidence indicates that Pteridophytes often separated under 'Ferns and Fern Allies', form a paraphyletic assemblage of groups, which represent distinct evolutionary lines and are lumped together for convenience. Recent genetic data has shown that Lycopodiophyta are only distantly related to other vascular plants, having radiated evolutionarily at the base of vascular plant clade, whereas Psilophyta and Equisetophyta are much closer to true ferns.

Classification of Smith et al.

A more recent classification of Smith et al. (2006), based on morphology as well as molecular data as such excludes Lycopodiophytes from Ferns. The classification recognizes 4 classes, 11 orders and 37 families under ferns. They estimated Peridophytes include nearly 10280 species, 1280 belonging to lycophytes and nearly 9000 to ferns:

- **Class: Psilotopsida**
  - **Order: Ophioglossales**
    - Family: **Ophioglossaceae**
  - **Order: Psilotales**
    - Family: **Psilotaceae**
- **Class: Equisetopsida**
  - **Order: Equisetales**
    - Family: **Equisetaceae**
- **Class: Marattiopsida**
  - **Order: Marratiales**
    - Family: Marratiaceae
- **Class: Polypodiopsida**
  - **Order: Osmundales**
    - Family: **Osmundaceae**
  - **Order: Hymenophyllales**
    - Family: Hymenophyllaceae
  - **Order: Gleicheniales**
    - Family: Gleicheniaceae
    - Family: Dipteridaceae
    - Family: Matoniaceae
  - **Order: Schizaeales**
    - Family: Lygodiaceae
    - Family: Anemiaceae
    - Family: Schizaeaceae
  - **Order: Salviniales**
    - Family: **Marsileaceae**
    - Family: **Salviniaceae**
  - **Order: Cyatheales**
    - Family: Thyrsopteridaceae
    - Family: Loxomataceae
    - Family: Culcitaceae
    - Family: Plagiogyriaceae
    - Family: Cibotiaceae
    - Family: **Cyatheaceae**
    - Family: Dicksoniaceae
    - Family: Metaxyaceae
  - **Order: Polypodiales**
    - Family: Lindsaeaceae
    - Family: Saccolomataceae
    - Family: Dennstaedtiaceae
    - Family: Pteridaceae
    - Family: **Aspleniaceae**
    - Family: Thelypteridaceae
    - Family: Woodsiaceae
    - Family: Blechnaceae
    - Family: Onocleaceae
    - Family: **Dryopteridaceae**
    - Family: Lomariopsidaceae
    - Family: Tectariaceae
    - Family: Oleandraceae
    - Family: Davalliaceae
    - Family: **Polypodiaceae**

Consensus Classification of Pteridophytes (Christenhusz and Chase, 2014. Ophioglossaceae shifted to Ferns. Total 24 families are recognized).

- **Division: Lycopodiophyta** (Lycopods)
  - **Subclass: Lycopodiidae**
    - **Order: Lycopodiales**
      - Family: **Lycopodiaceae**
    - **Order: Selaginellales**
      - **Family: Selaginellaceae**
    - **Order: Isoetales**
      - Family: **Isoetaceae**
- **Division: Polypodiophyta** (Ferns)
  - **Subclass: Equisetidae**
    - **Order: Equisetales**
      - Family: **Equisetaceae**
  - **Subclass: Ophioglossidae**
    - **Order: Ophioglossales**
      - Family: **Ophioglossaceae**
    - **Order: Psilotales**
      - Family: **Psilotaceae**
  - **Subclass: Marattiidae**
    - **Order: Marattiales**
      - Family: Marattiaceae
  - **Subclass: Polypodiidae**
    - **Order: Osmundales**
      - Family: **Osmundaceae**
    - **Order: Hymenophyllales**
      - Family: Hymenophyllaceae
    - **Order: Gleicheniales**
      - Family: Gleicheniaceae
      - Family: Dipteridaceae
      - Family: Matoniaceae
    - **Order: Schizaeales**
      - Family: Schizaeaceae
    - **Order: Salviniales**
      - Family: **Marsiliaceae**
      - Family: **Salviniaceae**
    - **Order: Cyatheales**
      - Family: **Cyatheaceae**

***Bold typed families are described and illustrated**

Pteridophyte Phylogeny Group (PPG)

Largely based on patterns of Angiosperm Phylogeny Group (APG), Smithsonian National Museum of Natural History has instituted **Pteridophyte Phylogeny Group (PPG)** in 2014, which aims to produce, and continually update, a community-derived classification for lycophytes and ferns-based on our understanding of phylogeny—at the family and genus levels.

PPG I (2016) proposed by Pteridophyte Phylogeny Group.

Class: Lycopodiopsida
- **A. Order: Lycopodiales**
  - 1. Family: **Lycopodiaceae**
- **B. Order: Isoetales**
  - 2. Family: **Isoëtaceae**
- **C. Order: Selaginellales**
  - 3. Family: **Selaginellaceae**

Class: Polypodiopsida
- **Subclass: Equisetidae**
  - **D. Order: Equisetales**
    - 4. Family: **Equisetaceae**
- **Subclass: Ophioglossidae**
  - **E. Order: Psilotales**
    - 5. Family: **Psilotaceae**
  - **F. Order: Ophioglossales**
    - 6. Family: **Ophioglossaceae**
- **Subclass: Marattiidae**
  - **G. Order: Marattiales**
    - 7. Family: Marattiaceae
- **Subclass Polypodiidae**
  - **H. Order: Osmundales**
    - 8. Family: **Osmundaceae**
  - **I. Order: Hymenophyllales**
    - 9. Family: Hymenophyllaceae
  - **J. Order: Gleicheniales**
    - 10. Family: Matoniaceae
    - 11. Family: Dipteridaceae
    - 12. Family: Gleicheniaceae
  - **K. Order: Schizaeales**
    - 13. Family: Lygodiaceae
    - 14. Family: Schizaeaceae
    - 15. Family: Anemiaceae
  - **L. Order: Salviniales**
    - 16. Family: **Salviniaceae**
    - 17. Family: **Marsileaceae**
    - 18. Family: Thyrsopteridaceae
    - 19. Family: Loxsomataceae
    - 20. Family: Culcitaceae
    - 21. Family: Plagiogyriaceae
    - 22. Family: Cibotiaceae
    - 23. Family: Metaxyaceae
    - 24. Family: Dicksoniaceae
    - 25. Family: **Cyatheaceae**
  - **N. Order: Polypodiales**
    - **Suborder: Saccolomatineae**
      - 26. Family: Saccolomataceae
    - **Suborder: Lindsaeineae**
      - 27. Family: Cystodiaceae
      - 28. Family: Lonchitidaceae
      - 29. Family: Lindsaeaceae
    - **Suborder: Pteridineae**
      - 30. Family: **Pteridaceae**
    - **Suborder: Dennstaedtiineae**
      - 31. Family: Dennstaedtiaceae
    - **Suborder: Aspleniineae**
      - 32. Family: Cystopteridaceae
      - 33. Family: Rhachidosoraceae
      - 34. Family: Diplaziopsidaceae
      - 35. Family: Desmophlebiaceae
      - 36. Family: Hemidictyaceae
      - 37. Family: **Aspleniaceae**
      - 38. Family: Woodsiaceae
      - 39. Family: Onocleaceae
      - 40. Family: Blechnaceae
      - 41. Family: Athyriaceae
      - 42. Family: Thelypteridaceae
    - **Suborder: Polypodiineae**
      - 43. Family: Didymochlaenaceae
      - 44. Family: Hypodematiaceae
      - 45. Family: **Dryopteridaceae**
      - 46. Family: Nephrolepidaceae
      - 47. Family: Lomariopsidaceae
      - 48. Family: Tectariaceae
      - 49. Family: Oleandraceae
      - 50. Family: Davalliaceae
      - 51. Family: **Polypodiaceae**

Bold typed families are described and illustrated.

The ordinal and familial classification of ferns by Smith et al. (2006), founded on the principle of monophyly, but also recognizing the desire to maintain established names, has generally been well received by the pteridological community. However, many important advances have since been made in our understanding of relationships. These changes, of course, are not accounted for in the Smith et al. (2006) classification. Furthermore, although this work provided a list of genera included in each family, it did not explicitly address generic classification.

The first version of this classification **PPG I** was published in 2016. It provides a modern, comprehensive classification for lycophytes and ferns, down to the genus level, utilizing a community-based approach, using monophyly as the primary criterion for the recognition of taxa, but also aim to preserve existing taxa and circumscriptions that are both widely accepted and consistent with our understanding of pteridophyte phylogeny. In total, this classification treats an estimated 11916 species (1338 lycophyte species and 10578 fern species) in 337 genera, 51 families, 14 orders, and two classes.

The classification focuses on extant lycophytes and ferns. Although some fossils could easily be accommodated, the phylogenetic affinities of most extinct plants are rather unclear. Many fossil taxa represent distinct evolutionary lineages and their inclusion in the PPG I classification would not only require revised circumscriptions, but almost certainly the recognition of new families, orders, and even classes. The classification presented is community-derived, and the taxa recognized have the support of most of the contributors.

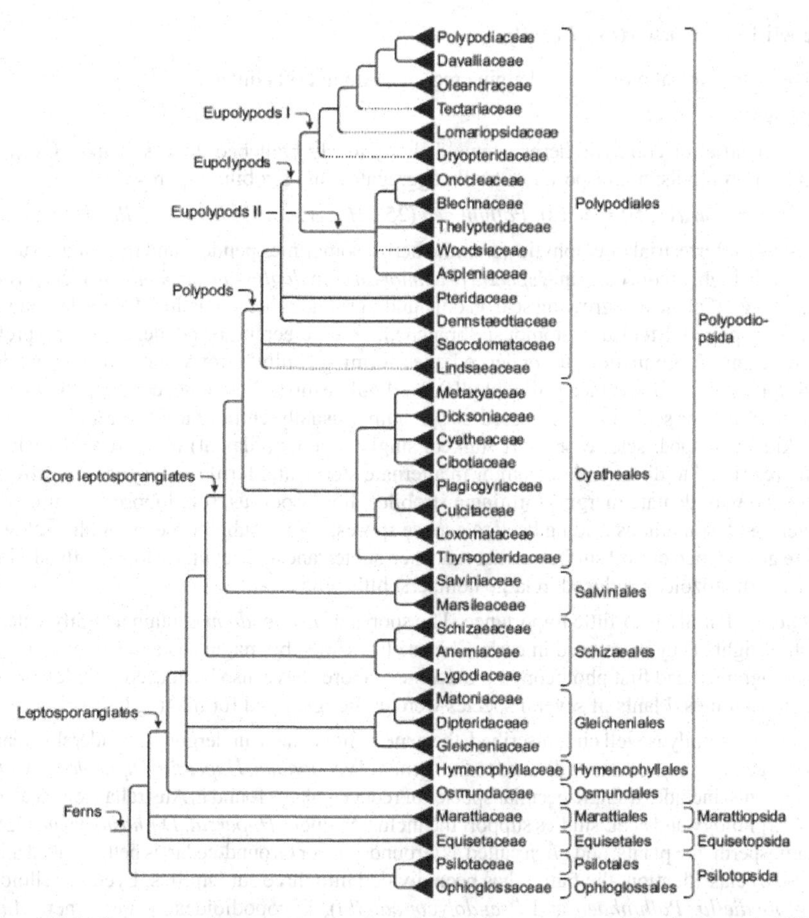

Figure 11.1: Consensus phylogeny representing relationships among ferns (after Smith et al., 2006).

The classification recognizes two classes with pteriphytes: **Lycopodiopsida** (lycophytes) and **Polypodiopsida** (ferns). These are distinct lineages within the tracheophyte tree of life, with ferns resolved as more closely related to seed plants than to lycophytes. Lycopodiopsida is a relatively smaller group divided into three orders: Lycopodiales (16 genera), Isoëtales and Selaginellales (single genus each). Polypodiopsida, on the other hand represents a large group with four subclasses: Equisetidae (horsetails); Ophioglossidae; Marattiidae; and Polypodiidae, last being the largest group with seven orders. Whereas the first six orders include 19 families, the last largest order Polypodiales includes 26 families distributed among 6 suborders.

PPG I classification is compatible with a scheme recently proposed by Ruggiero et al. (2015) who, however, treated lycophytes and ferns as distinct subphyla (Lycopodiophytina and Polypodiophytina), each with a single class (Lycopodiopsida and Polypodiopsida). They also agree in recognizing four subclasses within ferns. Pteridophyte classification has been exceedingly stable at the ordinal level over the past decade. Smith et al. (2006b) recognized 11 orders within ferns, one for each of four major eusporangiate lineages and seven within leptosporangiates. Christenhusz et al. (2011) and Christenhusz and Chase (2014) adopted these same 11 fern orders plus three orders for lycophytes (not treated by Smith et al., 2006b). Unlike these schemes, PPG I is much more comprehensive, listing genera also, recognize 337 pteridophyte genera, representing a 50% increase over the 223 genera included in the pre-molecular system of Kramer and Green (1990).

The PPG I classification does not provide species lists; however, it presents estimates of species number for each genus. The numbers are remarkably similar to the family-based sums provided by Christenhusz and Chase (2014), 1300 lycophytes and 10535 ferns.

PPG I classification has largely relied on two paths of resolution, when existing generic concepts are found to conflict with the results of molecular phylogenetic analyses. Using monophyly as criterion, resolution is achieved through **disintegration** (i.e., the "splitting" of a non-monophyletic genus into multiple monophyletic genera to maintain a nested genus); or **integration** (i.e., the "lumping" of a nested genus with the genus it renders non-monophyletic).

Lycopodiaceae Mirbel ***Club Moss Family***

Cosmopolitan, diverse in tropical montane and alpine regions, rare in arid climate.

16 genera, 388 species

Salient features: Terrestrial or epiphytic herbs, stems dichotomously branched; leaves simple, 1-veined, non-ligulate; sporangia in axils of sporophylls, homosporous, usually aggregated into strobilus; spores with 3-branched scar.

Major Genera: *Phlegmariurus* (250 species), *Palhinhaea* (25), *Huperzia* (25) *Lycopodiella* (15) and *Lycopodium* (15).

Description: Non-woody terrestrial or epiphytic perennial herbs, sometimes pendent and reaching up to 2 m, arising from creeping rhizome, rarely highly reduced as in *Huperzia drummondii* (*Phylloglossum drummondii*) scarcely exceeding 10 cm and all aerial parts dying off at the end growing season only underground tuber persisting. **Roots** dichotomously branched, adventitious, root hairs paired. **Stem** dichotomously branched, erect, creeping or pendent, slender, protostelic; unequal dichotomy often resulting in production of condensed axes forming 'bulbils' for vegetative propagation (*Lycopodium selago*). **Leaves** simple, small and 1-veined (microphylls), nonligulate, up to 2 cm long, covering the stem densely, spirally arranged or opposite, linear or scale-like, appressed or spreading, usually entire, rarely serrate (*Lycopodium serratum*). **Sporangia** large, kidney-shaped, sessile or short stalked, singly in axils (adaxial) of sporophylls which are similar to foliage leaves (and restricted to distal end of stem or in alternate sterile and fertile zones) or well differentiated (smaller than foliage leaves and with dentate margin) forming a strobilus, homosporous, subglobose to reniform, shortly stalked, opening by transverse slit, sometimes folding back to expose spores; spores subglobose or tetrahedral, with a 3-branched scar. **Gametophyte** green when on soil surface, nongreen when subterranean, irregularly lobed, often living up to 25 years; antheridia sunken, spermatozoids produced in large numbers, biflagellate.

Economic importance: Family is of little importance. The spores of *Lycopodium* contain a highly volatile oil and ignite rapidly into a flash of light. They were used in early Chinese fireworks, by magicians and sorcerers in Middle ages, and as flash in early photography, and first photocopying machines. Spores have also been used as industrial lubricants and in surgical gloves and condomes. Plants of several species were earlier gathered for making Christmas wreaths.

Phylogeny: Although the family is well circumscribed, the generic limits have undergone considerable readjustment. Often treated under a single genus *Lycopodium*, or divided to five genera *Lycopodium*, *Huperzia*, *Diphasiastrum*, *Lycopodiella* and *Phylloglossum*. Last genus includes a single peculiar species of reduced plants found in Australia New Zealand and Tasmania, but gametophyte morphology and *rbc*L studies support the inclusion under *Huperzia*. *Diphasiastrum*, a genus of nearly 20 species of low gymnosperm like plants, and often called as Ground-pine or Ground-cedar, is better placed under *Lycopodium*. In recent PPG I (2016) classification, the family has been divided into three subfamilies: Lycopodielloideae (four genera *Lateristachys, Lycopodiella, Palhinhaea* and *Pseudolycopodiella*), Lycopodioideae (nine genera *Austrolycopodium, Dendrolycopodium, Diphasiastrum, Diphasium, Lycopodiastrum, Lycopodium, Pseudodiphasium, Pseudolycopodium,* and *Spinulum*) and Huperzioideae (three genera *Huperzia, Phlegmariurus* and *Phylloglossum*). The family is very old in fossil record, dating back to 380 MYA, mostly dominated by lycopod trees.

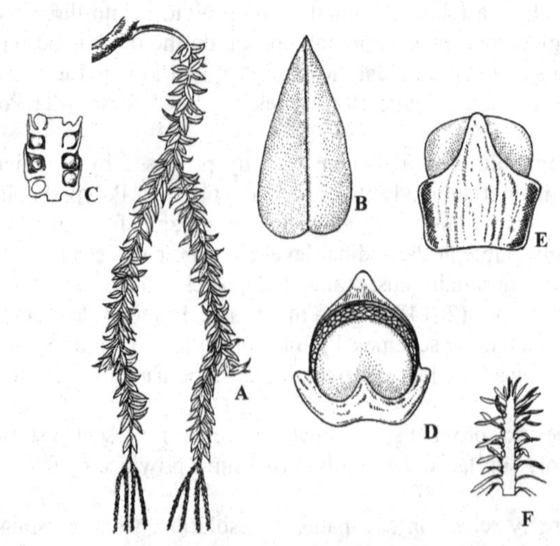

Figure 11.2: Lycopodiaceae. *Lycopodium phlegmaria.* **A:** Pendulous branch with terminal strobili; **B:** Vegetative leaf; **C:** Longitudinal section of strobilus; **D:** Sporophyll in adaxial view; **E:** Same in abaxial view; **F:** Vertical section of fetile branch of *L. lucidulum* showing sporangia in axils of unmodified sporophylls.

Selaginellaceae Wilk. *Spike Moss Family*

Worldwide, predominant in tropics, a few species extending to temperate and Arctic region.

1 genus, 700 species

Salient features: Terrestrial or epiphytic herbs, stems dichotomously branched, producing rhizophores; leaves simple, 1-veined, ligulate, dimorphic, often 4-ranked; sporangia in axils of sporophylls, heterosporous, aggregated into usually 4-angled strobilus.

Genera: Single genus *Selaginella* (700 species).

Description: Mainly terrestrial perennial plants, usually low-growing, mostly growing in moist habitats, creeping (*S. kraussiana*) or erect (*S. rupestris*), few in semiarid regions, rarely epiphytic. **Roots** dichotomously branched, arising from distal end of rhizophore, adventitious; rhizophores arising from stem. **Stem** dichotomously branched, with both creeping and short erect branches or with only erect branches, protostelic, with vessels. **Leaves** simple, small and 1 veined (microphylls), ligulate (ligule axillary or near leaf base; tongue- or fan-shaped with hyaline sheath at base), up to 1 cm long, covering the stem densely, spirally arranged, isomorphic (all similar) or dimorphic and 4-ranked, with two upper or dorsal rows of smaller leaves and two lower or lateral rows of larger leaves. **Sporangia** borne in axils of well-differentiated sporophylls, usually on 4-angled terminal strobili, heterosporous, microsporangia and megasporangia occurring on same (with megasporangia in upper part of strobilus, microsporangia in lower part—*S. helvetica* or in two opposite rows—*S. oregana*) or different strobili; microsporangium with more than 100 microspores about 20–60 µ in diameter; megasporangium with 4 large megaspores about 200–600 µ in diameter; spores tetrahedral with prominent triradiate mark and with characteristic ornamentation. **Gametophyte** unisexual, developed within respective spore walls; spermatozoids biflagellate, smallest among vascular plants. Several species especially those growing in dry climate can survive long periods of drought due to small leaves covered with thick cuticle, and branches curling up into a ball. Such plants revive fast with availability of water and are known as resurrection plants (*S. bryopteris* of India (regarded by some as Sanjeevani booti of Ramayana legend), and *S. lepidophylla* of Mexico and Texas).

Economic importance: Family is of little importance. Only a few species are grown as ornamentals.

Phylogeny: The family with single genus is well differentiated from Lycopodiaceae in ligulate leaves and heterosporous habit. From Isoetaceae, it is differentiated by smaller leaves and superficial sporangia, leaves being 2–100 cm long, onion-like and sporangia initially embedded in Isoetaceae. Leaf dimorphism is considered as an adaptation to poor light, as the species commonly inhabits forest floor. The family dates back to the Late Devonian-Early Carboniferous. Weststrand and Korall (2016) who estimate 750 species in the family, based on morphological and DNA sequence data from both chloroplast (rbcL) and single-copy nuclear gene data (pgiC and SQD1) using a Bayesian inference approach concluded that seven major clades are identified, which each can be uniquely diagnosed by a suite of morphological features.

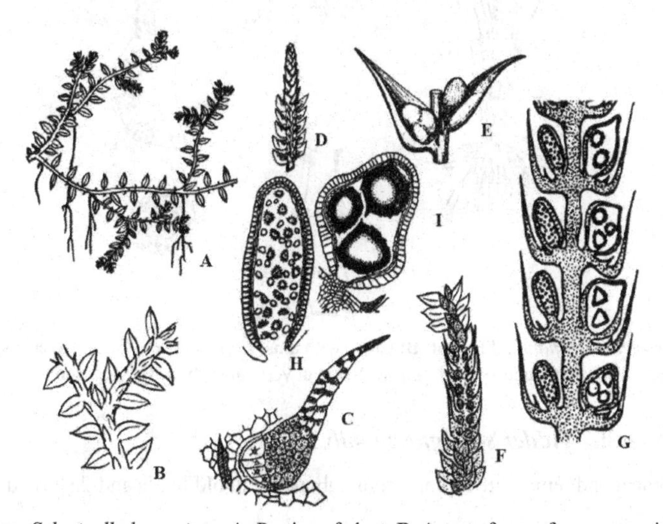

Figure 11.3: Selaginellaceae. *Selaginella kraussiana.* **A:** Portion of plant; **B:** A part of same from upper view showing arrangement of leaves; **C:** Vertical section of ligule; **D:** Portion of branch of *S. pallescens*; **E:** Small portion of same showing one megasporangium (left) and one microsporangium (right); **F:** Strobilus of *S. watsoni* proliferating at apex into vegetative shoot *S. oregana.* **G:** Vertical section of portion of strobilus; **H:** Vertical section of microsporangium; **I:** Vertical section of megasporangium.

Isoetaceae Reichenbach ***Quillwort Family***

Widely distributed in tropical and temperate Americas, Europe, Asia, Africa, Australia and New Zealand.

1 Genus, 150 species

Salient features: Terrestrial perennial herbs of marshy areas, stem short with secondary growth, leaves simple, long, quill-like, 1-veined, ligulate; sporangia borne singly, sunken at base of leaves, heterosporous.

Genera: Single genus *Isoetes* (150 species).

Description: Terrestrial tufted perennial plants usually found in marshy areas, often in periodically inundated pools. **Roots** firm, arising from grooves of lobed stem in radiating rows, unbranched or dichotomously branched, containing eccentric vascular strand and surrounding lacuna. **Stem** short, erect, cormose (corm-like), rarely rhizomatous, having secondary growth, protostelic (stele anchor-shaped with upturned lobes near base), lobed by a broad basal groove into 2–4 lobes, rough on sides due to sloughing off of cortical tissue. **Leaves** simple, linear, long, quill-like, resembling narrow-leaved species of *Allium*, up to 100 cm long, swollen at base, 1-veined (microphylls), containing 4 transversely septate longitudinal lacunae, a central collateral vascular strand and frequently several peripheral fibrous bundles, ligulate with ligule inserted above sporangium, hardened scales and phyllopodia surrounding the leaves; all leaves potential sporophylls. **Sporangia** solitary, borne adaxially embedded in cavity of swollen base of sporophylls, microsporophylls and megasporophylls usually borne in alternate cycles, sporangium covered partially or completely on adaxial side by velum; megasporangium with 50–300 megaspores; microsporangium with 0.15–1 million microspores; microspores elongate, about 45 μ long; megaspores 250–900 μ in diameter; spores set free by disintegration of sporangial wall, dehiscence zone not developed. **Gametophyte** unisexual, developed within respective spore walls (endosporic); microgametophyte 9-celled, antheridium single, spore wall cracking to release 4 multiflagellate spermatozoids each with terminal vesicle; megagametophyte with 1-several archegonia, often with rhizoids, exposed by cracking of spore wall.

Economic importance: None.

Phylogeny: The family with single genus is well differentiated from Lycopodiaceae in ligulate leaves and heterosporous habit. From Selaginellaceae, it is differentiated by much longer linear leaves, short erect cormose stem and sunken sporangia; leaves being less than 1 cm long, sporangia superficial and stem branched in Selaginellaceae. Fossil tree *Lepidodendron* is more closely related to *Isoetes* among the extant lycophytes.

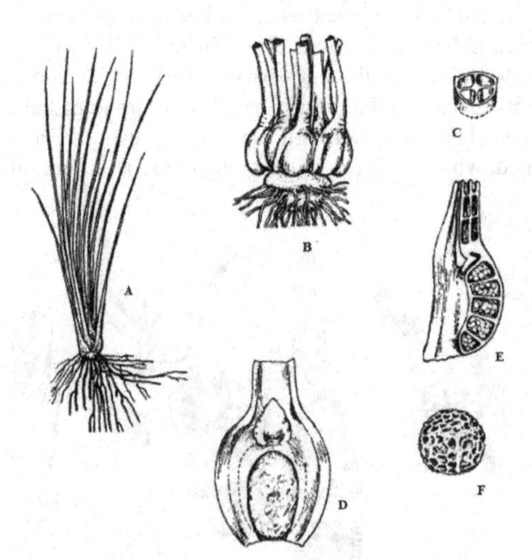

Figure 11.4: Isoetaceae. *Isoetes engelmanni.* **A:** Plant; **B:** Corm with attached leaf bases; **C:** Transverse section of leaf; **D:** Leaf base with ligule and sporangium; **E:** Same in vertical section; **F:** Megaspore.

Ophioglossaceae C. Agardh ***Adder's-tongue Family***

Widely distributed in tropical and temperate regions, more abundant in old fields and disturbed pastures.

10 Genera, 112 species

Salient features: Terrestrial perennial herbs with tubers or rhizomes, leaves with branched veins, fertile portion a simple or branched spike, arising from surface of leaf at the junction of petiole and blade, stipules present, sporangia homosporous, aggregated into sporophore, sporangium wall more than one cell thick, annulus absent, gametophytes subterranean.

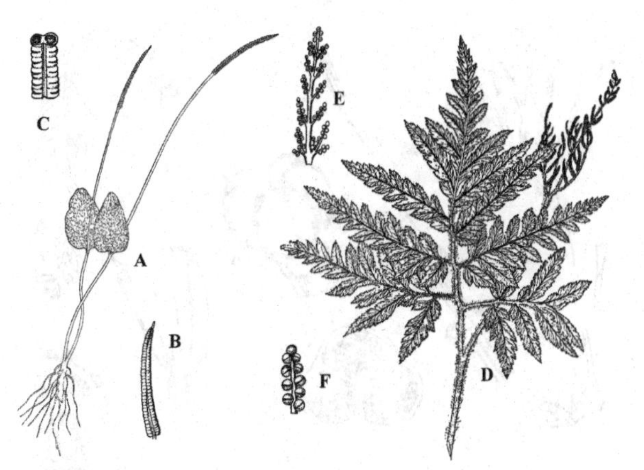

Figure 11.5: Ophioglossaceae. *Ophioglossum reticulatum.* **A:** Plant; **B:** Portion of fertile spike; **C:** Small portion of same enlarged. *Botrychium daucifolium*; **D:** Leaf with fertile branch; **E:** Portion of fertile branch; **F:** Small portion of same showing sporangia.

Major Genera: *Ophioglossum* (41 species), *Botrychium* (35), *Sceptridium* (25).

Description: Terrestrial perennial plants, very rarely epiphytic. **Roots** unbranched, adventitious, lacking root hairs. **Stem** short, tuberous or rhizomatous, aerial portion perishing after growing season, rarely evergreen (*Botrychium dissectum*, *B. multifidum*). **Leaves** simple (*Ophioglossum*) or more or less palmately compound (and looking like hand-*Cheiroglossa*) to many times pinnately compound (*Botrychium*), up to 50 cm long, unfolding lengthwise (conduplicate) and not circinate, with branched veins (euphyll or macrophyll), venation reticulate (*Ophioglossum*) or open dichotomous (*Botrychium*); single fertile branched (*Botrychium)* or simple (*Ophioglossum*) sporophore arising from surface of leaf (trophophore), more than one sporophore arising in *Cheiroglossa*; petiole fleshy with expanded sheathing base; stipules present, sheathing, persisting even after decay of leaves. **Sporangia** aggregated in fertile portion (sporophore), thick walled (eusporangiate-more than one celled thick), homosporous, not clustered in sori, separated (*Botrychium*) or forming synangia (*Ophioglossum*), exposed or embedded in spike-like sporophore, annulus absent; spores thousands per sporangium, chlorophyllous. **Gametophyte** subterranean, nongreen, mycorrhizal; antheridia and archegonia distributed over the surface of gametophyte; antheridia sunken producing multiflagellate spermatozoids in large numbers.

Economic importance: None.

Phylogeny: The family shares apomorphies of unbranched roots and absence of root hairs. The two may represent a transitional stage for total absence of these in Psilotaceae. The distinctness from other ferns and affinity with Psilotaceae is supported by DNA sequence data, although there is little morphological support. The sporophore represents a unique apomorphy of the family. The family represents highest chromosome number in plants ($n = 621 + 10$ fragments). *Mankyua chejuense* is a recently discovered genus and species found in a lowland swampy area on Cheju Island, off the southern coast of the Korean Peninsula (Sun et al., 2001). Its affinities with the Ophioglossaceae are fairly clear. It has creeping rhizome, alternate ternately compound leaf, open dichotomous venation and baculate spores like *Helminthostachs*; open dichotomous venation and horizontally dehiscent sporangia like *Botrychium*; fleshy spike, sunken horizontally dehiscent sporangia and vegetative leaf propagation like *Ophioglossum*; and nearly sessile sporophore like *Cheiroglossa*. *Mankyua* possesses a fertile spike (sporophore) originating dorsiventrally from the adaxial side of the vegetative trophophore, eusporangiate sporangia without annuli, and noncircinate leaf vernation. These three characters, especially the first, place it in the Ophioglossaceae. Also, recent DNA analyses confirm the placement of *Mankyua* in Ophioglossaceae (Sun, 2002). The family Ophioglossaceae is sometimes divided into two families: Ophioglossaceae and Botrychiaceae. The discovery of *Mankyua* unites the family. The mixture of character states presents in *Mankyua* combines the distinguishing states for each family, and thus abolishes the taxonomic boundaries between the two segregate families. PPG I has made significant changes recognising 4 subfamilies: Helminthostachyoideae (*Helminthostachys*), Mankyuoideae (*Mankyua*), Ophioglossoideae (*Cheiroglossa, Ophioderma, Ophioglossum* and *Rhizoglossum*) and 10 genera.

Psilotaceae Kanitz ***Whisk Fern Family***

Pantropical and warm tropical regions, mainly in Southeast Asia and South Pacific.

2 Genera, 17 species

Salient features: Perennial rhizomatous herbs without roots, stems dichotomously branched, leaves scale-like, 1-veined or veinless, 2–3 sporangia fused into thick walled synangium, homosporous, gametophyte free living, subterranean.

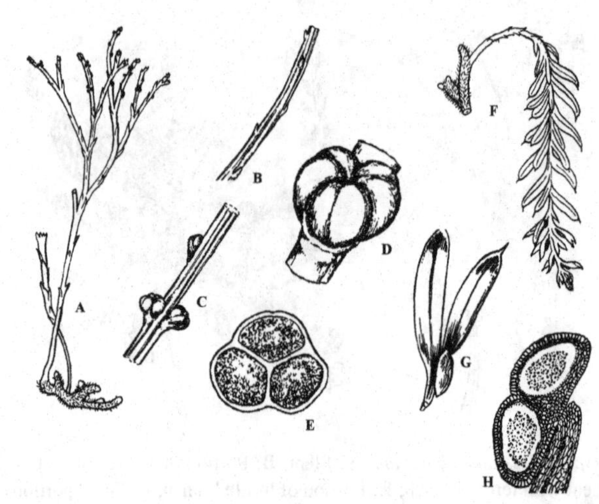

Figure 11.6: Psilotaceae. *Psilotum nudum.* **A:** Plant; **B:** Portion of sterile branch with leaves; **C:** Portion of fertile branch with 3-lobed synangia; **D:** Synangium; **E:** Synangium in cross section. *Tmesipteris tannensis.* **F:** Plant; **G:** Sporophyll; with 2-lobed synangium; **H:** Longitudinal section of synangium.

Genera: *Tmesipteris* (15 species) and *Psilotum* (2).

Description: Terrestrial or more commonly epiphytic perennial herbs, often pendulous. **Roots** absent, plant anchored by rhizome with rhizoids and mycorrhizal fungi; rhizome producing gemmae for vegetative reproduction in *Psilotum*. **Stem** erect, pendent or creeping; dichotomously branched, appearing like bundle of green forking sticks; vascular slender protostelic; branch tip flattened laterally and appearing leaf-like in *Tmesipteris*. **Leaves** (euphylls) scale-like, spirally arranged, 2-ranked, 1-veined (*Tmesipteris*) or veinless (*Psilotum*-termed enations) or nearly so, awl-shaped to lanceolate, simple or once forked. **Sporangia** homosporous with 2–3 celled thick wall, two or three together, fused into 2–3 lobed synangium (2-lobed in *Tmesipteris*, 3-lobed in *Psilotum*), yellowish at maturity, subtended by a forked appendage (Sporophyll); spores numerous, bean-shaped, pale, in tetrads. **Gametophyte** free living, subterranean or superficial, often with mycorrhizal fungi, irregularly branched; antheridia and archegonia in large numbers all over the gametophyte; spermatozoids spirally coiled, multiflagellate.

Economic importance: Family is of little importance, *Psilotum nudum* often grown as greenhouse plant.

Phylogeny: The phylogeny of the family is a matter of considerable speculation. Wettstein (1901) agreed with Engler and Prantl in regarding the group as advanced over Selaginellales and Isoetales. Eames (1936), however, considered the family as the most primitive of the extant vascular plants, lacking roots and true leaves. The recent morphological, chemical and molecular studies, however, support relationship with eusporangiate group Ophioglossales, and that the group lost roots secondarily. Recent genomic studies of Hidalgo et al. (2017) have shown that *Tmesipteris japonica* exhibits genomic gigantism with genome size of 1C = 150.61 pg, slightly smaller than largest genome size of 1C = 152.23 pg recorded in angiosperm species *Paris japonica* (Melanthiaceae). The addition of this new record doubles the range of genome size values so far encountered in ferns from 97.2-fold (0.77–74.84 pg/1C) to 196-fold (0.77–150.61 pg/1C). This finding emphasizes the importance of filling taxonomic gaps in our knowledge to uncover the full extent of genome size diversity across the different lineages of land plants.

Shen et al. (2018), on the basis of a large-scale phylogenomic analysis using high-quality transcriptome sequencing data, which covered 69 fern species from 38 families and 11 orders, concluded that Psilotales (whisk ferns), Ophioglossales (moonworts), and Marattiales (king ferns) form a monophyletic clade as ([Psilotales, Ophioglossales], Marattiales), which is sister to leptosporangiate ferns. The monophyletic origin of Psilotales, Ophioglossales, and Marattiales, which belong to eusporangiate ferns, is supported by the structure of sporangia.

Equisetaceae Michx. ***Horsetail Family***

Almost worldwide except Australia, New Zealand and Antarctica, in temperate and tropical climate.

1 Genus, 15 species

Salient features: Annual or perennial rhizomatous herbs with jointed ribbed aerial stems, leaves reduced with single vein, sporangia on peltate sporangiophores, homosporous, spore wall with elaters, gametophyte green, thallus shaped, spermatozoids multiflagellate.

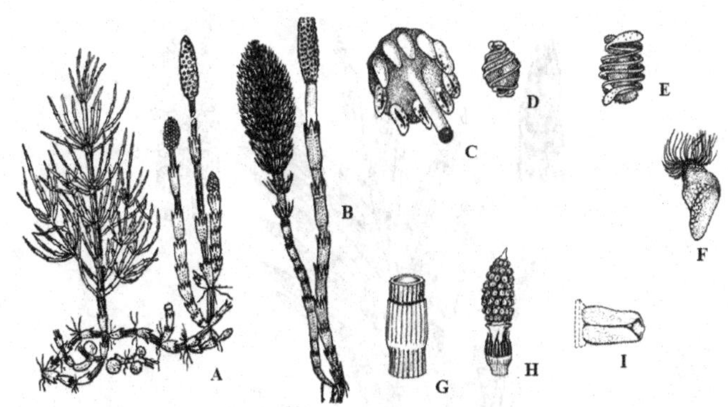

Figure 11.7: Equisetaceae. A: Plant of *Equisetum arvense* showing rhizome, vegetative branch and fertile branches. *E. telmateia.* **B:** Portion of plant with sterile and fertile branches; **C:** Sporangiophore viewed from below; **D–E:** Spore with coiled elaters; **F:** Antherozoid. *E. hyemale.* **G:** Node with sheath of leaves; **H:** Strobilus; **I:** Sporangium.

Genus: Single genus *Equisetum* (15 species).

Description: Terrestrial or aquatic annual or perennial herbs, sometimes evergreen, with subterranean much-branched rhizome. **Roots** slender, arising from horizontal subterranean rhizome. **Stem** subterranean as well as aerial, latter erect, green, unbranched or with whorled branches, with swollen jointed nodes, internodes longitudinally ribbed, with ridges and grooved outside, hollow with central canal, with additional smaller canals under the ridges; growth intercalary; outer surface covered with silica cells giving plant the texture of sand paper, hence the name 'scouring rush'. **Leaves** small, usually less than 1 cm, 1-veined, whorled and fused into a sheath, latter more or less swollen, each leaf corresponds to the ridge below, thus the number of leaves at each node as the number of ridges, the leaves of the successive nodes as well as ridges alternating. **Sporangia** large, lacking annulus, homosporous, hanging from lower surface of peltate sporangiophore, whorls of sporangiophores aggregated into strobilus terminating green branches or unbranched nongreen stems arising separately from rhizome; base of strobilus with collar of fused sterile appendages; sporangium with 2-layered wall; sporangiophore with slender stalk and hexagonal disc at distal end, 5–10 sporangia hanging from each disc; spores spherical, many thousand per sporangium, green, wrapped by 4–6 strap like elaters arising from outer wall, assisting in spore dispersal, wall with four apertures. **Gametophyte** pinhead sized, thallus shaped, green, developing near soil surface; antheridia and archegonia developing simultaneously; spermatozoids multiflagellate.

Economic importance: Family is of lesser importance. Silica covered stems of several species, especially *E. hyemale*, were used for scouring pots and pans, hence the common name scouring rush. *E. arvense* is reputed medicinal plant used traditionally against numerous conditions and many herbal products are sold on the market mainly against urinary and renal conditions, as well as skin, hair and nail remedies, potentially due to the species' high silica content. Closely related *E. palustre* contains toxic levels of pyridine alkaloid palustrine.

Phylogeny: The family is quite distinct in its jointed stems with ridges and grooves and hollow within, whorled leaves and peltate sporangiophores. Earlier the family was considered distinct from ferns, but the molecular data and morphological characters such as spermatozoids and root structure, support placement with ferns. The genus is divided into two subgenera: *Equisetum* (8 species) with branched stem and superficial stomata and *Hippochaete* (7 species) with unbranched stems and sunken stomata, sometimes recognized as distinct genera. Several studies have questioned the monophyly of subgenus *Equisetum*. Gullion (2007) based on original (atpB) and published (rbcL, trnL-trnF, rps4) sequence data to investigate the phylogeny of the genus, concluded that atpB sequences give an unusual topology, *Equisetum bogotense* is sister to subgenus *Hippochaete.* Study supported an early divergence of *E. bogotense* as sister to subgenus *Hippochaete* and the occurrence of two monophyletic clades (subgenus *Equisetum* minus *E. bogotense* and subgenus *Hippochaete*). Lagoudakis et al. (2015) based on DNA barcoding also concluded *Equisetum bogotense* as sister to the rest of the genus and not as a member of subgenus *Hippochaete*. The remainder of the genus is resolved into two major clades, each comprising seven species and corresponding to the two subgenera *Equisetum* and *Hippochaete.*

Shen et al. (2018), based on Large-scale genomic analysis proposed that ex-annulus sporangia, as in Equisetaceae, Psilotaceae, and Ophioglossaceae, is the ancestral state in ferns and that Equisetales is sister to rest of the ferns.

The family is represented in the fossil record as early as Devonian 408–360 MYA.

Osmundaceae Bercht. & J. Presl ***Royal Fern Family***

Worldwide except very cold climates and Pacific Islands.

6 Genera, 18 species

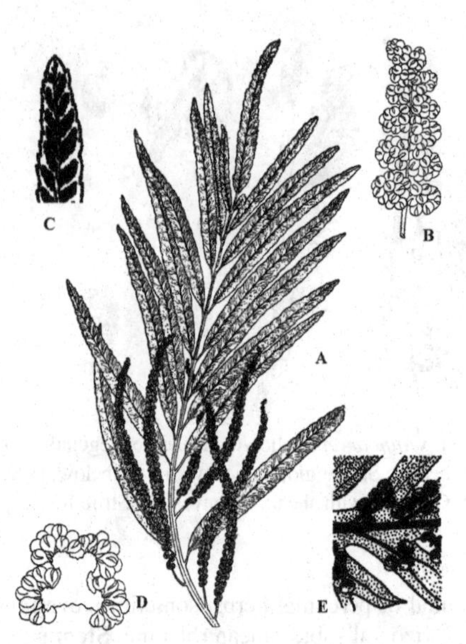

Figure 11.8: Osmundaceae. *Osmunda javanica* **A:** Portion of plant with sterile and fertile fronds; **B:** Portion of fertile frond; **C:** Fertile frond in cross section; **D:** Portion of fertile pinna of *Todea africana*; **E:** Portion of fertile pinna of *Leptopteris hymenophylloides.*

Salient features: Terrestrial plants with wiry roots, persistent stipe bases, dimorphic fronds, pinnate compound sterile fronds, sporangia with many spores, annulus shieldlike, gametophyte green, developing on soil surface.

Genera: *Leptopteris* (6 species), *Osmunda* (4), *Plenasium* (4) and *Todea* (2).

Description: Terrestrial plants common in wetlands and lowland forests, sometimes tree-like (*Leptopteris*). **Roots** wiry, adventitious, generally two below each leaf base. **Stem** erect to decumbent, branched, massive, often covered by persistent leaf bases, ectophloic siphonostele with ring of discrete xylem strands, latter often conduplicate or twice conduplicate in cross section. **Leaves** up to 2 m long, spirally arranged, usually dimorphic with distinct sterile and fertile fronds; sterile fronds green, once- thrice-pinnate compound, with expanded petiole base, circinate before unfolding, usually leathery, rarely filmy (*Leptopteris hymenophylloides*), usually covered with hairs especially when young; fertile fronds usually brown, much narrower; sometimes sterile and fertile segments present on same leaf. **Sporangia** large, shortly stalked, intermediate between eusporangiate and leptosporangiate ferns arising from single initial (leptosporangiate) or many initials (eusporangiate) and archesporial cell tetrahedra (leptosporangiate) or cubical (eusporangiate), separate or in loose clusters, borne on distinct fertile fronds (*Osmunda*) or on undersurface of foliage leaves along veins (crowded in *Todea*, sparse in *Leptopteris*), not forming sori, annulus poorly developed shield-like plate or broad horizontal band around the sporangium, homosporous, producing 128–512 spores, sporangia opening by apical slit; spores green, subglobose, with triradiate mark. **Gametophyte** large green, cordate, developing on soil surface; antheridia emergent producing up to 100 spermatozoids.

Economic importance: Family is of lesser importance. A few species like *Osmunda cinnamomea* (Cinnamon fern) and *O. regalis* (Royal fern) are grown as ornamentals.

Phylogeny: The family is considered sister to leptosporangiate ferns evidenced by numerous spores, rudimentary annulus, more than one celled thick wall, lack of sorus; position supported by *rbc*L sequence data. This conclusion is also supported by long fossil record, dating back to Permian 286–245 MYA.

Marsileaceae Mirbel ***Water-Clover Family***

Nearly worldwide in warm temperate and tropical areas.

3 Genera, 61 species

Salient features: Aquatic plants with floating long-petioled leaves, leaflets 2 to 4, sori enclosed in hard sporocarp, heterosporous, sporangia without annulus, megasporangium with one megaspore, microsporangium with 16–64 microspores.

Genera: *Marsilea* (55 species), *Pilularia* (5) and *Regnellidium* (1).

Description: Plants perennial, aquatic or rooted in mud with creeping rhizome, rarely xerophytic and developing underground tubers from rhizome (*M. hirsuta*); land forms with short internodes, branched roots, long petiole and stomata on both leaf surfaces; aquatic forms with long internodes, unbranched roots, flexible petiole and stomata mainly on the

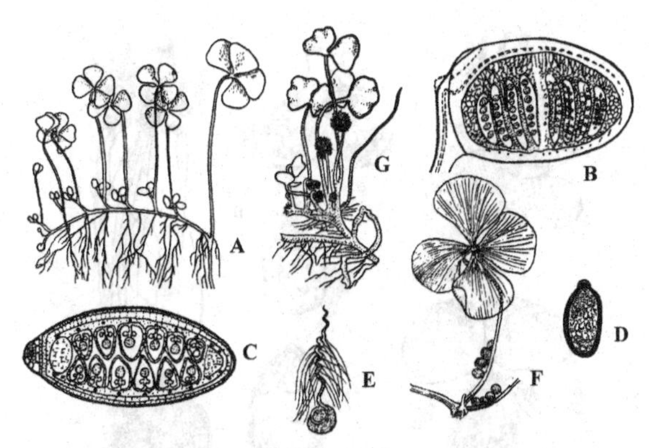

Figure 11.9: Marsileaceae. *Marsilea quadrifolia.* **A:** Plant with sporocarps; **B:** Vertical section of sporocarp; **C:** Sporocarp in longitudinal section; **D:** Spore; **E:** Spermatozoid; **F:** Portion of plant of *M. polycarpa*; **G:** Portion of plant of *Regnellidium diphyllum.*

upper surface. **Roots** arising from creeping rhizome, one or two at each node. **Stem** a slender rhizome, creeping, growing on soil surface or subterranean, often with hairs, dichotomously branched. **Leaves** floating or emergent, long petioled, blade filiform (*Pilularia*) or divided into 2 (*Regnellidium*) or 4 clover-like (*Marsilea*) leaflets, circinate before unfolding, leaflets folded together upwards until nearly mature, veins dichotomously branched but often fusing towards tips. **Sporangia** heterosporous, arranged in sori, latter without indusium, enclosed in hard pea-shaped, bean-shaped or subglobose sporocarps borne singly on short stalks near or at base of petioles, sometimes stalk branched bearing 2–3 sporocarps (*M. quadrifolia*), rarely several sporocarps on one petiole (*M. polycarpa*); each sporocarp with rows of sori along either side, with 2 (*M. aegyptiaca*) to 20 (*M. quadrifolia*) sori with large megasporangia along the crest and microsporangia along the sides; sporangia without annulus; megasporangium with one megaspore; microsporangium with 16–64 microspores; sporangia attached to a gelatinous ring-like structure called sorophore, that swells with water; microspore small globular, producing 16 spermatozoids, latter multiflagellate and corkscrew shaped with prominent vesicle; megagametophyte producing single archegonium.

Economic importance: Family is of little importance, with *Marsilea* species often grown as curiosity.

Phylogeny: The family is considered sister to leptosporangiate ferns due to presence of numerous spores, rudimentary annulus, more than one celled thick wall and lack of annulus. This conclusion is also supported by evidence *rbc*L sequence data and long fossil record, dating back to Permian 286–245 MYA. *Regnellidium* is exceptional non-flowering plant with latex tubes.

Salviniaceae Martynov ***Mosquito-fern Family***

Worldwide in tropical and temperate climates.

2 Genera, 21 species

Salient features: Aquatic free-floating plants, small simple leaf blades, veins branched, sporocarps flattened and soft, sporangia large, heterosporous, gametophyte endosporous.

Genera: *Salvinia* (12 species) and *Azolla* (9).

Description: Free-floating aquatic plants growing in lakes and ponds, often forming dense floating mats, with slender branching rhizome. **Roots** hanging down from zigzag stem (*Azolla*) or absent and represented by lower third row of leaves which are modified into root-like structures (*Salvinia*). **Stem** a rhizome, zigzag, horizontal, dichotomously branched, protostelic; stem fragile and readily breaks resulting in proliferous vegetative propagation, and in often covering entire water surface. **Leaves** simple, sessile, less than 15 mm long, rounded to oblong, entire; in *Azolla* imbricated in two rows harboring nitrogen-fixing cyanobacterium *Anabaena azollae* in leaf cavities; in whorls of three appearing in three rows in *Salvinia* with upper two rows floating and covered with hairs whereas the third lower row submerged and finely dissected into root-like structure (hairs septate unlike true root hairs); veins of leaf-blades branched, free (*Azolla*) or anastaomosing (*Salvinia*). **Sporangia** heterosporous, leptosporangiate, arranged in sori; In *Salvinia* megasporangia and microsporangia in different sporocarps (megasporocarp with up to 25 megasporangia; microsporocarp with numerous microsporangia); sporocarps soft; in *Azolla* sporangia are enclosed in indusium with either one megasporangium or several microsporangia; megasporangium with single megaspore, microsporangium with several microspores (all microspores surrounded by hardened tapetal cytoplasm are shed as a single mass known as massula); annulus absent; spores globose, trilete. **Gametophyte** endosporous, megagametophytes and microgametophytes protruding through sporangium wall;

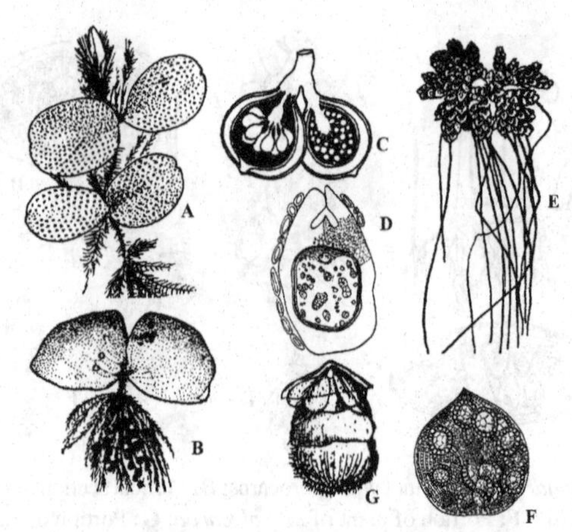

Figure 11.10: Salviniaceae. *Salvinia natans.* **A:** Plant; **B:** Two floating leaves and a submerged leaf. **C:** Sporocarps of *S. rotundifolia* with a megasporangium and a microsporangium; **D:** Vertical section megaspore. *Azolla microphylla.* **E:** Plant; **F:** Microsporocarp split open to show microsporangia; **G:** Megaspore.

megagametophytes floating on water surface with archegonia directed downward; microgametophytes remaining fixed to sporangium wall.

Economic importance: Both genera are invasive weeds in warm climates, often choking lakes, ponds and drains. *Salvinia* is frequently sold as aquarium plant. *Azolla* is often used as green manure, owing to the presence of nitrogen-fixing cyanobacterium.

Phylogeny: The family is sometimes split into two distinct families based on absence of roots and sori in distinct sporocarps in *Salvinia,* as against the presence of roots and absence of sporocarps in *Azolla.*

Cyatheaceae Kaulf. **Scaly *Tree-fern Family***

Tropical and subtropical montane forests and cloud forests.

3 Genera, 643 species

Salient features: Tall arborescent palm-like ferns with thick trunk, leaves covered with scales, large, pinnate to bipinnate, circinate before unfolding, homosporous, sporangia in sori, annulus continuous.

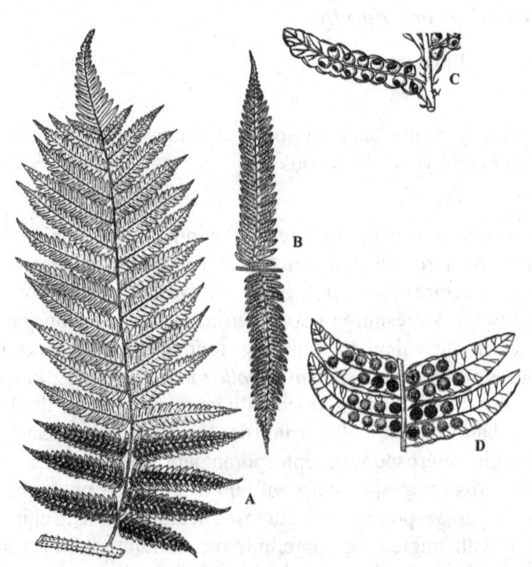

Figure 11.11: Cyatheaceae. *Cyathea spinulosa.* **A:** Portion of frond with fertile (below) and sterile (above) pinnules; **B:** Sterile pinnules enlarged; **C:** Portion of fertile frond of *Cyathea elegans*; **D:** Portion of fertile frond of *C. medullaris.*

Major Genera: *Alsophila* (275 species), *Cyathea* (265) and *Sphaeropteris* (103).

Description: Palm-like ferns with single erect arborescent trunk. **Stem** erect, usually unbranched, reaching 25 m in height, rarely decumbent or creeping, dictyostelic, stem and leaves covered with large and small scales. **Leaves** large often reaching 5 m in length, once to twice pinnate compound, with deeply pinnately lobed leaflets, rarely simple, circinate before unfolding; petioles with obvious, usually discontinuous pneumathodes in two lines; veins free or anastomosing. **Sporangia** homosporous, arranged in sori on abaxial surface of fronds, annulus continuous, not interrupted by sporangium stalk, oblique, allowing the sporangium to open horizontally; indusium completely covering the sorus or absent; spores usually 64 per sporangium, tetrahedral, trilete, variously ornamented. **Gametophyte** green, cordate.

Economic importance: Various members of the family are often grown as ornamentals, fibrous rhizomes often used as base for epiphytes in greenhouses.

Phylogeny: The family is well characterized by uniform chromosome number n = 69. *Hymenophyllopsis* with small distinct stems is quite distinct but shares the presence of scales with other members of the family. There is some molecular evidence supporting the inclusion of this genus in *Cyathea.*

Pteridaceae Ching *Maidenhair Fern Family*

Tropical and subtropical montane forests and cloud forests.

53 Genera, 1211 species

Salient features: Terrestrial or epiphytic, rhizomes covered with scales, leaves all similar, petiole base with persistent scales, sori near margin, indusium absent.

Major Genera: *Pteris* (250 species), *Adiantum* (225) and *Cheilanthes* (100).

Description: Terrestrial or epiphytic, rarely aquatic, growing in wide varieties of habitats such as deserts, ponds, cultivated areas, forest canopies to mangrove swamps. **Root** adventitious, arising from lower surface of rhizome. **Stem** represented by rhizome, short, creeping to erect, sometimes deeply penetrating, dichotomously branched, usually covered with scales, less often with hairs. **Leaves** all similar (monomorphic), dimorphic in few genera; petiole with persistent scales near base; blade simple, 1–6 pinnate or pedate; veins free or forking, variously anastomosing and forming a reticulate pattern without included veinlets. **Sporangia** homosporous, leptosporangiate, arranged in sori on abaxial surface near margin, forming a continuous band and protected by reflexed leaf margin forming false indusium (*Pteris*) or two-lipped indusium with thick upper lip and thin papery lower lip (*Pteridium*), true indusium being absent, or arranged along all leaf veins; sporangium stalk with 2–3 rows of cells; sporangia each with a vertical, interrupted annulus covering three fourth of sporangium, a strip of 4 cells below free end of annulus forms stomium, receptacles not or only obscurely raised; spores 64 per sporangium, globose or tetrahedral, trilete, variously ornamented. **Gametophyte** heart-shaped.

Economic importance: Some species of *Adiantum* and *Pteris* are grown as ornamentals.

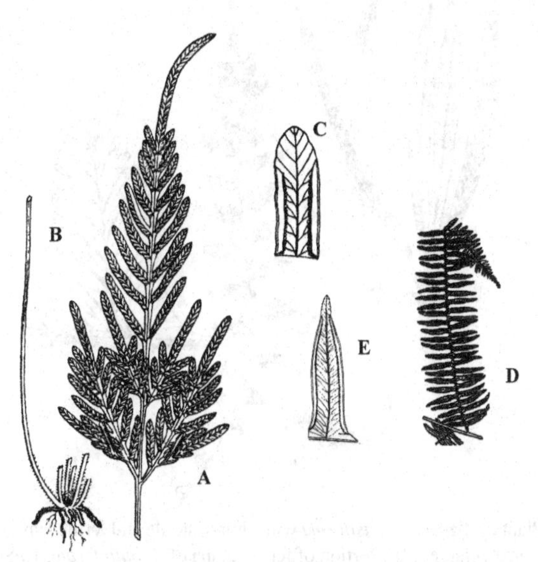

Figure 11.12: Pteridaceae. *Pteris griffithii.* **A:** Portion of frond; **B:** Petiole of same; **C:** Portion of a pinnule. *Pteris patens.* **D:** Portion of fertile frond; **E:** Portion of pinnule.

Phylogeny: The family shows a wide diversity due to its adaptation to diverse habitats. The family contains five clades that are recognised in PPG I (2016) as subfamilies Parkerioideae (*Acrostichum* and *Ceratopteris*), Cryptogrammoideae (*Coniogramme*, *Cryptogramma*, and *Llavea*), Pteridoideae (13 genera), Vittarioideae (12 genera), and Cheilanthoideae (23 genera). *Platyzoma*, sometimes recognized as an isolated family, is aberrant in chromosome base number ($x = 38$ as against 29 or 30 in the family) and in having dimorphic spores ("incipient heterospory") but nests with other genera of Pteridaceae, subfam. Pteridoideae. *Ceratopteris* is distinct from other members of the family in coarsely ridged spores with parallel striations; spores 32 or fewer per sporangium; sporangia with ill-defined annuli; aquatic habitat; $x = 38$, often placed in distinct family, but nests within Pteridaceae in all molecular analyses, and it appears to be sister to *Acrostichum*.

Aspleniaceae Newman ***Spleenwort Family***

Nearly worldwide, most diverse in tropics.

2 Genera, 730 species

Salient features: Terrestrial or epiphytic, rhizomes covered with scales, leaves all similar, sori elongate, indusia linear, laterally attached, sporangium stalk with one row of cells, spores bilateral.

Genera: *Asplenium* (700 species) and *Hymenasplenium* (30).

Description: Terrestrial or epiphytic perennials, some members growing on rocks (epipetric) in moist or wet forests. **Stem** represented by rhizome, erect or creeping, usually covered with scales especially towards apex, dictyostelic. **Leaves** all similar (monomorphic); petiole with scales near base, with C-shaped vascular strands fused distally back-to-back to form x-shape; blade simple to 5-pinnate, often with glandular hairs and a few linear scales; veins pinnate or forking, usually free, sometimes reticulate without included veinlets. **Sporangia** homosporous, arranged in sori on abaxial surface along veins, linear or curved; indusia linear, laterally attached; sporangium stalk with one row of cells; spores bilateral, reniform, monolete, with winged perine.

Economic importance: Some species of *Asplenium* such as *A. scolpendrium* (heart's tongue fern) are grown as ornamentals.

Phylogeny: The family is closely related to Blechnaceae, Onocleaceae and Thelypteridaceae. Some species have variously been removed to distinct genera to establish up to 10 genera in the family, although large number of species fit well under *Asplenium*. The segregate genera *Camptosorus* and *Loxoscaphe*, *Diellia* (endemic to Hawaii), *Pleurosorus*, *Phyllitis*, *Ceterach*, and *Thamnopteris* clearly nest within *Asplenium* s.l. *Hymenasplenium*, however, with a different chromosome base number than nearly all of the other segregates, as well as distinct root characters (Schneider et al., 2004), appears to represent the sister clade to the rest of the species in the family, and has been recognized as second genus within the family by PPG I (2016).

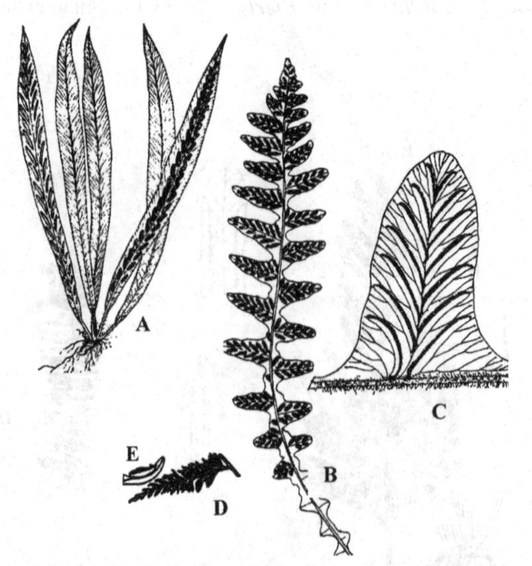

Figure 11.13: Aspleniaceae. A: Plant of *Asplenium ensiforme* with simple sterile and fertile leaves; **B:** Leaf of *A. alternans*; **C:** Portion of same enlarged; **D:** Portion of fertile pinna of *A. bulbiferum*; **E:** Sorus.

Dryopteridaceae Ching ***Wood Fern Family***

Nearly worldwide, most diverse in tropics with several representatives in temperate region.

26 Genera, 2115 species

Salient features: Mostly terrestrial, rarely epiphytic, rhizomes covered with scales towards apex, leaves all similar, petioles with persistent scales towards base, sori round, covering the leaf surface, indusia round-reniform shaped, sporangium stalk with three rows of cells, spores bilateral.

Major Genera: *Elaphoglossum* (600 species), *Polystichum* (500), *Dryopteris* (400), *Ctenitis* (125), *Megalastrum* (91), *Bolbitis* (80).

Description: Terrestrial, epiphytic or epipetric ferns. **Stem** represented by rhizome, erect, ascending or creeping, sometimes scandent or climbing, usually covered with non-clathrate scales especially towards apex, dictyostelic. **Leaves** all similar (monomorphic), rarely dimorphic; petiole with persistent scales near base, with numerous round vascular strands arranged in a ring; blade simple to 5-pinnate, sometimes scaly or glandular, rarely hairy; veins pinnate or forking, free to variously anastomosing, with or without included veinlets. **Sporangia** homosporous, arranged in sori on abaxial surface, sori rounded, closely spaced and covering the leaf surface; indusia round-reniform or peltate, rarely absent; sporangium stalk with 3 rows of cells; spore reniform monolete, perine winged.

Economic importance: None.

Phylogeny: Dryopteridaceae is well defined clade, except for three genera *Didymochlaena, Hypodematium* and *Leucostegia,* whose inclusion renders this family paraphyletic. *Hypodematium,* often removed to a distinct family Hypodematiaceae and *Leucostegia* is nearly always placed in Davalliaceae because of its similar indusia and sori terminal on the veins, but it differs from members of Davalliaceae in the terrestrial habit, the more strongly verrucate spores with rugulate perispore. Tsutsumi and Kato (2006) found support for a sister relationship between *Hypodematium* and *Leucostegia,* and also support for these as sister to the remaining Eupolypods. *Aenigmopteris* and *Dryopolystichum* have uncertain position.

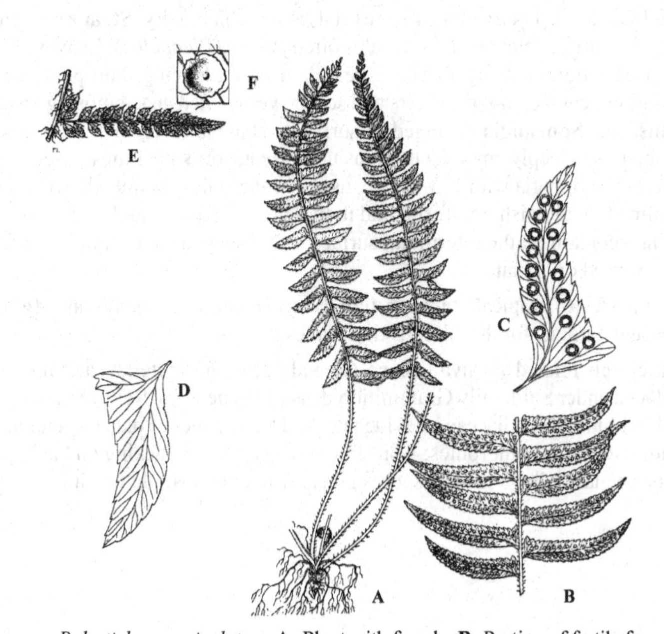

Figure 11.14: Dryopteridaceae. *Polystichum auriculatum.* **A:** Plant with fronds; **B:** Portion of fertile frond with sori; **C:** Fertile pinna enlarged; **D:** Sterile pinna enlarged; **E:** Portion of rachis of *P. setiferum* with fertile pinna; **F:** Sorus.

Polypodiaceae Bercht. & J. Presl. ***Polypod Family***

Widely distributed in tropics with few representatives in temperate region.

65 Genera, 1652 species

Salient features: Mostly epiphytic, rhizomes covered with scales, leaves all similar, usually simple, petioles without scales, sori round, indusia absent, sporangium stalk with 1–3 rows of cells.

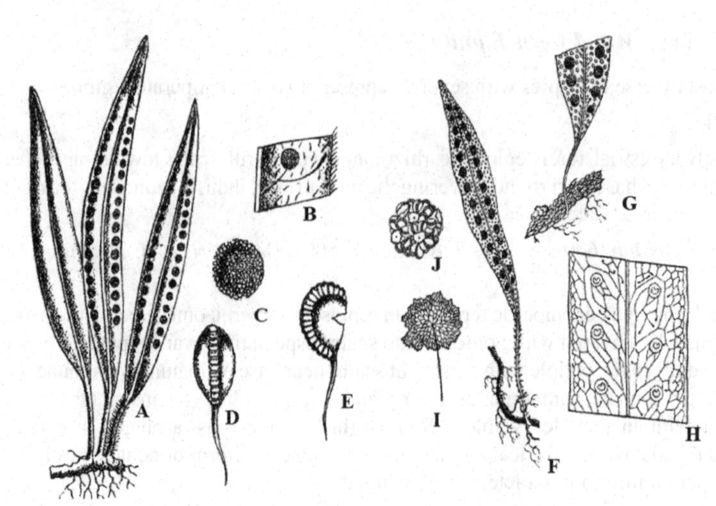

Figure 11.15: Polypodiaceae. *Polypodium wallii.* **A:** Plant with fronds; **B:** Portion of fertile frondwith sorus; **C:** Sorus; **D–E:** Sporangia in different views. *Pleopeltis lanceolata.* **F:** Portion of plant with fertile frond; **G:** Basal portion of fertile frond; **H:** Portion of leaf showing venation; **I:** Sorus; **J:** Peltate scale.

Major Genera: *Oreogrammitis* (156), *Pleopeltis* (90), *Prosaptia* (87), *Selliguea* (85), *Lepisorus* (80), *Ceradenia* (73), *Calymmodon* (65), *Pyrrosia* (52), *Aglaomorpha* (50), *Lellingeria* (49), *Grammitis* (43), *Tomophyllum* (42), *Polypodium* (40) and *Serpocaulon* (40).

Description: Mostly epiphytic and epipetric, a few terrestrial. **Roots** thick, wiry. **Stem** represented by rhizome, long- to short-creeping, dictyostelic, bearing scales and hairs, scales often peltate (*Pleopeltis*). **Leaves** all similar, usually thick and coriaceous, rarely dimorphic; petioles cleanly abscising near their bases, leaving short phyllopodia; blade mostly simple to pinnatifid or 1-pinnate; indument lacking or of hairs and scales; veins often anastomosing or reticulate, sometimes with included veinlets, or veins free. **Sporangia** arranged in sori; sori abaxial, rarely marginal, round to oblong or elliptic, occasionally elongate, sometimes deeply embedded; indusium absent, sori sometimes covered by caducous scales when young (*Lepisorus*, *Pleopeltis*); sporangia with 1–3-rowed, usually long stalks, frequently with paraphyses on sporangia or on receptacle; spores hyaline to yellowish, reniform, and monolete, or greenish and globose-tetrahedral, trilete (more or less triangular in aspect, and contain on their germinal surface a "Y" shaped suture called a **trilete** scar); perine various, usually thin, not strongly winged or cristate.

Economic importance: Species of tropical staghorn fern *Platycerium, Phlebidium* and *Aglaomorpha* are commonly cultivated in greenhouses and gardens for their interesting leaves.

Phylogeny: The family is closely related to Davalliaceae, Oleandraceae and Tectariaceae. Grammitid ferns often removed as Grammitidaceae but placed under Subfamily Grammitidoideae (33 genera and 911 species) as one of the six subfamilies in PPG I (2016), nest well within Polypodiaceae as evidenced by DNA sequence data (Schneider et al., 2004b), and share a large number of morphological synapomorphies: veins free (mostly); scales lacking on blades; setiform, often dark red-brown hairs on leaves; sporangial stalks 1-rowed; spores green, trilete; gametophytes ribbon-shaped.

Chapter 12

Families of Gymnosperms

Gymnosperms comprise a small group of seed plants characterized by naked seeds (gymno-naked, sperms-seeds) and absence of vessels (except Gnetopsids), endosperm formation independent of fertilization and commonly resulting in halploid endosperm (absence of double fertilization), absence of sieve tubes and companion cells. Group is represented by evergreen trees and shrubs, distributed worldwide and forming extensive forests in North America, Europe and Asia. They represent some of the largest (*Sequoiadendrod giganteum* of California), tallest (*Sequoia sempervirens* of California and Oregon) and longest living (*Pinus aristata*) organisms in the world.

Gymnosperms are woody trees or shrubs, herbaceous plants being absent from the group. The plants have well-developed tap root system, sometimes with symbiotic nitrogen fixing cyanobacterium (coralloid roots of *Cycas*) or mycorrhizae (*Pinus*). Vascular cylinder has xylem with tracheids with bordered pits and phloem with sieve cells. Leaves lack lateral veins but are compensated by transfusion tissue. Sporangia are heterosporous, microsporangia and megasporangia borne on microsporophylls and megasporophylls, respectively; latter often arranged in distinct cones. Each microsporangium produces numerous microspores arranged in tetrads, since each **microspore mother cell** undergoes meiosis to produce four haploid microspores. **Microspore** nucleus undergoes repeated divisions to form **male gametophyte**, which develops wall to become a pollen grain. The **megasporangium**, known as **ovule**, on the other hand, develops a single **megaspore mother cell**, surrounded by **nucellus** and **integument**, with an opening known as **micropyle**, at the end of integument. Of the four haploid megaspores resulting after meiosis, three degenerate, and only one **megaspore** is functional. Latter, after repeated nuclear divisions and wall formations produces a **female gametophyte** with several **archegonia**, consisting of an enlarged **egg** cell and two or four **neck cells**. Pollen grains of gymnosperms are carried by wind, land on micropyle and adhere to sticky fluid released by the female gametophyte. The pollen germinates to produce a pollen tube, that grows through nucellus and releases two sperms. One fuses with the egg to form zygote after fertilization. Latter develops into an embryo within matured ovule known as **seed**.

Gymnosperms have been recognized as group distinct from angiosperms since Robert Brown (1827) establish their identity. Four distinct groups of extant gymnosperms viz: Cycads, Conifers, Gnetopsids and monotypic *Ginkgo* have, however been treated differently by various authors. Chamberlain (1935) divided gymnosperms into two classes: Cycadophytes (Cycadofilicales, Bennettitales and Cycadales) and Coniferophytes (Cordaitales, Ginkgoales, Coniferales and Gnetales). Arnold (1948) separated Gnetalean members under a separate third class, raised to the level of divisions by Pant (1957) recognizing Cyacadophyta, Chlamydospermophyta and Coniferophyta. Cronquist et al. (1966) included gymnosperms under Pinophyta, divided into three subdivisions: Cycadicae (classes Lyginopteridatae, Cycadatae and Bennettitatae), Pinicae (classes Ginkgoatae and Pinatae) and Gneticae. Bierhorst (1971) divided gymnosperms into three classes: Cycadopsida (5 orders Pteridospermales, Caytoniales, Cycadeoidales, Pentoxylales and Glossopteridales), Coniferopsida (5 orders Cordaitales, Protopityales, Ginkgoales, Coniferales and Taxales) and Gnetopsida (3 orders Ephedrales, Gnetales and Welwitschiales). Taylor (1981) recognized 6 divisions within Gymnosperms:

Gnetales (sometimes treated under three separate orders: Gnetales, Ephedrales and Welwitschiales, and then collectively termed as Gnetopsids) are unique among gymnosperms in presence of vessels and occurrence of double fertilization in *Ephedra*, wherein one male nucleus fuses with ventral canal nucleus producing supernumerary embryos (and not endosperm of angiosperms); endosperm, however, remaining haploid. *Ephedra* also depicts flower-like reproductive structures and highly reduced female gametophyte. Early studies considered Gnetales to be more closely related (even sister) to angiosperms, making gymnosperms paraphyletic. Recent studies based on molecular evidence, however, point out affinities between Gnetales and conifers.

The monotypic genus *Ginkgo* (placed in monotypic Ginkgoales, Ginkgoaceae) is unique and apparently unrelated to other gymnosperms, and often termed as living fossil, retaining several primitive features like dichotomously veined fan-shaped leaves, motile sperms and lack of pollen tubes. The genus is also unique in having sex chromosomes: two *X* chromosomes in female plants and *XY* in male plants. Recent molecular studies (Qui et al., 2006; Wu et al., 2007) have pointed to close relationship with Cycadales, with which the genus shares features like dioecious habit, branched pollen tube growing away from the ovule, motile male gametes with several flagella and cell wall.

Recent treatments of Gymnosperms following the APG tradition, prefer to treat the four distinct groups of extant gymnosperms under distinct orders: Cycadales, Ginkgoales, Coniferales and Gnetales. Christenhusz et al. (2011) in recent classification of gymnosperms recognize 12 families of extant gymnosperms.

It is widely accepted that the gymnosperms originated in the late Carboniferous period, replacing the lycopsid rainforests of the tropical region. This appears to have been the result of a whole genome duplication event around 319 million years ago. Early characteristics of seed plants were evident in fossil progymnosperms of the late Devonian period around 383 million years ago.

Christenhusz et al. (2011) divide Gymnosperms into 4 subclasses: Cycadidae (one order and two families), Ginkgoidae (1 order, 1 family), Gnetidae (3 orders with one family each), and Pinidae (3 orders and 6 families). Christenhusz and Byng (2016) estimate 83 genera and 1079 known species in gymnosperms. Conifers constitute dominant group of living gymnosperms with nearly 79 genera and 629 species. Cycads form the next largest group with 10 genera and 337 species. Recent treatments of Gymnosperms following the APG tradition (APweb, version 14, 2017, updated December 2018), prefer to treat the four distinct groups of extant gymnosperms under distinct orders: Cycadales, Ginkgoales, Coniferales and Gnetales. The fossil groups which constitute bulk of gymnosperms such as Pteridosperms, Bennettitales, Caytoniales and Glossopterids have been excluded in this classification. The classification of Christenhusz et al. (2011) is followed in the present book.

Classification of Gymnosperms by Christenhusz et al. (2011)

Subclass I: Cycadidae
Order A: Cycadales
Family 1: **Cycadaceae** (*Cycas*)
Family 2: **Zamiaceae** (*Dioon, Bowenia, Macrozamia, Lepidozamia, Encephalartos, Stangeria, Ceratozamia, Microcycas, Zamia.*)

Subclass II: Ginkgoidae
Order B: Ginkgoales
Family 3: **Ginkgoaceae** (*Ginkgo*)

Subclass III: Gnetidae
Order C: Welwitschiales
Family 4: **Welwitschiaceae** (*Welwitschia*)
Order D: Gnetales
Family 5: **Gnetaceae** (*Gnetum*)
Order E: Ephedrales
Family 6: **Ephedraceae** (*Ephedra*)

Subclass IV: Pinidae
Order F: Pinales
Family 7: **Pinaceae** (*Cedrus, Pinus, Cathaya, Picea, Pseudotsuga, Larix, Pseudolarix, Tsuga, Nothotsuga, Keteleeria, Abies*)
Order G: Araucariales
Family 8: **Araucariaceae** (*Araucaria, Wollemia, Agathis*)
Family 9: **Podocarpaceae** (*Phyllocladus, Lepidothamnus, Prumnopitys, Sundacarpus, Halocarpus, Parasitaxus, Lagarostrobos, Manoao, Saxegothaea, Microcachrys, Pherosphaera, Acmopyle, Dacrycarpus, Dacrydium, Falcatifolium, Retrophyllum, Nageia, Afrocarpus, Podocarpus*)
Order H: Cupressales
Family 10: **Sciadopityaceae** (*Sciadopitys*)
Family 11: **Cupressaceae** (*Cunninghamia, Taiwania, Athrotaxis, Metasequoia, Sequoia, Sequoiadendron, Cryptomeria, Glyptostrobus, Taxodium, Papuacedrus, Austrocedrus, Libocedrus, Pilgerodendron, Widdringtonia, Diselma, Fitzroya, Callitris (incl. Actinostrobus*

and Neocallitropsis), Thujopsis, Thuja, Fokienia, Chamaecyparis, Callitropsis, Cupressus, Juniperus, Xanthocyparis, Calocedrus, Tetraclinis, Platycladus, Microbiota)

Family 12: **Taxaceae** (*Austrotaxus, Pseudotaxus, Taxus, Cephalotaxus, Amentotaxus, Torreya*)

Cycadaceae Pers. *Cycad Family*

Tropics and subtropics of Africa, Southeast Asia, Malaysia, Philippines and Polynesia.

1 Genus, 116 species

Salient features: Palm-like, unbranched stem, leaves fern-like, pinnately compound, thick, young pinnae circinate, with mid-vein but no side-veins, plants dioecious, megasporophylls leaf-like with ovules along margin, sperms motile.

Genus: Single genus *Cycas* (116 species).

Description: Palm-like plants with unbranched stem, rarely fern-like with underground rhizome (*C. siamensis*); bulbils often developing from axils of leaf bases; plants dioecious. **Roots** adventitious, large, fleshy; some roots near soil surface become inhabited with nitrogen fixing cyanobacteria, giving coralloid appearance (coralloid roots). **Stem** unbranched, covered with persistent leaf bases and scale leaves (cataphylls) in alternate spiral bands; outline irregular in section, large pith, numerous small vascular bundles in a ring, a wide cortex; parenchyma cells of cortex and pith containing a lot of starch; secondary growth initiated very early in life of plant but produced in small amount (manoxylic); mucilage canals abundant in stem; leaf traces forming girdle in stem. **Leaves** large, compound, forming a crown at the apex of stem, circinate when young, pinnate compound, spirally arranged, alternately with scale leaves; each leaflet with single mid-vein, side veins absent and compensated by transfusion tissue; leaf petiole with omega-like pattern of vascular bundles. **Microsporophylls** aggregated into compact large strobilus (cone), arranged spirally, hard, wedge-shaped with small sterile projection coiled at tip; microsporangia borne on abaxial surface, in groups of 3–5, surrounded by hairs; male strobilus terminal in position, becoming lateral due to lateral bud continuing the stem growth, axis becoming pseudopodial; pollen nonsaccate, with a single furrow; sperms motile. **Megasporophylls** spirally arranged, somewhat leaflike, not forming a strobilus, with 2–8 large ovules (up to 7 cm-largest in plant kingdom) along margin; upper sterile portion pinnate (*C. revoluta*) or reduced and serrated (*C. circinalis*); seeds elliptic or egg shaped, slightly flattened, fleshy, brightly colored: orange or red.

Economic importance: Several species notably *C. revoluta* and *C. circinalis* are commonly grown as ornamentals. Sago starch is obtained from stem of *C. circinalis* and other species. Seeds may also be utilized for yielding starch, but after the removal of toxins. Chamarrow people of Guam who suffer from fatal neurological disease apparently consume frugivorous

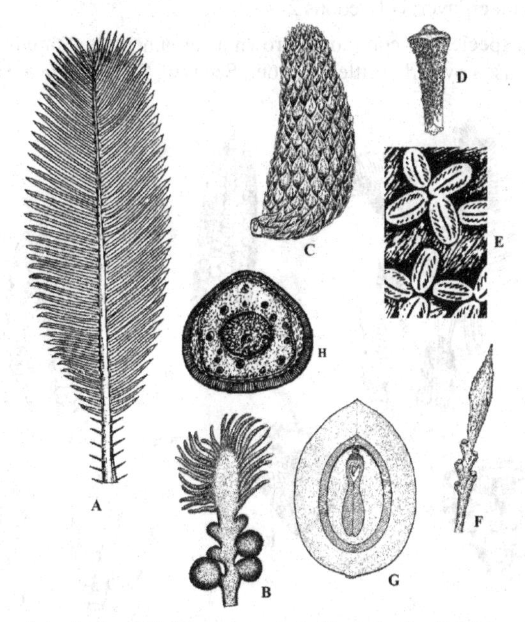

Figure 12.1: Cycadaceae. *Cycas revoluta.* **A:** Leaf; **B:** Megasporophyll. *C. circinalis.* **C:** Male cone; **D:** Microsporophyll; **E:** Portion of microsporophyll with sori; **F:** Megasporophyll; **G:** Longitudinal section of mature seed; **H:** Pollen grain.

bats known as flying foxes, who eat *Cycas* seeds. BMAA (b-methylamino-L-alanine), a neurotoxic non-protein amino acid, present in the seeds of *Cycas*, is produced by nitrogen fixing cyanobacteria in coralloid roots.

Phylogeny: The family was formerly circumscribed to include all genera now removed under Zamiaceae. The genus *Cycas* is distinct from all these genera in circinate young leaves, leaflets with midvein and no side veins, leaf-like megasporophylls not aggregated into strobilus and marginal ovules. Morphological and molecular studies have strongly supported *Cycas* to be sister genus to the rest of cycads. *Cycas* lineage may already have diverged from Zamiaceae by the Permian at least 250 million years before present (Hermsen et al., 2006) or as recently as 92 million years ago with diversification within the clade occurring ca 36 million years ago (Wink, 2006).

Zamiaceae Horianow ***Coontie Family***

Tropical to warm temperate regions of New World, Africa, and Australia.

9 Genera, 235 species

Salient features: Fern-like with rhizome or palm-like with unbranched stem, leaves fern-like, pinnately compound, thick, flat or conduplicate but not circinate, veins numerous, plants dioecious, megasporophylls reduced, forming strobilus, sperms motile.

Major Genera: *Encephalartos* (65 species), *Zamia* (77), *Macrozamia* (41), *Ceratozamia* (31), *Dioon* (15) and *Bowenia* (2).

Description: Fern-like plants with rhizome or Palm-like plants with unbranched stem; plants dioecious. **Roots** adventitious, large, fleshy; some roots near soil surface become inhabited with nitrogen fixing cyanobacteria, giving coralloid appearance (coralloid roots). **Stem** unbranched, covered with persistent leaf bases and scale leaves (cataphylls) in alternate spiral bands; outline irregular in section, large pith, numerous small vascular bundles in a ring, a wide cortex; parenchyma cells of cortex and pith containing a lot of starch; secondary growth initiated very early in life of plant, but produced in small amount (manoxylic); mucilage canals abundant in stem; leaf traces forming girdle in stem. **Leaves** large, thick, leathery, compound, forming a crown at the apex of stem, conduplicate or flat, not circinate when young (circinate in *Bowenia*), pinnate compound or bipinnate compound (*Bowenia*), spirally arranged, alternately with scale leaves; each leaflet with numerous parallel veins (there being no midvein); midvein present with dichotomous secondary veins in *Stangeria*; margin entire, dentate or with sharp spines; leaf petiole with omega-like pattern of vascular bundles; petiole and rachis with or without stout spines. **Microsporophylls** aggregated into compact strobilus (cone), arranged spirally, hard, wedge-shaped with minute sterile projection; microsporangia numerous, borne on abaxial surface, often clustered; pollen nonsaccate, with a single furrow; sperms motile. **Megasporophylls** aggregated into strobilus, densely crowded, symmetrical to asymmetrically peltate, valvate or imbricate, each with two ovules; strobili 1-several per plant, globose, ovoid or cylindrical, disintegrating at maturity; seeds large, 1–2 cm long, rounded in cross section, often brightly colored with fleshy outer layer and hard inner layer; cotyledons 2.

Economic importance: Several species are commonly grown as ornamentals. Starchy underground rhizome of *Zamia pumila* of tropical America were used by early settlers as flour. Seeds of *Dioon edule* are also ground into meal and eaten.

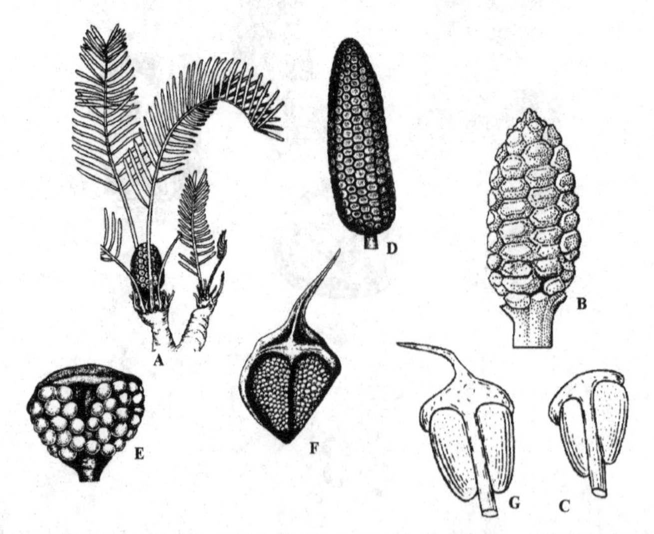

Figure 12.2: Zamiaceae. *Zamia floridana.* **A:** Plant with female cone; **B:** Female cone; **C:** Megasporophyll; **D:** Male cone; **E:** Microsporophyll. *Macrozamia.* **F:** Microsporophyll; **G:** Megasporophyll.

Several species are also used for the production of sago starch. Removal of toxic glycosides is however imperative before consumption. This is easily achieved through boiling.

Phylogeny: The family is distinct from Cycadaceae in presence of lateral veins, megasporophylls reduced and aggregated into strobili, absence of circinate vernation, and two reflexed ovules per megasporophyll. *Stangeria* is unique in the family with buds arising from roots, absence of scales and with presence of midvein and dichotomously branched lateral veins. It was removed into distinct family Stangeriaceae by Johnson (1959). *Bowenia,* similarly has bipinnate leaves, shows circinate vernation in leaflets and removed to Boweniaceae by Stevenson (1981), on subsequent morphological evidence included it under Stangeriaceae. Molecular studies, however, include both genera under Zamiaceae (Zgurski et al., 2008). The relationships of these genera, and other members of the family, however, are not clear.

Ginkgoaceae Engler *Maidenhair Tree Family*

Limited to remote mountain valleys of China. Possibly extinct in wild but planted in temples and gardens in many parts of the world, especially in temperate climate.

1 Genus, 1 species

Salient features: Deciduous tree with simple fan-like leaves, veins dichotomously branched, dioecious, ovules paired on long stalks, seed juicy with unpleasant smelling outer layer.

Genus: Single genus *Ginkgo* (1 species).

Description: Tall deciduous tree reaching 30 m. **Roots** diarch when young, older tetrarch or hexarch, containing VAM. **Stem** branched, crown asymmetric with curved branches attached to stout trunk; bark furrowed, grey; secondary growth profuse in long shoots forming broad wood zone (pycnoxylic), whereas short shoots show little of secondary growth (manoxylic). **Leaves** simple, fan-shaped, bilobed or entire, spirally arranged, turning bright yellow in autumn, widely placed on long shoots, crowded and appearing whorled on stout short shoots, latter arising from old long shoots; veins dichotomously branched, without midvein. Plants dioecious. **Microsporophylls** spirally arranged, aggregated in small loose strobili resembling angiosperm catkins, borne on short shoots; microsporophyll with short stalk and knob-like hump, bearing a pair of microsporangia; pollen tube not developed; sperms motile; pollination by wind. **Ovules** paired, on long stalks, on short shoots, hanging like cherries, naked; seeds one per stalk (other ovule not maturing), about 2.5 cm in diameter; outer coat fleshy, white-pink, unpleasant smelling (like rancid butter or human vomit); cotyledons 2–3.

Economic importance: The tree has long been grown as ornamental near temples and religious institutions in Eastern Asia. Male plants are commonly grown as they do not produce the unpleasant smell unlike female plants. More recently

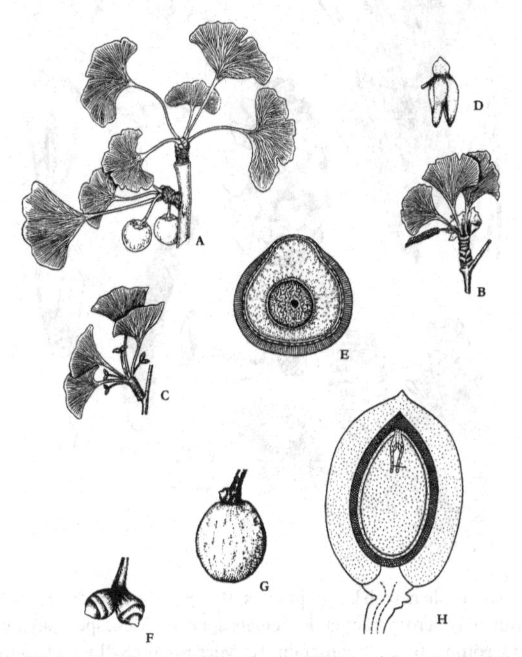

Figure 12.3: Ginkgoaceae. *Ginkgo biloba.* **A:** Shoot with seeds; **B:** Shoot with male strobilus; **C:** Shoot with paired ovules; **D:** Microsporophyll with two microsporangia; **E:** Pollen grain; **F:** Pair of ovules; **G:** Developing seed with aborted ovule near base; **H:** Longitudinal section of mature seed.

it has been planted in Canada, USA and Europe. It is relatively disease resistant and tolerates high air pollution. The seeds boiled or fried, are delicacy in some Chinese dishes.

Phylogeny: *Ginkgo biloba* is known in fossil record in Triassic and Jurassic periods, having appeared 200 million years ago, and reproductive structures seem to have changed little at least for last 120 m years, justifying its being called as "living fossil". It is not closely related to any extant group but shares motile sperms with cycads. Absence of pollen tube is another primitive feature. The short shoots are manoxylic (like Cycadales), whereas the long shoots are pycnoxylic (like conifers).

Ginkgo is unique in having sex chromosomes: *XY* male and *XX* female. The leaves resemble *Adiantum,* the Maidenhair fern, hence the name Maidenhair tree for *Ginkgo.*

Pinaceae Adanson ***Pine Family***

Distributed extensively in Northern Hemisphere, from warm temperate to arctic regions, extending to tree limit in mountains.

11 Genera, 228 species

Salient features: Evergreen resinous trees with linear to needle-like leaves, male and female cones distinct, seeds several in woody cones hidden by scales, ovuliferous scales imbricate, flat, distinct from bract scales, seeds winged, 2 per scale.

Major Genera: *Pinus* (100 species), *Abies* (40), *Picea* (40), *Larix* (10), *Tsuga* (10) and *Pseudotsuga* (5).

Description: Trees, rarely shrubs, evergreen, rarely deciduous (*Larix, Pseudolarix*), with strong smell from bark and leaves. **Roots** containing ectomycorrhiza. **Stem** branched, trunk elongate with whorled or opposite branches, rarely alternate; resin canals present in wood and leaves, crown pyramidal or spreading; wood pycnoxylic; p-plastids in sieve cells. **Leaves** simple, linear to needle-like, spiral but often appearing 2-ranked by twisting of leaf base to bring leaves into one plane, clustered in sheathed fascicles on short shoots of *Pinus*; sessile or short petioled; buds enclosed in bud scales. Plants monoecious. **Male cones** small, microsporophylls papery, spirally arranged; microsporangia abaxial, two

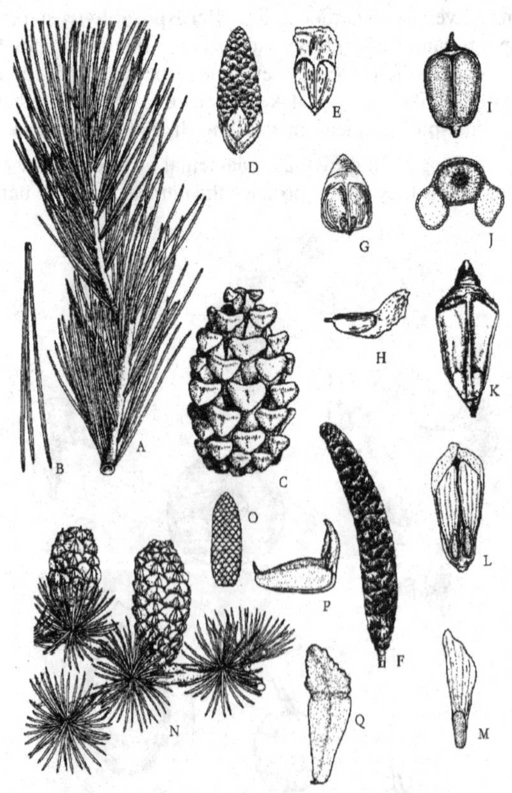

Figure 12.4: Pinaceae. *Pinus gerardiana.* **A:** Shoot with leaves in spurs; **B:** Spur with three needles; **C:** Mature female cone. *P. wallichiana*; **D:** Male cone; **E:** Microsporophyll with two microsporangia; **F:** Female cone; **G:** Megasporophyll with two ovules; **H:** Microsporophyll in lateral view. *P. roxburghii*; **I:** Microsporophyll; **J:** Pollen grain; **K:** Microsporophyll with two ovules; **L:** Microsporophyll with two seeds; **M:** Seed with wing. *Cedrus deodara*; **N:** Shoot with leaf clusters and mature female cones; **O:** Male cone; **P:** Microsporophyll in lateral view; **Q:** Microsporophyll in dorsal view.

per microsporophyll; pollen grains saccate with two saccae (saccae absent in *Larix*, *Pseudotsuga*). **Female cone** woody with persistent spirally arranged ovuliferous scales, each with two ovules, each scale in axil of but free from bract scale; bract longer or shorter than scales; ovules inverted, on adaxial surface of scale; archegonia few per ovule, seed with long terminal wing derived from scale; embryo straight, cotyledons 2–18, seeds shedding after elongation of cone axis, allowing scales to open; female cones taking 1–2 years to mature.

Economic importance: The family is the leading source of timber in the world. The wood of *Pinus* (Pines), *Pseudotsuga* (Douglas fir), *Picea* (Spruce), *Abies* (Fir), *Tsuga* (Hemlock) and *Cedrus* (Cedars) is extensively used for timber, fence posts, furniture and paper pulp. Many of these are also grown as ornamentals. Seeds of several species of *Pinus* (pinon pines or pine nuts) particularly *P. gerardiana* of W. Himalayas (Chilgoza; Neoza) are eaten as nuts. Rosin and turpentine are extracted from several species of pines.

Phylogeny: Pinaceae is a well-defined family. Numerous features such as inverted ovule, winged seed, woody cones, p-plastids in sieve cells, simple linear or acicular leaves and absence of biflavonoid compounds strongly support the monophyly of family. The family is sister to rest of the conifers as evidenced by morphological (Hart, 1987) and molecular evidence (Quinn et al., 2002).

The family is commonly divided into two subfamilies: Pinoideae (*Cathaya, Larix, Picea, Pinus* and *Pseudotsuga*) and Abietoideae (*Abies, Cedrus, Keteleria, Pseudolarix* and *Tsuga*). The separation is supported by data from chloroplast *matK*, mitochondrial *nad5* and nuclear *4CL* genes, although *Cedrus* is sister to rest of the family.

Cupressaceae Gray *Cypress Family*

Cosmopoliton distribution, mainly in warm and cold temperate climates, more abundant in Northern Hemisphere.

30 Genera, 149 species

Salient features: Trees or shrubs, leaves scale-like or needle-like, persistent on branches after dying, cone scales usually valvate, pollen grains nonsaccate, ovuliferous scale fused with bract scale, seeds with 2–3 lateral wings.

Major Genera: *Juniperus* (68 species), *Callitropis* (18), *Callitris* (14), *Cupressus* (12), *Chamaecyparis* (7), *Thuja* (5), *Taxodium* (3), *Sequoia* (1) and *Sequoiadendron* (1).

Description: Trees or shrubs with aromatic wood and foliage. **Stem** branched, trunk usually short with diffuse branches, bark fibrous, shedding in long strings, buds without bud scales; branches after shedding with age. **Leaves** simple, small, scale-like to needle-like, persistent, often closely appressed to branches or spreading and shedding with them, spiral but often appearing 2-ranked, opposite or whorled, with resin canals; often linear leaves on leading branches and scale-like

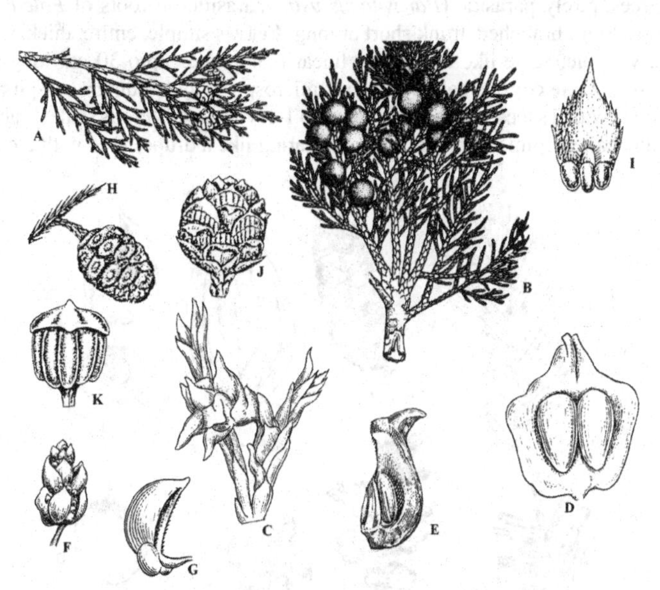

Figure 12.5: Cupressaceae. A: Shoot of *Cupressus torulosa* with female cones; **B:** Branch of *Juniperus indica* with berry-like female cones. *Thuja orientalis.* **C:** Curved branch with cone; **D:** Megasporophyll in abaxial view with two ovules; **E:** Megasporophyll in lateral view; **F:** Mature male cone; **G:** Side view of microsporophyll. *Sequoiadendron giganteum.* **H:** Portion of shoot with female cone; **I:** Microsporophyll; **J:** Female cone of *Taxodium distichum*; **K:** Megasporophyll of *Sequoia sempervirens* with several ovules.

on lateral branches; scale-like leaves often dimorphic: those towards base and top of branch flat, middle ones keeled and folded around branch. Plants monoecious, rarely dioecious (*Juniperus*). **Male cones** small, inconspicuous, microsporophylls spiral or opposite, microsporangia abaxial, 2–10 per microsporophyll; pollen grains nonsaccate. **Female cones** terminal or lateral on short branches; ovuliferous scale and bract scale fused, basally attached and flat or peltate, persistent, rarely deciduous (*Taxodium*), each with 1–20 ovules, inverted, on adaxial surface of scale, erect or inverted; archegonia clustered; seed with 2–3 short lateral wings derived from seed coat; embryo straight, cotyledons 2–15; cone maturing in 1–3 years, woody, rarely fleshy and berry-like (*Juniperus*).

Economic importance: The family is known for its ornamental shrubs and trees, particularly Species of *Cryptomeria* (*C. japonica*), *Cupressus*, *Thuja* (*T. orientalis*-arbor vitae) and *Juniperus*, the leading source of timber in the world. The wood of *Juniperus virginiana* (Eastern red cedar) is used for cedar chests, to line closets, for pencils and for shingles. Oil from the cones of *J. communis* is used to flavour gin. Juniper pollen is known to cause pulmonary allergies in humans.

Phylogeny: The family has often been considered distinct from Taxodiaceae (including *Taxodium*, *Sequoia*, *Sequoiadendron*, etc.) in either opposite and scale-like or whorled and linear leaves as against spiral and linear in latter (although *Metasequoia* has opposite, and *Athrotaxis* scale-like leaves). Page (1990) also suggested fundamental differences in reproductive structures. The two, however, share features like fused ovuliferous and bract scale, wings derived from seed coat, more than 2 microsporangia per microsporophyll, more than 2 seeds per scale, shedding of small branches, wingless pollen grains and clustered archegonia. Eckenwalder (1976) suggested the merger of two based on phenetic evidence. The merger of two families is also supported by molecular evidence (Quinn et al., 2002; Farjon, 2005). Unified monophyletic Cupressaceae perhaps arose from the paraphyletic assemblage "Taxodiaceae" which form a basal clade. The genera from Northern Hemisphere and Southern Hemisphere, however, form two distinct clades, cuppresoid clade and callitroid clade, respectively. The presence of several teeth on ovuliferous scales of *Cryptomeria* is perhaps reversal to plesiomorphic condition.

Podocarpaceae Endlicher ***Podocarp Family***

Tropical and subtropical regions of Southern Hemisphere, extending northwards to Japan, Central America and West Indies.

19 Genera, 187 species

Salient features: Shrubs or trees, leaves linear or broader, persistent, microsporophyll with two microsporangia, mature ovuliferous scale with one ovule, seeds surrounded by specialized scale called epimatium, bracts juicy, pollen saccate.

Major Genera: *Podocarpus* (100 species), *Dacrydium* (20), *Dacrycarpus* (9), *Afrocarpus* (6) and *Phyllocladus* (5).

Description: Shrubs or trees, rarely parasitic (*Parasitaxus ustus* parasitic on roots of *Falcatifolium taxoides*, another podocarp), slightly resinous. **Stem** branched, trunk short or long. **Leaves** simple, entire, thick, persistent, alternate, rarely opposite (*Microcachrys*), variable, scale-like to broadly linear (sometimes upto 30 cm long and 5 cm broad). Plants monoecious, rarely dioecious. **Male cones** small, cylindical; microsporophylls numerous, spirally arranged, each with 2 microsporangia, pollen grains with 2 saccae. **Female cones** with 1-several ovuliferous scales; each scale with single ovule, modified into juicy structure called epimetium, thus cone appearing like a drupe; seed with 2 cotyledons.

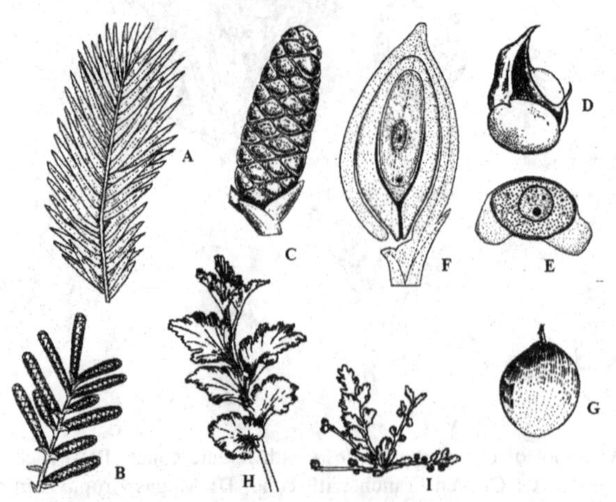

Figure 12.6: Podocarpaceae. *Podocarpus spicatus.* **A:** Vegetative shoot; **B:** Branch with male cones; **C:** Single cone showing arrangement of microsporophylls; **D:** Microsporophyll; **E:** Pollen grain; **F:** Vertical section of ovule; **G:** Mature seed. *Phyllocladus alpinus*; **H:** Branch with flattened phylloclads; **I:** Portion of branch with female cones.

Economic importance: Species of *Podocarpus* and *Dacrydium* are valuable sources of timber. *Podocarpus macrophyllus* is widely planted as ornamental.

Phylogeny: The family has been considered closer to Taxaceae with which it shares features of resinous plants, ovuliferous scale with solitary ovule and cone more or less fleshy. In two families, however, the nature of fleshy structure is different. It represents ovuliferous scale in Podocarpaceae, whereas it represents an aril, an outgrowth from the base of ovule, in Taxaceae (Quinn et al., 2002). Genus *Phyllocladus* with branches flattened into phylloclads looking like leaves (leaves reduced to scales) and with arillate seeds is often separated into distinct family Phyllocladaceae. Recent *rbcL* studies, however, point to their being sister groups (Quinn et al., 2002), or *Phyllocladus* embedded in Podocarpaceae (Wagstaff, 2004), latter conclusion also reached based on nuclear gene XDH (Peery et al., 2008).

Araucariaceae Henckel & W. Hochst ***Monkey-puzzle tree Family***

Nearly restricted to Southern Hemisphere mainly in S. E. Asia to Australia, New Zealand, and S. America.

3 Genera, 37 species

Salient features: Large trees with naked buds, highly resinous, leaves needle-like to lanceolate, persistent, shedding along with small branches, pollen grains nonsaccate, exine pitted, female cones woody, ovuliferous scale with single ovule.

Major Genera: *Araucaria* (18 species), *Agathis* (13) and *Wollemia* (1).

Description: Tall long-lived trees reaching up to 65 m in height and up to 6 m in diameter, highly resinous, usually symmetrical with conical crown, buds naked. **Stem** branched, trunk stout and thick, small branches shedding along with leaves. **Leaves** simple, entire, varying in shape from awl-shaped, scale-like, linear, oblong to elliptic, persistent, shedding on small branches, spiral (*Araucaria*) or opposite (*Agathis*). Plants monoecious (*Agathis*) or dioecious (*Araucaria*). **Male cones** small, cylindrical; microsporophylls numerous, spirally arranged, each with 4–20 microsporangia, pollen grains nonsaccate, exine pitted; sperms nonmotile. **Female cones** solitary, more or less erect, maturing in 2–3 years, disintegrating on tree, ovuliferous scales numerous, spirally arranged, flattened, linear to peltate, longer than and fused with bract scale, each scale with single ovule; ovule free from scale (*Araucaria*) or fused with it (*Agathis*); seed large, with (*Agathis*) or without (*Araucaria*) marginal wings, with 2 cotyledons sometimes deeply divided and appearing 4.

Economic importance: Species of *Araucaria* mainly *A. araucana* (Monkey-puzzle tree) with spectacular whorled branches and *A. heterophylla* (Norfolk Island pine) of Chile are prized ornamentals grown as avenue trees and house plants. Many species mainly *Agathis australis* (Kauri) with massive trees are utilized for timber.

Phylogeny: The family is distinct, restricted to the Southern Hemisphere, fossil record of Araucaria extending to Jurassic. Both genera are, *Agathis* and *Araucaria* are well separated, former with opposite broader leaves, monoecious habit, ovule free from scale, winged seeds and latter with spiral linear leaves, dioecious habit, ovule fused with scale, and wingless seeds. Monophyly of two genera is supported by *rbcL* sequence data.

Wollemia nobilis, discovered only in 1994 from Wollemi National Park in Australia was earlier known only from fossil record extending to 150 MYA. The species is represented by less than 50 trees and has unique dark brown and knobby bark described as "bubbling chocolate". It is also multi-trunked appearing as clumps of trunks.

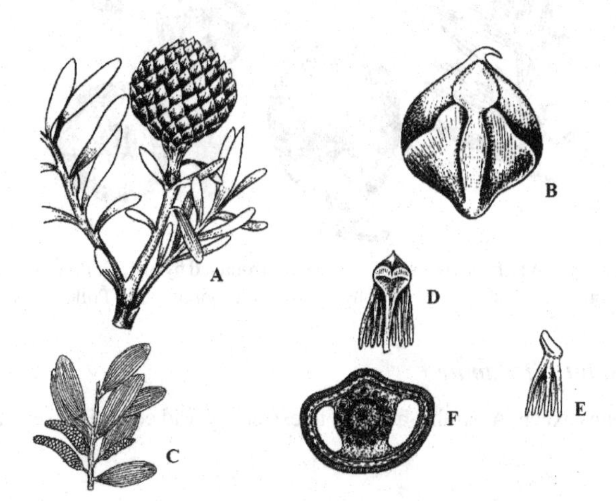

Figure 12.7: Araucariaceae. *Agathis australis.* **A:** Shoot with female cone; **B:** Megasporophyll; **C:** Shoot of *Agathis alba* with male cones. *Araucaria angustifolia*; **D–E:** Microsporophylls in different views; **F:** Pollen grain.

Taxaceae Bercht. & J. Presl. ***Yew Family***

Largely distributed in Northern Hemisphere, extending south to Guatemala and Java and Caledonia.

6 Genera, 30 species

Salient features: Shrubs to small trees, wood without resin canals, leaves linear, flattened, persistent, pollen grains nonsaccate, ovules solitary, not in cones, with fleshy aril at base surrounding the ovule fully at maturity and becoming bright red.

Genera: *Taxus* (10 species), *Cephalotaxus* (8), *Torreya* (5), *Amenotaxus* (5), *Pseudotaxus* (1) and *Austrotaxus* (1).

Description: Shrubs or moderately sized trees, not resinous or slightly resinous, fragrant or not. **Stem** much branched, wood without resin canals. **Leaves** simple, entire, linear, flattened, with abruptly tapering apex, persistent for several years, shedding singly, spiral, often twisted at base to appear 2-ranked, leaf base with decurrent petiole. Plants dioecious or monoecious. **Male cones** with 6–14 microsporophylls, each with 2–9 microsporangia radially arranged or on abaxial surface; pollen grains nonsaccate, sperms nonmotile. **Ovules** solitary, not forming cones, surrounded by fleshy aril, arising from base, aril brightly colored at maturity; seed with hard seed coat, cotyledons 2.

Economic importance: Various species of Taxus, especially *T. baccata* (English yew) and *T. cuspidata* (Oriental yew) are commonly planted as ornamentals. Wood of yew family has been popular since Middle Ages for making bows, owing to the presence of extra spiral thickenings on the xylem cells. Wood of *Taxus* is used in high grade furniture. The presence of taxol, a highly toxic alkaloid having antimitotic activity makes it potential agent for anticancer chemotherapeutic treatment.

Phylogeny: The family is unique in conifers in the absence of female cone, and in having solitary ovules, and has often been removed to a distict order Taxales. The family, however, shares similar embryology, wood anatomy, leaf and pollen morphology with rest of the conifers. The placement within Coniferales is supported by evidence from DNA studies (Chase et al., 1993; Price, 2003) and micromorphology (Anderson and Owens, 2003). Two distinct clades are established within the family: one including *Taxus*, *Austrotaxus* and *Pseudotaxus* (aril partly enclosing ovule, maturing in 6–8 months and mature seed 5–8 mm long) and the other including *Torreya*, *Cephalotaxus* and *Amenotaxus* (aril completely enclosing ovule, maturing in 18–20 months and mature seed 12–40 mm long). The two are sometimes placed in distinct families Taxaceae s. s. and Cephalotaxaceae, respectively. There is, however, strong molecular evidence to support their merger into broadly circumscribed Taxaceae (Price, 2003; Rai et al., 2008), as recognition of Cephalotaxaceae renders it para- or polyphyletic.

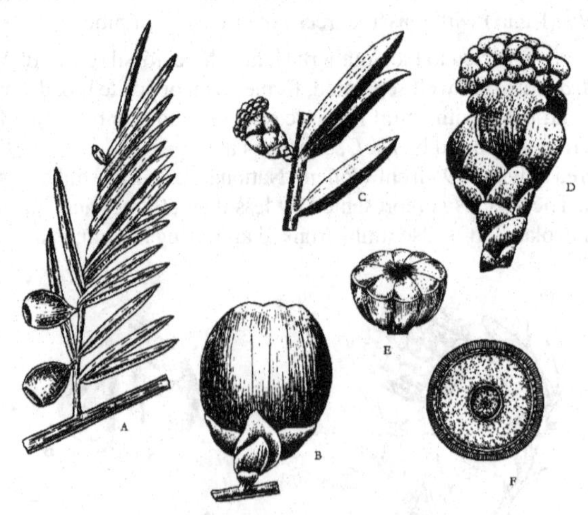

Figure 12.8: **Taxaceae.** *Taxus baccata.* **A:** Shoot with seeds; **B:** Seed surrounded by aril; **C:** Portion of branch with male cone; **D:** Male cone; **E:** Peltate microsporophyll with microsporangia; **F:** Pollen grain.

Ephedraceae Dumort ***Joint Fir Family***

Worldwide in temperate regions except Australia, adapted to extremely arid conditions, extending up to 4000 m in Andes and Himalayas.

1 Genus, 68 species

Salient features: Shrubs with jointed stems with clustered or whorled branches, vessels present, leaves scale-like, fused into sheath, microsporophylls stalked, pollen furrowed, nonsaccate, each ovule surrounded by pair of fused bracts, seed covered.

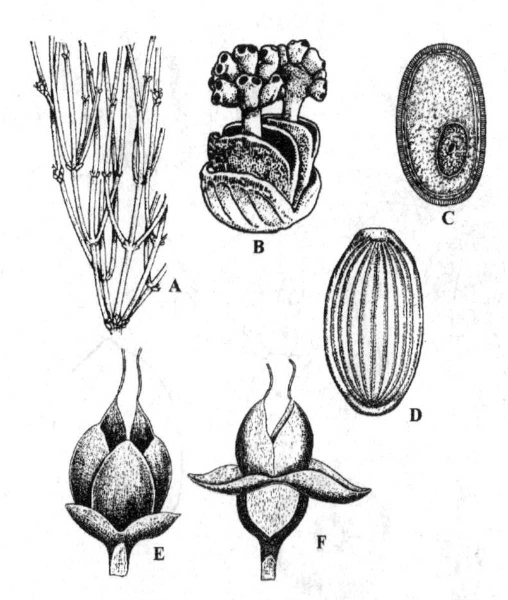

Figure 12.9: Ephedraceae. *Ephedra gerardiana.* **A:** Shoot with strobili; **B:** Male strobilus; **C:** Pollen grain; **D:** Surface of pollen grain with parallel ridges on exine; **E:** Female strobilus with bracts and two ovules; **F:** Same with seeds.

Genus: *Ephedra* (68 species).

Description: Small shrubs, often trailing, rarely climbing (*E. foliata*), very rarely almost tree-like reaching 30 cm in diameter and height of several metres (*E. triandra*). **Root** a tap root in seedling, persisting for long time, but gradually replaced by adventitious roots. **Stem** much branched, aerial stems arising from spreading rhizome; branches numerous, whorled or clustered, longitudinally grooved, distinctly jointed with long internodes, usually green and photosynthetic, horsetail like, wood with vessels. **Leaves** scale-like, opposite or in whorls of 3–4 leaves, fused at base into a sheath at each node, often shedding early, each leaf with two parallel veins; with axilliary buds; resin canals absent. Plants usually dioecious, rarely monoecious. **Male strobili** (inflorescence) in whorls of 1–10 at nodes in axil of scale leaf, each with 2–8 successive pairs of cupped bracts, lower one or two pairs of bracts sterile rest bearing solitary microsporangiate shoots (microsporophylls, flowers), each on short secondary axis (microsporangiophore) arising between each pair of fertile bracts and bearing two opposite scales (perianth) and into 2–10 microsporangia (stamens with filamented or sessile anthers), dehiscence by terminal pores; pollen furrowed, nonsaccate, inaperturate, exine shed on germination, pollen becoming naked. **Female strobili** opposite or in whorls of 3–4 at branch nodes, each with 2–10 successive pairs of bracts, uppermost fused to form a fleshy cup around the ovule borne on a stalk (stalk and ovule constitute female flower); ovule single or in pairs, with 2 integuments; seeds 1–2 per strobilus, surrounded by leathery yellow to dark brown cup; cotyledons 2.

Economic importance: Several species of *Ephedra* were used as beverage by early Mormon settlers, hence the name Mormon tea for the genus. The alkaloid drug ephedrine used as decongestant, treatment for cough and circulatory problems is obtained from several species especially *E. sinica* (ma huang), which has been used in China earlier than 2500 B.C.

Phylogeny: The genus *Ephedra* shown superficial resemblance to *Equisetum* and *Casuarina*, all three exhibiting "switch habit", owing to sheathed nodes and scale leaves. The presence of vessels, perianth-like bracts, extremely reduced gametophytes and the fusion of second male gamete with venter canal nucleus have been taken as angiosperm affinities, although the origin of vessels in two has been separate and several primitive angiosperms are vessel-less. It has been suggested that Gnetales are sister to a clade including all other seed plants based on studies of *rbc*L (Seider et al., 2002) and nuclear genes (Rydin et al., 2002), rendering gymnosperms as paraphyletic. The bulk of evidence in recent years, however, points to gymnosperms being monophyletic and Gnetales sister to Pinales. The binucleate sperm cells, basic proembryo structure, development of polyembryony, etc., of *Ephedra* agree with Pinales in general and perhaps Pinaceae in particular.

Gnetaceae Lindley ***Gnetum Family***

Distributed in tropical regions of W. Africa, Brazil, India and Southeast Asia.

1 Genus, 43 species

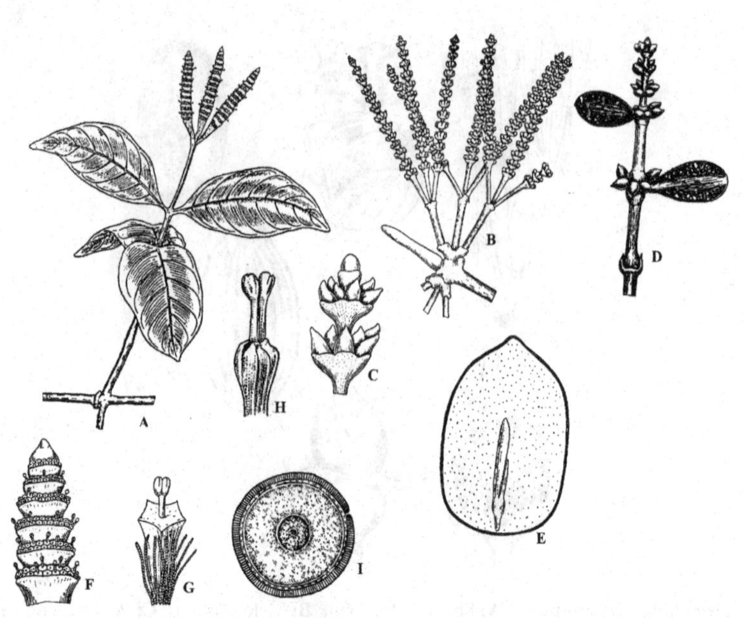

Figure 12.10: Gnetaceae. A: Shoot of *Gnetum latifolium* with opposite leaves and male strobili; **B:** Portion of branch with female strobili; **C:** Portion of female strobilus showing ovules; **D:** Portion of female strobilus with two mature seeds; **E:** Longitudinal section of seed; **F:** Portion of male strobilus of *G. ulva*; **G:** Single microsporophyll of same; **H:** Microsporophyll of *G. gnemon*; **I:** Pollen grain.

Salient features: Evergreen lianas, trees or shrubs, with vessels, leaves opposite, angiosperm-like, net-veined, microsporophylls stalked, with bracts, pollen spinose, seeds large, fleshy.

Genus: *Gnetum* (43 species).

Description: Usually liana, sometimes scadent shrubs (*G. contractum*) shrubs, rarely trees (*G. gnemon*). **Stem** climbing often reaching top of tall trees, rarely erect, sometimes with two types of shoots: long shoots and short shoots, rarely all shoots similar (*G. gnemon*), vessels present, phloem with companion cells, arising from cambium cells and not from mother-cell as in angiosperms. **Leaves** large, entire, oblong, elliptic or lanceolate, subsessile or short petiolate, opposite and decussate, net-veined, appearing like a dicot leaf, appearing only on short shoots in lianas, often reduced to scales on long shoots, stipules absent. Plants dioecious, rarely monoecious. **Male strobili** arising in axils of paired and decussate scale leaves, slender, elongate, with several (10–25) pairs of decussate bracts, bracts of each pair fused to form a cup known as cupule or collar; each bearing several microsporangiophores (microsporophylls, flowers) in 3–6 rings, with one ring of abortive female flowers or ovules above rings of male flowers; each microsporangiophore with two bracts (perianth) at base and two microsporangia (anthers) at top (one microsporangium in *G. gnemonoides*); perianth containing sclereids and latex tubes; pollen not striate, surface spinose. **Female strobili** with a pair of opposite sheathing bracts at base, followed by 5–6 whorls of ovules; each whorl with 4–10 ovules in a single ring above the collar, each ovule with two bracts forming perianth which persists in seed, a terminal ovule sometimes present on the strobilus; only few ovules maturing into seeds, others aborting and falling down; uniseriate hairs interspersed among the ovules; ovule with 2 integuments, subtended by perianth forming third outer envelope; perianth and outer integument with sclereids and laticiferous ducts; seed large, fleshy.

Economic importance: The tree species *G. gnemon* from Malaya, is widely cultivated as ornamental, and as food, with leaves and strobili cooked in coconut milk, and fibre for making ropes and fishing nets. The kernels of *G. ulva* yield an oil used for illumination and also as a massage in rheumatism. The plant of *G. montanum* is used as fish poison.

Phylogeny: The genus *Gnetum* along with other members of Gnetales share angiosperm features of presence of vessels, reduced gametophytes and presence of perianth like bracts. *Gnetum* appears more like angiosperms with net-veined leaves and presence of companion cells, although their origin is quite different. The reaction wood in *Gnetum* consists of gelatinous extra-xylary (reaction) fibers in the adaxial position—i.e., it is unique among seed plants (Tomlinson, 2003) and is unlike tension wood of angiosperms.

Figure 12.11: Selaginellaceae. A: *Selaginella martensii*. Equisetaceae. **B:** *Equisetum telemateia* var. *braunii*, vegetative shoots; **C:** Portion of stem enlarged. **Osmundaceae. D:** *Osmunda regalis*, portion of plant bearing sterile and fertile fronds; **E:** Portion of fertile frond enlarged; **F:** *Todea barbara*, plant with sterile and fertile fronds. **Blechnaceae.** *Woodwardia semicordata*. **G:** portion of leaf with sori; **H:** Small portion enlarged. **Dryopteridaceae.** *Cyrtomium falcatum*. **I:** Portion of leaf; **J:** small portion showing sori.

Figure 12.12: Cycadaceae. *Cycas circinalis*; **A:** Plant; **B:** Young cercinate leaves; **C:** Portion of mature leaf of *C. revoluta*; **Zamiaceae. D:** *Macrozamia communis*, apex with female cones; **E:** Plant with leaves and female cones; **F:** Plant of *Zamia integrifolia*; **G:** *Encephalartos trispinosus*, plant; **H:** *Dioon edule*, plant.

Figure 12.13: Ginkgoaceae. *Ginkgo biloba.* **A:** Branch with leaves; **B:** Portion enlarged. **Pinaceae. C:** *Cedrus deodara*, branch with leaves and female cones; **D:** Portion of bark; **E:** *Tsuga canadensis*, portion of plant with female cones. **Cupressaceae. F:** *Chamaecyparis lawsoniana*, portion of twig with female cones; **G:** *Sequoia sempervirens*, portion of branch enlarged; **H:** *Cryptomeria japonica*, portion of branch with female cones.

Figure 12.14: Podocarpaceae. A: *Podocarpus totara;* **B:** *Podocarpus gracilior*, branch enlarged; **C:** *Dacrydium cupressinum*; **D:** Twig of same showing leaves; **E:** Twig of *Phyllocladus trichmanoides.* **Taxaceae; F:** *Taxus baccata*, twig.**G:** *Torreya californica*, twig with seed enclosed in aril. **Ephedraceae; H:** *Ephedra americana*, plant with strobili.

Chapter 13

Major Families of Angiosperms

Flowering plants (Angiosperms) represent the most dominant group of plants on this planet with recent count of 369,400 (Kew Report, 2016) and 295,383 (Christenhusz and Byng, 2016) species of angiosperms. Earlier estimates were 253,300 species (Thorne, 2007) 315,000 (The Plant List version 1.1, 2013; total published names of vascular plants and bryophytes 1,064,035; accepted 350,700, incl ca 35,000 nonangiosperms), 400,000 (Edwards, 2010 on Physorg.com, 600,000 names of estimated 1 million names published to be deleted), depicting a wide variety of habit and growing in diverse habitats. The group represents tall evergreen trees often reaching more than 100 m in height (*Eucalyptus regnans*), tall deciduous trees (*Populus nigra, Platanus orientalis*), evergreen shrubs (species of *Rhododendron, Euonymus, Skimmia*), herbaceous plants more than a metre tall (*Zea mays, Saccharum officinarum, Silybum marianum*) or almost microscopic not more than 1–2 mm (*Wolfia arrhiza*). Several angiosperms with weeker stems may twine around the support (*Ipomoea, Convolvulus*), climb with the help of tendrils (*Cucurbita, Cucumis*), hooks (*Galium*) or thorns (*Bougainvillea*). The leaves may be as small just 1–2 mm (*Pilea microphylla*) or over 20 m long (*Raphia fainifera*). Broad rounded floating leaves of *Victoria amazonica* often reach 2 m in diameter and can support the weight of a child. Although flowering plants are largely green and photosynthetic, few grow as partial (*Viscum*) or total parasites on roots (*Balanophora*) or even leaves (*Arceuthobium*—growing on pinus leaves and being amongst the smallest angiosperms). A few, mainly the species of orchids (*Dendrobium* spp.) and *Tillandsia* grow as epiphytes.

As departure from third edition, the treatment of major families of Angiosperms presented in the following pages follows APG IV (2016), a continuously evolving classification of seed plants. An integration of the principal systems of classification also provided by indicating placement of the family in Principal recent systems. The view that the division of Angiosperms into monocots and dicots, renders the latter as paraphyletic, has firmly been established. Views are further consolidating to interpolate monocot taxa between primitive dicots and the more advanced ones, because it had long been suggested that monocots arose from primitive dicotyledons. The strong workforce of Angiosperm Phylogeny Group is attempting to establish monophyletic groups, more consistently up to the family level. The success above the family level is much short of any level of consistency for any meaningful application. From as many as 10 unplaced families in APG III (2009), revised APG IV (2016) has only single unplaced family Chloranthaceae under unplaced independant lineage Chloranthales, one unplaced genus (tentatively under Gesneriacea) and **COM** Clade of uncertain position with three orders Celastrales, Oxalidales and Malpighiales inserted among Rosids. The artificial clades above the order level are too arbitrary, and it may take a long time before the system may have a practical approach for the purpose for which the classifications are meant. Nevertheless, the placement of the family visa-vis all recent major systems of classification is compared, and latest phylogenetic position largely based on recent developments in serotaxonomy and molecular systematics discussed. It should be borne in mind that the same group names are not always comparable. Thus, whereas class names Magnoliopsida and Liliopsida refer to dicots and monocots in the classification systems of Cronquist and Takhtajan, the term Magnoliopsida was used for Angiosperms (for which Takhtajan and Cronquist had used the division name Magnoliophyta) by both Dahlgren and Thorne. Both of them had used the name Magnoliidae for dicots and Liliidae for monocots. However, Thorne has lately abandoned this distinction into traditional dicots and monocots, and instead used subclass names more or less comparable to Takhtajan and Cronquist (but without the super groups dicots and monocots). He has also taken the bold decision of separating the primitive dicots from the more advanced ones by inserting monocots in between, thus bringing the classification system much nearer to the Angiosperm Phylogeny Group but retaining the essence of hierarchical arrangement through consistent use of superorders and subclasses above the level of order.

Number of genera and species in each family are mostly based on Christenhusz and Byng (2016), based on families recognized by APG IV (2016). It is not under the scope of the present book to include all the families of angiosperms, but

Arrangement after APG IV (2016)
(For distribution of orders in Informal groups see Table 10.14 on page 267)

*** Listing all orders and families. Those families which are described and illustrated are indicated in bold type on the specified page.**

in addition to the major families, those which have been the subject of considerable phylogenetic interest over the recent years have been especially chosen for treatment.

Arrangement after APG IV (2016)
Basal Angiosperms:

Order 1. Amborellales
Family: **Amborellaceae**
Order 2. Nymphaeales
Family: Hydatellaceae
Family: **Cabombaceae**
Family: **Nymphaeaceae**
Order 3. Austrobaileyales
Family: **Austrobaileyaceae**
Family: Trimeniaceae
Family: **Schisandraceae**

Families in boldface are described in detail.

ANGIOSPERMS ROLL OF HONOR

Amborellaceae Pichon

1 genus, with a single species *Amborella trichopoda*

Endemic to Iceland of New Caledonia in South Pacific Ocean.

Placement:

	B & H	**Cronquist**	**Takhtajan**	**Dahlgren**	**Thorne**	**APG IV/(APweb)**
Informal Group						Basal Angiosperms/ (none)
Division		Magnoliophyta	Magnoliophyta			
Class	Dicotyledons	Magnoliopsida	Magnoliopsida	Magnoliopsida	Magnoliopsida	
Subclass	Monochlamydeae	Magnoliidae	Magnoliidae	Magnoliidae	Chloranthidae	
Series+/ Superorder	Microembyeae+		Nymphaeanae	Magnolianae	Chloranthanae	
Order		Laurales	Amborellales	Laurales	Chloranthales	Amborellales

B & H under family Monimiaceae. **B & H-** Bentham and Hooker, 1862–1883; Cronquist, 1988; **Takhtajan**, 2009; **Dahlgren- G. Dahlgren**, 1989; **Thorne**, 2007; **APG IV**, 2016; **APweb**, 2018 (Stevens).

Top left: *Victoria amazonica* (Nymphaeaceae), having large floating leaves often reaching 2 m in diameter and can often support the weight of a child. **Top right:** *Wolffia arrhiza* (Lemnaceae), the smallest known angiosperm, barely about 1 mm in size, visible as scum on the surface of water (photo Christian Fischer, CC BY-SA 3.0, https://commons.wikimedia.org/w/index.php?curid=398351). **Above left:** *Eucalyptus regnans* (Myrtaceae), the tallest tree, with recorded height of 97 m and girth of 7.5 m (Photo By Taylor—Own work, CC BY 2.5, https://commons.wikimedia.org/w/index.php?curid=1070068). **Above middle:** *Rafflesia arnoldii* (Rafflesiaceae), a bizarre plant with plant body no more than a fungus mycelium, yet producing largest sized flower sometimes reaching 1 m in diameter (photo courtesy Julie Barcelona, Manilla, Philippines). **Above left:** *Arabidopsis thaliana* (Brassicaceae), the guinea pig of plant kingdom, with most completely known genome (photo Alok Mahendroo).

Salient features: Shrubs lacking vessels, with simple alternate leaves, stipules lacking, nodes unilacunar, flowers unisexual, with multiseriate perianth, stamens numerous, pollen with granulate ektexine, carpel incompletely closed, fruit aggregate of drupes.

Genus: Only genus *Amborella* with 1 species.

Description: Sprawling shrub; wood with tracheids but no vessels, nodes unilacunar, primary medullary rays narrow, sieve-tube plastids S-type. **Leaves** evergreen, alternate, spiral to two-ranked, simple, entire to pinnately lobed, venation pinnate, stomata anomocytic, stipules absent, mesophyll without ethereal oils. **Inflorescence** cymose, plants dioecious. **Flowers** small, unisexual, hypanthium present. **Perianth** with 5–13 tepals, number more in staminate flowers than pistillate flowers, slightly united at base, spirally arranged, not differentiated into sepals and petals. **Androecium** with 12–22(–100) stamens, free, in 3–5 whorls, outer whorl adnate to tepals at base, anthers adnate, dehiscence longitudinal, introrse, microsporogenesis successive, pollen inaperturate to ulcerate, ektexine granulate, pistillate flower with 1–2 staminodes. **Gynoecium** with 5–8 carpels in a single whorl, free, ovary stalked, superior, carpel margins incompletely closed (unsealed at tip), stigma sessile with two expanded flanges, ovule 1, placentation marginal, ovule pendulous, hemianatropous, sessile. **Fruit** an aggregate of drupes with pock-marked stones, and pockets of resinous substances, seeds endospermic, embryo minute, cotyledons 2.

Economic importance: No economic value known.

Phylogeny: The family is unique in angiosperms in having granular ektexine, lacking tectum. This combined with the absence of vessels, partially closed carpels, places this family on the lowest branch of angiosperm family tree, having

Figure 13.1: Amborellaceae. *Amborella trichopoda.* **A:** A specimen growing at Arboretum of University of California, Santa Cruz (Manager of Arboretum Brett Hall is squatting besides the tree) (Photo Tim Stephens); **B:** Fully opened female flower. **C:** Close up of a branch; **D:** Male flower (photos A and C: courtesy University of California, Santa Cruz; B: photo courtesy Missouri Botanical Garden).

evolved from an unknown common ancestor of all angiosperms. The family has traditionally been placed under Laurales (Cronquist, Dahlgren, Takhtajan). The multigene analyses (Qui et al., 1999; Soltis et al., 1999; Zanis et al., 2002), support this family as sister to all extant angiosperms, with Nymphaeaceae as subsequent sister to the rest of angiosperms, which may ultimately result in placing these families in separate orders or in a common order. APG IV (2016) places this family at the begining of Angiosperms under Basal Angiosperms, under monotypic order Amborellales, a position earlier suggested by Takhtajan (2009), who placed it as first family of Magnollidae, under Nymphaeanae. APweb (2018) places similarly but without using informal group name Basal Angiosperms. Thorne had earlier (1999, 2000) placed Winteraceae at the beginning of Angiosperms (and the Magnoliidae), and Amborellaceae in the third suborder, shifted (2003) Amborellaceae, like APG schemes, to the beginning of Magnoliidae under order Chloranthales. He has further (2006, 2007) removed the order together with Nymphaeales under a distinct subclass Chloranthidae, superorder Chloranthanae.

Cabombaceae Richard ex A. Richard

2 genera, 6 species

America, India, Australia and tropical Africa.

Placement:

	B & H	**Cronquist**	**Takhtajan**	**Dahlgren**	**Thorne**	**APG IV/(APweb)**
Informal Group						Basal Angiosperms/ (none)
Division		Magnoliophyta	Magnoliophyta			
Class	Dicotyledons	Magnoliopsida	Magnoliopsida	Magnoliopsida	Magnoliopsida	
Subclass	Polypetalae	Magnoliidae	Magnoliidae	Magnoliidae	Chloranthidae	
Series+/ Superorder	Thalamiflorae+		Nymphaeanae	Chloranthanae	Nymphaeanae	
Order	Ranales	Nymphaeales	Nymphaeales	Nymphaeales	Nymphaeales	Nymphaeales

B & H under family Nymphaeaceae.

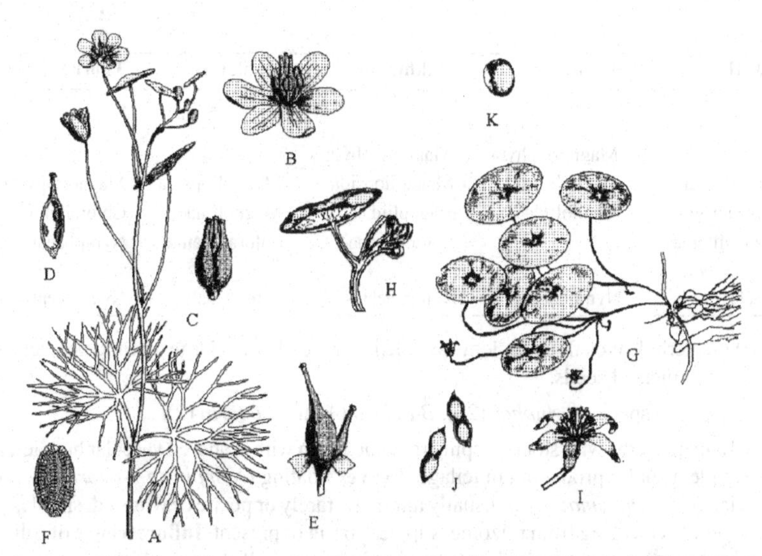

Figure 13.2: Cabombaceae. *Cabomba carolinaria.* **A:** Flowering branch with submerged dissected leaves and broad peltate floating leaves; **B:** Flower; **C:** Gynoecium with 3 free carpels; **D:** Longitudinal section of carpel; **E:** Fruit; **F:** Seed. *Brasenia schreberi.* **G:** Portion of plant with peltate leaves and small flowers; **H:** Submerged part of plant covered with thick jelly; **I:** Flower with three sepals and three petals, essentially similar; **J:** Two-seeded panduraeform nut-like fruit; **K:** Globose seed.

Salient features: Aquatic herbs, leaves floating, long-petioled, peltate, flowers large on long pedicels, stamens numerous, fruit spongy with several immersed seeds.

Genera included: *Cabomba* (5 species) and *Brasenia* (1).

Description: Perennial aquatic herbs, rhizomatous, secretary cavities present, vessels absent, sieve-tube plastids S-type. **Leaves** submerged, or submerged and floating, similar (*Brasenia*), or heterophyllous with dissected submerged leaves and entire floating leaves (*Cabomba*), alternate or opposite (submerged leaves of *Cabomba*), simple or compound, peltate, stipules absent, without sclerenchymatous idioblasts. **Inflorescence** with solitary axillary flowers. **Flowers** bisexual, 3-merous, cyclic, or partially acyclic. **Calyx** with 3 sepals, petaloid, in one whorl, free. **Corolla** with 3 petals, in one whorl, free, yellow, or purple, or white, clawed, or sessile. **Androecium** with 3–6 (*Cabomba*) stamens, or 12–18 (*Brasenia*), maturing centripetally, free, filaments slightly flattened, anthers bithecous, dehiscence by longitudinal slits, extrorse, pollen grains monosulcate, sometimes trichotomosulcate. **Gynoecium** with (2–)3–18 carpels, free, ovary superior, with a longitudinal stigmatic surface (*Brasenia*), or apically stigmatic (*Cabomba*), (1–)2(–3) ovuled, placentation parietal, ovules pendulous, anatropous, bitegmic, crassinucellate, outer integument not contributing to the micropyle. **Fruit** aggregate of follicles, many seeded spongy berry, sometimes indehiscent and nut-like (*Brasenia schreberi*); seeds endospermic, perisperm present, cotyledons 2.

Economic importance: None.

Phylogeny: The family is considered to be more strongly linked to monocots rather than Nymphaeaceae. The family has been located under superorder Nymphaeanae after Magnolianae but Takhtajan has finally taken it under distinct subclass Nymphaeidae. During the last two decades the family has been identified as a constituent of Paleoherb complex, forming the most primitive lineage among angiosperms. The paleoherb complex is characterized by scattered vascular bundles, absence of vascular cambium, leaves alternate, usually palmately veined, adventitious root system and lack of etherial cells. Judd et al. (2002) include genus *Cabomba* and *Brasenia* under Cabomboideae in Nymphaeaceae. APG IV and APweb (2018) separate them under Cabombaceae on the basis of trimerous flowers, with distinct sepals and petals, 2–3 free carpels, and fruit a follicle, included under Nymphaeales. Cabombaceae is closely related to Nymphaeaceae as evidenced by several studies. Taylor (2008), based on studies of vegetative and leaf architectural characters concluded that two families are monophyletic. *Pluricarpellatia*, known from the Early Cretaceous of Brazil probably to be placed in or near Cabombaceae (Doyle and Endress, 2014).

Nymphaeaceae Salisbury *Water Lily Family*

5 genera, 70 species (excluding Cabombaceae)

Throughout the world forming floating masses in freshwater habitats.

Placement:

	B & H	Cronquist	Takhtajan	Dahlgren	Thorne	APG IV/(APweb)
Informal Group						Basal Angiosperms/ (none)
Division		Magnoliophyta	Magnoliophyta			
Class	Dicotyledons	Magnoliopsida	Magnoliopsida	Magnoliopsida	Magnoliopsida	
Subclass	Polypetalae	Magnoliidae	Magnoliidae	Magnoliidae	Chloranthidae	
Series+/ Superorder	Thalamiflorae+		Nymphaeanae	Chloranthanae	Nymphaeanae	
Order	Ranales	Nymphaeales	Nymphaeales	Nymphaeales	Nymphaeales	Nymphaeales

Salient features: Aquatic herb, leaves floating, long-petioled, peltate, flowers large on long pedicels, stamens numerous, fruit spongy with several immersed seeds.

Major genera: *Nymphaea* (40 species), *Nuphar* (20), *Barcleya* (4) and *Victoria* (2).

Description: Perennial aquatic herbs with stout creeping rhizome. Stem with scattered vascular bundles, numerous air canals and laticifers. Hairs simple, usually producing mucilage. **Leaves** floating (*Nymphaea*, *Victoria*, etc.) or immersed, often very large (up to 2 m dia. in *Victoria amazonica*) usually alternate, rarely opposite or whorled, simple, cordate or orbicular, often peltate with long petiole emerging from rhizome, stipules absent or present. **Inflorescence** of solitary axillary flowers. **Flowers** floating or raised above water, bisexual, actinomorphic with spirally arranged stamens, hypogynous. **Calyx** with 4–12 sepals, free or connate, often petaloid. **Corolla** represented by staminodes, absent or many, free or connate at base, often passing into stamens. **Androecium** with many stamens, free, spirally arranged, filaments flattened sometimes poorly differentiated from anthers, sometimes adnate to petaloid staminodes, pollen grains usually monosulcate or inaperturate. **Gynoecium** with 3-many free or connate carpels with several locules and parietal placentation, unilocular with single or many ovules, stigmas often elongated and radiating from the disc, often surrounding a central bump, ovary superior (*Nuphar*), semi-inferior (*Nymphaea*), or inferior (*Euryale*). **Fruit** a spongy berry, rarely an aggregate of nuts or pods; seeds usually operculate, arillate, with small embryo, endosperm absent but with a conspicuous perisperm. Pollination by beetles, flies and bees. Flowers of *Victoria* and some species *Nymphaea* have starch-filled apical appendages of carpels as insect attraction, providing food, heat and characteristic smell. The fruit, on maturity, splits to separate individual segments (*Nuphar*) or ruptures under water so as to release seeds.

Economic importance: Species of *Nymphaea* (water lily), *Nuphar* (yellow water lily, spatterdock), and *Victoria* (Amazon lily) are ornamentals grown in ponds and lakes. The leaves of *Victoria amazonica* (Royal water lily) are so large that they can support the weight of a child. The seeds of *Victoria*, *Nymphaea* and *Euryale ferox* (Makhana) are often consumed.

Phylogeny: The family has been a subject of considerable discussion, often strongly linked with monocots, although traditionally classified with dicots. The family has been located under superorder Nymphaeanae after Magnolianae but Takhtajan has finally taken it under a distinct subclass Nymphaeidae. During the last decade, the family has been identified

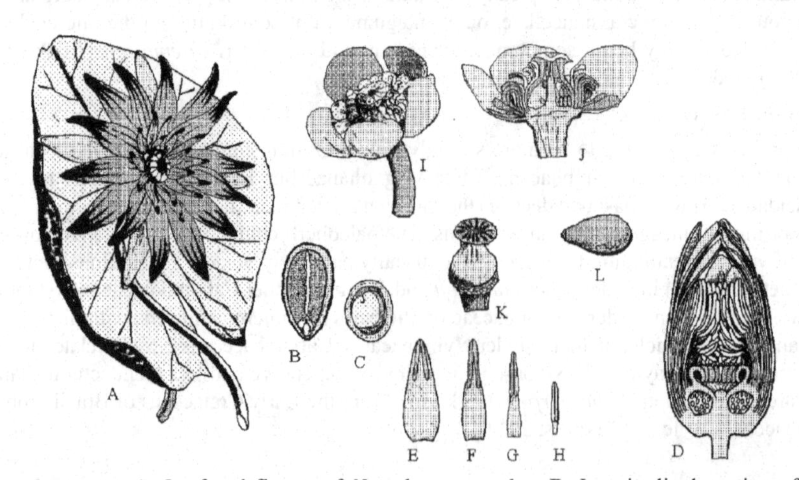

Figure 13.3: Nymphaeaceae. A: Leaf and flower of *Nymphaea coerulea*. **B:** Longitudinal section of seed of *N. alba*. **C:** Transverse section of seed of the same showing plumule lying in the cavity of one cotyledon. **D:** Longitudinal section of flower of *N. odorata*; **E-H:** Successively outer to inner stamens. *Nuphar* sp. **I:** Flower; **J:** Longitudinal **section** of flower; **K:** Gynoecium with stigmatic disc; **L:** Seed.

as a constituent of Paleoherb complex, forming the most primitive lineage among angiosperms. The paleoherb complex is characterized by scattered vascular bundles, absence of vascular cambium, leaves alternate, usually palmately veined, adventitious root system and lack of etherial cells. The family formerly also included genus *Nelumbo*, which has now been separated under family Nelumbonaceae because of distinct tricolpate pollen grains and absence of laticifers. Takhtajan places it under separate subclass Nelumbonidae, whereas in APG II classification it is removed under Tricolpates (Eudicots) clade. *Cabomba* and *Brasenia* earlier placed under Cabomboideae in Nymphaeaceae, in APG II, have been removed to Cabombaceae in APG IV and APweb, as was earlier done by Thorne (2000, 2003, 2006, 2007). Earlier placed after the Magnoloid complex, Thorne has finally shifted Cabombaceae and Nymphaeaceae in basal angiosperm clades as suggested by molecular studies (Qui et al., 2000; Soltis et al., 2000). Wikström et al. (2001) suggest an age for the Nymphaeales clade some 171–153 MYA, with divergence occurring 144–111 MYA. Yoo et al. (2005) based on molecular and morphological data concluded that Nymphaeales diversified into two major clades corresponding to Cabombaceae and Nymphaeaceae during the Eocene (44.6 ± 7.9 mya). The inclusion of Nuphar in Nymphaeaceae needs to be confirmed (Stevens, APweb, 2018), as such he places it under separate subfamily Nupharoideae, the rest placed under Nymphaeoideae.

Austrobaileyaceae (Croizat) Croizat

1 genus, single species *Austrobaileya scandens*
Native to Queensland, Australia.

Placement:

	B & H	Cronquist	Takhtajan	Dahlgren	Thorne	APG IV/(APweb)
Informal Group						Basal Angiosperms/ (none)
Division		Magnoliophyta	Magnoliophyta			
Class		Magnoliopsida	Magnoliopsida	Magnoliopsida	Magnoliopsida	
Subclass		Magnoliidae	Magnoliidae	Magnoliidae	Chloranthidae	
Series+/ Superorder			Nymphaeanae	Magnolianae	Chloranthanae	
Order	Not described then	Magnoliales	Austrobaileyales	Annonales	Chloranthales	Austrobaileyales

Salient features: Climbing shrubs with opposite leaves, flowers solitary in leaf axils, bisexual with numerous tepals gradually intergrading from sepals to petals, stamens numerous, laminar, inner modified into staminodes, carpels several, free, partially unsealed with bilobed style.

Major genera: Single genus with one species. Originally, two species *Austrobaileya maculata* and *A. scandens* were described, but they have now been combined into a single species under the latter name.

Description: Large climbing shrubs bearing essential oils, nodes unilacunar, with two traces, vessel end-walls scalariform, sieve-tube plastids S-type. **Leaves** evergreen, opposite to sub-opposite, leathery, petiolate, simple, entire, pinnately veined, stipules intrapetiolar, caducous, small, mesophyll with spherical etherial oil cells. **Inflorescence** with solitary axillary flowers. **Flowers** bisexual, bracteate, pedicellate, bracteolate. **Perianth** with tepals nearly sequentially intergrading from sepals to petals (9–)12(–14), free, imbricate. **Androecium** with 12–25 stamens, maturing centripetally, free, outer laminar, petaloid, fertile, inner gradually smaller, innermost reduced to staminodes, anthers adnate, dehiscence longitudinal, introrse, pollen grains monosulcate. **Gynoecium** with (6–)9(–12) carpels, free, spirally arranged, ovary superior, 8–14 ovuled, placentation marginal (biseriate), ovules collateral, anatropous, bitegmic, crassinucellate, style partially unsealed, bilobed. **Fruit** baccate, seeds with ruminate endosperm. Pollination by insects.

Economic importance: No economic value known.

Phylogeny: When first described by C. T. White (1933), it was considered to be belonging to Magnoliales, a placement also followed by Cronquist, but due to unique combination of characters, and for want of a better place, Hutchinson (1973) placed the family in Laurales, near Monimiaceae. According to Thorne (1996), the family Austrobaileyaceae is so intermediate between Magnoliales and Laurales, that Laurales should not be recognized as a separate order. He, accordingly, placed Magnoliaceae (and related families), and Lauraceae (and related families) under separate suborders Magnoliineae, and Laurineae, respectively. He later (1999, 2000) placed this family closer to Monimiaceae and Chloranthaceae under suborder Chloranthineae, before suborder Magnoliineae under Magnoliales. In his subsequent revision (2003), however, he recognized the suborders as independent orders, Chloranthales (upgraded Chloranthineae) being placed towards the beginning of Magnoliidae. In latest revision (2006, 2007), however he has shifted this family under a distinct Subclass Chloranthidae, superorder Chloranthanae, order Chloranthales, suborder Chloranthineae. Dahlgren (1989) placed it under the first order Annonales of Magnolianae. Takhtajan removed it to order Austrobaileyales within Magnolianae. APG II and

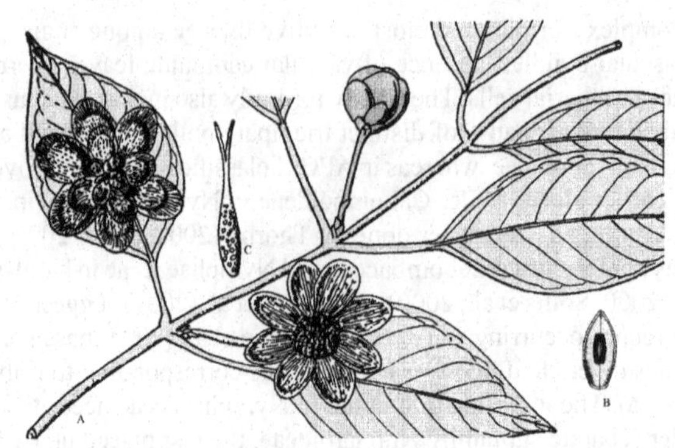

Figure 13.4: Austrobaileyaceae. *Austrobaileya scandens.* **A:** Flowering branch; **B:** Stamen, broad and petal-like; **C:** Carpel with bifid style.

APweb (2008) place this family together with Trimeniaceae, Schisandraceae and Illiciaceae under the order Austrobaileyales, placement adopted in APG IV and APweb, but prefering name Schisandraceae for Illiciaceae. The combination of these families under single order received 99 per cent bootstrap support in several multigene analyses (Soltis et al., 1999; Soltis et al., 2000). *Ondinea*, although with distinctive wind pollinated flowers, are derived from Nymphaea-type flowers (Löhne et al., 2009) and has been merged with *Nymphaea.*

Schisandraceae Blume *Nom. Cons.* ***Magnolia Vine Family***

(including **Illiciaceae** (DC) A. C. Smith ***Star Anis Family***)

3 genera, 85 species (92 as per APweb)

Southeast United States, West Indies, Mexico, China, Japan and Southeast Asia.

Placement:

	B & H	Cronquist	Takhtajan	Dahlgren	Thorne	APG IV/(APweb)
Informal Group						Basal Angiosperms/ (none)
Division		Magnoliophyta	Magnoliophyta			
Class	Dicotyledons	Magnoliopsida	Magnoliopsida	Magnoliopsida	Magnoliopsida	
Subclass	Polypetalae	Magnoliidae	Magnoliidae	Magnoliidae	Chloranthidae	
Series+/ Superorder	Thalamiflorae+		Nymphaeanae	Magnolianae	Chloranthanae	
Order	Ranales	Illiciales	Illiciales	Illiciales	Chloranthales	Austrobaileyales

B & H under family Magnoliaceae; **Cronquist, Takhtajan, Dahlgren** treated two families separate.

Salient features: Trees, shrubs or lianas with simple alternate leaves, stipules lacking, nodes unilacunar, flowers solitary with multiseriate perianth, stamens numerous, carpels free, in a single whorl, fruit a star-like aggregate of follicles.

Genera: *Illicium* (42 species), *Schisandra* (25) and *Kadsura* (18).

Description: Shrubs or small trees (*Illicium*) or llianas (*Schisandra, Kadsura*), containing aromatic terpenoids and branched sclereids. **Leaves** evergreen, alternate, often clustered at tips of branches, sometimes subverticillate, entire, gland-dotted, containing terpenoids, simple with reticulate venation, stipules absent. **Inflorescence** with solitary flowers, axillary or supra-axillary, rarely 2–3 together. **Flowers** bisexual (*Illicium*) or unisexual (other two), actinomorphic, hypogynous. **Perianth** with numerous tepals, in several whorls, sepals and petals not differentiated, outermost somewhat sepal-like, inner gradually becoming smaller and petal-like. **Androecium** with many stamens, free, spirally arranged, filaments short and thick, anthers basifixed, dehiscence longitudinal, connective extending beyond anther lobes, pollen tricolpate. **Gynoecium** with 5–20 free ascidiate carpels, in a single whorl, ovary superior with a single ovule or more, placentation basal, stigma extending down on style. **Fruit** a star-like aggregation of follicles (**follicetum,** *Illicium*) or berries, embryo minute, endosperm conspicuous, seeds glossy. Pollination primarily by flies. Dispersal of follicles by elastic opening, shooting out seeds.

Economic importance: The family is important for producing aromatic oils. Oil from the bark of *Illicium parviflorum* (yellow star anis) is used in flavourings. *I. verum* (star anis) and *I. anisatum* (Japanese star anis) are sources of anethole, used in dentistry, flavourings and perfumes.

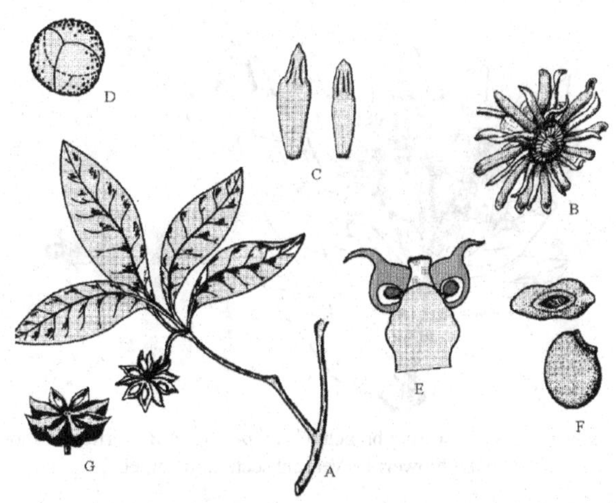

Figure 13.5: Schisandraceae. *Illicium floridanum*. **A:** Fruiting branch; **B:** Flower; **C:** Stamens with broad filaments, left with almost petaloid filament; **D:** Tricolpate pollen grain; **E:** Longitudinal section through carpels, all other floral parts removed; **F:** Two seeds in different view. **G:** A dehiscing follicetum of *I. anisatum*.

Phylogeny: The family is closely related to Winteraceae, and although the vessels are present, the elements are long, slender, angular, thin and greatly overlapping end walls with many scalariform perforation plates. The fruit is a primitive whorl of single-seeded follicles. Although the pollen grains are tricolpate, but their corpus morphology is different from eudicots. Loconte (1996) considers Illiciales among the most basal lineages of angiosperms.

The family has been traditionally placed in the Magnoloid complex under order Illiciales but has been removed in APG II and APweb along with Austrobaileyaceae, Schisandraceae and Monimiaceae into a separate order Austrobaileyales placed towards the beginning of angiosperms without any informal superclade. Eames (1961) considered Schisandraceae to be the closest family. The family Illiciaceae has been found to be very closely related to Austrobaileyaceae and Schisandraceae through multigene analyses (Soltis et al., 1999; Soltis et al., 2000) having received 99 per cent bootstrap support. APG II suggests optionally including Illiciaceae under Schisandraceae (because the latter is a priority name). Schisandraceae includes climbing or trailing shrubs. Thorne includes the two families under suborder Illicineae, which was earlier (1999) placed under order Magnoliales, but now (2003) shifted to Canellales (2003) and finally to Chloranthales under newly created Chloranthidae, Chloranthanae. APG III, APG IV and APweb have finally merged the two families under Austraobaileyales.

Arrangement after APG IV (2016)

Magnoliids:

Order 1. Canellales
- Family: Canellaceae
- Family: **Winteraceae**

Order 2. Piperales
- Family: **Saururaceae**
- Family: **Piperaceae**
- Family: Aristolochiaceae

Order 3. Magnoliales
- Family: Myrsticaceae
- Family: **Magnoliaceae**
- Family: **Degeneriaceae**
- Family: Himantandraceae
- Family: Eupomiaceae
- Family: **Annonaceae**

Order 4. Laurales
- Family: **Calycanthaceae**
- Family: Siparunaceae
- Family: Gomortegaceae
- Family: Atherospermataceae
- Family: Hernandiaceae
- Family: Monimiaceae
- Family: **Lauraceae**

Families in boldface are described in detail.

Winteraceae R. Br. ex Lindley *Winter's Bark Family*

5 genera, 65 species

Tropical, subtropical and temperate regions of Madagascar, South America, Mexico, Australia, New Caledonia and New Guinea.

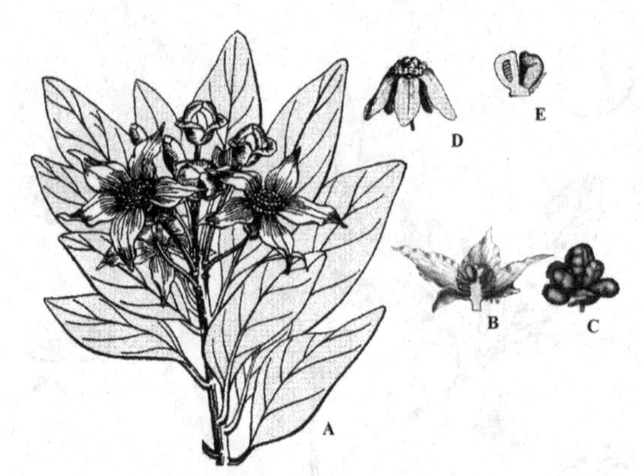

Figure 13.6: Winteraceae. *Drimys winteri.* **A:** Flowering branch of var. *punctata*; **B:** Vertical section of flower; **C:** fruits. *Tasmannia* sp. **D:** Flower; **E:** Vertical section of carpel.

Placement:

	B & H	Cronquist	Takhtajan	Dahlgren	Thorne	APG IV/(APweb)
Informal Group						Magnoliids
Division		Magnoliophyta	Magnoliophyta			
Class	Dicotyledons	Magnoliopsida	Magnoliopsida	Magnoliopsida	Magnoliopsida	
Subclass	Polypetalae	Magnoliidae	Magnoliidae	Magnoliidae	Chloranthidae	
Series+/ Superorder	Thalamiflorae+		Magnolianae	Magnolianae	Magnolianae	
Order	Ranales	Magnoliales	Canellales	Winterales	Canellales	Canellales

B & H under family Magnoliaceae.

Salient features: Trees and shrubs with simple alternate leaves with glaucous under surface, stipules lacking, nodes trilacunar, vessels absent, flower medium sized in cymes, stamens numerous with flattened filament, pollen grains in tetrads, stigma extending down on style and fruit a follicle.

Major genera: *Tasmannia* (40 species), *Drimys* (8), *Zygogynum* (8).

Description: Trees or shrubs lacking vessels and with narrow elongated tracheids, nodes trilacunar, sieve-tube plastids S-type. **Leaves** leathery, alternate, aromatic, gland-dotted, containing terpenoids, entire, simple with reticulate venation, under surface glaucous due to waxy coating, stipules absent. **Inflorescence** cymose or fasciculate, with medium-sized few flowers, solitary terminal in *Zygogynum*. **Flowers** usually bisexual, rarely polygamous, actinomorphic with spirally arranged stamens, hypogynous. **Calyx** with 2–6 sepals, free or connate at base (*Drimys*), valvate, sometimes falling off as cap. **Corolla** with 5 or many petals, 2- or more-seriate, mostly conspicuous in bud, imbricate. **Androecium** with many stamens, centrifugal, free, filaments flattened or almost laminar, poorly differentiated from anthers, anthers bithecous, dehiscing longitudinally, introrse, connective frequently extending beyond anthers, tapetum amoeboid or glandular, pollen uniporate, released in tetrads. **Gynoecium** with 1-many carpels, in a single whorl, usually free, sometimes slightly connate or syncarpous (*Zygogynum*), ovary superior with parietal placentation, ovules 1-many, anatropous, bitegmic, crassinucellate, stigma extending down on style or capitate, carpels sometimes partially unsealed (*Drimys*). **Fruit** an etaerio of berries or follicles, embryo minute, endosperm conspicuous. Pollination by small beetles (*Drimys*), flies and moths, some species wind pollinated (*Tasmannia*). Dispersal especially of berries is by vertebrates.

Economic importance: The bark of *Drimys winteri* (winter's bark) is of medicinal importance, used locally in South America as a tonic. It was also used once by mariners for scurvy prevention. Some other species also have medicinal uses. The fruits and seeds of *D. lanceolata* (mountain pepper) are used as pepper and allspice substitute.

Phylogeny: The family has gained considerable phylogenetic significance during the last three decades and has been regarded as the most primitive living family of angiosperms, and *Drimys* (according to Eames, 1961 the combination of characters suggests *Belliolum* as most primitive genus of the family) as the most primitive genus in the recent classifications of Thorne (pre-2003 versions) and Cronquist. Takhtajan also regarded this as a very primitive family but considered *Degeneria* (formerly under Winteraceae, but now removed to Degeneriaceae) to be the most primitive genus. The primitive position of *Drimys* is supported by the absence of vessels, narrow elongated tracheids, laminar stamens and more primitive beetle pollination. The fossil records of the family also go back to 100–140 years. Only Chloranthaceae is perhaps as old in fossil history of angiosperms. Thorne (1996) lists other primitive features of *Drimys* as alternate, entire, exstipulate leaves, pollen grains in tetrads, long cambial initials and tracheids, heterogenous rays, and poorly-organized pinnate venation, small medium sized flowers in cymes, style-less carpel, partly sealed stigmatic margins, and follicle fruit.

The position of Winteraceae at the base of angiosperms, however, has been refuted during the last few decades, largely due to emergence of the herbaceous origin hypothesis, and the results of cladistic studies largely based on molecular data. Young (1981) interpreted neoteny in the family with a series of reversals. It is also suggested that the family shares common ancestry with Illiciaceae (Doyle and Donoghue, 1993) and Amborellaceae (Loconte and Stevenson, 1991). Loconte (1996), on comparison of various hypotheses concluded that the tree based on Winteraceae hypothesis is two steps longer than one based on Calycanthaceae. The family has recently been placed along with Canellaceae in a separate order Canellales (APG II, APG IV, APweb), not at the beginning of angiosperms but after Amborellaceae, Chloranthaceae and Austrobaileyaceae. The sister-group relationship of Winteraceae and Canellaceae has received bootstrap support of 99 or 100 per cent in all recent multigene analyses (Qui et al., 1999; Soltis et al., 1999; Zanis et al., 2002, 2003). APweb (2018) recognizes two subfamilies: Taktajanioideae with single species *Takhtajania perrieri* and Winteroideae including the rest. Thomas et al. (2014) suggested that *Takhtajania* showed a Gondwanan distribution, rather unusual in flowering plants.

Saururaceae Martynov ***Lizard-tail Family***

4 genera, 6 species

Temperate or subtropical Coasts of North America and East Asia.

Placement:

	B & H	**Cronquist**	**Takhtajan**	**Dahlgren**	**Thorne**	**APG IV/(APweb)**
Informal Group						Magnoliids
Division		Magnoliophyta	Magnoliophyta			
Class	Dicotyledons	Magnoliopsida	Magnoliopsida	Magnoliopsida	Magnoliopsida	
Subclass	Monochlamydeae	Magnoliidae	Magnoliidae	Magnoliidae	Magnoliidae	
Series+/ Superorder	Microembryeae+		Piperanae	Nymphaeanae	Magnolianae	
Order		Piperales	Piperales	Piperales	Piperales	Piperales

B & H under family Piperaceae.

Salient features: Perennial herbs, leaves alternate, stipules adnate to petiole, flowers reduced, in dense spikes, colored bracts often surrounding the base of spike, looking like petals, the whole inflorescence looking like a flower, stamens 6, somewhat attached to carpels, carpels free or united, fruit a capsule.

Major genera: *Saururus* (2 species), *Gymnotheca* (2), *Houttuynia* (1) and *Anemopsis* (1).

Description: Perennial aromatic herbs bearing essential oils, rhizomatous, nodes 5-lacunar, or multilacunar, vascular bundles in one ring, vessel elements with oblique end-walls, scalariform, sieve-tube plastids S-type. **Leaves** alternate; spiral to distichous, petiolate, aromatic, simple, pinnately or palmately veined, stipules intrapetiolar and adnate to the petiole, stomata cyclocytic, mesophyll with spherical etherial oil cells. **Inflorescence** a raceme, or spike, often with involucral bracts resembling petals (*Houttuynia*, *Anemopsis*) and inflorescence appearing like a flower, or without involucral bracts (*Saururus*, *Gymnotheca*). **Flowers** small, regular; bisexual, cyclic **Perianth** absent. **Androecium** with 3, or 6, or 8 stamens,

Figure 13.7: Saururaceae. *Saururus cernuus.* **A:** Flowering branch with elongated spike. **B:** Flower with subtending bract; **C:** Vertical section of flower; **D:** Transverse section of fruit. **E:** Flowering portion of *Anemopsis californica* with showy involucre bracts below the spike, and the plant with basal leaves.

united with the gynoecium or not, free, 1 whorled (when 3), or 2 whorled (when six or eight), with slender filaments, anthers basifixed, dehiscence by longitudinal slits, extrorse to latrorse to introrse, tapetum glandular, pollen grains aperturate, or nonaperturate, usually monosulcate. **Gynoecium** with 3 or 4(–5) carpels, free or united, semicarpous in *Saururus* (the conduplicate carpels distinct above the connate base), superior (mostly), or inferior (*Anemopsis*); carpel in *Saururus* incompletely closed, style with decurrent stigma, (1–)2–4 ovuled, placentation dispersed (laminar-lateral), ovary 1 locular, Styles 3–4(–5); in the genera other than *Saururus* apical, stigmas 3–4(–5), placentation parietal, ovules in the single cavity 20–40(–50) (6–10 on each placenta), orthotropous to hemianatropous, bitegmic, tenuinucellate, or crassinucellate, outer integument contributing to the micropyle. **Fruit** an aggregate (*Saururus*), or not, indehiscent (*Saururus*) or dehiscent, fleshy, a capsule, or capsular-indehiscent; seeds scantily endospermic, perisperm present, embryo rudimentary.

Economic importance: *Houttuynia cordata* forms a good ground cover and is commonly cultivated. The leaves of this species are used as salad and for treating eye diseases in Vietnam. *Saururus chinensis* is also occasionally cultivated. The aromatic stoloniferous stock of *Anemopsis californica* was once fashioned into cylindrical necklace beads by American Indians, and hence the name Apache beads. The stock infused in water is also a reputed treatment for malaria and dysentery.

Phylogeny: The family is considered to be less specialized than Piperaceae in its free to united carpels and parietal placentation, and is believed to be belonging to paleoherb complex, early basal branch of angiosperms. Hickey and Taylor (1996), who proposed the herbaceous origin hypothesis, believe that the flowers of Piperaceae, as well as of *Anemopsis* and *Houttuynia*, arose through suppression of the system of inflorescence axes of gnetopsids and bracts to bring either a single distal and one more proximal pair of anthions together above the subtending bract of the second-order inflorescence axis. The four carpelled flowers of *Saururus* and *Gymnotheca* are the result of the reduction of an inflorescence axis to a penultimate and ultimate pair of anthions. Zeng et al. (2002) on the basis of matR gene studies concluded that 4 genera and six species of Saururaceae form a monophyletic group. *Circaeocarpus saurroides* C. Y. Wu earlier placed in Saururaceae is conspecific with *Zippelia begoniaefolia* Blume and belongs to Piperaceae. Two pairs of genera [*Saururus* + *Gymnotheca*] and [*Houttuynia* + *Anemopsis*] are supported in molecular analyses (Massoni et al., 2014), although not supported by morphological data.

Piperaceae Batsch ***Pepper Family***

5 genera, 3,700 species

Tropical and subtropical regions, mainly in rain forests.

Placement:

	B & H	Cronquist	Takhtajan	Dahlgren	Thorne	APG IV/(APweb)
Informal Group						Magnoliids
Division		Magnoliophyta	Magnoliophyta			
Class	Dicotyledons	Magnoliopsida	Magnoliopsida	Magnoliopsida	Magnoliopsida	
Subclass	Monochlamydeae	Magnoliidae	Magnoliidae	Magnoliidae	Magnoliidae	
Series+/ Superorder	Microembryeae+		Piperanae	Nymphaeanae	Magnolianae	
Order		Piperales	Piperales	Piperales	Piperales	Piperales

Salient features: Herbs, shrubs or climbers with jointed nodes, vascular bundles scattered, leaves alternate, petioles sheathing the nodes, flowers small in dense spikes, perianth absent, stamens 2–6, ovary with single ovule, embryo very small.

Major genera: *Piper* (2,000 species), *Peperomia* (1,600).

Description: Herbs, shrubs, or woody climbers, or small trees bearing essential oils, stems conspicuously jointed, nodes 3-lacunar to multilacunar, vascular bundles scattered, vessel-elements with scalariform or simple end-walls, sieve-tube plastids S-type. **Leaves** alternate, spiral, herbaceous or fleshy, simple, entire, pinnately or palmately veined, petiolate sheathing, stipules intrapetiolar, adnate to petiole, hydathodes commonly present, stomata cyclocytic or anisocytic, mesophyll with spherical ethereal oil cells. **Inflorescence** spadix or spike. **Flowers** bracteate; minute to small, usually bisexual, sometimes unisexual. **Perianth** absent. **Androecium** with 1–10 stamens, adnate to the base of ovary or not, free, often more or less monadelphous, staminodes often present, anthers bithecous (monothecous in *Peperomia*), dehiscence by longitudinal slits, extrorse, tapetum glandular, pollen grains monosulcate or nonaperturate. **Gynoecium** with 2–4 united carpels, or single carpel (*Peperomia*), ovary superior, unilocular, stigmas 1–5, placentation basal; ovule, ascending, orthotropous, bitegmic or unitegmic (*Peperomia*), crassinucellate. **Fruit** fleshy, usually a drupe; seeds scantily endospermic, perisperm copious, embryo minute.

Economic importance: *Piper nigrum* is the source of black and white pepper (ripe and unripe, respectively). The roots of *P. methystichum* are used for making the famous national beverage Kava in Polynesia. The leaves ('Paan' leaves) of *P.*

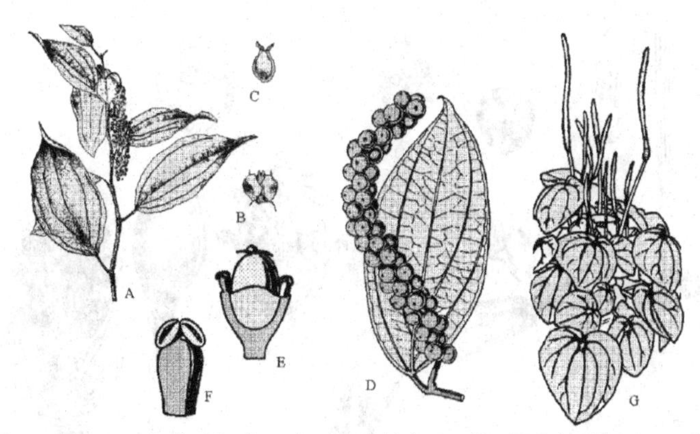

Figure 13.8: Piperaceae. *Piper guineense.* **A:** Fruiting branch with pendulous spike; **B:** Paired flowers and their bracts; **C:** Longitudinal section of female flower. *Piper nigrum.* **D:** Fruiting branch; **E:** Flower; **F:** Stamen. **G:** *Peperomia griseo-argenteum*, a cluster of flowering shoots.

betle (betel-vine) are used as masticatory in East Africa, India and Indonesia. Some species of *Peperomia* (*P. caperata, P. hederaefolia, P. magnoliaefolia*) are grown as ornamental foliage plants.

Phylogeny: Piperaceae, together with Saururaceae, constitute a monophyletic group often considered as order Piperales. Thorne had earlier (1999, 2000) placed these families under suborder Piperineae of the order Magnoliales, but has subsequently (2003, 2006, 2007) shifted them under Piperales together with Aristolochiaceae, Hydnoraceae and Lactoridaceae. Piperaceae, like Saururaceae, are also monophyletic (Tucker et al., 1993). *Peperomia* is considered to be the most derived member of the family with numerous apomorphies such as single carpel, monothecous anthers, unitegmic ovule, inaperturate pollen grains and succulent leaves, and is often removed to a distinct family Peperomiaceae. Thorne (2003, 2006, 2007) and Stevens (2008) place *Peperomia* under separate subfamily Peperomioideae, whereas the other 4 genera are placed under Piperoideae. APweb (2018) following Samain et al., 2008, recognizes three subfamilies: Verhuellioideae (*Verhuellia*), Zippelioideae (*Zippelia* and *Manekia*) and Piperoideae (*Piper* and *Peperomia*). *Peperomia*, formerly placed in a separated family Peperomiaceae is divided into 14 well-supported monophyletic subgenera by Frenzke et al. (2015).

Magnoliaceae A. L. de Jussieu ***Magnolia Family***

2 genera, 294 species

Warm temperate to tropical regions of Southeast, North and Central America, West Indies, Brazil, and East Asia.

Placement:

	B & H	Cronquist	Takhtajan	Dahlgren	Thorne	APG IV/(APweb)
Informal Group						Magnoliids
Division		Magnoliophyta	Magnoliophyta			
Class	Dicotyledons	Magnoliopsida	Magnoliopsida	Magnoliopsida	Magnoliopsida	
Subclass	Polypetalae	Magnoliidae	Magnoliidae	Magnoliidae	Magnoliidae	
Series+/ Superorder	Thalamiflorae+		Magnolianae	Magnolianae	Magnolianae	
Order	Ranales	Magnoliales	Magnoliales	Magnoliales	Magnoliales	Magnoliales

Salient features: Trees or shrubs with alternate simple leaves, stipules caducous, leaving a circular scar at the node, nodes multilacunar, flowers usually solitary, bisexual, large, floral parts numerous, spirally arranged on elongated thalamus, tepals gradually passing from outer sepals to inner petals, stamens laminar, carpels free, seed often suspended by thread like funiculus.

Genera: *Magnolia* (292 species) and *Liriodendron* (2).

Description: Trees or shrubs, nodes 5-lacunar or multilacunar, vessels-elements with scalariform ends, vessels without vestured pits, wood parenchyma apotracheal (terminal), sieve-tube plastids S-type, or P-type and S-type; when P-type, subtype Ib. **Leaves** evergreen or deciduous, alternate, spiral, petiolate, simple, dissected (*Liriodendron*), pinnatifid or entire, pinnately veined, or palmately veined, stipules large, sheathing, enclosing the terminal buds, caducous, leaving a ring-shaped scar at the node, stomata paracytic, or anomocytic, minor leaf veins without phloem transfer cells (*Magnolia*).

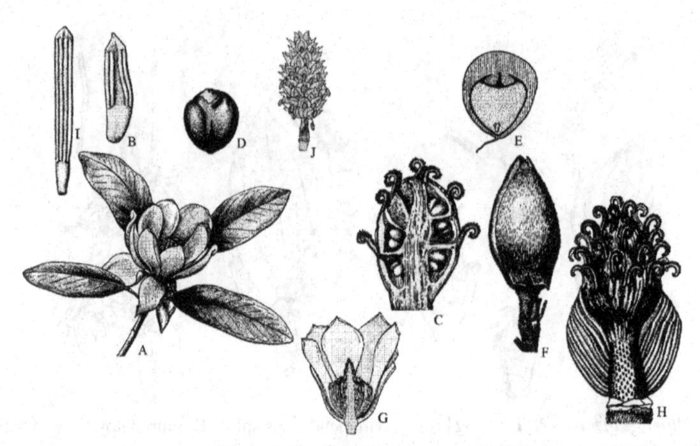

Figure 13.9: Magnoliaceae. *Magnolia virginiana.* **A:** Flowering branch with single terminal flower; **B:** Stamen, laminar and with apical sterile appendage; **C:** Longitudinal section of gynoecium, two anatropous ovules in each carpel; **D:** Seed with fleshy seed coat removed; **E:** Longitudinal section of seed showing fleshy seed coat, copious endosperm and small embryo. *M. grandiflora.* **F:** Flower bud; **G:** Vertical section of flower; **H:** Floral receptacle with half of the stamens removed; **I:** Anther; **J:** Dehisced fruit with arillate seeds hanging through thread-like funiculus.

Inflorescence with usually solitary terminal, or axillary flowers. **Flowers** bracteate (the bracts spathaceous); large, regular, bisexual. **Perianth** with 6–18 tepals, free, sequentially intergrading from sepals to petals, or petal-like usually spirally arranged, rarely 3–4 whorled, white, or cream, or pink, deciduous. **Androecium** with numerous (50–200) stamens, maturing centripetally, free, spirally arranged, all fertile, usually laminar (the four paired microsporangia embedded, the stamens often more or less strap-shaped), anthers adnate, dehiscence longitudinal, through slits or valves, extrorse (*Liriodendron*), or latrorse to introrse, bithecous, appendaged often by prolongation of the connective or unappendaged, pollen grains monosulcate. **Gynoecium** with (2–) 20–200 free carpels, ovary superior, carpel fully or incompletely closed, 2 (–20) ovuled, placentation marginal; ovules funicled, pendulous, biseriate (on the ventral suture), anatropous, bitegmic, crassinucellate; stigma extending down the style, but sometimes terminal. **Fruit** an aggregation of follicles or indehiscent samaras (*Liriodendron*), or united into fleshy syncarp (*Magnolia* subgenus *Yulania*, section *Michelia*, subsection *Aromadendron*); seeds endospermic, endosperm oily, seeds usually large, often with long thread-like funiculus. Pollination primarily by beetles. The fruits are primarily dispersed by animals, but the samaras of *Liriodendron* are wind dispersed.

Economic importance: Various species of *Magnolia* (*M. grandiflora, M. kobus, M. stellata, M. fuscata, M. champaca*—sapu, also source of timber) are grown as ornamentals. *Liriodendron tulipifera* (tulip tree or yellow poplar) is a valuable timber source in USA. Species of *Magnolia* (*M. hypoleuca* and others) also constitute sources of timber.

Phylogeny: The family was regarded as the most primitive of the extant angiosperms for several decades in the classification systems of Hallier (1905), Hutchinson (1926, 1973), and earlier versions of Cronquist and Takhtajan. The view was first challenged by Smith (1945), who considered that Magnoliaceae are relatively highly specialized both vegetatively and florally, casting some doubt on the assumption of the primitive nature of the family, and implying that groups such as Winteraceae, etc., may be at least as primitive. The status of Magnoliaceae as the most primitive family was strongly challenged by Carlquist (1969), Gottsberger (1974) and Thorne (1976), claiming Winteraceae to be the most primitive family. The primitive features of Magnoliaceae include spirally arranged floral parts, laminar stamens, fruit a follicle, longer and narrower vessel elements, monosulcate pollen grains and beetle pollination.

The family is considered to be monophyletic based on the support from *rbcL* and *ndhF* sequences (Qui et al., 1993; Kim et al., 2001). These studies, however, question the recognition of *Talauma*, *Michelia* and *Manglietia* as distinct genera, as it renders *Magnolia* as paraphyletic. Although *Liriodendron* is quite distinct, all other genera have been merged with *Magnolia* in the recent works. Figlar and Nootebroom (2004) divide the enlarged genus *Magnolia* into three subgenera: *Magnolia*, *Yulania* and *Gynopodium*. Two clades are distinguished within the family one represented by *Liriodendron*, and the other by rest of the genera. Judd et al. (2008), Stevens, APweb (2008, 2018) as such recognize only 2 genera *Magnolia* and *Liriodendron* within the family. Thorne (2003, 2006, 2007), places *Magnolia* and other 5 genera in subfamily Magnolioideae, whereas *Liriodendron* is placed in monogeneric Liriodendroideae.

The large genus *Magnolia* is divided by Magnolia Society into three subgenera (listed above), 12 sections and 13 subsections. Sima and Lu (2012), on the other hand, provide a reclassification of Magnolia in which it is divided into 16 genera placed in two tribes.

Degeneriaceae I. W. Bailey & A. C. Smith ***Degeneria Family***

1 genus, 2 species

Endemic to Fiji.

Placement:

	B & H	Cronquist	Takhtajan	Dahlgren	Thorne	APG IV/(APweb)
Informal Group						Magnoliids
Division		Magnoliophyta	Magnoliophyta			
Class		Magnoliopsida	Magnoliopsida	Magnoliopsida	Magnoliopsida	
Subclass		Magnoliidae	Magnoliidae	Magnoliidae	Magnoliidae	
Series+/ Superorder		Magnolianae	Magnolianae	Magnolianae		
Order	Not described then	Magnoliales	Magnoliales	Magnoliales	Magnoliales	Magnoliales

Salient features: Trees or shrubs with alternate simple leaves, stipules absent, nodes 5-lacunar, flowers usually solitary, bisexual, large, sepals and petals distinct, sepals 3, petals 12–18, stamens many, laminar, 3-veined, inner modified into staminodes, carpel single, incompletely sealed, fruit leathery with many seeds.

Genus: Single genus *Degeneria* with 2 species, *D. vitiensis* and *D. roseiflora*, endemic to Fiji.

Description: Large trees; bearing essential oils, nodes 5-lacunar, vessel-elements with oblique end walls, sieve-tube plastids P-type, pith with diaphragms. **Leaves** alternate, petiolate, non-sheathing, gland-dotted, aromatic, simple, entire, pinnately veined, exstipulate, stomata paracytic, mesophyll with spherical ethereal oil cells. **Inflorescence** with solitary pendulous flowers (supra) axillary. **Flowers** medium-sized to large, regular, polycyclic, thalamus shortly raised, sepals and petals distinct. **Calyx** with 3 sepals, 1 whorled, free, persistent. **Corolla** with 12–18 petals, larger than the sepals, 3–5 whorled, free, fleshy, deciduous, sessile. **Androecium** with about 30–50 stamens, maturing centripetally, free, 3–6 whorled, innermost 3–10 modified into staminodes; fertile stamens laminar, flattened, oblong, 3-veined; anthers bithecous, adnate, dehiscence longitudinal, with slits or valves, extrorse, tapetum glandular, pollen grains monosulcate. **Gynoecium** with single carpel, ovary superior, single chambered, carpel incompletely closed (largely unsealed at anthesis), style absent, stigma running nearly the entire length of carpel, placentation marginal, ovules 20–30, in two rows, long funicled, with a conspicuous funicular obturator; anatropous, bitegmic, crassinucellate, outer integument not contributing to the micropyle. **Fruit** leathery, with a hard exocarp, dehiscent, or indehiscent, 20–30 seeded; seeds flattened, more or less sculptured, with an orange-red sarcotesta, embryo well differentiated but small, cotyledons 3 (–4), copiously endospermic, endosperm ruminate, oily. Pollination by beetles.

Economic importance: No economic value known.

Phylogeny: The family was earlier included under Winteraceae and was considered closer to *Zygophyllum* by Hutchinson (1973). It is now treated to be an independent family, more closely allied to Magnoliaceae and Himantandraceae. Takhtajan (1987, 1997), considers Winteraceae and Degeneriaceae to be more primitive families, but is perhaps is the only one among the recent authors to consider Degeneriaceae as the most primitive family of extant angiosperms. The primitive features of the family include alternate simple leaves, numerous laminar stamens, partially closed carpel with stigma running the entire length of carpel, 3–4 cotyledons, and monosulcate pollen grains. Thorne (2003, 2006, 2007) and APG IV (2016) place Degeneriaceae between Magnoliaceae and Himantandraceae.

Figure 13.10: Degeneriaceae. *Degeneria vitiensis.* **A:** Tree growing in natural habitat in Fiji.; **B:** A branch with flowers; **C:** Stamen, laminar with undifferentiated filament and anther part; **D:** Transverse section of carpel; **E:** Fruit.

Annonaceae A. L. de Jussieu ***Annona or Pawpaw Family***

105 genera, 2500 species

Temperate and tropical regions of Eastern North America and Eastern Asia, and tropical South America. Mainly distributed in Old World tropics, in moist forests.

Placement:

	B & H	Cronquist	Takhtajan	Dahlgren	Thorne	APG IV/(APweb)
Informal Group						Magnoliids
Division		Magnoliophyta	Magnoliophyta			
Class	Dicotyledons	Magnoliopsida	Magnoliopsida	Magnoliopsida	Magnoliopsida	
Subclass	Polypetalae	Magnoliidae	Magnoliidae	Magnoliidae	Magnoliidae	
Series+/ Superorder	Thalamiflorae+		Magnolianae	Magnolianae	Magnolianae	
Order	Ranales	Magnoliales	Annonales	Annonales	Magnoliales	Magnoliales

Salient features: Trees or shrubs with alternate distichous leaves, stipules absent, leaves glaucous or with metallic sheen, flowers fragrant, flowers trimerous with numerous spirally arranged stamens, many carpels free, fruit an aggregate of berries, seed with ruminate endosperm.

Major genera: *Guateria* (250 species), *Xylopia* (150), *Uvaria* (100), *Artabotrys* (100), *Annona* (100), and *Polyalthia* (100).

Description: Trees, or shrubs, or lianas, bearing essential oils, nodes unilacunar, or bilacunar, vessel end-walls horizontal, simple, vessels without vestured pits, wood diffuse porous; partially storied, sieve-tube plastids P-type, subtype Ia, pith commonly with diaphragms **Leaves** evergreen, alternate, distichous, non-sheathing, simple, entire, pinnately veined, stipules absent, domatia recorded in 3 genera, stomata paracytic, secretory cavities containing oil, mucilage, or resin. **Inflorescence** with solitary flowers, or racemose. **Flowers** regular, cyclic, usually bisexual, rarely unisexual, thalamus sometimes elongated (*Mischogyne*) **Calyx** usually with 3 sepals, or 6 and 2-whorled, free, valvate. **Corolla** with 3–6 petals, 1–2 whorled, free, imbricate or valvate. **Androecium** with 25–100 stamens, maturing centripetally; free, all equal, spirally arranged, rarely 3–6 whorled, rarely outer forming staminodes (e.g., in *Uvaria)*, anthers bithecous, adnate, dehiscing by longitudinal slits or valves; extrorse, connective prolonged into appendage, tapetum glandular, pollen shed in aggregates (5 genera), or as single grains; when aggregated, in tetrads (usually), or in polyads (octads in *Trigynaea*). Pollen grains monosulcate or nonaperturate, or with two parallel furrows at the equator, or ulcerate. **Gynoecium** with 10–100 carpels, usually free, rarely united, placentation of free carpels basal, when syncarpous 1 locular, or 2–15 locular, parietal, or basal. Ovules 1–50, apotropous, with ventral raphe, bitegmic, crassinucellate, outer integument not contributing to the micropyle. **Fruit** fleshy, commonly an aggregate of berries; seeds endospermic, endosperm ruminate, oily. Pollination mostly by beetles. Dispersal especially of fleshy fruits by birds, mammals and turtles.

Economic importance: Many species of *Annona* are cultivated for their edible fruits: *A. squamosa* (sweet sop), *A. muricata* (sour sop), *A. reticulata* (custard apple), and *A. cherimola* (cherimoya). Flowers of *Cananga odorata* (ylang-ylang) and *Mkilua fragrans* are used in perfumes. The spicy fruits of West African *Xylopia aethiopica* are the so-called 'Negro pepper' used as a condiment, and those of *Monodora myristica* used as substitute for nutmeg.

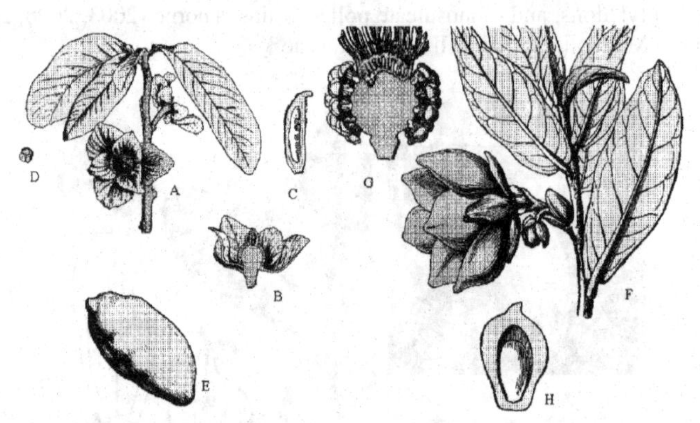

Figure 13.11: Annonaceae. *Asimina triloba.* **A:** Flowering branch bearing solitary flowers; **B:** Vertical section of flower; **C:** Longitudinal section of carpel showing ovules; **D:** Pollen grain; **E:** Fruit. *Annona furfuracea.* **F:** Flowering branch; **G:** Vertical section of receptacle showing male flowers towards the center and female flowers towards the periphery; **H:** Longitudinal section of carpel showing basal ovule.

Phylogeny: It is generally agreed upon that the family is derived from Magnoliaceous stock. Hutchinson placed the family under Annonales after Magnoliales, from which, according to him, they were clearly derived. The primitive features include spirally arranged numerous stamens and carpels, connective prolonged into an appendage. The sepals and petals are more advanced than Magnoliaceae. Most of the recent authors (except Dahlgren and Takhtajan, who place it under Annonales), however, include this family under Magnoliales (Stevens, 2008; Thorne, 2003, 2006, 2007). The genera with connate carpels and with fleshy berries are considered more advanced than those with free carpels. It is proposed that the diversification of Annonaceae may have occurred (84)82–57 mybp (Doyle et al., 2004; Scharaschkin and Doyle, 2005). Chatrou et al. (2012) divide family into 4 subfamilies: Anaxagoreoideae (single genus *Anaxagorea*), Ambavioideae (9 genera), Malmeoideae (8 tribes, 50 genera) and Annonoideae (7 tribes, 49 genera), followed in Aweb (2018). Annonoideae show more internal molecular divergence, hence the name long branch clade.

Calycanthaceae Lindley *Strawberry Shrub Family*

3 genera, 10 species (including Idiospermaceae)

Family with discontinuous distribution, found in North America, East Asia and Queensland.

Placement:

	B & H	Cronquist	Takhtajan	Dahlgren	Thorne	APG IV/(APweb)
Informal Group						Magnoliids
Division		Magnoliophyta	Magnoliophyta			
Class	Dicotyledons	Magnoliopsida	Magnoliopsida	Magnoliopsida	Magnoliopsida	
Subclass	Polypetalae	Magnoliidae	Magnoliidae	Magnoliidae	Magnoliidae	
Series+/ Superorder	Thalamiflorae+		Lauranae	Magnolianae	Magnolianae	
Order	Ranales	Laurales	Laurales	Laurales	Laurales	Laurales

Salient features: Shrubs with opposite simple leaves, stipules absent, flowers with numerous spirally arranged tepals, numerous stamens along the rim of cup-like receptacle, fruit single-seeded achene.

Major genera: *Chimonanthus* (6 species), *Calycanthus* (3 species) and *Idiospermum* (1).

Description: Small trees, or shrubs with aromatic bark, bearing essential oils, nodes unilacunar, vessel-elements with oblique end-walls, sieve-tube plastids P-type, subtype Ia. **Leaves** opposite, leathery, petiolate, gland-dotted, simple, entire, pinnately veined, stipules absent, stomata paracytic, hairs unicellular or absent, mesophyll with spherical ethereal oil cells. **Inflorescence** with solitary terminal flowers on specialized leafy short-shoots. **Flowers** medium-sized to large, regular, bisexual with spirally arranged floral parts, markedly perigynous. **Perianth** with 15–30 tepals, each with 3–4 veins, sequentially intergrading from sepals to petals, free, inserted along the rim of receptacle. **Androecium** with 15–55 stamens, maturing centripetally, free, spirally arranged at the top of the hypanthium, laminar or linear, bithecous, inner 10–25 modified into usually nectariferous staminodes, anthers adnate, dehiscence by longitudinal slits, extrorse, connective extended into an appendage, pollen grains 2(–3) aperturate, sulcate. **Gynoecium** with 5–45 free carpels, spirally arranged within the hypanthium, ovary superior, style distinct, stigma terminal, ovary 2 ovuled (upper often abortive), placentation marginal, ovules ascending, apotropous with ventral raphe, anatropous, bitegmic, crassinucellate, or pseudocrassinucellate, outer integument not contributing to the micropyle. **Fruit** an aggregate of achenes enclosed in the fleshy hypanthium; seeds nonendospermic, embryo well differentiated (large), cotyledons 2, spirally twisted. Pollination by insects, mainly beetles.

Economic importance: *Calycanthus floridus* (Carolina allspice) and *C. occidentalis* are grown as ornamental shrubs. Bark of *C. fertilis* and *C. floridus* yield medicinal extracts. *Chimonanthus praecox* (winter sweet) is widely cultivated and is one of the few species flowering in cold winter with snow around, the flowers used in Japan to make perfumes.

Phylogeny: The family is closely related to Magnoliaceae and Annonaceae in its numerous spirally arranged floral parts and free carpels. Hutchinson included the family under Rosales primarily because of perigynous flowers and free carpels, a position contested by several authors. Thorne (1996, 2000) regarded it to be closely related to Monimiaceae and placed under suborder Laurineae of Magnoliales, but subsequently (2003, 2006, 2007) under independent order Laurales. Loconte and Stevenson (1991), projected Calycanthaceae as basic angiosperms with a series of vegetative and reproductive angiosperm plesiomorphies such as shrub habit, unilacunar two-trace nodes, vessels with scalariform perforations, sieve-tube elements with starch inclusions, opposite leaves, strobilar flowers, leaf-like bracteopetals, poorly differentiated numerous spirally arranged tepals, and few ovulate carpels. Food bodies terminating the stamen connectives indicate beetle pollination. It is interesting to note that genus *Idiospermum* (which was recognized as new genus based on *Calycanthus australiensis* by S. T. Blake in 1972) was considered as the most primitive flowering plant by these authors. Blake had separated *Idiospermum* under distinct family Idiospermaceae, also recognized by Hutchinson and Cronquist, as distinct from Calycanthaceae. However, it is monotypic and shares many features with the other Calycanthaceae, although the alkaloids and the distribution

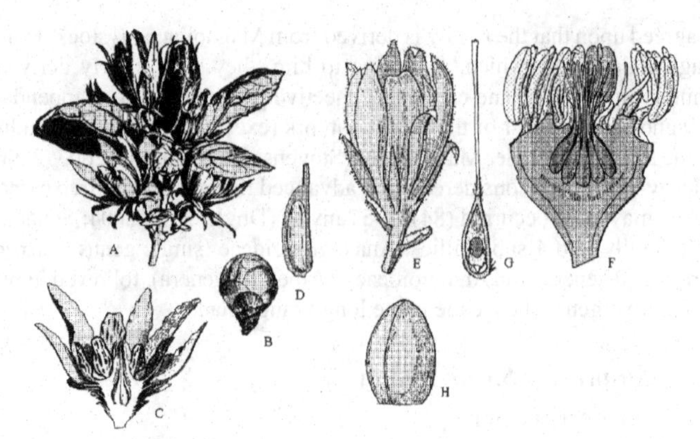

Figure 13.12: Calycanthaceae. A: Flowering branch of *Calycanthus laevigatus. Chimonanthus praecox.* **B:** Flower bud. **C:** Vertical section of flower; **D:** Longitudinal section of carpel. *Calycanthus floridus.* **E:** Flower; **F:** Vertical section of flower; **G:** Longitudinal section of carpel; **H:** Nut.

of xylem parenchyma differ in detail, as such has been merged with latter (APG IV, APweb). Endress (1983) had described 'In all respects, *Idiospermum* gives the impression of a strange living fossil'. Molecular studies suggest Calycanthaceae to be basal within Laurels, probably sister to all other Laurales (Doyle and Endress, 2000). APweb (2018) recognizes two subfamilies: Monotypic Idiospermoideae and Calycanthoideae (including rest).

Lauraceae Durande ***Laurel Family***

45 genera, 2850 species

Throughout tropical and subtropical regions of the world, primarily in rain forests of Southeast Asia and North America.

Placement:

	B & H	**Cronquist**	**Takhtajan**	**Dahlgren**	**Thorne**	**APG IV/(APweb)**
Informal Group						Magnoliids
Division		Magnoliophyta	Magnoliophyta			
Class	Dicotyledons	Magnoliopsida	Magnoliopsida	Magnoliopsida	Magnoliopsida	
Subclass	Monochlamydeae	Magnoliidae	Magnoliidae	Magnoliidae	Magnoliidae	
Series+/ Superorder	Daphnales+		Lauranae	Magnolianae	Magnolianae	
Order		Laurales	Laurales	Laurales	Laurales	Laurales

B & H as Laurineae.

Salient features: Aromatic trees or shrubs, leaves alternate, perianth small and undifferentiated, stamens in several whorls, fruit single seeded drupe or berry.

Major genera: *Litsea* (400 species), *Ocotea* (350), *Cinnamomum* (250), *Cryptocarya* (250), *Persea* (200), *Beilschmeidia* (150) and *Lindera* (100).

Description: Aromatic trees and shrubs, sometimes parasitic climbers (*Cassytha*) nodes unilacunar with two traces, vessel elements with scalariform or simple end-walls, without vestured pits, wood partially storied, sieve-tube plastids P-type, or S-type; when P-type, subtype Ib. **Leaves** nearly always evergreen, usually alternate and spiral, rarely opposite or whorled, leathery, petiolate, non-sheathing, gland-dotted, aromatic, simple, entire, sometimes lobed (*Sassafras*), pinnately veined, stipules absent, domatia (14 genera) represented by pits, pockets, or hair tufts, stomata, paracytic, hairs mostly unicellular, mesophyll usually with spherical etherial oil cells. **Inflorescence** cymose, or racemose, often umbelliform with involucral bracts, rarely solitary. **Flowers** small, often fragrant, regular, bisexual, rarely unisexual, usually 3-merous, cyclic, with well-developed hypanthium. **Tepals** usually 6, sometimes 4, free, (1–)2(–3) whorled, similar, sepaloid to petaloid, green, or white, or cream, or yellow. **Androecium** with (3–)9(–26) stamens, free, equal, or markedly unequal, (1–)3(–4) whorled, inner sometimes modified into staminodes, somewhat laminar to petaloid by expansion of the filament and connective, filaments appendaged or not, anthers bithecous, basifixed, dehiscence longitudinal by valves opening from base to apex, or dehiscing by pores (in *Hexapora*), usually introrse, sometimes extrorse, tapetum amoeboid (mostly), or glandular (in several genera), pollen grains nonaperturate, exine spiny. **Gynoecium** with 1 carpel, ovary usually superior, sometimes inferior (*Hypodaphnis*), style distinct with terminal stigma, placentation apical; ovules pendulous, apotropous, with dorsal

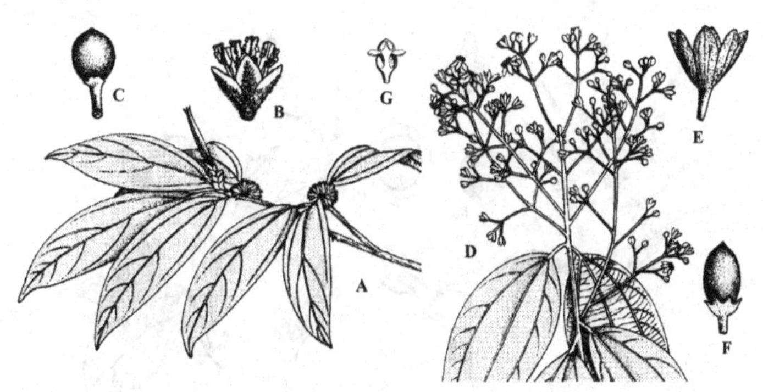

Figure 13.13: Lauraceae. *Litsea doshia.* **A:** Flowering branch with flowers in globose axillary clusters; **B:** Flower; **C:** Fruit. *Cinnamomum tamala.* **D:** Flowering branch with terminal paniculate inflorescence; **E:** Flower; **F:** Fruit; **G:** Anther dehiscing by valves (After Polunin and Stainton, Flowers of the Himalaya, 1984).

raphe, non-arillate, anatropous, bitegmic, crassinucellate, outer integument not contributing to the micropyle. **Fruit** fleshy, drupaceous, or baccate, enclosed in the fleshy receptacle, 1 seeded; seeds nonendospermic, embryo well differentiated, cotyledons massive, occasionally ruminate.

Economic importance: The family contributes some important spices from plants such as *Laurus nobilis* (bay leaves), *Cinnamomum verum* (cinnamon), *C. camphora* (camphor), and *Sassafras albidum* (sassafras). *Persea americana* (avocado) is an important tropical fruit. Aromatic oils are obtained from *Lindera* (benzoin) and *Sassafras*. Species of *Litsea*, and *Ocotea* yield fragrant woods used in cabinet-making.

Phylogeny: Lauraceae is generally considered to be a more specialized family placed closer to Monimiaceae and Calycanthaceae. The order Laurales is generally considered to belong to the magnoloid complex and represents an early divergent lineage. The derived apomorphies of Lauraceae and Monimiaceae include ovary with single carpel, and spinose pollen grains. The families also share pollen lacking aperture, stamens with paired appendages, and anthers opening by valves.

The family is traditionally divided into two subfamilies Cassythoideae (*Cassytha*) and Lauroideae (rest of genera). The latter is variously divided into 3 (Werff and Richter, 1996) or 5 tribes (Heywood, 1978). It has been suggested that the perianth in some Lauraceae may represent modified stamens (Chanderbali et al., 2004) as both the tepals and the stamens of *Persea* have three traces. *Hypodaphnis*, with an inferior ovary is considered to be sister to the rest of the family (Stevens, 2008).

Arrangement after APG IV (2016)
Independent lineage: unplaced to more inclusive clade
Order 1. Chloranthales
Family: **Chloranthaceae**

Chloranthaceae R. Br. Ex Lindl. *Chloranthus Family*

4 genera, 77 species

Tropical, subtropical and South Temperate regions.

Placement:

	B & H	Cronquist	Takhtajan	Dahlgren	Thorne	APG IV/(APweb)
Informal Group						Unplaced Independant Lineage/ (Mesangiosperms)
Division		Magnoliophyta	Magnoliophyta			
Class	Dicotyledons	Magnoliopsida	Magnoliopsida	Magnoliopsida	Magnoliopsida	
Subclass	Monochlamydeae	Magnoliidae	Magnoliidae	Magnoliidae	Chloranthidae	
Series+/ Superorder	Microembyeae+		Nymphaeanae	Magnolianae	Chloranthanae	
Order		Piperales	Chloranthales	Chloranthales	Chloranthales	Chloranthales

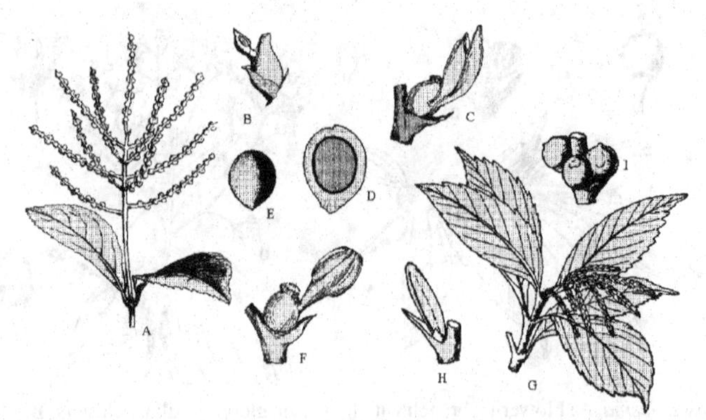

Figure 13.14: Chloranthaceae. A: Flowering branch of *Chloranthus inconspicuous*; **B:** Bisexual flower of *C. brachystachys*; **C:** Bisexual flower of *C. henryi* with single bract, trimerous stamen having bithecous (tetrasporangiate) middle anther and monothecous (bisporangiate) lateral anthers, and a single pistil with tufted stigma. *Sarcandra.* **D:** Transverse section of fruit of *S. chloranthoides*; **E:** Seed; **F:** Bisexual flower of *S. glabra*. *Ascarina lanceolata.* **G:** Flowering branch; **H:** Male flower and **I:** Fruit.

Salient features: Leaves aromatic, opposite, simple with connate petiole bases, stipules small, flowers small, lacking perianth, stamens 1–3, connate in a mass, carpel 1, ovary inferior with a single ovule, fruit a small drupe.

Major genera: *Hedyosmum* (40 species), *Ascarina* (18), *Chloranthus* (14), and *Sarcandra* (3).

Description: Herbs or evergreen shrubs or trees containing essential oils. Vessels absent in *Sarcandra* (vessels reported in roots, but not in stem), present in others but with long tapering elements, with scalariform perforation plates, nodes unilacunar or trilacunar, sieve-tube plastids S-type. **Leaves** aromatic, opposite, simple, usually serrate along margin, petioles often connate at base, stipules small, interpetiolar, leaf mesophyll with spherical etherial oil cells. **Inflorescence** a spike, panicle or capitate, ultimate inflorescence units cymose. **Flowers** usually unisexual or pseudobisexual due to coherence of male and female flowers, unisexual in *Ascarina* and *Hedyosmum* but bisexual in *Chloranthus* and *Sarcandra*, flowers minute, bracteate, actinomorphic. **Perianth** absent in male flowers, rudimentary and calyx-like and adnate to ovary in female flowers, latter sometimes completely naked (*Ascarina*) or enclosed by a cupular bract (*Hedyosmum*). **Androecium** with a single (*Sarcandra*) or three stamens connate into single mass with often bithecous middle anther and monothecous lateral anthers (*Chloranthus*), dehiscence longitudinal. **Gynoecium** with 1 carpel, ovary inferior, unilocular, ovule 1, orthotropous, pendulous, bitegmic, crassinucellate, placentation apical, style very short or absent. **Fruit** an ovoid or globose drupe, seed with abundant oily endosperm and minute embryo, perisperm present.

Economic importance: *Chloranthus glaberi* is grown as an ornamental shrub. The leaves of *C. officinalis* are used to make a drink in parts of Malaya and Indonesia. The infusion of flowers and leaves of *C. inconspicuous* are used to treat coughs and the flowers used to flavour tea in various parts of East Asia. Extract from the leaves of *Hedyosmum brasiliense* is used locally in tropical South America as tonic, to induce sweating and also to treat stomach complaints.

Phylogeny: The family is traditionally placed in the magnoloid complex under order Piperales (Cronquist), Chloranthales (Takhtajan, Dahlgren, Thorne). Donoghue and Doyle (1989) placed Chloranthaceae under Laurales, but this position is not supported by DNA-based cladistic analyses. Taylor and Hickey (1996) consider Chloranthaceae as the basic angiosperm family. The family shows several plesiomorphic characters such as flowers in an inflorescence, plants dioecious, carpels solitary, placentation apical, and fruit drupaceous with small seeds. The family is the oldest in the fossil record, the fossil genus *Clavitopollenites* being assigned to Chloranthaceae and closer to the genus *Ascarina*. The stems of *Sarcandra* are primitively vessel-less. In other genera vessels are primitive with long vessel-elements, tapered and with many barred perforation plates. The family is considered to be earliest to record wind pollination in angiosperms. Taylor and Hickey believe in the origin of Chloranthaceae from gnetopsids, hypothesizing that the ovule and the bract subtending the floral unit in Chloranthaceae are homologous with one of the terminal ovules and proanthophyll subtending the anthion (inflorescences unit) of gnetopsids. Chloranthaceae has undergone considerable reduction in its number of parts as well as general level of elaborateness. Thorne (1996) considered Trimeniaceae to be the closest relative of Chloranthaceae.

The position of this family in APG system is uncertain. The family is sister to magnoliids + eudicots in the six-gene compartmentalized analysis with 84 per cent bootstrap support (Zanis et al., 2003), but APG II prefers to keep the family unplaced at the beginning of angiosperms, without assigning it to any order. APweb places it under Chloranthales, before Magnoliids, but after Commelinids. Thorne had earlier (1999) placed family under Magnoliales under suborder Chloranthineae after Winterineae and Illicineae, but in subsequent revision (2003) placed it (along with Amborellaceae, Trimeniaceae and Austrobaileyaceae) under distinct order Chloranthales at the beginning of

Magnoliidae (first subclass of angiosperms), finally removed in 2006, 2007 together with Nymphaeales under a distinct subclass Chloranthidae, superorder Chloranthanae. APG IV places under Chloranthales, an Independent Unplaced Lineage after Magnoliids. APweb (2018) places it under Mesangiosperms before Magnoliids.

Arrangement after APG IV (2016)

Monocots:

Order 1. Acorales
- Family: **Acoraceae**

Order 2. Alismatales
- Family: **Araceae**
- Family: Tofieldiaceae
- Family: **Alismataceae**
- Family: **Butomaceae**
- Family: **Hydrocharitaceae**
- Family: Scheuchzeriaceae
- Family: Aponogetonaceae
- Family: Juncaginaceae
- Family: Maundiaceae
- Family: Zosteraceae
- Family: **Potamogetonaceae**
- Family: Posidonaceae
- Family: Ruppiaceae
- Family: Cymodociaceae

Order 3. Petrosaviales
- Family: Petrosaviaceae

Order 4. Dioscoreales
- Family: Nartheciaceae
- Family: Burmanniaceae
- Family: **Dioscoreaceae**

Order 5. Pandanales
- Family: Triuridaceae
- Family: Velloziaceae
- Family: Stemonaceae
- Family: Cyclanthaceae
- Family: **Pandanaceae**

Order 6. Liliales
- Family: Campynemataceae
- Family: Corsiaceae
- Family: Melanthiaceae
- Family: Petermanniaceae
- Family: Alstroemeriaceae
- Family: Colchicaceae
- Family: Philesiaceae
- Family: Ripogonaceae
- Family: **Smilacaceae**
- Family: **Liliaceae**

Order 7. Asparagales
- Family: **Orchidaceae**
- Family: Boryaceae
- Family: Blandfordiaceae
- Family: Asteliaceae
- Family: Lanariaceae
- Family: Hypoxidaceae
- Family: Doryanthaceae
- Family: Ixioliriaceae
- Family: Tecophilaeaceae
- Family: **Iridaceae**
- Family: Xeronemataceae
- Family: **Asphodelaceae**
- Family: **Amaryllidaceae**
- Family: **Asparagaceae**

Order 8. Arecales
- Family: Dasypogonaceae
- Family: **Arecaceae**

Order 9. Commelinales
- Family: Hanguanaceae
- Family: **Commelinaceae**
- Family: Philydraceae
- Family: Pontederiaceae
- Family: Haemodoraceae

Order 10. Zingiberales
- Family: Strelitziaceae
- Family: Lowiaceae
- Family: Heliconiaceae
- Family: **Musaceae**
- Family: **Cannaceae**
- Family: Marantaceae
- Family: Costaceae
- Family: **Zingiberaceae**

Order 11. Poales
- Family: Typhacea
- Family: Bromeliaceae
- Family: Rapateaceae
- Family: Xyridaceae
- Family: Eriocaulaceae
- Family: Mayacaceae
- Family: Thurniaceae
- Family: **Juncaceae**
- Family: **Cyperaceae**
- Family: Restoniaceae
- Family: Flagellariaceae
- Family: Joinvilleaceae
- Family: Ecdeiocoleaceae
- Family: **Poaceae**

Families in boldface are described in detail.

Acoraceae Martynov ***Sweet Flag Family***

1 genus, 2 species

North temperate region, Paleotropical. Frigid zone, temperate, and subtropical. Celebes and New Guinea, Eastern Asia to Norway approaching the Arctic circle, Central and Western North America.

Placement:

	B & H	**Cronquist**	**Takhtajan**	**Dahlgren**	**Thorne**	**APG IV/(APweb)**
Informal Group						Monocots
Division		Magnoliophyta	Magnoliophyta			
Class	Monocotyledons	Liliopsida	Liliopsida	Liliopsida	Magnoliopsida	
Subclass		Arecidae	Alismatiidae	Liliidae	Alismatidae	
Series+/ Superorder	Nudiflorae+		Aranae	Aranae	Acoranae	
Order		Arales	Acorales	Arales	Acorales	Acorales

B & H under family Aroideae.

Salient features: Marshy herbs, sweetly fragrant, with rhizomes, lacking oxalate crystals, inflorescence a spadix without spathe, flowers small, bisexual, tepals 6 in two whorls, stamens in two whorls, carpels 3, united, fruit a berry.

Genus: Single genus *Acorus* with 2 species, *A. calamus* and *A. gramineus*.

Description: Aromatic perennial marshy herbs with rhizomes, bearing essential oils, root xylem with vessels with scalariform end-walls. **Leaves** alternate, distichous, flat, sessile, sheathing, entire, ventralized isobifacial [oriented edge on to the stem, bases equitant], intravaginal squamules, parallel-veined, mesophyll with spherical etherial oil cells, lacking calcium oxalate crystals. **Inflorescence** scapigerous, spadix without spathe. **Flowers** ebracteate, small, bisexual, regular, 3-merous, cyclic. **Perianth** with 6 tepals, free, in two whorls, concave or hooded, similar, membranous. **Androecium** with 6 stamens, free, in two whorls, anthers basifixed, dehiscence by longitudinal slits, introrse, tapetum glandular, pollen grains monosulcate to subulcerate. **Gynoecium** with 3, rarely 2 or 4 carpels, united, ovary superior, 3-locular (rarely with 2 or 4 locules), placentation axile, ovules 2–4(–5) per locule, pendulous, orthotropous, bitegmic, ovary loculi with secretory trichomes. **Fruit** fleshy, berry; seeds endospermic, perisperm present, cotyledon 1, no double fertilization.

Economic importance: Dried rhizomes of *A. calamus* (rhizoma calami) have been used for a very long time and were already known to ancient Egyptians, Greeks and Romans. 'Oleum calami' is distilled from the rhizomes of *A. calamus*, for use in perfumery and medicine. The chopped rhizomes are still used as a tea to treat stomach ailments.

Phylogeny: The genus was earlier included under family Araceae. Hutchinson (1973) placed it under tribe Acorae along with *Gymnostachys*. Grayum (1987) justified removal of *Acorus* from Araceae, and this has been followed in all recent classifications. Whereas Dahlgren and Cronquist included the family under the order Arales, Takhtajan and Thorne took it under a separate order Acorales. Thorne (1999) removed the order from Aranae and placed it under Acoranae, along with order Nartheciales. He has now (2003, 2006, 2007) added Petrosaviales (family Petrosaviaceae; regarded unplaced in monocots in APG II) to Acoranae. Subsequent molecular analyses portray it, alone or with *Gymnostachys*, as the sister group of all other Monocotyledons. The rooting of Acoraceae as sister group of monocots is also supported by the studies of Chase et al. (2000); Soltis et al. (2000) and Fuse and Tamura (2000), based on multigene analyses. APG II onwards (APG III, IV) and APweb, accordingly place Acoraceae under a separate order Acorales at the beginning of monocots. Carlquist (2012) regards *Acorus* plant functionally vesselless, the tracheids being "pre-vessel" in morphology—although derived. Crown-group Acoraceae were dated to 19 ± 5.7 m.y. by Merckx et al. (2008), 52–4 m.y. by Mennes et al. (2013) and (50–)30(–15) or (10–)9 m.y. by Hertweck et al. (2015).

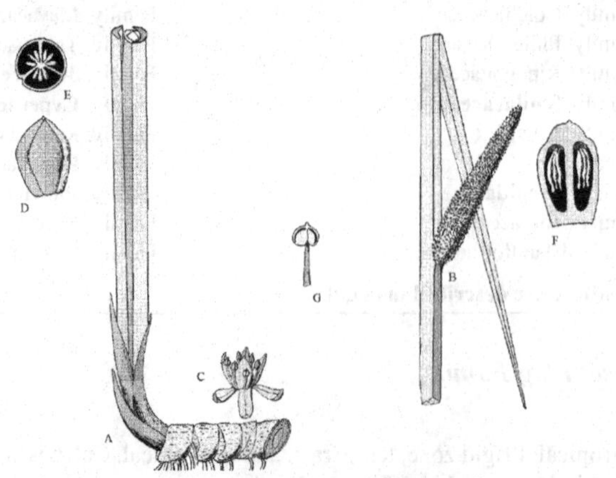

Figure 13.15: Acoraceae. *Acorus calamus.* **A:** Rhizome with basal leaves; **B:** Spadix with subtending leaf; **C:** Flower; **D:** Gynoecium; **E:** Transverse section of ovary; **F:** Longitudinal section of ovary; **G:** Stamen.

Araceae A. L. de Jussieu *Arum Family*

134 genera, 3,750 species

Throughout world but mainly in tropical and subtropical regions, very common in tropical forests and wetlands, a few species in temperate regions.

Placement:

	B & H	Cronquist	Takhtajan	Dahlgren	Thorne	APG IV/(APweb)
Informal Group						Monocots
Division		Magnoliophyta	Magnoliophyta			
Class	Monocotyledons	Liliopsida	Liliopsida	Liliopsida	Magnoliopsida	
Subclass		Arecidae	Alismatiidae	Liliidae	Alismatidae	
Series+/ Superorder	Nudiflorae+		Aranae	Aranae	Aranae	
Order		Arales	Arales	Arales	Arales	Alismatales

B & H as family Aroideae.

Salient features: Terrestrial erect or climbing or aquatic herbs with rhizomes or corms, leaves often large, mucilaginous, inflorescence scapigerous, a spadix subtended by a large spathe, flowers very small, reduced, usually unisexual, fruit a berry or utricle.

Major genera: *Anthurium* (900 species), *Philodendron* (500), *Arisaema* (150), *Amorphophalus* (100), *Pothos* (55), *Dieffenbachia* (40) and *Syngonium* (30).

Description: Terrestrial or aquatic herbs, sometimes epiphytic or climbing (*Pothos*, *Syngonium*), usually with rhizomes, or corms, sometimes free floating (*Pistia*), usually mucilaginous, raphide crystals of calcium oxalate, containing chemicals causing irritation of mouth (or temporary dumbness: *Dieffenbachia*, the dumbcane), vessels absent in stem and leaves, sieve-tube plastids P-type, subtype II; roots with vessels having scalariform end-walls, rarely with velamen. **Leaves** small to very large with sheathing base, alternate, spiral, or distichous, petiolate or sessile (*Pistia*), with parallel, pinnate or palmate venation, often cordate, or hastate, or sagittate; stipules absent; stomata paracytic, tetracytic, cyclocytic, or anomocytic. **Inflorescence** scapigerous, a spadix consisting of a dense spike subtended and enclosed by a large spathe (spathe absent in *Gymnostachys*, *Orontium*). **Flowers** very small, sessile, very rarely subsessile (*Pedicellarum*), unisexual, or bisexual, ebracteate, often fragrant, or malodorous; regular to very irregular. **Perianth** lacking or with 4–6 tepals (rarely 12), free or connate, usually in two whorls, green. **Androecium** with 1 (*Cryptocoryne*) –6(–12) stamens, free or with connate filaments, usually in two whorls, anthers basifixed, dehiscence by pores, longitudinal slits, or transversely, extrorse, tapetum amoeboid, pollen grains aperturate, or nonaperturate. **Gynoecium** with 2–3 carpels, rarely up to 8 carpels, united, ovary superior, usually unilocular, 1–5 ovuled, placentation apical, or marginal, rarely mutilocular with axile placentation, ovule orthotropous or anatropous, bitegmic, tenuinucellate (*Pistia*), or crassinucellate. **Fruit** usually a berry or drupe, rarely a utricle or capsule; seeds endospermic, or nonendospermic, cotyledon 1. Pollination mainly by insects, especially beetles, flies and bees. Dispersal of berries by birds and animals.

Economic importance: The family furnishes numerous horticultural ornamentals such as *Pothos*, *Alocasia*, *Arum*, *Dieffenbachia*, *Monstera*, *Philodendron*, *Zantedeschia* and *Syngonium*. *Epipremnum aureum* (money plant) is commonly grown as house plant. The corms of *Colocasia esculenta* (taro or dasheen), *Amorphophallus campanulatus* (Elephant-foot yam), *Cryptosperma* and *Xanthosoma* (tanier, yautia) and fruits of *Monstera* (Mexican breadfruit) are used as food.

Phylogeny: The family is considered to be monophyletic. Hutchinson believed it had been derived from the tribe Aspidistreae of Liliaceae. The bisexual flowers are considered to be more primitive, while those with unisexual flowers to be more highly evolved. Most recent classifications place this family under the order Arales along with Lemnaceae, Acoraceae having been removed to a separate order Acorales. Arales has also been placed under independent superorder Aranae in these classifications (Takhtajan shifts Aranae under separate subclass Aridae; Dahlgren and Cronquist also include Acoraceae under Arales). APG II onwards and APweb place Acoraceae under separate order but include Araceae along with several others under order Alismatales, Lemnaceae being merged with Araceae. The family is considered to be a fairly early divergent lineage within monocots and sister to remaining families of Alismatales. Thorne (2003) recognized 7 subfamilies under Araceae: Gymnostachyoideae (1 genus), Orontioideae (3 genera), Pothoideae (4 genera), Monsteroideae (12 genera), Lasioideae (10 genera), Calloideae (1 genus), and Aroideae (73 genera), treating Lemnaceae as a distinct family. APweb (2003, 2008) also recognises 7 subfamilies but Monsteroideae is merged with Pothoideae and Lemnoideae (based on Lemnaceae) included instead. APweb Version 14 (2018 update) reinstates Monsteroideae, merging Calloideae with Aroideae, thus recognizing 8 subfamilies. Thorne (2006, 2007) has followed APweb (2003), and included two additional subfamilies Philodendroideae (27 genera) and Schismatoglottidoideae (7 genera), both segregates from

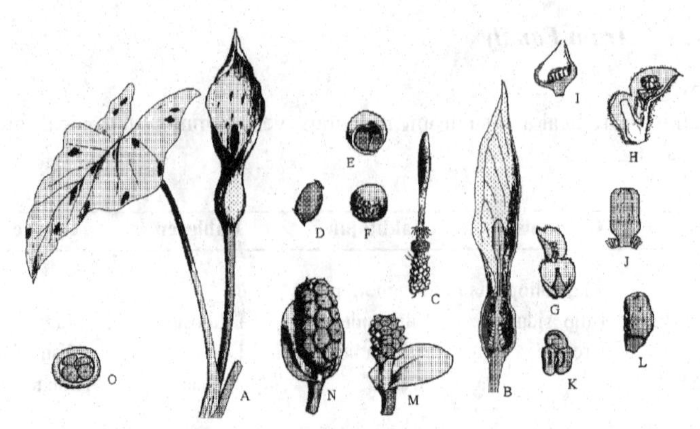

Figure 13.16: Araceae. *Arum maculatum.* **A:** Plant with flowering spadix; **B:** Vertical section of spadix and spathe; **C:** Spadix; **D:** Gynoecium; **E:** Fruit cut to show seeds; **F:** Seed. *Pistia stratiotes.* **G:** Inflorescence; **H:** Vertical section of inflorescence; **I:** Longitudinal section of ovary; **J:** Longitudinal section of orthotropous ovule with placental hairs; **K:** Portion of androecium; **L:** Seed. *Calla palustris.* **M:** Inflorescence; **N:** Mature fruits; **O:** Transverse section of carpel.

Aroideae (Keating, 2003a, 2003b, 2004a, 2004b), thus recognizing nine in all. Mayo et al. (2003) on the analysis of five plastid genes did not find a clear resolution of affinities within the family. However, the basal clade (Gymostachydoideae + Orontioideae) remains the same, and Lemnoideae are strongly supported as sister to the rest of the family.

Alismataceae Ventenat *Water Plantain Family*

16 genera, 135 species (including Limnocharitaceae)

Throughout world, plants of fresh water marshes, swamps, lakes, rivers and streams. Majority of the species found in the New World.

Placement:

	B & H	Cronquist	Takhtajan	Dahlgren	Thorne	APG IV/(APweb)
Informal Group						Monocots
Division		Magnoliophyta	Magnoliophyta			
Class	Monocotyledons	Liliopsida	Liliopsida	Liliopsida	Magnoliopsida	
Subclass		Alismatidae	Alismatidae	Liliidae	Alismatidae	
Series+/ Superorder	Apocarpae+		Alismatanae	Alismatanae	Alismatanae	
Order		Alismatales	Alismatales	Alismatales	Alismatales	Alismatales

B & H as family Alismaceae.

Salient features: Aquatic or marsh plants with laticifers, leaves petiolate with well-developed blade, inflorescence scapigerous, perianth in two whorls, differentiated into sepals and petals, stamens 6 to many, carpels 6 to many, free, ovule usually one per carpel, fruit an etaerio of achenes, embryo curved.

Major genera: *Echinodorus* (35 species), *Sagittaria* (25), *Alisma* (9) and *Burnatia* (3).

Description: Aquatic or marshy plant with basal leaves, rhizomatous, laticifers present with white latex, root xylem with vessels having scalariform to simple end-walls, vessels absent in stem and leaves. **Leaves** submerged and emergent, often heterophyllous, alternate, petiolate or sessile, sheathing, simple, pinnately, palmately, or parallel-veined, stomata paracytic or tetracytic, axillary scales present. **Inflorescence** scapigerous, paniculate, ultimate branches cymose or racemose, sometimes umbellate or even solitary, with or without involucral bracts. **Flowers** bracteate, bisexual or unisexual and monoecious, or dioecious (*Burnatia*), regular, trimerous, cyclic. **Perianth** differentiated into calyx and corolla. Sepals 3, free, imbricate. Petals 3, free, white, red, or pink. **Androecium** usually with 6 stamens, rarely more with branched outer stamens, free, anthers bithecous, dehiscence by longitudinal slits, extrorse, pollen grains usually 2–3 aperturate. **Gynoecium** with 3 carpels, rarely more, free, ovary superior, 1 ovuled, placentation basal, ovule anatropous, or amphitropous. **Fruit** etaerio of achenes, seeds nonendospermic, cotyledon 1, embryo strongly curved.

Economic importance: *Sagittaria sagittifolia* (arrowhead) is cultivated in China and Japan for its edible corms. Several species of *Sagittaria*, *Alisma* (water plantain), and *Echinodorus* (bur-heads) are cultivated as poolside plants and used as aquarium plants.

Figure 13.17: Alismataceae. *Alisma plantago-aquatica.* **A:** Basal part of plant with leaves; **B:** Inflorescence; **C:** Flower; **D:** Outer tepal; **E:** Inner tepal; **F:** Achene. *Sagittaria sagittifolia.* **G:** Plant with sagittate leaves and basal part of scape; **H:** Inflorescence; **I:** Male flower with petals removed; **J:** Petal; **K:** Stamens in different views; **L:** Achene; **M:** Carpel (After Sharma and Kachroo, Fl. Jammu, vol. 2, 1983).

Phylogeny: The family has been redefined (Pichon, 1946) to shift all genera with laticifers, petioled leaves with expanded blades, campylotropous ovules, and seeds with curved embryos, including several genera formerly included under Butomaceae. According to Judd et al. (2002) Butomaceae, Hydrocharitaceae and Alismataceae form one aquatic clade of Alismatales, supported by the apomorphies of perianth differentiated into sepals and petals, stamens more than six and carpels more than three, and the ovules scattered over the inner surface of locules. The genera with achenes (*Sagittaria*, *Alisma*, *Echinodorus*, etc.) may form a monophyletic group (Chase et al., 1993). The family is often regarded as primitive due to numerous stamens and carpels. The developmental and anatomical studies have, however, indicated that these numerous stamens are due to secondary increase, from ancestral condition of six stamens in two whorls. According to Hutchinson (1973) the family reminds of Ranunculaceae and but for solitary cotyledon and lack of endosperm, the genus *Ranalisma* might be well be placed in Ranunculaceae. According to Soros and Les (2002) *Echinodorus* is polyphyletic and evidently needs splitting. *Butomopsis* and *Limnocharis* were placed in distinct family Limnocharitaceae by Takhtajan, Cronquist and APG II (2003), but merged with Alismataceae in APG III (2009), APG IV (2016) and recent versions of APweb.

Butomaceae Richard *Flowering Rush Family*

1 genus, single species *Butomus umbellatus*

North temperate region, widespread in Asia and Europe, naturalized in tropical America.

Placement:

	B & H	Cronquist	Takhtajan	Dahlgren	Thorne	APG IV/(APweb)
Informal Group						Monocots
Division		Magnoliophyta	Magnoliophyta			
Class	Monocotyledons	Liliopsida	Liliopsida	Liliopsida	Magnoliopsida	
Subclass		Alismatidae	Alismatidae	Liliidae	Alismatidae	
Series+/ Superorder	Apocarpae+		Alismatanae	Alismatanae	Alismatanae	
Order		Alismatales	Hydrocharitales	Alismatales	Alismatales	Alismatales

B & H under family Alismaceae.

Salient features: Aquatic or marsh plants with linear triquetrous leaves, inflorescence scapigerous, umbellate cymes, perianth in two whorls, outer tinged green, stamens 9, carpels free, fruit etaerio of follicles.

Genus: *Butomus* (1 species).

Description: Aquatic or marshy plant with basal leaves, rhizomatous, secretory cavities present, with latex, root xylem with vessels with simple and scalariform end-walls, stem xylem without vessels, sieve-tube plastids P-type, subtype II. **Leaves**

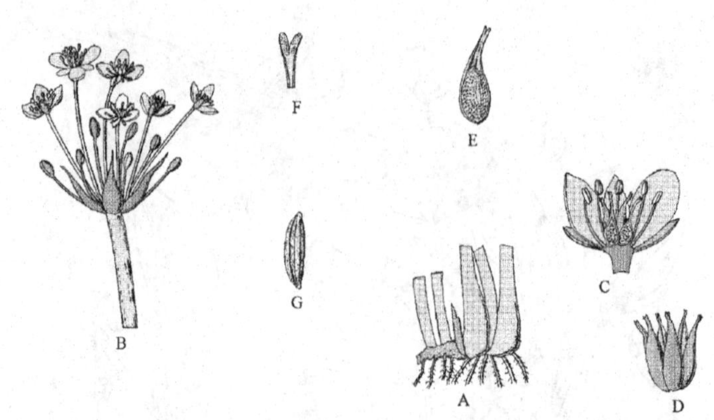

Figure 13.18: Butomaceae. *Butomus umbellatus*. **A:** Rhizome with basal leaves; **B:** Umbellate cymose inflorescence with involucral bracts; **C:** Vertical section of flower; **D:** Carpels; **E:** Longitudinal section of carpel showing scattered ovules; **F:** Stigmas; **G:** Seed.

emergent, alternate, distichous, petiolate or sessile, sheathing, simple, entire, linear, triquetrous, parallel-veined, stomata paracytic. **Inflorescence** scapigerous, umbellate cymes, involucral bracts three. **Flowers** medium-sized, on long pedicels, trimerous, regular, bisexual, cyclic. **Perianth** with 6 tepals, free, 2 whorled, whorls similar or outer sepaloid and inner petaloid, outer often tinged with green. **Androecium** with 9 stamens, 2 whorled (6 + 3), anthers basifixed, dehiscence by longitudinal slits, latrorse, pollen grains monosulcate. **Gynoecium** with 6 carpels, free or connate at base, ovary superior, carpel incompletely closed, style short, with ventral decurrent stigmatic region, ovules 20–100, placentation dispersed, stigma papillate. **Fruit** an aggregate of follicles, seeds nonendospermic, embryo straight (not curved), cotyledon 1.

Economic importance: Cultivated as an ornamental. The rhizomes are edible when baked.

Phylogeny: The genus was earlier placed under the family Alismaceae by Bentham and Hooker. Buchenau in Engler's '*Das Pflanzanreich*' (1903), recognized the family Butomaceae to include all members of Alismaceae (this original name of the family has now been replaced by Alismataceae) with numerous ovules and parietal placentation. Pichon (1946) redefined the circumscription of these two families to shift all genera with petioled leaves with expanded blades, campylotropous ovules, and seeds with curved embryos, to Alismaceae and retaining only genus *Butomus* under Butomaceae, a treatment followed in most of the recent publications. According to Hutchinson (1973), who retained a broader circumscription of the family, the gynoecium of Butomaceae represents probably the most ancient type of the monocotyledons, the free carpels recalling Helleboraceae, and the peculiar placentation of the ovules dispersed all over the surface of the carpel more ancient character than found in any herbaceous dicotyledon except in Cabombaceae, which is similar in this respect. Cabombaceae also have trimerous flowers and aquatic habit, and Butomaceae is separated only on the basis of a single cotyledon and absence of endosperm. The recent cladistic studies reveal this family to be closer to Hydrocharitaceae. According to Judd et al. (2002) Butomaceae, Hydrocharitaceae and Alismataceae form one aquatic clade of Alismatales, supported by the apomorphies of perianth differentiated into sepals and petals, stamens more than six and carpels more than three, and the ovules scattered over the inner surface of locules.

Hydrocharitaceae A. L. de Jussieu ***Tape Grass Family***

16 genera, 135 species

Throughout the world, mostly tropical and subtropical regions in fresh water and marine habitats.

Placement:

	B & H	**Cronquist**	**Takhtajan**	**Dahlgren**	**Thorne**	**APG IV/(APweb)**
Informal Group						Monocots
Division		Magnoliophyta	Magnoliophyta			
Class	Monocotyledons	Liliopsida	Liliopsida	Liliopsida	Magnoliopsida	
Subclass		Alismatidae	Alismatidae	Liliidae	Alismatidae	
Series+/ Superorder	Microspermae+		Alismatanae	Alismatanae	Alismatanae	
Order		Hydrocharitales	Hydrocharitales	Alismatales	Alismatales	Alismatales

B & H as family Hydrocharideae.

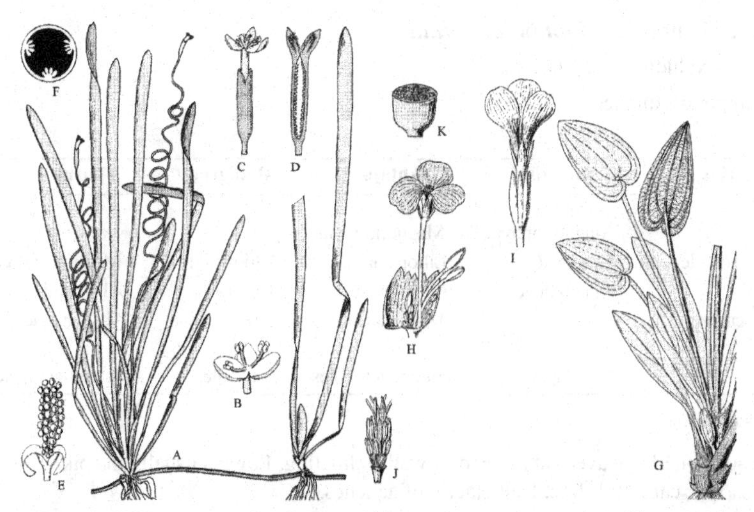

Figure 13.19: Hydrocharitaceae. *Vallisneria spiralis.* **A:** Plant with creeping stem, strap-shaped leaves and female flowers on coiled long pedicels; **B:** Male flower which detaches and floats on water; **C:** Female flower; **D:** Vertical section of female flower; **E:** Male inflorescence subtended by two connate bracts; **F:** Transverse section of ovary with parietal placentation. *Ottelia cordata.* **G:** Plant with leaves and emerging flower bud; **H:** Male inflorescence with spathe opened out; **I:** Female flower; **J:** Male flower with perianth removed, showing stamens and pistillodes; **K:** Transverse section of ovary.

Salient features: Aquatic fresh water or marine herbs, leaves submerged, usually ribbon-like, flowers subtended by paired bracts, male flowers often detaching and floating on water, carpels united, ovary inferior with many scattered ovules, fruit a capsule or berry.

Major genera: *Najas* (38 species), *Ottelia* (19), *Vallisneria* (8), *Elodea* (6), *Hydrocharis* (6), *Halophila* (4) and *Hydrilla* (1).

Description: Aquatic herbs, submerged or partly emergent, rooted in mud or unattached, in freshwater and marine habitats, annual or perennial. **Leaves** alternate (*Nechamandra*), opposite (*Elodea* some species) or whorled (*Hydrilla, Lagarosiphon*), in basal rosettes or cauline, simple, entire or serrate, with parallel or palmate venation, sheathing at base, small scales, at nodes inner to leaf base, stipules absent. **Inflorescence** with solitary flower (female) or short cymes (usually male), subtended by two often connate bracts. **Flowers** bisexual or unisexual, male flowers often disconnected and floating on water surface (*Vallisneria, Enhalus, Lagarosiphon*). **Perianth** often with distinct sepals and petals. Sepals 3, free, valvate, green. Petals 3, free, usually white, imbricate, sometimes lacking (*Thalassia, Halophila*). **Androecium** with 2 to 3 stamens, rarely more (*Egeria*), free with connate filaments, anthers bithecous, dehiscence by longitudinal slits, innermost sometimes modified into staminodes which often act as sails (*Lagarosiphon*), pollen grains monosulcate or inaperturate, sometimes united into thread-like chains (*Thalassia, Halophila*). **Gynoecium** with 3–6 (rarely 15), connate carpels, single in *Najas*, ovary inferior, unilocular, ovules many (1 in *Najas*), scattered over the surface, placentae often deeply intruded, styles often divided and twice the number of carpels, stigmas elongate and papillose. **Fruit** fleshy berry or dry capsule rupturing irregularly, nut in *Najas*; seeds without endosperm, embryo straight, cotyledon 1. Pollination in some (*Vallisneria, Enhalus*) by water, in others (*Egeria, Limnobium*) by insects. Dispersal by water or animals.

Economic importance: Species of many genera (*Hydrilla, Vallisneria, Elodea, Egeria*, etc.) are used as aquarium plants. Some species like *Hydrilla verticillata, Elodea canadensis* have become troublesome weeds in many parts of the world.

Phylogeny: The family, along with Butomaceae and Alismataceae, forms a well-defined clade, as indicated by cladistic analysis. Although monophyletic (Dahlgren and Rasmussen, 1983), the family is morphologically heterogenous and divided into 3 (Hutchinson, 1973; Thorne, 2003: Hydrocharitoideae, Thalassioideae and Halophiloideae) to 5 subfamilies (Dahlgren et al., 1985). Les et al. (1997) concluded that the family forms a well-defined lineage. Tanaka et al. (1997) suggest a series of quite well-supported nodes based on analysis of variation in two genes, the ultimate groupings recognized are similar to those of Les et al. (1997). The family Najadaceae, though distinct with single carpel, single ovule and superior ovary, has been included in Hydrocharitaceae by Thorne, APG II and APweb. It is possible *Najas* may be sister to the rest of Hydrocharitaceae, in which case it may probably be recognized as a separate family. APweb recognises 7 well defined groups (lineages) within Hydrocharitaceae as established by the studies of Les et al. (1997). Thorne (2006, 2007) has revised the classification of the family recognising 4 subfamilies: Hydrocharitoideae (2 genera), Stratiotoideae (1 genus), Anacharidoideae (7 genera) and Hydrilloideae (8 genera), followed in APweb (2018).

Potamogetonaceae Dumortier ***Pondweed Family***

6 genera, 110 species (Excluding Ruppiaceae)

Throughout world, in ponds, ditches and lakes.

Placement:

	B & H	Cronquist	Takhtajan	Dahlgren	Thorne	APG IV/(APweb)
Informal Group						Monocots
Division		Magnoliophyta	Magnoliophyta			
Class	Monocotyledons	Liliopsida	Liliopsida	Liliopsida	Magnoliopsida	
Subclass		Alismatidae	Alismatidae	Liliidae	Alismatidae	
Series+/ Superorder	Apocarpae+		Alismatanae	Alismatanae	Alismatanae	
Order		Najadales	Potamogetonales	Najadales	Potamogetonales	Alismatales

B & H under family Naiadaceae.

Salient features: Aquatic herbs, leaves submerged as well as floating, flowers usually in spikes, bisexual, perianth with 4 free tepals, stamens 1–4, carpels 4, free, fruit etaerio of achenes.

Major genera: *Potamogeton* (83 species), *Coleogeton* (6) and *Zanichellia* (5).

Description: Perennial or rarely annual fresh water herbs with rhizomes, stems mostly submerged, with reduced vascular bundles in a ring, with air cavities, tannins often present, root xylem with vessels having scalariform end-walls, stem without vessels, sieve-tube plastids P-type, subtype PII. **Leaves** submerged as well as floating, sheathing at the base, alternate or opposite (*Groenlandia*), simple, entire, **venation** parallel, submerged leaves thin, without cuticle and stomata, floating leaves thick, small scales present at nodes inner to the leaf sheath. **Inflorescence** terminal or axillary spike, often carried on a long peduncle, raised above water surface, peduncle surrounded by sheath at base. **Flowers** ebracteate, regular, bisexual, hypogynous, cyclic. **Perianth** with 4 tepal (often interpreted as appendages from connective of the anthers and thus perianth absent), free, fleshy, usually clawed. **Androecium** with 4 stamens, free, adnate to and opposite each tepal, anther sessile, dehiscence by longitudinal slits, pollen grains globose, inaperturate. Gynoecium with 4 free carpels, ovary superior with basal to apical placentation, ovule 1, campylotropous, bitegmic, crassinucellate, style short or lacking, truncate to capitate. **Fruit** an etaerio of achenes or drupes, seeds nonendospermic, with starch, cotyledon 1, embryo slightly curved. Pollination by wind, dispersal by water or animals.

Economic importance: The family is of little economic importance but biologically an important source of food for aquatic life. Many species of *Potamogeton* are troublesome weeds. Fleshy starchy rootstocks are sometimes used as food.

Phylogeny: The affinities of the family are not clear. The family is sometimes also interpreted to include *Ruppia* and/ or *Zannichelia*. The studies of Les et al. (1997) however, have shown that *Zannichelia* is rather weakly embedded in the

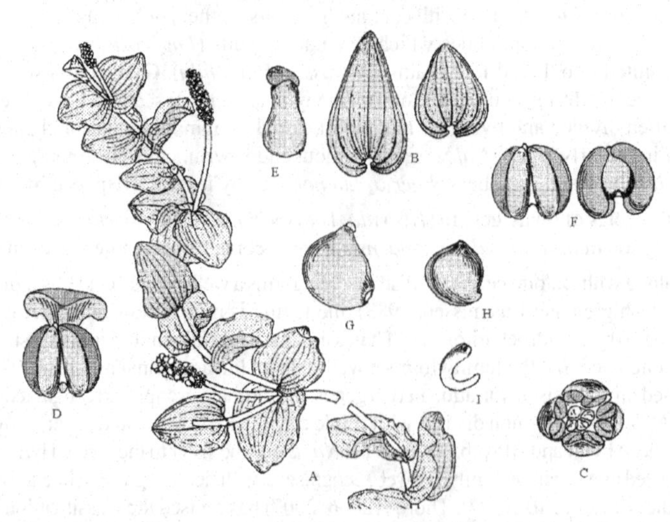

Figure 13.20: Potamogetonaceae. *Potamogeton perfoliatus.* **A:** Plant with flowering and fruiting inflorescences; **B:** Leaves; **C:** Flower; **D:** Tepal with attached stamen; **E:** Carpel; **F:** Anther in two different views; **G:** Fruit; **H:** Seed enclosed in hard endocarp; **I:** Seed (After Sharma and Kachroo, Fl. Jammu, vol. 2, 1983).

family, and the inclusion of *Ruppia* makes family biphyletic. *Potamogeton* itself is considered para- or polyphyletic (Les and Haynes, 1995). Uhl (1947) had earlier supported the segregation of *Zannichelia* into separate family by Hutchinson (1934). The tepals are often interpreted variously. Uhl suggested that so-called perianth parts are in fact individual bracts, subtending and adnate to stamens and the flower is fundamentally an inflorescence with staminate flowers (each represented by a monbracteate perianth) and naked female flowers, the view first proposed by Kunth (1841) and supported by Miki (1937). Most of the authors (Rendle, 1925; Watson and Dallwitz, 2000; Judd et al., 2008) consider tepals to represent appendages from the connective of the anther. As per Hutchinson (1973), these are normal perianth-segments on claws of which the extrorse anthers are sessile, a view also supported by Heywood (1978) and Woodland (1991). Hutchinson stressed that in the petaloid monocotyledons the stamens are always opposite the perianth-segments, and it is not a great step for those species of *Aponogeton* with more than one perianth-segment to *Potamogeton.* If the anther were introrse in *Potamogeton*, then the petal-like organ might be regarded as an outgrowth from the base of the connective, a very unusual feature indeed in any flowering plant. *Groenlandia* is sister to the rest of the family (Du et al., 2016). *Zannichellia* and relatives, which alone in the family commonly have flavone sulphates, is rather weakly embedded within Potamogetonaceae (Les et al., 1997) or sister to the rest of the family (Les and Tippery, 2013).

Dioscoreaceae R. Brown *Yam Family*

9 genera, 715 species

Mainly tropical and subtropical, few in the temperate region.

Placement:

	B & H	Cronquist	Takhtajan	Dahlgren	Thorne	APG IV/(APweb)
Informal Group						Monocots
Division		Magnoliophyta	Magnoliophyta			
Class	Monocotyledons	Liliopsida	Liliopsida	Liliopsida	Magnoliopsida	
Subclass		Liliidae	Liliidae	Liliidae	Liliidae	
Series+/ Superorder	Epigynae+		Dioscoreanae	Lilianae	Dioscoreanae	
Order		Liliales	Dioscoreales	Dioscoreales	Dioscoreales	Dioscoreales

Salient features: Woody or herbaceous climbers, leaves alternate, cordate, petiole with pulvinus at both ends, venation reticulate, inflorescence axillary racemes, spikes or umbels, flowers unisexual, fruit a capsule, seeds winged.

Major genera: *Dioscorea* (660 species), *Rajania* (17) and *Tacca* (16).

Description: Perennial herbaceous or woody climbers with tubers or rhizomes, a few dwarf shrubs, usually twining over the support, stem with vascular bundles in one or two rings. **Leaves** usually alternate, sometimes opposite (*Dioscorea alata*) simple, cordate, sometimes palmately lobed or compound (*D. pentaphylla*), petiolate, petiole with pulvinus both at base and above, sometimes with stipule like flanges on both sides, sometimes with superficial or internal glands containing nitrogen fixing bacteria, venation palmate, reticulate, stomata anomocytic, leaf axils often with bulbils. **Inflorescence** axillary panicles, racemes or spikes, flowers arranged singly or in 2–3 flowered clusters. **Flowers** usually unisexual (plants dioecious), small, sessile or rarely pedicellate, actinomorphic. **Perianth** with 6 tepals, in two whorls, free or connate at base into tube. **Androecium** with 6 stamens, in two whorls, attached to the base of perianth, 3 sometimes reduced to staminodes, filaments free or slightly connate, anthers bithecous, connective sometimes broad; pollen grains monosulcate or variously porate. **Gynoecium** with 3 carpels, united, ovary inferior, trilocular with axile placentation, ovules 2 (rarely many) in each locule, styles 3, free or connate. **Fruit** a 3-valved capsule or berry, rarely samara; seeds usually flattened and winged, with endosperm and small embryo, often with second scaly cotyledon, seed coat with yellow-brown to red pigments, and crystals.

Economic importance: Many species of *Dioscorea* are cultivated for starchy tubers (Yam). Some species are source of diosgenin, a steroidal sapogenin developed in recent years for its use in oral contraceptives.

Phylogeny: The family is often placed under order Dioscoreales. Dahlgren (1985) also included Smilacaceae and Trilliaceae and considered Dioscoreales to represent primitive monocots. The primitive position of the order (also advocated by Stevenson and Loconte, 1995), is not supported by cladistic analysis based on *rbcL* sequence and morphological data (Chase et al., 1995). Thorne has included Dioscoreaceae along with Taccaceae under order Taccales, Thorne who had earlier (1999, 2000) placed Taccales after Liliales under Lilianae, removed (2003) under distinct superorder Taccanae placed before Lilianae, but after Pandananae, which has been shifted to the beginning of Liliidae. In 2006, 2007 he has preferred the name Dioscoreanae for Taccanae and Dioscoreales for Taccales, as he has merged Taccaceae and Stenomeridaceae with Dioscoreaceae. Takhtajan took Taccaceae under monotypic order Taccales, keeping Dioscoreaceae

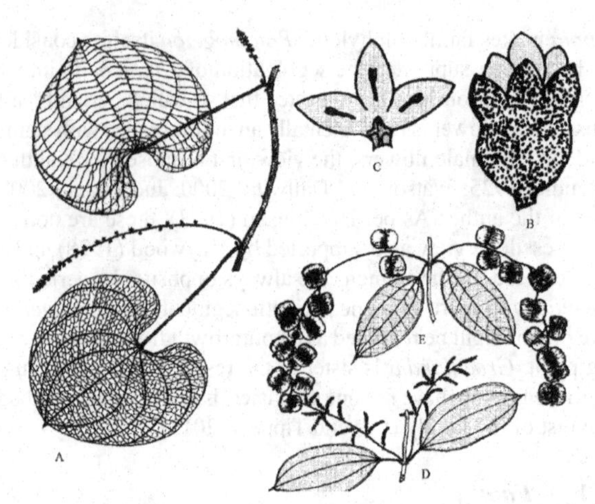

Figure 13.21: Dioscoreaceae. *Dioscorea esculenta.* **A:** Male plant with alternate leaves; **B:** Male flower; **C:** Male flower opened. *D. oppositifolia.* **D:** Male plant with opposite leaves and flowers; **E:** Female plant with fruits.

along with closely related families under Dioscoreales. Cronquist had earlier included it under broadly circumscribed Liliales. The circumscription of Dioscoreales has been narrowed in APG II onwards (incl. APG III, 2009; APG IV, 2016) and APweb to include only Nartheciaceae, Dioscoreaceae and Burmanniaceae, shifting Smilacaceae and Trilliaceae (latter under Melanthiaceae) to Liliales and Stemonaceae to Pandanales, but have merged Taccaceae, Stenomeridaceae and Trichopodaceae with Dioscoreaceae (both regarded as distinct families by Thorne). The narrowly circumscribed Dioscoreales is monophyletic as supported by morphological and *rbcL* sequence evidence (Chase et al., 1995), although the Placement of Nartheciaceae received poor support in the studies of Chase et al. (2000) and Caddick et al. (2002a, 2002b). *Trichopus* and *Avetra*, formerly under Trichpodaceae, has been merged with Dioscoreaceae.

Pandanaceae R. Brown ***Screw Pine Family***

5 genera, 982 species

Throughout tropics and subtropics of the Old World, mostly in coastal and marshy areas.

Placement:

	B & H	Cronquist	Takhtajan	Dahlgren	Thorne	APG IV/(APweb)
Informal Group						Monocots
Division		Magnoliophyta	Magnoliophyta			
Class	Monocotyledons	Liliopsida	Liliopsida	Liliopsida	Magnoliopsida	
Subclass		Arecidae	Aridae	Liliidae	Liliidae	
Series+/ Superorder	Nudiflorae+		Pandananae	Pandananae	Pandananae	
Order		Pandanales	Pandanales	Pandanales	Pandanales	Pandanales

Salient features: Large shrubs, trees or climbers with annual scars of leaf bases, bearing aerial roots, leaves 3-ranked, stiff, inflorescence a spadix with unisexual flowers, flowers naked, male flowers with many stamens, carpels many, free or united, fruit a berry or multilocular drupe, often aggregated into a cone appearing like pineapple.

Major genera: *Pandanus* (750 species), *Freycinetia* (123) and *Sararanga* (2).

Description: Trees, shrubs or climbing perennials (*Freycinetia*), supported by aerial roots, roots often penetrating supporting host (*Freycinetia*) or even absent (*Sararanga*), trunk bearing annual scars of leaf bases, stem and leaves also with xylem with scalariform end-walls. **Leaves** forming terminal crown, 3-ranked or 4-ranked, sometimes appearing spirally arranged due to twisting of stem, long, narrow, usually stiff or sword-like, sheathing at base, keeled, often spinose along margin and keel, sometimes even grass-like. **Inflorescence** a spadix subtended by a brightly colored spathe, and usually containing one type of flowers, male and female flowers being borne on separate plants (plants dioecious), spadix lacking in *Sararanga* and inflorescence paniculate. **Flowers** sessile, pedicellate in *Sararanga*, unisexual, without perianth, hypogynous. **Perianth** absent or vestigial, sometimes forming a short cupule (*Sararanga*). **Androecium** in male flower with numerous stamens, filaments free or connate, anthers erect, bithecous, basifixed, dehiscence by longitudinal slits, staminodes often present (*Freycinetia*) in female flower. **Gynoecium** in female flower with many carpels, free or connate, ovary superior, unilocular

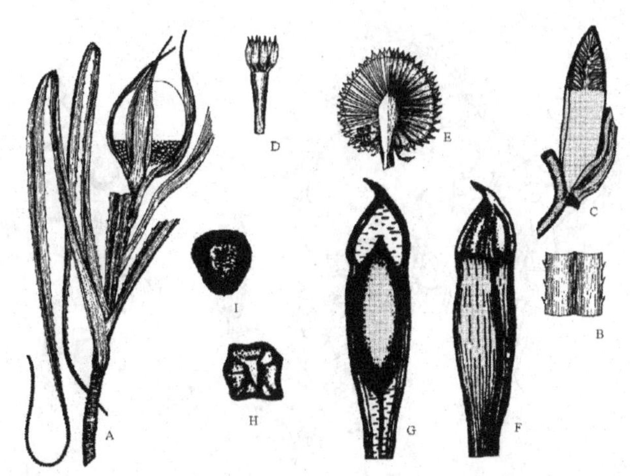

Figure 13.22: Pandanaceae. *Pandanus ceylonicus.* **A:** Plant with leaves and female inflorescence; **B:** Portion of leaf showing marginal prickles; **C:** Male inflorescence; **D:** Male flower with apiculate anthers; **E:** Female spadix split to show arrangement of flowers; **F:** Female flower with curved style; **G:** Longitudinal section of female flower; **H:** Drupe tipped by persistent style; **I:** Seed (After Dassanayake, Fl. Ceylone, vol. 3, 1981).

(if carpels free) or multilocular (if united), ovules 1 (*Pandanus*) to many (*Freycinetia*), anatropous, style short or absent, stigma nearly sessile, rudimentary ovary often present in male flower. **Fruit** a berry or multilocular drupe, often aggregated to form oblong or globose syncarps resembling a pineapple; seeds small with fleshy endosperm and minute embryo.

Economic importance: Several species of *Pandanus* (screw pine) are useful sources of food. *P. leram* (Nicobar breadfruit) produces a large globose fruit which is boiled in water. Other species also yielding edible fruits include *P. utilis* and *P. andamanensium*. The leaves of *P. odoratissimus* are used for thatching and weaving. Fibres made from aerial roots are used for cordages and brushes. The flowers of this species are used for popular Indian essence Kewra. The fragrant leaves of *P. odorus* in Malaya are used in poutpourris. *Freycinetia banksii* and *Pandanus veitchii* are used as ornamentals.

Phylogeny: Pandanaceae forms a well-defined clade included under Pandanales either singly (Thorne, Takhtajan, Dahlgren and Cronquist, Hutchinson), or together with Cyclanthaceae, Stemonaceae Velloziaceae, and Triuridaceae (APG II, APweb, continued in APG II, APG IV and latest update of APweb, 2018). The position of Cyclanthaceae does not show much departure as it has often been placed in the adjacent order. The placement of Triuridaceae within Pandanales is, however, interesting. Takhtajan places it under a separate subclass, Cronquist under Alismatidae, and Dahlgren under separate superorder Triuridanae. The placement of this family in Pandanales is supported by the studies of 18S rDNA (Chase et al., 2000). Thorne who had earlier (1999) placed Triuridaceae under Alismatidae has finally (2007) shifted it under Liliidae—> Pandananae—> Pananales, not recognizing superorder Triuridanae. He has also made a major change in shifting Pandananae to the beginning of Liliidae. He had earlier placed it towards the end of the subclass. Thorne (2003, 2007) recognizes two subfamilies Pandanoideae and Freycinetioideae. Nadaf and Zanan (2012: Indian species the focus) recovered the relationships [[*Benstonea* + *Pandanus*] [*Sararanga* [*Freycinetia* + *Martellidendron*]]].

Smilacaceae Ventenat ***Catbrier Family***

1 genus, 255 species

Mainly tropical and subtropical, extending into the temperate region.

Placement:

	B & H	**Cronquist**	**Takhtajan**	**Dahlgren**	**Thorne**	**APG IV/(APweb)**
Informal Group						Monocots
Division		Magnoliophyta	Magnoliophyta			
Class	Monocotyledons	Liliopsida	Liliopsida	Liliopsida	Magnoliopsida	
Subclass		Liliidae	Liliidae	Liliidae	Liliidae	
Series+/ Superorder	Coronariae+		Lilianae	Lilianae	Lilianae	
Order		Liliales	Smilacales	Dioscoreales	Liliales	Liliales

B & H under family Liliaceae.

Figure 13.23: Smilacaceae. *Smilax aspera.* **A:** Portion of plant with leaves, stipule tendrils and inflorescence; **B:** Male flower; **C:** Tepal with stamen; **D:** Female flower; **E:** Transverse section of ovary; **F:** Infructescence with berries.

Salient features: Woody or herbaceous climbers, climbing by stipular tendrils, stems sometimes prickly, leaves alternate, venation reticulate, inflorescence axillary racemes, spikes or umbels, flowers unisexual, fruit 1–3 seeded berry.

Genus: *Smilax* (255 species).

Description: Herbaceous or woody climbers with paired stipular tendrils, stem prickly, with underground rhizomes or tubers. **Leaves** alternate or opposite, mostly leathery, petiolate, 3-veined, stipules (or leaf sheath) developing into tendrils, venation reticulate. **Inflorescence** axillary raceme, spike or cyme. **Flowers** small, unisexual (plants dioecious), regular, hypogynous, trimerous. **Perianth** with 6 tepals, equal or subequal, in two whorls, free or united into a tube. **Androecium** with 6 stamens (rarely 3 or 9), free or united, anthers monothecous due to confluence of two locules, introrse, pollen grains inaperturate or monosulcate, staminodes present in female flower. Gynoecium with 3 united carpels, ovary superior, 3-locular, 1–2 ovules in each locule, placentation axile, ovules pendulous, orthotropous or semianatropous, stigmas 3. **Fruit** a berry with 1–3 seeds, embryo small, endosperm hard. Pollination by insects, dispersal of fruits by birds.

Economic importance: Several species of *Smilax* are the source of sarsaparilla, used for treating rheumatism, and other ailments. The dried rhizomes of *Smilax china* (China root) yield an extract used as a stimulant. Young stems and berries are sometimes used as food.

Phylogeny: Earlier included under Family Liliaceae (Bentham and Hooker; Engler and Prantl), it was separated as a distinct family by Hutchinson (1934, 1973), according to whom, the members are distinct from Liliaceae in habit, dioecious flowers and confluent anther loculi. He considered Smilacaceae to be considerably advanced from the general stock of the Liliaceae. Dahlgren et al. (1985) considered the family to be related to Dioscoreaceae and included it under Dioscoreales. The morphological studies (Conran, 1989) and *rbcL* sequences (Chase et al., 1993), however, support placement under Liliales (Cronquist, 1988; Thorn; APG II; APweb). Cronquist also included genera *Luzuriaga*, *Petermannia* and *Philesia* under the family Smilacaceae, but according to Chase et al., their inclusion makes Smilacaceae polyphyletic. The monophyly of the family (including only genera *Smilax, Heterosmilax* and *Ripogonum*) is supported by morphological as well as molecular evidence. APG II and APweb include *Ripogonum* under a distinct family Rhipogonaceae (leaves opposite, pollen reticulate), whereas Judd et al. (2002) include it (*Ripogonum*) under Smilacaceae. Thorne (2003) divided the family into two subfamilies Smilacoideae and Rhipogonoideae, latter has been elevated to a distinct family in 2006, 2007 revisions. Judd (1998) suggested merger of *Heterosmilax* with *Smilax*, since "the segregation of these two genera would result in a paraphyletic *Smilax sensu stricto* because no apomorphies are known that would unite the remaining members of the group, which are characterized by flowers with free tepals and six more or less distinct stamens". This proposal was further supported by recent molecular phylogenetic evidence (Cameron and Fu, 2006; Qi et al., 2012), which shows that *Heterosmilax* distributed in Eastern and Southeast Asia is monophyletic and embedded within *Smilax*, and as such Qui et al. (2013) proposed some new combinations under *Smilax*. APG IV (2016) and APweb (2018) accordingly recognize single genus *Smilax* under Smilacaceae, *Ripogonum* having already been segregated to Ripogonaceae.

Liliaceae M. Adanson ***Lily Family***

15 genera, 640 species

Widely distributed in the Northern Hemisphere, mainly in the temperate regions.

Placement:

	B & H	Cronquist	Takhtajan	Dahlgren	Thorne	APG IV/(APweb)
Informal Group						Monocots
Division		Magnoliophyta	Magnoliophyta			
Class	Monocotyledons	Liliopsida	Liliopsida	Liliopsida	Magnoliopsida	
Subclass		Liliidae	Liliidae	Liliidae	Liliidae	
Series+/ Superorder	Coronariae+		Lilianae	Lilianae	Lilianae	
Order		Liliales	Liliales	Liliales	Liliales	Liliales

Cronquist also included Amaryllidaceae, Asparagaceae, Alliaceae and several others under Liliaceae.

Salient features: Herbs with alternate or whorled leaves, base sheathing, flowers not subtended by spathaceous bracts, flowers bisexual, trimerous, perianth with 6 petaloid tepals, stamens 6, filaments free, carpels 3, united, ovary superior, placentation axile, fruit a capsule.

Major genera: *Fritillaria* (90 species), *Gagea* (80), *Tulipa* (80) and *Lilium* (75).

Description: Perennial herbs with underground bulb, generally with contractile roots. **Leaves** mostly basal, alternate or whorled, usually linear or strap shaped, simple, entire venation parallel, stipules absent. **Inflorescence** usually racemose (*Lilium*), sometimes solitary (*Tulipa*) or subumbellate (*Gagea*). **Flowers** showy, bisexual, actinomorphic, rarely zygomorphic, trimerous, hypogynous. **Perianth** with 6 tepals, in two whorls (outer representing sepals, inner petals), both whorls petaloid, often spotted or with lines, often united into tube, nectary at the base of tepal. **Androecium** with 6 stamens, in 2 whorls, epiphyllous, filaments free. **Gynoecium** with 3 carpels, united, ovary superior, trilocular with many ovules, placentation axile, styles simple with 3-lobed stigma. **Fruit** a loculicidal capsule, rarely a berry; seeds usually flat, with well-developed epidermis, seed coat not black, small embryo, endosperm copious. Pollination by insects, especially bees, wasps, butterflies. Seeds are dispersed by water or wind.

Economic importance: The family is important for its valuable ornamentals such as lily (*Lilium*), tulip (*Tulipa*) and *Fritillaria* (Fritillary).

Phylogeny: The circumscription of the family has undergone considerable reduction over the recent years. The genera formerly included under the family have now been removed to diverse families: *Colchicum* (Colchicaceae—with corm), *Trillium* (Trilliaceae—rhizome, leaves whorled, perianth with sepals and petals), *Allium* (Alliaceae—bulb, inflorescence umbellate, with spathaceous bracts, smell of onion, seeds black), *Asphodelus* and *Aloe* (Asphodelaceae—inflorescence racemose, seeds black, leaves succulent, often with colored sap), *Asparagus* (Asparagaceae—fruit a berry, leaves rudimentary, seeds black), *Ruscus* (Ruscaceae—leaves scarious, filaments connate), the last four were taken under a separate order Asparagales along with several other families, in the recent APG II, APG III, APG IV and APweb classifications. Thorne had earlier (1999, 2000) included these four families (along with other splitters from broadly circumscribed Liliaceae) under order Orchidales but has shifted (2003) these together with others to order Iridales, also bringing about

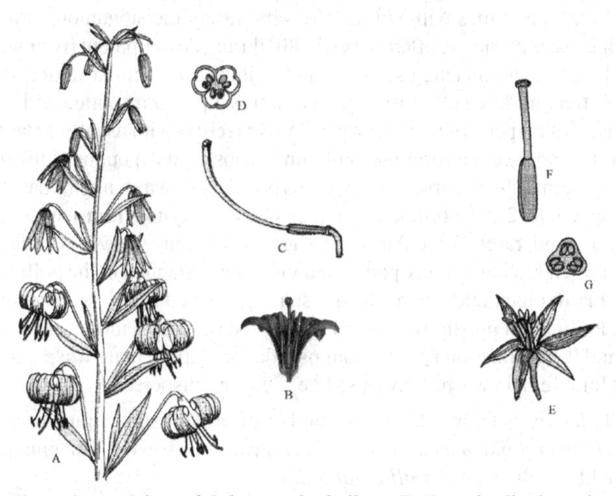

Figure 13.24: Liliaceae. A: Flowering portion of *Lilium polyphyllum*. **B:** Longitudinal section of flower of *L. canadense*. **C:** Gynoecium of *L. lancifolium*. **D:** Transverse section of ovary of *L. lancifolium*. *Gagea pseudoreticulata*. **E:** Flower; **F:** Gynoecium; **G:** Transverse section of ovary (A, after Polunin and Stainton, Fl. Himal., 1984).

certain changes in circumscription (Ruscaceae and Convallariaceae included under Asparagaceae). In 2006, 2007 revisions he has slightly enlarged the circumscription of family Liliaceae by merging Tricyrtidaceae and Calochortaceae but has divided the family into four subfamilies Medeoloideae (1 genus), Lilioideae (9 genera), Tricyrtidoideae (4 genera) and Calochortoideae (1 genus). The narrow circumscription of the family was first suggested by Dahlgren (1985), and forms a monophyletic group as confirmed by the cladistic studies of Chase et al. (1995a, 1995b). Cronquist, on the other hand, broadened the circumscription of the family, also including, in addition to the above families, large family Amaryllidaceae within Liliaceae. Peruzzi (2015) recognized six tribes within Lilaceae: Calochorteae, Lilieae, Medeoleae, Streptopeae, Tricyrtideae and Tulipeae. APweb (2018) recognizes three subfamilies: Lilioideae (2 tribes Medeoleae and Lilieae-which includes Tulipae of Peruzzi), Calochortoideae (including Tricyrtideae), Streptopoideae.

Orchidaceae A. L. de Jussieu *Orchid Family*

736 genera, 28,000 species (Overtakes Asteraceae in latest estimates of Christenhusz and Byng (2016), 26,000 species according to APWeb (2018) and now considered as largest family of Angiosperms).

Widely distributed, most common in moist tropical forests (where frequently epiphytic), also well distributed in subtropics and temperate regions.

Placement:

	B & H	Cronquist	Takhtajan	Dahlgren	Thorne	APG IV/(APweb)
Informal Group						Monocots
Division		Magnoliophyta	Magnoliophyta			
Class	Monocotyledons	Liliopsida	Liliopsida	Liliopsida	Magnoliopsida	
Subclass		Liliidae	Liliidae	Liliidae	Liliidae	
Series+/ Superorder	Microspermae+		Lilianae	Lilianae	Lilianae	
Order		Orchidales	Orchidales	Orchidales	Orchidales	Asparagales

Salient features: Herbaceous perennials, roots with velamen, leaves distichous, flowers trimerous, zygomorphic, corolla with 2 lateral petals and labellum, pollen in pollinia, ovary inferior, seeds minute.

Major genera: *Pleorothallus* (1100 species), *Bulbophyllum* (970), *Dendrobium* (900), *Epidendrum* (800), *Habeneria* (580), *Liparis* (320), *Malaxis* (280), *Oberonia* (280), *Calanthe* (200), *Vanilla* (100) and *Vanda* (60).

Description: Perennial herbs, terrestrial (*Malaxis*, *Orchis*), epiphytic (*Oberonia*, *Dendrobium*) or saprophytic (*Gastrodia*, *Epigonium*), rarely climbers (*Vanilla*), with rhizomes, tuberous roots, corms or rootstock, roots mycorrhizal, with multiseriate epidermis of dead cells known as velamen. Stems foliate or scapose, base often thickened to form pseudobulb, aerial roots present. **Leaves** usually alternate, distichous, rarely opposite, sometimes reduced to scales, often fleshy, simple, entire, sheathing at base, sheath closed and encircling stem, venation parallel, stipules absent, stomata tetracytic. **Inflorescence** racemose, spicate or paniculate, sometimes with solitary flowers, rarely cleistogamous. **Flowers** usually bisexual, very rarely unisexual, zygomorphic, usually showy, often twisted 180° during development (resupinate), **Perianth** differentiated into sepals and petals. Sepals 3, free or connate, usually petaloid, imbricate, similar or dorsal smaller, lateral more or less adnate to the ovary. Petals 3, free; middle petal forming labellum or lip, often spotted and variously colored, sometimes saccate or even spurred at base; lateral petals similar to sepals. **Androecium** with usually 1 stamen, sometimes 2 (*Apostasia*) or 3 (*Neuwiedia*), adnate to style and stigma forming a column (gynostemium) opposite the lip, anther sessile on column, bithecous, dehiscence by longitudinal slit, introrse; pollen grains powdery or waxy, agglutinated into pollinia, each pollinium with a sterile portion called caudicle, 2 to 8 pollinia formed in a flower. **Gynoecium** with 3 united carpels, ovary inferior, unilocular with parietal placentation, rarely 3-locular with axile placentation (*Apostasia*), stigmas 3, one often transformed into a sterile rostellum, latter often having a sticky pad called viscidium attached to the pollinia; ovules numerous, minute, anatropous, tenuinucellate. **Fruit** a loculicidal capsule or a sausage-shaped berry; seeds numerous, minute, embryo very minute, endosperm absent. Pollination mostly by insects such as bees, wasps, moths and butterflies. Flowers of *Ophrys* resemble the female wasp and the pollination results from pseudocopulation, male wasp attracted by the shape and smell of the flowers, mistaking it for a female wasp. Tiny dust-like seeds are dispersed by wind.

Economic importance: The family is known for large number of ornamentals reputed for their showy flowers mainly *Cattleya*, *Dendrobium*, *Cymbidium*, *Epidendrum*, *Vanda*, *Coelogyne* and *Brassia*. The only food product from the family is vanilla flavouring obtained from the fruits *Vanilla planifolia*.

Phylogeny: The family is generally considered as a natural group, the monophyly of the family supported by morphology and *rbcL* sequences (Dressler, 1993; Dahlgren et al., 1985). The family is commonly divided into three subfamilies:

Figure 13.25: Orchidaceae. A: *Cymbidium hookeranum* with leaves and flowers; **B:** *Eria muscicola,* plant with inflorescences. *Oberonia recurva*; **C:** Epiphytic plant with ensiform leaves and pendulous inflorescence; **D:** Flower; **E:** Floral parts separated showing from above downwards bract, three sepals, two lateral petals and anterior labellum. *Vanda tessellata*; **F:** Epiphytic plant with inflorescence; **G:** Floral parts separated; **H:** Pollinia from behind with gland and strap; **I:** Pollinia from front; **J:** Operculum from inside (After Dassanayake and Fosberg, Fl. Ceylone, vol. 2, 1981).

Apostasioideae, Cypripedioideae, and Orchidoideae. The former two include one tribe each, but the last one, which includes nearly 99 per cent of the orchid species is divided into 4 tribes. Apostasioideae are considered to be sister to the remaining orchids (Dressler, 1993; Cameron et al., 1999), and monophyletic as supported by vessel-elements with simple perforations and distinctive seed type. Cypripedioideae are usually considered clearly monophyletic (Judd et al., 1999) as supported by their saccate labellum, two functional stamens and absence of pollinia. Members of Orchidoideae share acute anther apex, soft stems and lack silica bodies. More recently however 5 subfamilies are recognized the other two being Vanilloideae and Epidendroideae (APweb, 2018). The recent studies have, however, put some uncertainty over the position of Cypripedioideae. They may group (albeit weakly) with Vanilloideae or be sister to Orchidaceae minus Apostasioideae (Stevens, 2003). Chase et al. (2015) in updated classification of family made substantial changes in Epidendroideae. Relationships within Vanilloideae are becoming fairly well resolved (Cameron and Berg, 2017).

Iridaceae A. L. de Jussieu *Iris Family*

66 genera, 2,244 species

Widely distributed, in tropical and temperate climates, mainly distributed in South Africa, the Mediterranean region, Central and South America.

Placement:

	B & H	Cronquist	Takhtajan	Dahlgren	Thorne	APG IV/(APweb)
Informal Group						Monocots
Division		Magnoliophyta	Magnoliophyta			
Class	Monocotyledons	Liliopsida	Liliopsida	Liliopsida	Magnoliopsida	
Subclass		Liliidae	Liliidae	Liliidae	Liliidae	
Series+/ Superorder	Epigynae+		Lilianae	Lilianae	Lilianae	
Order		Liliales	Iridales	Liliales	Iridales	Asparagales

Salient features: Herbaceous perennials, leaves equitant, leaf base sheathing, flowers bisexual, sepals petaloid, petals often spotted, ovary inferior, style petaloid, fruit a capsule.

Major genera: *Iris* (240 species), *Gladiolus* (230), *Moraea* (125), *Sisyrinchium* (100), *Crocus* (75), *Ixia* (45), *Freesia* (20) and *Tigridia* (12).

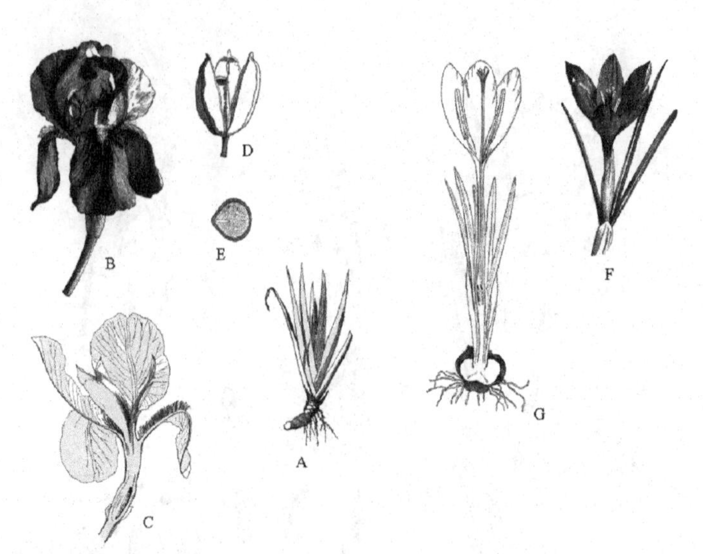

Figure 13.26: Iridaceae. *Iris germanica.* **A:** Rhizome and basal leaves; **B:** Flower; **C:** Vertical section of flower; **D:** Capsule dehiscing through valves; **E:** Longitudinal section of seed. *Crocus vernus.* **F:** Flower and leaves; **G:** Longitudinal section through entire plant.

Description: Perennial herbs with rhizomes (*Iris*), corms (*Gladiolus*) or bulbs, bundle sheaths with styloid crystals of calcium oxalate, tannins and terpenoids present. **Leaves** alternate, distichous, usually sessile, equitant (oriented edgewise to the stem), simple, entire, sheathing at base, venation parallel, stipules absent. **Inflorescence** a cyme, raceme, spike or panicle, sometimes solitary, commonly subtended and enclosed by one or more spathaceous bracts. **Flowers** bisexual, showy, actinomorphic (*Sisyrinchium*) or zygomorphic (*Gladiolus*), trimerous, epigynous. **Perianth** 6 in two whorls often differentiated into sepals and petals. Sepals 3, free or united, imbricate, sometimes deflexed and with a patch of hairs (bearded Irises). Petals 3, free (*Moraea*) or united (*Crocus*), adnate to the sepals forming a perianth tube, petals sometimes spotted, erect in bearded Irises (forming standard). **Androecium** with usually 3 stamens, rarely 2, filaments free or connate, sometimes adnate to perianth, anthers bithecous, dehiscence by longitudinal slits, extrorse, sometimes sticking to style branches. **Gynoecium** with 3 carpels, united, ovary inferior, rarely superior (*Isophysis*), 3-locular with axile placentation, rarely unilocular with parietal placentation (*Hermodactylus*), ovules few to numerous, anatropous or campylotropous, style three-lobed in upper part, sometimes petaloid. **Fruit** a loculicidal capsule, dehiscing by valves, usually with marked circular scar at tip; seeds often with aril, copious endosperm and small embryo, seed coat usually fleshy. Flowers are primarily insect pollinated, especially by bees and flies, some species by birds (*Rigidella*), a few being wind pollinated (*Dierama*). Seeds are dispersed by wind or water.

Economic importance: The family includes some of the most popular garden ornamentals such as *Gladiolus*, *Iris*, *Freesia*, *Sparaxis*, *Tigridia* and *Sisyrinchium*. The stigmas of *Crocus sativus* yield saffron, widely used as a coloring agent and for flavouring food stuffs. Orris root, from *Iris florentina* is used in making perfumes and cosmetics.

Phylogeny: The family is related to Liliaceae. The genus *Isophysis* with superior ovary is sometimes removed to a distinct family, but according to Hutchinson (1973), it is Iridaceous in all the characters except superior ovary, and hence included here only. The family is often divided into a number of tribes of which Sisyrincheae with free perianth, rhizome and undivided style branches is considered to be the most primitive (Hutchinson). Gladioleae and Antholyzeae are more advanced with zygomorphic perianth with curved tube and hood-like dorsal lobe. The position of family is uncertain. Whereas Hutchinson removed it to a distinct order Iridales, Takhtajan placed it under Orchidales, and Dahlgren and Cronquist under Liliales. In the recent classifications of APG, it is placed under a broadly circumscribed Asparagales primarily based on *rbcL* sequence (Chase et al., 1995a). The morphological studies, however (Chase et al., 1995b; Stevenson and Loconte) place the family within Liliales. The combined studies of the two, places the family under Asparagales. Rudall (2001) included an inferior ovary as a synapomorphy of the order, noting that in 'higher' Asparagales (reduced in APG II to only two families Alliaceae and Asparagaceae), there may well be a major reversal to superior ovaries. Thorne earlier (1999, 2000) included Iridaceae under order Orchidales but subsequently (2003, 2006, 2007) shifted it along with several other families to Iridales suborder Iridineae, restricting Orchidales to just 6 families. The family is traditionally divided into 4 subfamilies: Isophysidoideae (1 genus, 1 species); Nivenioideae (6, 92), Iridoideae (29: 890), and Crocoideae (29: 1032). Goldblatt et al. (2008), based on analyses of sequences of five plastid DNA regions, rbcL, rps4, trnL-F, matK, and rps16 concluded that Nivenioideae is paraphytetic and as such recognized 3 additional monogeneric subfamilies (total 7) Patersonioideae (1, 21), Geosiridaceae

(1, 1) and Aristeoideae (1, 55), restricting Nivenioideae to 3 genera, 15 species. They divide Iridioideae and and Crocoideae into 5 tribes each, a classification followed in APweb (2018).

Asphodelaceae A. L. de Jussieu *Aloe Family*

39 genera, 900 species

Temperate and tropical regions of the Old World, especially South Africa, usually in arid habitats.

Placement:

	B & H	Cronquist	Takhtajan	Dahlgren	Thorne	APG IV/(APweb)
Informal Group						Monocots
Division		Magnoliophyta	Magnoliophyta			
Class	Monocotyledons	Liliopsida	Liliopsida	Liliopsida	Magnoliopsida	
Subclass		Liliidae	Liliidae	Liliidae	Liliidae	
Series+/ Superorder	Coronarieae+		Lilianae	Lilianae	Lilianae	
Order		Liliales	Amaryllidales	Asparagales	Iridales	Asparagales

B & H, Cronquist under Liliaceae.

Salient features: Rhizomatous herbs or shrubs, leaves in rosettes, often succulent, vascular bundles in a ring surrounding mucilaginous central zone, inflorescence racemose, flowers bisexual, perianth not spotted, stamens 6, free, not adnate to tepals, ovary superior, nectaries in septa of the ovary, fruit a capsule, seeds black.

Major genera: *Aloe* (400 species), *Haworthia* (55), *Kniphofia* (50), *Bulbine* (50) and *Asphodelus* (12).

Description: Rhizomatous herbs (rarely bulbous), shrubs or trees often with anomalous secondary growth, anthraquinone present. **Leaves** in rosettes at base or tips of branches, simple, usually succulent, not fibrous, vascular bundles in a ring around mucilaginous parenchyma, phloem with a cap of aloine cells containing colored secretions, leaves sheathing at base, venation parallel, stipules absent. **Inflorescence** raceme, spike or panicle. **Flowers** usually bracteate, bisexual, hypogynous, often showy, trimerous. **Perianth** with 6 tepals, free or slightly connate, petaloid, not spotted. **Androecium** with 6 stamens, free, not adnate to tepals, bithecous, basifixed or dorsifixed, dehiscence longitudinal, introrse, pollen grains monosulcate. **Gynoecium** with 3 carpels, united, ovary superior, placentation axile, ovules many, orthotropous or anatropous, nectaries in septa of ovary, stigma discoid or 3-lobed. **Fruit** a loculicidal capsule, rarely a berry (*Kniphofia*); seeds flattened, black, usually with dry aril. Pollination by insects and birds. Seeds mainly dispersed by wind.

Economic importance: Several genera including *Aloe*, *Haworthia*, *Kniphofia* and *Gasteria* are grown as ornamentals. Several species of *Aloe* are used in cosmetics and as sources of medicine.

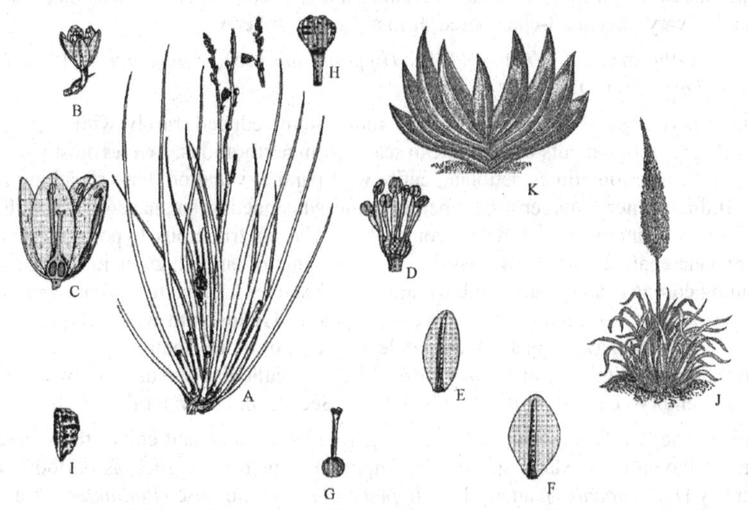

Figure 13.27: Asphodelaceae. *Asphodelus fistulosus.* **A:** Plant with scapigerous inflorescence; **B:** Flower; **C:** Vertical section of flower; **D:** Flower with perianth removed showing stamens and pistil; **E:** Outer tepal; **F:** Inner tepal; **G:** Pistil showing ovary, simple style and 3-lobed stigma; **H:** Capsule; **I:** Seed. **J:** *Eremurus olgae* with basal leaves and scapigerous inflorescence. **K:** *Gasteria verrucosa* with basal leaves (A–G, After Sharma and Kachroo, Fl. Jammu, 1983).

Phylogeny: The members of Asphodelaceae were earlier included under Liliaceae, but now placed separately. Dahlgren et al. (1985) being the authors of first major classification to recognize this and several other smaller families as indicated above. They recognized two subfamilies Asphodeloideae and Alooideae. The latter is clearly monophyletic as evidenced by apomorphies of leaves with central gelatinous zone with aloine layer and dimorphic karyotype. Asphodeloideae includes *Kniphofia*, which lacks aloine layer and has berry fruit, *Bulbine* which is closer to Alooideae, and *Asphodelus*. The recognition of Alooideae renders Asphodeloideae as such paraphyletic (Stevens, 2003). The family is included commonly under Asparagales (Dahlgren, Takhtajan, APG II, Judd et al., and APweb), but Thorne (2003, 2006, 2007) places it under order Iridales under suborder Asphodelineae. He like Dahlgren, recognizes the two subfamilies under Asphodelaceae. Treutlein et al. (2003) on the basis of chloroplast DNA sequences (*rbc*L, *mat*K) and from genomic finger-printing (ISSR) concluded that generic limits around *Aloe* are decidedly unsatisfactory. APG II (2003) suggested as an option the inclusion of Asphodelaceae, Xanthorrhoeaceae and Hemerocallidaceae in Xanthorrhoeaceae s.l., and this circumscription was adopted by APG III (2009), but because of nomenclatural issues the name has been changed to Asphodelaceae (APG IV 2016). Reveal et al. (2009) suggested recognition of these former families as subfamilies: Asphodeloideae, Hemerocallidoideae and Xanthorrhoeoideae, followed in APweb (2018). *Aloe* and its relatives were formerly placed in separate subfamily Alooideae, but its removal from Asphodelaoideae renders latter paraphyletic and necessitate the recognition of several other subfamilies.

Amaryllidaceae J. St.-Hilaire ***Amaryllis or Daffodil Family***

75 genera, 1600 species

Widely distributed, in tropical and temperate climates, frequently in semiarid habitats.

Placement:

	B & H	**Cronquist**	**Takhtajan**	**Dahlgren**	**Thorne**	**APG IV/(APweb)**
Informal Group						Monocots
Division		Magnoliophyta	Magnoliophyta			
Class	Monocotyledons	Liliopsida	Liliopsida	Liliopsida	Magnoliopsida	
Subclass		Liliidae	Liliidae	Liliidae	Liliidae	
Series+/ Superorder	Epigynae+		Lilianae	Lilianae	Lilianae	
Order		Liliales	Amaryllidales	Asparagales	Iridales	Asparagales

B & H as distinct family; **Cronquist** under family Liliaceae; **Takhtajan, Dahlgren, APweb** as distinct family; **Thorne** under family Alliaceae.

Salient features: Herbaceous perennials having bulbs, leaf base sheathing, inflorescence scapigerous, umbellate cyme, subtended by spathaceous bracts, flowers bisexual, perianth, not spotted, sometimes with staminal corona, stamens 6, often adnate to perianth, ovary inferior, 3-chambered, fruit a capsule or berry.

Major genera: *Allium* (499 species), *Crinum* (130), *Hippeastrum* (65), *Zephyranthes* (55), *Nothoscrodum* (35), *Hymenocallis* (48), and *Narcissus* (30).

Description: Perennial herbs with bulb and contractile roots, stem reduced, rarely with corm (*Milula*) or rhizome (*Agapanthus, Tulbaghia*), vessel-elements with simple to scalariform perforations, **Leaves** mostly basal, alternate, simple, cylindrical or flat, often fistular sometimes petiolate, entire with parallel venation, base sheathing and forming tunic of bulb, stipules absent. **Inflorescence** scapigerous, umbellate cyme, subtended by spathaceous bracts. **Flowers** ebracteate or ebracteate, bisexual, usually actinomorphic, rarely zygomorphic (*Gilliesia*) trimerous, hypogynous or epigynous. **Perianth** with 6 tepals, free or connate into a tube, in two whorls, petaloid, outer often with green midvein, sometimes with scale-like appendages forming corona (*Tulbaghia*). **Androecium** with 6 stamens, filaments free or connate and expanded into staminal corona (*Hymenocallis, Pancratium*), sometimes epitepalous, **Gynoecium** with 3 carpels, united, ovary superior (*Allium*) or inferior (*Crinum, Amaryllis*), placentation axile, ovules 2 or more, anatropous or campylotropous, nectaries in septa of ovary, style simple, stigma capitate to 3-lobed. **Fruit** a loculicidal capsule, rarely a berry; seeds globular or angular, seed coat black, embryo curved. Pollination by insects. Seeds wind or water dispersed.

Economic importance: The family is useful contributing garlic, onion, leek and chives the species of *Allium* used as important vegetables or flavourings. Many species are important ornamentals such as daffodils (*Narcissus*), swamp lily (*Crinum*), spider lily (*Hymenocallis*), amaryllis (*Hippeastrum*) and tuberose (*Polianthes*), some species of *Allium*, *Tulbaghia*, and *Gilliesia*.

Phylogeny: Formerly placed under distinct families Liliaceae or Alliaceae (Superior ovary) and Amaryllidaceae (inferior), the two have been brought together because of scapigerous habit, spathaceous bracts, umbellate inflorescence and presence of bulbs. Presence of alkaloides and rbcl sequences support this merger. Recent classifications of APG IV and APWeb divide family into three subfamilies Agapanthoideae, Allioideae and Amaryllidoideae.

Subfamily Allioideae (Syn: Family Alliaceae *Onion Family*)

13 genera, 791 species

Widely distributed, in tropical and temperate climates, frequently in semiarid habitats.

Placement:

	B & H	Cronquist	Takhtajan	Dahlgren	Thorne	APG IV/(APweb)
Informal Group						Monocots
Division		Magnoliophyta	Magnoliophyta			
Class	Monocotyledons	Liliopsida	Liliopsida	Liliopsida	Magnoliopsida	
Subclass		Liliidae	Liliidae	Liliidae	Liliidae	
Series+/ Superorder	Coronarieae+		Lilianae	Lilianae	Lilianae	
Order		Liliales	Amaryllidales	Asparagales	Iridales	Asparagales

B & H, Cronquist under family Liliaceae, **APG IV, APweb** under Amaryllidaceae.

Salient features: Herbaceous perennials with bulbs, latex present, smell of onion or garlic, leaf base sheathing, inflorescence scapigerous, umbellate cyme, subtended by spathaceous bracts, flowers bisexual, perianth not spotted, stamens 6, often adnate to perianth, ovary superior, fruit a capsule.

Major genera: *Allium* (500 species), *Nothoscordum* (35), *Tulbaghia* (22), *Gilliesia* (7) and *Miersia* (5).

Description: Perennial herbs with bulb and contractile roots, stem reduced, rarely with corm (*Milula*-now merged with *Allium*) or rhizome (*Tulbaghia*), vessel-elements with simple perforations, laticifers present, with onion or garlic scented sulphur compounds such as allyl sulphides, vinyl disulphide, etc. **Leaves** mostly basal, alternate, simple, cylindrical or flat, often fistular, entire with parallel venation, base sheathing. **Inflorescence** scapigerous, umbellate cyme, subtended by spathaceous bracts, enclosing flower buds, rarely spicate (formerly *Milula*). **Flowers** ebracteate, bisexual, usually actinomorphic, rarely zygomorphic (*Gilliesia*) trimerous, hypogynous, pedicels often unequal in size. **Perianth** with 6 tepals, free or connate at base, in two whorls, petaloid, outer often with green midvein, sometimes with scale-like appendages forming corona (*Tulbaghia*). **Androecium** with 6 stamens, filaments free or connate, sometimes epitepalous, sometimes with appendages, rarely 3 or 4 stamens without anthers, pollen grains monosulcate. **Gynoecium** with 3 carpels, united, ovary superior, placentation axile, ovules 2 or more, anatropous or campylotropous, nectaries in septa of ovary, style simple, stigma capitate to 3-lobed. **Fruit** a loculicidal capsule; seed coat black, embryo curved. Pollination by insects. Seeds wind or water dispersed.

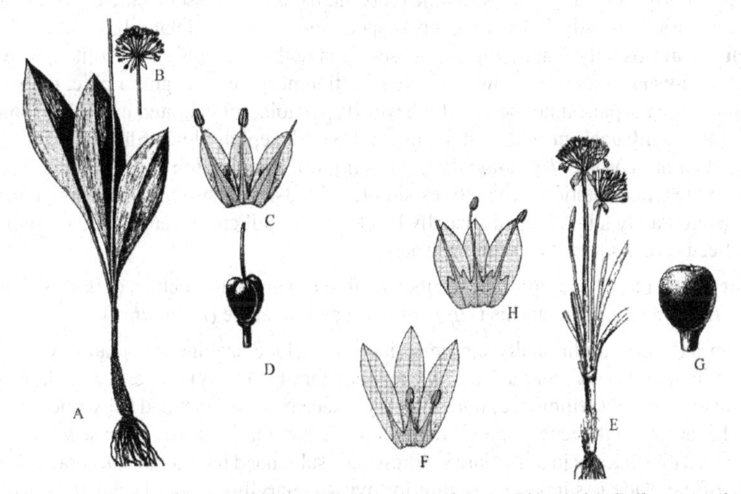

Figure 13.28: Amaryllidaceae, Subfamily **Allioideae.** *Allium victorialis.* **A:** Plant with bulb covered with reticulate fibres, broadly laceolate leaves and scape; **B:** Upper part of scape with inflorescence; **C:** Tepals with stamens; **D:** Capsule with long style. *A. humile.* **E:** Plant with scape and inflorescence; **F:** Tepals with stamens; **G:** Capsule. **H:** Tepal and stamens of *A. roylei* showing 2-toothed inner filaments.

Economic importance: The family is useful contributing garlic (*Allium sativum*), onion (*A. cepa*), leek (*A. porrum*), and chives (*A. schoenoprasum*), used as important vegetables or flavourings. Onion seeds are often used as substitute for nigella. A few species of *Allium*, *Tulbaghia*, and *Gilliesia* are cultivated as ornamentals.

Phylogeny: Originally included under Liliaceae, the first major shift was made by Hutchinson (1934) who abandoned the traditional separation of on the basis of superior (Liliaceae) or inferior (Amaryllidaceae) ovary and included *Allium* and its relatives with superior ovary under Amaryllidaceae, largely on the basis of spathaceous bracts. Cronquist had subsequently (1981, 1988) merged Amaryllidaceae with Liliaceae. In the recent years a number of distinct families have been segregated. The family is closely related to Amaryllidaceae in umbellate inflorescence subtended by spathaceous bracts, bulbs, and the presence of scape. Thorne had earlier (1999) considered Amaryllidaceae and Agapanthaceae as distinct families but has finally (2003, 2007) merged them in Alliaceae. The recent cladistic analysis have resulted in merging all families with umbellate inflorescence (Agapanthaceae, Amaryllidaceae and Alliaceae) under Amaryllidaceae, followed by APG IV and APweb who recognise three as subfamilies. The clade is characterized by bulbs, flavonols, saponins, laticifers, inflorescence scapigerous, umbellate, with scarious spathe, inflorescence bracts 2 (or more—external), pedicels not articulated, free or basally connate perianth, style long, and endosperm nuclear or helobial (Fay et al., 2000).

Subfamily Amaryllidoideae (Syn: Family Amaryllidaceae s.s.)

59 genera, 800 species

Widely distributed, in tropical and temperate climates, especially in South Africa, South America and the Mediterranean region.

Placement:

	B & H	Cronquist	Takhtajan	Dahlgren	Thorne	APG IV/(APweb)
Informal Group						Monocots
Division		Magnoliophyta	Magnoliophyta			
Class	Monocotyledons	Liliopsida	Liliopsida	Liliopsida	Magnoliopsida	
Subclass		Liliidae	Liliidae	Liliidae	Liliidae	
Series+/ Superorder	Epigynae+		Lilianae	Lilianae	Lilianae	
Order		Liliales	Amaryllidales	Asparagales	Iridales	Asparagales

Salient features: Herbaceous perennials having bulbs with contractile roots, leaf base sheathing, inflorescence scapigerous, umbellate cyme, subtended by spathaceous bracts, flowers bisexual, perianth, not spotted, sometimes with staminal corona, stamens 6, often adnate to perianth, ovary inferior, 3-chambered, fruit a capsule or berry.

Major genera: *Crinum* (130 species), *Hippeastrum* (65), *Zephyranthes* (55), *Hymenocallis* (48), and *Narcissus* (30).

Description: Perennial herbs having bulb with contractile roots, stem reduced, vessel-elements with scalariform perforations, **Leaves** mostly basal, alternate, mostly linear or strap shaped, sometimes petiolate, base sheathing, venation parallel, stipules absent. **Inflorescence** usually scapigerous, cymose, often umbellate clusters or solitary, flowers often subtended by spathaceous bracts. **Flowers** bracteate, showy, bisexual, actinomorphic or zygomorphic, epigynous. **Perianth** 6, in two whorls (outer representing sepals, inner petals), both whorls petaloid, often united into tube, sometimes with a corona on throat of perianth tube. **Androecium** with 6 stamens, in 2 whorls, epiphyllous, filaments free, sometimes expanded and connate forming staminal corona (*Hymenocallis*, *Pancratium*). **Gynoecium** with 3 united carpels, ovary inferior, trilocular with many ovules, placentation axile, styles simple with 3-lobed stigma, nectaries present in septa of ovary. **Fruit** a loculicidal capsule, rarely a berry; seeds usually black, with small curved embryo, endosperm fleshy. Pollination by insects and birds. Seeds are dispersed by wind or water.

Economic importance: The family is important for its valuable ornamentals such as daffodils (*Narcissus*), swamp lily (*Crinum*), spider lily (*Hymenocallis*), amaryllis (*Hippeastrum*) and tuberose (*Polianthes*).

Phylogeny: The group had been traditionally circumscribed to include scapigerous plants with spathaceous bracts in inflorescence and inferior ovary and regarded as independent family Amaryllidaceae. Hutchinson had also included genera with superior ovary (present Allioideae) under Amaryllidaceae. Cronquist had subsequently (1981, 1988) merged Amaryllidaceae with Liliaceae. In the recent years a number of distinct families have been segregated, as indicated above. The clade is closely related to Alliaceae in umbellate inflorescence subtended by spathaceous bracts, bulbs, and the presence of scape. Monophyly of the clade is supported by inferior ovary, amaryllid alkaloids and *rbc*L sequences (Chase et al., 1995a). Judd et al. (2002) recognize Amaryllidaceae as independent family. The recent cladistic analysis has resulted in optionally merging all families with umbellate inflorescence (Agapanthaceae, Amaryllidaceae and Alliaceae). Thorne

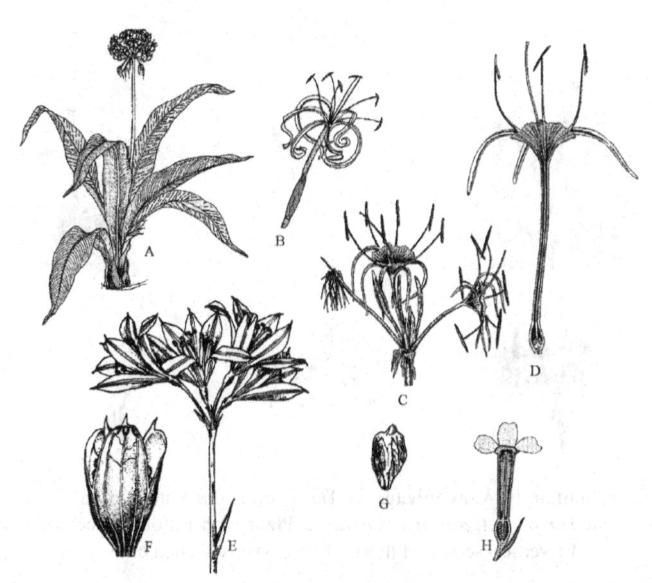

Figure 13.29: Amaryllidaceae, subfamily **Amaryllidoideae.** *Crinum asiaticum.* **A:** Plant with inflorescence; **B:** Flower with elongated perianth tube without corona. *Hymenocallis narcissifolia.* **C:** Inflorescence with a part of scape; **D:** Longitudinal section of flower showing staminal corona. *Ixiolirion tataricum* (now Ixioliriaceae). **E:** Inflorescence; **F:** Dehiscing capsule; **G:** Seed. **H:** Vertical section of flower of *Narcissus poeticus.*

(2003) earlier included it under Alliaceae in subfamily Amaryllidoideae, later (2006, 2007) prefering Narcissoideae. APG IV and APweb include it under Amaryllidaceae.

Asparagaceae A. L. de Jussieu *Asparagus Family*

(inc. Agavaceae S. L. Endlicher)

114 genera, 2900 species

Throughout tropics and subtropics, mainly in arid climate.

Placement:

	B & H	Cronquist	Takhtajan	Dahlgren	Thorne	APG IV/(APweb)
Informal Group						Monocots
Division		Magnoliophyta	Magnoliophyta			
Class	Monocotyledons	Liliopsida	Liliopsida	Liliopsida	Magnoliopsida	
Subclass		Liliidae	Liliidae	Liliidae	Liliidae	
Series+/ Superorder	Coronarieae/ Epigynae+		Lilianae	Lilianae	Lilianae	
Order		Liliales	Asparagales	Asparagales	Asparagales	Asparagales

B & H *Asparagus, Yucca* (and others with superior ovary) under Liliaceae (series Coronarieae), *Agave* (and others with inferior ovary) under Amaryllidaceae (series Epigynae).

Salient features: Large herbs, shrubs or trees with usually rosettes of leaves, leaves succulent with sharp spine at tip, fibrous, inflorescence paniculate, flowers bisexual, perianth 6, free or connate, stamens 6, often adnate to perianth, ovary inferior or superior, 3-chambered, nectaries in septa of ovary, fruit a capsule or berry, seed coat black, bimodal karyotype 5 large and 25 small chromosomes.

Major genera: *Asparagus* (300), *Agave* (240 species), *Chlorophytum* (150), *Dracaena* (120), *Sansevieria* (70), *Scilla* (70), *Anthericum* (65) and *Yucca* (40).

Description: Herbs, lianas (*Behnia*), shrubs or trees with basal or terminal rosettes of leaves, mostly rhizomatous, or bulbous (*Scilla*). **Leaves** alternate, simple, sometimes in rosettes (**Agave**) or two-ranked (*Behnia*), succulent or not (*Maianthemum*), margin entire or spinose-serrate, venation parallel, with thick tough fibres, base sheathing, sometimes reduced to scales subtending cladophylls (*Asparagus*), or scarious (*Aphyllanthes*). **Inflorescence** usually terminal raceme or panicle, sometimes thyrsoid (*Chlorophytum*), umbelate, paired, solitary or subspicate (*Anemarrhena*). **Flowers** usually

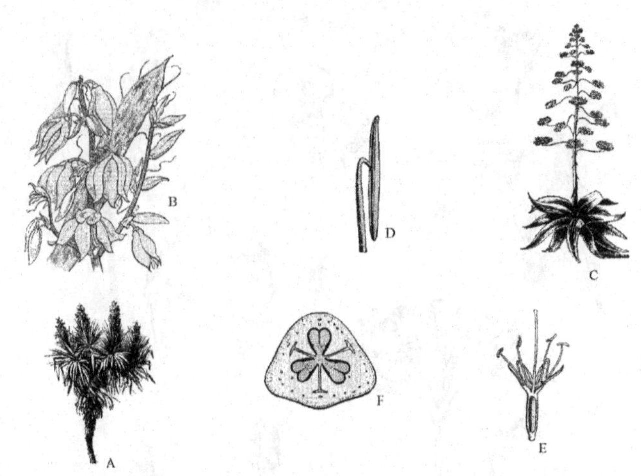

Figure 13.30: Asparagaceae, subfamily **Agavoideae. A:** *Yucca aloifolia* with a number of developing inflorescences. **B:** Part of inflorescence of *Y. filamentosa. Agave americana.* **C:** Plant with inflorescence; **D:** Stamen with versatile anther; **E:** Vertical section of flower; **F:** Transverse section of ovary.

bisexual, rarely unisexual (*Behnia*), actinomorphic, trimerous. **Perianth** with 6 tepals, free (*Asparagus, Yucca*) or connate into tube (*Agave*), petaloid, not spotted, usually white or yellow, sometimes blue (*Aphyllanthes*). **Androecium** with 6 stamens, longer than perianth (*Agave*) or shorter (*Furcraea*), free, anthers bithecous, basifixed (*Doryanthes*) or dorsifixed (*Beschorneria*), dehiscence by longitudinal slits, introrse. **Gynoecium** with 3 united carpels, ovary superior (*Asparagus, Yucca, Scilla*) or inferior (*Agave*), 3-locular with axile placentation; ovules many, anatropous, nectaries in septa of ovary, style short or long, stigma minute. **Fruit** a loculicidal capsule (*Yucca*) or baccate (*Asparagus*); seeds flat, globose or angular, black. Pollination by moths (*Yucca* by moth *Tegeticula*), others by bats (*Agave*, some species) or birds (*Beschorneria*). Seed dispersal by wind or animals.

Economic importance: Several species such as *Agave sisalana* (sisal hemp), *A. heteracantha* (Istle fibre or Mexican fibre), *A. morrisii* (Keratto fibre), are important sources of fibre. A few species of *Agave* are fermented to produce tequila and mescal. The species of both *Agave* and *Yucca* are used in the manufacture of oral contraceptives. Several species of *Agave, Yucca* and *Polianthes* (*P. tuberosa*-tube rose) are also used as ornamentals. Species of *Sansevieria* in addition to being used as ornamental are used in medicine and food additives. *Asparagus* stalks are commonly eaten as a vegetable. Roots, seeds, and extracts have been used as a treatment for various illnesses and as a diuretic.

Phylogeny: The members of the family were earlier placed in Liliaceae and Amaryllidaceae and were later removed to a separate family to include members with superior ovary (removed from Liliaceae) and inferior ovary (removed from Amaryllidaceae) representing advanced tribes in the respective families (Hutchinson, 1973), and lacking bulb, having arborescent habit, and inflorescence racemose (not an umbel). Hutchinson (1973) and Cronquist (1981, 1988) circumscribed the family Agavaceae more broadly also to include genera which have now been removed to Dracaenaceae, Nolinaceae, and Laxmanniaceae (incl. Lomandraceae). The APG II system of 2003 allowed two options as to the circumscription of the family: either Asparagaceae sensu lato ("in the wider sense") combining seven previously recognized families, or Asparagaceae sensu stricto ("in the strict sense") consisting of very few genera (notably *Asparagus*, also *Hemiphylacus*), but nevertheless totalling a few hundred species. APG III (2009) and APG IV (2016) adopted circumscription of Asaparagaceae including genera which were formerly placed in 8 separate families: Agavaceae, Hesperocallidaceae, Aphyllanthaceae, Asparagaceae, Themidaceae, Laxmanniaceae, Ruscaceae and Hyacinthaceae. Chase et al. (2009) suggested division of expanded Asparagaceae into 7 subfamilies: Agavoideae, Aphyllanthoideae, Asparagoideae, Brodiaeoideae, Lomandroideae, Nolinoideae and Scilloideae, which had earlier been regarded as forming well supported clade in Fay et al. (2000), also supported in subsequent studies of relationships in Chinese Asparagales as detailed in Chen et al. (2016) and the relationships in McLay and Bayly (2016). According to Stevens (2018, APweb) this is a highly unsatisfactory family. Nothing characterize it, and while some of the subfamilies have several distinctive apomorphies and are also easy to recognize, others are difficult to recognize. Although he follows Chase et al. (2009), but has also included familial names for each clade, in part because the roots of the two sets of names differ in over half the cases and both will be encountered in the literature. CAM photosynthetic pathway has evolved ca three times in Agaveae, a major CAM clade, and is found in nearly all species of Agave. Interestingly, succulent leaves with a three-dimensional venation system that are characteristic of the CAM species are also found in related C3 species, and the CAM-type morphology may have evolved before CAM photosynthesis itself (Heyduk et al., 2016). Bat pollination is common in the large genus *Agave* and its relatives (Fleming et al., 2009).

Arecaceae C. H. Schultz ***Palm Family***

(= **Palmae** A. L. de Jussieu)

181 genera, 2,600 species

Widespread in tropics of both hemispheres, a few in warm temperate regions.

Placement:

	B & H	Cronquist	Takhtajan	Dahlgren	Thorne	APG IV/(APweb)
Informal Group						Monocots
Division		Magnoliophyta	Magnoliophyta			
Class	Monocotyledons	Liliopsida	Liliopsida	Liliopsida	Magnoliopsida	
Subclass		Arecidae	Arecidae	Liliidae	Commelinidae	
Series+/ Superorder	Calycinae+		Arecanae	Arecanae	Arecanae	
Order		Arecales	Arecales	Arecales	Arecales	Arecales

B & H as family Palmae.

Salient features: Woody shrubs or trees, trunk with scars of fallen leaves, leaves large, fan-shaped or pinnately compound, with sheathing bases, inflorescence paniculate, spathes often present, flowers small.

Major genera: *Calamus* (350 species), *Bactris* (180), *Pinanga* (120), *Licuala* (105), *Daemonorops* (100), *Areca* (60) and *Phoenix* (17).

Description: Trees or shrubs with unbranched trunk, rarely branched (*Hyphaene*, *Nypa*), with prominent scars of fallen leaves, sometimes spiny due to modified leaves roots or exposed fibres, sometimes rhizomatous, tannins and polyphenols often present, vascular bundles with hard fibrous sheath, apical bud well protected by leaf sheaths, the longest unbranched stems, perhaps up to 200 m long in *Calamus manan*. **Leaves** alternate, usually forming a terminal crown, petiolate (petiole often with a flap called hastula at base), with pinnate (feather palms) or palmate (fan palms) segments, sometimes pinnately or twice pinnately compound, plicate (folded like a fan), blades rarely entire (*Licuala*); leaf segments folded V-shaped (induplicate) or inverted-V-shaped (reduplicate) in cross section; leaves sometimes very large sometimes over 20 m (*Raphia fainifera* with largest known leaf). **Inflorescence** axillary or terminal, often covered with spathes, a repeatedly branched panicle (*Calamus*) or almost spicate, largest inflorescence almost 7.5 m long in *Corypha umbraculifera*. **Flowers** bisexual (*Licuala*, *Livistona*) or unisexual with monoecious (*Reinhardtia*) or dioecious (*Borassus*, *Rhapis*) plants, flowers small, actinomorphic, usually sessile, trimerous, often with bracteoles connate below flowers. **Perianth** differentiated into sepals and petals, sometimes vestigial (*Nypa*). Sepals 3, free (*Arenga*) or connate (*Didymosperma*), usually imbricate. Petals 3, free or connate, usually valvate in male flower and imbricate in female flower (valvate in female flowers of *Arenga*). **Androecium** usually with six stamens in two whorls, sometimes numerous (*Reinhardtia*, *Howea*), rarely only 3 (*Nypa*), free, rarely with connate filaments (*Nypa*), anthers bithecous, basifixed or dorsifixed, rarely versatile, dehiscence by longitudinal slits; pollen grains usually monosulcate, smooth or echinulate. **Gynoecium** with usually 3 carpels, free or united, only 1 fertile in *Phoenix*, carpels sometimes many, ovary superior, placentation usually axile, rarely parietal (*Gronophyllum*), stigma usually terminal, sometimes lateral (*Heterospatha*) or basal (*Phloga*), ovules usually 1, rarely up to 3, orthotropous or anatropous. **Fruit** single seeded berry or drupe, exocarp often fibrous or covered with reflexed scales; seeds free or adhering to endocarp, endosperm present, embryo small.

Largest seed in angiosperms formed in double coconut (*Lodoicea maldivica*) measures up to 50 cm long and 15–30 kg.

Economic importance: The family is of great economic importance. Most useful member is Coconut palm (*Cocos nucifera*), with almost every part put to use. Mesocarp of the fruit is the source of coir fibre, the seed endosperm (copra) yielding coconut oil, and the leaves used in thatching, basket making and a variety of toys and decoration articles. Palm oil is extracted from *Elaeis guineensis*. Sago, a major source of carbohydrate food is obtained from *Metroxylon sagu* (sago palm) and some species of *Arenga* and *Caryota*. Palm wine (toddy) is obtained from species of *Borassus* and *Caryota*. Fibre is also extracted from many species of palms particularly belonging to *Raphia* (raffia), *Caryota* (kitul fibre) and *Leopoldinia* (Piassava fibre). Dates are obtained from Date palm (*Phoenix dactylifera*). Vegetable ivory is obtained from the seeds of ivory nut palm (*Phytelephas macrocarpa*) and was once used for buttons and as a substitute for real ivory. Waxes are obtained from *Copernicia* (carnauba wax) and *Ceroxylon*. Betel nut is obtained from *Areca catechu* of Africa and Southeast Asia. The family also contributes many ornamentals such as Royal palm (*Roystonea regia*), fishtail palm (*Caryota*), Chinese fan palm (*Livistona*), and cabbage palm (*Sabal*). Various species of *Calamus* are source of commercial cane used in furniture and polo sticks.

Phylogeny: Despite being very large and diverse, and often divided into numerous subgroups, the family is distinct, easily recognized and monophyletic. Uhl and Dransfield (1987) divided family into 6 subfamilies: Coryphoideae,

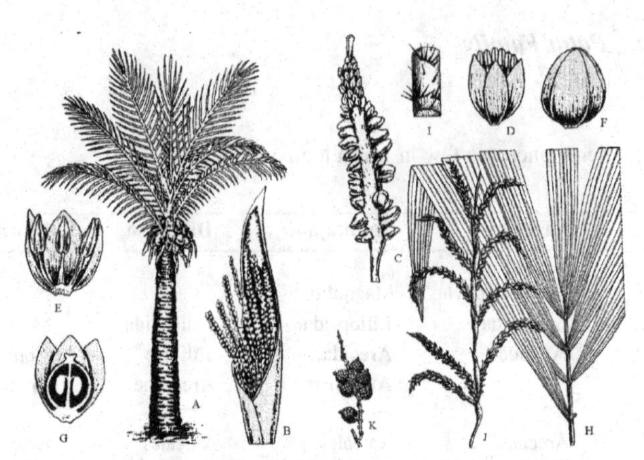

Figure 13.31: Arecaceae. *Cocos nucifera.* **A:** Habit; **B:** Inflorescence; **C:** Branch of inflorescence with female flowers towards base, male flowers towards the top; **D:** Male flower; **E:** Vertical section of male flower; **F:** Female flower; **G:** Vertical section of female flower. *Calamus pseudotenuis.* **H:** Vegetative branch; **I:** Portion of stem showing thorns; **J:** Male inflorescence; **K:** Female inflorescence.

Calamoideae, Nypoideae, Ceroxyloideae, Arecoideae and Phytelephantoideae. Uhl et al. (1995) carried out cladistic analysis of the family using morphological data as well as cpDNA restriction site analysis and found support for placement of *Nypa* (Nypoideae) a sister of rest of the palms. More recent studies of Asmussen et al. (2000) indicated that Nypoideae + Calamoideae (strong support) + the rest of the family (moderate support) form a basal trichotomy; other characters support these general relationships. However, other work suggests that details of the relationships of Nypoideae and Calamoideae to the rest of the family are unclear, and some morphological groupings are not supported by molecular data (Hahn, 2002). In a new phylogenetic classification of Arecaceae Dransfield et al. (2008) merged Phytelephantoideae with Ceroxyloideae (3 tribes Cyclospatheae, Ceroxyleae and Phytelepheae). The circumscription of the remaining five subfamilies has also changed, the most significant changes being the removal of tribe Caryoteae from Arecoideae and its placement within Coryphoideae, and the removal of Hyophorbeaefrom Ceroxyloideae and its placement (using the earlier published name Chamaedoreeae) within Arecoideae.

Commelinaceae R. Brown ***Spiderwort Family***

41 genera, 731 species

Widespread in tropical, subtropical and warm temperate regions.

Placement:

	B & H	**Cronquist**	**Takhtajan**	**Dahlgren**	**Thorne**	**APG IV/(APweb)**
Informal Group						Monocots/ (Commelinids)
Division		Magnoliophyta	Magnoliophyta			
Class	Monocotyledons	Liliopsida	Liliopsida	Liliopsida	Magnoliopsida	
Subclass		Commelinidae	Commelinidae	Liliidae	Commelinidae	
Series+/ Superorder	Coronarieae+		Commelinanae	Commelinanae	Commelinanae	
Order		Commelinales	Commelinales	Commelinales	Commelinales	Commelinales

Salient features: Herbs with succulent stems, nodes swollen, leaves with closed basal sheath, outer perianth whorl (sepals) green, inner (petals) colored, flowers bisexual, trimerous, in axils of spathaceous bracts, filaments usually hairy, ovary superior, 3-chambered.

Major genera: *Commelina* (170 species), *Tradescantia* (70), *Aneilema* (65), *Murdannia* (50), *Cyanotis* (50), *Dichorisandra* (30) and *Zebrina.*

Description: Annual or perennial herbs, rarely climbers (*Streptolirion*) with commonly succulent stems and swollen nodes, often with mucilage cells or canals containing raphides. **Leaves** alternate, simple, entire, flat or folded V-shaped in cross section, leaf sheath closed at base, venation parallel, stomata tetracytic, stipules absent. **Inflorescence** a helicoid cyme at the end of stem in leaf axil, sometimes solitary, subtended by spathaceous bracts. **Flowers** bisexual (rarely unisexual) actinomorphic (zygomorphic in *Commelina*), hypogynous. **Perianth** 6, in two whorls, outer representing sepals, inner

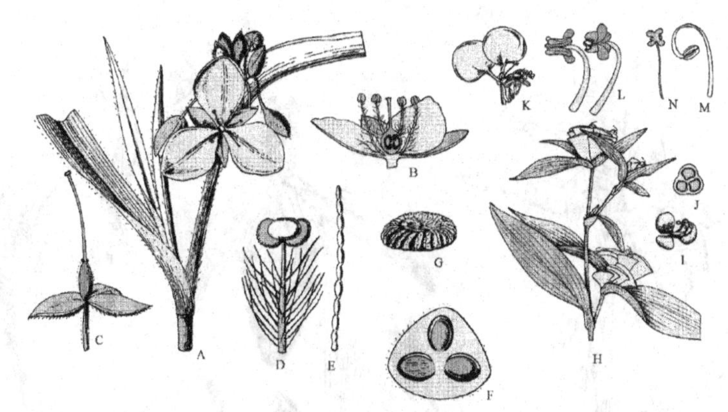

Figure 13.32: Commelinaceae. *Tradescantia virginiana.* **A:** Plant with flowers; **B:** Vertical section of flower; **C:** Flower with sepals and gynoecium; **D:** Stamen with hairy filament; **E:** Moniliform staminal hair; **F:** Transverse section of ovary showing one ovule in each chamber; **G:** Seed with aril. *Commelina paludosa.* **H:** Plant with flowers; **I:** Flower; **J:** Transverse section of ovary. *C. kurzii.* **K:** Flower; **L:** Stamen with large anther in different views; **M:** One of the lateral stamens; **N:** Staminode (A–E, after Hutchinson, Fam. Fl. Pl, ed. 3, 1973; H–J, after Polunin and Stainton, Fl. Himal. 1984).

petals. Sepals green and free. Petals colored (blue, violet or white) free (*Tradescantia, Rhoeo*) or connate into a tube (*Cyanotis, Zebrina*) or connivent (*Setcreasea*), perishing soon after anthesis, imbricate, crumpled in bud. **Androecium** with 6 stamens (a few often reduced to staminodes), in 2 whorls, filaments free, often hairy with simple or moniliform hairs, sometimes adnate to petals, connective often flattened, anthers bithecous, dehiscence by longitudinal slits, rarely by apical pores (*Dichorisanda*), pollen grains monosulcate. **Gynoecium** with 3 carpels, united, ovary superior, trilocular with 1-few orthotropous or anatropous ovules, placentation axile, styles simple with 3-lobed or capitate stigma. **Fruit** a loculicidal capsule, rarely a berry; seeds with aril, endosperm present, mealy.

Economic importance: The family is important for its valuable ornamentals such as dayflower (*Commelina*), spiderwort (*Tradescantia*), Moses-in-the-bulrushes (*Rhoeo*) and wandering Jew (*Zebrina*). In Africa, *Aneilema beninense* is used as a laxative. Leaf sap of *Floscopa scandens* is used in tropical Asia to treat inflammation of the eyes. The young shoots and leaves of *Tradescantia virginiana* and *Commelina clavata* are edible.

Phylogeny: The family is commonly divided into two subfamilies Tradescantoideae and Commelinoideae—well formed clades. The former is characterized by nonspiny pollen grains, medium to large chromosomes, actinomorphic flowers and moniliform hairs. Commelinoideae is characterized by spiny pollen, zygomorphic flowers, and filament hairs not moniliform. Monophyly of the family is supported by both morphological and molecular data (Evans et al., 2000). Thorne (2007) included the basal *Cartonema* and widely separated *Triceratella* in subfamily Cartonemoideae, merging the other two under Commelinoideae, but separated as tribes. Burns et al. (2011) based on DNA sequences from the plastid gene rbcL, provided a phylogenetic tree of the family recognising Cartenemoideae (including *Cartonema*) and Commelinoideae (two tribes Tradescantieae and Commelineae). Stevens (APweb, 2018) suggests inclusion of *Triceratella* from Zimbabwe under Cartonemoideae.

Musaceae A. L. de Jussieu ***Banana Family***

3 genera, 91 species

Mainly wet tropical lowlands from West Africa to Pacific (Southern Japan to Queensland).

Placement:

	B & H	**Cronquist**	**Takhtajan**	**Dahlgren**	**Thorne**	**APG IV/(APweb)**
Informal Group						Monocots/ (Commelinids)
Division		Magnoliophyta	Magnoliophyta			
Class	Monocotyledons	Liliopsida	Liliopsida	Liliopsida	Magnoliopsida	
Subclass		Zingiberidae	Commelinidae	Liliidae	Commelinidae	
Series+/ Superorder	Epigynae+		Zingiberanae	Zingiberanae	Commelinanae	
Order		Zingiberales	Zingiberales	Zingiberales	Cannales	Zingiberales

B & H under family Scitamineae.

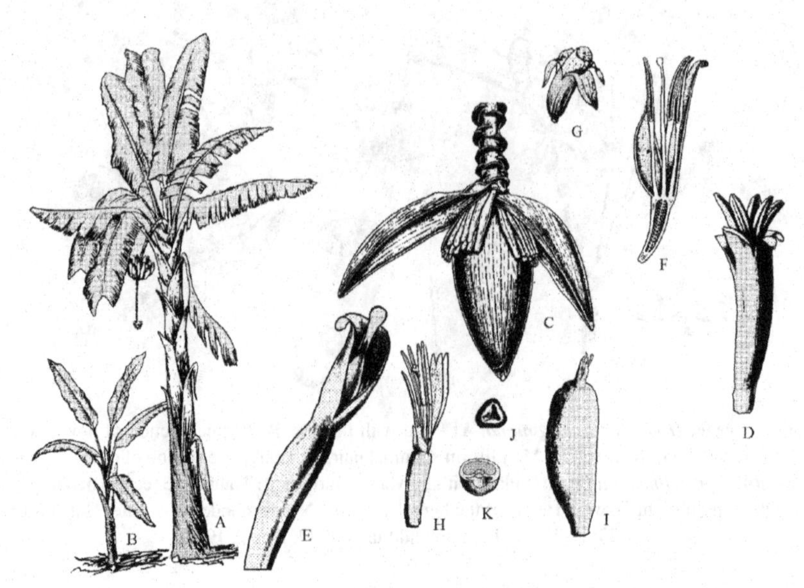

Figure 13.33: **Musaceae.** *Musa Paradisiaca* subsp. *sapientum* (A–C, F, G) **A:** Plant with inflorescence and split old leaves; **B:** Young plant; **C:** Apical portion of inflorescence; **D:** Male flower of *M. rubra*; **E:** Female flower of *M. rubra*; **F:** Vertical section through bisexual flower; **G:** Fruit partially opened to show edible berry sliced at top. *Ensete edule*. **H:** Bisexual flower; **I:** Fruit; **J:** Seed; **K:** Transverse section of seed showing pit of hilum.

Salient features: Large herbs with pseudostems formed by leaf sheaths, leaves large with thick midrib, parallel venation, flowers unisexual, inflorescence subtended by large spathaceous bracts, corolla 2-lipped, stamens 5 (sixth rudimentary), carpels 3, ovary inferior, 3-locular, ovules numerous, fruit fleshy berry with numerous small black seeds.

Major genera: *Musa* (33 species), *Ensete* (6) and *Musella* (1).

Description: Large usually tree-like perennial herbs with pseudostems formed from overlapping leaf sheaths, with laticifers, rhizomatous. **Leaves** large, spirally arranged, simple, entire, margin often torn and blade appearing pinnate, venation parallel with stout midrib, sheathing at base. **Inflorescence** a panicle-like cyme with one or more spathes, axis arising from basal rhizome and growing up through pseudostem. **Flowers** unisexual (plant monoecious), male within upper bracts, female in clusters within lower bracts. **Perianth** 6 in two whorls, petaloid. Sepals 3, adnate to 2 petals, narrowly tubular, soon splitting on one side, variously toothed at apex. Petals 3, somewhat 2-lipped, 2 adnate with sepals, 1 free. **Androecium** with 5 fertile stamens and 1 forming staminode, adnate to petals, filaments free, anthers linear, bithecous, dehiscence by longitudinal slits, pollen sticky. **Gynoecium** with 3 united carpels, ovary inferior, 3-locular, ovules many, placentation axile, style filiform, stigma 3-lobed. **Fruit** elongated berry containing numerous seeds, fruits forming compact bunches; seed with copious and small embryo.

Economic importance: Banana (*Musa paradisiaca* subsp. *sapientum*) is a staple food in many tropical countries. Manila hemp or Abaca obtained from fibres of *M. textilis* is used in making ropes and cordage. Inset or Abyssinian banana (*Ensete ventricosa*) is cultivated for its fibre and for food; the stem pulp and young shoots are eaten cooked. Some dwarf cultivars of *Musa* (*M. acuminata* 'Dwarf Cavendish') are often grown as greenhouse plants in temperate climates.

Phylogeny: The family is usually placed in Zingiberales (Cronquist, Dahlgren, Takhtajan, APG IV, APweb) along with Cannaceae, Zingiberaceae, Marantaceae and other closely related families. The genus *Heliconia*, earlier placed in this family has been removed to a separate family Heliconiaceae (Thorne, APG IV, APweb) or placed under Strelitziaceae (Heywood, 1978). Thorne (2003, 2007) prefers name Cannales over Zingiberales for the broadly circumscribed order, divided into 6 suborders. Fossil record has been found in Eocene of W. North America. Liu et al. (2010) and Li et al. (2010) discuss the phylogeny of the family, in which there are two main clades; the suckering *Musella* is derived from the non-suckering *Ensete* in the former, but in some analyses in the latter the two were sister taxa (see also Givnish et al., 2018).

Cannaceae A. L. de Jussieu ***Canna Family***

1 genus, 25 species

Mainly tropical and subtropical America, several species having naturalized in Asia and Africa.

Placement:

	B & H	Cronquist	Takhtajan	Dahlgren	Thorne	APG IV/(APweb)
Informal Group						Monocots/ (Commelinids)
Division		Magnoliophyta	Magnoliophyta			
Class	Monocotyledons	Liliopsida	Liliopsida	Liliopsida	Magnoliopsida	
Subclass		Zingiberidae	Commelinidae	Liliidae	Commelinidae	
Series+/ Superorder	Epigynae+		Zingiberanae	Zingiberanae	Commelinanae	
Order		Zingiberales	Zingiberales*	Zingiberales	Cannales	Zingiberales

B & H under family Scitamineae; ***Takhtajan** Zingiberales (Cannales).

Salient features: Herbaceous perennials, leaves broad, sheathing at base, flowers showy, bisexual, trimerous, perianth petaloid, stamens petaloid, only one fertile, ovary inferior, style flat with marginal stigma, fruit covered with warts.

Major genera: Single genus *Canna* (10 species).

Description: Perennial herbs with underground rhizome. Stem with mucilage canals. **Leaves** large, spirally arranged, broad, with distinct midrib containing air canals, alternate, spirally arranged, simple, venation parallel, petiole sheathing the stem, stipules and ligule absent. **Inflorescence** a terminal raceme, panicle, or spike of commonly 2-flowered cincinni; axis 3-angled in section with 3-ranked bracts, each bract associated with each 2-flowered (rarely 1-flowered) cincinnus. Flowers showy, bisexual, zygomorphic, epigynous, with a bract and a bracteole. **Perianth** 6, in two whorls (outer representing sepals, inner petals). Sepals 3 free, green or purple, persistent in fruit. Petals 3, connate at base and adnate to staminal column. **Androecium** with 6 stamens, in 2 whorls, connate and adnate to petals, three outer modified into petaloid imbricate staminodes, of 3 inner 2 modified into petaloid staminodes and the third with one anther lobe fertile and other modified into petaloid staminode, pollen grains inaperturate. **Gynoecium** with 3 united carpels, ovary inferior, trilocular with many ovules, placentation axile, styles simple, petaloid with marginal stigma. **Fruit** a capsule covered with warts, dehiscing by collapse of the pericarp; seed spherical, black, with tuft of hairs (modified aril) with straight embryo, endosperm hard, perisperm present. Most species self pollinated. Seeds often dispersed by water.

Economic importance: Various species of *Canna* especially *C. indica* and various hybrids are grown as garden ornamentals. The starch from rhizome of *C. edulis* (Queensland Arrowroot) is used as diet for infants as it is easily digestible.

Phylogeny: The family is closely related with other families such as Zingiberaceae, Musaceae, Marantaceae and Strelitziaceae, usually placed under order Zingiberales, differing from Zingiberaceae in lacking ligule. Thorne (2003, 2006, 2007) prefers name Cannales for the order, placing the family Cannaceae under suborder Cannineae. Takhtajan (1997) has narrowly circumscribed Cannales, to include Cannaceae and Marantaceae (suborder Cannineae of Thorne),

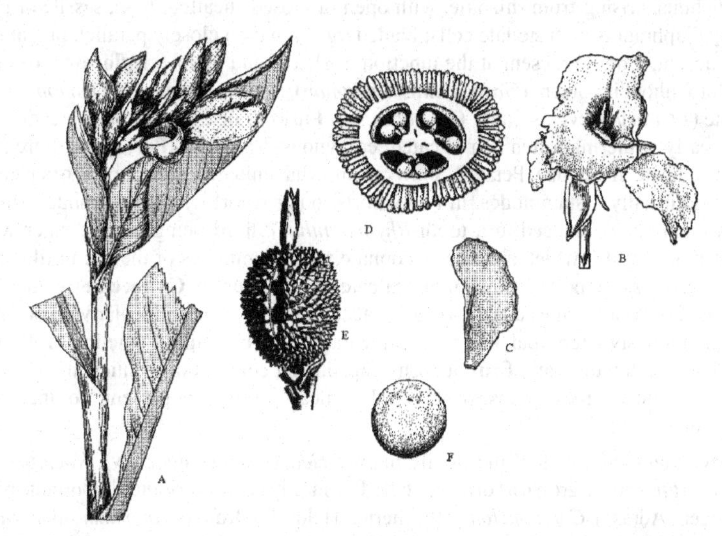

Figure 13.34: Cannaceae. *Canna indica.* **A:** Plant with leaves and inflorescence; **B:** Open flower showing petaloid staminodes, half anther and tip of style; **C:** Fertile stamen with half petaloid staminode and half anther; **D:** Transverse section of ovary showing axile placentation and ovary wall covered with warts; **E:** Dehiscing capsule covered with warts; **F:** Seed.

Zingiberales restricted to include only Zingiberaceae and Costaceae. Lowiaceae is removed to Lowiales and Musaceae together Heliconiaceae and Strelitziaceae to Musales. All 4 orders are placed under superorder Zingiberanae.

Grootjen and Bouman (1988) described a pachychalaza in Cannaceae, with mitosis occurring during ovule development in the chalaza and basal part of the nucellus. This is unlike the other Zingiberalean families. The family is monophyletic, as supported by DNA and morphology (Kress, 1990, 1995). The nature of androecium is unclear. Miao et al. (2014a, b) suggested that in *Canna indica* the fertile ½ stamen represents two primordia, one member of the outer whorl (the fertile bit) and one member of the inner whorl (the petaloid bit); the labellum there consists of another member of the outer whorl and another member of the inner whorl. On the other hand, Almeida et al. (2013) thought that both the fertile and petaloid parts were produced by a single half anther. ABC-type floral genes have very broad expression patterns across the various floral organs (Almeida et al., 2013).

Zingiberaceae Lindley ***Ginger Family***

50 genera, 1,600 species

Widespread in tropics mainly under forest shade and wetlands, chiefly distributed in Indomalaysia.

Placement:

	B & H	Cronquist	Takhtajan	Dahlgren	Thorne	APG IV/(APweb)
Informal Group						Monocots/ (Commelinids)
Division		Magnoliophyta	Magnoliophyta			
Class	Monocotyledons	Liliopsida	Liliopsida	Liliopsida	Magnoliopsida	
Subclass		Zingiberidae	Commelinidae	Liliidae	Commelinidae	
Series+/ Superorder	Epigynae+		Zingiberanae	Zingiberanae	Commelinanae	
Order		Zingiberales	Zingiberales	Zingiberales	Cannales	Zingiberales

B & H under family Scitamineae.

Salient features: Perennial rhizomatous aromatic herbs, leaves alternate, distichous, sheathing at base, flowers bisexual, zygomorphic, perianth 6 in two whorls, one petal often larger than others, fertile stamen 1, staminodes 3 or 4, petaloid, two staminodes forming a lip or labellum, carpels 3, united, ovary inferior, placentation axile, fruit a capsule or berry.

Major genera: *Alpinia* (165 species), *Amomum* (130), *Zingiber* (95), *Globba* (65), *Curcuma* (55), *Kaempheria* (65) and *Hedychium* (66).

Description: Perennial rhizomatous herbs, often with tuberous roots, aromatic, containing etherial oils, terpenes and phenyl-propanoid compounds, aerial stems short, usually leafless, rarely foliate, vessels present in roots as well as stem. **Leaves** alternate, distichous, arising from rhizome, with open or closed sheath at base, sessile or petiolate; petiole with air canals separated by diaphragms with stellate cells; blade large, venation closely parallel, pinnate, diverging obliquely from midrib, stipules absent, a ligule present at the junction of sheath and petiole. **Inflorescence** usually surrounded by involucre (*Geanthus*) or without involucre (*Gastrochilus, Amomum*), a dense spicate head (*Amomum*) or cyme, sometimes racemose or paniculate (*Elettaria*), even solitary (*Monocostus*). **Flowers** bisexual, often subtended by a sheathing bract, usually zygomorphic, early withering, often complicated, epigynous, trimerous. **Perianth** differentiated into sepals and petals. Sepals 3, green, connate into a tube. Petals 3, showy, somewhat united, posterior petal often enlarged. **Androecium** with 1 fertile stamen and usually 4 staminodes, in two whorls; outer whorl with 2 staminodes often fused to form 2–3 lobed lip or labellum (sometimes reduced to a tooth-*Rhynchanthus*), third being absent; inner whorl with one fertile stamen and two smaller staminodes which are free or connate with staminodes of the lip; fertile stamen with bithecous anther, grooved and grasping style, pollen grains monosulcate or inaperturate. **Gynoecium** with 3 united carpels, ovary inferior, 3-locular, rarely 2-locular, ovules usually many, placentation axile, or unilocular with parietal (*Globba*) or basal (*Haplochorema*) placentation, style terminal, undivided, free or grasped by anther, sometimes 2-lipped or dentate, stigma funnel-shaped, nectaries 2, on top of ovary. **Fruit** a fleshy capsule, indehiscent or loculicidal, rarely berry; seeds globose or angular, with large aril, endosperm copious, white, hard or mealy, perisperm present. Pollination by insects or birds. Fruits are dispersed by birds.

Economic importance: Many members of the family mainly *Hedychium* (ginger lily), *Kaempheria*, *Costus*, *Nicolaia* (torch ginger) and *Alpinia* are widely grown as ornamentals. The family also contributes important spices from the rhizome *Zingiber officinale* (ginger, 'Adrak'), *Curcuma longa* (turmeric, 'Haldi'), or fruits of *Amomum subulatum* (Bengal cardamon) and *Elettaria cardamomum* (Malabar cardamon). East Indian arrowroot is obtained from tubers of *Curcuma angustifolia*. A perfumed powder abir is obtained from the rhizomes of *Hedychium spicatum*. Spice Melegueta pepper is obtained from

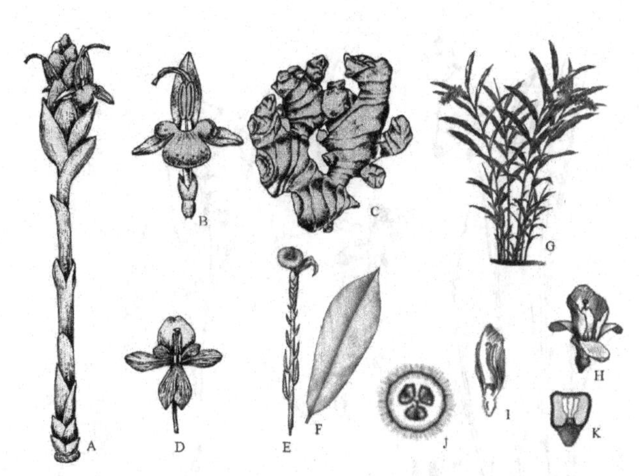

Figure 13.35: Zingiberaceae. *Zingiber officinale*. **A:** Plant with inflorescence; **B:** Flower; **C:** Rhizome. **D:** Flower of *Roscoea alpina*. *Aframomum laurentii*. **E:** Inflorescence; **F:** Leaf. *Alpinia nutans*. **G:** Plant with inflorescence; **H:** Flower; **I:** Vertical section of flower; **J:** Transverse section of fruit. **K:** Longitudinal section of seed of *Amomum*.

Aframomum melegueta. Rhizomes of *Alpinia zedoaria* are also sources of spice, tonic and perfume (zeodary), whereas those of *A. galanga* are used in medicine and flavouring (galangal).

Phylogeny: The family forms a monophyletic group along with other families included under Zingiberales as indicated above. Genus *Costus* sometimes included under a separate subfamily within Zingiberaceae has been removed to a distinct family Costaceae. Monophyly of the family is supported by morphology (Kress, 1990) and DNA information (Kress, 1995). The family was divided by Loesener (1930) into 2 subfamilies Zingiberoideae and Costoideae (latter with four tribes Hedychieae, Globbeae and Zingibereae). Hutchinson (1934, 1973) treated the four groups as four tribes (fourth being Zingibereae). Kress et al. (2001, 2002) has redefined the classification of the family, recognizing four subfamilies: distributing most genera of the three tribes (Costoideae has already been removed as distinct family) among two subfamilies Alpinoideae and Zingiberoideae, and recognizing two additional subfamilies Siphonochiloideae (single genus *Siphonochilus*) and Tamijioideae (single monotypic genus—*Tamijia*). Two genes analyses by these authors indicated strong support for Siphonochiloideae being sister to other three, and Tamijioideae to the other two Alpinioideae + Zingiberoideae. APWeb (2018) while recognising these four subfamilies divides Alpinoideae into two tribes Alpinieae and Riedelieae and Zingiberoideae into two tribes Zingibereae and Globbeae. Basic relationships in the family are [Siphonochiloideae [Tamijioideae [Alpinioideae + Zingiberoideae]]]; all clades have strong support (Harris et al., 2006). *Alpinia*, *Etlingera*, and *Amomum* are all more or less strongly para/polyphyletic.

Juncaceae A. L. de Jussieu ***Rush Family***

6 genera, 345 species

Worldwide, mostly in cold temperate and montane regions, usually in damp habitats.

Placement:

	B & H	Cronquist	Takhtajan	Dahlgren	Thorne	APG IV/(APweb)
Informal Group						Monocots/ (Commelinids)
Division		Magnoliophyta	Magnoliophyta			
Class	Monocotyledons	Liliopsida	Liliopsida	Liliopsida	Magnoliopsida	
Subclass		Arecidae	Commelinidae	Liliidae	Commelinidae	
Series+/ Superorder	Calycinae+		Juncanae	Commelinanae	Commelinanae	
Order		Juncales	Juncales	Cyperales	Juncales	Poales

Salient features: Tufted herbs, leaves grass-like, sometimes reduced to basal sheath, perianth 6, in two whorls, stamens 6, pollen in tetrads, ovary superior, placentation axile, stigmas 3, fruit a capsule.

Major genera: *Juncus* (260 species), *Luzula* (65), *Oxychloe* (7) and *Distichia* (3).

Figure 13.36: Juncaceae. *Juncus articulatus*. **A:** A plant with inflorescences; **B:** Flower; **C:** Perianth and stamens; **D:** Gynoecium; **E:** Capsule; **F:** Seeds. *Luzula albida*. **G:** Plant with inflorescence; **H:** Flower; **I:** Vertical section of flower; **J:** Gynoecium; **K:** Dehiscing fruit; **L:** Seed; **M:** Longitudinal section of seed. **N:** Flower of *Juncus sphacelatus*.

Description: Perennial or annual tufted herbs, often with rhizomes, stems cylindrical and solid, usually foliate only at base. **Leaves** alternate, mostly basal, 3-ranked, rarely distichous (*Distichia*), cylindrical or flat, sheathing at base or reduced only to sheath, sheath open or closed, blade grass-like, entire, venation parallel, stipules and ligule absent. **Inflorescence** with cymes clustered in heads or forming panicles, corymbs or even solitary (*Andesia*). **Flowers** usually bisexual, sometimes unisexual (*Distichia*) and plants dioecious, rarely monoecious (*Rostkovia*), actinomorphic, very small, trimerous. **Perianth** with 6 tepals, in two whorls, free, green brown or black, rarely scarious, inner sometimes smaller (*Marsippospermum*). **Androecium** with usually 6 stamens, rarely 3 (*Voladeria*), opposite the tepals, free, anthers bilocular, basifixed, dehiscence by longitudinal slits, introrse, pollen in tetrads, monoporate. **Gynoecium** with 3 united carpels, ovary superior, placentation axile, sometimes parietal (*Marsippospermum*), ovules many, styles 3 or 1, stigmas 3. **Fruit** a loculicidal capsule; seeds spherical or flat, sometimes pointed and spindle-shaped (*Marsippospermum*), with small straight embryo, endosperm present. Wind pollinated.

Economic importance: Family is not of much commercial use. Split rushes used in basket making are taken from stems of *Juncus effusus* (soft rush) and *J. squarosus* (heath rush). Juncio, used in binding, is derived from *Juncus maritimus* (sea rush). A few species of *Juncus* and *Luzula* are grown as ornamentals.

Phylogeny: Juncaceae are closely related to Liliaceae (Hutchinson, 1973; Heywood, 1978) representing reduced forms derived from that stock. *Prionium* which was earlier considered to be the most primitive genus of the family Juncaceae, linking it to Liliaceae, has now been removed to a distinct family Prioniaceae (Thorne, 1999, 2003), or under Thurniaceae (APG II; Apweb, 2008; Judd et al., 2008; Thorne, 2006, 2007). Takhtajan (1997) places *Prionium* under Juncaceae. Restionaceae, which have become totally dioecious, form the closest relatives of Juncaceae. The family is connected to Cyperaceae through genus *Oreobolus*, the most primitive genus of that family. Muasya et al. (1998) suggest that *Oxychloe* (Juncaceae) is sister to Cyperaceae, with moderate support, other Juncaceae are also basal and paraphyletic, but with poor support, while *Prionium* is sister to the whole clade, with good support. A study by Plunkett et al. (1995) placed *Oxychloe* within Cyperaceae. The relationships of the latter genus in particular are still unclear. According to the studies of Bremer (2002) Thurniaceae (including *Prionium*) are sister to Juncaceae plus Cyperaceae (*Oxychloe* not included) with strong support. Záveská Drábková and Vlcek (2009) found that *Juncus trifidus* and *J. monanthos* were separate from other *Juncus* and sister to the rest of the family (recognized as *Oreojuncus*—Záveská Drábková and Kirschner, 2013).

Cyperaceae A. L. de Jussieu ***Sedge Family***

90 genera, 5,500 species

Worldwide, mostly in cold temperate regions, usually in damp habitats.

Placement:

	B & H	Cronquist	Takhtajan	Dahlgren	Thorne	APG IV/(APweb)
Informal Group						Monocots/ (Commelinids)
Division		Magnoliophyta	Magnoliophyta			
Class	Monocotyledons	Liliopsida	Liliopsida	Liliopsida	Magnoliopsida	
Subclass		Arecidae	Commelinidae	Liliidae	Commelinidae	
Series+/ Superorder	Glumaceae+		Juncanae	Commelinanae	Commelinanae	
Order		Cyperales	Juncales	Cyperales	Juncales	Poales

Salient features: Herbs, stems often 3-angled, solid, leaves 3-ranked, containing silica bodies, sheaths closed, ligule absent, glumes present, flower subtended by a single bract, lodicules absent, perianth represented by bristles, scale or absent, ovary superior with single ovule, fruit a nut.

Major genera: *Carex* (1800 species), *Cyperus* (580), *Fimbristylis* (290), *Scirpus* (280), *Rhynchospora* (240), *Scleria* (200) and *Eleocharis* (190).

Description: Annual or perennial herbs, usually rhizomatous, stems mostly 3-angled, solid. **Leaves** alternate, 3-ranked, often crowded at the base of stem, simple, grass-like, with silica bodies, entire or serrulate, venation parallel, stipules and ligule absent, sheath closed, stomata with dumbbell-shaped guard cells. **Inflorescence** consisting of small spikes (sometimes called spikelet but different from spikelet of grasses which has two basal glumes, and each floret enclosed in a lemma and a palea) each often subtended by a bract (prophyll) and bearing (on the axis called rachilla) spirally arranged (*Cladium*) or distichous (*Cyperus*) bracts (glumes), each subtending one flower; small spikes (spikelets) aggregated in spikes, panicles or even umbels, the whole inflorescence subtended by one or more usually leaf-like involucral bracts. **Flowers** very small, bisexual (*Cyperus*, *Scirpus*) or unisexual (*Scleria*), subtended by bract (glume), female flower often with second bract surrounding the pistil and forming sac-like perigonium. **Perianth** represented by bristles, sometimes scales (*Oreobolus*, *Lipocarpha*), or even absent (*Bulbostylis, Scirpus*). **Androecium** with 3 stamens, sometimes more (6 in *Arthrostylis*; 12–22 in *Evandra*), free, anthers bithecous, basifixed, oblong or linear, dehiscence by longitudinal slits, pollen grains uniporate, in pseudomonad (out of the four microspores, three degenerate and form the part of fourth fertile forming pollen grain). **Gynoecium** with 2 (*Kyllinga*) or 3 (*Cyperus*) united carpels, ovary superior, unilocular, ovule 1, anatropous, placentation basal, style with 2–3 branches, stigmas 2 or 3. Fruit a nut (often called achene but the latter is strictly derived from a single carpel), sometimes enclosed in a utricle (*Carex*), with often persistent style and associated with persistent perianth bristles, bifacial or trigonous; seed erect, embryo small, endosperm conspicuous, mealy or fleshy.

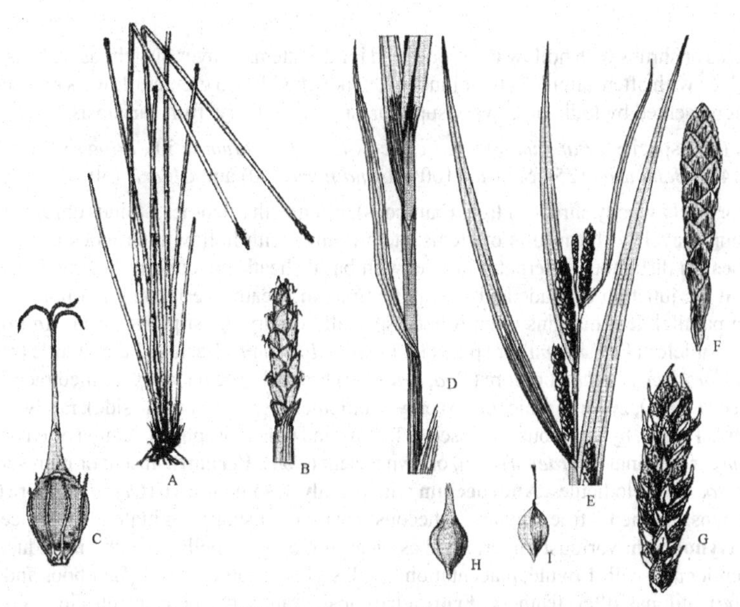

Figure 13.37: Cyperaceae. *Eleocharis lankana.* **A:** Plant in flower; **B:** Inflorescence; **C:** Fruiting gynoecium with hypogynous bristles. *Carex ligulata.* **D:** Lower part of plant; **E:** Part of inflorescence; **F:** Male spike; **G:** Female spike; **H:** Utricle; **I:** Nut.

Economic importance: Various species of Cyperaceae are useful in different ways. Stems of *Cyperus papyrus* (papyrus or paper weed) were much used in ancient times for making paper and is now commonly grown as an ornamental. The stems of *Cladium effusum* (saw grass) are also source of cheap paper. Stems and leaves of *Carex brizoides* and *Lepironia mucronata* are used for packing and basket work. Underground organs of *Cyperus esculentus* (tigernut, Zulu nut or rush nut), *Scirpus tuberosus* and *Eleocharis tuberosa* (matai, Chinese water chest nut) are used as food. The stems of *Scirpus totara* are used for making canoes and rafts and those of *S. lacustris* for basketwork, mats and chair seats.

Phylogeny: The family is considered to be monophyletic, connected to Juncaceae through genus *Oreobolus*, the most primitive genus (Hutchinson, 1973). Muasya et al. (1998) suggest that *Oxychloe* (Juncaceae) is sister to Cyperaceae, with moderate support. Plunkett et al. (1995) placed *Oxychloe* within Cyperaceae. The relationships of the latter genus in particular are still unclear. According to the studies of Bremer (2002) Thurniaceae (including *Prionium*) are sister to Juncaceae plus Cyperaceae, with strong support. He did not include *Oxychloe* under Cyperaceae. According to him Cyperaceae, Juncaceae and Thurniaceae form a well defined cyperid clade. The family is traditionally divided into 3 subfamilies (Engler): Scirpoideae, Rhynchosporoideae and Caricoideae. APWeb (2018) recognises 2 subfamilies Mapanioideae and Cyperoideae, latter further subdivided into 7 tribes Cypereae, Scirpeae, Sclerieae, Rhynchosporeae, Schoeneae, Dulichieae and Cariceae. Simpson et al. (2003) based on the pollen and plastid DNA sequence data concluded that Mapanioideae are sister to the rest of the family.

Poaceae Barnhart ***Grass Family***

(= Gramineae A. L. de Jussieu**)**

780 genera, 12000 species (Fourth largest family after Orchidaceae, Asteraceae, and Fabaceae)

Worldwide, distributed from poles to equator and from mountain peaks to sea level, in all types of climates and habitats.

Placement:

	B & H	Cronquist	Takhtajan	Dahlgren	Thorne	APG IV/(APweb)
Informal Group						Monocots/ (Commelinids)
Division		Magnoliophyta	Magnoliophyta			
Class	Monocotyledons	Liliopsida	Liliopsida	Liliopsida	Magnoliopsida	
Subclass		Arecidae	Commelinidae	Liliidae	Commelinidae	
Series+/ Superorder	Glumaceae+		Poanae	Commelinanae	Commelinanae	
Order		Cyperales	Poales	Poales	Poales	Poales

B & H as Gramineae.

Salient features: Herbs or shrubs with hollow internodes and jointed stems, leaves distichous with distinct sheath enclosing the stem and linear blade with often a ligule at their junction, spikelet with two glumes, flowers reduced, enclosed in lemma and palea, perianth represented by lodicules, ovary superior, stigma feathery, fruit caryopsis.

Major genera: *Poa* (500 species), *Panicum* (450), *Festuca* (430), *Paspalum* (350), *Stipa* (300), *Bromus* (160), *Elymus* (150), *Sporobolus* (140), *Bambusa* (125), *Setaria* (100), *Arundinaria* (50) and *Chloris* (50).

Description: Herbs or rarely woody shrubs or trees (bamboos), often with rhizomes, stolons or runners, frequently tillering (branching from ground level) to form tufts of stems, stem (culm) with hollow internodes and jointed swollen nodes, with silica bodies. **Leaves** distichous, alternate, simple, with basal sheath surrounding internode and free linear blade, a ligule often present at the junction of blade and sheath, margins of sheath overlapping but not fused, sometimes united into a tube, venation parallel, leaf margins often rolled especially on drying, stipules absent. **Inflorescence** of spikelets arranged in racemes, panicles (*Poa, Avena*) or spikes (*Triticum, Hordeum*). Each spikelet with 2 (rarely 1 as in *Monera*) glumes enclosing 1 (*Hordeum, Nardus*) or more (*Poa, Triticum*) florets borne on an axis called rachilla, usually in 2 rows. **Flowers** small, reduced (floret), zygomorphic (due to only 2 lodicules displaced on one side), rarely actinomorphic, usually bisexual rarely unisexual (*Zea*), hypogynous, enclosed in lemma and palea (prophyll), lemma often bearing dorsal (*Avena*), subterminal (*Triticum*) or terminal (*Hordeum*) awn, or awn absent (*Poa*). **Perianth** absent or represented by 2 (rarely 3, as in *Bambusa* and *Streptochaeta*) lodicules. **Androecium** with usually 3, sometimes 6 (*Oryza*) or more (*Arundinaria*), rarely 1–2 (Leptureae) stamens, filaments free, anthers bithecous, basifixed, usually sagittate, dehiscence longitudinal, pollen grains monoporate. **Gynoecium** variously interpreted as bicarpellary, tricarpellary (with one reduced style), syncarpous or monocarpellary, unilocular with 1 ovule, placentation basal, styles 2, sometimes 3 (Bamboos and *Streptochaeta*), very rarely 1 (*Anomochloa*), stigmas often feathery. **Fruit** a caryopsis, rarely nut berry or utricle; seed fused with pericarp, embryo straight, endosperm starchy.

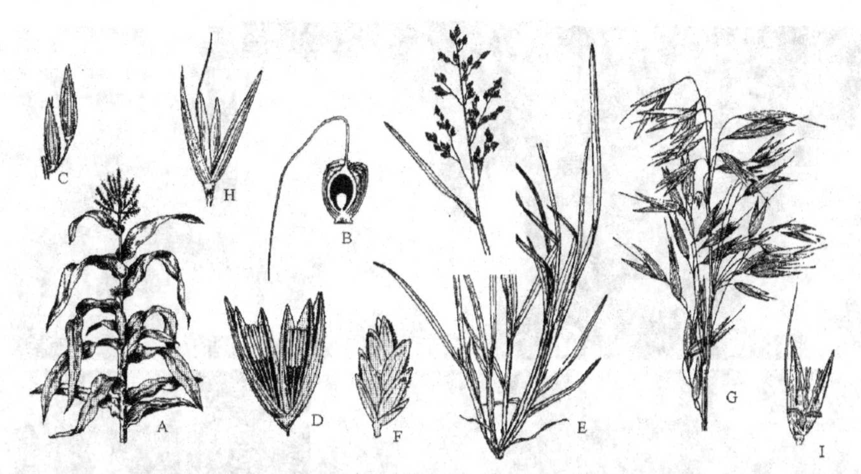

Figure 13.38: Poaceae. *Zea mays.* **A:** Plant with terminal male inflorescence and axillary female inflorescence (Cob); **B:** Vertical section of female spikelet; **C:** Paired male spikelets; **D:** Male spikelet opened to show two fertile florets. *Poa annua.* **E:** Plant in flower; **F:** Spikelet. *Avena sativa.* **G:** Inflorescence; **H:** Spikelet opened; **I:** Fertile floret with awned lemma.

Economic importance: The family is of great economic importance, being a source of important cereals such as rice (*Oryza sativa*), wheat (*Triticum aestivum*) and corn or maize (*Zea mays*). The family also includes other food crops such as barley (*Hordeum vulgare*), pearl millet (*Pennisetum glaucum*), oats (*Avena sativa*), rye (*Secale cereale*) and sorghum (*Sorghum vulgare*). Grasses such as *Cynodon*, *Axonopus* and *Agrostis* are extensively used in lawns and turfs. *Andropogon*, *Agropyron*, and *Phleum* are major forage grasses. Sugarcane (*Saccharum officinarum*) is the major source of commercial sugar. Bamboos are employed in big way in construction work, wickerwork and thatching in different parts of the world. Young bamboo shoots are used as food and often pickled. Lemon grass (*Cymbopogon*) leaves are distilled to yield essential oil for imparting citronella scent. Grains of *Coix lacryma-jobi* (Job's tears) are use as necklace beads. Roots of *Vetiveria zizanoides* (vetivar grass) are used for making fragrant cooling pads and extraction of vetiver oil.

Phylogeny: Although a very large assemblage Poaceae are easily recognized and form a monophyletic group, as supported by morphology (lodicules, spikelets with glumes, lemma and palea, fruit caryopsis) and DNA characters (*rbcL* and *ndhF* sequences). Cronquist (1988) places Poaceae and Cyperaceae under the same order Cyperales, but similar morphology of two is believed to be due to convergent evolution, Cyperaceae being more closely related to Juncaceae (Judd et al., 1999). The studies of Bremer (2002), using *rbcL* and taq analyses found strong support for Cyperaceae, Juncaceae, and Thurniaceae forming cyperid clade and Poaceae along with other families forming a graminoid clade.

The nature of gynoecium in this family has been a matter of controversy. Most early authors including Haeckel (1883), Rendle (1930) and Diels (1936) considered it to consist of a single carpel terminated by 2–3 branched stigma. Lotsy (1911), Weatherwax (1929) and Arber (1934) considered that it represents tricarpellary ovary having evolved from an ovary with parietal placentation, a view supported by studies on floral anatomy (Belk, 1939). Others believe that gynoecium consists of 2–3 carpels (depending on the number of stigmas visible; Cronquist, 1988; Woodland, 1991).

The family is variously classified by different authors. Hutchinson (1973) recognized two subfamilies Pooideae (with 24 tribes) and Panicoideae (with 3 tribes). Heywood (1978) recognized 6 subfamilies (Bambusoideae, Centostecoideae, Arundinoideae, Chloridoideae, Panicoideae and Pooideae), further subdivided to include 50 tribes. Of these subfamilies, Centostecoideae occupies an isolated position although related to both Bambusoideae and Panicoideae and includes broad-leaved herbs with single- to several-flowered spikelets. Studies of Clark et al. (1995) and Soreng and Davis (1998) suggest that Arundinoideae, Chloridoideae and Panicoideae form a well supported clade (often called PACC clade) based on embryological and DNA data. Arundinoideae as generally defined are not monophyletic, and many of their members such as *Aristida*, *Phragmites*, etc., are spread over in other two subfamilies. Chloridoideae and Panicoideae are generally found to be monophyletic. Stevens (Apweb, 2003) lists 12 subfamilies under Poaceae: Anomochlooideae, Pharoideae, Puelioideae, Panicoideae, Arundinoideae, Centothecoideae, Chloridoideae, Aristidoideae, Danthonioideae (six forming **PACCAD** clade), Bambusoideae, Ehrhartoid-eae, Pooideae (**BEP** clade), prefering names Oryzoideae and Micrairoideae in place of Centothecoideae and Ehrhartoideae in APweb (2018). There is great diversity in the morphology and biochemistry of C4 photosynthesis in the family (Kellogg, 2000). Studies based on gene expression (Ambrose et al., 2000) indicate that the palea and perhaps even lemma are calycine in nature and the lodicules are corolline.

Figure 13.39: Amborellaceae. *Amborella trichopoda.* **A:** Plant; **B:** Flower; **Chloranthaceae**. **C:** *Chloranthus glaber* plant in flower; **Calycanthaceae. D:** *Calycanthus occidentalis*, plant in flower; **Nymphaeaceae. E:** *Nuphar polysepala*, plant; **F:** Flower of *N. lutea*; **G:** *Nymphaea odorata*, plant in flower; **H:** Flower enlarged.

Figure 13.40: Magnoliaceae. *Magnolia grandiflora.* **A:** Tree in flower; **B:** Flower; **C:** Fruit; **D:** Leaf of *Liriodendron chinense.* **Degeneriaceae.** *Degeneria vitiensis.* **E:** Plant; **F:** Fruit; **G:** Flower. **Lauraceae. H:** *Laurus nobilis* plant in flower; **I:** Plant of *Neolitsea sericea.*

Figure 13.41: Araceae. A: *Amorphophallus titanum*, unopened inflorescence; **B:** Part of fruiting inflorescence; **C:** *Epipremnum aureum*, plant; **D:** *Anthurium andreanum*, inflorescence; **Alismataceae. E:** *Sagittaria sagittifolia*, plant; **Hydrocharitaceae. F:** *Hydrilla verticillata*, plant; **Liliaceae. G:** *Tulipa* cultivar, flower; **H:** Stamens and pistil enlarged; **I:** *Lilium* sp., plant in flower; **J:** Flower enlarged.

Figure 13.42: Iridaceae. A: *Iris* sp., plant; **B:** Flower enlarged; **C:** *Iris germanica*, flower. **Asphodelaceae. D:** *Kniphofia thomsoni*, plant; **E:** Part of inflorescence; **F:** *Asphodelus fistulosus*, plant; **G:** Flower and fruit; **Amaryllidaceae**. **H:** *Clivia chrysanthifolia*, plant; **I:** Part of inflorescence; **J:** *Allium cepa*, plant; **K:** Inflorescence enlarged; **L:** *Agapanthus praecox*, inflorescence; **M:** Basal part with leaves and scapes.

Figure 13.43: Hyacinthaceae. A: *Eucomis autumnalis* plant with inflorescence; **B:** *E. bicolor*, inflorescence. **Agavaceae. C:** *Yucca rupicola*, plant with inflorescence; **D:** Part of inflorescence; **E:** *Agave parrii*, plant; **F:** *A. wightii*, inflorescence. **Asparagaceae. G:** *Ophiopogon planiscaposus*, plant; **H:** *Ruscus aculeatus*, plant; **I:** Flower; **J:** *Asparagus racemosus*, fruit. **Nolinaceae. K:** *Nolina recurvata*, swollen base; **L:** *N. nelsoni*, plant.

Figure 13.44: Arecaceae. A: *Roystonea regia*, plant; **B:** Trunk with leaf scars; **C:** *Parajubaea coccoides*, portion of plant with fruits; **D:** *Caryota urens*, plant; **E:** Inflorescence; **F:** Part in Fruit. **Musaceae. G:** *Musa paradisiaca* subsp. *sapientum*, plant with inflorescence; **H:** Inflorescence with fruits. **Commelinaceae. I:** *Rhoeo discolor*, plant; **J:** *Tradescantia pallida*, plant; **K:** Flower. **Cyperaceae. L:** *Cyperus alternifolius*, plant; **M:** Part of spike enlaged; **N:** *Scirpoides holoschoenus*, plant. **Poaceae. O:** *Triticum aestivum*, plant; **P:** Spike; **Q:** Portion of spike of *Pennisetum glaucum;* **R:** *Avena sativa*, plant with inflorescence; **S:** *Zea mays*, male inflorescence.

Arrangement after APG IV (2016)
Probable sister to Eudicots:
Order 1. Ceratophyllales
Family: **Ceratophyllaceae**

Ceratophyllaceae S. F. Gray ***Hornwort Family***

1 genus, 6 species

Widespread, forming floating masses in fresh water bodies.

Placement:

	B & H	Cronquist	Takhtajan	Dahlgren	Thorne	APG IV/(APweb)
Informal Group						Probable sister to Eudicots/ (Commelinids)
Division		Magnoliophyta	Magnoliophyta			
Class	Dicotyledons	Magnoliopsida	Magnoliopsida	Magnoliopsida	Magnoliopsida	
Subclass	Monochlamydeae	Magnoliidae	Magnoliidae	Magnoliidae	Chloranthidae	
Series+/ Superorder	Ordines anomali+		Nymphaeanae	Nymphaeaneae	Nymphaeanae	
Order		Nymphaeales	Ceratophyllales	Nymphaeales	Nymphaeales	Ceratophyllales

Salient features: Submerged aquatic herbs, roots absent, leaves whorled, often dichotomously dissected, flowers minute, unisexual, perianth with 7-numerous bract-like segments, stamens 10 to numerous, anthers with prolonged connective, carpel 1, ovary superior, placentation apical, fruit an achene with 2 or more projections.

Major genera: Single genus *Ceratophyllum* (6 species).

Description: Submerged aquatic herbs often forming floating masses, rootless but sometimes with colorless root-like branches anchoring the plant; stems branched but with never more than one branch at one node, with single vascular strand with central air canal surrounded by starch-containing cells, with tannins. **Leaves** whorled, 3–10 at each node, once to four times dichotomously dissected, ultimate leaf-segments with two rows of minute teeth and tipped by two bristles, stomata and cuticle absent, stipules absent. **Inflorescence** with solitary axillary flowers, usually one flower in a whorl of leaves. **Flowers** unisexual (plants monoecious), male and female flowers usually on alternate nodes, actinomorphic, very small. **Perianth** with 7 to numerous tepals, linear, bract-like, slightly connate at base. **Androecium** with numerous

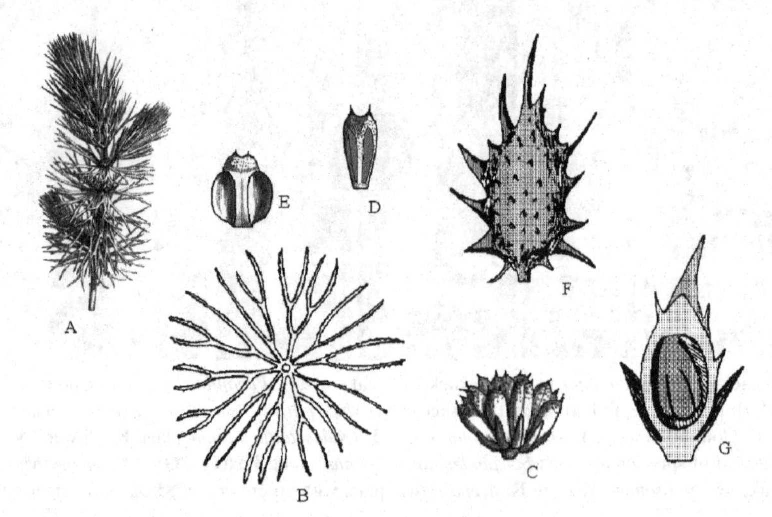

Figure 13.45: Ceratophyllaceae. *Ceratophyllum submersum*. **A:** Portion of plant; **B:** Whorl of leaves at node spread out to show dichotomous segments; **C:** Male flower with perianth and several sessile stamens; **D:** Young stamen with connective produced into two prominent teeth; **E:** Dehiscing stamen; **F:** Fruit with persistent style and spiniscent projections; **G:** Longitudinal section of fruit with pendulous seed.

stamens, filaments indistinct, anther oblong-linear, bithecous, dehiscence longitudinal, extrorse, connective prolonged beyond anthers into two prominent often colored teeth, staminodes absent in female flower; pollen grains inaperturate, exine reduced, pollen tubes branched. **Gynoecium** with 1 carpel, ovary superior, unilocular with 1 ovule, placentation apical, ovule pendulous, style continuous with the ovary, stigma extending along one side of style. **Fruit** a nut tipped by persistent spine-like style and often with two or more projections; seed with straight embryo, endosperm absent.

Economic importance: Floating masses provide protection to fish fry. The masses also support bilharzia-carrying snails and malaria- or filaria-carrying mosquito larvae. Fruits and foliage form food for migratory waterfowl. It is sometimes troublesome, choking waterways.

Phylogeny: The phylogeny of the family has been a matter of great speculation. Bentham and Hooker placed this family along with others of uncertain affinities under Ordines anomali. It is usually considered to be related to Nymphaeaceae (Lawrence, 1951; Heywood, 1978—both include *Nelumbo* under Nymphaeaceae) particularly genus *Nelumbo*, which has now been removed to a distinct family Nelumbonaceae. Cronquist (1988) placed all the three families under the same order Nymphaeales. Dahlgren (1989) included Nelumbonaceae under order Nelumbonales of superorder Magnolianae, whereas Nymphaeaceae and Ceratophyllaceae were placed under order Nymphaeales of superorder Nymphaeanae. Takhtajan (1997) removed Nelumbonaceae under distinct order Nelumbonales, distinct superorder Nelumbonanae and even a distinct subclass Nelumbonidae. Ceratophyllaceae and Nymphaeaceae are placed under subclass Nymphaeidae but separate superorders Ceratophyllanae and Nymphaeanae, under respective orders Ceratophyllales and Nymphaeales. In last revision in 2009, he included it under Magnoliidae, Nymphaeanae, Ceratophyllales. Thorne (2003) placed Nymphaeaceae and Cabombaceae under Nymphaeales, superorder Nymphaeanae, under subclass Magnoliidae, subsequently (2006) shifting to Chloranthidae. Ceratophyllaceae is taken closer to Nelumbonaceae under Ranunculanae of subclass Ranunculidae but under separate orders Ceratophyllales and Nelumbonales, respectively. Latest 2007 revision showed a major shift of Ceratophyllaceae to Chloranthidae under order Nymphaeales.

The family Ceratophyllaceae has attracted a lot of interest with morphological and fossil evidence (Les et al., 1991) and molecular evidence (Chase et al., 1993) suggesting basal placement in angiosperms. Hickey and Taylor (1996), however, suggested that the aquatic plant with highly reduced vegetative body and pollen wall, tenuinucellate unitegmic ovules and problematic fossil record is a poor candidate for basal-most position. Ceratophyllaceae is possibly sister to monocots (e.g., Graham and Olmstead, 2000; Zanis et al., 2002; Whitlock et al., 2002), and accordingly in APG II and Apweb, placed in distinct order Ceratophyllales before monocots and Chloranthales, respectively without any supraordinal informal group. More recent studies of low-copy nuclear genes by Zhang et al. (2012) supported a possible **CCMM** clade including Chloranthaceae, Ceratophyllaceae, Magnoliids and Monocots, in apparent divergence from results based on plastid genomes that placed Ceratophyllaceae and Monocots as successive sisters of Eudicots (Moore et al., 2007, 2010). APG III (2009), APG IV (2016) as such place Ceratophyllaceae under Ceratophyllales before and probable sister to Eudicots. APweb (2018) also places it before Eudicots but under and at the end of Commelinids.

Arrangement after APG IV (2016)

Eudicots:

Order 1. Ranunculales
- Family: Eupteleaceae
- Family: **Papaveraceae**
- Family: Circaeasteraceae
- Family: Lardizabalaceae
- Family: Menispermaceae
- Family: **Berberidaceae**
- Family: **Ranunculaceae**

Order 2. Proteales
- Family: Sabiaceae
- Family: Nelumbonaceae
- Family: Platanaceae
- Family: Proteaceae

Order 3. Trochodendrales
- Family: Trochodendraceae

Order 4. Buxales
- Family: Buxaceae

Families in boldface are described in detail.

Papaveraceae A. L. de Jussieu *Poppy Family*

42 genera, 775 species

Widely distributed, primarily in temperate regions of the Northern Hemisphere, also in Southern Africa and Eastern Australia.

Placement:

	B & H	Cronquist	Takhtajan	Dahlgren	Thorne	APG IV/(APweb)
Informal Group						Eudicots
Division		Magnoliophyta	Magnoliophyta			
Class	Dicotyledons	Magnoliopsida	Magnoliopsida	Magnoliopsida	Magnoliopsida	
Subclass	Polypetalae	Magnoliidae	Ranunculidae	Magnoliidae	Ranunculidae	
Series+/ Superorder	Thalamiflorae+		Ranunculanae	Ranunculanae	Ranunculanae	
Order	Parietales	Papaverales	Papaverales	Papaverales	Ranunculales	Ranunculales

Salient features: Herb, sap usually milky or colored, flowers bisexual, sepals caducous, petals crumpled in bud, stamens numerous in several whorls, ovary superior, unilocular, fruit a capsule.

Major Genera: *Corydalis* (400 species), *Papaver* (80), *Fumaria* (55), *Meconopsis (5)*, *Argemone* (30) and *Eschscholzia* (10).

Description: Annual or perennial herbs, rarely soft-wooded shrubs (*Dendromecon*), or small trees (*Bocconia*), vascular bundles often in several rings, white or colored latex. Hairs simple, sometimes barbellate (*Cathcartia*). **Leaves** usually alternate, floral leaves sometimes subopposite (*Platystemon*), simple, often much dissected, sometimes entire (*Dendromecon*) or spinose (*Argemone*), venation reticulate, stipules absent. **Inflorescence** usually with solitary flowers, scapigerous in *Sanguinaria*, racemose in *Eomecon*, paniculate in *Bocconia*. **Flowers** bisexual, actinomorphic, sometimes zygomorphic (*Fumaria, Corydalis*). **Calyx** with 2 sepals, sometimes 3, caducous or calyptrate, free, usually enclosing bud. **Corolla** with usually 4 petals, sometimes 6 or even 8–12 (*Sanguinaria*), free, usually in two whorls, two outer sometimes saccate or spurred containing nectary (*Fumaria, Corydalis*), inner sometimes connivent at tip (*Fumaria*), imbricate, often crumpled in bud, absent in *Bocconia*. **Androecium** with numerous stamens (*Papaver*), sometimes 4 and opposite the petals (*Corydalis*) or 6 in two bundles of 3 each (*Dicentra*), anthers bithecous (in *Fumaria* of the 6 stamens 2 are with bithecous anthers, 4 with monothecous anthers), dehiscence longitudinal, pollen grains tricolpate to polyporate. **Gynoecium** with usually 2 united carpels, sometimes loosely united and becoming free in fruit (*Platystemon*), ovary superior, unilocular with parietal placentation, sometimes becoming multilocular due to intrusion of placentae, ovules numerous, sometimes 1 (*Baccopia*), anatropous, stigma discoid or lobed, sometimes capitate. **Fruit** a capsule dehiscing by valves or splitting into 1-seeded segments, sometimes nut (*Fumaria*); seeds small, sometimes with aril, embryo minute, endosperm copious, fleshy or oily. Pollination usually by insects, rarely wind (*Bocconia*). Seeds are dispersed by explosive opening of capsules, those with aril often by ants.

Economic importance: Many species of *Papaver* (poppy), *Eschscholzia* (Californian poppy), *Argemone* (Prickly poppy), *Corydalis* (harlequin), *Sanguinaria* (blood-root) and *Dicentra* (Dutchman's breeches, bleeding heart) are grown as ornamentals. Opium poppy (*Papaver somniferum*) is the most valuable member yielding opium (obtained from the latex of capsules) and its derivatives heroin, morphine and codeine. Seeds of this species do not contain opium and as such are used in baking, and also yield a drying oil. Seeds of *Glaucium flavum* and *Argemone mexicana* also yield oils used in the manufacture of soaps.

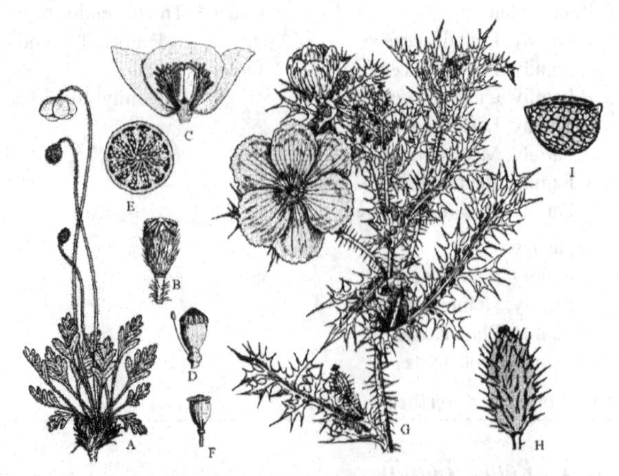

Figure 13.46: Papaveraceae. *Papaver nudicaule*. **A:** Plant with flowers; **B:** Fruit with bristly hairs. *P. rhoeas*. **C:** Vertical section of flower; **D:** Gynoecium with one stamen still attached, others having shed; **E:** Transverse section of ovary with intruded placentae; **F:** Fruit with glabrous surface and broad stigmatic disc. **G:** *Argemone ochroleuca* with flowers and a fruit towards the base with conspicuous style; **H:** Fruit of *A. mexicana* with sessile style; **I:** Seed of *A. mexicana* (A–B after Polunin and Stainton, Fl. Himal., 1984).

Phylogeny: The family is considered to be related backwards to Helleboraceae (Hutchinson, 1973) but with syncarpous gynoecium and parietal placentation, and very clearly forwards to Brassicaceae, which also has parietal placentation but with false septum. Genera with zygomorphic flowers, and with saccate or spurred petals are sometimes treated under distinct family Fumariaceae (Hutchinson, 1926, 1973; Lawrence, 1951; Cronquist, 1988; Dahlgren, 1989 and Takhtajan) but morphological and nucleotide sequence data supported the monophyly of the family including these genera, which are better placed under subfamily Fumarioideae (Thorne, 2003; Judd et al., 2002). APG II also optionally include Fumariaceae under Papaveraceae. There is, however, difference of opinion regarding basal genera. Loconte et al. (1995) proposed Platystemonoideae (*Platystemon* and relatives) with numerous slightly fused carpels and free stigmas as the basal clade. Hoot et al. (1997) on the other hand, on the basis of morphology and nucleotide sequence, regarded *Pteridophyllum* as sister to the remaining genera. This monotypic genus has been removed to a distinct family Pteridophyllaceae by Takhtajan (1997, 2009), Thorne (2003, 2006, 2007) and APG II (2003), but merged back in APG III (2009), APG IV (2016) and APweb (2018). Thorne (2007) divides Papaveraceae into 5 subfamilies: Papaveroideae, Eschscholzioideae, Chelidonioideae, Hypecoideae and Fumarioideae. APweb (2018) recognises 3 subfamilies: Papaveroideae (4 tribes Papaverae, Chelidonieae (Chelidonioideae of Thorne), Eschscholtzieae (Eschscholzioideae of Thorne) and Platystemoneae) Hypecoöideae, and Fumarioideae, being unsure about position *Pteridophyllum racemosum.* Earlier Pérez-Gutiérrez et al. (2015) and Sauquet et al. (2015: matk sequence suspect) all found its position to be unclear.

The family was earlier placed closer to Brassicaceae and Capparaceae, due to the parietal placentation, but has now been shifted closer to (or under) Ranunculales, the shift supported by chemical evidence—absence of glucosinolates and the presence of alkaloid benzylisoquinolene.

Berberidaceae Durande *Barberry Family*

14 genera, 700 species

Widespread chiefly in North Temperate regions and the Andes of South America.

Placement:

	B & H	Cronquist	Takhtajan	Dahlgren	Thorne	APG IV/(APweb)
Informal Group						Eudicots
Division		Magnoliophyta	Magnoliophyta			
Class	Dicotyledons	Magnoliopsida	Liliopsida	Magnoliopsida	Magnoliopsida	
Subclass	Polypetalae	Magnoliidae	Ranunculidae	Magnoliidae	Ranunculidae	
Series+/ Superorder	Thalamiflorae+		Ranunculanae	Ranunculanae	Ranunculanae	
Order	Ranales	Ranunculales	Berberidales	Ranunculales	Ranunculales	Ranunculales

Salient features: Herbs or shrubs, stipules absent, flowers bisexual, sepals and petals similar, stamens 6, outer whorl opposite the petals, anthers dehiscing by valves, carpel 1, ovary superior, fruit a berry.

Major genera: *Berberis* (540 species), *Mahonia* (60), *Epimedium* (55), *Podophyllum* (12), *Jeffersonia* (2) and *Nandina* (1).

Description: Perennial herbs or shrubs, rarely small trees, stem sometimes with scattered vascular bundles, wood usually colored yellow by berberine (an isoquinolene alkaloid). **Leaves** usually alternate, rarely opposite (*Podophyllum*), simple (*Berberis*), or palmately lobed (*Podophyllum*), or pinnate compound (*Mahonia*), rarely 2–3 times pinnately compound (*Nandina*), leaves of longer shoots sometimes modified into spines (*Berberis*), leaves entire or spinose-serrate, venation pinnate or palmate, reticulate, stipules absent. **Inflorescence** a raceme, panicle (*Nandina*) or even solitary (*Jeffersonia*). **Flowers** bisexual, actinomorphic, hypogynous. **Calyx** with 3 to 6 sepals, free, imbricate, green (*Podophyllum*) or petaloid (*Berberis*), rarely absent (*Achlys*). **Corolla** with 3–6 petals, sometimes more, free, inner whorl often in the form of petaliferous nectaries, rarely absent (*Achlys*). **Androecium** with anthers bithecous, dehiscence by longitudinal valves opening from base upwards, sometimes by usually 6 stamens, opposite the petals, sometimes upto 18 (*Podophyllum*) or reduced to 4 (*Epimedium*), longitudinal slits (*Nandina, Podophyllum*), pollen grains usually tricolpate. **Gynoecium** with 1 carpel, ovary superior, unilocular with many ovules, sometimes with 1 ovule (*Nandina*), anatropous, placentation parietal or basal, style very short, stigma almost sessile, sometimes 3-lobed. **Fruit** usually a berry, rarely a dehiscent capsule (*Jeffersonia*), or an achene (*Achlys*); seeds with small embryo, endosperm copious, sometimes with aril. Pollination by insects. Dispersal by birds or animals. Bladder-like capsule of *Leontice* dispersed by wind. In *Caulophyllum* the fleshy blue seeds burst through the ovary wall and develop in completely exposed state.

Economic importance: Many species of *Berberis* (*B. buxifolia, B. darwinii*), *Mahonia* (*M. aquifolium*) and *Nandina* (*N. domestica*) are commonly grown as ornamentals. The rhizomes of *Podophyllum hexandrum* (May apple) yield a resin which is used as a purgative and incorporated in many laxative pills.

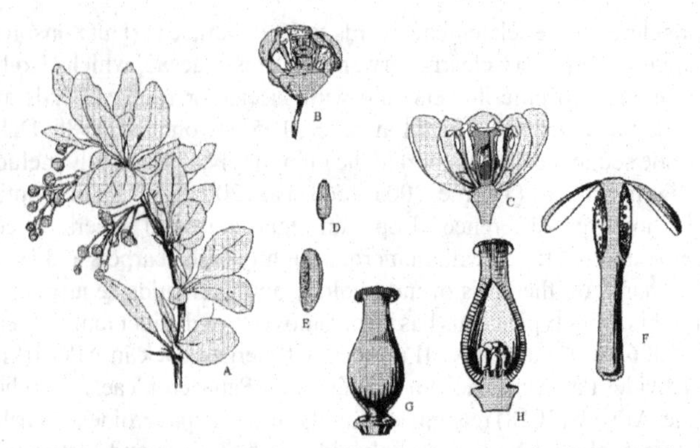

Figure 13.47: Berberidaceae. *Berberis vulgaris.* **A:** Twig with leaves, flowers and spines; **B:** Flower; **C:** Longitudinal section of flower; **D:** Fruit; **E:** Seed. *B. stenophylla.* **F:** Stamen with anther dehiscing by two valves; **G:** Ovary; **H:** Longitudinal section of ovary (F–H after Hutchinson, 1973).

Phylogeny: The family includes genera which are quite distinct from one another. Chapman (1936) based on carpellary anatomy proposed that Berberidaceae and Ranunculaceae arose by parallel evolution from a proranalian complex, and also doubted whether any existing families may be related as the immediate predecessors of Berberidaceae. She also demonstrated that single carpel of this family arose from ancestors having 3 carpels with axile placentation, and that two carpels were suppressed, and their placentae moved towards one side of the ovary, and the locules lost by compression, resulting in a unilocular condition. According to Kim and Jansen (1998) the gynoecia of the $n = 6$ clade alone (*Epimedium, Podophyllum, Jeffersonia*) being derived from two carpels. Hutchinson (1973) separated the genera included here under three families: Berberidaceae (including woody genera *Berberis* and *Mahonia* in which anthers open by flaps), Nandinaceae (single woody genus *Nandina* with 2–3 times pinnate compound leaves and anthers opening by slits) and Podophyllaceae (including herbaceous genera). Interestingly, whereas former two were included under order Berberidales, the last family was included under Ranales along with Ranunculaceae, Nymphaeaceae, Ceratophyllaceae, etc. Takhtajan (2009) also segregated Ranzaniaceae and included all the 4 families under order Berberidales. The recent classifications treat them under the same family Berberidaceae which is considered to be monophyletic as supported by morphology and DNA data. The family is placed under order Ranunculales along with Ranunculaceae and other related families in most of the recent systems. *Nandina* is considered to be sister to rest of the family and often included under separate subfamily Nandinoideae, and rest of the genera under Berberidoideae. A number of distinct clades are recognized within Berberidoideae (Loconte, 1993). *Leontice, Gymnospermum* and *Caulophyllum* are characterized by petaliferous nectaries (staminodes), pollen with reticulate sculpturing and basal placentation. Similarly, *Epimedium, Vancouveria* and *Jeffersonia* are distinct in the sense that large fleshy blue seed develops in an exposed condition. Thorne (2003, 2006, 2007) recognizes 4 subfamilies under Berberidaceae: Nandinoideae, Berberidoideae, Leonticoideae and Podophylloideae. APweb (2008) recognized only two, monogeneric Nandinoideae and Berberidoideae including rest of the genera. In Version 14 of APweb (updated 2018), Podophylloideae is recognized as third subfamily (including Leonticoideae of Thorne).

Ranunculaceae M. Adanson ***Buttercup or Crowfoot Family***

43 genera, 2,346 species

Primarily in temperate and boreal regions of the Northern Hemisphere.

Placement:

	B & H	Cronquist	Takhtajan	Dahlgren	Thorne	APG IV/(APweb)
Informal Group						Eudicots
Division		Magnoliophyta	Magnoliophyta			
Class	Dicotyledons	Magnoliopsida	Magnoliopsida	Magnoliopsida	Magnoliopsida	
Subclass	Polypetalae	Magnoliidae	Ranunculidae	Magnoliidae	Ranunculidae	
Series+/ Superorder	Thalamiflorae+		Ranunculanae	Ranunculanae	Ranunculanae	
Order	Ranales	Ranunculales	Ranunculales	Ranunculales	Ranunculales	Ranunculales

Salient features: Herbs, leaves with sheathing base, blade often divided, flowers bisexual, petals with nectary, stamens and carpels numerous, free and spirally arranged, ovary superior, fruit a follicle or achene.

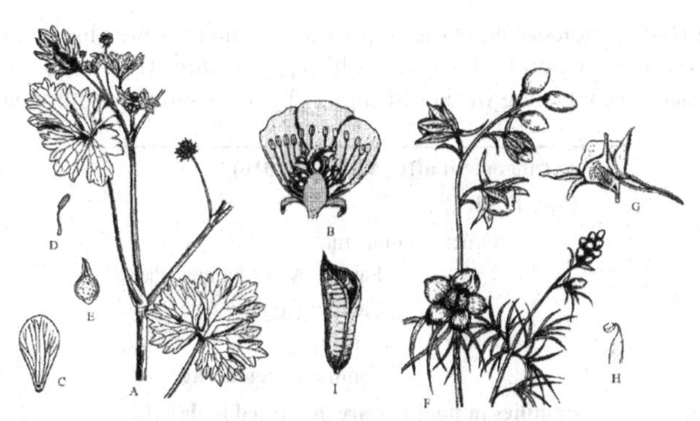

Figure 13.48: Ranunculaceae. *Ranunculus muricatus.* **A:** A portion of plant with flowers and fruits; **B:** Vertical section of flower; **C:** Petal with nectary; **D:** Stamen; **E:** Achene. *Delphinium ajacis.* **F:** A branch with young inflorescence and an expanded inflorescence; **G:** Vertical section of flower; **H:** Stamen; **I:** Dehiscing follicle (A–E, after Sharma and Kachroo, Fl. Jammu, 1983).

Major genera: *Ranunculus* (400 species), *Clematis* (200), *Delphinium* (250), *Aconitum* (245), *Anemone* (150) and *Thalictrum* (100).

Description: Mostly herbs, sometimes woody climbers (*Clematis*), or shrubs (*Xanthorhiza*). Stem with scattered or several rings of vascular bundles. Hairs simple. **Leaves** usually alternate (opposite in *Clematis*), undivided (*Caltha*) palmately lobed (*Ranunculus*) or compound (*Clematis*), stipules absent (present in *Thalictrum*). Tendrils for support may sometimes be formed from petiole (*Clematis*) or terminal leaflet (*Naravelia*). **Inflorescence** of solitary flowers (*Anemone*) or cymose, sometimes racemes (*Delphinium*) or panicles (*Clematis natans*). **Flowers** bracteate (*Clematis*) or ebracteate (*Anemone*) bisexual (unisexual in *Thalictrum*), actinomorphic (zygomorphic in *Delphinium*) with spirally arranged stamens and carpels, hypogynous. **Calyx** with 5 (4 in *Clematis*) or many sepals, free, one (*Delphinium*) or all five (*Aquilegia*) sepals often produced into spur at base. **Corolla** with 5 or many (*Helleborus*) petals, free, often with nectaries or represented only by nectaries (*Delphinium*), sometimes produced into spur which enters the spur formed by sepal, sometimes perianth is not differentiated (*Anemone*, *Helleborus*) into sepals and petals. **Androecium** with many stamens, free, spirally arranged, anthers often extrorse, dehiscence longitudinal. **Gynoecium** with single (subgenus *Consolida* of *Delphinium*) or many free carpels (syncarpous in *Nigella*), unilocular (multilocular in *Nigella*) with single (*Ranunculus*) or many (*Delphinium*) ovules, placentation marginal or basal, rarely axile (*Nigella*), ovary superior, style 1, sometimes feathery (*Clematis*), stigma 1. **Fruit** an achene (*Ranunculus*), follicle (*Delphinium*), berry (*Actaea*) or rarely a capsule (*Nigella*); seed with small embryo, endosperm present. Pollination usually by insects. *Clematis* and *Anemone*, which lack nectaries are pollinated by pollen-gathering insects. *Ranunculus*, *Delphinium*, etc., with nectaries by usually bees. Some species of *Thalictrum* are wind pollinated. Achenes may be provided with hairs for wind dispersal (*Clematis*), with tubercles or hooked spines for dispersal by animals (*Ranunculus*). Berries of *Actaea* are mainly dispersed by birds.

Economic importance: *Delphinium* (Larkspur), *Anemone* (windflower), *Aquilegia* (columbine), *Ranunculus* (buttercup), and *Helleborus* (hellebore) are grown as ornamentals. *Aconitum napellus* yields aconite, whereas *A. ferox* is source of bikh poison. Roots of *Hydrastis* (removed by Takhtajan to Hydrastidaceae) are used for stomach ailments. Seeds of *Nigella sativa* (Nigella, black seed, 'Kalonji') are used as flavouring, medicinally to treat asthma, bronchitis and rheumatism. Thymoquinone extracted from the seeds of this species have recently been found to be useful in treatment of cancer.

Phylogeny: The family is largely considered to be a monophyletic group as supported by morphology and molecular evidence. *Hydrastis*, with 3-merous perianth, vessels with scalariform perforations, ovule with two integuments, and fleshy follicles occupies a unique basal position along with *Glaucidium*, as evidenced by molecular data. Both these genera were removed by Takhtajan (1997) into distinct families Hydrastidaceae and Glaucidiaceae, under Hydrastidales and Glaucidiales, respectively, subsequently (2009) shifting Hydrastidaceae to Ranunculales. Thorne (2003) includes Glaucidiaceae under Paeoniales, but Hydrastidaceae near Ranunculaceae under Ranunculales. Studies based on cpDNA restriction sites and sequence data (Hoot, 1995) suggest that these two genera along with other genera placed in Thalictroideae form basal paraphyletic group, thus justifying retaining all these genera within Ranunculaceae. These basal genera retain plesiomorphies such as presence of berberine, yellow creeping rhizomes, small hairs and small chromosomes, linking them to Berberidaceae. The separation of follicle bearing genera under Helleboraceae by Hutchinson is rejected by the evidence from floral anatomy. The reduction in the number of ovules per carpel and the evolution of achenes has occurred several times within the family. The separation is also negated by nucleotide sequences (Hoot, 1995). The petals with nectary are often considered to represent petaliferous nectaries, the petals being absent. According to Erbar et al. (1999) they are interpreted as being derived from stamens, and that stamens are secondarily spiral. Thorne (2003, 2006) divided family Ranunculaceae into 3 subfamilies: Coptidoideae, Isopyroideae (Thalictroideae in 2007 revision) and Ranunculoideae. Stevens (APweb,

2006) recognized 5, adding Hydrastidoideae and Glaucidioideae, division retained in version 14 (updated 2018). Wang et al. (2009) found strong molecular support for the relationships [*Glaucidium* [*Hydrastis* + rest of Ranunculaceae]], that for [*Hydrastis* + rest of Ranunculaceae] being weakened slightly by the addition of morphological data.

Arrangement after APG IV (2016)

Core Eudicots:

Order 1. Gunnerales

Family: Myrothamnaceae

Family: Gunneraceae

Order 2. Dilleniales

Family: **Dilleniaceae**

Families in boldface are described in detail.

Dilleniaceae Salisbury ***Dillenia Family***

11 genera, 430 species

Mostly found in Tropics and Subtropics and most of Australia.

Placement:

	B & H	Cronquist	Takhtajan	Dahlgren	Thorne	APG IV/(APweb)
Informal Group						Core Eudicots/ (Pentapetalae)
Division		Magnoliophyta	Magnoliophyta			
Class	Dicotyledons	Magnoliopsida	Magnoliopsida	Magnoliopsida	Magnoliopsida	
Subclass	Polypetalae	Dilleniidae	Dilleniidae	Magnoliidae	Caryophyllidae	
Series+/ Superorder	Thalamiflorae+		Dillenianae	Theanae	Dillenianae	
Order	Ranales	Dilleniales	Dilleniales	Dilleniales	Dilleniales	Dilleniales

Salient features: Mostly woody lianas, shrubs or trees, leaves often rough with toothed margin, flowers symmetrical with overlapping sepals, petals crumpled in bud, numerous stamens, free carpels, endosperm copious.

Major genera: *Hibbertia* (175 species), *Dillenia* (60), *Tetracera* (44), *Doliocarpus* (40), and *Davilla* (25).

Description: Woody lianas, shrubs or trees, rarely herbs (*Pachynema*), with distinctive flavanoides; leaves spiral, exstipulate, simple, rarely pinnately divided, surface scabrid, margin toothed, sometimes entire and greatly reduced and winged (*Hibbertia*), lateral veins parallel; flowers often terminal, single or in racemes or panicles, with articulated pedicel, mostly bisexual, rarely unisexual, actinomorphic; sepals mostly 3–5, sometimes more, imbricate, mostly persistent and leathery, sometimes enlarged in fruit (*Dillenia*); petals 3–5, mostly yellow, sometimes white or red, imbricate, crumpled in bud; stamens often many, centrifugal, free or united in fascicles, anthers basifixed with longitudinal dehiscence or with apical pores, flowers of *Didesmandra aspera* have two bundles of stamens on the functionally upper side of the flower, in each there is a single fertile stamen longer than the rest, on the other hand monosymmetric flowers of *Schumacheria* have only a single staminal bundle in which all stamens are about the same length; carpels usually many, free or slightly fused at base, ovary superior, styles free, stigma capitate or peltate, ovules 1 or more, anatropous; fruit a follicle, berry or capsule, seeds with copious endosperm, embryo small, usually arillate, straight; seed dispersal by birds and monkeys (*Dillenia*), myrmecochorous in *Hibbertia*.

Economic importance: A few species of *Dillenia* are useful for their timber and as a source of tannin. *D. indica*, a tree native to Southeast Asia but widely planted elsewhere, is valued for its scented flowers and lemon-flavored fruits used in jellies and curries. Fruits of other species of the genus have similar uses. Several species of *Hibbertia* are grown as ornamentals, especially *H. scandens*, a woody vine with yellow ill-smelling flowers, which is grown mainly in warm areas or in greenhouses.

Phylogeny: The position Dilleniaceae is quite uncertain. Formerly placed in Ranales complex closer to Ranunculaceae and Magnoliaceae, it has more recently been grouped under separate Subclass Dilleniidae, superorder Dillenianae, order Dilleniales. Analysis of plastid loci rbcL, infA, rps4, and the rpl16 intron using maximum parsimony and Bayesian methods resolve *Tetracera*, the only pantropical genus in the family, as sister to all other Dilleniaceae (Horn, 2009). APG III (2009) considered it unplaced under Core Eudicots near Caryophyllales but shifted it to Core Eudicots order Dilleniales in APG IV (2016). APWeb (2018) is uncertain about position of this family (under order Dilleniales) but places it closer to Saxifragates and Vitales under informal group Pentapetalae. The family is divided by APWeb into 4 subfamilies: Delimoideae (lianas,

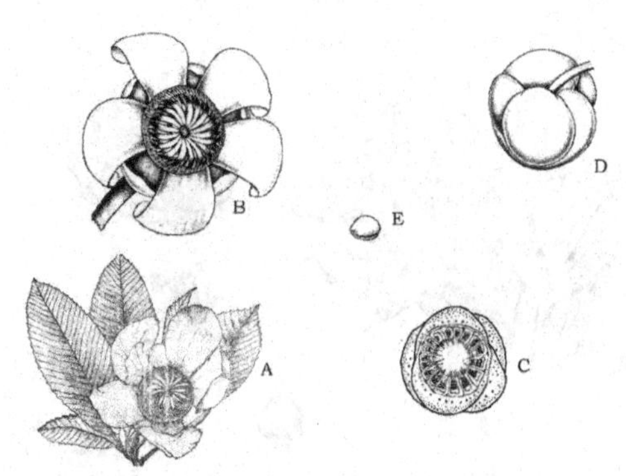

Figure 13.49: Dilleniaceae. *Dillenia indica.* **A:** Flowering branch; **B:** Flower; **C:** Section of fruit surrounded by calyx; **D:** Fruit; **E:** Seed.

vessel elements with simple perforation plates, stomata paracytic, plant functionally dioecious, filament apex expanded, stigma peltate; 1 genus *Tetracera*) Doliocarpoideae (lianas, small vessel elements with scalariform perforation plates, carpels 1–3, stigma peltate-infundibular, ovules 2 per carpel, calyx thin accrescent in fruit; 5 genera), Hibbertioideae (leaves reduced, stem photosynthetic, leaf margin entire, tertiary venation not scalariform, leaf base not sheathing, ovule 1 per carpel; 1 genus *Hibbertia*) and Dillenioideae (leaves deciduous, blade pinnate in *Acrotrema*, carpels often connate with a central receptacular cone, ovule one per carpel; 4 genera). Dilleniaceae may be an ancient clade that expresses some phylogenetic relation between the higher Eudicots and the rather more primitive groups. It is estimated that the clade diverged around 115 million years ago in Mid Cretaceous, but the crown group was formed much later—only 52 million years before the present.

Arrangement after APG IV (2016)
Superrosids:

Order 1. Saxifragales
- Family: Peridiscaceae
- Family: **Paeoniaceae**
- Family: Altingiaceae
- Family: Hamamelidaceae
- Family: Cercidiphyllaceae
- Family: Daphniphyllaceae
- Family: Iteaceae
- Family: Grossulariaceae
- Family: **Saxifragaceae**
- Family: Crassulaceae
- Family: Aphanopetalaceae
- Family: Tetracarpaeaceae
- Family: Penthoraceae
- Family: Haloragaceae
- Family: Cynomoriaceae

Families in boldface are described in detail.

Paeoniaceae Rafinesque *Peony Family*

1 genus, 33 species

Mainly temperate regions of Asia and Europe, also in Northwest America.

Placement:

	B & H	**Cronquist**	**Takhtajan**	**Dahlgren**	**Thorne**	**APG IV/(APweb)**
Informal Group						Superrosids/ (Pentapetalae)
Division		Magnoliophyta	Magnoliophyta			
Class	Dicotyledons	Magnoliopsida	Liliopsida	Magnoliopsida	Magnoliopsida	
Subclass	Polypetalae	Dilleniidae	Ranunculidae	Magnoliidae	Ranunculidae	
Series+/ Superorder	Thalamiflorae+		Ranunculanae	Theanae	Ranunculanae	
Order	Ranales	Dilleniales	Paeoniales	Paeoniales	Paeoniales	Saxifragales

B & H under family Ranunculaceae.

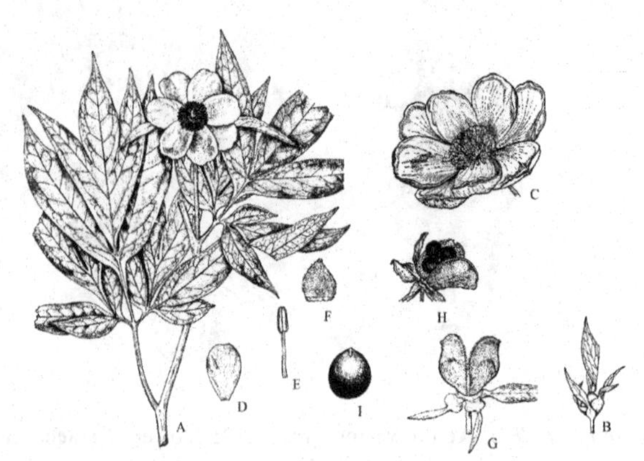

Figure 13.50: Paeoniaceae. *Paeonia emodi.* **A:** Branch with flower; **B:** Flower bud with leafy bracts; **C:** Flower magnified to show numerous stamens; **D:** Petal; **E:** Stamen; **F:** Ovary covered with hairs; **G:** Follicle splitting; **H:** Follicle dehisced to expose seeds; **I:** Seed with aril.

Salient features: Perennial rhizomatous herbs or shrubs, leaves alternate, compound or lobed, without stipules, flowers large, bisexual, sepals 5, green and leathery, petals 5–10, colored, stamens many, centrifugally arranged, carpels 5, free, ovules many, fruit etaerio of follicles.

Genus: Single genus *Paeonia* (33 species).

Description: Perennial herbs or soft shrubs, with tubers or rhizomes, stem base covered with scale-like sheaths. **Leaves** alternate, petiolate, pinnately to ternately compound to highly dissected or lobed, stipules absent. **Inflorescence** usually with solitary flowers with leafy bracts at base. **Flowers** Large, showy, bisexual, hypogynous, almost globular in appearance. **Calyx** with 5 sepals, free, green, unequal, imbricate, subfoliaceous, persistent. **Corolla** with 5 petals, sometimes 6–10, large, free, orbicular, subequal, imbricate. **Androecium** with numerous stamens, centrifugal, attached to fleshy disc present around the carpels, free, spirally arranged, bithecous, basifixed, dehiscence by longitudinal slits, extrorse. **Gynoecium** with 5 carpels, sometimes up to 2, borne on fleshy disc, free, fleshy, ovary superior, unilocular, ovule 2-many, placentation marginal, stigma sessile, thick, falcate, 2-lipped. **Fruit** an etaerio of leathery follicles, dehiscing by adaxial suture, seeds globose, with aril, red turning black at maturity, with prominent umbilicus, embryo small, endosperm copious.

Economic importance: The family contributes many ornamentals cultivated for attractive flowers. The flowers of *Paeonia officinalis* may reach 15 cm in diameter.

Phylogeny: The genus *Paeonia* was once included under family Ranunculaceae, from which, however, it is distinct in having 5 large chromosomes, centrifugal (and not centripetal) stamens, persistent sepals, disc and seeds with aril. The separation of the genus into a distinct family was first advocated by Worsdell (1908) based on anatomical evidence. Corner (1946) considered the centrifugal development of stamens of considerable importance in phylogeny and advocated placing Paeoniaceae near Dilleniaceae, a placement followed by Cronquist (1981, 1988), but not supported by Hutchinson, who in 1969 placed Paeoniaceae before Helleboraceae under order Ranales. The placement of Paeoniaceae near Dilleniaceae is also contradicted by difference in gynoecial development, nectary morphology (Stevens in Apweb, 2008). Hutchinson considered *Paeonia* to be a link between the Magnoliaceae and Helleboraceae, but much more closely related to latter. *Paeonia* was linked with moderate support to the Crassulaceae clade, or, more weakly, with the Crassulaceae + Saxifragaceae clades in some analyses in Fishbein et al. (2001) and accordingly placed under order Saxifragales (Core Eudicots) by APG II and APweb. Paeoniaceae and another monogeneric family Glaucidiaceae are often considered related. They are together placed in the same order Paeoniales by Dahlgren (1989) and Thorne (2003, 2006, 2007), whereas as Takhtajan (1997, 2009), places them under two adjacent orders Paeoniales and Glaucidiales under superorder Ranunculanae of subclass Ranunculidae. Mabberley (1997) includes *Glaucidium* in Paeoniaceae. Hoot et al. (1998) included *Glaucidium* and *Hydrastis* under Ranunculaceae, being sister to rest of the family, a treatment followed in APG II, APG III (2009), APG IV (2016) and APweb. Within Saxifragales relationships other than the Saxifragaceae/Crassulaceae clade ("S.-C. clade") have been unclear for sometime, and even now many of the deeper nodes remain poorly supported (Soltis et al., 2013). According to studies of Jian et al. (2006) and Soltis et al. (2011) Paeoniaceae are sister to the S.-C. clade.

Saxifragaceae A. L. de Jussieu *Saxifrage Family*

33 genera, 525 species

Widespread but best represented in the Northern Hemisphere, mainly in the temperate and arctic climate.

Placement:

	B & H	**Cronquist**	**Takhtajan**	**Dahlgren**	**Thorne**	**APG IV/(APweb)**
Informal Group						Superrosids/ (Pentapetalae)
Division		Magnoliophyta	Magnoliophyta			
Class	Dicotyledons	Magnoliopsida	Magnoliopsida	Magnoliopsida	Magnoliopsida	
Subclass	Polypetalae	Rosidae	Rosidae	Magnoliidae	Hamamelididae	
Series+/ Superorder	Calyciflorae+		Rosanae	Rosanae	Hamamelidanae	
Order	Rosales	Rosales	Saxifragales	Saxifragales	Saxifragales	Saxifragales

B & H as Saxifrageae.

Salient features: Perennial herbs, leaves alternate, gland-toothed, stipules absent, flowers actinomorphic, usually perigynous, sepals and petals 5 each, stamens 5 to 10, carpel 2, united, ovary superior, placentation axile, fruit a capsule.

Major genera: *Saxifraga* (310 species), *Heuchera* (50), *Chrysoplenium* (45), *Mitella* (18), *Astilbe* (18) and *Bergenia* (6).

Description: Perennial herbs, vessel elements with simple perforations, often with tannins, sometimes cyanogenic. **Leaves** alternate, usually in basal rosette, simple or pinnately or palmately compound, venation pinnate or palmate, reticulate, stipules absent or represented by expanded margins of petiole base. **Inflorescence** racemose or cymose, rarely with solitary flowers. **Flowers** bisexual, rarely unisexual (plants monoecious or dioecious), actinomorphic, rarely zygomorphic, usually perigynous with distinct hypanthium, rarely epigynous. **Calyx** usually with 5 sepals, rarely 4, free or connate, often persistent. **Corolla** usually with 5 petals, free, often clawed, imbricate or convolute, sometimes reduced or absent. **Androecium** with 5 to 10 stamens, free, anthers bithecous, dehiscence longitudinal, pollen grains tricolporate. **Gynoecium** with usually 2 carpels, rarely up to 5, united, free or adnate to hypanthium, ovary superior or inferior, placentation axile or parietal, ovules numerous, styles free, stigmas free, capitate. **Fruit** a septicidal capsule or follicle; seed with small straight embryo surrounded by endosperm. Pollination mainly by insects. Seeds dispersed by wind or passing animals.

Economic importance: The family has little economic importance with a few genera *Saxifraga*, and *Astilbe* grown in rock gardens or perennial borders.

Phylogeny: The family was earlier broadly circumscribed to include genera, which have now been separated to different families such as Grossulariaceae (*Ribes*), Hydrangeaceae (*Hydrangea*—separated by Thorne (2003) to Asteridae—> Cornanae—> Hydrangeales; Asterids—> Cornales in APG II), Parnassiaceae (*Parnassia*—under Rosidae—> Celastranae—> Celastrales by Thorne; Eurosids I—> Celastrales by APG II and APweb), etc. Hydrangeaceae are woody, tenuinucellate and unitegmic and related to Asterids. Similarly, separation of Parnassiaceae is in agreement with data from floral anatomy (Bensel and Palser 1975b, c). The family has long been considered as closely related to Rosaceae.

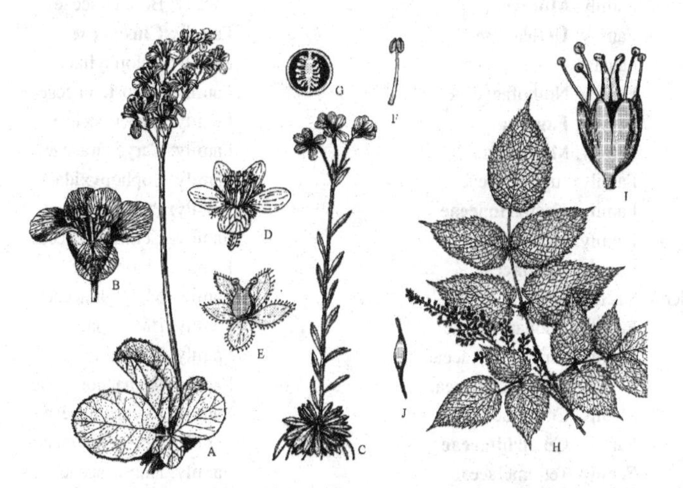

Figure 13.51: Saxifragaceae. *Bergenia ciliata.* **A:** Plant with basal leaves and corymbose panicle carried on a long scape; **B:** Flower with petals distinctly larger than nearly cup-shaped calyx. *Saxifraga flagellaris.* **C:** Plant with thick stolons, small leaves and few flowers; **D:** Flower with almost free sepals, both pedicel and sepals glandular. **E:** Flower with calyx and gynoecium, petals and stamens removed; **F:** Stamen; **G:** Transverse section of ovary with axile placentation. *Astilbe rivularis.* **H:** Portion of bipinnate leaf and a paniculate inflorescence alongside; **I:** Flower lacking petals, with 5 stamens and 2 carpels; **J:** Seed, tailed at both ends.

Astilbe of Saxifragaceae can be confused with *Aruncus* of Rosaceae but the former quite often have opposite leaves, their carpels are usually two and connate at the base, and their stamens are fewer. These resemblances are mainly superficial. There are two major clades in Saxifragaceae, *Saxifraga* s. str. and the *Heuchera* clade, members of the latter containing the bulk of the floral variation in the family (Soltis et al., 2001). Generic limits are unclear; hybridization is extensive and there are various combinations of chloroplast and nuclear genomes. For example, the chloroplast genome of *Tellima* is also found in *Mitella* (Soltis et al., 1993). However, the unitegmic *Darmera* with scapigerous inflorescence is properly to be retained in Saxifragaceae (Gornall, 1989). Thorne had earlier placed Saxifragales under Rosidae but has subsequently (2003, 2007) shifted it to newly created subclass Hamamelididae. The family is monophyletic as evidenced by data from cpDNA restriction sites, *rbcL, matK,* and 18S sequences and morphology. The members, in addition share an *rpl2* intron deletion. Recent studies have shown that genera like *Saxifraga* and *Mitella* are not monophyletic. In addition, hybridization often causes taxonomic problems. Thorne (2006, 2007) recognizes 2 subfamilies: Astilboideae and Saxifragoideae, APWeb (2018) preferring name Heucheroideae for the former.

Arrangement after APG IV (2016)

Rosids:

Order 1. Vitales
- Family: Vitaceae

Order 2. Zygophyllales
- Family: **Zygophyllaceae**
- Family: Krameriaceae

Order 3. Fabales
- Family: Quillajaceae
- Family: **Fabaceae**
- Family: Surianaceae
- Family: Polygalaceae

Order 4. Rosales
- Family: **Rosaceae**
- Family: Barbeyaceae
- Family: Dirachmaceae
- Family: Elaeagnaceae
- Family: **Rhamnaceae**
- Family: **Ulmaceae**
- Family: Cannabaceae
- Family: **Moraceae**
- Family: **Urticaceae**

Order 5. Fagales
- Family: Nothofagaceae
- Family: **Fagaceae**
- Family: Myricaceae
- Family: Juglandaceae
- Family: **Casuarinaceae**
- Family: Ticodendraceae
- Family: **Betulaceae**

Order 6. Cucurbitales
- Family: Apodanthaceae
- Family: Anisophylleaceae
- Family: Corynocarpaceae
- Family: Coriariaceae
- Family: **Cucurbitaceae**
- Family: Tetramelaceae
- Family: Datiscaceae
- Family: Begoniaceae

COM Clade:

Order 1. Celastrales
- Family: Lepidobotryaceae
- Family: **Celastraceae**

Order 2. Oxalidales
- Family: Huaceae
- Family: Connaraceae
- Family: **Oxalidaceae**
- Family: Cunoniaceae
- Family: Elaeocarpaceae
- Family: Cephalotaceae
- Family: Brunelliaceae

Order 3. Malpighiales
- Family: Pandaceae
- Family: Irvingiaceae
- Family: Ctenolophonaceae
- Family: Rhizophoraceae
- Family: Erythroxylaceae
- Family: Ochnaceae
- Family: Bonnetiaceae
- Family: **Clusiaceae**
- Family: Calophyllaceae
- Family: Podostemaceae
- Family: Hypericaceae
- Family: Caryocaraceae
- Family: Lophopyxidaceae
- Family: Putranjivaceae
- Family: Centroplacaceae
- Family: Elatinaceae
- Family: Malpighiaceae
- Family: Balanopaceae
- Family: Trigoniaceae
- Family: Dichapetalaceae
- Family: Euphroniaceae
- Family: Chrysobalanaceae
- Family: Humiriaceae

(...Malpighiales Continued)

Families in boldface are described in detail.

Arrangement after APG IV (2016)

Order 3. Malpighiales (continued)
- Family: Achariaceae
- Family: **Violaceae**
- Family: Goupiaceae
- Family: Passifloraceae
- Family: Lacistemataceae
- Family: **Salicaceae**
- Family: Peraceae
- Family: **Rafflesiaceae**
- Family: **Euphorbiaceae**
- Family: Linaceae
- Family: Ixonanthaceae
- Family: Picrodendraceae
- Family: Phyllanthaceae

Rosids (continued):

Order 7. Geraniales
- Family: **Geraniaceae**
- Family: Francoaceae

Order 8. Myrtales
- Family: Combretaceae
- Family: **Lythraceae**
- Family: **Onagraceae**
- Family: Vochysiaceae
- Family: **Myrtaceae**
- Family: Melastomataceae
- Family: Crypteroniaceae
- Family: Alzateaceae
- Family: Penaeaceae

Order 9. Crossosomatales
- Family: Aphloiaceae Takht.
- Family: Geissolomataceae
- Family: Strasburgeriaceae
- Family: Staphyleaceae
- Family: Guamatelaceae
- Family: Stachyuraceae
- Family: Crossosomataceae

Order 10. Picramiales
- Family: Picramiaceae

Order 11. Huerteales
- Family: Gerrardinaceae
- Family: Petenaeaceae
- Family: Tapisciaceae
- Family: Dipentodontaceae

Order 12. Sapindales
- Family: Biebersteiniaceae
- Family: Nitrariaceae
- Family: Kirkiaceae
- Family: Burseraceae
- Family: **Anacardiaceae**
- Family: **Sapindaceae**
- Family: **Rutaceae**
- Family: Simaroubaceae
- Family: **Meliaceae**

Order 13. Malvales
- Family: Cytinaceae
- Family: Muntingiaceae
- Family: Neuradaceae
- Family: **Malvaceae**
- Family: Sphaerosepalaceae
- Family: Thymelaeaceae
- Family: Bixaceae
- Family: Cistaceae
- Family: Sarcolaenaceae
- Family: **Dipterocarpaceae**

Order 14. Brassicales
- Family: Akaniaceae
- Family: Tropaeolaceae
- Family: Moringaceae
- Family: Caricaceae
- Family: Limnanthaceae
- Family: Setchellanthaceae
- Family: Koeberliniaceae
- Family: Bataceae
- Family: Salvadoraceae
- Family: Emblingiaceae
- Family: Tovariaceae
- Family: Pentadiplandraceae
- Family: Gyrostemonaceae
- Family: Resedaceae
- Family: **Capparaceae**
- Family: **Cleomaceae**
- Family: **Brassicaceae**

Families in boldface are described in detail.

Zygophyllaceae R. Brown ***Creosote Bush Family***

22 genera, 285 species

Widespread in tropics and subtropics, mainly in arid regions.

Placement:

	B & H	Cronquist	Takhtajan	Dahlgren	Thorne	APG IV/(APweb)
Informal Group						Rosids/(Rosids/ Fabid/Rosid I)
Division		Magnoliophyta	Magnoliophyta			
Class	Dicotyledons	Magnoliopsida	Magnoliopsida	Magnoliopsida	Magnoliopsida	
Subclass	Polypetalae	Rosidae	Rosidae	Magnoliidae	Rosidae	
Series+/ Superorder	Disciflorae+		Geranianae	Rutanae	Geranianae	
Order	Geraniales	Sapindales	Zygophyllales	Geraniales	Zygophyllales	Zygophyllales

Salient features: Leaves usually opposite, pinnate compound, stipules paired persistent, flowers with disc, stamen with gland or appendage at base, ovary 4–5 locular, syle 1.

Major genera: *Roepera* (60 species), *Zygophyllum* (50), *Fagonia* (40), *Tetraena* (40), *Tribulus* (20), *Balanites* (20), *Guaiacum* (6), and *Larrea* (5).

Description: Herbs (*Tribulus*), shrubs or trees (*Guaiacum*) with often jointed nodes, xylem elements arranged in horizontally aligned tier, with steroidal or triterpenoid saponins and alkaloids. **Leaves** opposite, 2-ranked, rarely alternate, usually paripinnate, rarely simple or 2-foliate, strongly resinous, leaflets entire, venation reticulate, pinnate or palmate, stipules paired, commonly spiny. **Inflorescence** cymose, sometimes reduced to single flower. **Flowers** bisexual, rarely unisexual (*Neoleuderitzia*), actinomorphic, rarely zygomorphic, hypogynous, usually pentamerous. **Calyx** with 5 sepals, rarely 4, free or slightly connate at base. **Corolla** with 5 petals, rarely 4 or absent, free, often clawed, imbricate. **Androecium** with 10 stamens, rarely 15, usually in whorls of 5, outer whorl opposite the petals, free, each filament with a gland or appendage at base, anthers bithecous, basifixed, dehiscence by longitudinal slits, pollen grains usually tricolporate. **Gynoecium** with 5 united carpels, rarely 2–6, ovary superior, usually furrowed or winged, placentation axile, locules as many, ovules one to many in each locule, pendulous, anatropous or orthotropous, style 1, short, stigma capitate or lobed, nectar disc present at the base of ovary. **Fruit** a usually a capsule, septicidal or loculicidal, rarely schizocarpic, berry or drupe, sometimes winged; seeds usually with aril, embryo curved or straight, endosperm usually absent or scanty. Pollination by insects. Arillate seeds (*Guaiacum*) are dispersed by birds, schizocarpic winged fruits by wind and spiny fruits (*Tribulus*) by exozoochory.

Economic importance: The family is of minor economic importance. Wood of *Guaiacum officinale* (lignum vitae), being the strongest and heaviest wood, is highly prized timber in tropical Central America and West Indies. The tree also yields medicinal resin guaiacum, once used to treat syphilis. Species of *Bulsenia* (*B. arborea*: Maracaibo lignum vitae; *B. sarmienti*: Paraguay lignum vitae) yield valuable timber and perfume oil. Seeds of *Peganum harmala* are the source of dye turkey red. *Tribulus terrestris* is a troublesome weed whose spines on fruit are similar to sharp iron caltrops once used in battlefields to stab the feet of men and horses. They also often puncture cycle

Figure 13.52: Zygophyllaceae. *Tribulus terrestris.* **A:** Part of plant with flowers and fruits; **B:** Flower; **C:** Vertical section of flower; **D:** Flower with sepals and petals removed; **E:** Ovary covered with hairs, short style and stigma; **F:** Transverse section of ovary; **G:** Fruit; **H:** One of the cocci enlarged showing sharp spines (A–B, D–E, after Sharma and Kachroo, Fl. Jammu, 1983).

tyres; hence the names caltrops, puncture vine and goat head for the weed. Species of *Zygophyllum* are used as spices: buds of *Z. fabago* used in sauces and fruits of *Z. coccinium* as substitute for black pepper.

Phylogeny: The family is usually placed under order Geraniales, although more recently shifted to order Zygophyllales (Takhtajan, 1997; Thorne, 2007; Stevens, 2008). The family is considered to be monophyletic after the exclusion of a few genera to Peganaceae and Nitrariaceae (Thorne to order Rutales) or Nitrariaceae (Sapindales of Eurosids II by APG II). Monophyly of Zygophyllaceae supported by morphology and DNA characters. The family is sister to Krameriaceae as supported by *rbc*L sequences (Soltis et al., 1998; Savolainen et al., 2000). The family is divided into 5 subfamilies: Morkillioideae, Tribuloideae, Seetzenioideae, Larreoideae and Zygophylloideae, a classification suggested by Sheahan and Chase (2000) and followed by APweb (2018). *Balanites* is very different from other Zygophyllaceae in floral, vegetative and seed anatomy, although tentatively included under Tribuloideae. Hilu et al. (2003) reported *Larrea* to be weakly associated with Fabaceae in their *rbc*L analysis; they note that the possession of anthroquinones is a possible synapomorphy between Zygophyllaceae and the N-fixing clade Fabaceae. Bellstedt et al. (2008) suggested that repeated radiations from the horn of Africa to southern Africa and Asia and back lead to the present distribution of the taxa in the subfamily Zygophylloideae. Genus *Zygophyllum,* previously contained nearly 100 species which did not form a monophyletic group and as such was split to establish *Roepera*, *Tetraena* and few more genera.

Fabaceae Lindley ***Bean or Pea Family***

(= **Leguminosae** A. L. de Jussieu)

765 genera, 19,580 species (Third largest family after Orchidaceae and Asteraceae) according to estimate of LPWG (2017). Cosmopolitan in distribution, primarily in warm temperate regions.

Placement:

	B & H	**Cronquist**	**Takhtajan**	**Dahlgren**	**Thorne**	**APG IV/(APweb)**
Informal Group						Rosids/(N-Fixing clade)
Division		Magnoliophyta	Magnoliophyta			
Class	Dicotyledons	Magnoliopsida	Magnoliopsida	Magnoliopsida	Angiospermae	
Subclass	Polypetalae	Rosidae	Rosidae	Magnoliidae	Rosidae	
Series+/ Superorder	Calyciflorae+		Fabanae		Rutanae	Rosanae
Order	Rosales	Fabales	Fabales	Fabales	Fabales	Fabales

B & H as Leguminosae **Takhtajan, Thorne, APG IV** and **APweb** as Fabaceae.

Cronquist and **Dahlgren** recognize three independent families Fabaceae, Caesalpiniaceae and Mimosaceae, thus restricting the name Fabaceae to include only papilionaceous (members of Papilionoideae) members, for which the alternate name is Papilionaceae and not Leguminosae.

This large family has traditionally been divided into three subfamilies Papilionoideae (Faboideae), Caesalpinioideae and Mimosoideae. These have been recognized as independent families Fabaceae (Papilionaceae), Caesalpiniaceae and Mimosaceae in several recent systems of classification, a trend that tends to be reversing in last decade or so. It must be noted that name Fabaceae is valid for family sensu lato as well as for Papilionoideae upgraded as family. Leguminosae is the alternate name only for former whereas Papilionaceae is the alternate name for latter. Common features of the family include leaves usually compound with pulvinate base, odd sepal anterior, flowers perigynous, carpel 1 with marginal placentation and fruit commonly a pod or lomentum. Revision of the family by The Legume Phylogeny Working Group (LPWG) in 2017 brought about major changes recognizing 6 subfamilies: Cercidoideae (12 genera, 335 species), Detarioideae (84, 760), Duparquetioideae (1, 1), Dialioideae (17, 85), recircumscribed Caesalpinioideae (148, 4400, incl. Mimosoideae) and Papilionoideae (503, 14000), adopted by APweb (2018), preferring Faboideae for Papilionoideae.

Subfamily Faboideae DC. (=**Papilionoideae** L. ex A. DC.)

B & H as Papilionoideae **Takhtajan, Thorne, APG III** and **APweb** as Faboideae
Cronquist and **Dahlgren** as family Fabaceae (Papilionaceae).

503 genera, 14,800 species

Cosmopolitan in distribution, primarily in warm temperate regions.

Salient features: Leaves usually pinnate compound with pulvinate base, flowers zygomorphic, corolla papilionaceous corolla, sepals united, odd sepal anterior, stamens 10, usually diadelphous (1+ (9)), carpel 1, ovary superior, fruit a pod.

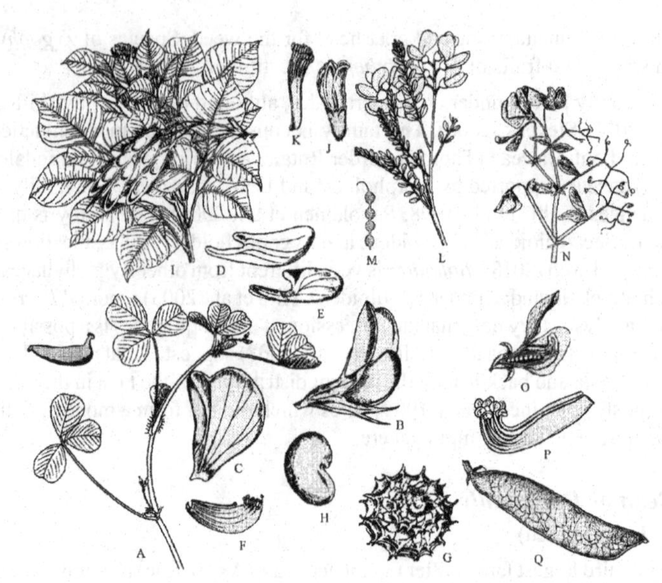

Figure 13.53: Fabaceae, subfamily **Faboideae.** *Medicago polymorpha.* **A:** Portion of plant with trifoliate leaves, laciniate stipules and few flowered axillary clusters on long peduncles; **B:** Flower; **C:** Standard; **D:** Wing; **E:** Keel; **F:** Androecium with diadelphous (1 free, 9 with united filaments) stamens; **G:** Fruit covered with tubercles; **H:** Seed. *Dalbergia sissoo.* **I:** Flowering shoot with a fruiting twig; **J:** Flower; **K:** Androecium with 9 monadelphous stamens. *Sophora mollis.* **L:** Branch with flowers; **M:** Moniliform pod. *Lathyrus odoratus.* **N:** Portion of a flowering branch, upper leaflets modified into tendrils; **O:** Vertical section of flower; **P:** Diadelphous andoecium; **Q:** Pod.

Major genera: *Astragalus* (2000 species), *Indigofera* (700), *Crotalaria* (600), *Desmodium* (400), *Tephrosia* (400), *Trifolium* (300), *Dalbergia* (200), *Lathyrus* (150), *Milletia* (100) and *Lotus* (100).

Description: Trees (*Dalbergia, Erythrina*), shrubs (*Tephrosia, Alhagi, Indigofera*) or herbs (*Medicago, Melilotus*), sometimes woody climbers (*Wisteria*), commonly with root nodules. **Leaves** alternate, pinnately (*Pisum, Vicia*) or palmately compound (*Trifolium*), sometimes simple (*Alysicarpus, Alhagi*), whole leaf (*Lathyrus aphaca*) or upper leaflets (*Vicia, Pisum*) sometimes modified into tendrils, leaf base (sometimes also the base of leaflets) pulvinate, stipules present. **Inflorescence** racemose, in racemes, heads (*Trifolium*) or spikes (*Ononis*), sometimes in clusters (*Lotus, Caragana*). **Flowers** bracteate (bracts often caducous), bisexual, zygomorphic, perigynous. **Calyx** with 5 sepals, more or less united, usually campanulate, odd sepal anterior. **Corolla** with 5 petals, free, papilionaceous consisting of *a posterior* standard or vexillum, two lateral wings or alae and two anterior petals fused along margin to form keel or carina which encloses stamens and pistil, posterior petal outermost. **Androecium** with 10 stamens, diadelphous (1 posterior free and filaments of nine fused into a tube which is open posteriorly), sometimes 5+5 as in *Smithia*, rarely monadelphous (*Ononis*), or free (*Sophora, Thermopsis*) anthers bithecous, dehiscence longitudinal. **Gynoecium** with a single carpel, unilocular with many ovules, placentation marginal, ovary superior, style single, curved. **Fruit** a legume or pod, rarely a lomentum (*Desmodium*), sometimes indehiscent (*Melilotus*), rarely spirally coiled (*Medicago*); seeds 1-many, seed coat hard, endosperm minute or absent, food reserves in cotyledons. Pollination primarily by insects, mostly bees. Dispersal is commonly by wind, but often exozoochorus (*Medicago*), or by mammals (*Tamarindus*).

Economic importance: The subfamily is of major economic importance, ranking second to Poaceae. It is the source of several pulse crops such as kidney bean (*Phaseolus vulgaris*), green gram (*P. aureus*), black gram (*P. mungo*), lentil (*Lens esculenta*), chick pea (*Cicer arietinum*), pea (*Pisum sativum*) and pigeon pea (*Cajanus cajan*). Soybean (*Glycine max*) and peanut (*Arachis hypogaea*) yield oil and high-protein food. Indigo dye is obtained from *Indigofera tinctoria*. The seeds of *Abrus precatorius* are used in necklaces and rosaries but are extremely poisonous and can be fatal if ingested. The important fodder plants include alfalfa (*Medicago sativa*) and clover (*Trifolium*). Common ornamentals include lupin (*Lupinus*), sweet pea (*Lathyrus odoratus*), Wisteria (*Wisteria*), *Laburnum*, coral tree (*Erythrina*), false acacia (*Robinia*) and broom (*Cytisus*).

Subfamily Caesalpinioideae DC.

B & H, Takhtajan, Thorne, APG III and **APweb** as Caesalpinioideae
Cronquist and **Dahlgren** as family Caesalpiniaceae.

150 genera, 2,700 species (excluding now merged Mimosoideae)

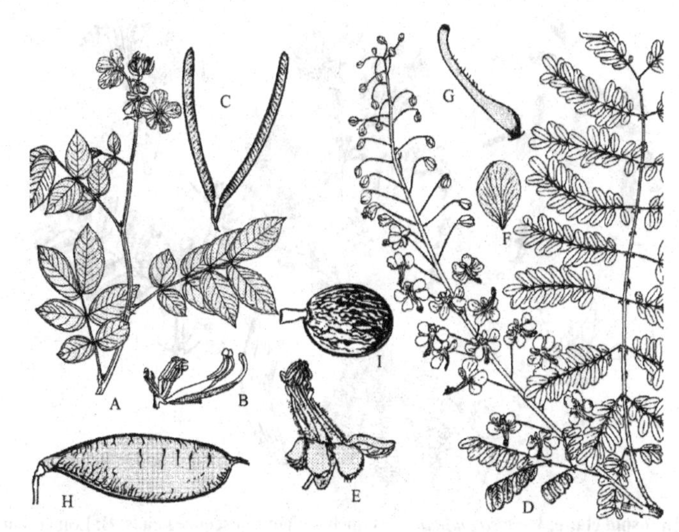

Figure 13.54: Fabaceae, subfamily **Caesalpinioideae.** *Cassia occidentalis.* **A:** Portion of plant with flowers and paripinnate leaves; **B:** Flower with sepals and petals removed, showing gynoecium and stamens of three different sizes; **C:** A pair of pods. *Caesalpinia decapetala.* **D:** Portion of plant with bipinnate leaves and racemose inflorescence; **E:** Flower; **F:** One of the four large petals; **G:** Gynoecium; **H:** Pod; **I:** Seed.

Distributed mainly in tropics and subtropics, a few species in the temperate regions.

Salient features: Trees, shrubs or herbs, leaves usually pinnate compound with pulvinate base, flowers zygomorphic corolla not papilionaceous, posterior petal innermost, sepals free, odd sepal anterior, stamens 10, usually free, in two whorls, ovary superior, carpel 1, fruit a pod.

Major genera: *Chamaecrista* (260 species), *Senna* (250), *Caesalpinia* (120) and *Cassia* (30).

Description: Trees (*Delonix*), shrubs (*Cassia occidentalis*) or herbs (*Cassia obtusa*), rarely woody climbers (*Pterolobium*). **Leaves** alternate, pinnately or palmately compound, sometimes simple (*Bauhinia*), leaf base (sometimes also the base of leaflets) pulvinate, stipules present. **Inflorescence** racemose, in racemes or spikes (*Dimorphandra*). **Flowers** bracteate (bracts usually caducous) bisexual, zygomorphic, perigynous. **Calyx** with 5 sepals, rarely 4 (*Amherstia*), free or rarely connate (*Bauhinia*), odd sepal anterior. **Corolla** with 5 petals, free, not papilionaceous, posterior petal innermost. **Androecium** with 10 stamens, sometimes lesser, rarely more, free, sometimes unequal in size (*Cassia*), anthers bithecous, dehiscence longitudinal or by apical pores. **Gynoecium** with a single carpel, unilocular with many ovules, placentation marginal, ovary superior, style single, curved. **Fruit** a legume or pod, rarely a lomentum; seeds 1-many, seed coat hard, endosperm minute or absent, food reserves in cotyledons.

Economic importance: The subfamily includes several ornamentals such as pride of Barbados (*Caesalpinia pulcherrima*), paulo verde (*Parkinsonia*), red bud (*Cercis canadensis*), Gul-mohar (*Delonix regia*), and several species of *Cassia* and *Senna*. Many species of *Senna* are cultivated for leaves that yield drug senna. The heartwood of *Haematoxylon campechianum* (logwood) yields the dye hematoxylin.

Mimosoid clade of Caesalpinioideae (= Mimosoideae DC)

B & H, Takhtajan, Thorne, APG III and **APweb** as Mimosoideae

Cronquist and **Dahlgren** as family Mimosaceae. Mostly merged with Caesalpinioideae in LPWG Classification, 2017; APWeb, 2018.

40 genera, 2,500 species

Distributed mainly in tropical and subtropical regions.

Salient features: Trees, shrubs or herbs, leaves usually pinnate compound with pulvinate base, flowers actinomorphic, corolla not papilionaceous, petals valvate, sepals united, odd sepal anterior, stamens 4-many, free or connate, filaments often long exserted and showy, ovary superior, carpel 1, fruit a pod or lomentum.

Major genera: *Acacia* (1300 species), *Mimosa* (500), *Inga* (250), *Pithecellobium* (170), *Calliandra* (150) and *Albizia* (150).

Description: Trees (*Acacia, Albizia*), shrubs (*Calliandra*) or herbs (*Mimosa pudica*), rarely climbers (*Entada*), or aquatic plants (*Neptunia*). **Leaves** alternate, pinnately or palmately compound, sometimes simple, leaf base (sometimes also the

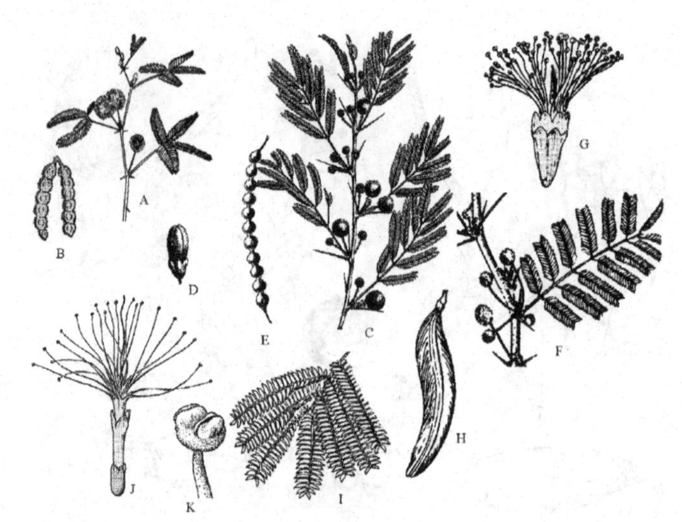

Figure 13.55: Fabaceae, Mimosoid clade. *Mimosa pudica.* **A:** Branch with inflorescence heads; **B:** Lomentum fruits constricted between and splitting into 1-seeded segments. *Acacia nilotica.* **C:** Branch with long spines and inflorescence heads; **D:** Flower bud; **E:** Moniliform pod. *A. farnesiana.* **F:** Portion of a branch with spines, leaf and inflorescence heads; **G:** Flower with numerous stamens; **H:** Pod. *Albizia julibrissin.* **I:** Part of a bipinnate leaf; **J:** Flower with monadelphous stamens; **K:** Part of a stamen showing anther.

base of leaflets) pulvinate, petiole sometimes modified into phyllode (*Acacia auriculiformis*), stipules present, sometimes spiny and hollow inside sheltering ants (*Acacia sphaerocephala*), leaves of *Mimosa pudica* sensitive to touch and showing sleeping movements. **Inflorescence** racemose, in racemes (*Adenanthera*) or spikes (*Prosopis*), sometimes in cymose heads (*Mimosa, Acacia*). **Flowers** small, bracteate (bracts usually caducous), sessile, or short-pedicelled, bisexual, actinomorphic, perigynous. **Calyx** with 5 sepals (4 in *Mimosa*), connate, odd sepal anterior, usually valvate, teeth small. **Corolla** with 5 petals (4 in *Mimosa*), free or united (*Acacia, Albizia*), valvate. **Androecium** with 4-many (4 in *Mimosa*, 10 in *Prosopis*, numerous in *Acacia* and *Albizia*) stamens, free (*Acacia, Prosopis*) or filaments connate (*Albizia*), anthers bithecous, dehiscence longitudinal, filaments long and anthers usually exserted. **Gynoecium** with a single carpel, unilocular with many ovules, placentation marginal, ovary superior, style single, curved. Fruit a legume or lomentum (*Mimosa, Acacia*); seeds 1-many, seed coat hard, endosperm minute or absent.

Economic importance: The subfamily is of lesser economic importance. Sensitive plant touch-me-not (*Mimosa pudica*) is grown as a curiosity. Various species of *Acacia* (*A. senegal, A. stenocarpa*) yield gum arabic. The pods and seeds of mesquite (*Prosopis juliflora*) are used as animal feed, wood in cooking. Wood of *Xylia* is hard and used in ship building. *Calliandra, Dichrostachys* are grown as ornamentals, *Pithecellobium* as a useful hedge plant.

Phylogeny of Fabaceae: The family is commonly circumscribed to include all the three subfamilies. Hutchinson as early as 1926 had recognized these as independent families Fabaceae, Caesalpiniaceae and Mimosaceae, a position that he maintained even in his last revision in 1973, regarding Caesalpiniaceae as the most primitive of the three, Mimosaceae relatively advanced and Fabaceae to be the climax group. The trend was followed and maintained in their latest classifications by Cronquist (1988) and Dahlgren (1989). Takhtajan who also began with the same treatment, has in his last three versions (1987, 1997, 2009) included all the three under broadly circumscribed Fabaceae, giving these three the rank of subfamily. Thorne has consistently included all the three subfamilies under broadly circumscribed Fabaceae, a position also justified by APG II. In his latest revision Thorne (2006, 2007) placed Fabaceae, Surianaceae, Polygalaceae and Quillajaceae in separate order Fabales (under Rosanae), a treatment similar to APG III and APweb. Affinities with Rutales have been supported based on wood anatomy and embryology (Thorne, 1992).

Recognition of broadly circumscribed Fabaceae is supported by its monophyly as evidenced by common morphological features, and the results of *rbcL* sequence data (Chappill, 1994; Doyle, 1994). Studies also indicated that Caesalpinioideae are paraphyletic with some genera more closely related to Mimosoideae, and others to Faboideae than they are to one another. It is now established that *Swartzia* and *Sophora* (and relatives) represent basal clades of Faboideae lack a 50 kb inversion in the *trnL* intron that is found in other members of the subfamily. Studies of Doyle et al. (2001) and Bruneau et al. (2001) suggest that *Cercis* and *Bauhinia* are basal in Fabaceae and as such discussed under distinct group Cercideae in APweb, characterized by simple leaves, sometimes bilobed; vestured pits, which they lack, are also absent in Cassieae. The flowers of *Cercis* are only superficially similar to those of Faboideae (Tucker, 2002) and removed to distinct subfamily Cercidoideae along with *Bauhinia* and 10 other genera in LPWG scheme (2017).

Mimosoideae are largely monophyletic, Faboideae are monophyletic, Caesalpinioideae are paraphyletic and basal. Wojciechowski et al. (2003) on basis of studies on sequences of the plastid *matK* gene note than non-protein amino acids seem to have originated once in this clade. Fabaceae s. l. are often referred to their own order, as in both Cronquist (1981) and Takhtajan (1997), former placing it closer to Rosales and latter closer to Sapindales. They can be confused with Connaraceae (Oxalidales), although the latter lack stipules, their flowers are radially symmetrical and have stamens of two distinctly different lengths, and their gynoecium is frequently multicarpellate. However, in both the RP122 chloroplast gene has moved to the nucleus! Also, the ovaries of both have adaxial furrows (cf. the ventral slit: Matthews and Endress, 2002). A complete reorganization of family by LPWG (2017) merges Mimosoideae with Caesalpinioideae but excludes several genera to create 4 more subfamilies, thus recognizing a total of 6 subfamilies as indicated earlier.

Rosaceae A. L. de Jussieu ***Rose Family***

91 genera, 2,950 species

Widespread but best represented in the Northern Hemisphere, mainly in the temperate and arctic climate.

Placement:

	B & H	Cronquist	Takhtajan	Dahlgren	Thorne	APG IV/(APweb)
Informal Group						Rosids/(N-Fixing clade)
Division		Magnoliophyta	Magnoliophyta			
Class	Dicotyledons	Magnoliopsida	Magnoliopsida	Magnoliopsida	Magnoliopsida	
Subclass	Polypetalae	Rosidae	Rosidae	Magnoliidae	Rosidae	
Series+/ Superorder	Calyciflorae+		Rosanae	Rosanae	Rosanae	
Order	Rosales	Rosales	Rosales	Rosales	Rosales	Rosales

Salient features: Herbs shrubs or trees, leaves usually serrate, stipules conspicuous, flowers actinomorphic, usually perigynous and with hypanthium, sepals and petals 5 each, petals usually clawed, well-developed nectary on hypanthium or base of stamens, stamens numerous, carpel single or numerous and free, rarely united, fruit usually fleshy.

Major genera: *Rubus* (750 species), *Potentilla* (500), *Prunus* (430), *Crataegus* (240), *Cotoneaster* (230), *Sorbus* (230), *Rosa* (225), *Alchemilla* (220), *Spiraea* (100), *Pyrus* (60), *Malus* (55) *Geum* (40) and *Fragaria* (15).

Description: Herbs (*Alchemilla, Fragaria*), shrubs (*Rosa, Rubus*) or trees (*Prunus, Malus, Pyrus*), rarely climbing (some species of *Rosa*), sometimes with runners (*Fragaria*), often with prickles and thorns, without latex, nodes trilacunar, rarely unilacunar. **Leaves** alternate, rarely opposite (*Rhodotypos*), simple (*Malus, Prunus*), palmately compound (*Fragaria*) or pinnate compound (*Sorbaria*), leaf blade often with gland-tipped teeth, usually serrate, venation pinnate or palmate, reticulate, stipules present, often adnate to petiole. **Inflorescence** with solitary flowers (some species of *Rosa*), racemes (*Padus*), panicles or cymose umbels (*Spiraea*), sometimes corymbs (*Crataegus*), rarely catkin-like (*Poterium*). **Flowers** bisexual, rarely unisexual (*Poterium*; plants monoecious or dioecious), actinomorphic, rarely zygomorphic (*Parinarium*), usually perigynous with distinct hypanthium (flat, cup-shaped or cylindrical); hypanthium free from or adnate to carpels, often enlarging in fruit, with nectar ring on inside, rarely epigynous (*Malus*). **Calyx** usually with 5 sepals, united at base, sometimes with 3–5 **epicalyx** (*Fragaria*) on outside, often persistent. **Corolla** usually with 5 petals, free, often clawed, imbricate. **Androecium** with numerous stamens, free or fused at base with nectar disc (*Chrysobalanus*), 4 in *Sanguisorba*, 2 in *Parastemon urophylla*, anthers bithecous, rarely monothecous (*Alchemilla*), dehiscence longitudinal, pollen grains tricolporate. **Gynoecium** with 1 (*Prunus*), 2–3 (*Crataegus*) to many carpels (*Rosa*), usually free, rarely connate (*Crataegus, Pyrus*), sometimes adnate to hypanthium, ovary superior or inferior, usually unilocular, ovules 1, 2 or more, unitegmic or bitegmic, crassinucellate, placentation basal, lateral or apical, rarely axile (*Pyrus*). **Fruit** a follicle (*Spiraea*), achene (*Rosa*), drupe (*Prunus*), pome (*Malus*), or aggregate (etaerio of achenes in *Potentilla*, etaerio of drupes in *Rubus*); seed with straight embryo, without endosperm. Pollination mainly by insects. Dispersal by birds, animals or wind.

Economic importance: The family is largely known for its temperate fruits: apple (*Malus domestica*), pear (*Pyrus*), plums (*Prunus*-several species), cherries (*Prunus avium, P. cerasus*), peaches (*Prunus persica*), almonds (*Prunus dulcis*), apricots (*Prunus armeniaca*), strawberry (*Fragaria vesca*), loquots (*Eriobotrya*), raspberries (*Rubus*), quince (*Cydonia*), etc. Popular ornamentals include species of *Rosa* (rose) *Rubus* (raspberry), *Chaenomeles* (flowering quince), *Potentilla* (cinquefoil), *Geum* (avens), *Cotoneaster*, *Crataegus* (hawthorn), *Pyracantha* (firethorn), and *Sorbus* (mountain ash). Flowers of *Rosa damascena* are used for extracting attar of roses. The bark of *Quillaja* (soap-bark tree) contains saponin used as substitute for soap in cleaning textiles, and also yield tannin. Bark of *Moquilla utilis* (pottery tree) of Amazon is used in making heat-resistant pots. The wood of *Prunus serotina* is used for making furniture and cabinets. Several species are also valuable sources of timber.

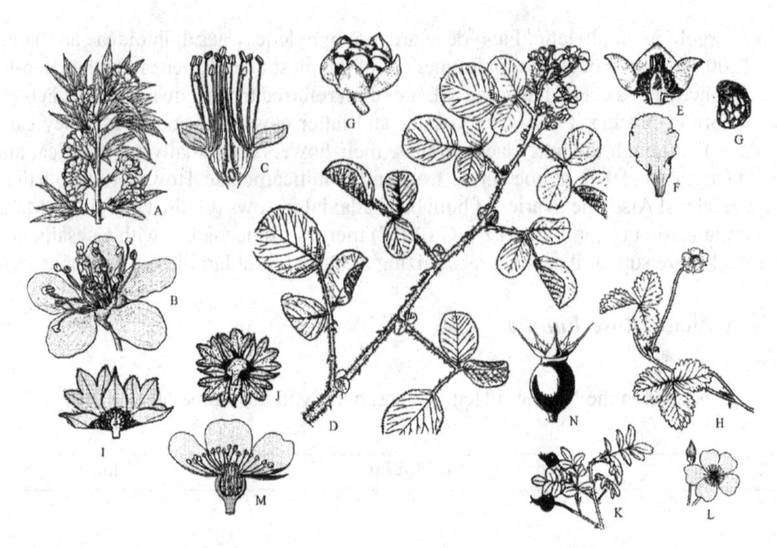

Figure 13.56: Rosaceae. *Prunus domestica.* **A:** Portion of a flowering twig; **B:** Flower from above; **C:** Vertical section of flower, petals removed. *Rubus ellipticus.* **D:** Branch with terminal inflorescence; **E:** Vertical section of flower with petals removed; **F:** Petal; **G:** Fruit covered with persistent calyx. *Duchesnia indica.* **H:** Portion of a branch with trifoliate leaves and flower; **I:** Vertical section of flower with petals removed, 3-lobed bracteoles (epicalyx) present outside calyx; **J:** Calyx and 5 3-lobed bracteoles. *Rosa pimpinellifolia.* **K:** Branch with fruits; **L:** Flower and bud; **M:** Vertical section of flower showing cup shaped hypanthium and numerous free carpels; **N:** Fruit (hip) enclosing achenes and with persistent calyx.

Phylogeny: In spite of great morphological diversity the family Rosaceae is a well-recognized group whose monophyly has been supported by *rbcL* sequences (Morgan et al., 1994). More than 27 family names have been proposed for groups of different genera taken out from Rosaceae, but according to Hutchinson (1973) if one or two tribes of the family are taken out, at least 18 or 19 should follow suit, and the Rosaceae would be reduced to the genus *Rosa* only. He like most recent authors follows a broader circumscription of the family but does not recognize separation of Chrysobalanaceae and Neuradaceae (established as distinct in 12th edition of the Engler's *Syllabus* published in 1964). These two last families have been recognized as distinct in all major classifications. Cronquist places them together with Rosaceae under Rosales. Dahlgren places Chrysobalanaceae under Theanae—> Theales, but Neuradaceae along with Rosaceae in Rosales. Takhtajan places Neuradaceae in Rosales along with Rosaceae, but Chrysobalanaceae in distinct order Chrysobalanales. Thorne (1999) shifted both families from Rosidae to Dilleniidae, Chrysobalanaceae under Dillenianae—> Dilleniales and Neuradaceae under Malvanae—> Malvales. In latest revision (2007), however, he has abolished Dilleniidae and brought all the three families under Rosidae but placing Rosaceae under Rosidae—> Rosanae—> Rosales, Neuradaceae under Malvidae—> Malvanae—> Malvales—> Cistineae and Chrysobalanaceae under Rosidae—> Podostemanae—> Euphorbiales. APG III and APweb have shifted Chrysobalanaceae to Eurosids I—> Malpighiales and Neuradaceae to Eurosids II—> Malvales, retaining Rosaceae in Eurosids I—> Rosales. The family has often been considered closely related to Saxifragaceae and Crassulaceae but the *rbcL* data identify Ulmaceae, Celtidaceae, Moraceae, Urticaceae and Rhamnaceae as sister groups (Savolainen et al., 2000a). Usually 4 subfamilies are recognized within Rosaceae: Maloideae (fruit a pome), Amygdaloideae (syn: Prunoideae; fruit a drupe, carpel 1, nectaries on petiole and lamina), Rosoideae (fruit achenes or drupelets) and Spiraeoideae (follicle or capsule). Although Rosoideae and Maloideae are reasonable clades, little can yet be said of larger patterns of relationship in the rest of the family (Potter et al., 2002). *Porteranthus* is sister to Maloideae; *Gillenia* is sister to that whole clade (Potter et al., 2002; Evans et al., 2002a,b). The position of Dryadeae (incl. *Cercocarpus*, *Dryas*, *Cowania* and *Chamaebatia*) included in Rosoideae is uncertain, they lack phragmidiaceous rusts; their roots are associated with N-fixing *Frankia* and their fruits are achenes with hairy styles. They are rather basal (Potter et al., 2002; Evans et al., 2002). APWeb (2018) recognizes three subfamilies: Dryadoideae, Rosoideae (Six tribes) and Amygdaloideae (11 tribes including Maloideae (placed in tribe Malinae) and Spiraeoideae (placed in tribe Spiraeeae)). Xiang et al. (2016) found that Dryadoideae were sister to the rest of the family, and they thought that other positions were unlikely.

Rhamnaceae A. L. de Jussieu ***Buckthorn Family***

53 genera, 875 species

Distributed worldwide but more common tropical and subtropical regions.

Placement:

	B & H	Cronquist	Takhtajan	Dahlgren	Thorne	APG IV/(APweb)
Informal Group						Rosids/(N-Fixing clade)
Division		Magnoliophyta	Magnoliophyta			
Class	Dicotyledons	Magnoliopsida	Magnoliopsida	Magnoliopsida	Magnoliopsida	
Subclass	Polypetalae	Rosidae	Rosidae	Magnoliidae	Malvidae	
Series+/ Superorder	Disciflorae+		Rutanae	Rutanae	Malvanae	
Order	Rhamnales	Rhamnales	Sapindales	Sapindales	Rhamnales	Rosales

Salient features: Trees and shrubs, leaves toothed with strong secondary veins, stipulate, flowers perigynous, petals strongly concave, stamens opposite the petals, hypanthium with nectary inside, ovules on basal placentas.

Major genera: *Rhamnus* (150 species), *Phylica* (150), *Ziziphus* (100), *Ceanothus* (40), *Gouania* (35), *Colubrina* (15), *Berchemia* (10) and *Sageretica* (10).

Description: Erect or climbing shrubs, trees, rarely herbs, sometimes thorny, sometimes associated with nitrogen-fixing Actinomycetes bacteria, stems often modified into thorns, tendrils or hooks. **Leaves** usually opposite, sometimes alternate, simple, toothed, venation reticulate with strong secondary veins, stipules usually present and often modifies into spines, leaves sometimes rudimentary. **Inflorescence** axillary corymb or cymose clusters, rarely solitary. **Flowers** small, usually inconspicuous, actinomorphic, bisexual, rarely unisexual, perigynous, hypanthium present. **Calyx** with 5 sepals, rarely 4, free or united, lobes valvate. **Corolla** with 5 petals, rarely 4, free, sometimes absent, often concave and hooded (cucullate) over anthers, usually clawed. **Androecium** with as many stamens as petals and opposite them, arising from outside disc that lines the rim of hypanthium, anthers bithecous, dehiscence longitudinal, anthers with sterile tip formed by extension of connective, pollen grains tricolpate or triporate. **Gynoecium** with 2–4 united carpels, rarely 5, locules as many, ovary superior or partly inferior (*Gouania*) due to adnation with disc, ovule 1 in each locule on basal placenta, pendulus, anatropous, style one, often lobed or cleft. **Fruit** a drupe with 1-many endocarps, capsule or samaroid nut; seed large, usually straight, sometimes curved, without or with scanty endosperm.

Economic importance: The family yields important fruits from species of *Ziziphus* (*Z. jujuba*, the jujube or Chinese date; *Z. mauritiana*, the Indian jujube; *Z. lotus*, lotus fruit), also used to make jelly-like candy. Plants of the family yielding dyes include *Rhamnus cathartica* (green dye from sap), *R. tinctoria* (yellow dye from fruits), *R. chlorophora* (Chinese green indigo from bark). Several species are also used medicinally: fruits of *R. cathartica* and *R. purshiana* are strongly laxative; extract of *Gouania* bark are used as wood dressing in Africa; *Ventilago oblongifolia* is used to treat cholera in Malaya. Many species of *Ceanothus* (tea bush) with beautiful panicles of blue, pink or white flowers are grown as ornamentals. Other members used as ornamental include *Hovenia* (raisin tree), *Berchemia* (supplejack), *Poliurus* (Jerusalem thorn) and *Reynosia*.

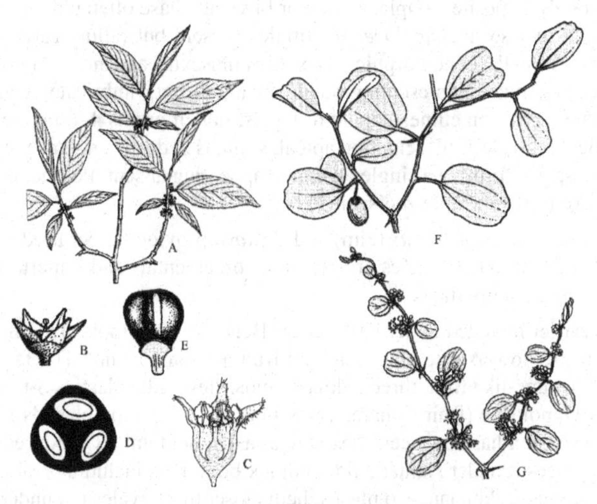

Figure 13.57: **Rhamnaceae.** *Rhamnus purpurea*. **A:** Branch with flowers; **B:** Flower; **C:** Vertical section of flower; **D:** Transverse section of ovary; **E:** Fruit. **F:** Fruiting shoot of *Ziziphus mauritiana*. **G:** Flowering shoot of *Z. nummularia* (A, B and E, after Polunin and Stainton, Fl. Himal., 1984; F and G, after Maheshwari Illus. Fl. Del., 1966).

Phylogeny: Family is often placed closer to Vitaceae, in the same order Rhamnales (Rendle, Cronquist, Hutchinson but he also included Elaeagnaceae and Heteropyxidaceae). Thorne (2006) retains Rhamnaceae, Elaeagnaceae, Dirachmaceae and Barbeyaceae under Rhamnales, but shifts Vitaceae to independent order Vitales under Malvanae. APG-II includes Vitaceae unplaced in Rosids, whereas APWeb included it under Vitales but towards end of Core Eudicots. Takhtajan (1997) also separated two families under separate orders. The family Rhamnaceae shows affinities with Rosales, under which it is placed in both APG-II and APweb. There are three main clades in the family are recognized by APWeb, the rhamnoids, which include *Maesops* and *Ventilago* (three tribes), the ziziphoids (five tribes), which include most of the rest of the family, and the ampeloziziphoids (three tribes). The four families included under the order by Thorne (2006) form a well-defined clade (Sytsma et al., 2002), with dense curly hairs on abaxial surface of leaf, a possible synapomorphy. Rhamnaceae is sister to this clade. Richardson et al. (2000) proposed a phylogenetic classification of Rhamnaceae dividing it into clade-based tribes: Ampeloziziphеae, Bathiorhamneae, Colletieae, Doerpfeldieae, Gouanieae, Maesopsideae, Paliureae, Phyliceae, Pomaderreae, Rhamneae and Ventilagineae. The classification is followed by Sun et al. (2016) in their dense taxon sampling analysis of Rosidae, atleast 12 genera having uncertain position. They also found Rhamnaceae + (Elaeagnaceae + Barbeyaceae) + Dirachmaceae, with 56%, 62%, and 84% BS support form a well-defined clade within Rosales.

Ulmaceae Mirbel ***Elm Family***

7 genera, 45 species

Mainly distributed in temperate region, also tropics and subtropics.

Placement:

	B & H	**Cronquist**	**Takhtajan**	**Dahlgren**	**Thorne**	**APG IV/(APweb)**
Informal Group						Rosids/(N-Fixing clade)
Division		Magnoliophyta	Magnoliophyta			
Class	Dicotyledons	Magnoliopsida	Magnoliopsida	Magnoliopsida	Magnoliopsida	
Subclass	Monochlamydeae	Hamamelidae	Dilleniidae	Magnoliidae	Malvidae	
Series+/ Superorder	Unisexuales+		Malvanae	Malvanae	Malvanae	
Order		Urticales	Urticales	Urticales	Urticales	Rosales

B & H under family Urticaceae.

Salient features: Trees or shrubs with watery sap, sieve-tube plastids P-type, leaves simple, serrate or biserrate, vascular bundles entering teeth, leaf base oblique, venation pinnate, flowers often bisexual, fruit a winged samara or drupe.

Major genera: *Ulmus* (20 species), *Ampelocera* (12), *Zelkova* (6), *Phyllostylon* (3) and *Planera* (1).

Description: Shrubs or trees without laticifers, often with tannins, cystoliths present, branching profusely and often spreading. **Leaves** alternate, rarely opposite, simple, serrate or biserrate, base often oblique, venation pinnate, reticulate, vascular bundles entering teeth (not so in *Ampelocera*), stipules present but falling early. **Inflorescence** consisting of axillary cymose clusters. **Flowers** small, actinomorphic, bisexual or unisexual and monoecious, hypogynous or perigynous. **Perianth** with 4–9 tepals, free or connate, representing sepals (petals absent), imbricate. **Androecium** with 4–9 stamens, as many as tepals and opposite them, sometimes adnate to tepals, pollen grains 4–6-porate. **Gynoecium** with 2 united carpels, ovary superior, unilocular, ovule 1, placentation apical, stigmas 2, decurrent on style. **Fruit** a nut or samara; seed flat with straight embryo, endosperm forming a single layer and appearing absent. Pollination by wind. Winged fruits are also dispersed by wind, nut-like fruits of *Planera* dispersed by water.

Economic importance: Various species of *Ulmus* (elm) and *Zelkova* provide timber used for furniture, posts and under water pillings. *Ulmus americana* and other species are grown as ornamentals and important shade trees. Mucilaginous inner bark of *U. rubra* has medicinal importance.

Phylogeny: The family was earlier included under Urticaceae (Bentham and Hooker) but now separated due to veins of leaves running directly into teeth, flowers often bisexual and fruit a samara or nut. The family was earlier divided into two subfamilies: Celtidoideae (drupe-like fruit, three palmate veins, sieve tube plastids S-type; style with single vascular bundle, embryo curved) and Ulmoideae (fruit samara, veins pinnate; sieve tube plastids P-type, style with 3 vascular bundles, embryo straight). The former has now been separated as a distinct family Celtidaceae. The family is often placed in order Urticales. Cronquist places it under Hamamelid complex but others including Dahlgren and Thorne place them along with other Malvanean groups. Takhtajan also places them closer to Malvales but under superorder Urticanae. APG classifications place them closer to Rosaceae, and Rhamnaceae under order Rosales, Rosaceae being considered sister to rest of the families. Poorly known genus *Ampelocera,* although with smooth hairs and ascending, deserves placement in

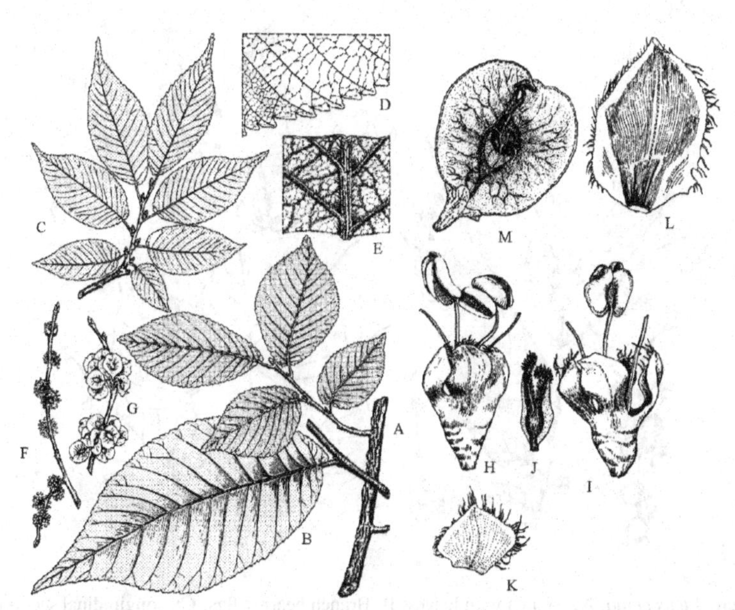

Figure 13.58: Ulmaceae. *Ulmus chumlia.* **A:** Short shoot, lower surface & corky bark; **B:** Leaf of coppice shoot, upper surface; **C:** Normal adult short shoot, upper surface; **D:** Leaf margin; **E:** Indumentum of midrib portion, lower surface; **F:** Flowering shoot; **G:** Fruiting shoot; **H:** Flower; **I:** Slightly older flower; **J:** Gynoecium; **K:** Bract; **L:** Inner bud scale; **M:** Mature winged fruit (After Melville and Heybroek, 1971).

Ulmaceae (APweb, 2018). A clade including *Ampelocera, Phyllostylon* and *Holoptelea* is sister to the rest of the family (Neubig et al., 2012b; Sun et al., 2016).

Moraceae Link ***Mulberry Family***

38 genera, 1,180 species

Distributed mainly in tropics and subtropics with some species in temperate regions.

Placement:

	B & H	Cronquist	Takhtajan	Dahlgren	Thorne	APG IV/(APweb)
Informal Group						Rosids/(N-Fixing clade)
Division		Magnoliophyta	Magnoliophyta			
Class	Dicotyledons	Magnoliopsida	Magnoliopsida	Magnoliopsida	Magnoliopsida	
Subclass	Monochlamydeae	Hamamelidae	Dilleniidae	Magnoliidae	Malvidae	
Series+/ Superorder	Unisexuales+		Malvanae	Malvanae	Malvanae	
Order		Urticales	Urticales	Urticales	Urticales	Rosales

B & H under family Urticaceae.

Salient features: Trees and shrubs with milky latex, leaves alternate, flowers unisexual, small, carpels usually 2, ovary superior, single chambered, ovule 1.

Major genera: *Ficus* (600 species), *Dorstenia* (110), *Artocarpus* (50), *Morus* (15), *Maclura* (12) and *Broussonetia* (8).

Description: Trees or shrubs, sometimes lianas, rarely epiphytic in early stage (strangling species of *Ficus*), often with milky latex distributed in all parenchymatous tissues, cystoliths present, usually globose, tannins often present. **Leaves** alternate (rarely opposite), usually distichous, simple with entire or lobed margin, with pinnate or palmate reticulate venation, stipules present and leaving a circular scar when shed. **Inflorescence** of various types, erect or pendulous (catkin), spike (*Morus*), hypanthodium (*Ficus*), or raceme. **Flowers** small, unisexual (monoecious or dioecious), actinomorphic, hypogynous. **Perianth** usually with 4–6 tepals (representing sepals, petals absent), free or united, often persisting and becoming fleshy in fruit, sometimes absent. **Androecium** with 4–6 (as many as tepals) stamens, opposite the tepals, filaments free, incurved in bud or straight, anthers bithecous or monothecous, dehiscence longitudinal, pollen grains multiporate or with 2–4 pores.

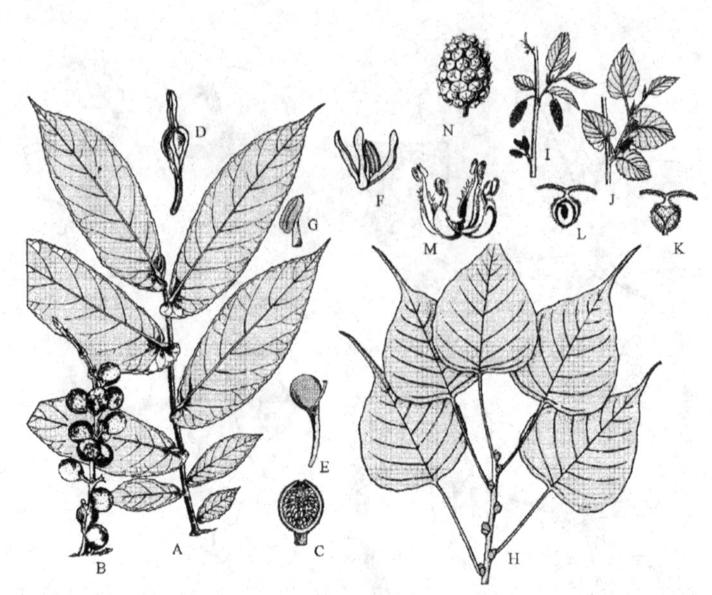

Figure 13.59: Moraceae. *Ficus cunia.* **A:** Branch with leaves; **B:** Branch bearing figs; **C:** Longitudinal section of hypanthodium (fig, receptacle); **D:** Female flower; **E:** Gynoecium; **F:** Male flower with single stamen; **G:** Stamen. **H:** Twig of *Ficus religiosa. Morus alba.* **I:** Male branch; **J:** Female branch; **K:** Female flower with closely appressed perianth; **L:** Longitudinal section of female flower; **M:** Male flower with four tepals and 4 stamens; **N:** Multiple fruit (Sorosis).

Gynoecium with 2 united carpels, ovary superior, unilocular, ovule 1, anatropous to campylotropous, placentation apical, styles usually 2. **Fruit** usually a multiple fruit sorosis (*Morus*), syconium (syconus; *Ficus*), sometimes etaerio of drupes or a berry; seed with curved or straight embryo, endosperm present or absent.

Economic importance: The family is important for its fruits such as mulberry (*Morus alba, M. nigra*), fig (*Ficus carica*) and breadfruit (*Artocarpus altilis*). Fruits of *Artocarpus heterophyllus* ('kathal') are cooked as vegetable, whereas those of *A. lakoocha* ('dheon') are pickled. Leaves of *Morus* are also used for raring silkworms. Various species of *Ficus* including *F. elastica* (Indian rubber tree or rubber plant) are grown as ornamentals.

Phylogeny: The family was earlier placed in Urticaceae (Bentham and Hooker) but now considered distinct in woody habit with milky latex, 2 carpels, ovary with single apical ovule and usually curved embryo. Cronquist places Urticales (including Urticaceae and related families) under Hamamelid complex but others including Dahlgren and Thorne place them along with other Malvanean groups. Takhtajan also places them closer to Malvales but under superorder Urticanae. APG classifications place them closer to Rosaceae, and Rhamnaceae under order Rosales, Rosaceae being considered sister to rest of the families. *Cecropia* and related genera earlier included under Moraceae (Hutchinson and earlier authors), were separated under Cecropiaceae by APG (1998), Thorne (1999, 2000) and Judd et al. (1999, 2002) are intermediate between Moraceae and Urticaceae, but closer to Urticaceae in restriction of laticifers to bark, basal ovule, straight embryo and with one carpel aborted (pseudomonomerous). The family Cecropiaceae has appropriately been merged with Urticaceae by APG II (2003), APweb (2003) and Thorne (2003). The family Moraceae as narrowly circumscribed here is monophyletic as supported by rbcL sequences (Sytsma et al., 1996). The reduction of one carpel is also indicated in slightly or strongly unequal styles in *Artocarpus, Dorstenia* and *Ficus*. A complete loss of one of the two styles probably occurred in common ancestor of Cecropiaceae + Urticaceae clade. Clement and Weiblen (2009) recognised following six tribes in Moraceae: Artocarpeae, Moreae, Maclureae, Dorstenieae, Ficeae and Castilleae. Some of the old and paraphyletic Moreae with their incurved stamens are now placed in Maclureae, Castilleae are sister to Ficeae, and relationships between the tribes are on the whole strongly supported (Clement and Weiblen, 2009). This classification is followed in APweb (2018) infrageneric classification of *Ficus* was discussed by Pederneiras et al. (2015).

Interestingly, The straightening stamens and reflexing tepals of *Morus alba* are reported to show the fastest movement of any plant parts known, over half the speed of sound (Taylor et al., 2006).

Urticaceae A. L. de Jussieu *Nettle Family*

53 genera, 2625 species

Widespread in tropics and temperate climates, poorly represented in Australia.

Placement:

	B & H	Cronquist	Takhtajan	Dahlgren	Thorne	APG IV/(APweb)
Informal Group						Rosids/(N-Fixing clade)
Division		Magnoliophyta	Magnoliophyta			
Class	Dicotyledons	Magnoliopsida	Magnoliopsida	Magnoliopsida	Magnoliopsida	
Subclass	Monochlamydeae	Hamamelidae	Dilleniidae	Magnoliidae	Malvidae	
Series+/ Superorder	Unisexuales+		Malvanae	Malvanae	Malvanae	
Order		Urticales	Urticales	Urticales	Urticales	Rosales

Salient features: Usually herbs with stinging hairs, leaves with stipules, flowers small, unisexual, tepals and stamens usually 4 each, carpel 1, style 1, fruit achene or fleshy drupe.

Major genera: *Pilea* (370 species), *Elatostema* (170), *Boehmeria* (80), *Cecropia* (60), *Urtica* (50), *Parieteria* (30) and *Laportea* (20).

Description: Usually herbs, rarely trees or shrubs, sometimes climbers, with often milky latex restricted to bark or reduced with clear sap, cystoliths present, usually elongated, tannins often present, hairs simple, usually stinging. **Leaves** alternate or opposite, usually distichous, simple with entire or lobed margin, with pinnate or palmate venation, reticulate, stipules present, leaf base cordate or asymmetrical. **Inflorescence** cymose or heads, sometimes with solitary flowers. **Flowers** small, unisexual (monoecious or dioecious), actinomorphic, hypogynous. **Perianth** usually with 4 tepals (representing sepals, petals absent), rarely only 3 or up to 6, free or united, imbricate or valvate. **Androecium** with 4–5 (as many as tepals) stamens, opposite the tepals, filaments free, incurved in bud, reflexed at anthesis, anthers bithecous, dehiscence longitudinal, pollen grains multiporate or with 2–3 pores. **Gynoecium** with single carpel (actually 2 but with one reduced: pseudomonomerous), ovary superior, unilocular, ovule 1, orthotropous, placentation basal, style 1, stigmas 1 or 2, extending on style or capitate. **Fruit** usually an achene, embryo straight, endosperm sometimes lacking.

Economic importance: In addition to being a noxious weed, *Urtica dioica* (common stinging nettle) yields silky bast fibre. Fibre is also extracted on the commercial scale from *Boehmeria nivea* (ramie or china grass). Species of *Pilea* and *Soleirolia* (baby's tears) provide important ornamentals.

Phylogeny: The family was earlier broadly circumscribed (Bentham and Hooker) to include families which have now been separated as Moraceae, Ulmaceae, Celtidaceae, etc. The family is now circumscribed to include mainly herbaceous species with elongate cystoliths, laticifers restricted to the bark, with clear sap, incurved stamens, pseudomonomerous

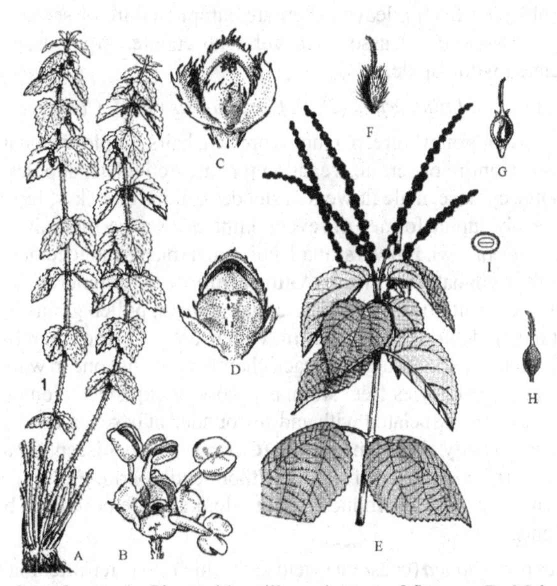

Figure 13.60: Urticaceae. *Urtica hyperborea.* **A:** Plant with axillary clusters of flowers; **B:** Male flower with four tepals and four stamens; **C:** Female flower with unequal tepals; **D:** Achene surrounded by persistent perianth. *Boehmeria platyphylla.* **E:** Plant with interrupted spikes; **F:** Female flower with bristly perianth and hairy style; **G:** Vertical section of female flower; **H:** Gynoecium; **I:** Transverse section of achene.

gynoecium and basal ovule. Thorne (1999) and the APG classifications (APG, 1998; Judd et al., 1999, 2002) also included in Urticaceae the genus *Poikilospermum*, formerly placed in Cecropiaceae. Cronquist places Urticales (including Urticaceae and related families) under Hamamelid complex but others including Dahlgren and Thorne place them along with other Malvanean groups. Takhtajan also places them closer to Malvales but under superorder Urticanae. APG classifications place them closer to Rosaceae, and Rhamnaceae under order Rosales, Rosaceae being considered sister to rest of the families. Single carpel in the family has been derived through abortion of the second carpel, as borne out by the aborted vascular bundles in the ovary of *Urtica* and *Laportea*. The basal placentation has similarly been derived from apical placentation of Moraceae. This is inferred from the fact that in *Boehmeria cylindrica* the vascular bundle supplying the ovule ascends the carpel wall for a short distance and then reverses direction to enter the ovule at the base of the ovary. The family Cecropiaceae, which was formerly recognized (APG, 1998; Judd et al., 1999, 2002; Thorne, 1999) as distinct family has finally been merged with Urticaceae (Thorne, 2003, 2007; APG II, 2003; APG III, 2009; APG, 2016; APweb), though Takhtajan in his last revision (2009) retained Cecropiaceae. Wu et al. (2013) based on studies of multiple loci of three genomes concluded that Urticaceae includes four well defined clades recognized as tribes (Wu et al., 2013; Wu et al., 2015): Cecropieae, Boehmerieae, Elatostemateae and Urticeae. APweb (2018) removes [*Leucosyke* + *Maoutia*] from Cecropieae. Position of *Metatrophis* is uncertain.

Fagaceae Dumortier ***Oak Family***

8 genera, 927 species

Widespread in tropical and temperate regions of the Northern Hemisphere.

Placement:

	B & H	**Cronquist**	**Takhtajan**	**Dahlgren**	**Thorne**	**APG IV/(APweb)**
Informal Group						Rosids/(N-Fixing clade)
Division		Magnoliophyta	Magnoliophyta			
Class	Dicotyledons	Magnoliopsida	Magnoliopsida	Magnoliopsida	Magnoliopsida	
Subclass	Monochlamydeae	Hamamelidae	Hamamelididae	Magnoliidae	Hamamelididae	
Series+/ Superorder	Unisexuales+		Hamamelidanae	Rosanae	Hamamelidanae	
Order		Fagales	Fagales	Fagales	Betulales	Fagales

B & H under Cupuliferae.

Salient features: Trees or shrubs with tannin, leaves alternate, simple, entire or serrate, stipules present, inflorescence cymose, female flowers usually in groups of 1–3, associated with scaly cupule, carpels usually 3, ovary inferior, placentation axile, fruit a nut, closely associated with cupule.

Major genera: *Quercus* (430 species), *Lithocarpus* (280), *Castanopsis* (100), *Castanea* (12) and *Fagus* (8).

Description: Trees or shrubs, deciduous or evergreen, tannins present, hairs simple or stellate, sometimes glandular. **Leaves** simple, alternate, sometimes lobed, entire or serrate, venation pinnate, reticulate, stipules present, early deciduous, often narrowly triangular. **Inflorescence** cymose, male flowers in slender catkins or spikes, female flowers solitary or in groups of up to three, associated with a scaly cupule formed of several imbricate scales, male and female flowers sometimes in the same inflorescence (*Castanea*, *Lithocarpus*). **Flowers** small, unisexual (plants monoecious), actinomorphic. **Perianth** with 4–6 tepals, reduced, free or slightly connate, imbricate. **Androecium** with 4-numerous stamens, filaments free, filiform, anthers erect, bithecous, loculi often contiguous, dehiscence longitudinal, pollen grains usually tricolporate or tricolpate, staminodes sometimes present in female flower. **Gynoecium** with usually 3 carpels, rarely upto 12, united, ovary inferior, locules as many as carpels, placentation axile, ovules 2 in each chamber but only one in whole ovary developing, pendulous, bitegmic, outer integument vascularized, styles free, stigmas porose or expanded along upper side of style, fertilization porogamous. **Fruit** a nut (acorn) closely associated with and surrounded at base (*Quercus*) or completely (*Castanea*) with cupule, cupule often hardened and woody, sometimes spiny (*Castanea*), indehiscent (*Quercus*) or dehiscent by splitting of cupule like pericarp into valves (*Castanea*); seed single, without endosperm. Flowers of *Fagus* and *Quercus* are wind pollinated, those of *Castanea* and *Castanopsis* produce strong odour and are pollinated by flies, beetles and bees. Fruits are dispersed by birds and rodents.

Economic importance: Species of *Castanea* (chestnut) yield nuts which are eaten after roasting, but have a very short shelf life, turning rancid within a few days. Fruits of some species of *Quercus* (oak) and *Fagus* (beech) are also occasionally eaten. Cork is made from the bark of *Quercus suber*. Wood of several species is a source of timber used for construction, furniture, barrels and cabinetry. Several species of *Quercus*, *Fagus*, *Castanea* and *Castanopsis* are grown as ornamentals.

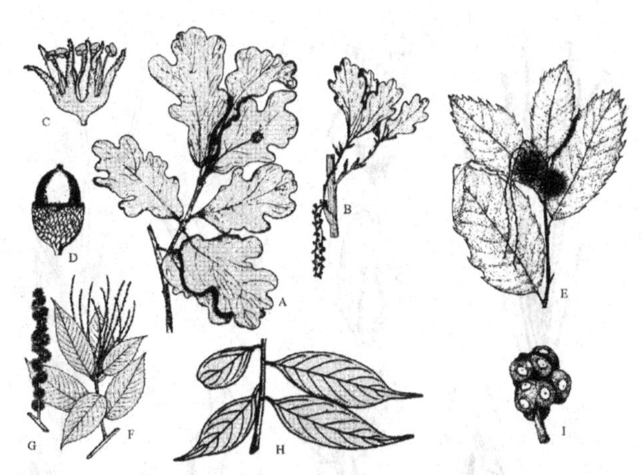

Figure 13.61: Fagaceae. *Quercus robur*. **A:** Branch with lobed leaves and long peduncled female flower; **B:** Young shoot with male catkin; **C:** Male flower; **D:** Fruit enclosed upto nearly half by cupule. **E:** Branch of *Castanea sativa* with long spikes each bearing single female flower at base and numerous male flowers above, cupule spiny. *Castanopsis indica*. **F:** Branch with several spikes; **G:** Fruiting spike. *Lithocarpus pachyphylla*. **H:** Portion of branch with leaves; **I:** Portion of fruiting branch with nuts in groups of three.

Phylogeny: The family is closely related to Betulaceae and the two are usually placed under the same order, although Takhtajan places only Fagaceae and Nothofagaceae under Fagales and separates Betulaceae and others under Corylales. The family is monophyletic as supported by morphology, cpDNA restriction sites (Manos et al., 1993) and *matK* sequences (Manos and Steele, 1997). *Castanea*, *Lithocarpus* and *Chrysolepis* have retained numerous plesiomorphic morphological characters such as monoecious inflorescences, perianth better developed, exserted stamens and minute stigmas. The nature of cupule has been a subject of considerable discussion. It is generally regarded to represent a cymose inflorescence in which outer axes of the cyme are modified into cupule valves which bear scales or spines (Manos et al., 2001). The cupules of *Nothofagus* (which was earlier placed (Hutchinson, 1973; Cronquist, 1981) under Fagaceae but now separated under Nothofagaceae) are composed of clustered bracts and stipules and not homologous with the cupule of Fagaceae. Heywood (1977) recognized three subfamilies under Fagaceae: Fagoideae, Quercoideae and Castanoideae. *Trigonobalanus* is considered sister to the rest of family (excluding *Fagus*) and as such removed together with *Colombobalanus* and *Formanodendron* under fourth subfamily Trigonobalanoideae by Thorne. Thorne had earlier (1999) placed Betulales under Rosidae—> Rosanae but has subsequently (2003, 2006, 2007) shifted it to Hamamelididae—> Hamamelidanae. APG II and AP web prefer order name Fagales placed under the clade Eurosids I, shifted to Rosids in APG IV (2016). *Fagus* is sister to all other Fagaceae (Sauquet et al., 2012), and as such placed in separate subfamily Fagoideae, all other genera under Quercoideae (including Castanoideae and Trigonobalanoideae) thus recognizing only two subfamilies in recent treatments (Manos et al., 2008; Xiang et al., 2014; APweb, 2018).

Casuarinaceae R. Brown ***She-Oak Family***

4 genera, 91 species

Widespread in Southeast Asia and Australia, naturalized in the coastal regions of tropical and subtropical Africa and America.

Placement:

	B & H	**Cronquist**	**Takhtajan**	**Dahlgren**	**Thorne**	**APG IV/(APweb)**
Informal Group						Rosids/(N-Fixing clade)
Division		Magnoliophyta	Magnoliophyta			
Class	Dicotyledons	Magnoliopsida	Magnoliopsida	Magnoliopsida	Magnoliopsida	
Subclass	Monochlamydeae	Hamamelidae	Hamamelididae	Magnoliidae	Hamamelididae	
Series+/ Superorder	Unisexuales+		Hamamelidanae	Rosanae	Hamamelidanae	
Order		Casuarinales	Casuarinales	Casuarinales	Betulales	Fagales

Salient features: Usually trees with jointed stems, appearing like conifers, leaves scale-like, whorled at nodes, inflorescence catkin, flowers unisexual, subtended by bracts, perianth absent, stamen 1, carpels 2, united, ovary superior, fruit a samara, fruits aggregated like cones.

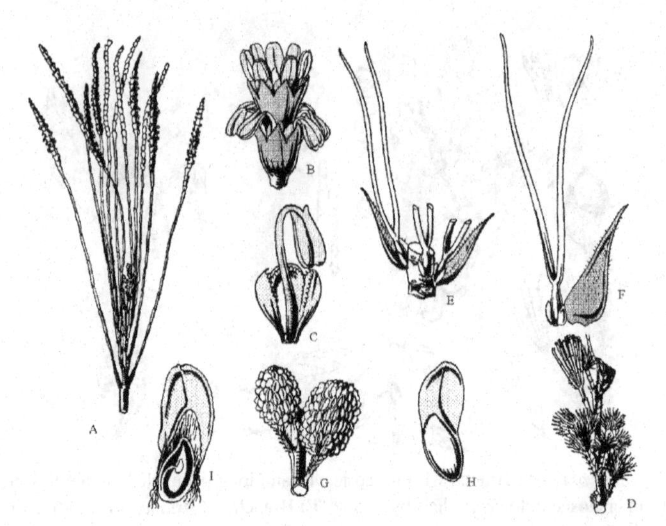

Figure 13.62: Casuarinaceae. *Casuarina suberba.* **A:** Branch with male inflorescences; **B:** Portion of male inflorescence; **C:** Male flower with single stamen; **D:** Portion of branch with female inflorescences; **E:** Part of female inflorescence showing 3 flowers; **F:** Female flower with bract, 2 small bracteoles and gynoecium with 2 long stylar branches; **G:** Fruits; **H:** Seed with broad wing; **I:** Longitudinal section of seed.

Major genera: *Allocasuarina* (50 species), *Casuarina* (25), *Gymnostoma* (14) and *Ceuthostoma* (2), often combined into single genus *Casuarina.*

Description: Trees or shrubs with a weeping habit due to long slender branches, stems jointed with circular sheath at nodes (switch habit), branches grooved, photosynthetic, sometimes aromatic (*Allocasuarina*), sieve-tube plastids S-type, nodes unilacunar, roots with nodules containing nitrogen-fixing bacteria, tannins present. **Leaves** whorled (4–20 in a whorl), scale-like, connate forming a toothed sheath at each node, stipules absent. **Inflorescence** forming catkins at tips of lateral branches. **Flowers** small, unisexual (plants monoecious or dioecious), actinomorphic, solitary in axil of each bract of inflorescence, associated with 2 bracteoles. **Perianth** absent in female flower, sometimes represented by 1–2 vestigial scales in male flower (often interpreted as inner bracteoles). **Androecium** with single stamen, anthers bithecous, incurved in bud, dehiscence longitudinal, pollen grains usually triporate. **Gynoecium** with 2 united carpels, ovary superior, bilocular with axile placentation, one often reduced and ovary appearing unilocular, ovules 2 but only one developing, orthotropous, bitegmic, crassinucellate, style short with 2 linear branches. **Fruits** crowded into cones with persistent bracts, fruit an indehiscent samara associated with 2 woody bracteoles which open like a capsule; seed with straight embryo, without endosperm. Wind pollinated. Fruits are also dispersed by wind.

Economic importance: The wood of several species is extremely hard and valued for furniture making. *Casuarina equisetifolia* (red beefwood) is most widely cultivated as ornamental tree.

Phylogeny: The family is monophyletic, as are the four genera recognized independently or combined into *Casuarina.* The family is considered to be a part of the Hamamelid complex, now included along with the broadly circumscribed Rosalean complex Rosanae (Dahlgren) or Rosidae (Thorne, 1999 under order Casuarinales; subsequently shifted to Hamamelididae—> Hamamelidanae—> Betulales in 2003, 2006 and 2007). APG II and APweb include the family under Fagales (under Eurosids; under Rosids in APG IV, 2016; APweb, 2018 under N-Fixing Clade), shifting Hamamelidaceae and some other families of the complex to Saxifragales. Fagales are the core of the old Englerian, Amentiferae which have since been demolished, several members shifted to otherwise entirely unrelated groups within the Eudicots (Qiu et al., 1998). The family Casuarinaceae was once considered to be the most primitive among dicots (Engler and Prantl) derived from Ephedraceae. The studies of wood anatomy and floral anatomy have shown that it is sufficiently advanced, having undergone considerable reduction in floral features and vegetative morphology. The genus was split into four genera indicated above (Johnson and Wilson, 1993). *Gymnostoma* is sister to the rest of the family and has many plesiomorphous features (both carpels fertile, with 2 ovules in each carpel) (Steane et al., 2003; Sun et al., 2016).

Betulaceae S. F. Gray ***Birch Family***

6 genera, 167 species

Widespread in temperate and boreal regions, *Alnus* being distributed in South America in Andes and Argentina.

Placement:

	B & H	Cronquist	Takhtajan	Dahlgren	Thorne	APG IV/(APweb)
Informal Group						Rosids/(N-Fixing clade)
Division		Magnoliophyta	Magnoliophyta			
Class	Dicotyledons	Magnoliopsida	Magnoliopsida	Magnoliopsida	Magnoliopsida	
Subclass	Monochlamydeae	Hamamelidae	Hamamelididae	Magnoliidae	Hamamelididae	
Series+/ Superorder	Unisexuales+		Hamamelidanae	Rosanae	Hamamelidanae	
Order		Fagales	Betulales	Fagales	Betulales	Fagales

Salient features: Trees or shrubs with tannin, bark sometimes exfoliating in thin layers, leaves alternate, simple, doubly serrate, stipules present, inflorescence a catkin, male and female inflorescences distinct, tepals 2–4, stamens 2–4, perianth absent in female flower, carpels usually 2, placentation axile, fruit a nut, surrounded by fused bract and bracteoles.

Major genera: *Betula* (55 species), *Alnus* (30), *Carpinus* (28), *Corylus* (15), *Ostrya* (10) and *Ostryopsis* (2).

Description: Trees or shrubs, deciduous, tannins present, bark smooth or scaly, with prominent horizontal lenticels, sometimes exfoliating in thin layers, hairs simple, glandular or peltate. **Leaves** simple, alternate, doubly serrate, venation pinnate, reticulate, secondary veins running into serrations, stipules present. **Inflorescence** a catkin, with separate male and female inflorescences but plant monoecious, male inflorescence usually pendulous, female short and erect, flowers borne singly at each node of catkin or in cymose cluster of 2–3, adhering to involucre of bracts and bracteoles. In *Alnus,* female inflorescence has each cymose cluster with 2 flowers associated with 1 bract, two secondary bracteoles and 2 tertiary bracteoles all connate into woody persistent involucre. In *Betula,* there are 3 female flowers in the cluster with 1 bract, 2 bracteoles, all fused into a 3-lobed 'bract' or involucre. **Flowers** small, unisexual (plants monoecious), actinomorphic. **Perianth** with usually 2–4 tepals, rarely 1 or up to 6, reduced, free, imbricate, absent in male (Coryloideae) or female flower (Betuloideae). **Androecium** with 2 (*Betula*) or 4 (*Alnus*) stamens, rarely 1 or up to 12 (Coryloideae), sometimes appearing many due to close association of three flowers, filaments free or connate at base, anthers bithecous, loculi distinct or contiguous, dehiscence longitudinal, pollen grains usually 2-multiporate, staminodes absent in female flower. **Gynoecium** with 2 united carpels, ovary inferior, bilocular, placentation axile, ovules 2 in each chamber but only one in whole ovary developing, pendulous, unitegmic, styles free, cylindrical, stigma running along adaxial side of style, pistillode absent in male flower. **Fruit** a single-seeded nut or 2-winged samara often with persistent styles, involucre of bract and bracteoles deciduous or persistent, scaly and woody or enlarged and foliaceous, sometimes bladder-like (*Ostrya*); seed solitary pendulous, embryo straight, cotyledons large, endosperm absent. Flowers are wind pollinated and emerge before leaves. Winged fruits of *Betula* and *Alnus* are dispersed by wind. Large nuts of *Corylus* are dispersed by rodents.

Economic importance: Papery bark of *Betula utilis* (birch) was used as a writing surface (bhojpatra) in place of paper in ancient Vedic manuscripts; also used for roofing and umbrella covers. *B. lutea* and *B. lenta* are important hardwoods in North America providing wood used for plywood, boxes and turnery. *Alnus rubra* (alder) provides a valuable timber

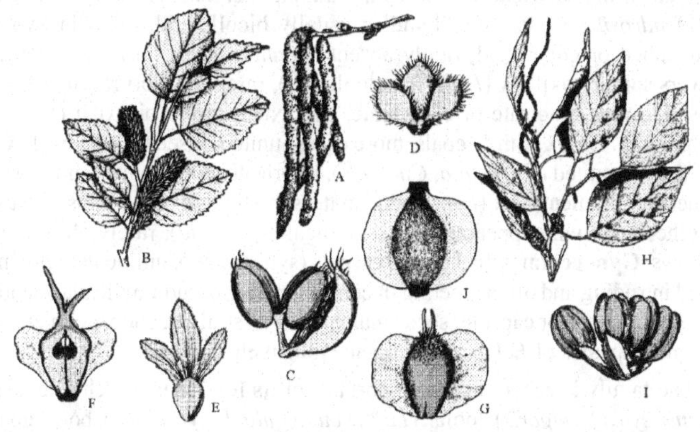

Figure 13.63: Betulaceae. *Betula utilis.* **A:** Portion of branch with male catkins, appearing before or along with leaves; **B:** Branch with female spikes; **C:** Single stamen (second one removed) with bracteole and forked filament separating anthers; **D:** Bract and lateral bracteoles of male flower; **E:** Fused bract and bracteoles of female flower; **F:** Young winged carpel; **G:** Nut with 2 wings and persistent styles. *Alnus nitida.* **H:** Branch with slender male catkins towards top and ovoid female spikes lower down; **I:** Male flower with 4 tepals and 4 stamens; **J:** Nut with 2 wings (A–B, after Polunin and Stainton, Fl. Himal., 1984).

which is a good imitation of mahogany. Nuts such as hazelnuts, filberts from species of *Corylus* are edible. Many species of *Betula, Alnus, Corylus, Ostrya* are grown as ornamentals.

Phylogeny: The family is closely related to Fagaceae and the two are usually placed under the same order, although Takhtajan places only Fagaceae and Nothofagaceae under Fagales and has separated Betulaceae and others under Corylales, Takhtajan (2009) suggesting both Betulales and Corylales as alternate names. The family is usually divided into two subfamilies: Betuloideae (male flowers with perianth, female lacking perianth, stamens 2 or 4, involucre scaly or woody, fruit 2-winged samara) and Coryloideae (male flower without perianth, female with perianth, stamens usually more than 3, involucre foliaceous, nut not flattened). Nuclear ribosomal ITS and *rbcL* sequences also support these two subfamilies (Chen et al., 1999). Hutchinson treated them as distinct families, also proposing that as the female flowers of Betulaceae lack perianth, the ovary is superior, and that of Corylaceae with perianth and ovary is inferior, a contention not supported by other authors. Both groups are monophyletic, although monophyly of *Ostrya* and *Carpinus* is doubtful (Yoo and Wen, 2002). Li et al. (2004) suggested that Betuloideae were paraphyletic, with *Alnus* and *Betula* being successively sister to the rest of the family. Forest et al. (2005), analyzing variation in ITS and the 5S spacer, recovered the two subfamilies above as monophyletic. The ITS phylogeny of *Betula* provided by Wang et al. (2016) suggests that there has been hybridization, even between species now placed in separate subgenera. The orientation of stamens or carpels in the flower may change during development in *Betula* and other taxa (Lin et al., 2010), while Endress (2008) discussed the structural lability of carpellate flowers of *Carpinus betulus*. The ovary of *Corylus* is not always obviously inferior.

Cucurbitaceae A. L. de Jussieu *Cucurbit or Gourd Family*

95 genera, 965 species

Mainly distributed in tropics and subtropics, in temperate regions often found in cultivation.

Placement:

	B & H	Cronquist	Takhtajan	Dahlgren	Thorne	APG IV/(APweb)
Informal Group						Rosids/(N-Fixing clade)
Division		Magnoliophyta	Magnoliophyta			
Class	Dicotyledons	Magnoliopsida	Magnoliopsida	Magnoliopsida	Magnoliopsida	
Subclass	Polypetalae	Dilleniidae	Dilleniidae	Magnoliidae	Rosidae	
Series+/ Superorder	Calyciflorae+		Violanae	Violanae	Violanae	
Order	Passiflorales	Violales	Cucurbitales	Cucurbitales	Violales	Cucurbitales

Salient features: Tendril climbing plants, leaves palmately veined, flowers unisexual, stamens 5, variously united, carpels usually 3, united, ovary inferior, fruit a berry or pepo.

Major Genera: *Cayaponia* (60 species), *Momordica* (45), *Gurania* (40), *Sicyos* (40), *Cucumis* (30) and *Cucurbita* (27).

Description: Climbing annuals with coiled tendrils, sometimes trailing (*Ecballium*), rarely xerophytic shrubs (*Acanthosicyos horrida*) or even trees (*Dendrosicyos*), vascular bundles usually bicollateral, often in two rings. **Leaves** alternate, simple, palmately veined, lobed or compound, rarely absent (*Acanthosicyos horrida*), stipules absent. **Inflorescence** cymose (*Bryonia*) or flowers solitary axillary (*Luffa* female flower), rarely in short racemes (*Luffa* male flowers), plants monoecious or dioecious. **Flowers** bracteate or ebracteate, unisexual, rarely bisexual (*Schizopepon*), actinomorphic, epigynous, with long hypanthium. **Calyx** with 5 sepals, more or less united, fused to ovary wall. **Corolla** with 5 petals, free (*Luffa, Lagenaria, Benincasa*) or united (*Cucurbita, Cucumis*), imbricate, commonly yellow or white. **Androecium** with 5 stamens, anthers monothecous, filaments free (*Luffa*) or connate, sometimes 4 of these fused in two pairs thus two stamens bithecous and third monothecous giving appearance of 3 stamens (*Coccinia*), rarely all five fused (*Cucurbita*), pollen grains with 3 to many furrows. **Gynoecium** with 3 united carpels (syncarpous), unilocular with many ovules, placentation parietal, placentae enlarged intruding and often meeting in centre forming pseudo-axile placentation, ovary inferior, styles simple or 3-partite. **Fruit** a berry, pepo or capsule; seeds many, embryo straight, endosperm absent. Pollination mostly by insects. Dispersal by animals, capsules of *Echinocystis* open explosively.

Economic importance: The family is economically important for its food plants such as cucumber (*Cucumis sativus*), watermelon (*Citrulus lanatus, syn. C. vulgaris*), loofah (*Luffa acutangula, L. cylindrica*), bottle gourd (*Lageneria siceraria*), melon (*Cucumis melo*) and red pumpkin (*Cucurbita maxima*). The dried fruit of *Luffa* yields bathroom sponge loofah. Species of *Bryonia, Cucumis, Momordica* are of medicinal importance.

Phylogeny: The family was earlier considered closely related to Passifloraceae and included under the same order. Hutchinson (1973) placed them under separate orders, Cucurbitales derived from Passiflorales through formation of

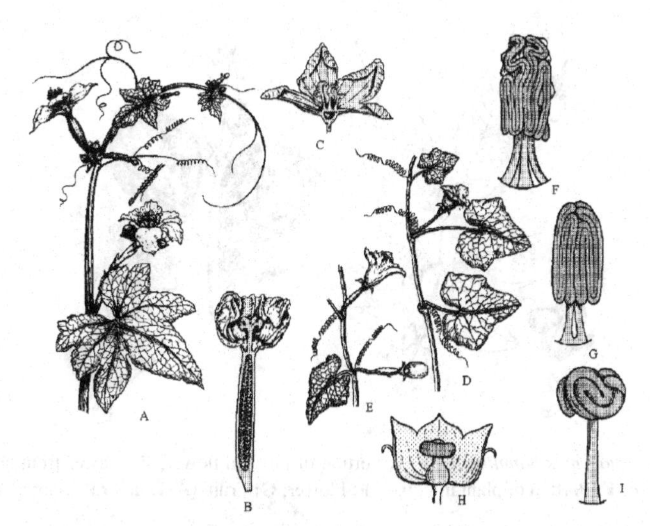

Figure 13.64: Cucurbitaceae. *Luffa cylindrica.* **A:** Branch with male flowers on peduncle towards the base and solitary axillary female flowers towards the top; **B:** Vertical section of female flower; **C:** Vertical section of male flower. *Coccinia cordifolia.* **D:** Branch with male flower; **E:** Branch with female flowers. **Stamen types. F:** *Lagenaria* with 3 stamens, 2 with bithecous anthers and 1 with monothecous anther; **G:** *Cucurbita* with anthers united into a column; **H:** *Cyclanthera* with anthers united into 2 rings running around the top; **I:** *Sicyos* with filaments as well as anthers united.

unisexual flowers, parietal placentation, inferior ovary, and modification of stamens. The separation is followed by Takhtajan, Dahlgren, and APG group. Cronquist and Thorne preferred to retain these and other families under the same order Violales. Thorne (1999) placed Cucurbitaceae along with Begoniaceae and Datiscaceae under a separate suborder Begoniineae. Subsequently (2003, 2006, 2007) he has added Tetramelaceae (earlier included under Datiscaceae). It is interesting to note that order Cucurbitales of APG IV and APweb include the same four families along with a few more, Cucurbitaceae and Begoniaceae share the apomorphies of inferior ovary, strongly intruded placentae and imperfect flowers. Monophyly of Cucurbitales is supported by serological data and *rbcL* sequences. Cucurbitaceae is easily recognized and monophyletic, but of the two subfamilies commonly recognized only Cucurbitoideae is monophyletic, Nhandiroboideae (Zanonioideae) being paraphyletic. Renner et al. (2002) from the molecular studies multiple chloroplast loci. P. 169, concluded that Nhandiroboideae form an unresolved basal group. Schaefer and Renner (2011) divided Cucurbitaceae into 15 tribes. Renner and Schaefer (2016) based on Ribosomal RNA studies developed a phylogenetic scheme of the family.

COM Clade

Celastraceae R. Brown ***Spindle-Tree Family***

96 genera, 1,350 species

Widely spread, mainly in tropical and subtropical regions, a few species in temperate regions.

Placement:

	B & H	Cronquist	Takhtajan	Dahlgren	Thorne	APG IV/(APweb)
Informal Group						COM Clade/ (Rosids/Fabid/ Rosid I)
Division		Magnoliophyta	Magnoliophyta			
Class	Dicotyledons	Magnoliopsida	Magnoliopsida	Magnoliopsida	Magnoliopsida	
Subclass	Polypetalae	Rosidae	Rosidae	Magnoliidae	Rosidae	
Series+/ Superorder	Disciflorae+		Celastranae	Rutanae	Celastranae	
Order	Celastrales	Celastrales	Celastrales	Celastrales	Celastrales	Celastrales

Salient features: Trees, shrubs or climbers, leaves simple often leathery, stipules usually small, falling early or absent, flowers small, greenish, pentamerous, in clusters, ovary superior with large fleshy disc at base fused with the base of stamens, carpels 2–5, seeds with brightly colored coat or aril.

Figure 13.65: Celastraceae. *Euonymus hamiltonianus.* **A:** Portion of plant in flower; **B:** Flower from above; **C:** Flower from base; **D:** Fruit. *Celastrus paniculatus.* **E:** Portion of plant in flower; **F:** Flower; **G:** Fruit. (A–D, after Polunin and Stainton, Fl. Himal, 1984).

Major genera: *Maytenus* (200 species), *Salacia* (150), *Euonymus* (130), *Hippocratea* (120), *Cassine* (60) and *Crossopetalum* (50).

Description: Trees, shrubs or climbers (*Hippocratea*), stem smooth or with spines, with or without laticifers, juice not colored, nodes unilacunar, vessels with simple or scalariform end-walls, vestured pits absent. **Leaves** alternate (*Maytenus*) or opposite (*Euonymus*), simple, usually leathery, serrate, pinnately veined, stipules small and caducous or absent. **Inflorescence** of flat-topped axillary or terminal clusters, rarely solitary or racemose. **Flowers** small, greenish or greenish-white, regular, bisexual or unisexual, hypogynous perigynous or epigynous. **Calyx** with usually 4–5 small sepals, distinct or connate at base, rarely 3, imbricate, rarely valvate. **Corolla** with 4–5 free petals, rarely 3, somewhat similar to sepals, rarely absent. **Androecium** with usually 3–5 stamens rarely many (*Plagiopteron*), free, often attached at base to enlarged disc, rarely connate at base, anthers bithecous, dehiscence longitudinal, pollen grains aperturate or colporate. **Gynoecium** with 2–5 united carpels, ovary superior, sometimes inferior (*Empleuridium*) due to elarged disc, placentation axile with 2–6 ovules in each chamber, style short, terminal, stigma capitate or lobed, dry type, non-papillate. **Fruit** a berry, drupe, capsule or samara; seed usually surrounded by brightly colored aril, embryo large and straight, endosperm present.

Economic importance: Economically the family is of lesser importance, a few used as ornamentals. Climber *Celastrus scandens* is grown for its attractive colored fruits and seeds. Various species of *Euonymus* are grown for their attractive foliage, *E. japonicus* with shiny leathery leaves very popular as hedge plant along pathways in temperate climate. The toxic alkaloid maytansine (from *Maytenus*), when delivered by antibodies, may have application in treating colon cancers.

Phylogeny: The family was earlier placed closer to Rhamnaceae, latter removed along with a few other families to Rhamnales by Hutchinson (1973). Dahlgren (1989) placed Stakhousiaceae, Lophopyxidaceae, Cardiopteridaceae, Corynocarpaceae and Celastraceae under Celastales under superorder Rutaneae. Both Takhtajan (1997) and Thorne (2003) placed Celastales under superorder Celastanae of Rosidae. Thorne had earlier placed this superorder as fifth after Rosanae, subsequently (2006, 2007) bringing it to the begining of Rosidae. Celastales of Thorne (2006) included 4 families Celastaceae, Parnassiaceae, Lepidobotryaceae and Huaceae. Zhang and Simmons (2006) found that Huaceae were sister to Oxalidales, with quite strong support (jacknife values over 80%); they suggest that Huaceae should be included in Oxalidales. Thorne (2007) and Stevens (APweb, 2008), as such exclude Huaceae from the order Celastrales. Whereas Stevens kept, it unplaced within Eurosids I, Thorne shifted it under order Oxalidales, followed by Stevens (2018). Thorne (2007) recognizes 4 subfamilies with Celastraceae: Celastroideae, Hippocrateoideae, Macgregorioideae and Stackhousioideae (Siphonodontoideae recognized as 5th subfamily in 2006 version having been merged with Celastroideae). Stevens (2018) believes these subfamily groups to be not well delimited and divides family into two groups: *Parnassia* et al. and the rest. The family appears to be more uniform with the removal of *Bhesa* to Malpighiales (Zhang and Simmons, 2006) and *Perrottetia* to Huerteales (Crossosomatales of Thorne) near *Tapiscia* (Simmons in Matthews and Endress, 2005b). Sun et al. (2016) found *Mortonia* to be sister to the rest of the family, with [*Parnassia* + *Lepuropetalum*] and [*Zinowiewia* [*Quetzalia* + *Microtropis*]] together forming a clade.

Oxalidaceae R. Brown ***Wood Sorrel Family***

6 genera, 570 species

Distributed mainly in tropical and subtropical regions, a few species in temperate regions.

Placement:

	B & H	Cronquist	Takhtajan	Dahlgren	Thorne	APG IV/(APweb)
Informal Group						COM Clade/ (Rosids/Fabid/ Rosid I)
Division		Magnoliophyta	Magnoliophyta			
Class	Dicotyledons	Magnoliopsida	Magnoliopsida	Magnoliopsida	Magnoliopsida	
Subclass	Polypetalae	Rosidae	Rosidae	Magnoliidae	Rosidae	
Series+/ Superorder	Disciflorae+		Geranianae	Rutanae	Oxalidanae	
Order	Geraniales	Geraniales	Oxalidales	Linales	Oxalidales	Oxalidales

B & H under family Geraniaceae.

Salient features: Herbs or shrubs, leaves usually compound, with a sour taste, leaflets pulvinate at base, entire, stipules usually absent, flowers pentamerous, heterostylus, petals clawed, stamens united, outer stamens shorter, styles 5, seeds with conspicuous endosperm, arillate.

Major genera: *Oxalis* (500 species), *Biophytum* (50) and *Averrhoa* (12).

Description: Herbs with bulbous tubers or fleshy rhizome, or shrubs, rarely trees, often with soluble and crystalline oxalates. **Leaves** alternate or all basal, pinnately (*Biophytum*) or palmately compound or trifoliate (some species of *Oxalis*), rarely replaced by phyllode (*Oxalis bupleurifolia*, petioles forming phyllodes), leaflets often folding in cold or at night, entire, often emarginate, with pinnate or palmate reticulate venation, leaflets often with prominent pulvinus, stipules small or absent. **Inflorescence** cymose umbel, rarely solitary. **Flowers** bisexual, actinomorphic, usually heterostylus, sometimes cleistogamous and apetalous (*Oxalis acetosella*). **Calyx** with 5 sepals, free, green, persistent. **Corolla** with 5 petals, free or connate at base, often clawed, usually convolute, absent in cleistogamous flowers. **Androecium** with 10 stamens, usually in two whorls, usually connate at base, outer filaments usually shorter than inner, anthers bithecous, dehiscence by longitudinal slits, pollen grains tricolpate or triporate, nectar glands at the base of filaments or alternating with petals. **Gynoecium** with 5 united carpels, rarely free (*Biophytum*), ovary superior, placentation axile, 1 or more ovules in each loculus, styles 5, free, persistent, stigmas capitate or shortly divided. **Fruit** a loculicidal capsule or berry, often angled; seeds usually with an aril, embryo straight, endosperm copious, testa often elastic turning inside out and ejecting seed. **Pollination** by insects, heterostyly resulting in outcrossing. Mostly self-dispersed by explosive inversion of testa and aril.

Economic importance: The family is of little importance. The tubers of *Oxalis tuberosa* (oca) are eaten in Andean South America, and that of *O. crenata* boiled and eaten in Peru. The leaves of *O. acetosella* are sometimes used as salad. The bulbous stem of O. *pes-caprae* (Bermuda buttercup) are sometimes used as vegetable in France and North Africa. *Averrhoa carambola* (carambola or star fruit) is cultivated widely for its edible fruit.

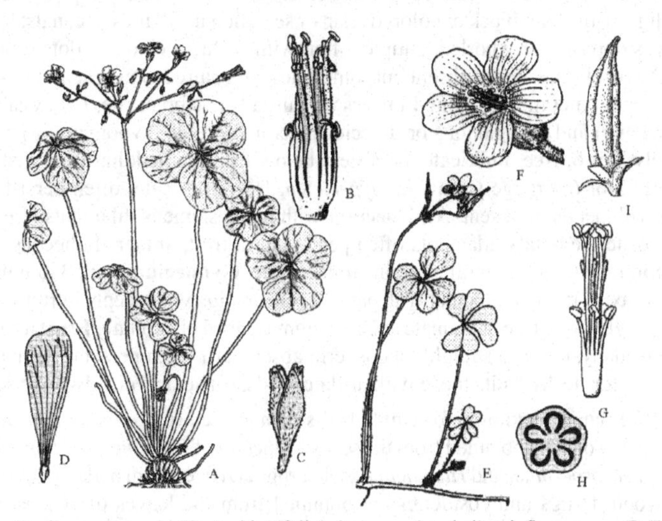

Figure 13.66: Oxalidaceae. *Oxalis martiana*. **A:** Plant with trifoliate leaves and umbellate inflorescence; **B:** Flower with calyx and corolla removed; **C:** Calyx; **D:** Petal. *O. corniculata*. **E:** Portion of plant rooting at nodes and umbellate inflorescence; **F:** Flower; **G:** Flower with calyx and corolla removed; **H:** Transverse section of ovary; **I:** Fruit with persistent calyx (A–D, after Sharma and Kachroo, 1983).

Phylogeny: The family was earlier included under Geraniaceae (Bentham and Hooker) but now separated into a distinct family and separable by 5 distinct styles, possession of arillate seeds and absence of stipules. Phylogenetic studies based on *rbcL* sequences (Chase et al., 1993) indicate that Oxalidaceae are more closely related to Cunoniaceae and Cephalotaceae (and related families) and included under Oxalidales (Judd et al., APG II, APweb), distinct from Geraniales. The family is also related to Linaceae with it was placed by Dahlgren under Linales. Woody genera including *Averrhoa* are sometimes placed in a distinct family but are better placed here (APweb, 2018). The genus *Hypseocharis* with united style, included here by Hutchinson and Cronquist has been shifted (optionally APG II) to Geraniaceae (APG IV, 2016; APweb), and to a distinct family Hypseocharitaceae by Takhtajan (2009). Thorne who had earlier (1999) included *Hypseocharis* under Geraniaceae and placed both Oxalidaceae and Geraniaceae closer together under Dilleniidae—> Geranianae—> Geraniales has subsequently (2003, 2006, 2007) placed Oxalidaceae under Rosidae—> Oxalidanae—> Oxalidales whereas Hypseocharitaceae and Geraniaceae are placed under Rosidae—> Geranianae—> Geraniales, far removed from Oxalidaceae.

Clusiaceae Lindley ***St. John's Wort Family***

19 genera, 1010 species (Including Hypericaceae)

Distributed widely, mainly in moist tropics, some in temperate regions of New as well Old World. *Hypericum* and *Triadenum* distributed in temperate regions.

Placement:

	B & H	**Cronquist**	**Takhtajan**	**Dahlgren**	**Thorne**	**APG IV/(APweb)**
Informal Group						COM Clade/ (Rosids/Fabid/ Rosid I)
Division		Magnoliophyta	Magnoliophyta			
Class	Dicotyledons	Magnoliopsida	Magnoliopsida	Magnoliopsida	Magnoliopsida	
Subclass	Polypetalae	Dilleniidae	Dilleniidae	Magnoliidae	Rosidae	
Series+/ Superorder	Thalamiflorae+		Theanae	Theanae	Podostemanae	
Order	Guttiferales	Theales	Hypericales	Theales	Podostemales	Malpighiales

Salient features: Leaves opposite or whorled, dotted with resin or secretary cavities or canals, margin entire, stipules absent, sepals and petals free, stamens many in bundles, ovary 3–5 chambered, superior, styles free, stigma papillate, fruit a capsule or berry.

Major Genera: *Hypericum* (350 species), *Garcinia* (210), *Calophyllum* (180), *Clusia* (150), *Kayea* (7), *Mammea* (65), *Vismia* (55), *Chrysochlamys* (55), *Kielmeyera* (50), *Harungana* (50) and *Triadenum* (10).

Description: Herbs (*Hypericum*), shrubs or trees (*Garcinia*), rarely as woody lianas or epiphytes (some species of *Clusia*) behaving like strangling figs, with clear, black or colored resin or secretion in cavities or canals; hairs simple, multicellular, sometimes stellate. **Leaves** opposite or whorled, simple, often with pellucid or black dots (punctate) or canals, margin entire, unicostate reticulate venation, stipules absent but sometimes with paired glands at nodes. **Inflorescence** of terminal cymes or with solitary flowers, sometimes thyrsoid. **Flowers** with usually two bracteoles below calyx, bisexual (*Hypericum*) or unisexual (*Clusia, Garcinia*) and polygamous or dioecious, actinomorphic, hypogynous, pentamerous. **Calyx** with 5 sepals, sometimes 4, rarely 3 or 6, free, imbricate, persistent below fruit or shedding from mature fruit. **Corolla** with 5 petals, sometimes 4, rarely 3 or 6, orange-yellow in *Hypericum*, free, imbricate, often persisting as withered remains, nectar glands alternating with petals or absent. **Androecium** with many stamens, filaments free or united in 3–5 bundles (rarely 6–8 bundles), opposite the petals, filaments often persisting in fruit, anthers bithecous, dehiscence longitudinal, staminodes often present in female flower, pollen grains tricolporate. **Gynoecium** with 3–5 united carpels, rarely many, ovary superior, placentation axile, rarely unilocular with parietal placentation with deeply intruded placentas, ovules 2-many in each locule, anatropous, styles 3–5, free or connate at base, stigmas lobed or capitate. **Fruit** usually a capsule, sometimes berry or drupaceous; seeds many, embryo straight, endosperm absent, aril often present. Pollination mostly by bees and wasps. Dispersal by animals for fleshy fruits those with arillate seeds, capsular fruits disperse seeds by wind or water.

Economic importance: The family is known for edible fruits mangosteen (*Garcinia mangostana*) and mammey apple (*Mammea americana*). Fats and oils are obtained from the seeds of species of *Calophyllum*, *Pentadesma*, etc. The species of *Harungana*, *Calophyllum*, *Psorospermum* and *Harungana* yield drugs and dyes from bark. Species of *Clusia* and *Hypericum* yield hard and durable wood. Drugs and cosmetics are obtained from the leaves of *Hypericum* spp. and *Harungana madagascariensis*, and flowers of *Mesua ferrea*. Gums and pigments are extracted from the stems of *Garcinia* (source of gamboge) and *Clusia* (source of healing gums). Species of *Vismia* with showy flowers are often grown as ornamentals.

Figure 13.67: Clusiaceae. *Hypericum calycinum.* **A:** Branch with terminal flower. *H. lobbii* **B:** Portion of branch showing one flower; **C:** Androecium. **D:** Flower with petals and stamens removed; **E:** Transverse section of ovary. *Clusia purpurea.* **F:** Branch with inflorescence; **G:** Flower. *Garcinia mangostana.* **H:** Branch with fruit; **I:** flower; **J:** Vertical section of flower; **K:** Fruit with rind from top removed. *Hypericum myrtifolium.* **L:** Portion of branch with flowers; **M:** Flower top view, enlarged. (L–M, after Goodfrey and Wooten, Aq. Wetland Pl. SE US, vol. 2, 1981; H–K, after Bailey, Man. Cult. Pl., 1949).

Phylogeny: The family Hypericaceae was treated as distinct from Guttiferae (Clusiaceae) by Bentham and Hooker (1862). Engler and Prantl (1887) combined the two families under Clusiaceae, a treatment followed by Heywood (1978) and Cronquist (1988). Hutchinson (1973) justified separation of Hypericaceae based on constantly bisexual flowers and gland-dotted leaves, as against unisexual flowers, close veins and secretary canals. He argued that 'Hypericaceae is fairly well circumscribed, and there seems little to be gained, in these days of smaller family concepts, by including them in Clusiaceae (Guttiferae), as in Engler and Prantl system'. The two are combined in treatments of Judd et al. (2002, 2008; under Clusiaceae) and Thorne (1999, 2000 and 2007 as Clusiaceae, 2003, 2006; prefered priority name Hypericaceae). The affinities within the group are not clearly resolved as suggested by the studies of Chase et al. (2002) and Gustafsson et al. (2002). Thorne (2003, 2006), placed Bonnetiaceae, Hypericaceae, Elatinaceae and Podostemaceae under order Hypericales, prefering name Podostemales in 2007 revision. Thorne (2007) divides Clusiaceae into five subfamilies Kielmeyeroideae, Calophylloideae, Clusioideae, Chrysopioideae and Hypericioideae. APG II, APG IV (2016) and APweb treat the two as distinct families, latter earlier (2008) recognizing two subfamilies Kielmeyeroideae and Clusioideae (including Calophylloideae), but in latest Version 14 (updated 2018) recognizes three tribes: Clusieae, Garcinieae and Symphonieae under Clusiaceae and three: Vismieae, Cratoxyleae and Hypericeae under Hypericaceae. Broadly circumscribed family (under name Clusiaceae or Hypericaceae) is assumed to be monophyletic based on anatomical and chemical evidence. Clusieae are a well-supported clade clearly sister to the rest of the family (Ruhfel et al., 2016).

Violaceae Batsch ***Violet Family***

31 genera, 980 species

Distributed widely, mainly in temperate regions.

Placement:

	B & H	**Cronquist**	**Takhtajan**	**Dahlgren**	**Thorne**	**APG IV/(APweb)**
Informal Group						COM Clade/ (Rosids/Fabid/ Rosid I)
Division		Magnoliophyta	Magnoliophyta			
Class	Dicotyledons	Magnoliopsida	Magnoliopsida	Magnoliopsida	Magnoliopsida	
Subclass	Polypetalae	Dilleniidae	Dilleniidae	Magnoliidae	Rosidae	
Series+/ Superorder	Thalamiflorae+		Violanae	Violanae	Violanae	
Order	Parietales	Violales	Violales	Violales	Violales	Malpighiales

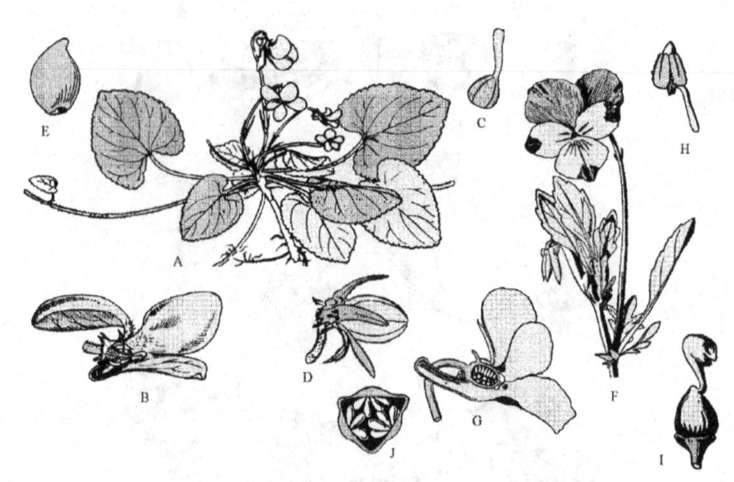

Figure 13.68: Violaceae. *Viola canescens.* **A:** Plant with flowers; **B:** Vertical section of flower; **C:** Gynoecium; **D:** Fruit with persistent calyx; **E:** Seed. *V. tricolor.* **F:** Portion of plant with flowers; **G:** Vertical section of flower showing spurred lower petal; **H:** Stamen with spurred anther; **I:** Ovary with style and enlarged stigma; **J:** Transverse section of ovary with parietal placentation.

Salient features: Herbs, leaves serrate, stipules present, flowers zygomorphic, bisexual, petals 5, anterior spurred, anthers with spur-like nectaries, carpels 3, united, placentation parietal, fruit a loculicidal capsule, seeds large dispersed explosively as the fruit wall closes round them and squeezes them out.

Major genera: *Viola* (450 species), *Rinorea* (280), *Hybanthus* (110), *Anchietia* (8) and *Leonia* (6).

Description: Herbs (*Viola*) shrubs (*Rinorea*) or trees (*Rinorea maingayi*), rarely climbers (*Anchietia*) with often saponins or alkaloids. **Leaves** alternate rarely opposite (*Hybanthus*), mostly basal, simple, sometimes lobed, entire to serrate, venation pinnate or palmate, veins often conspicuous, stipules present, sometimes foliaceous (*Viola*). **Inflorescence** usually with solitary axillary flower, sometimes in racemes or spikes. **Flowers** bisexual, rarely unisexual, actinomorphic (*Rinorea*) or zygomorphic (*Viola*), hypogynous, pentamerous, sometimes cleistogamous. **Calyx** with 5 sepals, usually free, sometimes slightly connate to form a ring around ovary, imbricate, persistent. **Corolla** with 5 petals, free, imbricate or convolute, unequal, anterior usually largest and spurred or saccate. **Androecium** with 5 stamens, filaments short, free or slightly connate at base, anthers erect, somewhat connivent forming a ring around ovary, 2 anterior anthers often with spur-like nectaries, connective often with triangular appendage, dehiscence longitudinal, introrse, pollen grains usually tricolpate. **Gynoecium** with 3 united carpels, carpels rarely 2–5 (*Melicystus*), ovary superior, unilocular with parietal placentation, ovules many, anatropous, style 1, stigma often expanded but with small receptive region, sometimes lobed. **Fruit** a loculicidal capsule; seeds with straight embryo, endosperm and aril present. Pollination by insects, attracted by nectar in the spur. Seeds are often dispersed explosively as the fruit wall closes round them and squeezes them out. Ants also disperse seeds, attracted by oily aril.

Economic importance: The family is mainly known for ornamental pansy flowers (*Viola*) and green-violet (*Hybanthus*). Flowers of *Viola odorata* is largely grown in France for essential oil used in the manufacture of perfumes, flavourings and toiletries. The flowers are also preserved in sugar ('banafsha'). *Hybanthus ipecacuanha* has been used as substitute for true ipecac (*Psychotria ipecacuanha*) as emetic. Roots of *Anchietia salutaris* are used as an emetic and to treat sore throats and lymphatic tuberculosis. The roots of *Corynostylis hybanthus* are used as an emetic.

Phylogeny: The family is clearly defined and uniformly placed in most classifications under Violanae of Dilleniid complex. Hutchinson who had largely separated dicotyledons into woody (Lignosae) and herbaceous (Herbaceae) lineages, had especially chosen to justify the position of largely herbaceous *Viola* in the predominantly woody clade, and considered the herbaceous habit in this genus to be derived from woody ancestors. APG (latest APG IV, 2016) and APweb (latest version 14, updated 2018), however place this family in a broadly circumscribed order Malpighiales. Two tribes are commonly recognized: Rinoreae with mainly actinomorphic flowers and Violeae with zygomorphic flowers. Although the family is clearly monophyletic, rbcL sequences suggest that neither tribe is monophyletic. Violaceae are weakly associated with Acharaciaceae (and Goupiaceae, Lacistemataceae and Ctenolophonaceae) in Chase et al. (2002). Thorne, who had earlier (1999, 2000) placed Violanae under subclass Dilleniidae shifted it to Rosidae, Dilleniidae having been dismantled. He (2006, 2007) recognizes three subfamilies Violoideae, Leonioideae and Fusispermoideae under Violaceae, APweb (2018) recognizing only two, Leonioideae being merged with Violoideae. Wahlert et al. (2014) and Flicker and Ballard (2015) suggest splitting genera *Hybanthus* and *Rinorea.*

Salicaceae Mirbel ***Willow Family***

56 genera, 1220 species

Distributed widely, mainly in north temperate to arctic regions, in moist open habitats.

Placement:

	B & H	Cronquist	Takhtajan	Dahlgren	Thorne	APG IV/(APweb)
Informal Group						COM Clade/ (Rosids/Fabid/ Rosid I)
Division		Magnoliophyta	Magnoliophyta			
Class	Dicotyledons	Magnoliopsida	Magnoliopsida	Magnoliopsida	Magnoliopsida	
Subclass	Monochlamydeae	Dilleniidae	Dilleniidae	Magnoliidae	Rosidae	
Series+/ Superorder	Ordines anomali+		Violanae	Violanae	Violanae	
Order		Salicales	Salicales	Salicales	Violales	Malpighiales

Salient features: Deciduous trees and shrubs, leaves with salicoid teeth, stipules conspicuous, flowers unisexual, inflorescence a catkin, flowers naked, carpels 2, ovules many, seeds with hairs.

Major genera: *Salix* (445 species; incl. *Chosenia*), *Casearia* (180), *Homalium* (150), *Xylosma* (85), *Scolopia* (40) and *Populus* (40).

Description: Deciduous trees and shrubs containing phenolic heterosides salicin and populin, containing tannins, tension wood with multilayered cell walls. **Leaves** alternate, simple, serrate to dentate, teeth salicoid (vein entering tooth and associated with glandular seta), venation pinnate to palmate, reticulate, stipules present, sometimes foliaceous and persistent. **Inflorescence** erect or pendulous catkins, on short branches. **Flowers** unisexual (plants dioecious), actinomorphic, reduced, usually subtended by hairy bracts. **Calyx** reduced to a glandular disc (*Populus*) or 1 to 2 fringed nectar gland (*Salix*). **Corolla** absent. **Androecium** with 2 to numerous stamens, filaments free or slightly connate at base, anthers bithecous, dehiscence by longitudinal slits, pollen grains usually tricolpate or triporate, rarely inaperturate. **Gynoecium** with 2–4 carpels, united, ovary superior, unilocular with parietal placentation or with 2–4 basal placentas, ovules many, unitegmic, styles 2–4, stigmas 2–4, capitate, often expanded and lobed. **Fruit** a loculicidal capsule; seeds with a basal tuft of hairs, endosperm scanty or absent. Pollination by wind, flowers of *salix* are pollinated by insects attracted by nectar. Seeds are often dispersed by wind aided by hairs.

Economic importance: The family is important for several species grown as ornamentals, usually avenue trees. Cricket bats and polo balls are usually made from willow (*Salix* spp.) wood. Twigs of willow are commonly used in basket making. The bark of *Salix* contains salicylic acid, which reduces swellings and fever, and is constituent of aspirin.

Phylogeny: The affinities of this family were not known to Bentham and Hooker who placed it along with other uncertain families under Ordines anomali of Monochlamydeae. The reduced flowers of Salicaceae (and other members of Amentiferae) were considered to represent primitive dicots by Engler (1892) and Rendle (1904, 1930). Fisher (1928) on the basis of extensive studies concluded that the simplicity of flowers is largely due to extreme reduction and not a representation of archaic features, and that flowers in the ancestral form possessed a perianth of 1 or 2 series, which is now represented by a cupule-like gland. Hutchinson (1926), placed the family under Hamamelidales, treating it as the most primitive within the group. Hjelmquist (1948) believed the cup or finger like gland in the flower was formed by the reduction of an undifferentiated bracteal envelope and that it is not quite appropriate to designate them as perianth. He also separated the family under order Salicales. This treatment has been followed by Hutchinson (1973; placing Salicales after Hamamelidales), Cronquist (1988; placing Salicales under Dilleniidae after Violales and not in Hamamelidae), Dahlgren (1989; under Violanae after Violales and Cucurbitales), Takhtajan (1997; Dilleniidae-Violanae after Violales, Passiflorales and Caricales). Thorne (1999) placed it under Dilleniidae—> Violanae—> Violales after Violaceae and Flacourtiaceae, subsequently (2003, 2006, 2007) placing it under Rosidae owing to the abolition of Dilleniidae. APG (APG II onwards) and APweb include this family under Malpighiales.

The family is clearly monophyletic, having affinities with Flacourtiaceae, which also exhibit salicoid teeth, presence of salicin and apetalous flowers in some genera. Molecular data also support close affinities. The family Salicaceae as such has been broadly circumscribed in APG II (onwards) and APweb (55 genera, 1,010 species) to include larger part of Flacourtiaceae and smaller diverse families such as Bembiciaceae, Homaliaceae, Poliothyrsidaceae, Prockiaceae, Samydaceae and Scyphostegiaceae. Broadly circumscribed Salicaceae is defined by leaves with salicoid teeth, cocarcinogens and flowers in which sepals and petals if present are equal in number, those with sepals and petals not equal shifted to Achariaceae. The genus *Casearia* (formerly in Flacourtiaceae), which may lack salicoid leaf teeth and has apetalous flowers with the disc on the basal-adaxial surface of the calyx, is sister to the rest of Salicaceae, although this position is weakly

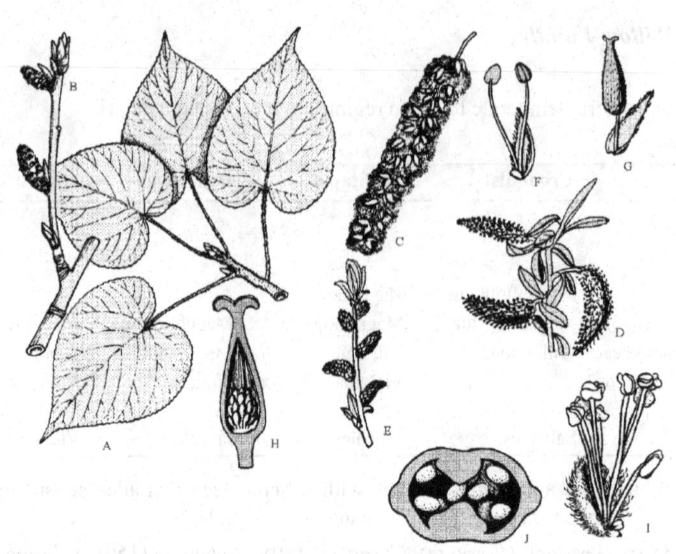

Figure 13.69: Salicaceae. *Populus cilliata.* **A:** Portion of a vegetative branch; **B:** Branch with male catkins; **C:** Fruiting female catkin. *Salix alba.* **D:** Portion of a shoot with female catkins; **E:** Portion of shoot with male catkins; **F:** Male flower with hairy bract and 2 stamens; **G:** Female flower with bract and stipitate ovary; **H:** Longitudinal section of ovary with basal placentas. *S. caroliniana.* **I:** Male flower with bract, 2 nectar glands and many stamens; **J:** Transverse section of ovary with parietal placentation.

supported by *rbcL* (Chase et al., 2002) but strongly supported by data based on three genes (Soltis et al., 2000a). Stevens (Apweb version 14, updated 2018) divides Salicaceae into three subfamilies: Samydoideae (13 genera), Scyphostegioideae (2 genera) and Salicoideae (6 tribes, 40 genera).

Rafflesiaceae Dumortier ***Rafflesia Family***

3 genera, 25 species

Southeast Asia, from India to Indonesia.

Placement:

	B & H	Cronquist	Takhtajan	Dahlgren	Thorne	APG IV/(APweb)
Informal Group						COM Clade/ (Rosids/Fabid/ Rosid I)
Division		Magnoliophyta	Magnoliophyta			
Class	Dicotyledons	Magnoliopsida	Magnoliopsida	Magnoliopsida	Magnoliopsida	
Subclass	Monochlamydeae	Rosidae	Magnoliidae	Magnoliidae	Malvidae	
Series+/ Superorder	Multiovulatae Terrestres+		Rafflesianae	Magnolianae	Rafflesianae	
Order		Rafflesiales	Rafflesiales	Rafflesiales	Rafflesiales	Malpighiales

B & H as family Cytinaceae.

Salient features: Parasitic on stem and roots, plant body filamentous like a fungal mycelium, flowers usually unisexual, with fleshy petaloid calyx, stamens in a column, ovary inferior, carpels fused, placentation parietal, fruit fleshy.

Major genera: *Rafflesia* (21 species), *Sapria* (2) and *Rhizanthes* (2).

Description: Total parasites on stems and roots of angiosperms, vegetative part filamentous, like fungal mycelium, rootless, permeating the host tissues, with only the flowers or the flowering stems emerging from the host tissue, xylem without vessels. **Leaves** much reduced, present at the bases of flowering stems, or beneath the flower, or absent, alternate, opposite, or whorled, of membranous scales, stomata absent. **Inflorescence** with solitary flowers. **Flowers** small to very large (*Rafflesia arnoldii*, with the largest known flowers in angiosperms, up to 1 m in diameter), regular, usually unisexual, cyclic. **Perianth** with tepal green or petaloid, 4, or 5(–10), free, or united into

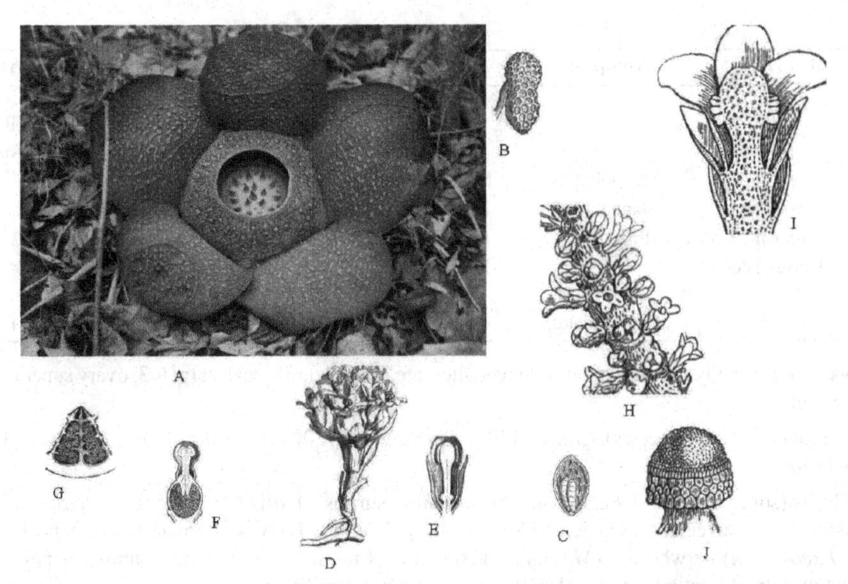

Figure 13.70: Rafflesiaceae (A–C). A: Fully opened flower of *Rafflesia speciosa* (photo courtesy Julie Barcelona, Manilla, Philippines). **B:** Seed of *R. arnoldii*. **C:** Partial section of seed showing undivided embryo. **Cytinaceae** (D–G). *Cytinus hypocistis*. **D:** Plant in flower; **E:** Vertical section of male flower; **F:** Vertical section of female flower; **G:** Portion of transverse section of ovary. **Apodanthaceae** (H–J) *Pilostyles berterii*. **H:** Host twig with flowers of *Pilostyles* emerging out; **I:** Vertical section of male flower; **J:** A head of stamens.

tube, usually fleshy, imbricate, rarely valvate. **Androecium** with 5–100 stamens, united with the gynoecium, free or filaments united into a tube round the stylar column, 1 whorled, filaments slender, or reduced, anthers monothecous or bithecous, dehiscing by longitudinal slits, pores, or transversely, pollen grains usually nonaperturate. **Gynoecium** with 4–8 carpels, ovary inferior, unilocular with 4–14 parietal placentas, or 3–10(–20) locular by deep intrusion of the placentas (*Rafflesia*), ovules 50–100 per locule, non-arillate, hemianatropous to anatropous, bitegmic, tenuinucellate, united, style expanded into an often large, complex disk, with stigmatal projections. **Fruit** usually fleshy berry, seeds endospermic, minute, embryo rudimentary.

Economic importance: None.

Phylogeny: The family is closely related to Hydnoraceae, often placed in Magnoloid complex under a distinct order Rafflesiales. Cronquist, however, placed this order under Rosidae. The family is often considered closer to Aristolochiaceae because of similar perianth. According to Judd et al. (2002), Rafflesiaceae (also Balanophoraceae and Hydnoraceae) look so different from other flowering plants that no one has been sure where to place them. Hydnoraceae appear to belong in Piperales, the other two, according to them are apparently dicots, but are no more precisely placed than that. The recent cladistic studies, however, place Hydnoraceae and Aristolochiaceae under Piperales based on multigene analyses. The position of Rafflesiaceae was uncertain in APG II, but placed under Malpighiales in APG IV (2016) and APweb (2018). Nickrent (2002) considers this family to be closer to Malvales. Thorne had earlier (1999) broadly circumscribed the family Rafflesiaceae, divided into 4 subfamilies: Mitrastemonoideae (*Mitrastemon*-flowers bisexual and solitary, ovary superior), Cytinoideae (*Cytinus* and *Bdallophyton*-flowers unisexual, in racemes, stamens in one ring, ovary inferior with 8–14 placentas), Apodanthoideae (*Apodanthes*, *Pilostyles* and *Berlinianche*-flowers small, unisexual, solitary, stamens in 2 or 4 rings, ovary inferior with 4 placentas or 1 continuous placenta) and Rafflesioideae (Flowers solitary and unisexual, large, stamens in 1 ring, ovary inferior with many irregular chambers). These have now (2003) been recognized as independent families Mitrastemonaceae, Cytinaceae, Apodanthaceae and Rafflesiaceae, respectively. APG II and APweb also recognize them as independent families but unplaced towards the end of angiosperms, shifting them to diverse orders in APG IV (2016) and APweb (2018): Mitrastemonaceae in Ericales, Cytinaceae in Malvales and Apodanthaceae in Cucurbitales. The family placement of *Bdallophyton* was uncertain but, has now been shifted to Cytinaceae.

Euphorbiaceae A. L. de Jussieu ***Spurge Family***

266 genera, 8,302 species (including Phyllanthaceae 57 genera, 2,050 species)

Distributed widely in tropical and subtropical regions, with few species in temperate regions.

Placement:

	B & H	Cronquist	Takhtajan	Dahlgren	Thorne	APG IV/(APweb)
Informal Group						COM Clade/ (Rosids/Fabid/ Rosid I)
Division		Magnoliophyta	Magnoliophyta			
Class	Dicotyledons	Magnoliopsida	Magnoliopsida	Magnoliopsida	Magnoliopsida	
Subclass	Monochlamydeae	Rosidae	Dilleniidae	Magnoliidae	Rosidae	
Series+/ Superorder	Unisexuales+		Euphorbianae	Malvanae	Podostemanae	
Order		Euphorbiales	Euphorbiales	Euphorbiales	Euphorbiales	Malpighiales

Salient features: Plants usually with milky latex, leaves alternate, flowers unisexual, carpels 3, ovary superior, 3-chambered, ovule with a caruncle.

Major genera: *Euphorbia* (2420 species), *Croton* (1300), *Phyllanthus* (1200), *Acalypha* (430), *Glochidion* (300), *Antidesma* (140), *Manihot* (160) and *Jatropha* (140).

Description: Herbs (some species of *Euphorbia, Phyllanthus*) shrubs (*Acalypha*) or trees (*Hevea*) with often milky or colored latex, sometimes succulent and cactus-like, usually poisonous. **Leaves** alternate rarely opposite (some species of *Euphorbia*; *Excoecaria*) or whorled (*Mischodon*), sometimes modified into spines, simple or palmate compound, venation pinnate or palmate, reticulate, stipules present, sometimes modified into spines (*Euphorbia milii*) or glandular, rarely absent. **Inflorescence** of various types, commonly a cup shaped cyathium (*Euphorbia*) having a cup-shaped involucre with usually 5 nectaries along the rim and enclosing numerous male flowers (arranged in scorpioid cymes, without perianth and represented by a single stamen) and single female flower in the centre; sometimes a raceme (*Croton*) or panicle (*Ricinus*). **Flowers** unisexual (monoecious or dioecious), actinomorphic, hypogynous. **Perianth** usually with 5 tepals (representing sepals, petals absent), rarely 6 in two whorls (*Phyllanthus*) or absent (*Euphorbia*), petals usually absent but present in *Jatropha* and *Aleurites*, free or connate. **Androecium** with 1 stamen (*Euphorbia*), 3 with fused filaments (*Phyllanthus*), 5 (*Bridelia*) or many (*Trewia*), sometimes polyadelphous (or with repeatedly branched filaments) as in *Ricinus*, anthers bithecous (sometimes monothecous in *Ricinus* due to splitting of filament), dehiscence longitudinal. **Gynoecium** with 3 united carpels, rarely 4-many, ovary superior, trilocular with 1–2 ovules in each chamber, placentation axile, styles usually 3. **Fruit** a schizocarpic capsule, a regma (*Ricinus*), rarely a berry or drupe (*Bridelia*); seed often with conspicuous fleshy outgrowth called caruncle, embryo curved or straight, endosperm abundant or absent.

Economic importance: The family includes several valuable plants. *Hevea brasiliensis* (Para rubber tree) is the source of natural rubber. Rubber is also obtained from *Manihot glaziovii* (ceara rubber). Thick roots of *Manihot esculenta* (cassava or tapioca) are important source of starch in tropical regions. The leaves of *Cnidoscolus chayamansa* are used as vegetable. The fruits of *Antidesma bunias* are also edible. *Aleurites moluccana* (candlenut tree) and *A. fordii* (Tung tree) are sources of oils used in the manufacture of paints and varnishes. Oil similar to tung is also obtained from the species of *Vernicia*. Castor oil obtained from *Ricinus communis* is used as purgative. The common ornamentals include *Euphorbia pulcherrima, E. milii, Acalypha hispida, Jatropha integerrima* (syn: *J. panduraefolia*) and *Codiaeum variegatum*. The fruit of *Phyllanthus emblica* ('amla') is very rich source of vitamin C. The greasy tallow surrounding the seeds of *Sapium sebiferum* (Chinese tallow tree) is used for making soaps and candles.

Phylogeny: The family was earlier broadly circumscribed (Bentham and Hooker) to include genera which have now been separated under Buxaceae. Earlier considered related to Euphorbiaceae the family Buxaceae has been far removed to Sapindales (Engler and Prantl), Hamamelidales (Hutchinson), Buxales (Takhtajan: under Caryophyllidae—> Buxanae), or Balanopales (Thorne: under Rosidae—> Rosanae near Hamamelidales), Proteales (Judd et al., under core tricolpates), Buxales (APweb) or unplaced at the beginning of Eudicots (APG II). Cronquist is the only recent author to include Buxaceae next to Euphorbiaceae under Euphorbiales (Rosidae). The genus *Ricinus* is sometimes included under a separate family Ricinaceae but is more appropriately included under Euphorbiaceae. Webster (1967, 1994), who studied this family extensively recognized five subfamilies: Phyllanthoideae, Oldfieldioideae, Acalyphoideae, Crotonoideae and Euphorbioideae. These five are also recognized by Thorne (1999, 2003). The former two based on evidence from *rbcL* sequences have been separated into a distinct family Phyllanthaceae by APG IV and APweb, as they do not seem to form a clade with other members of Euphorbiaceae. *Putranjiva, Lingelsheimia* and *Drypetes* have been removed to Putranjivaceae, and *Paradrypetes* shifted to Rhizophoraceae. Rest of the Euphorbiaceae including last three subfamilies form a well-defined clade with single ovule in each chamber. Thorne (2003, 2007) has recognized Putranjivaceae as distinct family (including *Drypetes*), followed in recent treatments. APWeb (2018) recognises 4 subfamilies: Cheilosoideae, Acalyphoideae, Crotonoideae and Euphorbioideae. Although Acalyphoideae in the old sense are paraphyletic, the great bulk of the subfamily is included in a strongly-supported clade, Acalyphoideae s. str., and Cervantes et al. (2016) suggest

Figure 13.71: Euphorbiaceae. *Euphorbia milii.* **A:** Branch with umbellate cyathia and spines; **B:** Vertical section of cyathium to depict showy scarlet bracts, single female flower and numerous male flowers, and nectaries along the rim of cyathium. *E. hirta.* **C:** Portion of plant showing opposite leaves and cyathia in heads; **D:** Cyathium with female flower protruding out and only 4 nectaries, showy bracts absent; **E:** Vertical section of cyathium. *Phyllanthus fraternus.* **F:** Portion of plant with flowers; **G:** Male flower with monadelphous stamens; **H:** Female flower; **I:** Vertical section of female flower. *Croton bonplandianum.* **J:** Portion of plant with flowers and fruits; **K:** Male flower with many stamens; **L:** Female flower; **M:** Vertical section of female flower.

that *Erismanthus* is sister to all other Acalyphoideae examined. There is no support for a monophyletic Crotonoideae s.l., although there are number of distinctive features.

Rosids (continued)

Geraniaceae A. L. de Jussieu *Geranium Family*

5 genera, 830 species

Widespread mainly in temperate and subtropical regions.

Placement:

	B & H	Cronquist	Takhtajan	Dahlgren	Thorne	APG IV/(APweb)
Informal Group						Rosids/(Malvid/ Rosid II)
Division		Magnoliophyta	Magnoliophyta			
Class	Dicotyledons	Magnoliopsida	Magnoliopsida	Magnoliopsida	Magnoliopsida	
Subclass	Polypetalae	Rosidae	Rosidae	Magnoliidae	Rosidae	
Series+/ Superorder	Disciflorae+		Rutanae	Rutanae	Geranianae	
Order	Geraniales	Geraniales	Geraniales	Geraniales	Geraniales	Geraniales

Salient features: Usually herbs, stems swollen at nodes, leaves usually deeply lobed, stipules conspicuous, flowers pentamerous, petals clawed, stamens united, styles 1, fruit with elastic dehiscent schizocarps that curl on the beak, aril absent.

Major genera: *Geranium* (430 species), *Pelargonium* (280), *Erodium* (80) and *Monsonia* (40).

Description: Usually herbs, rarely undershrubs, sometimes aromatic (*Pelargonium*), stems swollen at nodes, usually with stalked glandular hairs. **Leaves** alternate or opposite, simple or palmately lobed, or compound, venation palmate, reticulate, stipules conspicuous. **Inflorescence** cymose umbel, rarely solitary. **Flowers** bisexual, actinomorphic, rarely zygomorphic (*Pelargonium*), hypogynous, pentamerous. **Calyx** with 5 sepals, free, green, persistent, sometimes spurred (*Pelargonium*). **Corolla** with 5 petals, rarely 4 or absent, free, often clawed, imbricate, nectar glands alternating with petals or absent. **Androecium** with 10 (*Geranium*) or 15 (*Monsonia*) stamens, rarely 5 (other 5 sterile-*Erodium*), usually

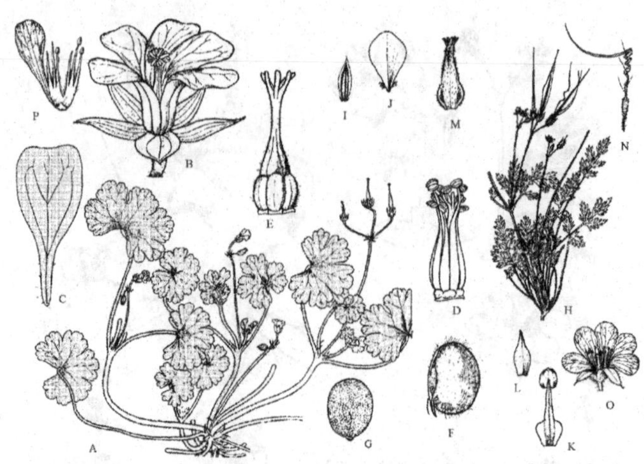

Figure 13.72: Geraniaceae. *Geranium rotundifolium*. **A:** Plant with palmately lobed leaves and umbellate inflorescence; **B:** Flower; **C:** Petal; **D:** Flower with sepals and petals removed to show androecium and gynoecium; **E:** Gynoecium with 5 carpels; **F:** One segment of capsule; **G:** Seed. *Erodium cicutarium*. **H:** Plant with pinnate leaves and umbellate inflorescence; **I:** Sepal; **J:** Petal; **K:** Stamen; **L:** Staminode; **M:** Gynoecium; **N:** One mericarp with long coiled beak. *Monsonia senegalensis*. **O:** Flower having 15 stamens; **P:** Portion of flower to show stamens with filaments united in groups of 3, the androecium being pentadelphous.

connate at base, sometimes pentadelphous (*Monsonia*), rarely free, anthers bithecous, dehiscence by longitudinal slits, pollen grains tricolpate or triporate. **Gynoecium** with 5 united carpels, ovary superior, usually lobed, placentation axile, ovules usually 2 in each loculus, anatropous or campylotropous, style 1, slender and beak-like. **Fruit** a capsular dehiscent schizocarp with 5, 1-seeded segments that separate elastically from central column, and often opening to release seeds (*Geranium*); seeds usually without aril, pendulous, embryo curved, endosperm usually absent or scanty. Pollination by insects. Mostly self dispersed by explosive opening of schizocarps throwing seeds several metres away.

Economic importance: The family is known for *Pelargonium* (often marketed as Geranium), grown as ornamental in pots and also for geranium oil extracted from the leaves and shoots of mainly *P. odoratissimum*. Species of *Geranium* (cranebill) and *Erodium* (storckbill) are also grown as ornamentals. The persistent dry style of *Erodium,* which is hygroscopic, is often used to indicate changes in humidity.

Phylogeny: The family is consistently placed under order Geraniales, sometimes along with Oxalidaceae. The recent DNA based studies (Chase et al., 1993), however, suggest that it is related to Crossosomataceae, Staphyleaceae, in a narrowed circumscribed order. Geraniaceae are well-defined monophyletic group based on *rbcL* sequences and loss of intron in the plastid gene *rpl16* (Price and Palmer, 1993). *Hypseocharis*, with capsular fruits and formerly placed under Oxalidaceae is sister to rest of the family. Takhtajan places it under a distinct family Hypseocharitaceae. Thorne who had earlier (1999) included both Geraniaceae and Oxalidaceae adjacent to each other under Geraniales, has subsequently (2003, 2006, 2007) shifted Oxalidaceae under distinct superorder Oxalidanae, order Oxalidales. He has also removed *Hypseocharis* to a distinct family Hypseocharitaceae, placed next to Geraniaceae. APG II optionally include Hypseocharitaceae under Geraniaceae, firmly established in APG IV (2016). APweb treats *Hypseocharis* as distinct group within Geraniaceae.

Lythraceae J. St.-Hilaire ***Loosestrife Family***

27 genera, 620 species (including Trapaceae)

Widely distributed mainly in tropics, more widespread in America, a few herbaceous species in temperate regions.

Placement:

	B & H	**Cronquist**	**Takhtajan**	**Dahlgren**	**Thorne**	**APG IV/(APweb)**
Informal Group						Rosids/(Malvid/ Rosid II)
Division		Magnoliophyta	Magnoliophyta			
Class	Dicotyledons	Magnoliopsida	Magnoliopsida	Magnoliopsida	Magnoliopsida	
Subclass	Polypetalae	Rosidae	Rosidae	Magnoliidae	Rosidae	
Series+/ Superorder	Calyciflorae+		Myrtanae	Myrtanae	Myrtanae	
Order	Myrtales	Myrtales	Myrtales	Myrtales	Myrtales	Myrtales

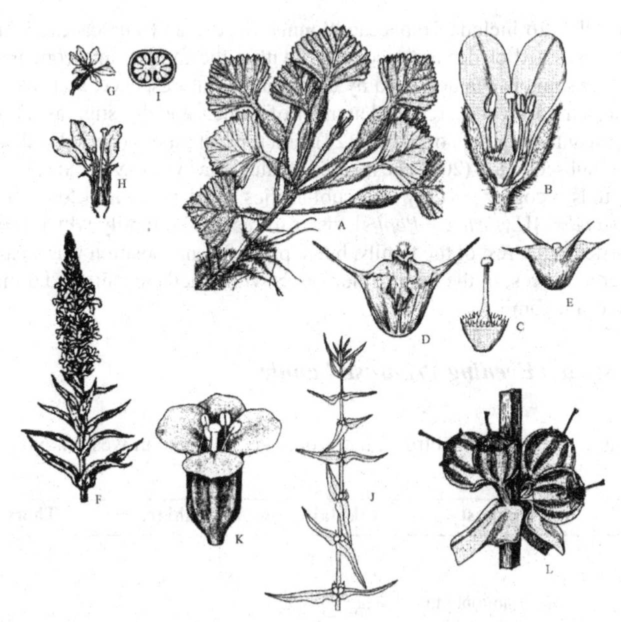

Figure 13.73: Lythraceae. *Trapa bispinosa.* **A:** Portion of plant with leaves (showing swollen petioles) and flowers; **B:** Flower with front sepals and petals removed; **C:** Pistil with disc; **D:** Fruit. **E:** Fruit of *T. natans. Lythrum salicaria* **F:** Branch with flowers; **G:** Flower; **H:** Vertical section of flower; **I:** Transverse section of ovary. *Ammania coccinia.* **J:** Branch with flowers and fruits; **K:** Flower; **L:** Cluster of fruits at node (F–I, after Bailey, Man. Cult. Pl., 1949. J–L, after Godfrey and Wooten, Aq. wetland Pl. SE US, 1981).

Salient features: Bark flaky, leaves opposite, simple, entire without gland dots, flowers in racemes or panicles, bisexual, hypanthium present and ribbed, sepals valvate, petals crumpled, stamens unequal in length, in two whorls, ovary superior, fruit dry, indehiscent or dehiscent capsule.

Major genera: *Cuphea* (280 species), *Diplusodon* (70), *Lagerstroemia* (55), *Nesaea* (50), *Rotala* (45), *Lythrum* (35), *Ammania* (20), and *Trapa* (3, sometimes split into up to 30).

Description: Herbs, shrubs or trees (with flaky bark), rarely spinescent (*Lawsonia*), rarely aquatic herbs (*Trapa*). **Leaves** opposite or whorled, rarely alternate, simple, entire, toothed and with swollen petiole in *Trapa*, venation pinnate, stipules absent or represented by minute hairs. **Inflorescence** of solitary flowers or raceme, panicle or cymose. **Flowers** bisexual, actinomorphic, rarely zygomorphic (*Cuphea*), perigynous with well developed ribbed hypanthium, epicalyx of connate pair of bracts sometimes present below hypanthium (*Lythrum*). **Calyx** with 4–8 sepals, free or connate, valvate, often thick. **Corolla** with usually 4–8 petals, free, imbricate, usually attached along the inner rim of hypanthium, crumpled in bud and wrinkled at maturity, sometimes lacking (*Peplis*, *Rotala*). **Androecium** with usually twice as many stamens as petals, in two whorls, outer whorl alternating with petals, sometimes only in one whorl, usually unequal in length, filaments free, anthers bithecous, dorsifixed, dehiscence longitudinal, introrse. **Gynoecium** with usually 2–6 united carpels, ovary superior, locules as many as carpels, rarely unilocular, placentation axile, septa sometimes incomplete and disappearing in upper part of ovary, ovules 2-several, anatropous, ascending, nectaries at base of hypanthium, style simple, stigma discoid or capitate. **Fruit** a capsule, indehiscent or dehiscent, rarely berry; seeds usually flattened or winged, seed coat often with hairs becoming mucilaginous on wetting, embryo straight, endosperm lacking. pollination by bees, beetles, flies or birds (*Cuphea*), sometimes by bats (*Sonneratia*). Cleistogamy prevalent in *Peplis* and *Ammania*. Dispersal of seeds occurs through wind or water.

Economic importance: The family is known for various ornamentals such as Crepe myrtle (*Lagerstroemia*), Mexican heather (*Cuphea*) and loosestrife (*Lythrum*). Migonette tree (*Lawsonia inermis*) is the source of henna, obtained from leaves. Leaves of *Woodfordia fruticosa* yield red color, and the bark of *Lafoensia pacari* a red dye. Fleshy seeds of Pomegranate (*Punica granatum*) are edible and also used as condiment after drying and powdering.

Phylogeny: The family is commonly placed under Myrtales although Hutchinson included it under order Lythrales. The genus *Trapa* formerly included in Onagraceae was separated to Trapaceae but has now been shifted to Lythraceae (APG IV, 2016; APweb, 2018; Thorne, 2006, 2007). Onagraceae and Lythraceae share features of tannins scarce, soluble oxalate present, wood with vessels in groups, petiole bundle arcuate, inflorescence racemose and clawed petals. The family

Lythraceae broadly circumscribed to include Trapaceae, Sonneratiaceae and Punicaceae is monophyletic as supported by *rbc*L sequences. Two well defined clades are recognised within the family: one containing Sonneratia, Duabanga, Punica, Lagerstroemia and Lawsonia is characterized by determinate inflorescence, and wet stigmas; second containing the remaining genera racemose inflorescence, reduced number of carpels and dry stigmas. Thorne (2006) recognised five subfamilies: first Lythroideae containing bulk of genera (27), and the rest monogeneric Duabangoideae, Sonneratioideae, Punicoideae and Trapoideae. Subsequently (2007) he merged Punicoideae with Lythroideae and combined the other three under Lagerstroemioideae, thus recognizing only two subfamilies. Graham et al. (2005) found maximum parsimony support for the topology [*Decodon* [[*Lythrum* + *Peplis*], sister to rest of the family which forms two clades. Chen et al. (2016) found *Rotala* to be sister to the rest of the family, but support for this position was weak]. Stevens (APweb, 2018) suggests, if *Decodon* is sister to the rest of the family, then seeds with mucilage hairs and 6-merous flowers are probably synapomorphies for the rest of the family.

Onagraceae A. L. de Jussieu ***Evening Primrose Family***

22 genera, 656 species

Widely distributed mainly in temperate and subtropical regions, very diverse in western North America.

Placement:

	B & H	Cronquist	Takhtajan	Dahlgren	Thorne	APG IV/(APweb)
Informal Group						Rosids/(Malvid/ Rosid II)
Division		Magnoliophyta	Magnoliophyta			
Class	Dicotyledons	Magnoliopsida	Magnoliopsida	Magnoliopsida	Magnoliopsida	
Subclass	Polypetalae	Rosidae	Rosidae	Magnoliidae	Rosidae	
Series+/ Superorder	Calyciflorae+		Myrtanae	Myrtanae	Myrtanae	
Order	Myrtales	Myrtales	Myrtales	Myrtales	Myrtales	Myrtales

Salient features: Herbs and shrubs, leaves simple, flowers usually 4-merous, sepals, petals and stamens inserted on rim of hypanthium, carpels 4, united, placentation axile, ovary inferior.

Major genera: *Epilobium* (180 species), *Oenothera* (120), *Fuchsia* (110), *Ludwigia* (80), *Camisonia* (60), *Clarkia* (45), *Gaura* (18, sometimes merged with *Oenothera*), *Lopezia* (17) and *Circaea* (12). Although *Oenothera* is the type genus of the family, the name Oenotheraceae Warming, 1879 is antedated by Onagraceae, 1829 and adopted by A. L. de Jussieu.

Description: Usually herbs, sometimes shrubs (*Fuchsia*), rarely aquatic herbs (*Jussiaea*) or trees (*Hauya*), raphides present, stems with internal phloem, often with epidermal oil cells. **Leaves** alternate, opposite or whorled, simple, rarely pinnate, entire or toothed, sometimes lobed, venation pinnate, stipules usually absent, if present caducous (*Fuchsia, Circaea*). **Inflorescence** of solitary flowers in leaf axils, sometimes spike or raceme, rarely panicle (*Fuchsia*). **Flowers** bisexual, actinomorphic, rarely zygomorphic (*Lopezia*), epigynous with well developed hypanthium often prolonged above ovary. **Calyx** with usually 4 sepals, rarely 2–7, free, rarely connate, valvate, sometimes petaloid, caducous, rarely persistent (*Ludwigia*). **Corolla** with usually 4 petals, rarely 2–7, free, sometimes clawed, imbricate, convolute or rarely valvate, rarely absent. **Androecium** with usually 4 stamens, mostly as many as petals, sometimes twice as many, rarely only one fertile and one staminode (*Lopezia*), filaments free, inserted on inner rim of hypanthium, anthers bithecous, sometimes with cross partitions, dehiscence logitudinal; pollen grains usually triporate, sometimes tricolporate or biporate, with paracrystalline beaded outer exine, associated with viscin threads, which help pollen to adher together. **Gynoecium** with usually 4 united carpels, rarely 2 or 5, ovary inferior, usually 4 chambered with axile placentation, septa sometimes incomplete, or with parietal placentation, ovules 1-many in each locule, anatropous, with monosporic 4-nucleate megagametophyte (*Oenothera*-type), nectary near or at base of hypanthium, style slender, stigma capitate or 4-lobed. **Fruit** a loculicidal capsule, rarely berry (*Fuchsia*), bristly 1–2 seeded nutlet (*Circaea*), or 1-seeded nut (*Gaura*); seeds commonly with hairy tufts (*Epilobium*) or wings (*Hauya*), rarely smooth, embryo straight, endosperm lacking. Pollination by bees, moths, flies and birds. Dispersal of winged and hairy-tuft seeds by wind, fleshy fruits of *Fuchsia* by birds, and hooked fruits of *Circaea* by exozoochory. *Epilobium* commonly colonizes burned areas, hence the name fireweed. *Oenothera* flowers often open in late afternoon and thus known as Evening primrose.

Economic importance: The family is known for showy flowers. Species of *Oenothera* (Evening pirmrose) and *Clarkia* are grown as ornamentals in flower beds. *Fuchsia* shrubs are grown in greenhouses or in open warm regions.

Phylogeny: The family is commonly placed under Myrtales although Hutchinson (1973) included it under order Lythrales. The genus *Trapa* formerly included in this family was separated to Trapaceae (Cronquist, 1988; Dahlgren, 1989; Takhtajan, 1997), but has now been shifted to Lythraceae (APG IV, 2016; APweb, 2018; Thorne, 2006, 2007).

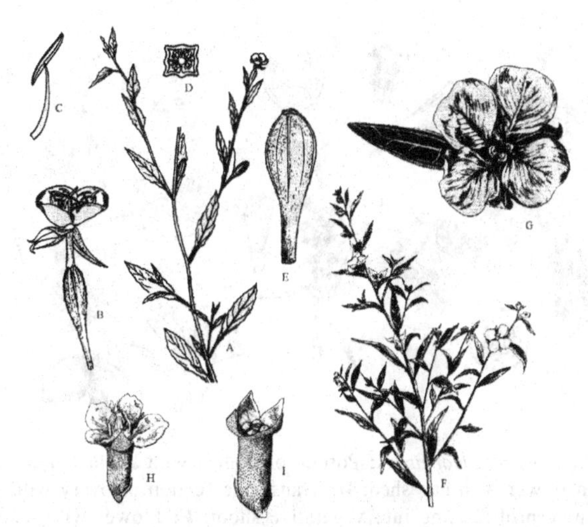

Figure 13.74: Onagraceae. *Oenothera rosea.* **A:** Portion of plant with flowers and fruits; **B:** Flower; **C:** Stamen; **D:** Transverse section of ovary; **E:** Fruit. *Ludwigia alternifolia* **F:** Branch with flowers and fruits; **G:** Flower from above. *L. linearis* **H:** Flower, side view; **I:** Fruit (A–E, after Sharma and Kachroo, Fl. Jammu, 1983. F–I, after Godfrey and Wooten, Aq. wetland Pl. SE US, 1981).

Onagraceae and Lythraceae share features of tannins scarce, soluble oxalate present, wood with vessels in groups, petiole bundle arcuate, inflorescence racemose and clawed petals. Pentamerous members *Decodon* and *Ludwigia* are sister to the respective families Lythraceae and Onagraceae. Two well defined subfamilies, Ludwigioideae (4–5 merous flowers, hypanthium absent, pollen in tetrads, stigma capitate) and Onagroideae (flowers 4-merous, hypanthium long, stigma divided) are commonly recognized in APweb (2008, 2018), although it is conventional to divide the family into number of tribes. Raimann (1893) recognised 8 tribes within the family: Jussieae, Epilobieae, Hauyeae, Onagreae, Gaureae, Fuchsieae and Circeae. Gaureae is often included under Onagreae. Despite intensive morphological and molecular studies of Onagraceae, relationships within the family are not fully understood. Levin et al. (2003) based on parsimony and maximum likelihood analyses with *rbcL* and *ndhF* sequence data for 24 taxa representing all 17 Onagraceae genera and two outgroup Lythraceae found strong support for monophyly of Onagraceae, with *Ludwigia* as the basal lineage and a sister-taxon relationship between *Megacorax* and *Lopezia.* Most relationships within Onagreae are weakly resolved, suggesting a rapid diversification of this group in western North America. Neither *Camissonia* nor *Oenothera* appears to be monophyletic. The study also showed that the small genus *Gongylocarpus* previously included in tribe Onagreae is strongly supported as sister to the rest of Onagreae + Epilobieae, and should be placed in its own tribe, Gongylocarpeae. Subsequent studies of Levin et al. (2004) on two biggest tribes based on DNA sequence data from one nuclear region (ITS) and two chloroplast regions *trnL-trnF* and *rps16* strongly suggest that tribe gongylocarpeae is sister to tribes Epilobieae + Onagreae, both of which are monophyletic. Within Onagreae, *Camissonia* seems to be broadly paraphyletic, and *Oenothera* is also paraphyletic. Within *Ludwigia*, taxa with five and those with ten stamens form separate clades (Barber et al., 2008), although support for the 10-stamen clade was strong only in Bayesian analyses (Liu et al., 2017).

Myrtaceae A. L. de Jussieu ***Myrtle Family***

132 genera, 5,950 species

Mainly distributed in tropics and subtropics, abundant in Australia.

Placement:

	B & H	**Cronquist**	**Takhtajan**	**Dahlgren**	**Thorne**	**APG IV/(APweb)**
Informal Group						Rosids/(Malvid/ Rosid II)
Division		Magnoliophyta	Magnoliophyta			
Class	Dicotyledons	Magnoliopsida	Magnoliopsida	Magnoliopsida	Magnoliopsida	
Subclass	Polypetalae	Rosidae	Rosidae	Magnoliidae	Rosidae	
Series+/ Superorder	Calyciflorae+		Myrtanae	Myrtanae	Myrtanae	
Order	Myrtales	Myrtales	Myrtales	Myrtales	Myrtales	Myrtales

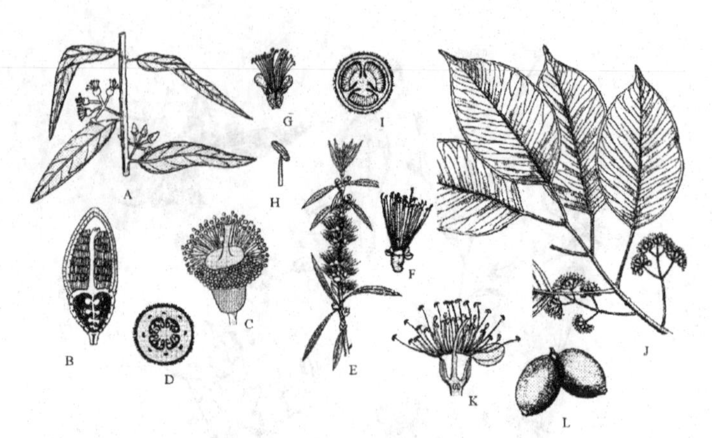

Figure 13.75: Myrtaceae. *Eucalyptus tereticornis.* **A:** Portion of branch with axillary umbellate inflorescences; **B:** Vertical section of flower bud; **C:** Open flower with cap shed; **D:** Transverse section of ovary with 4 axile placentas. *Callistemon lanceolatus.* **E:** Branch with spike proliferating into vegetative shoot; **F:** Flower with long exserted numerous stamens; **G:** Vertical section of flower; **H:** Stamen with dorsal fixation; **I:** Transverse section of ovary with 3 axile placentas. *Syzygium cuminii.* **J:** Branch with inflorescences borne on peduncles; **K:** Vertical section of flower; **L:** Fruits.

Salient features: Shrubs or trees, bark flaky, leaves gland-dotted, entire, inframarginal venation, stamens numerous, ovary inferior often united with hypanthium.

Major genera: *Eugenia* (1,115 species), *Syzygium* (1,045), *Eucalyptus* (800), *Myrcia* (800), *Corymbia* (115), *Psidium* (100), *Melaleuca* (250) and *Callistemon* (25).

Description: Evergreen Shrubs (*Myrtus*) or large trees (*Eucalyptus*) often with flaky bark, terpenes present. **Leaves** alternate (*Barringtonia, Callistemon*), opposite (*Eugenia*) or whorled, simple, entire, gland-dotted, usually coriaceous, venation often inframarginal, stipules absent. **Inflorescence** cymose (umbellate cyme in *Eucalyptus*) or racemose (*Barringtonia*), flowers sometimes solitary (*Psidium*), or in spikes (*Callistemon*—proliferating into vegetative shoots, giving appearance of a bottle-brush). **Flowers** bracteate (*Eugenia*) or ebracteate (*Eucalyptus*), bisexual, actinomorphic, epigynous (sometimes perigynous). **Calyx** with 4–5 sepals, more or less connate into a tube, imbricate, sometimes united into a cap (calyptra or operculum) which drops off as flower opens. **Corolla** with 4 (*Eugenia*) to 5 (*Psidium*) petals, (rarely absent), usually fugacious, free, rarely united with calyx to form cap like operculum (*Eucalyptus*) that falls off as the flower opens. **Androecium** with many stamens, filaments free or slightly connate at base (*Callistemon*), attached higher up on hypanthium, anthers bithecous, dehiscence longitudinal or by apical pores, pollen grains usually tricolpate with fused furrows **Gynoecium** with 2–5 united carpels (syncarpous), multilocular (locules as many as carpels) with 2-many ovules, placentation axile, rarely parietal (*Rhodamnia*) with intruded placentae, ovary inferior, or semi-inferior (*Melaleuca*) style long with capitate stigma. **Fruit** a fleshy berry (*Eugenia*), drupe (*Barringtonia*) or capsule (*Eucalyptus*), rarely one-seeded nut (*Calycothrix*); seeds 1-many, embryo curved or twisted, endosperm absent.

Economic importance: The family is the source of important oils such as eucalyptus oil (*Eucalyptus*) used as flavouring and inhalant, clove oil (*Syzygium aromaticum*) used as clearing agent and in tooth aches and oil of bay rum (*Pimenta racemosa*). *Callistemon* is commonly grown as ornamental with its beautiful bottle brush like inflorescence (hence the name). Guava fruit is obtained from *Psidium guajava*. Clove and allspice (*Pimenta dioica*) include important spices. Fruits of *Syzygium cuminii* (jambolan; 'jamun') are edible and grown in India and China.

Phylogeny: The family presents least taxonomic conflicts, almost universally placed under Myrtales under Rosids (whether rosid clade, Rasanae, or Rosidae depending upon the nomenclature followed by different authors). Monophyly of the family, together with the morphological data, is evidenced by molecular analysis through *rbcL* (Conti, 1994), *matK* (Wilson et al., 1996), and *ndhF* (Sytsma et al., 1998) sequences. *Heteropyxis* and *Psiloxylon* are basal taxa with perigynous flowers and stamens in two whorls. The family is closely related to Rosaceae, and probably the order Myrtales is derived from Rosales. The family is traditionally divided into two subfamilies: Leptospermoideae (leaves spiral to opposite; fruit dry, dehiscent) and Myrtoideae (polyhydroxy alkaloids common; leaves opposite; terpenoid-containing glands in the apex of the connective, stigma dry; fruit fleshy, indehiscent). The latter are largely derived. Leptospermoideae are basal and paraphyletic (Wilson et al., 2001; Salywon et al., 2002) as evidenced by molecular and morphological data. Genus *Syzygium*, sometimes included under *Eugenia*, represents an independent acquisition of the fleshy fruit from that in *Eugenia* and the bulk of Myrtoideae. Family was traditionally divided into two subfamilies: Leptospermoideae (leaves mostly alternate, fruit dry) and Myrtoideae (leaves opposite, fruit fleshy), a distinction which is untenable according to

Wilson (2011), who recognizes two subfamilies Psiloxyloideae (4 species in 2 genera) and Myrtoideae (15 tribes including Leptospermeae). The two genera *Heteropyxis* and *Psiloxylon* of Psiloxyloideae were earlier placed in distinct families, but now merged with Myrtaceae and are presently believed to be the earliest surviving lineages of Myrtaceae. Vasconcelos et al. (2017) suggested that within tribe *Myrtastrum rufopunctatum* could be sister to all the rest of the tribe.

Anacardiaceae R. Brown ***Cashew Family***

83 genera, 860 species

Distributed mainly in tropics but several species extending to north temperate regions of Asia, Europe and America.

Placement:

	B & H	**Cronquist**	**Takhtajan**	**Dahlgren**	**Thorne**	**APG IV/(APweb)**
Informal Group						Rosids/(Malvid/ Rosid II)
Division		Magnoliophyta	Magnoliophyta			
Class	Dicotyledons	Magnoliopsida	Magnoliopsida	Magnoliopsida	Magnoliopsida	
Subclass	Polypetalae	Rosidae	Rosidae	Magnoliidae	Malvidae	
Series+/ Superorder	Disciflorae+		Rutanae	Rutanae	Rutanae	
Order	Sapindales	Sapindales	Burserales	Rutales	Rutales	Sapindales

Salient features: Trees or shrubs with resin canals, alternate exstipulate leaves, flowers pentamerous, stamens inserted at the base of a disc, ovary unilocular, ovule one, fruit commonly a drupe.

Major genera: *Rhus* (100 species), *Mangifera* (68), *Semecarpus* (50), *Lannea* (40), *Toxicodendron* (30) and *Schinus* (30).

Description: Trees, shrubs or lianas, rarely perennial herbs, resin canals in bark, bigger veins of leaves and in parenchymatous tissue; resin clear when fresh but drying black and often causing dermatitis (*Toxicodendron*, some species of *Rhus*). **Leaves** alternate, rarely opposite (*Dobinea*), usually pinnate compound (*Rhus, Schinus*), rarely simple (*Mangifera*), entire or serrate, venation pinnate, stipules absent or vestigial. **Inflorescence** paniculate, axillary or terminal, usually thyrse with cymose branches. **Flowers** bracteate, bisexual or unisexual due to reduction of stamens or carpels, receptacle often swollen and fleshy, actinomorphic, small, hypogynous. **Calyx** with 5 sepals (rarely 3–7), free or connate at base, imbricate, rarely much enlarged (*Parishia*). **Corolla** with 5 petals (rarely 3–7), free, imbricate, rarely much enlarged (*Swintonia*). **Androecium** with 5–10 stamens, sometimes more or reduced to single fertile stamen (*Mangifera, Anacardium*), filaments usually glabrous, distinct, rarely connate at base, arising along outer (most genera) or inner margin (*Mangifera*) of rim of the disc, dehiscence longitudinal, pollen grains tricolporate or triporate. **Gynoecium** with usually 3 carpels, sometimes 5 (*Buchanania*) or only 1 (*Mangifera*), united, rarely free (*Buchanania*), ovary superior, locule usually 1 with basal or apical placentation, rarely multilocular with 1 ovule in each locule, styles 1–3, free or connate, stigma capitate, disc present between stamens and petals. **Fruit** a drupe with resinous mesocarp, rarely berry or nut (*Anacardium*), or samara (*Loxopterygium*); seed with straight or curved embryo, endosperm absent or scanty. Pollination by insects, dioecious habit promoting outcrossing. Fruits are dispersed by birds and mammals.

Economic importance: The family contributes important fruits such as mango (*Mangifera indica*), yellow mombin or Hog plum (*Spondias mombin*), Indian Hog plum (*S. indica*), Red mombin or Jamaica apple (*S. purpurea*), Otaheite apple (*S. cytherea*) and Kaffir plum (*Harpephyllum caffrum*). Nuts of Cashew (*Anacardium occidentale*) and Pistachio (*Pistacia vera*) are eaten after roasting. Species of *Rhus* and *Toxicodendron* cause dermatitis due to the phenolic compound 3-n-pentadecycatechol and should be touched with care. Lacquer is obtained from Varnish tree (*Toxicodendron vernicifera*) and Mastic tree (*Pistacia lentiscus*). Important ornamentals are contributed by the species of *Cotinus* (smoke tree), *Rhus* (Sumac) and *Schinus* (Brazilian pepper). Commercial supply of tannins is obtained from quebracho (*Schinopsis lorentzii*), Sicilian sumac (*Rhus coriaria*) and species of *Cotinus* and *Pistacia*. The first turpentine used by artists came from terebinth tree (*Pistacia terebinthus*).

Phylogeny: Anacardiaceae is closely related to Burseraceae (Thorne, 2006 places two families separately under suborder Anarcardiineae) as supported by rbcl sequences (Gadek et al., 1996). The two share resin canals and biflavones. The family is usually placed under Sapindales but shifted by Thorne under Rutales. Two clades (subfamilies) are recognised within the family (Thorne; APweb): Anacardioideae and Spondoideae, latter having retained many plesiomorphic features such as five carpels, multilocular ovary and fruit with thick endocarp having lignified and irregularly oriented sclereids, and sister rest of the family, i.e., Anacardioideae (Aguilar-Ortigosa and Sosa, 2004). Anacardioideae clade has 3 carpels, unilocular ovary with apical placenta and endocarp with regularly arranged cells. wind-pollinated taxa of this subfamily, however, do not form a single group (Pell and Mitchell, 2007). *Buchanania* in some analyses is quite well supported as sister to Anacardioideae (Aguilar-Ortigosa and Sosa, 2004; Wannan, 2007), consistent with its chemistry, endocarp anatomy and

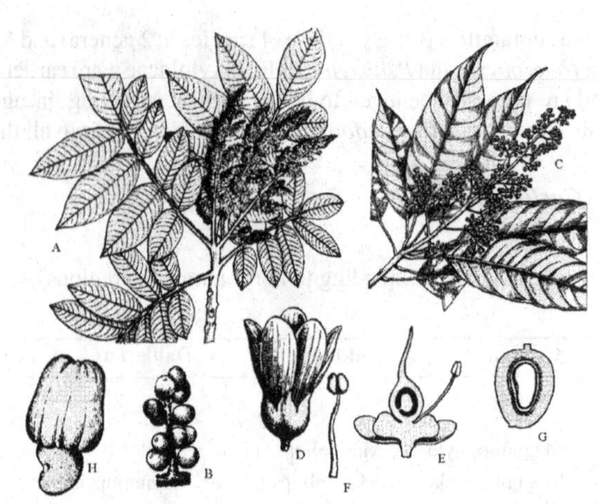

Figure 13.76: Anacardiaceae. *Rhus wallichii.* **A:** Branch with inflorescence; **B:** Portion of inflorescence in fruit. *Mangifera indica.* **C:** Branch with inflorescence; **D:** Flower; **E:** Vertical section of flower showing conspicuous disc; **F:** Single stamen. **G:** Longitudinal section of fruit, reduced view. **H:** Fruit of *Anacardium occidentale* (A–B, after Polunin and Stainton, Fl. Himal., 1984).

carpel number. Phylogeny of the family has been studied by Pell (2004) who covers the morphology of the whole family and Mitchell et al. (2006), who focus more on Spondoideae. *Rhus* and *Toxicodendron*, are distinct in former having red glandular-pubescent fruits and latter glabrous greenish or white ones. These are often combined, but the resultant genus won't be monophyletic. Spondiadoideae may be polyphyletic (Pell et al., 2011) or Paraphyletic (Weeks et al., 2014). Chen et al. (2016) found *Spondias*, *Dracontomelon*, and *Buchanania* to be in the same clade and sister to the rest of the family with moderate support.

Sapindaceae A. L. de Jussieu *Soapberry Family*

142 genera, 1860 species

Distributed mainly in tropical and subtropical regions, a few genera in temperate regions.

Placement:

	B & H	Cronquist	Takhtajan	Dahlgren	Thorne	APG IV/(APweb)
Informal Group						Rosids/(Malvid/ Rosid II)
Division		Magnoliophyta	Magnoliophyta			
Class	Dicotyledons	Magnoliopsida	Magnoliopsida	Magnoliopsida	Magnoliopsida	
Subclass	Polypetalae	Rosidae	Rosidae	Magnoliidae	Malvidae	
Series+/ Superorder	Disciflorae+		Rutanae	Rutanae	Rutanae	
Order	Sapindales	Sapindales	Sapindales	Sapindales	Rutales	Sapindales

Salient features: Usually trees or shrubs, leaves alternate, pinnately-compound or palmately-compound, strongly swollen petiole base, inflorescences paniculate, flowers often borne in congested groups along the axis, flowers small and conspicuously hairy inside, nectar disc between petals and stamens, fruit with only one or two seeds in each chamber, often deeply lobed, seeds with aril.

Major genera: *Serjania* (200 species), *Paullinia* (140), *Acer* (100), *Allophyllus* (95), *Dodonaea* (60), *Sapindus* (18), *Aesculus* (13), *Cardiospermum* (12), *Koelreuteria* (10) and *Litchi* (2).

Description: Shrubs or trees, herbaceous or woody lianas with tendrils (*Serjania*), rarely herbs (*Cardiospermum*), often with tannins, usually with triterpenoid saponins in secretary cells. **Leaves** alternate, rarely opposite (*Velenzuelia, Acer*), once or twice pinnate, sometimes palmately compound (*Aesculus*) trifoliate (*Billia*) or simple (*Litchi*), leaflets entire or serrate, venation pinnate, reticulate, base of petiole strongly swollen, stipules absent, rarely present (*Urvillea, Serjania*). **Inflorescence** usually cymose, aggregated in panicles, often quite congested along the axis. **Flowers** unisexual (plants monoecious, dioecious or polygamous), actinomorphic or zygomorphic (*Cardiospermum*). **Calyx** with usually 4 or 5 sepals, free or united. **Corolla** with 4 or 5 petals, usually as many as sepals, sometimes absent (*Dodonaea*), free, usually clawed, with basal appendages inside, imbricate. **Androecium** with usually *4* (*Glenniea*)-10 stamens, rarely more (*Deinbollia*), usually borne on a nectar

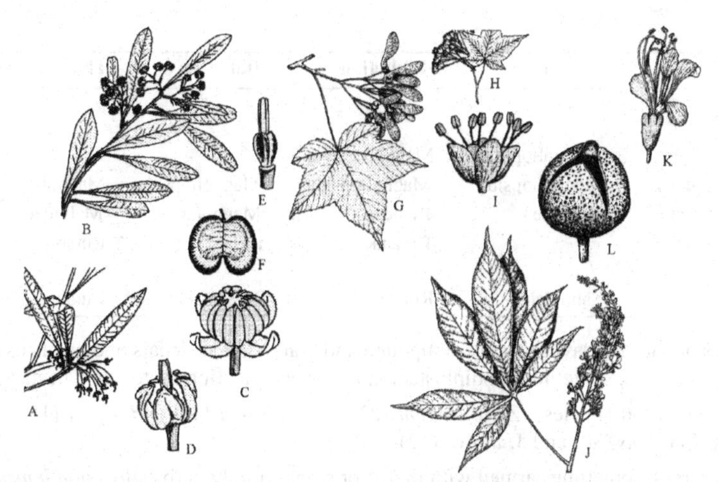

Figure 13.77: Sapindaceae. *Dodonaea viscosa.* **A:** Flowering twig of female plant; **B:** Flowering twig of male plant; **C:** Male flower with 5 sepals and 8 stamens, petals absent; **D:** Female flower with sepals and gynoecium; **E:** Gynoecium. **F:** Fruit with 2 wings. *Acer caesium.* **G:** Portion of a fruiting branch; **H:** Portion of flowering branch; **I:** Flower. *Aesculus indica.* **J:** Portion of flowering branch with palmately compound leaf; **K:** Flower with long-clawed petals and exserted stamens; **L:** Capsule opening by 3 valves (G–I after Fl. Himal., 1984).

disc present between petals and stamens, filaments free and usually hairy, anthers bithecous, dehiscence by longitudinal slits, pollen grains tricolpate, furrows often fused. **Gynoecium** with 2–3 united carpels, rarely upto 6, ovary superior, placentation axile, 1 or 2 ovules in each loculus, ovule orthotropous or anatropous, funiculus lacking, and ovule broadly attached to obturator (projection from placenta), style 1, stigma usually lobed. **Fruit** a drupe, berry, capsule (3-winged-*Bridgesia*), samara (*Acer*) or schizocarp; seeds often with an aril, embryo curved, endosperm absent. Pollination by birds and insects, *Dodonaea* and some species of *Acer* being wind pollinated. Dispersal often by birds attracted by aril, but inflated fruits and winged fruits are often dispersed by wind.

Economic importance: The family is important primarily for its fruits: *Litchi sinensis* (Litchi, lychy), *Nephelium lappaceum* (rambutan) and *Euphoria* (logan). The aril of the fruits of *Blightia sapida* (akee), a native of West Africa are also eaten cooked, tasting like scrambled egg, but poisonous if eaten when unripe. The fruits of different species of *Sapindus* (soapberry) are often used as a natural soap due to the presence of saponins. *Paullinia cupana* is the source of drink guarana, popular in Brazil. *Schleichera trijuga* is the source of macassar oil, used in ointments and for illumination. The species of *Aesculus* have various medicinal uses, and extracts from some have been used by North American Indians to stupefy fish. The family also contributes several ornamentals such as *Koelreuteria* (goldenrain tree), *Cardiospermum* (balloon vine), *Xanthoceras*, *Acer* (maple) and *Aesculus* (horse chestnut). The trees of maple are prized for their beautiful foliage and spectacular autumn colors. *Acer saccharum* (sugar maple) and some other species yield maple sugar.

Phylogeny: Sapindaceae is sometimes narrowly circumscribed to exclude Hippocastanaceae and Acerceae (Hutchinson, Takhtajan, Cronquist and Dahlgren), but their separation leads to paraphyletic Sapindaceae (Judd et al., 1994). The family is as such broadly circumscribed to include both Aceraceae and Hippocastanaceae (Thorne, Judd et al., APG IV and APweb). Monophyly of the family is supported by morphology and *rbcL* sequences. *Xanthoceras*, with simply 5-merous, polysymmetric flowers and complex, golden nectaries borne outside the eight stamens, is sister to rest of Sapindaceae, and the genera included in Aceraceae and Hippocastanaceae are monophyletic sister taxa (Savolainen et al., 2000). Sapindaceae are chemically similar to Leguminosae, and both have compound leaves, but they are unlikely to be immediately related. Thorne (1999, 2003) recognizes 5 subfamilies: Dodonaeoideae (*Dodonaea*), Koelreuterioideae, Sapindoideae, Hippocastanoideae (*Aesculus*, *Billia*) and Aceroideae (*Acer*, *Dipteronia*). APweb (2018) recognizes Four: Xanthoceratoideae (single species *Xanthoceras sorbifolium*), Hippocastanoideae (including Aceroideae), Dodonaeoideae and Sapindoideae (including Koelreuterioideae). Buerki et al. (2009, 2010) had earlier suggested that *Xanthoceras*, with 5-merous, polysymmetric flowers, eight stamens, ovules arranged in parallel (see also *Magonia*), and golden nectaries borne outside the stamens and opposite the sepals, might be sister to all other Sapindaceae.

Rutaceae A. L. de Jussieu *Citrus or Rue Family*

148 genera, 2,070 species

Distributed in warm temperate and tropical regions with the greatest diversity in Australia and South Africa.

Placement:

	B & H	Cronquist	Takhtajan	Dahlgren	Thorne	APG IV/(APweb)
Informal Group						Rosids/(Malvid/ Rosid II)
Division		Magnoliophyta	Magnoliophyta			
Class	Dicotyledons	Magnoliopsida	Magnoliopsida	Magnoliopsida	Magnoliopsida	
Subclass	Polypetalae	Rosidae	Rosidae	Magnoliidae	Malvidae	
Series+/ Superorder	Disciflorae+		Rutanae	Rutanae	Rutanae	
Order	Geraniales	Sapindales	Rutales	Rutales	Rutales	Sapindales

Salient features: Trees or shrubs, leaves usually compound and gland dotted, sepals and petals usually 5 each, stamens 5 or more free or polyadelphous, sometimes obdiplostemonous, ovary superior, seated on a nectary disc, fruit a berry.

Major Genera: *Melicope* (300 species), *Zanthoxylum* (225), *Agathosma* (180), *Boronia* (150), *Haplophyllum* (66), *Glycosmis* (50), *Ruta* (60) *Citrus* (30) and *Murraya* (12).

Description: Shrubs or trees, sometimes armed with thorns or spines, rarely herbs (*Boenninghausenia*), often aromatic, containing alkaloids and phenolic compounds. **Leaves** alternate, rarely opposite (*Evodia*), usually pinnate compound, sometimes unifoliate due to reduction of lower two leaflets (*Citrus*), less frequently simple (*Evodia*), gland dotted, stipules absent. **Inflorescence** cymose or flowers solitary (*Triphasia*), rarely racemose (*Atlantia*). **Flowers** ebracteate, bisexual or rarely unisexual (*Zanthoxylum*), actinomorphic or rarely zygomorphic (*Dictamus*), hypogynous. **Calyx** with 4–5 sepals, rarely 3 (*Lunasia*) free or more or less united, gland dotted. **Corolla** with 4–5 petals, rarely 3 (*Triphasia*) free, rarely united (*Correa*) valvate or imbricate, sometimes absent. **Androecium** with 8–10 rarely 5 (*Skimmia*), or many (*Citrus*) stamens, free (*Murraya*), or polyadelphous (*Citrus*), rarely monadelphous (*Atlantia*) sometimes obdiplostemonous, anthers bithecous, dehiscence longitudinal, pollen grains 3–6-colpate. **Gynoecium** with 2–5 united carpels (syncarpous), rare monocarpellary (*Teclea*), sometimes ovaries free (*Zanthoxylum*) and only styles united, multilocular (locules as many as carpels) with 1-many ovules, placentation axile, rarely parietal (*Feronia*), ovary superior and lobed, style 1, stigma small. **Fruit** a berry (*Murraya*), drupe (*Spathelia*), hesperidium (*Citrus*), samara (*Ptelea*), capsule (*Ruta*) or follicle (*Zanthoxylum*); seeds 1-many, embryo curved or straight, endosperm absent or present. Pollination mainly by insects, chiefly bees and flies. Dispersal usually by animals, rarely birds or even wind (*Ptelea*).

Economic importance: The family is important for its citrus fruits such as lemon (*Citrus limon*), lime (*C. aurantifolia*), citron (*C. medica*), bitter orange (*C. x aurantium*), Pomelo, Shaddak (*C. maxima*) orange, sweet orange, 'malta' (*C. sinensis*), tangerine or 'santra' (*C. reticulata*), mousambi (*C. limetta, C. limon* 'Limetta'), kumquat (*Citrus*

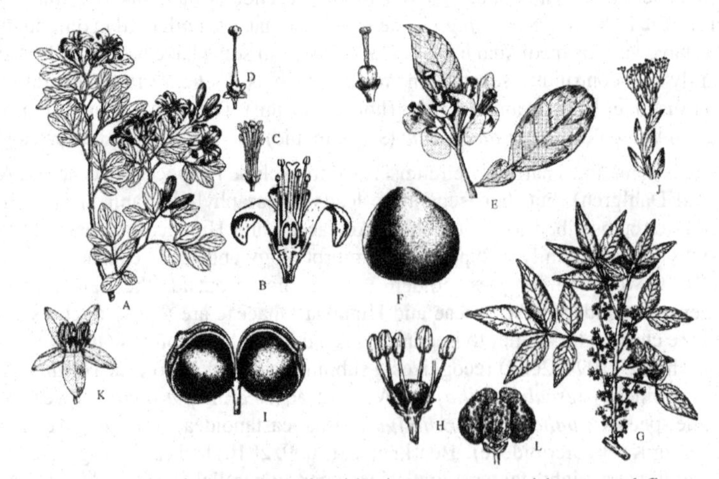

Figure 13.78: Rutaceae. *Murraya paniculata.* **A:** Branch with pinnate compound leaves and flowers; **B:** Vertical section of flower; **C:** Flower with petals removed showing 10 stamens in 2 whorls; **D:** Gynoecium with nectary at base. *Citrus paradisi.* **E:** Portion of branch with unifoliate leaf having broadly winged petiole and cluster of flowers with polyadelphous stamens; **F:** Fruit. *Zanthoxylum armatum.* **G:** Branch with pinnate compound leaves, spines and inflorescences; **H:** Male flower with 6 free stamens and abortive ovary, the sepals are small and petals absent; **I:** Schizocarpic fruit splitting into 2 segments. *Haplophyllum acutifolium.* **J:** Portion of plant with flowers; **K:** Flower with large petals, stamens with flattened filaments; **L:** Capsule covered with glands and deeply 5-lobed.

japonica, syn: *Fortunella japonica*) and grapefruit (*C. paradisi*). *Aegle marmelos* (Bel tree), and *Casimiroa* (white zapote) are grown for fruits. *Murraya paniculata* is cultivated as an ornamental shrub, whereas *M. koenigii* (Curry leaf) is cultivated for its curry leaves. Leaves of *Skimmia laureola* are burnt in order to purify air. *Ruta* (rue), *Zanthoxylum* (toothache tree), and *Casimiroa* are medicinal. *Boenninghausenia* ('Pisu-mar-buti') is used as an insecticidal.

Phylogeny: The family is related to Sapindaceae, Simaroubaceae, and Meliaceae. Although the family presents a variety of fruit types, it is a well circumscribed monophyletic taxon characterized by oil cavities appearing as pellucid dots, supported by data from *rbcL* and *atpB* sequences. Subfamilies are characterized by carpel number, extent of fusion and fruit type. Whereas Dahlgren (1989) and Takhtajan (1997) preferred to separate order Sapindales from order Rutales, others like Cronquist (1988), Thorne (1999, 2003), and APG II prefer to merge the two. Cronquist and APG II (also APG IV) use name Sapindales whereas Thorne prefered the priority name Rutales. Thorne, however, places Sapindaceae and Rutaceae under separate suborders. Earlier Hutchinson (1973) had also separated the two orders, also separating Meliaceae under Meliales, and placing between the two orders. The affinities of Meliaceae, Rutaceae and Sapindaceae have been long recognized. Thorne recognizes 3 subfamilies: Rutoideae, Aurantioideae and Cneoroidea. APWeb (2018) recognises 4 subfamilies: Cneoroideae, Amyridoideae, Rutoideae, Aurantioideae. Rutaceae are relatively young, and distributions are unlikely to be much affected by continental drift. Recent work suggests that the basic relationships in the rest of the family are [Amyridoideae [Aurantioideae + Rutoideae]] (Groppo et al., 2012). *Harrisonia* which has often been linked with Simaroubaceae, lacks quassinoids of that family, and is better placed in Rutaceae.

Meliaceae A. L. de Jussieu ***Mahogany Family***

53 genera, 600 species

Distributed mainly in tropical and subtropical regions.

Placement:

	B & H	**Cronquist**	**Takhtajan**	**Dahlgren**	**Thorne**	**APG IV/(APweb)**
Informal Group						Rosids/(Malvid/ Rosid II)
Division		Magnoliophyta	Magnoliophyta			
Class	Dicotyledons	Magnoliopsida	Magnoliopsida	Magnoliopsida	Magnoliopsida	
Subclass	Polypetalae	Rosidae	Rosidae	Magnoliidae	Malvidae	
Series+/ Superorder	Disciflorae+		Rutanae	Rutanae	Rutanae	
Order	Geraniales	Sapindales	Rutales	Rutales	Rutales	Sapindales

Salient features: Trees or shrubs containing bitter triterpenoid compounds, leaves alternate, pinnate compound, flowers unisexual, sepals 4–5, petals 4–5, stamens with connate filaments, ovary with axile placentation, stigma capitate, seeds dry and winged.

Major genera: *Aglaia* (95 species), *Trichilia* (60), *Turraea* (60), *Dysoxylum* (58), *Guarea* (32), *Toona* (15), *Melia* (15), *Cedrela* (6) and *Azadirachta* (2).

Description: Shrubs or trees, rarely herbs (*Naregamia*), commonly producing bitter triterpenoid compounds, usually with scattered secretary cells, wood sometimes yellow (*Chloroxylon*) or red (*Cedrela*), nodes pentalacunar, vessels with simple end-walls. **Leaves** alternate, once (*Azadirachta*) or twice pinnate (*Melia*), sometimes trifoliate (*Sandoricum*) or unifoliate (*Turraea*), venation pinnate, reticulate, stipules absent. **Inflorescence** axillary or terminal panicles, usually cymose. **Flowers** ebracteate, bisexual or rarely unisexual (*Amoora*), actinomorphic, trimerous to pentamerous, cyclic. **Calyx** with 3 (*Amoora*), 4 (*Dysoxylum*) or 5 (*Melia, Cedrela*) sepals, free or united (*Amoora, Melia*), valvate or imbricate, green. **Corolla** with 3–5 petals, usually as many as sepals, free, rarely united (*Munronia*), imbricate or valvate. **Androecium** with 3 (some species of *Amoora*), 4–6 (*Cedrela*), 5 (*Aglaia*) or upto 12 (*Melia*), free (*Walsura*) or monadelphous (*Melia*), usually inserted on a nectariferous disc, anthers bithecous, dorsifixed or versatile, introrse, dehiscence longitudinal, pollen grains 2- to 5-colporate, sometimes in tetrads (*Dysoxylum championii*). **Gynoecium** with 2–6 united carpels, ovary superior, 2–5 chambered with 1 or 2 (or more in *Swietenia*) ovules in each loculus, ovule orthotropous or anatropous (*Dysoxylum*), placentation axile, style 1, stigma capitate. **Fruit** a drupe, (*Melia azedarach*), berry (*Walsura*), or a capsule (*Amoora*); seeds winged or with an aril (Melioideae), embryo curved or straight, endosperm present or absent.

Economic importance: The family is highly prized for its true mahogany woods: *Swietenia mahogani* of the West Indies; *Entandrophragm*, *Khaya* and *Lovoa* of Africa; *Cedrela odorata* and *Toona* of Australia. These are renowned for their excellent color, working properties and finish. Oils for soap-making are extracted from the seeds of *Trichilia emetica* in Uganda. The oil from Malayan *Chisocheton macrophyllus* has been used as an illuminant. The flowers of *Aglaia odorata* are used in the East for flavouring tea. Species of *Melia, Aglaia, Chisocheton* and *Turraea* are grown as ornamentals.

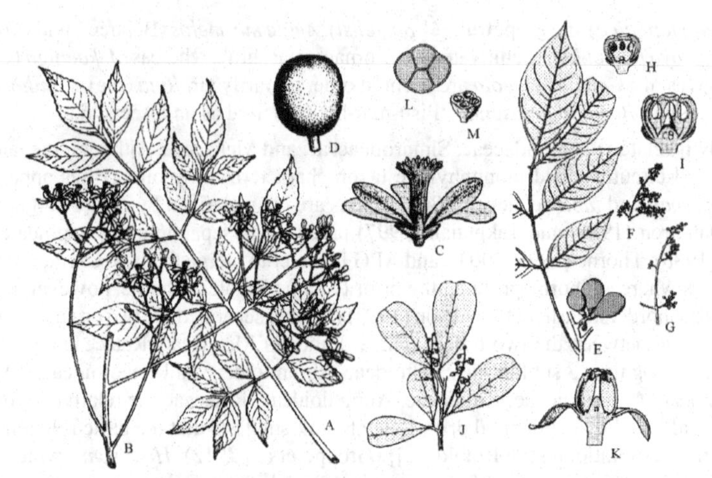

Figure 13.79: Meliaceae. *Melia azedarach*. **A:** Bipinnate leaf; **B:** Inflorescence; **C:** Flower with long staminal tube; **D:** Drupe. *Aglaia apiocarpa*. **E:** Fruiting branch with pinnate compound leaf; **F:** Portion of male inflorescence; **G:** Female inflorescence; **H:** Vertical section of male flower; **I:** Vertical section of female flower. *Dysoxylum championii*. **J:** Flowering branch; **K:** Vertical section of male flower; **L:** Tetrad of pollen grains; **M:** Transverse section of ovary.

Azadirachta indica (neem) from India has gained considerable importance in the recent years as a bioinsecticide and a component of several medicines and also tooth pastes. The tree, for a long time, has been grown as a shade tree and twigs used to brush teeth (datun).

Phylogeny: The family is generally included closer to Rutaceae in broadly circumscribed Geraniales (Bentham and Hooker, Bessey) or Sapindales (Cronquist, APG II (also APG IV), APweb) or narrowly circumscribed Rutales (Takhtajan, Dahlgren). Thorne (1999, 2003) combines Rutales with Sapindales but prefers name Rutales for the broadly circumscribed order. Hutchinson (1926, 1973) segregated the family to a distinct order Meliales primarily based on leaves usually not gland-dotted and stamens connate. The family is distinct and monophyletic as supported by morphology and *rbcL* sequences (Gadek et al., 1996). Two subfamilies are commonly recognized: Melioideae (seeds not winged, naked buds) and Swietenioideae (flattened or winged seeds, and scaly buds). Thorne had earlier (1999) merged these two under Melioideae and recognized two more Quivisianthoideae (single genus *Quivisiantha*) and Capuronianthoideae (single genus *Capuronianthus*). In later revision (2003) he resurrected Swietenioideae, thus recognizing four subfamilies. The genus *Cedrela* is distinct in its free stamens and erect petals, included by thorne under Swietenioideae. Apweb (2018) recognizes two subfamilies: Melioideae (including Capuronianthoideae) and Cedreloideae (including Quivisianthoideae).

Malvaceae A. L. de Jussieu *Mallow Family*

244 genera, 4,225 species

Distributed in tropical and temperate climates, mainly in the South American tropics.

Placement:

	B & H	**Cronquist**	**Takhtajan**	**Dahlgren**	**Thorne**	**APG IV/(APweb)**
Informal Group						Rosids/(Malvid/ Rosid II)
Division		Magnoliophyta	Magnoliophyta			
Class	Dicotyledons	Magnoliopsida	Magnoliopsida	Magnoliopsida	Magnoliopsida	
Subclass	Polypetalae	Dilleniidae	Dilleniidae	Magnoliidae	Malvidae	
Series+/ Superorder	Thalamiflorae+		Malvanae	Malvanae	Malvanae	
Order	Malvales	Malvales	Malvales	Malvales	Malvales	Malvales

Thorne under suborder Malvineae; Malvaceae, includes Sterculiaceae and Bombacaceae and truncated Tiliaceae (2 genera) but excludes Grewiaceae (majority genera of former Tiliaceae); **APG IV, APweb** Malvaceae includes Tiliaceae, Sterculiaceae and Bombacacea.

Family Malvaceae as understood in APG IV (2016) and updated APweb (2018) includes four traditional families: Malvaceae, Bombacaceae, Sterculiaceae and Tiliaceae, which form a natural group sharing common characters of mostly

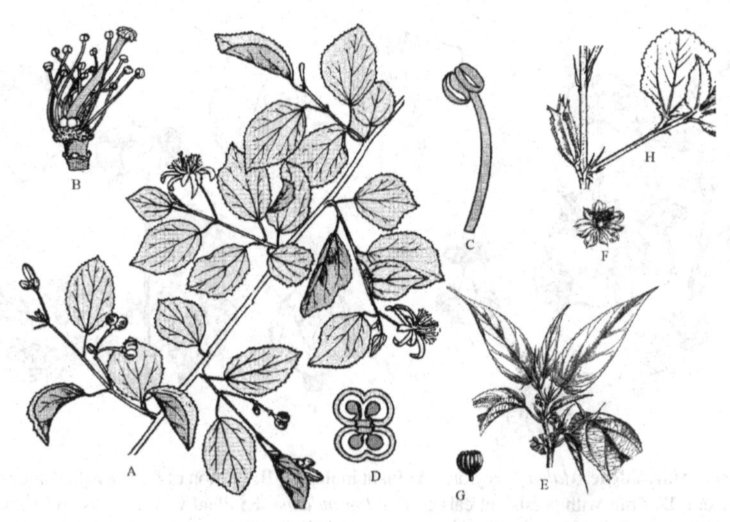

Figure 13.80: Malvaceae, Grewioideae. *Grewia tenax.* **A:** Portion of plant with flowers and fruits; **B:** Flower with sepals and petals removed; **C:** Stamen; **D:** Transverse section of ovary. *Corchorus capsularis.* **E:** Portion of plant with flowers and fruits; **F:** Flower from above; **G:** Fruit. **H:** *Corchorus aestuans*, portion of plant with fruit.

woody plants with cyclopropenoid fatty acids, sieve tubes with non-dispersive protein bodies, hairs stellate/lepidote, leaves spiral or two-ranked, lamina margins entire or toothed, single vein running to the non-glandular apex, secondary venation palmate, inflorescence cymose, calyx valvate, petals imbricate, stamens 5 to many, at least basally connate, extrose, tapetum amoeboid, carpels 3–5 to many, united, style usually 5-branched, ovules 1 to many per carpel, fruit usually a capsule, less commonly berry or schizocarp, endosperm often starchy. The family is divided into 9 subfamilies by APweb (2018): Grewioideae (2 tribes), Byttnerioideae (2 tribes), Tilioideae, Dombeyoideae, Brownlowioideae, Helicteroideae, Sterculioideae, Malvoideae (3 tribes) and Bombacoideae (3 tribes). Two Grewioideae and Malvoideae are described here.

Subfamily Grewioideae Hochreutiner

(= Grewiaceae (Dippel) Thorne)

B & H, Cronquist, Takhtajan, and **Dahlgren** under family Tiliaceae. **Thorne** as separate family under suborder Malvineae, **APG IV and APweb** do not recognize Tiliaceae or Grewiaceae as separate family, merge with Malvaceae.

25 genera, 770 species

Widely distributed in tropics and subtropics.

Salient features: Shrubs or trees, leaves with asymmetrical base, pubescence of branched hairs, stamens numerous with free or united filaments, anthers bithecous, carpels five or more, ovary superior, placentation axile.

Major Genera: *Grewia* (290 species), *Triumfetta* (150), *Microcos* (60) and *Corchorus* (50).

Description: Shrubs or trees, rarely herbs (*Corchorus, Triumfetta*). **Leaves** alternate, simple, deciduous, bases asymmetrical, pubescence of branched hairs, stipules present. **Inflorescence** cymose, usually in small clusters in leaf axils. **Flowers** bisexual, rarely unisexual, actinomorphic, hypogynous. **Calyx** with 3–5 sepals, free or connate, valvate. Corolla with 3–5 petals, free, imbricate or valvate, sometimes with glandular hairs at their bases, rarely absent. **Androecium** with many stamens, sometimes only 5 (*Triumfetta pentandra*), filaments free or united into groups of 5 or 10 (polyadelphous), adnate to base of petals, anthers bithecous, dehiscence longitudinal, or by apical pores. **Gynoecium** with 2-many, united carpels, multilocular (locules as many as carpels) with many ovules, placentation axile, ovary superior, style single, stigma lobed or capitate. **Fruit** a capsule or fleshy. Seeds 1-many, embryo straight, endosperm present.

Economic importance: Jute is obtained from stem fibres of *Corchorus capsularis* and *C. olitorius*. Leaves of *C. olitorius* are used for food in many eastern Mediterranean countries.

Subfamily Malvoideae Burnett

78 genera, 1670 species

Distributed in tropical and temperate climates, mainly in the South American tropics.

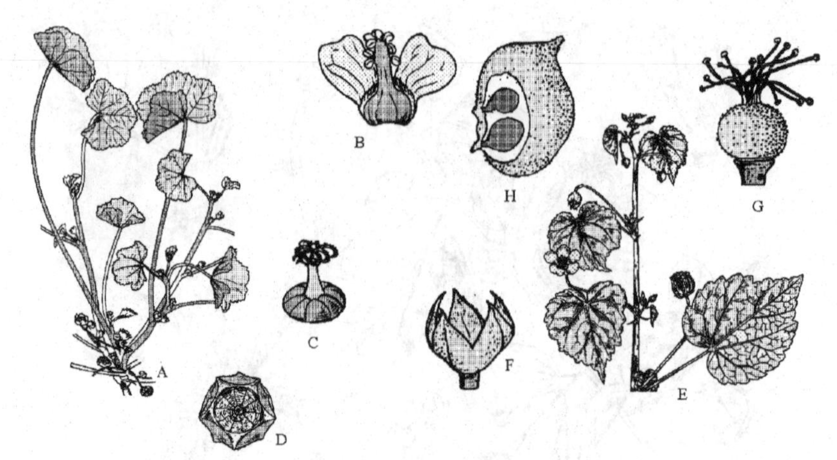

Figure 13.81: Malvaceae, Malvoideae. *Malva parviflora*. **A:** Plant in flower; **B:** Portion of flower with 2 petals and longitudinally split androecium; **C:** Gynoecium; **D:** Fruit with persistent calyx. *Abutilon indicum*. **E:** Plant with flowers and fruits on long peduncles; **F:** Calyx; **G:** Gynoecium with several carpels; **H:** One fruiting carpel split to show seeds.

Salient features: Herbs and shrubs with stellate pubescence, often mucilaginous, leaves palmately veined, stipules prominent, flowers usually with epicalyx, stamens numerous with united filaments, anthers monothecous, carpels five or more, ovary superior, placentation axile.

Major genera: *Hibiscus* (300 species), *Sida* (200), *Pavonia* (200), *Abutilon* (100).

Description: Herbs or shrubs, rarely small trees (*Thespesia*). Plants often mucilaginous. **Leaves** alternate, simple, sometimes palmately lobed (*Gossypium*), palmately veined, pubescence stellate or of peltate scales, stipules present. Inflorescence cymose (*Pavonia*) or flowers solitary axillary. **Flowers** bracteate (*Abutilon*) or ebracteate (*Hibiscus*) bisexual, actinomorphic, hypogynous. **Calyx** with 5 sepals, more or less united, often subtended by **epicalyx** (bracteoles), epicalyx 3 (*Malva*), 5–8 (*Althaea*) or absent (*Sida*). **Corolla** with 5 petals, free, imbricate, often adnate at base to staminal tube. **Androecium** with many stamens, filaments united into a tube (monadelphous), epipetalous, anthers monothecous, dehiscence transverse, pollen grains large with spinous exine, triporate or multiporate, tricolpate in *Abutilon*. **Gynoecium** with 2-many (usually 5) united carpels (syncarpous), multilocular (locules as many as carpels) with many ovules, placentation axile, ovary superior, styles branched above, stigmas as many as carpels or twice as many (*Malvaviscus*). **Fruit** a loculicidal capsule or schizocarp (*Malva*), follicles (*Sterculia*), rarely a berry (*Malvaviscus*); seeds 1-many, embryo curved, endosperm absent. Flowers are insect pollinated, nectar usually produced by inner surface of calyx. Dispersal may occur by wind, water, or animals.

Economic importance: The family is represented by several ornamentals such as China rose (*Hibiscus rosa-sinensis*), kenaf (*Hibiscus cannabinus*), hollyhock (*Alcea rosea*) and rose of Sharon (*Hibiscus syriacus*), Young fruits of okra (*Abelmoschus esculentus;* 'bhindi') are used as vegetable. Cotton is obtained from different species of *Gossypium*.

Phylogeny of Malvaceae: The family has been considered quite distinct based on monadelphous stamens with monothecous anthers, though it had been considered quite closer to Tiliaceae, Bombacaceae and Sterculiaceae by Cronquist (1988) and Takhtajan (1997). These families share the features of presence of stellate hairs, mucilaginous cells, pericycle strands above phloem, similar size and pitting of vessels, and the distribution of xylem parenchyma. According to Judd et al. (1999, 2002) the traditional distinctions between these families are arbitrary and inconsistent, and the merger of four would form a monophyletic Malvaceae. They, however, concede that genera such as *Grewia, Corchorus, Triumfetta*, etc., form a clade which has lost calyx fusion, also suggesting that Grewioideae and Byttnerioideae form distinct clades within Malvaceae. Traditional Tiliaceae was circumscribed by free stamens and bithecous anthers. Thorne (1999, 2000), obviously had kept Tiliaceae distinct, merging the other two families with Malvaceae. Recent molecular evidence (Alverson et al., 1998) suggests that half-anthers of the traditional Malvaceae are transversely septate bithecous anthers that are strongly connate. Earlier Hutchinson (1973) had proposed that the monothecous anthers arose from splitting (**chorisis**) of the filaments. Restriction site analysis of cpDNA has established that genera with loculicidal capsules and numerous seeds (*Hibiscus, Gossypium*), form a basal paraphyletic complex. Genera with schizocarpic fruits, more than five carpels, and ovules one or two per carpel depict synapomorphies. Thorne who had earlier recognized Tiliaceae as distinct family has finally (2003) shifted *Tilia* and *Craigia* to Malvaceae under Tilioideae, the remaining genera of family Tiliaceae being put under new family Grewiaceae. He recognizes 7 subfamilies under Malvaceae, establishing Grewioideae and Byttnerioideae as independent families Grewiaceae and Byttneriaceae, respectively. APweb (2018) recognizes following 9 subfamilies

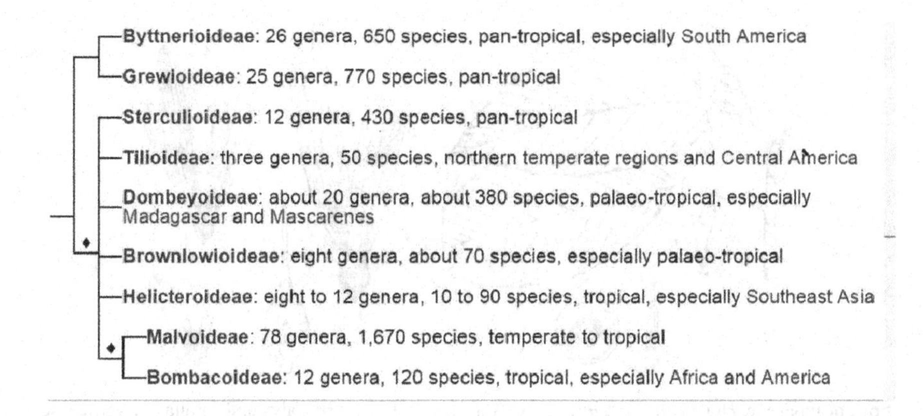

under the broadly circumscribed Malvaceae: Grewioideae, Byttnerioideae, Tilioideae, Dombeyoideae, Brownlowioideae, Helicteroideae, Sterculioideae, Malvoideae and Bombacoideae. [Grewioideae + Byttnerioideae] are probably sister to the rest of the family (Richardson et al., 2015; Chen et al., 2016). Brief summary of relationships, size and distribution of different subfamilies is presented below.

Dipterocarpaceae Blume ***Meranti Family***

17 genera, 550 species

Distributed mainly in tropical Asia and Indomalaysia, also represented in Africa and South America.

Placement:

	B & H	Cronquist	Takhtajan	Dahlgren	Thorne	APG IV/(APweb)
Informal Group						Rosids/(Malvid/ Rosid II)
Division		Magnoliophyta	Magnoliophyta			
Class	Dicotyledons	Magnoliopsida	Magnoliopsida	Magnoliopsida	Magnoliopsida	
Subclass	Polypetalae	Dilleniidae	Dilleniidae	Magnoliidae	Malvidae	
Series+/ Superorder	Thalamiflorae+		Malvanae	Malvanae	Malvanae	
Order	Guttiferales	Theales	Malvales	Malvales	Malvales	Malvales

B & H as Dipterocarpeae. **Thorne** under suborder Cistineae.

Salient features: Small or large trees with buttressed bases, leaves evergreen, alternate, often with domatia, flowers perigynous or epigynous, in racemes or panicles, sepals becoming winged in fruit, petals 5, often leathery, anthers with sterile tips, carpels 3, fruit a winged nut.

Major genera: *Shorea* (150 species), *Hopea* (110), *Dipterocarpus* (80), *Vatica* (60) and *Monotes* (26).

Description: Small or large trees, often buttressed at the base, trunk very long and smooth, branched at top with cauliflower-shaped crown, usually with special resin canals exuding aromatic dammar from wounds, nodes trilacunar or pentalacunar, roots with ectomycorrhiza. **Leaves** alternate, distichous, coriaceous, simple, evergreen, covered with fasciculate or stellate hairs, stipules present and frequently containing domatia housing insects, usually early shedding. **Inflorescence** racemose, axillary or terminal racemes or panicles. **Flowers** bisexual, actinomorphic, often showy, fragrant, hypogynous. **Calyx** with 5 sepals, free or slightly connate, sometimes enlarged and winged in fruit. **Corolla** with 5 petals, free or connate at base, spirally twisted, often leathery. **Androecium** with 5-numerous stamens, filaments free or connate at base, anthers bithecous, dorsifixed (Monotoideae), or basifixed (Dipterocarpoideae), dehiscence longitudinal, anthers with sterile tip formed by extension of connective, pollen grains tricolpate or triporate. **Gynoecium** with 3 united carpels, ovary superior or partly inferior (*Anisoptera*), 3-locular with 2 ovules in each chamber, placentation axile, ovules pendulus, anatropous, bitegmic, crassinucellate, only one ovule develops further. **Fruit** a single seeded nut with winged and membranous calyx; seeds without endosperm, cotyledons often twisted, enclosing radicle.

Economic importance: Many species of *Dipterocarpus*, *Shorea*, *Hopea* and *Vatica* usually grow together in tropical rain forests and are principal sources of hardwood timber. The wood is pale in color and in great demand for plywood and block wood. Dammar resin obtained from the tree is used for special varnishes.

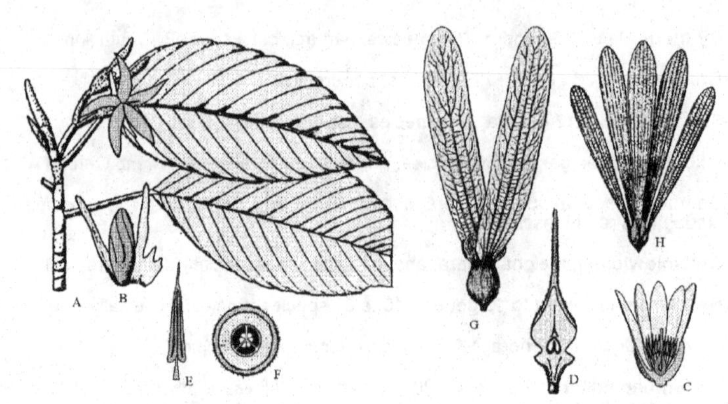

Figure 13.82: Dipterocarpaceae. *Dipterocarpus trinervis.* **A:** Branch with flower; **B:** Calyx and corolla; **C:** Vertical section of flower; **D:** Longitudinal section of ovary; **E:** Stamen with sterile tip above anther; **F:** Transverse section of ovary. **G:** Fruit of *D. pilosus* with two long wings. **H:** Fruit of *Parashorea stellata* with five wings.

Phylogeny: The family is related to Ochnaceae, Elaeocarpaceae, Grewiaceae and other members of Malvales. Cronquist considers it closer to Guttiferae and Theaceae in addition to Ochnaceae. The family is usually divided into 3 subfamilies: Monotoideae, Pakaraimaeoideae and Dipterocarpoideae. Molecular studies of Kubitzki and Chase (2002) have shown that Sarcolaenaceae, Cistaceae and Dipterocarpaceae form a well-defined clade having plant with secretory canals, calyx imbricate, two outer members often different from the rest, filaments not articulated, ovules both anatropous and Atropous; exotegmen curved inwards in chalazal region, and there is a strong case for merging former two in Dipterocarpaceae. Phylogenetic studies on family Dipterocarpaceae based on morphological and *rbcL* sequence data (Dayanandan et al., 1999) have shown that Monotoideae and Pakaraimaeoideae are cladistically basal, representing primitive members of the family. Thorne (2003) had earlier included the family (and the order Malvanae) under superorder Rosanae, but subsequently (2006) shifted to Malvanae. Recent evidence has shown that *Pakaraimaea* is sister to Cistaceae, and as such removed from Dipterocarpaceae (Kubitzki and Chase, 2002; Horn et al., 2016), leaving two subfamilies in Monotoideae and Dipterocarpoideae in the family (APweb, 2018). Based on genetic study, Ducousso et al. (2004) found that the Asian dipterocarps share a common ancestor with the Sarcolaenaceae, a tree family endemic to Madagascar. This suggests that ancestor of the Dipterocarps originated in the southern supercontinent of Gondwana, and that the common ancestor of the Asian dipterocarps and the Sarcolaenaceae was found in the India-Madagascar-Seychelles land mass millions of years ago, and were carried northward by India, which later collided with Asia and allowed the dipterocarps to spread across Southeast Asia and Malaysia.

Capparaceae A. L. de Jussieu ***Caper Family***

30 genera, 324 species

Widespread in tropical and subtropical regions.

Placement:

	B & H	**Cronquist**	**Takhtajan**	**Dahlgren**	**Thorne**	**APG IV/(APweb)**
Informal Group						Rosids/(Malvid/ Rosid II)
Division		Magnoliophyta	Magnoliophyta			
Class	Dicotyledons	Magnoliopsida	Magnoliopsida	Magnoliopsida	Magnoliopsida	
Subclass	Polypetalae	Dilleniidae	Dilleniidae	Magnoliidae	Malvidae	
Series+/ Superorder	Thalamiflorae+		Capparanae	Violanae	Capparanae	
Order	Parietales	Capparales	Capparales	Capparales	Capparales	Brassicales

B & H as Capparidaceae, Cronquist Thorne and Takhtajan as Capparaceae.

APG II merged with Brassicaceae, **APG IV** and **APWeb,** however, reinstated the family.

Salient features: Shrubs or trees, sepals and petals 4 each, free, stamens many, ovary unilocular with parietal placentation, superior, sometimes with gynophore, fruit a capsule or berry.

Major Genera: *Capparis* (250 species), *Maerua* (100), *Boscia* (37), *Cadaba* (30) and *Crateva* (20 species).

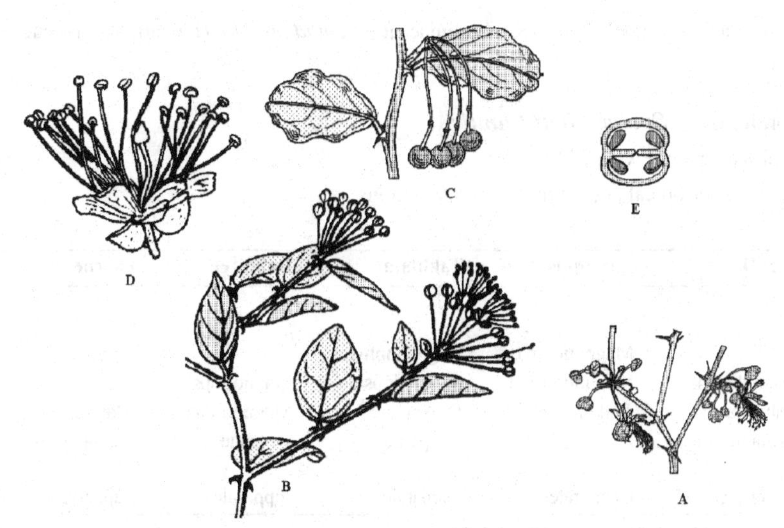

Figure 13.83: Capparaceae. A: *Capparis decidua* with flowers, leaves absent on branches. *C. sepiaria.* **B:** Portion of flowering branch; **C:** Small portion of fruiting branch; **D:** Flower with numerous stamens. **E:** Transverse section of ovary with intruded placentae.

Description: Shrubs (*Capparis*), rarely trees (*Crateva*) or climbers (*Maerua*). **Leaves** alternate, rarely opposite, simple, stipules present, sometimes reduced to glands or spines (*Capparis*). **Inflorescence** typically racemose, corymbose (*Crateva*), or in umbels (*Capparis*). **Flowers** bracteate (bracts often leafy), actinomorphic or zygomorphic (*Capparis*), bisexual, rarely unisexual or polygamous (*Crateva*), hypogynous, thalamus often prolonged into andro-gynophore. **Calyx** with 4 sepals, rarely 2–8, free or connate (*Maerua*), in two whorls, sometimes in one whorl. **Corolla** with 4 petals, cruciform (arranged in a cross), rarely 8 or even lacking (*Maerua*), clawed. **Androecium** with 4 or more stamens, free, often arising from androphore (lower portion of andro-gynophore), dehiscence longitudinal, nectaries often present near base of stamens. **Gynoecium** with 2–12 united carpels (syncarpous), unilocular with one-many ovules, replum absent, placentation parietal, ovary superior, often on a gynophore (upper part of andro-gynophore), style 1, stigma capitate or bilobed. **Fruit** a berry, capsule, drupe, or nut, often stalked; seeds 1-many, embryo curved, endosperm usually absent.

Economic importance: The family contributes a few ornamentals such as *Capparis* and *Crateva*. The fruit of *Capparis decidua* ('dela') is pickled and also given to heart patients. The dried floral buds of *C. spinosa* are called capers and are used in seasoning.

Phylogeny: Heterogeneity of broadly circumscribed Capparaceae was long recognized. Hutchinson (1973), based on extensive studies at Kew concluded that the family consisted of two distinct groups which are not really phylogenetically related. True Capparids, according to him are woody plants with indehiscent fruits, without a replum, and fairly closely related to Flacourtiaceae, whereas *Cleome* and its relatives are herbs with dehiscent fruits with replum, as in family Brassicaceae. This view was confirmed by morphological studies (Judd et al., 1994) and *rbcL* sequences (Rodman et al., 1993) as mentioned under Brassicaceae, ultimately leading to the merger of Capparaceae with Brassicaceae in APG classifications. Although position was kept up in the APG III but was pointed out that 'resurrection of Capparaceae and Cleomaceae may be appropriate in the future'. This change in position has largely been on account of the results of the studies Hall, Sytsma and Iltis (2002), who on the basis of chloroplast DNA sequence data concluded that the three, form distinct strongly supported monophyletic groups, as is also supported by morphological data. According to Puri (1950), bicarpellary syncarpous unilocular ovary with parietal placentation is derived from tetracarpellary condition with axile placentation. Thorne who had earlier (1999) recognized Capparaceae (also including *Cleome* and its relatives) and Brassicaceae as distinct families, has subsequently (2003, 2007) separated *Cleome* and relatives under distinct family Cleomaceae, as suggested in APG II. APG IV (2016) and APWeb (2018), recognize three as distinct families. *Neothorelia* formerly in Capparaceae has been excluded along with *Stixis*, *Borthwickia* and shifted to Resedaceae in APG IV (2016) and APWeb (2018). Earlier Su et al. (2012) shifted *Stixis* and *Borthwickia* into a distinct family Borthwickiaceae based on pollen and molecular data. The small-flowered *Dipterygium*, placed in Capparaceae in a subfamily by itself by Kers (2002), is to be included here; it is well embedded in the family Cleomaceae, in a clade with some other Old World taxa (Patchell et al., 2014). Tamboli et al. (2018) based on molecular studies concluded that Indian *Capparis* are more closely related to Old World taxa and have connections with African, Australian and Eastern Asian species and that Africa as the ancestral region for both Old World and New World *Capparis*. A few Genera formerly included in Capparaceae have been segregated into distinct families: *Koeberlinia* (Koeberliniaceae), *Pentadiplandra* (Pentadiplandraceae) and *Setchellanthus* (Setchellanthaceae).

Cleomaceae Horaninow ***Spider Plant Family***

25 genera, 270 species (as per APweb, 2018)

Widespread in tropical, subtropical and warm temperate regions.

Placement:

	B & H	Cronquist	Takhtajan	Dahlgren	Thorne	APG IV/(APweb)
Informal Group						Rosids/Malvids/ Rosid II
Division		Magnoliophyta	Magnoliophyta			
Class	Dicotyledons	Magnoliopsida	Magnoliopsida	Magnoliopsida	Magnoliopsida	
Subclass	Polypetalae	Dilleniidae	Dilleniidae	Magnoliidae	Malvidae	
Series+/ Superorder	Thalamiflorae+		Capparanae	Violanae	Capparanae	
Order	Parietales	Capparales	Capparales	Capparales	Capparales	Brassicales

B & H under Capparidaceae **Cronquist, Dahlgren** and **Takhtajan** under Capparaceae.

APG IV and APweb distinct family.

Salient features: Herbs, sepals and petals 4 each, free, stamens many, ovary unilocular with parietal placentation, superior, sometimes with gynophore, replum present, fruit a capsule or follicle.

Major Genera: *Cleome* (210 species), *Sieruela* (40), *Podandrogyne* (10) and *Polanisia* (7).

Description: Annual or perennial Herbs. **Leaves** alternate, rarely opposite, simple or palmately compound, stipules present. **Inflorescence** typically racemose, corymbose (*Cleome*). **Flowers** bracteate, bracts often leafy, actinomorphic, bisexual, hypogynous, thalamus often prolonged into andro-gynophore. **Calyx** with 4 sepals, rarely 2–8, free, in two whorls, sometimes in one whorl. **Corolla** with 4 petals, cruciform (arranged in a cross), clawed. **Androecium** with 4 or more stamens, free, often arising from androphore (lower portion of andro-gynophore), dehiscence longitudinal, nectaries often present near base of stamens. **Gynoecium** with 2–12 united carpels (syncarpous), unilocular (usually bilocular due to false septum and with distinct replum) with one-many ovules, placentation parietal, ovary superior, often on a gynophore (upper part of androgynophore), style 1, stigma capitate or bilobed. **Fruit** a capsule or siliqua, often stalked; seeds 1-many, embryo curved, endosperm usually absent.

Figure 13.84: Cleomaceae. *Cleome gynandra.* **A:** Lower part of the plant with palmately compound leaves; **B:** Inflorescence with flowers having conspicuous androgynophore; **C:** Sepal; **D:** Petal; **E:** Stamen; **F:** Gynoecium with distinct gynophore; **G:** Transverse section of ovary with parietal placentation; **H:** Seed.

Economic importance: The family contributes a few ornamentals such as *Cleome* and *Polanisia*. The decoction of *Cleome chelidonii* is used to cure scabies.

Phylogeny: Members of the family are generally included under family Capparaceae. Hutchinson (1973) based on extensive studies at Kew concluded that *Cleome* and its relatives are distinct from capparids. APG II classification included Capparaceae (including Cleomaceae) under Brassicaceae, but it was pointed out that 'resurrection of Capparaceae and Cleomaceae may be appropriate in the future'. This change in position has largely been on account of the results of the studies Hall, Sytsma and Iltis (2002), who based on chloroplast DNA sequence data, concluded that the three, form distinct strongly supported monophyletic groups, as is also supported by morphological data. Thorne who had earlier (1999) included this family under Capparaceae, has subsequently (2003) separated *Cleome* and relatives under distinct family Cleomaceae, as suggested in APG II, and continued in APG IV (2016) and APWeb (2018). *Cleomella* is sister to the rest of the family.

Brassicaceae Burnett *Mustard Family*

(= Cruciferae A. L. de Jussieu)

328 genera, 3,628 species

A cosmopolitan family mainly distributed in North Temperate Zone, particularly the Mediterranean region.

Placement:

	B & H	Cronquist	Takhtajan	Dahlgren	Thorne	APG IV/(APweb)
Informal Group						Rosids/(Malvids/ Rosid II)
Division		Magnoliophyta	Magnoliophyta			
Class	Dicotyledons	Magnoliopsida	Magnoliopsida	Magnoliopsida	Magnoliopsida	
Subclass	Polypetalae	Dilleniidae	Dilleniidae	Magnoliidae	Malvidae	
Series+/ Superorder	Thalamiflorae+		Capparanae	Violanae	Capparanae	
Order	Parietales	Capparales	Capparales	Capparales	Capparales	Brassicales

B & H as Cruciferae, others as Brassicaceae.

Salient features: Herbs, sap watery, sepals and petals 4 each, free, stamens tetradynamous, ovary with false septum and a thickened placental rim called replum, ovary superior, placentation parietal, fruit a siliqua or silicula.

Major Genera: *Draba* (350 species), *Erysimum* (180), *Lepidium* (170), *Cardamine* (160), *Arabis* (160), *Alyssum* (150), *Sisymbrium* (90) and *Brassica* (50).

Description: Annual, biennial or perennial herbs (rarely undershrubs: *Farsetia*) with watery sap, containing glucosinolates (mustard oils) and with myrosin cells. Hairs simple, branched, stellate or peltate. **Leaves** alternate or in basal rosettes, simple, often dissected, rarely pinnate compound (*Nasturtium officinale*) sometimes bearing bulbils in axil (*Dentaria bulbifera*) or leaf surface (*Cardamine pratensis*), stipules absent. **Inflorescence** typically racemose, corymbose raceme, or flat-topped corymb (*Iberis*), *Cardamine* also produces subterranean cleistogamous flowers. **Flowers** ebracteate, rarely bracteate (*Nasturtium montanum*), bisexual, actinomorphic or rarely zygomorphic (*Iberis*), hypogynous (perigynous in *Lepidium*). **Calyx** with 4 sepals, free, in two whorls, sepals of lateral pair sometimes saccate at base, green or slightly petaloid. **Corolla** with 4 petals, cruciform (arranged in a cross), clawed, sometimes absent in *Coronopus* and *Lepidium*. **Androecium** with 6 stamens (2 in *Coronopus*, 4 in *Cardamine hirsuta*, 16 in *Megacarpaea*), free, tetradynamous (2 short 4 long), dehiscence longitudinal, nectaries often present near base of stamens, pollen grains tricolporate or tricolpate. **Gynoecium** with two united (thus pistil single) carpels (syncarpous), rarely carpels 3 (*Lepidium*) or 4 (*Tetrapoma*), unilocular but becoming bilocular due to false septum that is surrounded by a thick placental rim called replum, ovules many, rarely single, placentation parietal, ovary superior, gynophore distinct, style 1, stigmas 2. **Fruit** a siliqua (long: length thrice width or more) or silicula (short: length less than thrice width), at dehiscence valves break away from below upward leaving seeds appressed to false septum, fruit moniliform lomentum on *Raphanus*; seed with large embryo, endosperm scant or absent. Pollination by insects, failure of cross pollination may result in self pollination. Seeds are usually dispersed by wind.

Economic importance: The family contributes several food plants such as radish (*Raphanus sativus*), cabbage (*Brassica oleracea* var. *capitata*), cauliflower (*B. oleracea* var. *botrytis*), Brussels sprouts (*B. oleracea* var. *gemmifera*), kohlrabi (*B. oleracea* var. *caulorapa*) and turnip (*B. rapa*). Seeds of *B. campestris* yield cooking oil those of black mustard (*B. nigra*) are used as condiment. Woad was formerly used a blue dye obtained from leaves of *Isatis tinctoria*. Common ornamentals include stock (*Matthiola*), candy tuft (*Iberis amara*), alyssum (*Alyssum*), wall flower (*Erysimum*) and sweet alyssum (*Lobularia*).

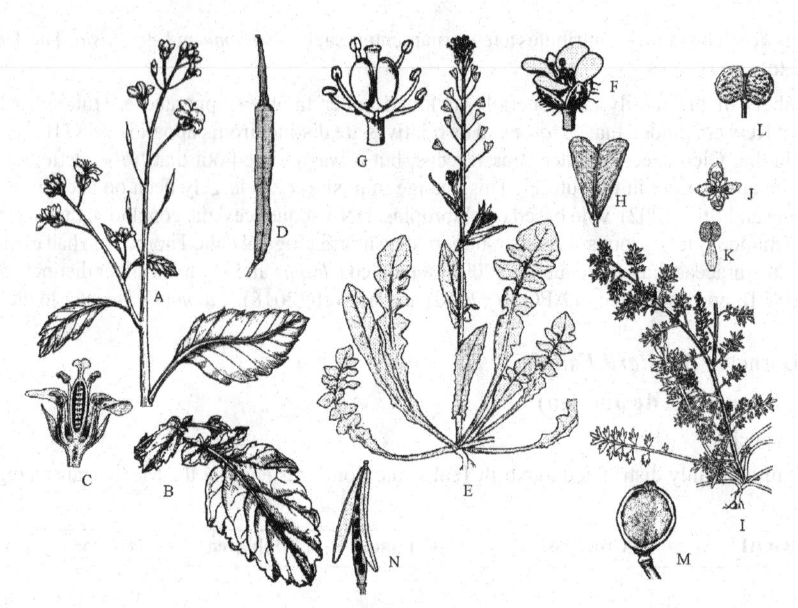

Figure 13.85: Brassicaceae. *Brassica campestris.* **A:** Upper part of plant with inflorescence; **B:** Lower leaf; **C:** Vertical section of flower; **D:** Siliqua with persistent style forming a long beak. *Capsella bursa-pastoris.* **E:** Plant with inflorescence; **F:** Flower; **G:** Flower with sepals and petals removed; **H:** Silicula with apical notch having persistent style, fruit flattened at right angles to the septum and as such replum appearing as vertical rim. *Coronopus didymus.* **I:** Plant with highly dissected leaves and axillary racemes; **J:** Flower from above showing minute petals and 2 stamens; **K:** Stamen; **L:** Silicula, deeply bilobed and prominent replum. **M:** Silicula of *Lobularia maritima* flattened parallel to the false septum and as such replum forming a ring around the fruit. **N:** Siliqua of *Brassica nigra* dehisced with valves separating and seeds attached to false septum.

Phylogeny: The family is regarded as monophyletic, supported by evidence from morphology (gynophore, exserted stamens), glucosinolates, dilated cisternae in endoplasmic reticulum and *rbcL* sequences. The order Brassicales (others prefer Capparales) had long been treated as a well-defined group, with Brassicaceae and Capparaceae considered to be fairly close as suggested by evidence from morphology, dilated cisternae, but have been treated as distinct largely because of several stamens and very long gynophore in Capparaceae.

Judd et al. (1994) based on morphological studies, and Rodman et al. (1993) on the basis rbcL sequences, concluded that out of the traditional Capparaceae, Capparoideae and Cleomoideae do not form a monophyletic group, as also concluded earlier by Hutchinson (1973). Capparoideae according to these authors form basal paraphyletic group within Brassicaceae. Cleomoideae and Brassicoideae (traditional Brassicaceae) form monophyletic group based on synapomorphies of herbaceous character, replum in fruit and rbcL sequences. The merger of Capparaceae with Brassicaceae avoids arbitrarily delimited paraphyletic taxa, and thus forms monophyletic group with broadened circumscription. The two have been merged in APG II and APweb classifications. APweb recognizes 3 subfamilies under broadly circumscribed Brassicaceae: Capparoideae, Cleomoideae and Brassicoideae. It is pertinent to note that Thorne (1999), who has been updating his classification in light of recent advances, has preferred to retain Brassicaceae and Capparaceae as distinct families, also separating Cleomaceae in the latest revision (2003, 2007), thus recognizing three subfamilies as independent families, a practice that has been followed by APG III, APG IV and recent version of APWeb (2018) According to the recent studies of Soltis et al. (2000) and Hall et al. (2002), Brassicaceae (Brassicoideae) and Cleomaceae (Cleomoideae) are more closely related and form a monophyletic group based on synapomorphies of herbaceous habit, rbcL sequences and presence of replum. It is interesting to record that although Hutchinson had indicated heterogeneity within Capparaceae and reasoned that *Cleome* and its relatives were much closer to Brassicaceae, he had placed Capparaceae and Brassicaceae in two distinct orders Capparales (in his diagram he used name Capparidales) and Brassicales, even further separating them under Lignosae and Herbaceae respectively, based on distinction of woody and herbaceous habits.

The family has been classified several times with different criteria, mostly based on fruit type and embryo curvature, but mostly unsatisfactory. Al-Shehbaz et al. (2006) recognised 25 monophyletic tribes based on well-supported clades, the number was increased to 49 (Al-Shehbaz, 2012) 321 genera, and 3660 species. Of these, 20 genera and 34 species remain to be assigned to tribes.

Figure 13.86: Paeoniaceae. A: *Paeonia suffruticosa*, plant. **Berberidaceae. B:** *Nandina domestica*, flowering branch; **C:** Branch with fruits. **Papaveraceae. D:** *Bocconia glaucifolia*, plant; **E:** Flower; **F:** *Papaver orientale*, plant; **G:** Flower; **H:** *Romneya coulteri*, portion of plant; **I:** Stamens and stigma.

Figure 13.87: Ranunculaceae. A: *Anemone occidentalis*, plant; **B:** Fruit; **C:** *Cimicifuga heracleifolia*, plant; **D:** *Clematis viticella*, plant; **E:** Central part of flower. **F:** *Caltha leptosepala*, plant; **G:** *Thalictrum polyganum*, plant; **H:** *Delphinium ajacis*; **I:** Flower; **J:** *Ranunculus muricatus*, plant; **K:** Flower; **L:** *R. sceleratus*, flower; **M:** Plant; **N:** *Helleborus argutifolius*, plant; **O:** *H. orientalis*, flower. **P:** *Nigella damascena*, plant.

Figure 13.88: Grossulariaceae. A: *Ribes menziesii* var. *leptosmum*, plant with fruits; **B:** Fruit. **C:** *Ribes sanguineum* var. *glutinosum*, fruit. **Fagaceae. D:** *Cyclobalanopsis glauca*, portion of trunk; **E:** Vegetative branches. **F:** *Lithocarpus densiflorus*, plant in flower. **Nothofagaceae. G:** *Nothofagus obliqua*, plant with fruits. **Betulaceae.** *Betula utilis.* **H:** Bark; **I:** Branch with Fruit; **J:** Fruiting inflorescence; Branch of *Corylus colurna*.

Figure 13.89: Celastraceae. A: *Euonymus grandiflorus*, plant; **B:** Flowers. **Violaceae. C:** *Viola tricolor*, plant; **D:** Flower. **Cucurbitaceae. E:** *Luffa cylindrica*, plant; **F:** Female flower; **G:** Young fruit. **Begoniaceae. H:** *Begonia foliosa*, plant; **I:** *B. sempervirens*, flower; **J:** *Begonia* 'Gene Daniels', flowers.

Figure 13.90: Clusiaceae. A: *Hypericum androsaemum*, plant; **B:** *H. monogynum*, plant; **C:** *Hypericum calycinum*, flower; **Euphorbiaceae. D:** *Jatropha integerrima*, plant; **E:** Female flower with fruit; **F:** Male flowers **G:** *Ricinus communis*, plant; **H:** Male Flower; **I:** Female flower; **J:** *Euphorbia milii*, plant; **K:** Cyathia with showy bracts; **L:** *E. pulcherrima*, plant; **M:** *Phyllanthus emblica*, branch in flower; **Oxalidaceae. N:** *Oxalis spiralis*, plant; **O:** Flower.

Figure 13.91: Rafflesiaceae. A: *Rafflesia speciosa*, flower. **Brassicaceae. B:** *Coronopus didymus*, plant; **C:** *Cheiranthus cheiri*, plant; **D:** *Matthiola incana*, plant; **E:** *Brassica oleracea*, flowering branch; **F:** Fruits; **G:** *B. campestris*, flowering branch; **H:** Fruits; **I:** Flowers; **J:** *Iberis amara*, flowering branch; **K:** *Cakile maritima*, flowering branch.

Figure 13.92: Fabaceae. A: *Lupinus arboreus*, plant; **B:** *Clitoria ternatea*, flower; **C:** *Parkinsonia aculeata*, plant; **D:** *Poinciana pulcherrima*, plant; **E:** Flower; **F:** *Saraca asoka*, plant with inflorescence; **G:** *Senna candolleana*, plant; **H:** Flower; **I:** *Calliandra haematocephala*, plant; **J:** Fruit; **K:** *Leucaena leucocephata*, plant; **L:** Fruits.

Figure 13.93: Geraniaceae. A: *Pelargonium zonale*, plant; **B:** Flower. **Rosaceae. C:** *Prunus campanulata*, plant; **D:** *Chaenomeles vilmoriana*, plant; **E:** Flower; **F:** *Cotoneaster microphyllus*, fruit; **G:** *Fragaria vesca*, plant; **H:** Flower; **I:** *Potentilla fruticosa*, plant; **J:** *Rubus trifidus*, flower.

Figure 13.94: Myrtaceae. A: *Lophostemon confertus*, plant; **B:** Flower; **C:** *Myrceugenella apiculata*, branch; **D:** *Metrosideros excelsa*, inflorescence; **Lythraceae. E:** *Cuphea* sp., flowering branch; **F:** Flower; **G:** *Lagerstroemia speciosa*, flowering twig. **Onagraceae. H:** *Fuchsia hatschbachii*, flowering branch; **I:** Flower; **J:** *Fuchsia microphylla*, flower.

Figure 13.95: Sapindaceae. A: *Acer japonicum,* branch; **B:** *A. griseum*, bark; **C:** Fruit; **D:** *Aesculus californica*, flowering branch; **E:** Flowers; **F:** *Ungnadia speciosa*, fruiting branch; **G:** Fruit.

Figure 13.96: Rutaceae. A: *Citrus limon*, flowering branch; **B:** Fruit; **C:** *Murraya paniculata*, flowering branch; **D:** Flower. **E:** *Citrus medica*, flower; **F:** *Ravenia spectabilis*, plant in flower; **G:** Flower. **H:** *Murraya koenigii*, flowering branch; **I:** Flowers. **Anacardiaceae. J:** *Mangifera indica*, flowering branch; **K:** Portion of inflorescence enlarged; **L:** Female flower. **Meliaceae. M:** *Melia azedarach*, flowering branch. **N:** Flower; **O:** Fruits.

Figure 13.97: Malvaceae. A: *Hibiscus rosasinensis*, flowering branch; **B:** stamens and stigmas; **C:** *Malvaviscus arboreus*, flowering branch; **D:** Flower; **E:** *Lavatera assurgentiflora*, flowering branch; **F:** Flower; **G:** *Gossypium hirsutum*, flowering branch; **H:** Flower; **I:** Fruit. **Rhamnaceae. J:** *Colletia paradoxa*, plant. **Moraceae. K:** *Morus alba*, plant with male catkins; **L:** Male catkin enlarged; **M:** Plant with female inflorescences. **N:** *Ficus religiosa*, twig with young hypanthodia.

Arrangement after APG IV (2016)

Superasterids:

Order 1. Berberidopsidales
- Family: Aextoxicaceae
- Family: Berberidopsidaceae

Order 2. Santalales
- Family: Olacaceae
- Family: Opiliaceae
- Family: Balanophoraceae
- Family: Santalaceae
- Family: Misodendraceae
- Family: Schoepfiaceae
- Family: Loranthaceae

Order 3. Caryophyllales
- Family: Frankeniaceae
- Family: Tamaricaceae
- Family: Plumbaginaceae
- Family: **Polygonaceae**
- Family: **Droseraceae**
- Family: Nepenthaceae
- Family: Drosophyllaceae
- Family: Dioncophyllaceae
- Family: Ancistrocladaceae
- Family: Rhabdodendraceae
- Family: Simmondsiaceae
- Family: Physenaceae
- Family: Asteropeiaceae
- Family: Macarthuriaceae
- Family: Microteaceae
- Family: **Caryophyllaceae**
- Family: Achatocarpaceae
- Family: **Amaranthaceae**
- Family: Stegnospermataceae
- Family: Limeaceae
- Family: Lophiocarpaceae
- Family: Kewaceae
- Family: Barbeuiaceae
- Family: Gisekiaceae
- Family: **Aizoaceae**
- Family: Phytolaccaceae
- Family: Petiveriaceae
- Family: Sarcobataceae
- Family: **Nyctaginaceae**
- Family: Molluginaceae
- Family: Montiaceae
- Family: Didiereaceae
- Family: Basellaceae
- Family: Halophytaceae
- Family: Talinaceae
- Family: **Portulacaceae**
- Family: Anacampserotaceae
- Family: **Cactaceae**

Families in boldface are described in detail.

Polygonaceae Durande *Buckwheat Family*

48 genera, 1,200 species

Distributed mainly in all temperate parts of the Northern Hemisphere, a few species in tropics, arctic region and the Southern Hemisphere.

Placement:

	B & H	Cronquist	Takhtajan	Dahlgren	Thorne	APG IV/(APweb)
Informal Group						Superasterids/ (Malvid/Rosid II)
Division		Magnoliophyta	Magnoliophyta			
Class	Dicotyledons	Magnoliopsida	Magnoliopsida	Magnoliopsida	Magnoliopsida	
Subclass	Monochlamydeae	Caryophyllidae	Caryophyllidae	Magnoliidae	Caryophyllidae	
Series+/ Superorder	Curvembryeae+		Polygonanae	Polygonanae	Caryophyllanae	
Order		Polygonales	Polygonales	Polygonales	Polygonales	Caryophyllales

Salient features: Mostly herbs with swollen nodes, stipules forming ochrea at nodes, flowers in spikes, heads or panicles, perianth usually petaloid, stamens 3–8, carpels 3, united, ovule solitary, fruit a nut.

Major genera: *Eriogonum* (250 species), *Rumex* (200), *Persicaria* (150), *Coccoloba* (120), *Polygonum* (60), *Rheum* (50) and *Fagopyrum* (15).

Description: Annual or perennial herbs, shrubs, small trees (*Triplaris*) or climbers with tendrils (*Antigonon*), with swollen nodes, usually with tannins, without laticifers, nodes pentalacunar or multilacunar, sieve-tube plastids S-type. **Leaves** usually alternate, rarely opposite (*Pterostegia*) or whorled (*Eriogonum*), sometimes reduced (*Coccoloba*) simple, usually entire, venation pinnate, reticulate, stipules connate to form ochrea at node (ochrea absent in *Eriogonum*). **Inflorescence** in axillary cymose clusters or forming spikes, heads or panicles. **Flowers** bisexual, rarely unisexual (*Rumex*), actinomorphic, hypogynous, showy (*Antigonon*) or inconspicuous (*Rumex*). **Perianth** with 6 tepals, in two whorls, usually petaloid,

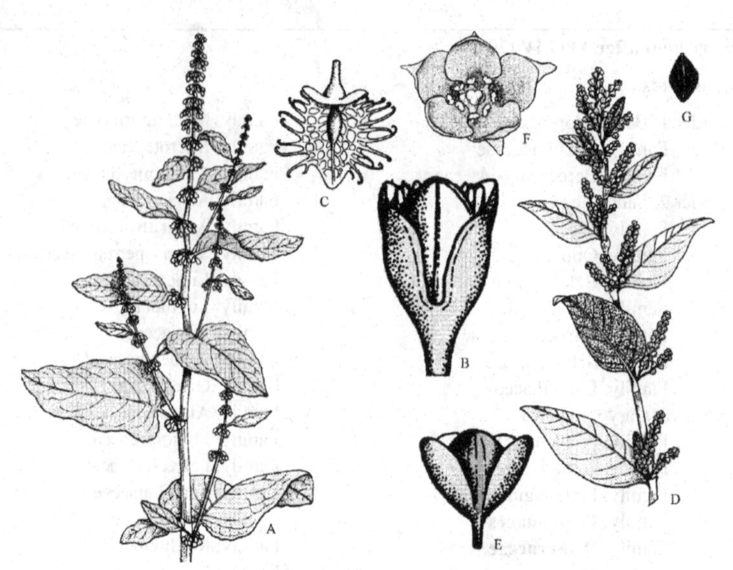

Figure 13.98: Polygonaceae. *Rumex nepalensis.* **A:** Portion of plant in flower; **B:** Flower with partially emergent anthers; **C:** Fruit with broad wings and hooked teeth. *Polygonum tortuosum.* **D:** Portion of plant in flowers; **E:** Flower with included stamens; **F:** Flower from top; **G:** Seed (After Polunin and Stainton, Fl. Himal., 1984).

sometimes 5 due to fusion of 2 tepals (*Polygonum*), free or slightly connate, imbricate, persistent, inner whorl often enlarged in fruit and tubercled (*Rumex*) or not (*Oxyria*), perianth often 4 in two whorls (*Oxyria*). **Androecium** with usually 6 stamens, 8 in *Fagopyrum*, 9 in *Rheum*, filaments free or slightly connate, anthers bithecous, dehiscence longitudinal, pollen grains tricolpate to polyporate. **Gynoecium** with 2–3 united carpels, ovary superior, unilocular with 1 ovule, orthotropous, placentation basal, sometimes partially divided by false septa, styles 2–3, a nectary surrounding the base of ovary, or paired glands associated with filaments. **Fruit** a trigonous or bifacial nut; seed with straight or curved embryo, endosperm copious, mealy. Pollination mostly by bees and flies. Fruits are usually dispersed by wind or water.

Economic importance: A few species of the family are of economic importance. Buckwheat (*Fagopyrum*) is an important source of food (millets) in some areas. The petioles of rhubarb (*Rheum rhaponticum*) are used as salad. Species of *Coccoloba*, *Antigonon*, *Muehlenbeckia*, and *Polygonum* are often cultivated as ornamentals. Fruits of *Coccoloba* are often used to make jellies. Leaves of *Rumex acetosa* and *R. crispus* are eaten as vegetables.

Phylogeny: The basic floral pattern of the family according to Laubengayer (1937) is trimerous and whorled and the apparent spiral condition in some members is actually whorled as shown by anatomical study. In members with 5 tepals, one outer and one inner tepal are fused. Laubengayer believed the family to be allied and more advanced than Caryophyllaceae and the seemingly basal placentation of the family is derived from free central placentation, the funiculus representing the greatly reduced free central placenta. Studies of Lamb Frye and Kron (2003) suggest that five petals is basic condition in the family.

The family is easily recognized and clearly monophyletic. The family is considered closer to Plumbaginaceae and according to Williams et al. (1994) although no plumbagin had been reported from the family, other quinones were found. Two subfamilies are commonly recognized in Polygonaceae: Polygonoideae with spiral leaves with sheathing ochrea and Eriogonoideae with opposite leaves and more or less cymose and involucrate inflorescence. Thorne (1999, 2003) recognized third subfamily Coccoloboideae characterized by reduced leaves, stems flattened and photosynthetic, for which, however, he subsequently (2006, 2007) preferred the name Brunnichioideae. The members of Eriogonoideae have generally 6 tepals and may form a basal paraphyletic complex (Cuénoud et al., 2002; Lamb Frye and Kron, 2003), thus justifying merger with polygonoideae. Sanchez et al. (2009) using chloroplast data placed *Symmeria* as sister to the whole of the rest of the family and *Afrobrunnichia* sister to Eriogonoideae, Burke and Sanchez (2011) as such, recognized three subfamilies: Eriogonoideae, Polygonoideae and Symmerioideae, six tribes within Eriogonoideae (Brunnicheae, Coccolobeae, Leptogoneae, Triplarideae, Gymnopodieae and Eriogoneae), Leptogoneae and Gymnopodieae being new additions. Schuster et al. (2015) found that the African *Oxygonum* is sister to other Polygonoideae, and then genera like *Persicaria* and *Bistorta* make up Persicarieae, and they are sister to the remainder of the subfamily. APweb (2018) places *Afrobrunnichia*, without assigning any subfamily.

Droseraceae Salisbury ***Sundew Family***

3 genera, 180 species

Widely distributed, mainly in marshy places low in nutrients, more commonly represented in Australia.

Placement:

	B & H	Cronquist	Takhtajan	Dahlgren	Thorne	APG IV/(APweb)
Informal Group						Superasterids/ (Malvid/Rosid II)
Division		Magnoliophyta	Magnoliophyta			
Class	Dicotyledons	Magnoliopsida	Magnoliopsida	Magnoliopsida	Magnoliopsida	
Subclass	Polypetalae	Dilleniidae	Dilleniidae	Magnoliidae	Caryophyllidae	
Series+/ Superorder	Thalamiflorae+		Nepenthanae	Theanae	Caryophyllanae	
Order	Parietales	Nepenthales	Nepenthales	Theales	Caryophyllales	Caryophyllales

B & H under Sarraceniaceae; **Dahlgren** under Nepenthaceae.

Salient features: Insectivorous plants, leaves in rosettes, covered with sticky insect-catching hairs, cercinate in bud, flowers bisexual, pentamerous, inflorescence determinate, stamens 5-many, carpels 3–5, fruit a capsule.

Major genera: *Drosera* (178 species), *Dionaea* (1) and *Aldrovanda* (1).

Description: Insectivorous herbs of bogs and wetlands, or submerged aquatic plant (*Aldrovanda*), annual or perennial. **Leaves** usually in basal rosette, upper alternate, circinate in bud, simple, blade modified into hinged trap (*Dionaea*) or covered with sticky glandular hairs (often called tentacles) for trapping insects (*Drosera*), venation obscure, stipules absent or present, petiole winged in *Dionaea*. **Inflorescence** scapigerous, determinate, cymose, often appearing like umbel, raceme or panicle, rarely solitary. **Flowers** bisexual, actinomorphic, hypogynous, pentamerous. **Calyx** with 5 sepals, rarely 4, slightly connate at base, imbricate, persistent. **Corolla** with 5 petals, rarely 4, free, convolute, mostly white or pinkish. **Androecium** with usually 5 stamens (*Drosera*), 10–20 in *Dionaea*, free or slightly connate at base, anthers bithecous, dehiscence longitudinal, pollen grains triporate to multiporate, released in tetrads. **Gynoecium** with 3 united carpels, rarely upto 5, ovary superior, unilocular with 3-many ovules, placentation parietal or basal, style single (*Dionaea*) or 3–5, each divided up to the base thus appearing 6–10 in number. **Fruit** a loculicidal capsule; seed small, with straight embryo, endosperm present, crystalline-granular, seed variously reticulated or ornamented. Pollination usually by insects, protoandy resulting in outcrossing, but selfing may result when flowers close by the end of the day. Small seeds are dispersed by wind or water. Detached leaves and inflorescences may also produce new plants.

The hinged leaves of *Dionaea* have two halves, each with marginal sensitive bristles. When an insect touches the bristles, the two halves close, entrapping the insect. The small glands on the leaf surface secrete enzymes, which digest

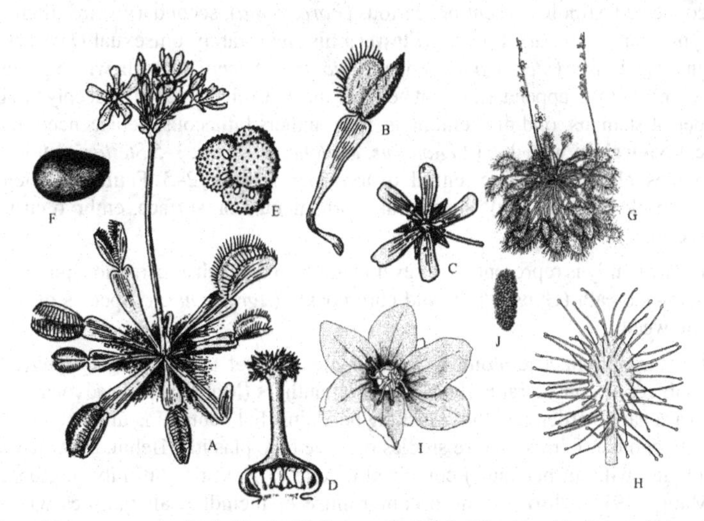

Figure 13.99: Droseraceae. *Dionaea muscipula.* **A:** Plant with flowers; **B:** Leaf with winged petiole and hinged leaf blade with marginal bristles; **C:** Flower; **D:** Ovary cut open to show ovules; **E:** Pollen tetrad; **F:** Seed. *Drosera intermedia.* **G:** Plant with flowers; **H:** Leaf blade with glandular hairs; **I:** Flower, top view; **J:** Seed covered with papillae (After Godfrey and Wooten, Aq. Wetland Pl SE US, 1981).

the insect. In *Drosera*, the gland-tipped sticky hairs when stimulated, entangle the insect by bending inwards, pressing the insect against the leaf blade.

Economic importance: The family is of little economic importance. Venus flytrap (*Dionaea muscipula*) and various species of *Drosera* (Sundew) are grown as novelties. Leaves of *Drosera* yield a violet dye but is no longer of commercial importance.

Phylogeny: The family was earlier included under Sarraceniaceae (Bentham and Hooker) but is now recognized independently. Wettstein (1907) placed Droseraceae in order parietales near Violaceae and Ochnaceae on the basis of parietal placentation. Insectivorous habit and aquatic habit exhibit homoplasy with family Lentibulariaceae. Earlier the family also included 4th subwoody genus *Drosophyllum* (being distinct in leaves not curling to envelop insects, different pollen grains and trichome anatomy), which has now been separated into a distinct family Drosophyllaceae (APG-II on wards; Thorne, 2006, 2007; APweb), being sister to Dioncophyllaceae + Ancistrocladaceae. Within Droseraceae, *Dionaea* and *Aldrovanda* with snap-trap leaves and n = 6 may be sister to rest of taxa.

Caryophyllaceae A. L. de Jussieu ***Pink Family***

96 genera, 2,415 species

Distributed mainly in all temperate parts of the world.

Placement:

	B & H	**Cronquist**	**Takhtajan**	**Dahlgren**	**Thorne**	**APG IV/(APweb)**
Informal Group						Superasterids/ (Malvids/Rosid II)
Division		Magnoliophyta	Magnoliophyta			
Class	Dicotyledons	Magnoliopsida	Magnoliopsida	Magnoliopsida	Magnoliopsida	
Subclass	Polypetalae	Caryophyllidae	Caryophyllidae	Magnoliidae	Caryophyllidae	
Series+/ Superorder	Thalamiflorae+		Caryophyllanae	Caryophyllanae	Caryophyllanae	
Order	Caryophyllineae	Caryophyllales	Caryophyllales	Capparales	Caryophyllales	Caryophyllales

Salient features: Herbs with swollen nodes, leaves opposite, inflorescence usually a dichasial cyme, corolla caryophyllaceous, stamens ten or lesser, obdiplostemonous, ovary unilocular with free central placentation, superior, fruit a capsule opening by valves or teeth.

Major genera: *Silene* (700 species), *Dianthus* (300), *Arenaria* (200), *Gypsophila* (150), *Minuartia* (150), *Stellaria* (150) and *Cerastium* (100).

Description: Annual or perennial herbs with swollen nodes, anthocyanins present. **Leaves** opposite, simple, bases of opposite leaves often connected, stipules absent or scarious (*Paronychia*), secondary veins often obscure. **Inflorescence** typically a dichasial cyme, rarely solitary flowers. **Flowers** bisexual, rarely unisexual (*Lychnis alba*), actinomorphic, hypogynous. **Calyx** with 5 sepals, free (*Stellaria*) or connate (*Dianthus*, *Silene*). **Corolla** with 5 petals, usually differentiated into a distinct claw and a limb, with an appendaged joint between the two, often notched or deeply bilobed at tip. **Androecium** with 10 or lesser number of stamens, obdiplostemonous, free, anthers bithecous, dehiscence longitudinal, pollen grains tricolpate to polyporate. **Gynoecium** with 2–5 (2 *Dianthus*, 3 *Silene*, 4 *Sagina*, 3–5 *Stellaria*) united carpels (syncarpous), unilocular with many ovules, placentation free central, ovary superior, styles 2–5. **Fruit** a loculicidal capsule opening by valves or teeth, rarely a utricle (*Paronychia*). Seeds many, ornamented on surface, embryo curved, endosperm absent, often replaced by perisperm.

Economic importance: The family is represented by several ornamentals such as carnation, pinks, sweet william (different species of *Dianthus*), baby's breath (*Gypsophila*) and corn cockle (*Agrostemma*). Species of *Arenaria*, *Cerastium* and *Stellaria* are troublesome weeds.

Phylogeny: The family Caryophyllaceae along with other members of the order Caryophyllales (Centrospermae of Engler) have been the subject of considerable debate, several authors (Mabry, 1963) advocating separation of betalain containing families into a separate order, and those lacking betalains (but containing anthocyanins as in Caryophyllaceae and Plumbaginaceae) into another. Ultrastructure studies of sieve tube plastids (Behnke, 1975, 1977, 1983) showed that all members (those with and without betalains) contained unique PIII plastids, affinity reinforced by studies of DNA/RNA hybridization (Mabry, 1975), throwing up a compromise of including all families within the same order, but separate suborders, a trend being followed by Takhtajan upto 1987 but finally discarded in 1997. Thorne (2006, 2007) has established 4 suborders within Caryophyllales, Caryophyllaceae being placed under monotypic

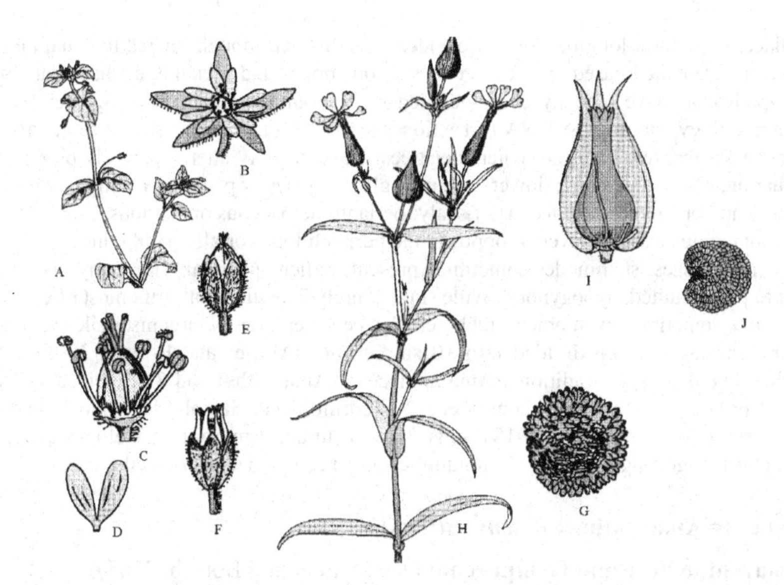

Figure 13.100: Caryophyllaceae. *Stellaria media.* **A:** A part of plant in flower; **B:** Flower showing hairy sepals and deeply bilobed petals; **C:** Flower with sepals and petals removed to show stamens and pistil; **D:** Bilobed petal; **E:** Mature capsule with persistent calyx; **F:** Capsule dehiscing through valves; **G:** Seed. *Silene conoidea.* **H:** A portion of plant in flower; **I:** Capsule with half of calyx removed to show dehiscence through teeth and remnants of petals and stamens; **J:** Seed (After Sharma and Kachroo, Fl. Jammu, 1983).

suborder Caryophyllineae. Centrospermae is a classic case of proving a point that no single character should be relied upon, and conclusions should be drawn only after the findings are reinforced by studies from other fields.

The family forms a well-defined monophyletic clade, as evidenced by morphology, and *rbcL* sequence. True petals are lacking in the family, and in most cases outer 4 to 5 stamens are transformed into petals. It was traditionally divided into three subfamilies: Silenoideae (or Caryophylloideae-stipules absent, sepals united into a tube), Alsinoideae (stipules absent, sepals free) and Paronychioideae (stipules present, petals free or united), a classification not supported by recent studies. Harbaugh et al. (2010) based on studies three chloroplast gene sequences (matK, trnL-F, and rps16) reveal abandon three subfamilies and instead recognize 11 tribes: Paronychioideae replaced by four tribes Corrigioleae, Paronychieae, Polycarpaeae and Sperguleae; Caryophylloideae replaced by two tribes Caryophylleae and Sileneae; and Alsinoideae with some members from Caryophylloideae replaced by 5 tribes Sclerantheae, Sagineae, Alsineae, Arenarieae and Eremogonieae. Greenberg and Donoghue (2011) based on Ribosomal DNA and chloroplast studies concluded that Paronychioideae is paraphyletic at the base of Caryophyllaceae, Alsinoideae and Caryophylloideae together form a clade, within which neither subfamily is monophyletic. They supported tribal classification of Harbaugh et al., proposing additional clade name Plurcaryophyllaceae.

Amaranthaceae M. Adanson ***Amaranth Family***

183 genera and 2,500 species

Cosmopolitan, mainly tropical, centered in Africa and America.

Placement:

	B & H	**Cronquist**	**Takhtajan**	**Dahlgren**	**Thorne**	**APG IV/(APweb)**
Informal Group						Superasterids/ (Malvids/Rosid II)
Division		Magnoliophyta	Magnoliophyta			
Class	Dicotyledons	Magnoliopsida	Magnoliopsida	Magnoliopsida	Magnoliopsida	
Subclass	Monochlamydeae	Caryophyllidae	Caryophyllidae	Magnoliidae	Caryophyllidae	
Series+/ Superorder	Curvembryeae+		Caryophyllanae	Caryophyllanae	Caryophyllanae	
Order		Caryophyllales	Caryophyllales	Caryophyllales	Caryophyllales	Caryophyllales

Family Chenopodiaceae has for a long time been considered as distinct, though related to Amaranthaceae. Traditional Amaranthaceae was differentiated based on presence of scarious bracts and perianth, connate stamens and presence of staminodes. Chenopodiaceae have recently been included in Amaranthaceae and two together form a monophyletic as supported by morphology, chloroplast DNA restriction sites and *rbcL* sequences. The reorganized Amaranthaceae now includes herbs or shrubs, few vines, succulents or trees; leaves mostly alternate, rarely opposite, exsipulate, with entire or toothed margin, sometimes scaly; flowers occur singly, in cymes or panicles, mostly bisexual, rarely unisexual, actinomorphic, bracts and bracteoles herbaceous or scaly; perianth herbaceous or scarious, usually 5, somewhat united; stamens usually 5, sometimes lesser or even 1, opposite the perianth lobes or alternating, inserted on hypogynous disc, anthers with 2 to 4 pollen sacs, staminodes sometimes present, pollen spherical with many pores; carpels usually 2, sometimes single or up to 6, united, hypogynous, ovule single, rarely 2, ovule basal; fruit most often single seeded utricle with persistent perianth, sometimes even bracts and bracteoles persistent, rarely circumscissile capsule or a berry.

Family Amaranthaceae is often divided into 10 subfamilies (Muller and Borsch, 2005) 2 (Amaranthoideae and Gomphrenoideae) belonging to traditional Amaranthaceae (Amaranths) and 8 (Betoideae, Camphorosmoideae, Chenopodioideae, Corispermoideae, Polycnemoideae, Salicornioideae, Salsoloideae and Suaedoideae) to former Chenopodiaceae (Chenopods). APweb (2018) adds 3 more under Amaranths: Celosioideae, Aervoideae and Achyranthoideae), thus recognising a total of 13 subfamilies under reorganized Amaranthaceae.

Amaranth complex (= Amaranthaceae *sensu stricto*)

(Subfamilies Amaranthoideae and Gomphrenoideae (Muller and Borsch, 2005))

(Celosioideae, Amaranthoideae, Aervoideae, Achyranthoideae, Gomphrenoideae (APweb, 2018))

79 genera, 1,000 species

Cosmopolitan, mainly tropical, centered in Africa and America.

Salient features: Herbs or small shrubs, stipules absent, flowers small often greenish, subtended by scarious or papery bracts, perianth papery, stamens opposite perianth lobes, slightly connate at base, staminodes present, carpels 2–3, ovary superior, fruit a capsule or utricle or nutlet, enclosed in persistent perianth, embryo curved.

Major genera: *Gomphrena* (120 species), *Alternanthera* (100), *Celosia* (65) and *Iresine* (80), *Amaranthus* (60).

Description: Herbs or small shrubs, very rarely climbing, often with swollen nodes, nodes unilacunar, vascular bundles in concentric rings, included phloem usually present, sieve-tube plastids PIII-A type, containing betalains instead of anthocyanins. **Leaves** alternate or opposite, herbaceous, sometimes aggregated at base (*Ptilotus*), petiolate to sessile, simple, entire, stipules

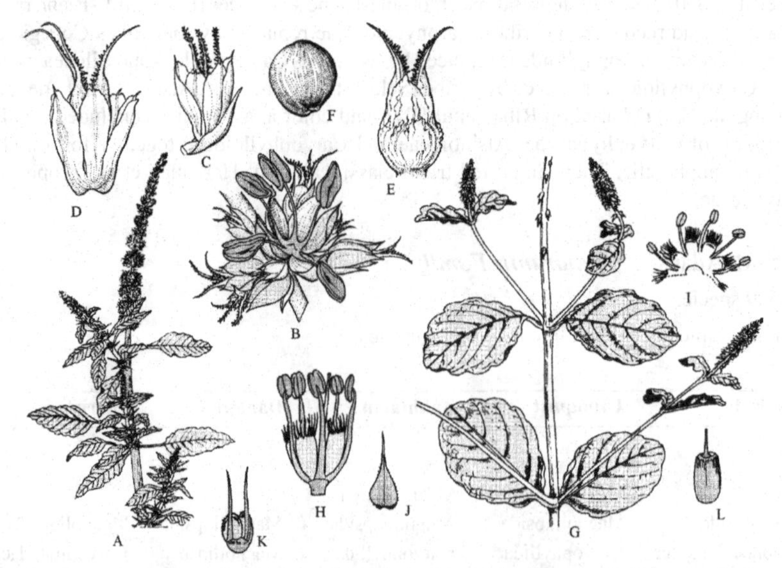

Figure 13.101: Amaranthaceae. *Amaranthus spinosus.* **A:** Part of plant in flower; **B:** Cymose cluster with one male and several female flowers; **C:** Female flower with 3 carpels; **D:** Mature fruit of same with enlarged persistent perianth; **E:** Mature utricle developed from flower with 2 carpels, perianth removed; **F:** Seed. *Achyranthes aspera.* **G:** Part of plant in flower; **H:** Flower with bract and perianth removed; **I:** Androecium showing stamens and staminodes; **J:** Bract; **K:** Bracteoles; **L:** Utricle with persistent style (A, G–L, after Sharma and Kachroo, Fl. Jammu, 1983).

absent. **Inflorescence** cymose, spikes or panicles, with conspicuous persistent bracts and bracteoles. **Flowers** small, greenish, bisexual (rarely unisexual), actinomorphic, hypogynous, cyclic. **Perianth** (represented by sepals, petals absent) with 3–5 free or united tepals, usually persistent, sometimes accrescent (*Ptilotus*) in fruit, usually dry and scarious. **Androecium** with 5 stamens, rarely 3 or even 6–10, opposite the tepals, filaments slightly connate at base, often adnate to tepals, anthers inflexed in bud, bithecous (*Amaranthus*) or monothecous (*Gomphrena*), dehiscence longitudinal, pollen grains multiporate, spinulose, staminodes often present, usually 1–3. **Gynoecium** with 2–3 united carpels, ovary superior, unilocular, ovule usually 1, placentation basal, rarely many (*Celosia*), styles 1–3. **Fruit** a circumscissile capsule, or nut or utricle (when enclosed in membranous perianth); seed lens shaped with curved or spiral embryo, endosperm absent, perisperm present.

Economic importance: The group includes several ornamentals such as *Celosia* (Cockscomb), *Amaranthus* (amaranth), *Gomphrena* (globe amaranth) and *Iresine* (bloodleaf). Species of *Alternanthera* and *Tilanthera* are grown as edge plants and have ornamental leaves. Seeds and leaves of several species of *Amaranthus* are edible, as are also the leaves of *Alternanthera sessilis*.

Chenopod complex (= Chenopodiaceae Ventenat)

(Subfamilies Betoideae, Camphorosmoideae, Chenopodioideae, Corispermoideae, Polycnemoideae, Salicornioideae, Salsoloideae and Suaedoideae (Muller and Borsch, 2005; APWeb, 2018))

104 genera, 1,500 species

Widely distributed in temperate and tropical climates but common in arid and semiarid saline habitats.

Salient features: Herbs or small shrubs usually in saline habitats, often covered with white bloom, stipules absent, flowers small often greenish, perianth herbaceous, stamens opposite perianth lobes, all fertile and similar, carpels 2, ovary superior, fruit a nut enclosed in persistent perianth, embryo curved.

Major genera: *Atriplex* (300 species), *Salsola* (120), *Chenopodium* (105), *Suaeda* (100) and *Salicornia* (35).

Description: Herbs or small shrubs, rarely small trees (*Haloxylon*), usually in saline habitats, sometimes succulent (*Salicornia*), often covered with whitish bloom, nodes unilacunar, vascular bundles in concentric rings, included phloem usually present, sieve-tube plastids PIII-C type, containing betalains instead of anthocyanins. Leaves minute to large, alternate, rarely opposite (*Salicornia*, *Nitrophila*), petiolate to sessile, simple, entire or variously lobed, sometimes fleshy or reduced to scales, stipules absent. **Inflorescence** cymose, spikes or panicles, sometimes catkins. **Flowers** small, greenish, bisexual, rarely unisexual and plants dioecious (*Grayia*) or monoecious, actinomorphic, hypogynous. **Perianth** (represented by sepals, petals absent) with 2–5 united tepals, rarely free (*Salsola*) usually persistent and accrescent in fruit, and appendaged with tubercles, spines or wings, sometimes absent. **Androecium** with 5 stamens, rarely 3, opposite the perianth lobes, filaments free, anthers inflexed in bud, bithecous, dehiscence longitudinal, pollen grains multiporate, spinulose. **Gynoecium** with 2 carpels, united, rarely carpels upto 5, ovary superior, unilocular, ovule 1,

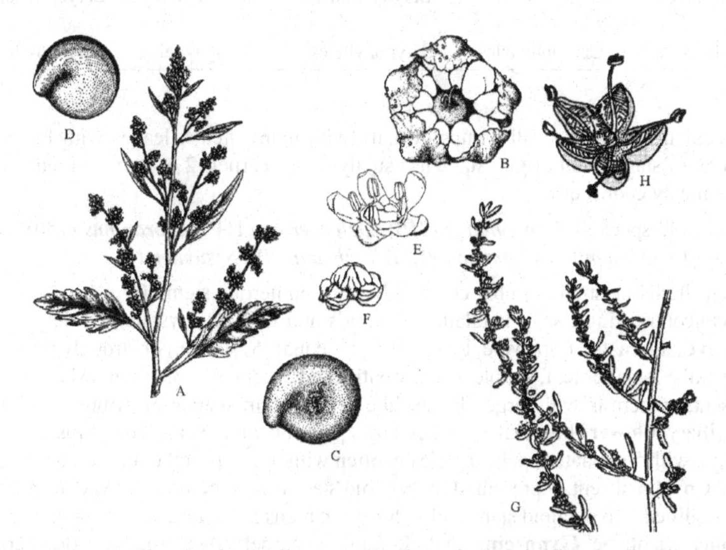

Figure 13.102: Chenopodiaceae. *Chenopodium album*. **A:** Portion of plant in flower; **B:** Flower partially opened with stamens still included; **C:** Fruit from above; **D:** Seed. *Beta vulgaris*. **E:** Flower; **F:** cluster of Fruits. *Suaeda maritima*. **G:** Portion of plant in flower; **H:** Flower.

placentation basal, styles 2 (rarely upto 5). **Fruit** a nut or utricle (when enclosed in membranous perianth); seed lens shaped with curved or spiral embryo, endosperm absent, perisperm present.

Economic importance: The family includes a few food plants such as beet (*Beta vulgaris;* root vegetable mainly for salad and a source of sugar), beet leaf (*B. vulgaris* ssp. *maritima*) used as leafy vegetable, often confused with spinach, spinach (*Spinacea oleracea*) and lambs quarters (*Chenopodium album;* bathoo in Hindi). *Chenopodium ambrosioides* is source of wormseed used as a vermifuge. Seeds and leaves of *C. quinoa* are eaten by Peruvians and Andes.

Phylogeny of Amaranthaceae: The traditional Amaranthaceae is closely related to Chenopodiaceae (which is placed within Amaranthaceae in APG classifications) but differentiated in scarious bracts and perianth, connate stamens and presence of staminodes. Chenopodiaceae on the other hand has herbaceous perianth, all fertile stamens equal in length, and free filaments Hutchinson (1926, 1973) believes the family to have evolved from caryophyllaceous ancestors. Cuénoud et al. (2002) found Amaranthaceae s. str. to be monophyletic, with very strong (97 per cent) support. The family, in the broader sense (including Chenopodiaceae), is monophyletic as supported by morphology, chloroplast DNA restriction sites and *rbcL* sequences. The separation of Chenopodiaceae and Amaranthaceae seems to be arbitrary. Others like Pratt et al. (2001) consider Amaranthaceae to be polyphyletic. The separation of these two families leads to paraphyletic Chenopodiaceae (Downie et al., 1997; Rodman, 1994; Pratt et al., 2001). Cuénoud et al. (2002) found Chenopodiaceae were perhaps monophyletic, but the branch collapsed in a strict consensus tree; the sampling was moderately good, but only one gene-*matK*—was sequenced and analyzed. *Sarcobatus* has long been acknowledged an anomalous member of this family, e.g., by Bentham and Hooker (1880), who presented it as a monogeneric tribe. Behnke (1997) proposes raising it to family rank (accepted by APG IV, and APWeb). Within Amaranthaceae s. str.—at least some flowers are imperfect—*Bosea* and *Charpentiera* were successively sister to the rest, but Amaranthoideae, Amarantheae and Amarathineae were paraphyletic (Ogundipe and Chase, 2009).

Aizoaceae Martynov *Stone Plant Family*

121 genera, 1,900 species

Widely distributed in tropical and subtropical regions, mainly in arid and coastal regions of South Africa.

Placement:

	B & H	Cronquist	Takhtajan	Dahlgren	Thorne	APG IV/(APweb)
Informal Group						Superasterids/ (Malvids/Rosid II)
Division		Magnoliophyta	Magnoliophyta			
Class	Dicotyledons	Magnoliopsida	Magnoliopsida	Magnoliopsida	Magnoliopsida	
Subclass	Polypetalae	Caryophyllidae	Caryophyllidae	Magnoliidae	Caryophyllidae	
Series+/ Superorder	Calyciflorae+		Caryophyllanae	Caryophyllanae	Caryophyllanae	
Order	Ficoidales	Caryophyllales	Caryophyllales	Caryophyllales	Caryophyllales	Caryophyllales

B & H as Ficoideae.

Salient features: Succulent herbs or small shrubs, usually with many stems, leaves with large bladder-like cells in epidermis, petals several of staminodal origin, stamens usually many, carpels 2 or more, united, fruit a capsule, embryo curved, surrounded by mealy endosperm.

Major genera: *Ruscia* (300 species), *Conophytum* (250), *Delosperma* (150), *Lapranthus* (150), *Drosanthemum* (100), *Antimima* (60), *Lithops* (40), *Mesembryanthemum* (30), *Trianthema* (20), *Sesuvium* (8).

Description: Succulent herbs or small shrubs, commonly with numerous stems from base, sometimes mat-forming, vascular bundles in concentric rings, with betalains, alkaloids and raphide crystals of calcium oxalate, usually with CAM metabolism. **Leaves** alternate or opposite, leaves of a pair equal (*Sesuvium*) or strongly unequal (*Trianthema*) with branches arising from axil of smaller leaf, simple, usually entire and succulent, veins somewhat obscure, stipules scarious or fringed, rarely absent, epidermis with large bladder-like cells. **Inflorescence** of cymose axillary cymes, in pairs or solitary, terminal or axillary. **Flowers** bisexual, subtended by a pair of laciniate bracts often fused with calyx tube, regular, with hypanthium. **Calyx** with 5 connate sepals, imbricate, often with horny protuberance on back below tip, persistent in fruit, margin scarious. **Corolla** absent, represented by petaloid staminodes, numerous. **Androecium** with 5 to numerous stamens, many outer modified into petaloid staminodes, fertile stamens 5 or more, free or with connate filaments, arising from hypanthium, pollen tricolpate. **Gynoecium** with 2–5 united carpels, ovary superior or inferior, 2–5 locular, with axile, parietal or basal placentation, ovules 1 to many, anatropous to campylotropous, disc usually present. **Fruit** a capsule, loculicidal, septicidal, or circumscissile, sometimes berry, rarely a nut; seed with large curved embryo, sometimes with

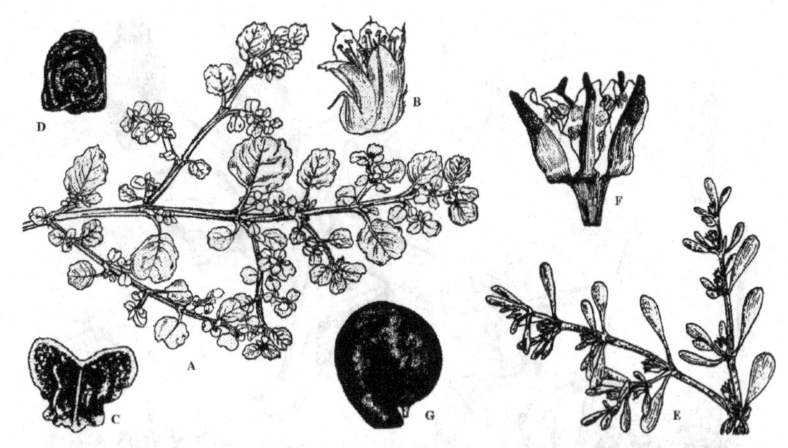

Figure 13.103: **Aizoaceae.** *Trianthema portulacastrum.* **A:** Portion of plant with flowers; **B:** Flower with two bracts and horned calyx lobes; **C:** Capsule; **D:** Seed with papilae. *Sesuvium verrucosum.* **E:** Portion of plant with flowers. **F:** Flower with horny calyx. **G:** Seed of *S. maritimum* (A, B after Maheshwari, Fl. Delhi, 1963; D–G after Godfrey & Wooten, Aq. Wetland Pl. SE US, Vol. 2, 1981).

aril, endosperm absent, replaced by perisperm. Flowers are pollinated by bees, wasps, butterflies and beetles. Seeds are dispersed by wind or water.

Economic importance: Species of *Mesembryanthemum* (ice plant), *Lapranthus*, *Dorotheanthus* and *Carpobrotus* are grown as garden ornamentals. Some like *Lithops* (stone plant) and *Titanopsis* are grown as curiosities. *Tetragonia* is used as vegetable. Some species help stabilize sand dunes and road banks.

Phylogeny: The family Aizoaceae as understood now includes former families Mesembryanthemaceae, Sesuviaceae and Tetragoniaceae. Reorganized family is monophyletic, represented by distinct clades, often recognised as four subfamilies Aizooideae, Mesembryanthemoideae, Sesuvioideae and Ruschioideae (Klak et al., 2004; Stevens, 2007; Thorne, 2006). Members with numerous petaloid staminodes, placed in Mesembryanthemoideae and Ruschioideae form a monophyletic group (Hartmann, 1993). Genus *Mesembryanthemum* was previously circumscribed to include more than 1000 species, but has subsequently been split into numerous genera, a bulk of genera shifted to Ruschioideae. There have been further attempts to divide the genus further into much smaller genera, an attempt that would result in many poorly characterised genera. A detailed phylogenetic analysis of the family by Klak et al. (2007) has resulted in better resolution affinities within the subfamily Mesembryanthemoideae. Klak et al. (2017) removed *Acrosanthes* into fifth new subfamily Acrosanthoideae, an isolated sister-lineage to the Mesembryanthemoideae + Ruschioideae. Basal, shortly stipitate ovules and a xerochastic, parchment-like capsule are synapomorphies for the Acrosanthoideae. The largest subfamily Ruschioideae with 111 genera and 1500 species has been further subdivided into three tribes: Apatesieae, Dorotheantheae and the rest Core Ruschioideae.

Nyctaginaceae A. L. de Jussieu *Four O'Clock Family*

31 genera, 400 species

Widely distributed in tropical and subtropical regions, mainly in the New World.

Placement:

	B & H	Cronquist	Takhtajan	Dahlgren	Thorne	APG IV/(APweb)
Informal Group						Superasterids/ (Malvids/Rosid II)
Division		Magnoliophyta	Magnoliophyta			
Class	Dicotyledons	Magnoliopsida	Magnoliopsida	Magnoliopsida	Magnoliopsida	
Subclass	Monochlamydeae	Caryophyllidae	Caryophyllidae	Magnoliidae	Caryophyllidae	
Series+/ Superorder	Curvembryeae+		Caryophyllanae	Caryophyllanae	Caryophyllanae	
Order		Caryophyllales	Caryophyllales	Caryophyllales	Caryophyllales	Caryophyllales

Salient features: Swollen nodes, wood quickly turning reddish-brown when cut, leaves usually opposite, sometimes unequal, without stipules, vascular bundles in concentric rings, flowers bisexual with conspicuous bracts, bracts often enlarged, perianth 5 with long tube, stamens usually 5, carpel single, ovule single, basal, fruit an achene or nut.

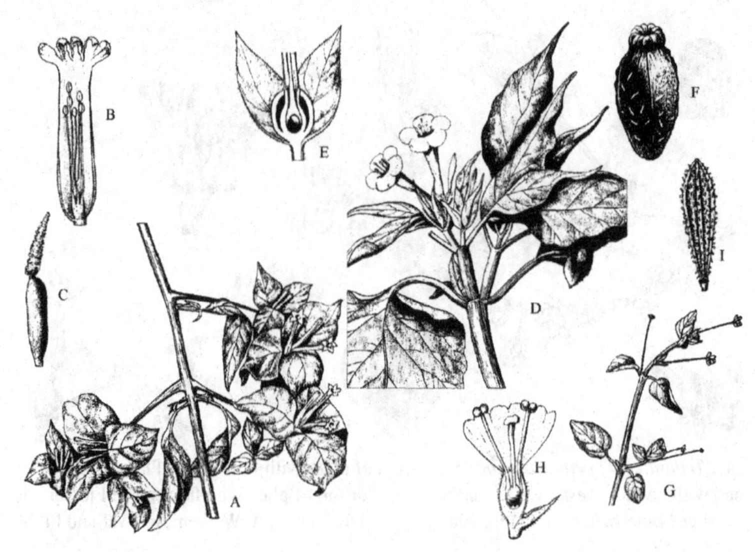

Figure 13.104: Nyctaginaceae. *Bougainvillea glabra.* **A:** Portion of plant with flowers; **C:** Carpel; **B:** Longitudinal section of flower of *B. spectabilis* showing perianth and stamens. *Mirabilis jalapa.* **D:** A portion of branch with flowers; **E:** Longitudinal section of ovary with subtending bracts; **F:** Anthocarp. *Boerhavia repens.* **G:** Portion of branch with flowers and fruits; **H:** Longitudinal section of flower to show perianth, stamens and ovary; **I:** Anthocarp.

Major genera: *Neea* (80 species), *Guapira* (70), *Mirabilis* (45), *Pisonia* (40), *Abronia* (30), *Boerhavia* (20) *and Bougainvillea.*

Description: Herbs (*Boerhavia, Mirabilis*) often with swollen nodes, shrubs (*Pisonia*) or woody climbers (*Boungainvillea*), rarely small trees (*Pisonia alba*), usually with concentric rings of vascular bundles, containing betalains and raphide crystals of calcium oxalate, woody oxidising, i.e., turning orange to reddish-brown when cut. **Roots** branched taproot, sometimes thick and tuberous (*Mirabilis*) or fusiform (*Boerhavia*). **Leaves** opposite, leaves of a pair equal or unequal, simple, usually entire, net-veined, without stipules. **Inflorescence** with usually cymose clusters, usually dichasial, subsequently monochasial. **Flowers** bisexual, complete, actinomorphic, subtended by an involucre of 3 (*Bougainvillea*) to 5 (*Mirabilis*) bracts which are often enlarged and colored (*Bougainvillea*), sometimes reduced to small teeth (*Boerhavia*), hypogynous, flowers rarely unisexual (*Pisonia*). **Perianth** with 5 united tepals, campanulate (*Boerhavia*) or more commonly tubular, lower part of tube persistent and surrounding fruit (and known as anthocarp), distal part usually petal-like and falling off. **Androecium** with 2–5 (*Mirabilis*), 5–10 (*Bougainvillea*), or more, filaments free or connate, equal or unequal, anthers bithecous, basifixed or dorsifixed, dehiscence longitudinal or by lateral slits, pollen grains tricolpate or polyporate. **Gynoecium** with single carpel, ovary superior, but often appearing inferior due to closely associated persistent part of perianth tube, placentation basal, ovule single, anatropous or campylotropous, style long and filiform, stigma capitate, nectar disc present. **Fruit** an achene or nut, usually enclosed in leathery or fleshy persistent perianth tube, latter winged or ribbed, often covered with glandular hairs; seed small with unequal cotyledons, curved embryo, endosperm absent, replaced by perisperm. Flowers are pollinated by bees, butterflies, moths and birds. Dispersal in species with fleshy perianth occurs by birds, those with glands and hooked hairs by exozoochory.

Economic importance: Species *Bougainvillea* are commonly grown as hedges and for covering walls and fences. Species of *Mirabilis* (Four O'Clock) are grown as garden ornamentals. *Boerhavia repens* is used as medicinal plant as a diuretic. *Fisonia aculeata* is used as hedge plant.

Phylogeny: Hiemerl (1934) divided the family into 5 tribes: first four with glabrous ovary and stamens connate at base, and the fifth Leacastereae with hairy ovary and distinct stamens. Of the first four, Mirabileae has straight embryo and large cotyledons, Pisoneae with shrubby habit, Boldoeae with herbaceous habit and alternate leaves, and Colignoneae with herbaceous habit and opposite leaves; all three have curved embryo. The family is closely related to Phytolaccaceae in anatomical features, and ovary with single carpel. The close affinities of the two families together with genus *Delosperma* (Aizoaceae) were confirmed by the studies of Soltis et al. (2000). Monophyly of the family found moderate support in the studies of Douglas and Manos (2007). Douglas and Spellenberg (2010) recognized 7 tribes: Leacastereae, Boldoeae, Colignoneae, Bougainvilleeae, Pisoneae, Nyctagineae and Caribeeae. APweb (2018) considers monotypic Caribeeae with only species *Caribea litoralis* as unplaced, The South American Leucastereae and Mexican-Central American Boldoeae are successively sister taxa to the remainder of the family, positions that have moderate to strong support. Within the remainder of the family a North American xerophytic clade has very strong support. Here Bougainvilleeae and Pisonieae (with minor additions) form a clade, while Abronieae are embedded in a highly paraphyletic Nyctagineae plus Boerhavieae complex, all three included in Nyctagineae above (Douglas and Manos, 2007). Latter also found the relationships [Nyctaginaceae [Sarcobataceae [Phytolaccaceae + Petiveriaceae]] and found only moderate support for the monophyly of Nyctaginaceae and vanishing little support for the monophyly of Phytolaccaceae (including Sarcobataceae). Some Nyctagineae have pollen grains ca 200 µm long, about the largest in angiosperms outside the aquatic Cymodoceaceae. Stevens in APweb version 14 (updated, November 2018) recognized 5 subfamilies Sesuvioideae, Sesuvioideae, Acrosanthoideae, Acrosanthoideae and Acrosanthoideae (with 4 tribes: Apatesieae, Dorotheantheae, Delospermeae and Ruschieae).

Portulacaceae A. L. de Jussieu *Purslane Family*

1 genus, 115 species

Widely distributed in tropical and temperate regions, mainly North and South America.

Placement:

	B & H	Cronquist	Takhtajan	Dahlgren	Thorne	APG IV/(APweb)
Informal Group						Superasterids/ (Malvids/Rosid II)
Division		Magnoliophyta	Magnoliophyta			
Class	Dicotyledons	Magnoliopsida	Magnoliopsida	Magnoliopsida	Magnoliopsida	
Subclass	Polypetalae	Caryophyllidae	Caryophyllidae	Magnoliidae	Caryophyllidae	
Series+/ Superorder	Thalamiflorae+		Caryophyllanae	Caryophyllanae	Caryophyllanae	
Order	Caryophyllineae	Caryophyllales	Caryophyllales	Caryophyllales	Caryophyllales	Caryophyllales

Salient features: Usually succulent herbs, mucilaginous, roots somewhat tuberous, leaves simple, flowers in cymes or solitary, bisexual, sepals 2, carpels united with unilocular ovary, fruit a capsule, embryo curved.

Genus: *Portulaca* (115 species).

Description: Annual herbs, somewhat succulents, Mucilage cells very common, containing betalains, often exhibiting CAM metabolism, hairs usually simple. **Stems** erect or prostrate, herbaceous. **Leaves** opposite, simple, usually fleshy, entire, often clustered at ends of branches, stipules scarious or setose, rarely absent. **Inflorescence** cymose with few or of solitary flowers. **Flowers** showy, bisexual, usually actinomorphic, with short or elongated hypanthium. **Calyx** of 2 sepals, antero-posterior, green, free or united at base. **Corolla** with usually 5 petals, rarely 4 or 6, free, rarely united at

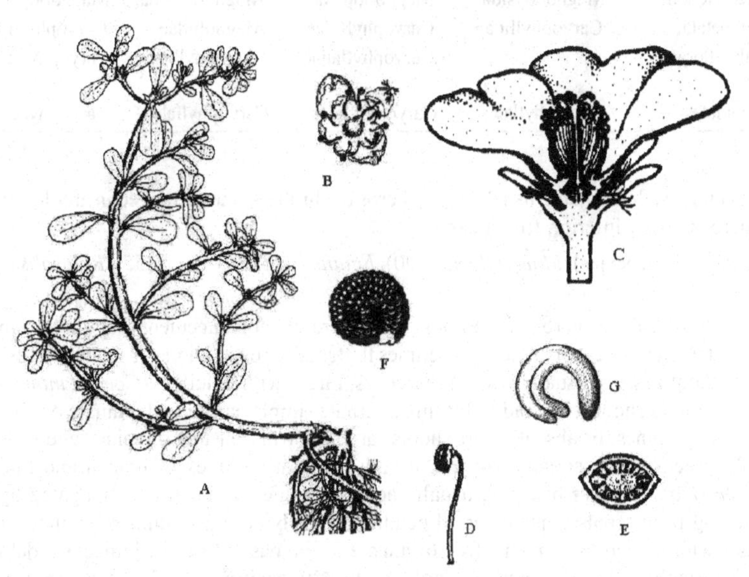

Figure 13.105: Portulacaceae. *Portulaca oleracea.* **A:** Portion of plant with flowers; **B:** Flowers seen from above; **C:** Flower enlarged in vertical section; **D:** Stamen; **E:** Transverse section of ovary with free central placentation; **F:** Seed; **G:** Embryo.

base, imbricate, falling early. **Androecium** with usually as many stamens as petals, opposite petals, filaments free from petals or epipetalous, dehiscence longitudinal, pollen tricolpate, polycolpate or polyporate. **Gynoecium** with 2 to 3 (rarely more) united carpels, ovary superior or half-inferior, single chambered, with single basal ovule or several ovules on free-central placenta attached at the base of ovary, style simple or split above, stigma minute. **Fruit** a loculicidal or circumscissile capsule; seeds lens-shaped, smooth, shining, embryo curved, endosperm absent, perisperm present, aril sometimes developed. Pollination by insects like bees, flies and beetles, flowers opening briefly in full sunlight. Seeds with aril dispersed by ants, smaller ones by wind or water.

Economic importance: The family is of little economic importance. Purslane (*Portulaca oleracea*) commonly growing wild is frequently cultivated as pot herb. Rose moss (*Portulaca grandiflora*), wingpod purslane (*P. umbraticola*) are grown as ornamentals.

Phylogeny: The family Portulacaceae has traditionally been considered closely related to Caryophyllaceae and Basellaceae, although the presence of betalains has often taken this family away from Caryophyllaceae. Phylogeny of Portulacaceae has been a matter of considerable speculation. The separation of Basellaceae and Didiereaceae is supported by morphological data. Portulacaceae is considered as more closely related to Cactaceae (although separation is supported by ITS sequence data) and Thorne (2006) has accordingly shifted family under suborder Cactineae. Applequist and Wallace (2001) on the basis analysis of chloroplast gene *ndhF* in Portulacaceae, Basellaceae, Cactaceae, and Didiereaceae concluded that the group forms a monophyletic group with two major clades. The first included *Portulaca, Anacampseros* and its relatives, much of *Talinum, Talinella*, and Cactaceae; the second, weakly supported, included the remaining genera of Portulacaceae, Basellaceae, and Didiereaceae. The separation of these families from Portulaceae renders it paraphyletic. Subsequent studies of these authors (2006) resulted in Stevens (2007) placing only genus *Portulaca* in the family Portulacaceae, separating 10 genera (incl. *Montia, Lewisia and Phemeranthus*) to Montiaceae and two (*Talinum, Talinella*) to Talinaceae, Monotypic Porlulacaceae is followed in APG IV and latest version of APweb (2018).

Cactaceae A. L. de Jussieu *Cactus Family*

127 genera, 1,750 species

Mainly in arid climate, in deserts regions of North and South America, several species introduced in Africa, India and Australia.

Placement:

	B & H	Cronquist	Takhtajan	Dahlgren	Thorne	APG IV/(APweb)
Informal Group						Superasterids/ (Malvids/Rosid II)
Division		Magnoliophyta	Magnoliophyta			
Class	Dicotyledons	Magnoliopsida	Magnoliopsida	Magnoliopsida	Magnoliopsida	
Subclass	Polypetalae	Caryophyllidae	Caryophyllidae	Magnoliidae	Caryophyllidae	
Series+/ Superorder	Calyciflorae+		Caryophyllanae	Caryophyllanae	Caryophyllanae	
Order	Ficoidales	Caryophyllales	Caryophyllales	Caryophyllales	Caryophyllales	Caryophyllales

B & H as Cacteae.

Salient features: Succulents, fleshy habit, usually spiny herbs or shrubs, spines arranged in areoles, flowers solitary, petals many, stamens numerous, ovary inferior, fruit a berry.

Major Genera: *Opuntia* (250 species), *Mammillaria* (190), *Echinopsis* (75), *Cereus* (55), *Rhipsalis* (50) and *Cleistocactus* (50).

Description: Spiny stem succulents, herbs, sometimes tree-like, rarely non-succulent (but with fleshy leaves-*Pereskia*) or epiphytic (*Rhipsalis*), stem cylindrical or angled, sometimes flattened, or even jointed, usually photosynthetic, usually with vessels, sometimes without vessels, usually without laticifers, rarely with laticifers (*Coryphantha*), plastids PIII-A type. **Leaves** usually borne on long shoots and readily falling, alternate simple, entire with pinnate or obscure venation, leaves sometimes represented by spines, or absent, short shoots (areoles) with clusters of spines and tufts of hairs (glochids); stipules absent. **Inflorescence** with solitary flowers, usually sunk at the apex of branch and thus appearing axillary, rarely in clusters (*Pereskia*). **Flowers** bisexual, usually actinomorphic, with short or elongated hypanthium. **Perianth** sequentially intergrading from sepals to petals or all petaloid, spirally arranged, numerous, innermost slightly coherent at base. **Androecium** with numerous stamens, free or adnate to the base of petals, bithecous, dehiscence longitudinal, introrse, pollen grains tricolpate to polycolpate or polyporate. **Gynoecium** with 2 to numerous carpels, united, ovary

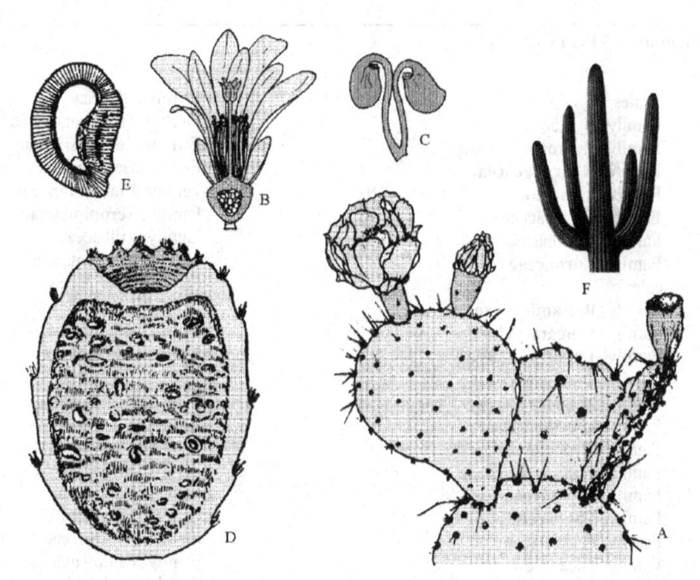

Figure 13.106: Cactaceae. *Opuntia rafinesqui.* **A:** Portion of plant with flowers and spines; **B:** Vertical section of flower showing many stamens, inferior ovary and parietal placentation; **C:** Ovules with long funiculus; **D:** Longitudinal section of fruit; **E:** Longitudinal section of seed. **F:** *Carnegiea gigantea* with characteristic branched habit and ribbed stem.

inferior, rarely semi-inferior (*Pereskia*) or even superior (some species of *Pereskia*), unilocular with numerous ovules, placentation parietal, sometimes divided by false septa or nearly basal (*Pereskia*), stigmas 2 to numerous, spreading, ovules campylotropous, bitegmic, crassinucellate. **Fruit** a berry, often covered with spines and/or glochids; seeds numerous, immersed in pulp, testa often black, endosperm usually absent, embryo usually curved. Pollination by insects, birds or bats. Berries are dispersed by animals or birds.

Economic importance: The family is known for large number of ornamentals (cacti) such as *Opuntia* (prickly pear), *Mammillaria* (pincushion cactus), *Cereus* (hedge cactus), *Echinopsis* (sea-urchin cactus), *Epiphyllum* (orchid cactus), *Schlumbergera* (Christmas cactus) and *Rhipsalis* (mistletoe cactus). Fruits of several species of *Opuntia* are eaten raw or made into jams or syrups. Spines of cacti are often used as gramophone needles. *Lophophora* contains mescaline alkaloids and is hallucinogenic. Cochineal dye is derived from small insects living on members of this family.

Phylogeny: The family is unique in combining unspecialized floral characters with highly advanced vegetative organs. The family is commonly divided into three subfamilies: Pereskioideae, Opuntioideae and Cactoideae (Heywood, 1978). Thorne (1999, 2003, 2006, 2007) places the two genera of Pereskoideae under two separate subfamilies Pereskioideae (*Pereskia*) and Maihuenioideae (*Maihuenia*), thus recognizing 4 subfamilies in all. *Pereskia* retains several plesiomorphic features such as non-succulent stems, well-developed persistent leaves, cymose inflorescence, superior ovary (in some species) with basal placentation and is sister to rest of the Cactaceae. The other two subfamilies form a well-defined clade with solitary flowers sunken into stem apices, inferior ovary and parietal placentation. Opuntioideae are monophyletic based on synapomorphies of presence of glochids on the areoles, seeds coat with bony aril and cpDNA characters. Cactoideae similarly has monophyly supported by extreme reduction of leaves and a deletion of the *rpoCl* intron in chloroplast genome (Wallace and Gibson, 2002). The affinities of Cactaceae have largely remained uncertain. It is now considered to be closely related to Portulacaceae, Phytolaccaceae, Basellaceae, Halophytaceae, Didiereaceae and Aizoaceae. Phylogenetic relationships within Cactaceae are still rather unclear, with chloroplast and nuclear genes sometimes suggesting different major clades. A study by Nyffeler (2002) found rather weak support for the subfamilies and perhaps rather distressingly no clear monophyly for the basal *Pereskia*. The distinctive *Blossfeldia* was sister to the other Cactoideae. Edwards et al. (2005) confirm that *Pereskia* s.l. is probably paraphyletic, which allows them to shed new light on the evolution of the cactus habit.

Detailed classification of Cactaceae was provided by Nyffler and Eggli (2010), recognizing 4 subfamilies: Pereskioideae (single genus *Pereskia*), Maihuenioideae (single genus *Maihuenia*), Opuntioideae (2 tribes, 17 genera, 3 of uncertain position) and Cactoideae (6 tribes, 112 genera, 3 of uncertain position). Stevens (APweb, 2018) recognizes 5th Leuenbergioideae, proposed earlier by Mayta and Molinari (2015) based on *Leuenbergia,* earlier placed under genus *Pereskia.*

Arrangement after APG IV (2016)

Asterids:

Order 1. Cornales
- Family: Nyssaceae
- Family: Hydrostachyaceae
- Family: **Hydrangeaceae**
- Family: Loasaceae
- Family: Curtisiaceae
- Family: Grubbiaceae
- Family: **Cornaceae**

Order 2. Ericales
- Family: **Balsaminaceae**
- Family: Marcgraviaceae
- Family: Tetrameristaceae
- Family: Fouquieriaceae
- Family: **Polemoniaceae**
- Family: Lecythidaceae
- Family: Sladeniaceae
- Family: Pentaphylacaceae
- Family: **Sapotaceae**
- Family: **Ebenaceae**
- Family: **Primulaceae**
- Family: Theaceae
- Family: Symplocaceae
- Family: Diapensiaceae
- Family: Styracaceae
- Family: Sarraceniaceae
- Family: Roridulaceae
- Family: Actinidiaceae
- Family: Clethraceae
- Family: Cyrillaceae
- Family: **Ericaceae**
- Family: Mitrastemonaceae

Order 3. Icacinales
- Family: Oncothecaceae
- Family: Icacinaceae

Order 4. Metteniusales
- Family: Metteniusaceae

Order 5. Garryales
- Family: Eucommiaceae
- Family: Garryaceae

Order 6. Gentianales
- Family: **Rubiaceae**
- Family: Gentianaceae
- Family: Loganiaceae
- Family: Gelsemiaceae
- Family: **Apocynaceae**

Order 7. Boraginales
- Family: **Boraginaceae**

Order 8. Vahliales
- Family: Vahliaceae

Order 9. Solanales
- Family: **Convolvulaceae**
- Family: **Solanaceae**
- Family: Montiniaceae
- Family: Sphenocleaceae
- Family: Hydroleaceae

Order 10. Lamiales
- Family: Plocospermataceae
- Family: Carlemanniaceae
- Family: Oleaceae
- Family: Tetrachondraceae
- Family: Calceolariaceae
- Family: Gesneriaceae
- Family: **Plantaginaceae**
- Family: **Scrophulariaceae**
- Family: Stilbaceae
- Family: Linderniaceae
- Family: Byblidaceae
- Family: Martyniaceae
- Family: Pedaliaceae
- Family: **Acanthaceae**
- Family: **Bignoniaceae**
- Family: Lentibulariaceae
- Family: Schlegeliaceae
- Family: Thomandersiaceae
- Family: **Verbenaceae**
- Family: **Lamiaceae**
- Family: Mazaceae
- Family: Phrymaceae
- Family: Paulowniaceae
- Family: Orobanchaceae

Order 11. Aquifoliales
- Family: Stemonuraceae
- Family: Cardiopteridaceae
- Family: Phyllonomaceae
- Family: Helwingiaceae
- Family: **Aquifoliaceae**

Order 12. Asterales
- Family: Rousseaceae
- Family: Campanulaceae
- Family: Pentaphragmataceae
- Family: Stylidiaceae
- Family: Alseuosmiaceae
- Family: Phellinaceae
- Family: Argophyllaceae
- Family: Menyanthaceae
- Family: Goodeniaceae
- Family: Calyceraceae
- Family: **Asteraceae**

Order 13. Escalloniales
- Family: Escalloniaceae

Order 14. Bruniales
- Family: Columelliaceae
- Family: Bruniaceae

Order 15. Paracryphiales
- Family: Paracryphiaceae

Order 16. Dipsacales
- Family: **Adoxaceae**
- Family: Dipsacaceae

Order 17. Apiales
- Family: Pennantiaceae
- Family: Torricelliaceae
- Family: Griseliniaceae
- Family: Pittosporaceae
- Family: **Araliaceae**
- Family: Myodocarpaceae
- Family: **Apiaceae**

Families in boldface are described in detail.

Hydrangeaceae Dumortier *Hydrangea Family*

9 genera, 223 species

Mainly distributed in Northern Hemisphere from Himalayas to Japan to North America and tropical Africa.

Placement:

	B & H	Cronquist	Takhtajan	Dahlgren	Thorne	APG IV/(APweb)
Informal Group						Asterids
Division		Magnoliophyta	Magnoliophyta			
Class	Dicotyledons	Magnoliopsida	Magnoliopsida	Magnoliopsida	Magnoliopsida	
Subclass	Polypetalae	Rosidae	Cornidae	Magnoliidae	Asteridae	
Series+/ Superorder	Calyciflorae+		Cornanae	Cornanae	Cornanae	
Order	Rosales	Rosales	Hydrangeales	Cornales	Cornales	Cornales

B & H under Saxifragaceae.

Salient features: Mostly shrubs, leaves simple, usually opposite, stipules absent, flowers bisexual, sepals often enlarged and petaloid, ovary inferior or semi-inferior, placentation axile or deeply intruded parietal, nectary present at top of ovary, fruit a capsule.

Major genera: *Philadelphus* (65 species), *Deutzia* (60), *Hydrangea* (60), *Dichroa* (13) and *Fendlera* (4).

Description: Herbs (*Cordiandra*), soft-wooded shrubs (*Hydrangea*) or rarely small trees or climbers (*Decumeria*), often with tannins and Iridoids and raphide crystals. **Leaves** usually opposite, rarely alternate (*Cordiandra*), simple, sometimes lobed, usually deciduous, rarely evergreen (*Pileostegia*), venation pinnate or palmate, reticulate, stipules absent. **Inflorescence** terminal racemes, cymes or corymbose, rarely solitary flowers. **Flowers** usually bisexual, sometimes unisexual (*Broussaisia*, polygamo-dioecious) outer usually sterile with enlarged petaloid sepals, actinomorphic, perigynous or epigynous. **Calyx** with 4–5 sepals, united, calyx tube often adnate to ovary, sepals of outer sterile flowers often large and petaloid. **Corolla** with 4–5 petals, free, imbricate, convolute, rarely valvate (*Platycrater*), usually white. **Androecium** with many stamens, sometimes 8–10 (*Hydrangea*), free or slightly connate at base, anthers bithecous, basifixed or dorsifixed, filaments often lobed or toothed, connective sometimes appendaged at tip (*Fendlera*), pollen grains tricolpate or triporate. **Gynoecium** with 2–7 united carpels, ovary semi-inferior (*Dichroa, Broussaisia*), inferior (*Philadelphus, Deutzia, Hydrangea*) or superior (*Jamesia*), 1–7 locular, ovules numerous, placentation axile or parietal with deeply intruded placentas, styles free, rarely united (*Carpenteria*), stigmas free, nectar disc usually present at top of ovary. **Fruit** usually a loculicidal (Hydrangeae) or septicidal (Philadelpheae) capsule, rarely a berry (*Dichroa*); seeds numerous, small, sometimes winged, with fleshy endosperm and straight embryo. Pollination by insects aided by epigynous disc. Small seeds are dispersed by wind.

Economic importance: The family is known for ornamental shrubs with showy flowers including *Hydrangea, Decumeria* (climbing hydrangea), *Schizophragma* (hydrangea vine), *Philadelphus* (mock orange) and *Deutzia*. Some species of *Hydrangea* are used as source of hydrangin, a compound used in medicine.

Phylogeny: The family was earlier included under Saxifragaceae (Bentham and Hooker; Engler and Prantl). It was separated as distinct family by Hutchinson (1927), who also treated Philadelphaceae as distinct family in his *The Genera of Flowering*

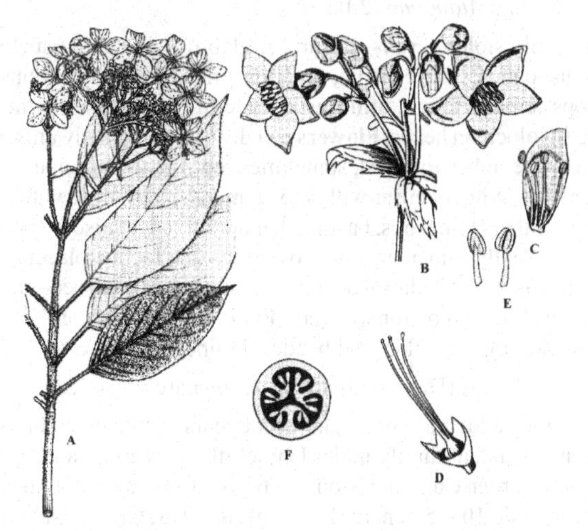

Figure 13.107: Hydrangeaceae. *Hydrangea heteromalla.* **A:** Branch with corymbose inflorescence with outer sterile flowers having 4 enlarged petaloid sepals. *Kirengeshoma palmata.* **B:** Terminal part of flowering branch; **C:** Petal with adnate stamens; **D:** Flower with stamens and petals removed; **E:** Stamens with dorsifixed anthers; **F:** Transverse section of ovary (B–F after Hutchinson, 1973).

Plants (1964) followed up in his last revision of classification (1973). Philadelphaceae was merged with Hydrangeaceae in classifications of Cronquist (1988) retaining the family in Rosales (placed under Rosidae) along with Saxifragaceae. The merger of Philadelphaceae with Hydrangeaceae is supported by morphology and DNA characteristics (Albach et al., 2001; Soltis et al., 1995; Hufford, 1997) and the treatment has been followed in recent classifications. Dahlgren (1983, 1989) shifted the family away from Rosanae to superorder Cornanae. Takhtajan (1997) placed family under separate subclass Cornidae (superorder Cornanae, order Hydrangeales), while Saxifragaceae was retained in Rosidae. The recent classifications of APG II, APG IV (2016), APweb and Thorne (2003, 2007) place family under Asterid complex (Thorne in Asteridae—> Cornanae—> Cornales; APG classifications in Asterids—> Cornales). The separation of Hydrangeaceae away from Saxifragaceae has been supported by cladistic analysis, indicating that the two are distantly related, and Hydrangeaceae is closer to Cornaceae.

The family Hydrangeaceae is divided into two subfamilies: Jamesioideae (2 genera *Jamesia* and *Fendlera*) and Hydrangeoideae (rest of genera). Hufford et al. (2001) based on analysis of sequences of *mat*K and their combination with *rbc*L and evidence from morphology concluded that Jamesioideae may be sister to the rest of the family. They also divide Hydrangeoideae into two tribes Hydrangeae (conspicuous sterile marginal flowers, valvate petals, loculicidal capsule) and Philadelpheae (sterile flowers absent, petals imbricate, capsules septicidal). De Smet et al. (2015) based on sequencing four chloroplast regions and ITS for an extensive set of taxa, proposed revised classification of Hydrangeae to include all satellite genera into single genus *Hydrangea*. Ohba and Akiyama (2016), on the other hand prefer several genera with narrower limits.

Cornaceae Dumortier ***Dogwood Family***

2 genera, 85 species (including Alangiaceae DC.)

Widely spread, mainly in north temperate regions.

Placement:

	B & H	Cronquist	Takhtajan	Dahlgren	Thorne	APG IV/(APweb)
Informal Group						Asterids
Division		Magnoliophyta	Magnoliophyta			
Class	Dicotyledons	Magnoliopsida	Magnoliopsida	Magnoliopsida	Magnoliopsida	
Subclass	Polypetalae	Rosidae	Asteridae	Magnoliidae	Asteridae	
Series+/ Superorder	Calyciflorae+		Cornanae	Cornanae	Cornanae	
Order	Umbellales	Cornales	Cornales	Cornales	Cornales	Cornales

Salient features: Mostly shrubs, leaves simple, usually opposite, stipules absent, flowers bisexual, sepals often enlarged and petaloid, ovary inferior or semi-inferior, placentation axile or deeply intruded parietal, nectary present at top of ovary, fruit a capsule.

Major genera: *Cornus* (65 species) and *Alangium* (20).

Description: Trees and shrubs, rarely stoloniferous subshrubs, glabrous or hairy, usually with iridoids. **Leaves** usually opposite, sometimes alternate or distichous, simple, entire, venation pinnate, secondary veins usually smooth arching towards margin or forming a series of loops, stipules absent. **Inflorescence** terminal branched cymes or heads, usually subtended by large showy bract, often forming involucre in heads. **Flowers** usually bisexual, rarely unisexual, actinomorphic, epigynous. **Calyx** with 4–5 sepals, united, valvate, lobes rounded, sometimes represented by small teeth, rarely absent. **Corolla** with 4–5 petals, free, imbricate, or valvate. **Androecium** with 4–5 stamens, alternating with petals, rarely up to 10, filaments free, arising from the edge of disc, anthers bithecous, basifixed or dorsifixed, dehiscing laterally. **Gynoecium** with 2 united carpels, rarely 1–4, ovary inferior, usually 2-locular, single ovule in each locule, placentation axile, axis lacking vascular bundles and the ovules attached to vascular bundles that arch over the top of each septum (apical axile placentation), style simple, stigma capitate or lobed, disc present on top of ovary. **Fruit** usually a drupe, 1–2 seeded, ridged or winged; seeds small, endosperm present. Pollination by bees, flies and beetles. Drupes are dispersed by birds and mammals.

Economic importance: Species of *Cornus* (Dogwood) and *Alangium* are grown as ornamental trees and shrubs.

Phylogeny: The family has undergone a lot realignment in recent years. Bentham and Hooker placed it under Araliaceae, but subsequently recognized as independent family under Umbelliflorae (Engler and Prantl). Hutchinson (1948) shifted Cornaceae along with Araliaceae to order Cunoniales on the basis of woody habit and stem anatomy. The family was earlier broadly circumscribed to include 10–15 genera (Hutchinson, 1973; Cronquist, 1988), but was split into a number of families (Takhtajan, 1997) such as Davidiaceae, Nyssaceae, Mastixiaceae, Curtisiaceae, Cornaceae and Alangiaceae. APG-II recognized only Curtisiaceae and Cornaceae, Nyssaceae optionally merged with latter. Thorne (2006, 2007) recognized monogeneric Cornaceae, Alangiaceae, whereas rest of the genera are united under Nyssaceae, divided into

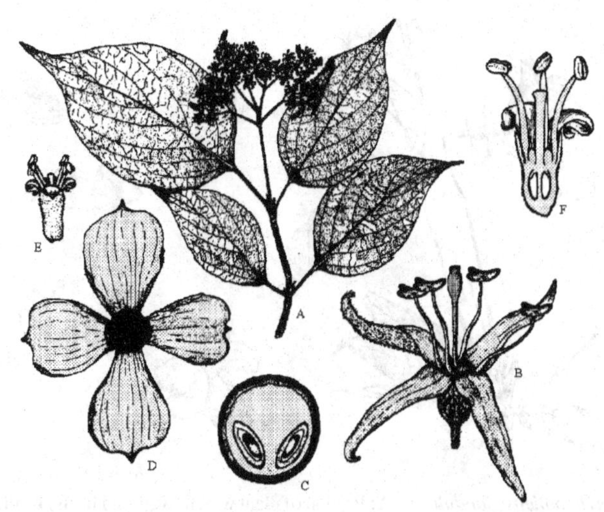

Figure 13.108: Cornaceae. *Cornus macrophylla.* **A:** Portion of plant with flowers; **B:** Flowertop, view; **C:** Transverse section of fruit. **D:** Flower head of *C. capitata* subtended by four bracts. *C. stolonifera.* **E:** Flower; **F:** Vertical section of flower (A–D, after Nasir and Ali, Fl. W. Pak.No. 88, 1975; E–F, Bailey, Man. Cult. Pl., 1949).

subfamilies Davidioideae, Nyssoideae and Mastixioideae. APweb (2007) united Alangiaceae and Cornaceae, thus including two genera under Cornaceae, followed up in APG IV (2016) and APweb Version 14 (2018 update). Analysis of a combined matK and rbcL sequence data set by Xiang et al. (1998) established *Cornus-Alangium* as distinct clade within Cornales, results confirmed by 26S rRNA and combined 26S rDNA-*mat*K-*rbc*L sequence data of Fan and Xiang (2003) establishing Grubbiaceae (*Grubbia* and *Curtisia*), Cornaceae (*Cornus* and *Alangium*) and Nyssaceae (*Nyssa, Davidia, Camptotheca, Mastixia* and *Diplopanax*) as distinct clades; Hydrangiaceae and Loasaceae forming independent clade.

Balsaminaceae Bercht. & J. Presl. ***Balsam Family***

2 genera, 1001 species

Widely distributed but more common in subtropics and tropics of Africa and Asia, and temperate regions New and Old world.

Placement:

	B & H	Cronquist	Takhtajan	Dahlgren	Thorne	APG IV/(APweb)
Informal Group						Asterids
Division		Magnoliophyta	Magnoliophyta			
Class	Dicotyledons	Magnoliopsida	Magnoliopsida	Magnoliopsida	Magnoliopsida	
Subclass	Polypetalae	Rosidae	Dilleniidae	Magnoliidae	Asteridae	
Series+/ Superorder	Disciflorae+		Ericanae	Rutanae	Ericanae	
Order	Geraniales	Geraniales	Balsaminales	Balsaminales	Balsaminales	Ericales

B & H under Geraniaceae.

Salient features: Somewhat succulent herbs, flowers bisexual, zygomorphic, spurred, anthers connate, filaments closely surrounding ovary, succulent capsule with elastic dehiscence.

Genera: *Impatiens* (1000 species) and *Hydrocera* (1).

Description: Herbs, rarely evergreen subshrubs, often with fleshy translucent stems, sometimes aquatic, rarely epiphytes. **Leaves** alternate, opposite or in whorls of three, simple, usually toothed, stipules as paired glands or absent. **Inflorescence** of solitary flowers, cymes, racemes or panicles, on axillary peduncles. **Flowers** bisexual, spurred, nodding, often resupinate, hypogynous, pentamerous, some flowers cleistogamous and self-pollinating. **Calyx** with usually 3 sepals, sometimes up to 5, free, petaloid, imbricate, posterior (lower in mature flower) largest and with nectar spur or pouch on back. **Corolla** with 5 petals, free, or lateral pairs connate and appear as 3 petals, lower pair larger than upper. **Androecium** with 5 stamens, filaments flattened, short, closely covering the ovary, free below, connate towards top with connate anthers (syngenesious) forming lid (calyptra) over the ovary, anthers bithecous, dehiscence longitudinal. **Gynoecium** with 5 united carpels, ovary

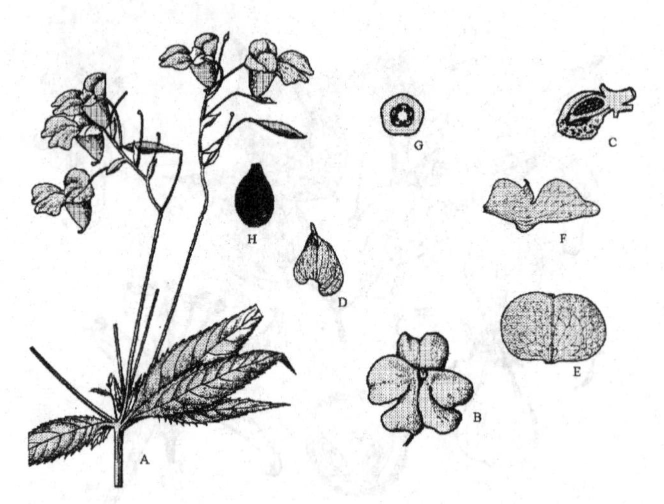

Figure 13.109: Balsaminaceae. *Impatiens glandulifera.* **A:** Portion of plant with flowers and fruits; **B:** Flower, top view; **C:** Longitudinal section of flower with sepals and petals removed; **D:** Lateral sepal; **E:** anterior petal; **F:** Lateral fused petals; **G:** Transverse section of ovary; **H:** Seed.

superior, 5-locular, ovules 3-many in each locule, anatropous, pendulous, placentation axile, style simple, often very short, stigmas 1–5. **Fruit** a fleshy 5-valved capsule, dehiscing explosively, valves coiling elastically and on splitting the tension forcibly distributing the seeds, rarely a berry-like drupe; seeds with straight embryo, endosperm absent. Pollination by insects. Capsules dehiscing explosively by autochory or when touched slightly, hence the name 'touch me not'.

Economic importance: The family is known for its ornamental herbs with showy flowers. Garden balsam (*Impatiens balsamina*), Himalayan balsam (*I. glandulifera*), Busy lizzie (*I. holstii*) and several hybrid cultivars are widely grown in temperate as well as tropical climates.

Phylogeny: The family was formerly placed under Geraniaceae, but differs in zygomorphic flowers, spurred sepal and connate anthers. It is often allied to Tropaeolaceae, but the spur in latter family is derived from receptacular tissue and not calyx as in Balsaminaceae. Balsaminaceae, Marcgraviaceae and Tetrameristaceae form a well-defined clade and probably sister to rest of Ericales. Monophyly of these three families is well supported by the studies of Nandi et al. (1998), Soltis et al. (2000) and Geuten et al. (2004). Latter authors concluding that Balsaminaceae and Marcgraviaceae are sister taxa, although there are no obvious synapomorphies. Balsaminaceae show the combination of leucoanthocyanins and raphides, rarely seen in herbs (Fischer, 2004). The family is vegetatively very uniform although florally diverse and duplication and probable subfunctionalisation of the class B *DEF* gene has occurred in this clade (Janssens et al., 2006; Geuten et al., 2006). *Hydrocera* and *Impatiens* are clearly sister taxa (Yuan et al., 2004; Janssens et al., 2006). Species of *Impatiens* with five sepals are scattered through the genus, so that condition is apparently at least sometimes derived.

Yu et al. (2015), based on analysis 46 morphological characters, nuclear ribosomal ITS and plastid *atpB-rbcL* and *trnL-F*, concluded that species with three-colpate pollen and four carpels form a monophyletic group. They divided Impatiens into 2 subgenera: *Clavicarpa* and *Impatiens.*

Polemoniaceae Bromehead ***Phlox Family***

20 genera, 360 species

Widely distributed, more common in temperate climate, especially western North America.

Placement:

	B & H	**Cronquist**	**Takhtajan**	**Dahlgren**	**Thorne**	**APG IV/(APweb)**
Informal Group						Asterids
Division		Magnoliophyta	Magnoliophyta			
Class	Dicotyledons	Magnoliopsida	Magnoliopsida	Magnoliopsida	Magnoliopsida	
Subclass	Gamopetalae	Asteridae	Lamiidae	Magnoliidae	Asteridae	
Series+/ Superorder	Bicarpellatae+		Solananae	Solananae	Ericanae	
Order	Polemoniales	Solanales	Polemoniales	Solanales	Polemoniales	Ericales

Salient features: Mostly herbs, leaves alternate, stipules absent, sepals 5, united, petals 5, united, stamens 5, epipetalous, carpels 3, united, placentation axile, ovules many, style 1, stigmas 3, fruit a capsule.

Major genera: *Gilia* (110 species), *Phlox* (75), *Polemonium* (40), *Limnanthus* (40), *Ipomopsis* (25), *Collomia* (15) and *Cantua* (12).

Description: Annual or perennial herbs, rarely shrubs or trees (*Cantua*) or climbers (*Cobaea*), often with glandular hairs, nodes unilacunar. **Leaves** usually alternate, rarely opposite or whorled (*Gymnosteris*), usually simple, sometimes dissected or pinnate compound (*Polemonium*), venation reticulate, stipules absent. **Inflorescence** terminal or axillary, often crowded into corymbs or heads, rarely solitary. **Flowers** bisexual, actinomorphic, rarely zygomorphic, hypogynous, usually showy. **Calyx** with 5 sepals, united, green. **Corolla** with 5 petals, united, often with a narrow tube, lobes plicate or convolute. **Androecium** with 5 stamens, adnate to corolla tube (epipetalous; inserted), alternating with lobes, filaments free, anthers bithecous, dehiscence longitudinal, pollen grains 4-many aperturate, colpate or porate. **Gynoecium** with 3 united carpels, rarely 2, ovary superior, seated on a nectary disc, 3-locular, rarely 2-locular, ovules 1 or more, unitegmic, placentation axile, style single, elongated, branched above, stigmas 3, rarely 2. **Fruit** a loculicidal capsule; seed with straight or curved embryo, endosperm copious, seed-coat mucilaginous when moistened. Pollination mostly by bees and flies. Seeds are dispersed aided by mucilaginous coat, sometimes by wind or water.

Economic importance: The family is known for its ornamentals *Phlox* (*P. drummondii* being most popular), *Gilia* and *Polemonium* with showy flowers.

Phylogeny: The family is considered to be closely related to Hydrophyllaceae, the two having closer affinities with group consisting of Solanaceae, Nolanaceae and Convolvulaceae. Hutchinson, who placed Cuscutaceae under the order Polemoniales, considered it to be most highly evolved within the order. Thorne (1999) shifted this family to Dilleniidae (Dillenianae) under separate order Polemoniales (in accordance with the findings of DNA sequence studies), retaining other families under Lamiidae, order Solanales. Takhtajan placed all families including Polemoniaceae under Lamiidae, superorder Solananae, but under separate orders. Judd et al. (1999) had placed Polemoniaceae under Solanales because of actinomorphic flowers, and united plicate corolla. The studies of Porter and Johnson (1998), based on morphology and DNA sequences, however, indicate that the family belongs to Ericales. The family has accordingly been shifted to Ericales under Aterids in Judd et al. (2002), APG IV (2016) and APweb.

Thorne (2003), who has brought about major realignments, and abolished Dilleniidae, distributing its members among various subclasses, has also placed Polemoniaceae under Ericales (superorder Ericanae) under Asteridae. Traditionally two subfamilies Cobaeoideae and Polemonioideae are recognized. *Acanthogilia* with dimorphic leaves and short shoots has been placed in its own subfamily (Porter et al., 2000). Thorne (2006, 2007) accordingly recognizes 3 subfamilies Cobaeoideae, Acanthogilioideae and Polemonioideae, followed by APweb (2018). Studies based on chloroplast gene *ndhF* (Prather et al., 2000), however, indicate that the genus *Acanthogilia* may be basal in the *Cobaea* lineage. Studies of Porter and Johnson (1998) have also indicated that woody tropical genera of Cobaeoideae form a paraphyletic basal complex, and the herbaceous genera of Polemonioideae mainly *Ipomopsis*, *Linanthus*, *Polemonium*, *Phlox* and *Gila* constitute a monophyletic group.

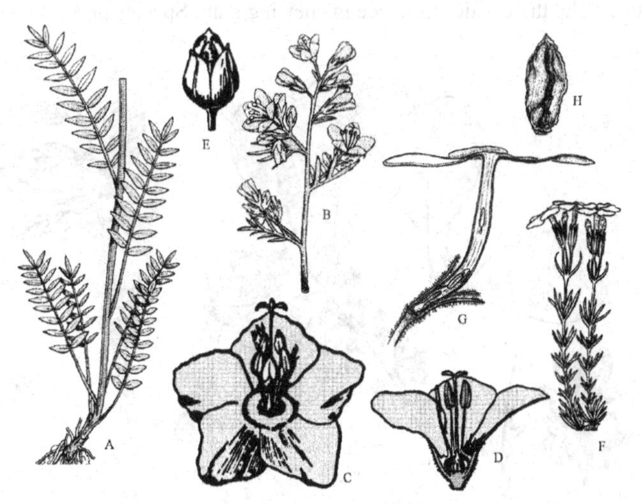

Figure 13.110: Polemoniaceae. *Polemonium caeruleum*. **A:** Basal part of plant; **B:** Upper leaves and inflorescence; **C:** Flower from above; **D:** Vertical section of flower; **E:** Fruit enclosed in persistent calyx. *Phlox nivalis*. **F:** Plant in flower; **G:** Vertical section of flower; **H:** Seed.

Sapotaceae A. L. de Jussieu *Sapodilla Family*

52 genera, 1250 species

Widely distributed in tropics of New World as well as Old World, especially in wet lowland forests, with a few species in temperate regions.

Placement:

	B & H	Cronquist	Takhtajan	Dahlgren	Thorne	APG IV/(APweb)
Informal Group						Asterids
Division		Magnoliophyta	Magnoliophyta			
Class	Dicotyledons	Magnoliopsida	Magnoliopsida	Magnoliopsida	Magnoliopsida	
Subclass	Gamopetalae	Dilleniidae	Dilleniidae	Magnoliidae	Asteridae	
Series+/ Superorder	Heteromerae+		Primulanae	Primulanae	Ericanae	
Order	Ebenales	Ebenales	Sapotales	Ebenales	Sapotales	Ericales

Salient features: Mostly trees with milky latex, small terminal naked buds with appressed T-shaped hairs, leaves simple, coriaceous, bisexual flowers, multilocular superior ovary, single ovule in each chamber, unitegmic ovules, seeds with shiny testa.

Major genera: *Pouteria* (325 species), *Palaquium* (110), *Planchonella* (100), *Madhuca* (110), *Sideroxylon* (75), *Chrysophyllum* (75), *Manilkara* (70) and *Mimusops* (60).

Description: Mostly trees, rarely shrubs (*Reptonia*), with milky latex, with sympodial branches or thorns, silica bodies, triterpenoids and cyanogenic compounds often present, small naked buds, brownish hairs, latter T-shaped but one branch often reduced. **Leaves** alternate, sometimes clustered at shoot apices, simple, entire, coriaceous, venation reticulate, stipules usually absent, rarely present (*Madhuca*), fresh petioles bottle-shaped. **Inflorescence** axillary cymes, rarely terminal cymes (*Madhuca*) or with solitary flowers. **Flowers** bisexual, ebracteate (*Manilkara*) or bracteate (*Madhuca*), actinomorphic, hypogynous. **Calyx** with 4–12 sepals, free or connate at base, sometimes in two whorls or spirally arranged, imbricate, persistent. **Corolla** with 4–12, as many as sepals, united, sometimes with paired petaloid appendages and as such corolla lobes appear 18–24 in number, usually imbricate, rarely valvate. **Androecium** with usually 8–16 stamens, in two or 3 whorls but only inner whorl fertile opposite petals, others reduced to staminodes, epipetalous, anthers bithecous, basifixed or dorsifixed, dehiscence longitudinal, pollen grains tricolpate or tetracolpate. **Gynoecium** with 2-many united carpels, ovary superior, locules many, placentation axile, each locule with 1 ovule, ovule anatropous with single integument, style single, protruding, stigma capitate or lobed. **Fruit** a berry with often leathery bony layer; seed with hard shiny testa and large hilum, endosperm usually fleshy, rarely absent. Pollination by insects. Dispersal of berries by birds and animals.

Economic importance: The family provides several delicious tropical fruits like sapodilla (*Manilkara zapota*), mamey sapote (*Pouteria sapota*), eggfruit or yellow sapote (*P. campechiana*) and star apple (*Chrysophyllum cainito*). Latex of *Manilkara zapata* provides chicle, the elastic substance in chewing gum. Species of *Palaquium* provide gutta-percha, a

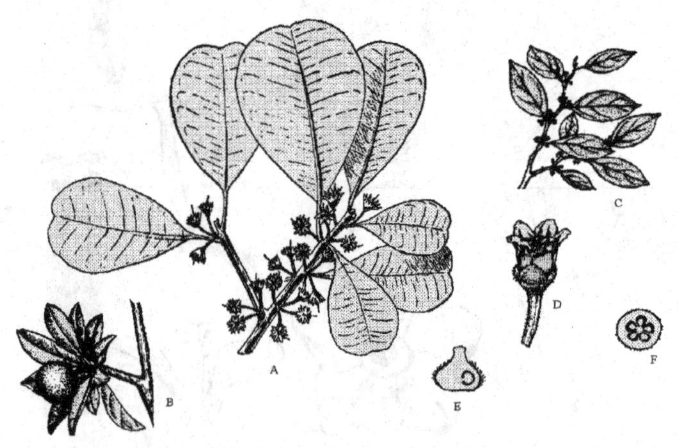

Figure 13.111: Sapotaceae. A: Flowering branch of *Mimusops hexandra.* **B:** Fruiting branch of *Pouteria sapota. Chrysophyllum olivaeforme.* **C:** Flowering branch; **D:** Flower; **E:** Vertical section of ovary; **F:** Transverse section of ovary (A, after Brandis, Ind. Trees, 1918; B–F, after Bailey, Man. Cult. Pl., 1949).

latex substance used in golf balls and submarine telephone cables as insulation and in dental stoppings. Several species are used for timber in Malaya. The fruits of Miracle berry (*Synsepalum dulcificum*) when eaten causes subsequently eaten sour fruits to taste sweet, due to the presence of glycoprotein with trailing carbohydrate chains called Miraculin, which bind to the taste buds of tongue.

Phylogeny: Family is closely related to Ebenaceae but distinct in the presence of milky latex, bisexual flowers, ovules singly in chambers and with single integument. The family is closer to Hoplestigmataceae in single ovule in chamber and single integument (Lawrence, 1951), but the latter family has been shifted to Lamiidae by Thorne (2006, 2007), whereas APG-II places *Hoplestigma* among unplaced genera, and APweb (2007) the family as unplaced eudicot but shifted to Cordiaceae in APG IV (2016) and APweb (2018). The studies of Pennington (1991) from the study of corolla lobes and stamens concluded that whereas genera having same number of stamens as corolla lobes constitute a well-defined clade, those with twice as many stamens form a heterogenous complex, which is probably paraphyletic and basal. Thorne (2006) and APweb (2018) recognize three subfamilies in Sapotaceae: Sarcospermatoideae, Sapotoideae and Chrysophylloideae. Above three clades are supported by combined morphological molecular analysis of Swenson and Anderberg (2005). They also concluded that the staminodes common in Chrysophylloideae, but derived within the clade, are perhaps not immediately comparable with the staminodes of other members of the family; the former are outside the staminal whorl, the latter in the same whorl as the stamens. *Sarcosperma* is regarded as sister to rest of family. *Eberhardtia* may be sister to the remainder (Chen et al., 2016; Rose et al., 2018).

Ebenaceae Gürcke *Ebony Family*

4 genera, 553 (–800) species

Widely distributed in tropics and subtropics, with a few species in temperate regions of North America and Australia.

Placement:

	B & H	Cronquist	Takhtajan	Dahlgren	Thorne	APG IV/(APweb)
Informal Group						Asterids
Division		Magnoliophyta	Magnoliophyta			
Class	Dicotyledons	Magnoliopsida	Magnoliopsida	Magnoliopsida	Magnoliopsida	
Subclass	Gamopetalae	Dilleniidae	Dilleniidae	Magnoliidae	Asteridae	
Series+/ Superorder	Heteromerae+		Primulanae	Primulanae	Ericanae	
Order	Ebenales	Ebenales	Styracales	Ebenales	Sapotales	Ericales

Salient features: Trees and shrubs without milky latex, bark blackish, leaves distichous, simple, unisexual flowers, multilocular superior ovary, bitegmic ovules in pairs.

Major genera: *Diospyros* (500 species), *Euclea* (20) and *Lissocarpa* (8).

Description: Shrubs and trees with black charcoal-like bark, heartwood black, red or green, lacking milky latex, buds covered with adpressed hairs. **Leaves** alternate, distichous, simple, entire, coriaceous, venation reticulate, lower surface with dark colored glands, turning blackish on drying due to presence of napthoquinones, stipules absent. **Inflorescence** axillary cymes or with solitary flowers; plants usually monoecious, male flowers in larger numbers. **Flowers** unisexual, actinomorphic, male flowers with pistillode, female with staminodes, hypogynous, two bracteoles below flower in *Lissocarpa*. **Calyx** with 3–7 sepals, united, persistent and often enlarged in fruit (accrescent). **Corolla** with 3–7 petals, united, urceolate, lobes imbricate and contorted, coriaceous. **Androecium** with same number of stamens as petals or twice as many, and as many whorls, free or united in pairs, epipetalous or free from petals, anthers bithecous, dehiscence longitudinal, introrse, rarely by apical pores. **Gynoecium** with 3–8 united carpels, ovary superior, rarely inferior (*Lissocarpa*), locules many, placentation axile, each locule with usually 2 ovules attached from the top but each ovule separated by partition, thus ovary with twice as many chambers and apical-axile placentation, ovule pendulous, anatropous, integuments two, styles anatropous to campylotropous, styles and stigmas 3–8, styles free or connate at base. **Fruit** a berry with often enlarged calyx; seed with straight embryo, endosperm copious, hard and irregularly grooved or ridged.

Economic importance: The family is important source of durable timber Macassar ebony (*Diospyros ebenum*) and Black ebony (*D. reticulata*). Other species of the genus are sources of common fruits such as Japanese persimon (*D. kaki*), American persimon (*D. virginiana*), and Date plum (*D. lotus*).

Phylogeny: Family is closely related to Sapotaceae but distinct in the absence of milky latex, unisexual flowers, multilocular superior ovary, ovules in pairs and with 2 integuments. From Styracaceae it is distinct by unisexual flowers and septate ovary. *Lissocarpa* (with large bracteoles, corolla with 8-lobed corona, connate filaments, anther with prolonged connective and inferior ovary) removed to a distinct unassigned family Lissocarpaceae in APG (1998), merged with Ebenaceae

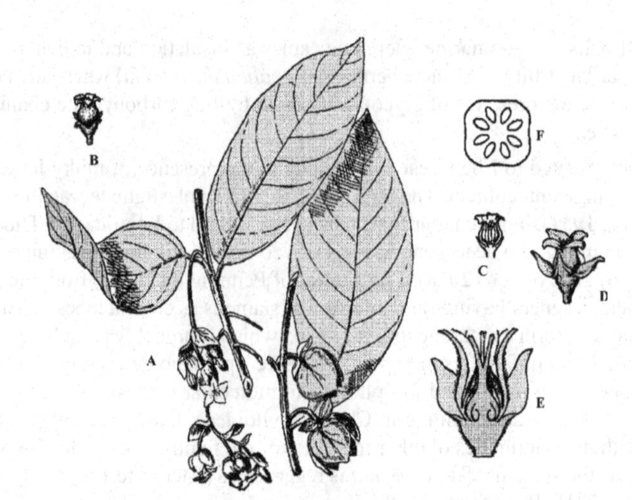

Figure 13.112: Ebenaceae. *Diospyros paniculata.* **A:** Flowering branch. *D. virginiana.* **B:** Staminate flower; **C:** Same in vertical section; **D:** Pistillate flower; **E:** Same enlarged in vertical section; **F:** Transverse section of ovary (A, after Brandis, Ind. Trees, 1918; B–F, after Bailey, Man. Cult. Pl., 1949).

in APG-II (2003), Thorne (2006, 2007) and APweb (2007). Latter places it under subfamily Lissocarpoideae, other two genera under Ebenoideae. Savolainen et al. (2000) suggested removal of *Lissocarpa* to Rutaceae, but rbcL studies supported it as sister to Ebenaceae s. str. (Berry et al., 2001; Rose et al., 2018). Being distinct from rest of Ebenaceae, it was placed under subfamily Lissocarpoideae by Wallnöfer and Halbritter (2003), all rest of genera under Ebenoideae, a placement followed in APweb (2018). *Euclea* and *Royena* (recognized as 4th genus within Ebenaceae by APWeb) were sister to *Diospyros*, and within latter there were a number of well-supported clades, although relationships between them are unclear (Duangjai et al., 2006).

Primulaceae Batsch ex Borkh. *Primrose Family*

53 genera, 5790 species

Largely distributed in north temperate regions, mainly in the Mediterranean region, Alps, and Asia Minor.

Placement:

	B & H	Cronquist	Takhtajan	Dahlgren	Thorne	APG IV/(APweb)
Informal Group						Asterids
Division		Magnoliophyta	Magnoliophyta			
Class	Dicotyledons	Magnoliopsida	Magnoliopsida	Magnoliopsida	Magnoliopsida	
Subclass	Gamopetalae	Dilleniidae	Dilleniidae	Magnoliidae	Asteridae	
Series+/ Superorder	Heteromerae+		Primulanae	Primulanae	Ericanae	
Order	Primulales	Primulales	Primulales	Primulales	Primulales	Ericales

Salient features: Herbs, leaves spiral, opposite or whorled or basal, petals united, ovary superior, stamens opposite the petals, carpels more than 2, placentation free-central, seeds numerous, more or less immersed in placenta, stigma not lobed.

Major genera: *Primula* (500 species), *Ardisia* (450), *Lysimachia* (200), *Myrsine* (155), *Maesa* (150), *Androsace* (150), *Oncostemum* (110), *Embelia* (100), *Dodecantheon* (50), *Anagallis* (28) and *Cyclamen* (15).

Description: Perennial herbs, usually with sympodial rhizomes (*Primula*) or tubers (*Cyclamen*) rarely annuals (*Anagallis*) or subshrubs, sometimes aquatic (*Hottonia*), nodes unilacunar, sieve-tube plastids S-type. **Leaves** opposite, whorled or alternate, sometimes all basal, simple, sometimes dissected (*Hottonia*), venation reticulate, stipules absent, rarely present (*Coris*). **Inflorescence** with solitary axillary flowers (*Anagallis*), to paniculate (*Lysimachia*) or umbellate (*Primula*), often scapigerous (*Primula*). **Flowers** bisexual, actinomorphic, rarely zygomorphic (*Coris*), hypogynous, rarely partly epigynous (*Samolus*), usually pentamerous. **Calyx** with 5 sepals, rarely 6 (*Lysimachia*) or even 9 (*Trientalis*), united, inflated or tubular, imbricate or twisted. **Corolla** with 5 petals, rarely 4 (*Centunculus*), 6 (*Lysimachia*) or 9 (*Trientalis*), or absent (*Glaux*), united, the tube often short, rotate (*Anagallis*) or tubular (*Primula*), lobes imbricate or twisted. **Androecium** with 5 stamens (rarely 4 or 6, depending on the number of petals), free, opposite the petals, epipetalous, anthers bithecous, dehiscence

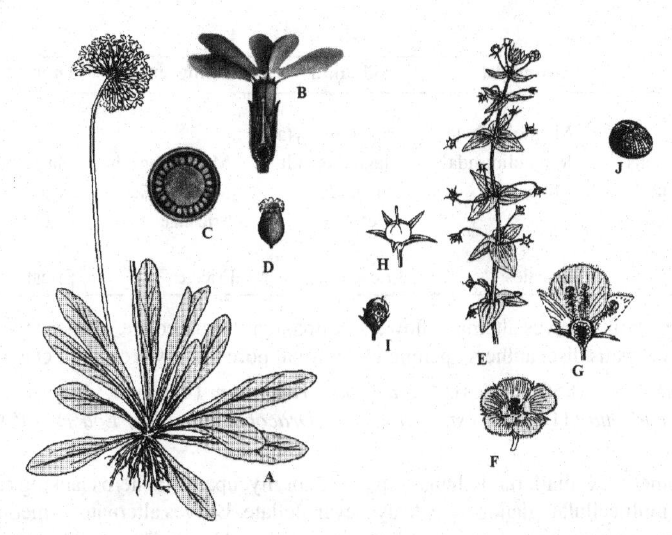

Figure 13.113: Primulaceae. *Primula longiscapa.* **A:** Plant with basal rosette of leaves and scapigerous inflorescence; **B:** Vertical section of flower to show long corolla tube and epipetalous stamens; **C:** Transverse section of ovary with free-central placentation; **D:** Fruit dehiscing by recurved apical teeth. *Anagallis arvensis.* **E:** Part of plant with opposite and whorled leaves and axillary flowers; **F:** Flower from above; **G:** Vertical section of flower; **H:** Fruit with persistent calyx and style; **I:** Pyxidium fruit dehiscing by terminal cap; **J:** Seed.

longitudinal, sometimes with apical pores, sometimes with staminodes alternating the petals. **Gynoecium** with 5 united carpels, ovary superior or half-inferior, unilocular, ovules many, anatropous to campylotropous, placentation free central, style simple, stigma capitate or minute, heterostyly is prevalent in the genus *Primula*. Fruit a capsule, variously dehiscent, pyxidium in *Anagallis*, opening by a cap like cover; seeds with straight embryo, endosperm present, sometimes with aril. Pollination by various insects. Small seeds are often dispersed by wind or water, some by ants attracted by the oily aril.

Economic importance: The family is important for several ornamental species of *Primula* and *Cyclamen*. *Anagallis arvensis* is of medicinal importance.

Phylogeny: The family is well defined and usually placed under Primulales along with other groups of Dillenianae or Dilleniidae (whichever is recognized). In some genera such as *Samolus*, there are five staminodes alternating with petals, in addition to 5 normal stamens opposite the petals, suggesting that antipetalous condition has resulted during course of evolution from the loss of the outer whorl (represented in some genera by staminodes). The family is mostly placed closer to Myrsinaceae. Hutchinson, however, advocated that the two are not related and their free central placentation and stamens opposite the petals are due to parallel evolution. He considers Primulaceae to have evolved from Caryophyllaceae with Portulacaceae being a connecting link. Over the recent years, there have been attempts to shift *Anagallis*, *Lysimachia* and other genera to Myrsinaceae, separation of *Maesa* from Myrsinaceae into a distinct family Maesaceae and shifting of *Samolus* from Primulaceae to Theophrastaceae (Anderberg et al., 2000, 2001). On the basis of phylogenetic analysis based on DNA sequences of *rbc*L and *ndh*F Kallesjo et al. (2000) concluded that genera of tribe Lysimachieae (*Anagallis, Cyclamen, Gaux, Lysimachia* and *Trientalis*) as also the genera *Coris* and *Ardisiandra* should be placed within expanded Myrsinaceae, restricting Primulaceae to herbaceous members with campanulate corolla and capsule fruits, a treatment followed in APweb. Judd et al. (2002) and Thorne (2003), however, have followed broader circumscription of family Primulaceae retaining these genera. Recent studies have indicated resemblances of the family Primulaceae with Ericalean complex under Asterids. As such Thorne in his recent revision (2003, 2006, 2007) has shifted Primulales to Asteridae (under Ericanae), and Judd et al. (2002), APG II (2003), APG IV (2016) and APweb have placed the family under order Ericales of Asterids.

The family is divided into four subfamilies: Maesoideae, Theophrastoideae (tribes Samoleae and Theophrasteae), Primuloideae and Myrsinoideae.

Ericaceae A. L. de Jussieu *Heath Family*

124 genera, 4,245 species

Widely distributed throughout temperate and subtropical regions, and to some extent in subarctic and alpine regions, mainly on acidic soils.

Placement:

	B & H	Cronquist	Takhtajan	Dahlgren	Thorne	APG IV/(APweb)
Informal Group						Asterids
Division		Magnoliophyta	Magnoliophyta			
Class	Dicotyledons	Magnoliopsida	Magnoliopsida	Magnoliopsida	Magnoliopsida	
Subclass	Gamopetalae	Dilleniidae	Dilleniidae	Magnoliidae	Asteridae	
Series+/ Superorder	Heteromerae+		Ericanae	Ericanae	Ericanae	
Order	Ericales	Ericales	Ericales	Ericales	Ericales	Ericales

Salient features: Mainly shrubs, leaves alternate, flowers campanulate to urceolate, stamens twice the number of corolla lobes, arising from nectariferous disc, anthers opening by terminal pores, ovary 4 or more chambered.

Major genera: *Rhododendron* (850 species), *Erica* (600), *Vaccinium* (450 incl. *Agapetes*: 90), *Gaultheria* (160), *Leucopogon* (150), *Cavendishia* (110), *Arctostaphylos* (70), *Dracophyllum* (50), *Epacris* (45), *Pyrola* (20), *Cassiope* (12) and *Monotropa* (5).

Description: Shrubs, sometimes small, rarely lianas, epiphytes or mycoparasitic herbs lacking chlorophyll (*Monotropa*). trichomes unicellular or multicellular, glandular or scaly, never stellate. **Leaves** alternate, sometimes opposite or whorled, simple, entire or serrate, often coriaceous, persistent, sometimes reduced to needles or scales, margin sometimes revolute, venation reticulate, usually pinnate, sometimes palmate, blade sometimes reduced to mycoparasite, stipules absent. **Inflorescence** with solitary axillary flowers, cymes, racemes or panicles. **Flowers** bisexual, rarely unisexual (*Empetrum*), actinomorphic, sometimes slightly zygomorphic (*Rhododendron*), hypogynous. **Calyx** with 5 sepals, rarely 4–7, distinct or slightly connate at base, persistent. **Corolla** with 5 petals, rarely 4–7, united, corolla tubular, campanulate or urceolate, lobes short, imbricate or convolute. **Androecium** with 5 stamens sometimes twice the number of corolla lobes, arising from nectar disc, free, filaments flattened at base, sometimes connate at base (*Vaccinium*), straight or S-curved, free from corolla or epipetalous, anthers bithecous, sometimes with paired appendages (spurs) near base, dehiscence by apical pores, rarely small slits near tip, pollen in tetrads. **Gynoecium** with 5 united carpels, rarely 2–10, ovary usually superior, rarely half-inferior (*Gaultheria*) or inferior (*Vaccinium*), disc present, placentation axile or parietal with deeply intruded placentae, unilocular, ovules many, anatropous, style simple, conical or filiform, rarely split above, stigma simple. **Fruit** a capsule, berry (*Vaccinium, Cavendishia*) or drupe (*Arctostaphylos, Styphelia*); seeds small, embryo straight, endosperm present, seed coat thin. Pollination mostly by bees and wasps, aided by nectar. Viscin threads present in *Rhododendron* and related genera help insects to pull out pollen tetrads. Capsule fruits are dispersed by wind, species with berries or drupes are usually dispersed by birds.

Economic importance: The family provides edible fruits blueberry (*Vaccinium* spp.), cranberry (*V. macrocarpum*) and huckleberry (*Gaylussacia;* sometimes included in *Vaccinium*). Several species of *Rhododendron* (rhododendrons, azaleas), *Calluna* (heather), *Kalmia* (*K. latifolia*, mountain laurel), *Erica* (heath), *Oxydendrum* (sourwood), *Arbutus* (madrone) and *Leucothoe* (fetterbush) are widely grown as ornamentals. Oil of wintergreen (methyle salicylate) is obtained from *Gaultheria procumbens*. Foliage of *Gaultheria shallon* is sold as 'lemon leaf'.

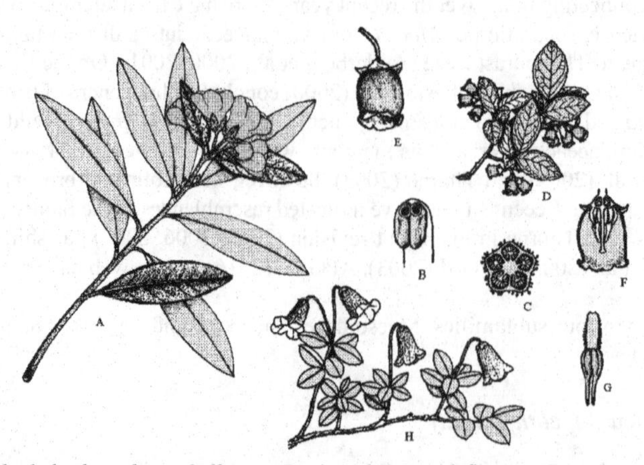

Figure 13.114: Ericaceae. *Rhododendron glaucophyllum.* **A:** Portion of plant with flowers; **B:** Anther; **C:** Transverse section of ovary; *Vaccinium vacillans.* **D:** Branch with flowers; **E:** Flower; **F:** Vertical section of flower; **G:** Anther. **H:** Flowering branch of *Lyonia villosa* (A, H, after Polunin and Stainton, Fl. East. Himal., 1984; rest, after Bailey, Man. Cult. Pl., 1949).

Phylogeny: The broadly circumscribed Ericaceae includes Empetraceae, Epacridaceae, Monotropaceae, Pyrolaceae and Vacciniaceae, sometimes recognized as independent families, or included under two families Ericaceae (ovary superior, fruit capsule) and Vacciniaceae (ovary semi-inferior or inferior, fruit a berry). Segregation these families would render Ericaceae as paraphyletic. Monophyly of broadly circumscribed Ericaceae is supported by evidence from morphology, *rbc*L and 18S rDNA sequences (Kron, 1996; Soltis et al., 1997). Engler and Diels (1936) included four subfamilies: Rhododendroideae (septicidal capsule, seed winged or ribbed, anthers without appendages), Arbutoideae (fruit berry or loculicidal capsule, seed not winged, anthers appendaged, ovary superior), Vaccinioideae (ovary inferior, rest similar to Arbutoideae) and Ericoideae (fruit loculicidal capsule or nut, ovary superior, seeds not winged, calyx persistent, anthers with apically spreading lobes). Thorne (2006) recognized eight subfamilies, merging Rhododendoideae with Ericoideae, and adding Enkianthoideae, Monotropoideae, Cassiopoideae, Styphelioideae and Empetroideae, subsequently (2007) adding ninth Harrimanelloideae segregated from Cassiopoideae. APweb (2008) also recognizes 8, but includes Empetroideae under Ericoideae, recognizing Harrimanelloideae as additional subfamily, version 14 of APweb (updated 2018) recognizes 9 subfamilies: Enkianthoideae, newly added Pyroloideae, Monotropoideae, Arbutoideae, Cassiopoideae, Ericoideae (5 tribes), Harrimanelloideae, Epacridioideae (7 tribes) and Vaccinioideae. *Monotropa uniflora* has smallest embryo, just two-celled (Olson, 1991).

Rubiaceae A. L. de Jussieu *Madder Family*

590 genera, 13,620 species

Worldwide in distribution, but mainly distributed in the tropics and subtropics, especially the woody members.

Placement:

	B & H	Cronquist	Takhtajan	Dahlgren	Thorne	APG IV/(APweb)
Informal Group						Asterids/ (Euasterids/ Lamiid/Asterid I)
Division		Magnoliophyta	Magnoliophyta			
Class	Dicotyledons	Magnoliopsida	Magnoliopsida	Magnoliopsida	Magnoliopsida	
Subclass	Gamopetalae	Asteridae	Lamiidae	Magnoliidae	Lamiidae	
Series+/ Superorder	Inferae+		Lamianae		Loasanae	Lamianae
Order	Rubiales	Rubiales	Rubiales	Gentianales	Rubiales	Gentianales

Salient features: Mainly shrubs and trees, leaves opposite or whorled, often turning blackish when dry, stipules interpetiolar, colleters present in leaf axils, inflorescence cymose, flowers pentamerous, stamens 5, ovary inferior.

Major genera: *Psychotria* (1450 species), *Galium* (410), *Ixora* (370), *Pavetta* (360), *Hedyotis* (360), *Tarenia* (350), *Randia* (240), *Gardenia* (240) and *Mussaenda* (190).

Description: Trees (*Adina, Neolamarckia*) or shrubs (*Ixora, Gardenia*), rarely herbs (*Galium*), sometimes climbing (*Rubia*) with hooked hairs, rarely epiphytic (*Myrmecodia*) with large swellings on roots inhabiting ants, usually with Iridoids, raphide crystals common. **Leaves** opposite, with interpetiolar stipules which often become as large as leaves and thus forming whorled arrangement of leaves, simple, entire, often turning blackish when dry, with colleters in leaf axils. **Inflorescence** cymose, sometimes capitate (*Adina*), or solitary (*Gardenia*). **Flowers** bisexual, actinomorphic, rarely zygomorphic (*Posoqueria*) epigynous, sometimes dimorphic (*Randia*). **Calyx** with 4–5 sepals, adnate to ovary, 5-lobed, lobes often very small, one sometimes enlarged and brightly colored (*Mussaenda*). Corolla with 4–5 petals, (rarely 8–10), united, tubular, rotate or funnel-shaped, valvate, imbricate or twisted. **Androecium** with 4–5 stamens, free, epipetalous, anthers bithecous, dehiscence longitudinal, introrse, pollen grains usually tricolporate. **Gynoecium** with 2 (rarely 1-many) united carpels, ovary inferior, rarely superior (*Pugama*) or semi-inferior (*Synaptantha*), bilocular (rarely 1-many locules) with 1-many ovules in each chamber, placentation axile (rarely apical or basal), nectar disc usually present above the ovary, style slender, stigma capitate or lobed. **Fruit** a berry, capsule, drupe or schizocarp; seeds 1-many, with small embryo, curved or straight, endosperm present or absent.

Economic importance: The family is economically important for being the source of coffee, quinine and many ornamentals. Coffee is obtained from roasted seeds of *Coffea arabica* and *C. canephora*. Quinine, a remedy for malaria is derived from several species of *Cinchona*. Madder (*Rubia tinctoria*) was formerly cultivated for its red dye alizarin. Important ornamentals include *Gardenia, Ixora, Hamelia, Neolamarckia* (cadamb tree) and *Mussaenda*.

Phylogeny: Rubiaceae form a well-defined group which is clearly monophyletic as supported by morphology (Bremer and Struwe, 1992) and *rbcL* sequences (Bremer et al., 1995). Affinities of the family lie with Gentianales (Dahlgren, APG

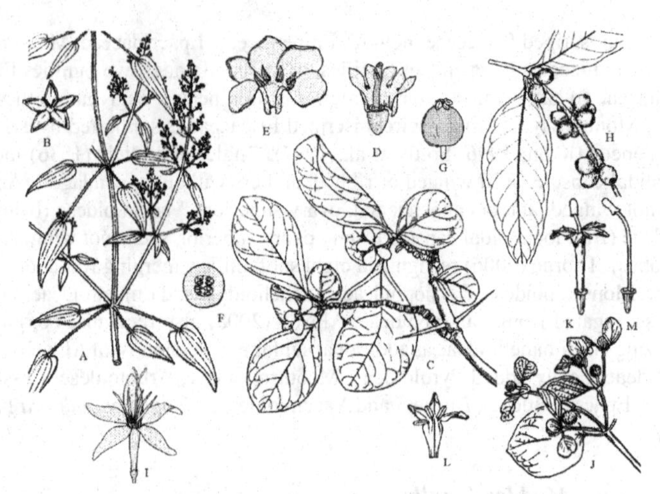

Figure 13.115: Rubiaceae. *Rubia manjith.* **A:** Portion of plant with axillary inflorescences; **B:** Flower. *Randia spinosa.* **C:** Twig showing spines and flowers; **D:** Flower with corolla removed and calyx opened to show gynoecium; **E:** Corolla opened to show epipetalous stamens; **F:** Transverse section of ovary; **G:** Fruit. *Coffea arabica.* **H:** Portion of twig with fruits; **I:** Flower. *Mitragyna parvifolia.* **J:** Twig with globose inflorescences. **K:** Flower with mitraeform (head-gear) stigma; **L:** Corolla opened to show androecium; **M:** Flower with corolla removed to show style and stigma.

II, APweb; Thorne places under Rubiales also containing Gentianaceae and related families; Takhtajan under Rubiales next to Gentianales under Lamiidae—> Gentiananae) or Dipsacales (Cronquist—next to Dipsacales towards the end of Asteridae, Gentianales towards the beginning), both having opposite leaves and 2 carpels. The separation exclusively based on ovary being inferior or superior, is slowly being abandoned as has also been done in the case of certain monocots such as Alliaceae, Agavaceae and Amaryllidaceae. The family is commonly divided into three subfamilies (Thorne, APweb): Cinchonoideae (mainly woody, raphides absent, seeds with endosperm, heterostyly absent), Ixoroideae (woody, raphides absent, pollination plunger-mechanism as in Asteraceae), Rubioideae (mainly herbaceous, raphides present in leaves, seeds with endosperm, heterostyly common). Molecular data (Fay et al., 2000a) provide support for including *Dialypetalanthus* (formerly placed under Dialypetalanthaceae-Thorne, 1999) in Rubiaceae, under subfamily Ixoroideae (Thorne, 2003). Recent molecular studies based on *trnL*-F and cpDNA data (Rova et al., 2002) and broadly-based molecular data involving several taxa (Bremer et al., 1999) suggest that Cinchonoideae and Ixoroideae are sister taxa. Robbrecht and Manen (2006) proposed merging the two, thus recognizing only two subfamilies: Rubioideae and Cinchonoideae, followed in APweb (2018). Rydin et al. (2017) based on mitochondrial genomic data and plastid data found conflicting results of tribal distribution in subfamilies.

Apocynaceae A. L. de Jussieu ***Dogbane Family***

366 genera, 5,100 species (including Asclepiadaceae Borkh.)
Mostly tropical and subtropical with a few species in temperate regions.

Placement:

	B & H	Cronquist	Takhtajan	Dahlgren	Thorne	APG IV/(APweb)
Informal Group						Asterids/ (Euasterids/ Lamiid/Asterid I)
Division		Magnoliophyta	Magnoliophyta			
Class	Dicotyledons	Magnoliopsida	Magnoliopsida	Magnoliopsida	Magnoliopsida	
Subclass	Gamopetalae	Asteridae	Lamiidae	Magnoliidae	Lamiidae	
Series+/ Superorder	Bicarpellatae+		Lamianae	Loasanae	Lamianae	
Order	Gentianales	Gentianales	Rubiales	Gentianales	Rubiales	Gentianales

Salient features: Herbs shrubs or climbers, latex milky, leaves opposite or whorled, throat of corolla tube with scales, pollinia absent, ovary superior, fruit a follicle, seed with a tuft of hairs (Asclepiadoideae, formerly Asclepiadaceae separated by pollinia, anthers adnate to stigmatic disc, stigmas united into gynostegium).

Figure 13.116: Apocynaceae. Subfamily Rauvolfioideae. *Catharanthus roseus.* **A:** Portion of plant with flowers and fruits; **B:** Flower bud showing twisted corolla with a long tube; **C:** Vertical section of flower from corolla throat showing free epipetalous stamens and calyptrate stigma; **D:** Anther with dorsal fixation; **E:** Transverse section of flower passing through ovary showing sepals, corolla tube 2 lateral nectaries and 2 free ovaries; **F:** Pair of follicles; **G:** Seed. *Nerium oleander.* **H:** Branch with whorled leaves and terminal inflorescence; **I:** Corolla opened to show corona of scales and anthers with tailed appendages forming single twisted hairy appendage; **J:** Pair of follicles. *Thevetia nerifolia.* **K:** Branch with subopposite and alternate leaves and large funnel-shaped flowers; **L:** Drupe fruit. *Rauvolfia serpentina.* **M:** Portion of plant with inflorescences in flower and fruit; **N:** Flower buds with twisted corolla; **O:** Seed.

Major genera: *Asclepias* (220 species), *Tabernaemontana* (220), *Cynanchum* (200), *Ceropegia* (140), *Hoya* (140), *Rauvolfia* (105), *Ervatamia* (80), *Allamanda* (15) and *Catharanthus* (5).

Description: Perennial herbs (*Catharanthus*), vines (*Cryptostegia, Daemia*), shrubs (*Calotropis, Nerium*), rarely trees (*Alstonia*), often fleshy (*Hoya*) or cactus-like (*Stapelia*), latex usually milky, Iridoids often present. **Leaves** simple, reduced or absent in some succulent species, opposite (*Calotropis, Catharanthus*) or whorled (*Nerium*), simple, entire, venation pinnate, reticulate, stipules absent, colleters often present at the base of petiole. **Inflorescence** dichasial or monochasial cyme, racemose or umbellate (*Calotropis*), sometimes solitary (*Vinca*), or axillary cymose pairs (*Catharanthus*). **Flowers** bisexual, actinomorphic, hypogynous, pentamerous, often coronate. **Calyx** with 5 sepals, distinct or basally connate, imbricate or valvate, often with glands (colleters) at base. **Corolla** with 5 united petals, the tube often short campanulate (*Calotropis*), salver-shaped (*Catharanthus*) or funnel-shaped (*Thevetia*) and with contorted or valvate lobes. **Corona** of usually 5 scales or appendages arising from corolla throat (corolline corona: *Nerium, Cryptostegia*), or from stamens (staminal corona: *Calotropis, Asclepias*), coronal appendages nectariferous. **Androecium** with 5 stamens, filaments free (Apocynaceae *sensu str.*) or connate (Asclepiadoideae except *Cryptostegia*) anthers free with separate pollen grains (Apocynaceae *sensu str.*). In Asclepiadoideae anthers adherent to the stigmatic area forming a 5 angled disc gynostegium (gynandrium), pollen agglutinated within the anther sacs to form waxy pollinia (corpusculum or gland joining together two pollinia one each from adjacent anthers with the help of caudicles to form translator, an adaptation for insect pollination); pollen grain tricolporate or biporate or triporate. **Gynoecium** bicarpellary of two apically united carpels, unilocular ovaries, marginal placentation, ovules 2 or more, unitegmic, styles 2, stigma 1, calyptrate (*Catharanthus*), dumb-bell shaped (*Nerium*), or 5 lobed and fused with anthers to form gynostegium (*Calotropis*), carpels sometimes united by ovaries with axile placentation (*Thevetia, Allamanda, Carissa*). **Fruit** etaerio of 2 follicles (*Nerium, Calotropis*), sometimes drupe (*Thevetia*), capsule or berry; seeds usually numerous, flattened and comose with long silky hairs. Pollination by insects, helped by special translators in Asclepiadoideae. Dispersal mostly by wind, aided by hairs.

Economic importance: *Nerium* (oleander), *Catharanthus* (Madagascar periwinkle), *Asclepias* (milkweed), *Hoya* (wax plant), *Stapelia* (carrion flower), *Plumeria* (frangipani) and *Tabernaemontana* are grown as ornamentals. *Nerium* and *Thevetia* are poisonous (can be fatal). Roots of *Rauvolfia serpentina* yield reserpine used as tranquillizer for patients suffering from schizophrenia and hypertension. *Catharanthus* provides antileukaemic drugs. Latex from *Plumeria* used for healing toothache. Low quality down is obtained from seeds of several species. Pitcher like leaves and roots of *Dischidia* chewed with betel. Rubbervine (*Cryptostegia*) is also as caoutchouc or rubber source. Tubers of *Ceropegia* are edible. Stem fibre of *Calotropis* and *Leptadaenia* is used for cordage. *Asclepias* is a livestock poison.

Phylogeny: For a long time Apocynaceae was considered to be closely related but distinct from Asclepiadaceae, latter with pollinia, gynostegium and usually staminal corona (Bentham and Hooker, Engler and Prantl, Hutchinson, Cronquist,

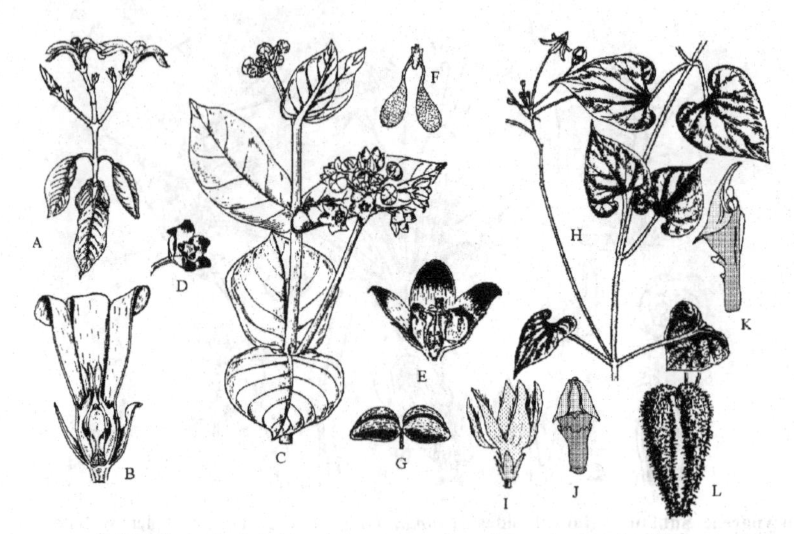

Figure 13.117: Apocynaceae, Subfamily **Asclepiadoideae.** *Cryptostegia grandiflora.* **A:** Branch with terminal inflorescence; **B:** Vertical section of flower with corolloin corona and gynostegium. *Calotropis procera.* **C:** Portion of plant with umbellate inflorescences on axillary peduncles; **D:** Flower with purple-tipped corolla; **E:** Vertical section of flower with staminal corona and broad gynostegium and free ovaries; **F:** Translator with 2 pollinia joined by caudicles to common corpusculum; **G:** Pair of follicles. *Pergularia daemia.* **H:** Portion of plant with inflorescence; **I:** Flower; **J:** Flower with calyx removed and corolla lobes cut to show corona and staminal tube; **K:** Stamen and corona in side view; **L:** Pair of follicles covered with bristles.

Dahlgren). The family Asclepiadaceae was merged with Apocynaceae by Thorne (1983) and practice was followed by Takhtajan (1987, 1997), Judd et al. (2002), APG II and APweb. The separation of Asclepiadaceae as distinct family would lead to paraphyletic Apocynaceae (Judd et al., 1994; Endress et al., 1996). The family Apocynaceae is appropriately divided into 5 subfamilies (Thorne, 2000, 2003, 2007): Rauvolfioideae (Plumerioideae), Apocynoideae, Periplocoideae, Secamonoideae and Asclepiadoideae. The generic limits are not clearly resolved. According to Sennblad and Bremer (2002), both Rauvolfioideae and Apocynoideae may be quite wildly paraphyletic. The position of the Periplocoideae as sister to Secamonoideae + Asclepiadoideae is also uncertain (Potgeiter and Albert, 2001; Sennblad and Bremer, 2002). The family is usually placed in Gentianales, but Thorne has merged this order with broadly circumscribed Rubiales. Dahlgren and APG IV (2016) classifications prefer the name Gentianales for the broadly circumscribed order. Endress et al. (2014) divide family into 5 subfamilies, 25 tribes, 49 subtribes and 366 genera, describing or validating several new tribes and subtribes.

Boraginaceae A. L. de Jussieu ***Borage Family***

135 genera, 2,535 species

Widely distributed in temperate, tropical and subtropical regions.

Placement:

	B & H	**Cronquist**	**Takhtajan**	**Dahlgren**	**Thorne**	**APG IV/(APweb)**
Informal Group						Asterids/ (Euasterids/ Lamiid/Asterid I)
Division		Magnoliophyta	Magnoliophyta			
Class	Dicotyledons	Magnoliopsida	Magnoliopsida	Magnoliopsida	Magnoliopsida	
Subclass	Gamopetalae	Asteridae	Lamiidae	Magnoliidae	Lamiidae	
Series+/ Superorder	Bicarpellatae+		Lamianae	Solananae	Solananae	
Order	Polemoniales	Lamiales	Boraginales	Boraginales	Solanales	Boraginales

Salient features: Bristly herbs, stems cylindrical, leaves alternate, inflorescence helicoid cymes, flowers pentamerous, actinomorphic, carpels 2, ovary 4-lobed, style gynobasic, fruit with 4 nutlets.

Major genera: *Cordia* (300 species), *Heliotropium* (250), *Tournefortia* (240), *Onosma* (140), *Myosotis* (90), *Cynoglossum* (75) and *Ehretia* (75).

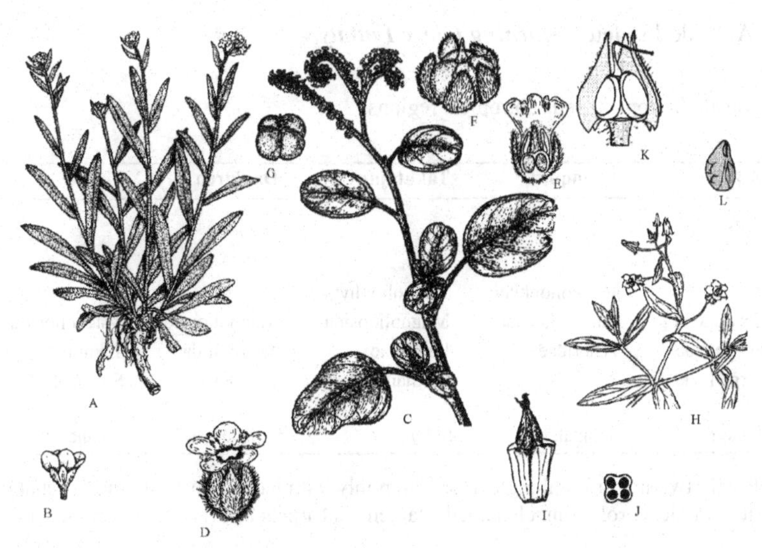

Figure 13.118: Boraginaceae. *Cynoglossum glochidiatum.* **A:** Plant with terminal inflorescences; **B:** Flower. *Heliotropium eichwaldii.* **C:** Portion of plant with terminal helicoid cymes; **D:** Flower with bristly calyx; **E:** Vertical section of flower; **F:** Fruit with persistent calyx; **G:** Fruit. *Trichodesma indicum.* **H:** Portion of plant in flower; **I:** Flower with corolla cut away to show androecium; **J:** Transverse section of ovary, 4-locular due to false septum; **K:** Fruit with two of the sepals removed; **L:** Seed.

Description: Herbs, shrubs or trees (*Cordia*), sometimes lianas, inner phloem lacking, hairs with basal cystolith and often calcified or silicified and as such plants bristly, rough to touch. **Leaves** alternate, simple, entire, venation pinnate, reticulate, stipules absent. **Inflorescence** usually of helicoid cymes, rarely scorpioid. **Flowers** bisexual, actinomorphic, rarely zygomorphic (*Echium*), hypogynous, pentamerous. **Calyx** with 5 sepals, free or slightly connate at base, persistent. **Corolla** with 5 petals, united, rotate, tubular or funnel shaped, usually plicate. **Androecium** with 5 stamens, epipetalous, inserted in corolla tube, filaments free, bithecous, anthers introrse, dehiscence longitudinal, pollen grains tricolporate or multiporate, filaments often with nectar discs at base. **Gynoecium** with 2 united carpels, ovary superior, deeply 4-lobed, bilocular, axile placentation, becoming 4-locular due to false septa, style 1, terminal or gynobasic, stigma 1 or bilobed, capitate or truncate, ovary seated on a nectary. **Fruit** a drupe with one 4-seeded, two 2-seeded or four 1-seeded pits or schizocarp with four 1-seeded nutlets; seeds with embryo straight or curved. Pollination mostly by insects. Drupaceous fruits are dispersed by birds, whereas corky ones (*Argusia, Cordia*) are carried away by water.

Economic importance: Several species of *Heliotropium* (heliotrope), *Mertensia* (virgin bluebells), *Myosotis* (forget-me-not), *Cordia* (Geiger tree), *Cynoglossum* (hound's tongue) and *Pulmonaria* (lungwort) are grown as ornamentals. Several species such as *Borago officinalis* (borage), *Symphytum officinalis* (comphrey) and *Lithospermum* spp. (pucoon) have been used as medicinal herbs. *Alkanna tinctoria* (alkanet) is a source of red dye used to stain wood and marble and to color medicines, wines and cosmetics.

Phylogeny: The family is closely related to Solanaceae, Convolvulaceae and Polemoniaceae with which it shares alternate leaves and actinomorphic flowers and mostly included under Boraginales next to Solanales (Dahlgren) or Solanales, Convolvulales and Polemoniales (Takhtajan), or under Solanales (Thorne). The family also shows close affinities with Lamiaceae and Verbenaceae in having gynobasic style, 4 lobed ovary becoming 4-locular by false septum and usually schizocarpic fruit. Cronquist accordingly places Boraginaceae under Lamiales closer to Lamiaceae and Verbenaceae. The position of both Vahliaceae and Boraginaceae in Euasterids I is uncertain. *Vahlia* is placed sister to Lamiales, but with only 63 per cent bootstrap support (Albach et al., 2001), or is associated more specifically with Boraginaceae (Lundberg, 2001). Relationships between Gentianales, Lamiales and Solanales are also unclear (Albach et al., 2001). APweb recognizes 6 groups within Boraginaceae: four subfamilies Boraginoideae, Heliotropioideae, Cordioideae and Ehretioideae and two family groups Hydrophyllaceae and Lennoaceae. Thorne (2003, 2006, 2007) treats last two as independent families and recognizes 5 subfamilies under Boraginaceae, adding Wellstedioideae as fifth subfamily.

Chacón et al. (2016) based on analysis of three plastid markers and taxon sampling with four outgroup and 170 ingroup species from 73 genera, proposed revised classification of Boraginaceae recognizing 3 subfamilies: Echiochiloideae (3 genera), Boraginoideae (2 tribes, 2 subtribes) and Cynoglossoideae (8 tribes, 6 subtribes). Earlier versions of APG, APG II (2003) and APweb included the family unplaced under Asterids I, subsequently assigned to Boraginales in APG IV (2016) and APweb (2017).

Convolvulaceae A. L. de Jussieu ***Morning Glory Family***

59 genera, 1,830 species

Widely distributed, mostly in tropical and subtropical regions.

Placement:

	B & H	Cronquist	Takhtajan	Dahlgren	Thorne	APG IV/(APweb)
Informal Group						Asterids/ (Euasterids/ Lamiid/Asterid I)
Division		Magnoliophyta	Magnoliophyta			
Class	Dicotyledons	Magnoliopsida	Magnoliopsida	Magnoliopsida	Magnoliopsida	
Subclass	Gamopetalae	Asteridae	Lamiidae	Magnoliidae	Lamiidae	
Series+/ Superorder	Bicarpellatae+		Lamianae	Solananae	Solananae	
Order	Polemoniales	Solanales	Solanales	Solanales	Solanales	Solanales

Salient features: Usually twining or climbing herbs, commonly with latex, leaves alternate, venation palmate, stipules absent, flowers actinomorphic, corolla funnel-shaped, stamens 5, carpels 2, ovary superior, 2-chambered, ovules 1 or 2, fruit a capsule.

Major genera: *Ipomoea* (550 species), *Convolvulus* (240), *Cuscuta* (140), *Jacquemontia* (110), *Evolvulus* (95) and *Calystegia* (25).

Description: Twining or climbing herbs, often rhizomatous, latex usually present, sometimes parasitic (*Cuscuta*), rarely tree (*Humbertia*), vascular bundles with both outer and inner phloem, sometimes with alkaloids, branching usually sympodial. **Leaves** alternate, simple, rarely lobed or compound, sometimes absent (*Cuscuta*), venation palmate, reticulate, stipules absent. **Inflorescence** cymose or with solitary flowers. **Flowers** bisexual, actinomorphic, rarely zygomorphic (*Humbertia*), hypogynous. **Calyx** with 5 sepals, free or slightly connate at base, persistent. **Corolla** with 5 petals, united, funnel-shaped, usually plicate. **Androecium** with 5 stamens, epipetalous, inserted in corolla tube, often unequal, filaments free, bithecous, anthers introrse, dehiscence longitudinal or by apical pores, pollen grains tricolpate or multiporate. **Gynoecium** with 2 united carpels, ovary superior, entire or deeply bilobed, bilocular, axile placentation, style 1, terminal or gynobasic, stigma bilobed, capitate or linear, ovary seated on a nectary. **Fruit** a capsule; seeds 1 or 2 in each chamber, embryo straight or curved, cotyledons folded. Pollination mostly by insects.

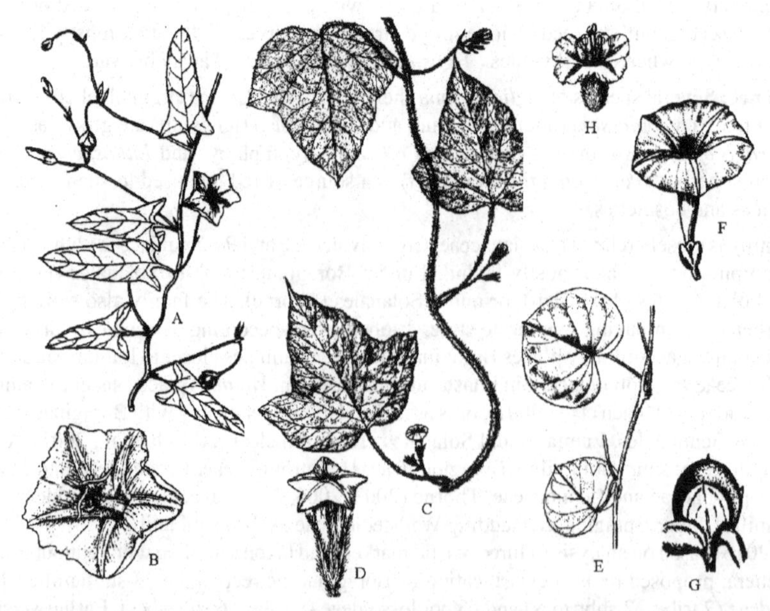

Figure 13.119: Convolvulaceae. *Convolvulus arvensis*. **A:** Branch with flowers and fruits; **B:** Flower from above. *Ipomoea arachnosperma*. **C:** Branch with flowers; **D:** Flower. *Rivea hypocrateriformis*. **E:** Portion of plant with leaves; **F:** Flower; **G:** Fruit with persistent calyx. **H:** Flower of *Seddera latifolia*.

Economic importance: *Ipomoea batatas* (sweet potato) is important for its edible roots. Important ornamentals include *Ipomoea* (morning glory), *Porana* (Christmas vine) and *Dichondra* (ponyfoot). Roots of *Convolvulus scammonia* (scammony) and of *Ipomoea purga* (jalap) yield a drug used medicinally as cathartic.

Phylogeny: The family is closely related to Solanaceae, Boraginaceae and Polemoniaceae. Cuscutaceae and Dichondraceae, sometimes recognized as distinct families, are better placed in Convolvulaceae, their separation leading to paraphyly of Convolvulaceae. Monophyly of the family is supported by morphological characters. Thorne (1999, 2000) recognized 4 subfamilies: Humbertioideae (*Humbertia;* tree), Dichondroideae (2 genera *Dichondra, Falkia*), Convolvuloideae and Cuscutoideae (*Cuscuta;* without leaves), subsequently (2003) merging Dichondroideae in Convolvuloideae, but restored in 2006 and 2007 revisions. On the basis of DNA sequences of multiple chloroplast loci, Stefanovic et al. (2002) concluded that Poranae (including *Porana,* itself polyphyletic) and Erycibeae successively form basal clades in Convolvuloideae. The basal Poraneae have foliaceous bracts and fruits that are utriculate. *Erycibe* (Erycibeae) has sessile stigmas. They provided a detailed tribal classification in 2003, recognizing 12 tribes: Ipomoeeae, Merremieae, Convolvuleae, Aniseieae, Cuscuteae, Jacquemontieae, Maripeae, Cresseae, Dichondreae, Erycibeae, Cardiochlamyeae and Humbertieae. APweb earlier (2007) recognized only two subfamilies Humbertioideae and Convolvuloideae (including both Cuscutoideae and Dichondroideae), increasing to six in Version 14 (updated 2018): Humbertioideae, Cardiochlamyeae, Eryciboideae, Cuscutoideae, Convolvuloideae (3 tribes) and Dichondroideae (4 tribes). The distinctive *Humbertia* is sister to the rest of the family (García et al., 2014).

Solanaceae A. L. de Jussieu *Nightshade* or *Potato Family*

100 genera, 2,600 species

Cosmopolitan in distribution, found both in temperate and tropical climates with largest concentration in Central and South America.

Placement:

	B & H	Cronquist	Takhtajan	Dahlgren	Thorne	APG IV/(APweb)
Informal Group						Asterids/ (Euasterids/ Lamiid/Asterid I)
Division		Magnoliophyta	Magnoliophyta			
Class	Dicotyledons	Magnoliopsida	Magnoliopsida	Magnoliopsida	Magnoliopsida	
Subclass	Gamopetalae	Asteridae	Lamiidae	Magnoliidae	Lamiidae	
Series+/ Superorder	Bicarpellatae+		Lamianae	Solananae	Solananae	
Order	Polemoniales	Solanales	Solanales	Solanales	Solanales	Solanales

Salient features: Leaves alternate, stipules absent, flowers actinomorphic, stamens 5, carpels 2, ovary superior, 2-chambered, placenta swollen, septum oblique, ovules numerous, fruit a berry or capsule.

Major genera: *Solanum* (1350 species), *Lycianthus* (190), *Cestrum* (160), *Nicotiana* (110), *Physalis* (95), *Lycium* (90), *Capsicum* (50), *Hyoscyamus* (25) and *Datura* (10).

Description: Herbs, shrubs (*Brunfelsia, Cestrum*) or small trees (*Solanum verbascifolia; Dunalia*), rarely lianas, often poisonous, sometimes with prickles, underground tubers in *Solanum tuberosum* (potato), vascular bundles with both outer and inner phloem. **Leaves** alternate, simple, rarely pinnately compound (potato), stipules absent, paired leaves adjacent on the stem are common. **Inflorescence** cymose (*Solanum*) or of solitary flowers (*Datura*). **Flowers** bisexual, actinomorphic, hypogynous. **Calyx** with 5 sepals, united, persistent, sometimes enlarged and swollen in fruit (*Withania, Physalis*). **Corolla** with 5 petals, united, rotate (*Solanum*) or tubular (*Cestrum*), rarely funnel shaped (*Datura*) or bilabiate (*Schizanthus*). **Androecium** with 5 stamens epipetalous, inserted in corolla tube, filaments free, bithecous, anthers introrse, dehiscence longitudinal or by apical pores. **Gynoecium** with 2 carpels, rarely 3–5 (*Nicandra*), united, ovary superior, bilocular, axile placentation, placenta swollen, septum oblique, ovary often further divided by false septa, style 1, rarely gynobasic (*Nolana*), stigma bilobed, ovary seated on a nectary. Fruit berry or capsule (*Datura*); seeds many, embryo straight, endosperm present. Pollination mostly by insects. Dispersal mostly by birds.

Economic importance: The family includes a number of food plants such as tomato (*Lycopersicon esculentum*), potato (*Solanum tuberosum*), egg plant or brinjal (*S. melongena*), ground cherry (*Physalis peruviana*). Peppers (*Capsicum annuum*) are used both as a food source (young) and spices (ripe). Many poisonous species are important drug plants such as *Atropa belladona* (atropine), *Hyoscyamus niger* (henbane-hypnotic drug), *Datura stramonium* (stramonium) and *Mandragora officinarum*. Tobacco (*Nicotiana tabacum and N. rustica*) contains toxic alkaloid nicotine and is grown for chewing, smoking

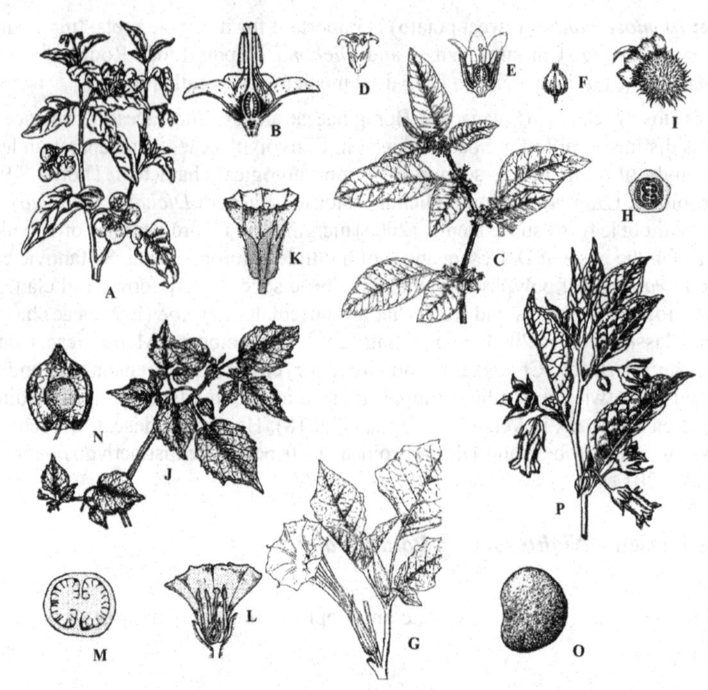

Figure 13.120: Solanaceae. *Solanum nigrum.* **A:** Branch with extra-axillary rhipidium infloresences and fruits; **B:** Vertical section of flower. *Withania somnifera.* **C:** Branch with axillary cymose clusters of flowers; **D:** Flower; **E:** Vertical section of flower with bell-shaped corolla; **F:** Fruit enclosed in enlarged urceolate calyx. *Datura inoxia.* **G:** Branch with axillary funnel-shaped flowers; **H:** Transverse section of ovary, tetralocular due to false septum; **I:** Capsule covered with tubercles and basal persistent portion of calyx. *Physalis minima.* **J:** Portion of plant with flowers; **K:** Flower; **L:** Vertical section of flower; **M:** Transverse section of ovary with swollen placentae; **N:** Fruit with inflated calyx removed from one side; **O:** Seed; **P:** Flowering branch of *Atropa belladona.*

and snuff. Some ornamental genera include *Brunfelsia* (lady-of-the-night; yesterday-today-and tomorrow), *Cestrum* ('Rat ki Rani'; night blooming jessamine), *Petunia, Physalis* (ground cherry) and *Solanum* (nightshade).

Phylogeny: The family is closely related to Scrophulariaceae from which it is differentiated in vascular bundles having outer and inner phloem, actinomorphic flowers and oblique septum of the ovary. *Schizanthus* with zygomorphic flowers is borderline genus. The family also has close affinities with Convolvulaceae, Boraginaceae and Gesneriaceae. Nolanaceae with gynobasic style and lobed ovary has been merged with Solanaceae. The following 7 subfamilies are recognized (Olmstead et al., 1999; APweb, 2003): Schwenckioideae (pericycle fibres present, stamens 4, didynamous, or 3 staminodes; embryo straight, short), Schizanthoideae (pericycle fibres absent, flowers zygomorphic, anterior petals connate, forming a keel, stamens 2, staminodes 3, embryo curved), Goetzeoideae (Fruit often a drupe, embryo curved: Takhtajan as family Goetzeaceae), Cestroideae (pericyclic fibres present, staminodes 4 or 5, often of two lengths), Petunioideae (flowers bisymmetric, embryo slightly curved), Solanoideae (seeds flattened, embryo curved, often coiled), and Nicotianoideae (Cork superficial pericyclic fibres present or absent, stamens 4 or 5, of two lengths, embryo straight or curved). The grouping (Petunioideae (Solanoideae + Nicotianoideae)) is well supported, although the relationships between the more basal branches have only weak support, but *Schwenkia* is probably sister to the rest of the family (Olmstead et al., 1999). Family Sclerophylacaceae has been variously included under Solanaceae (Hutchinson, Cronquist, APG II), treated as distinct family (Takhtajan, Dahlgren), or considered unplaced (APweb) has been treated as subfamily Sclerophylacoideae of Solanaceae by Thorne (2003), who recognizes Browallioideae, Solanoideae and Goetzeoideae as other three subfamilies. APweb (2018) following Olmstead et al. (2008) recognises 8 subfamilies: Schizanthoideae, Goetzeoideae, Duckeodendroideae, Cestroideae, Schwenkieae, Petunioideae, Nicotianoideae and Solanoideae.

Plantaginaceae A. L. de Jussieu ***Snapdragon Family***

94 genera, 1,900 species

Widely distributed from temperate to tropical regions, more diverse in temperate regions.

Placement:

	B & H	Cronquist	Takhtajan	Dahlgren	Thorne	APG IV/(APweb)
Informal Group						Asterids/ (Euasterids/ Lamiid/Asterid I)
Division		Magnoliophyta	Magnoliophyta			
Class	Dicotyledons	Magnoliopsida	Magnoliopsida	Magnoliopsida	Magnoliopsida	
Subclass	Gamopetalae	Asteridae	Lamiidae	Magnoliidae	Lamiidae	
Series+/ Superorder	Bicarpellatae+		Lamianae	Lamianae	Lamianae	
Order	Personales	Scrophulariales	Lamiales	Lamiales	Lamiales	Lamiales

B & H, Cronquist, Takhtajan and **Dahlgren** under Scrophulariaceae.

Salient features: Leaves alternate or opposite, stipules absent, flowers zygomorphic, stamens 4 or 2, anther opening by 2 distinct slits, anthers more or less sagittate at base, carpels 2, ovary superior, 2-chambered, many, fruit a capsule.

Major genera: *Veronica* (450 species), *Penstemon* (275), *Plantago* (210), *Linaria* (150), *Bacopa* (55), *Russelia* (50) and *Callitriche* (30).

Description: Herbs or small shrubs, rarely climbers (*Antirrhinum cirrhosum*), often with phenolic glycosides and triterpenoid saponins, and sometimes with cardiac glycosides, hairs usually simple, when glandular with short discoid head lacking vertical partitions. **Leaves** alternate or opposite, rarely whorled (*Russelia*), simple, entire or dentate, venation pinnate, reticulate, stipules absent. **Inflorescence** racemose: racemes or spikes. **Flowers** bisexual, zygomorphic, hypogynous. **Calyx** with 5 sepals, rarely 4 (*Veronica*), connate, persistent. **Corolla** with 5 petals, rarely 4 (due to fusion of 2 petals as in *Veronica*), united, usually bilabiate, sometimes with nectar sac or spur, lower lip sometimes with a bulge obscuring the throat (personate), lobes imbricate or valvate. **Androecium** with usually 4 stamens, didynamous, fifth stamen sometimes present as a staminode (*Penstemon*), rarely 2 (*Veronica*), epipetalous, inserted in corolla tube, filaments free, anther bilocular, locules distinct, opening by two longitudinal slits, pollen sacs divergent (anther sagittate), pollen grains tricolporate. **Gynoecium** with 2 united carpels, rarely only 1 carpel developed (*Globularia*), ovary superior, bilocular, ovules several in each chamber, rarely 1 or 2 (*Globularia*) unitegmic, axile placentation, style 1, stigma bilobed, ovary seated on a nectary. **Fruit** a septicidal capsule; seed angular or winged, with curved or straight embryo, endosperm present. Pollination by insects. Seeds or nutlets dispersed by wind.

Economic importance: The family contributes several ornamentals such as *Digitalis* (Foxglove), *Mimulus* (monkey flower), *Antirrhinum* (snapdragon), *Penstemon* (beardtongue), *Veronica* (speedwell) and *Russelia* (firecracker plant).

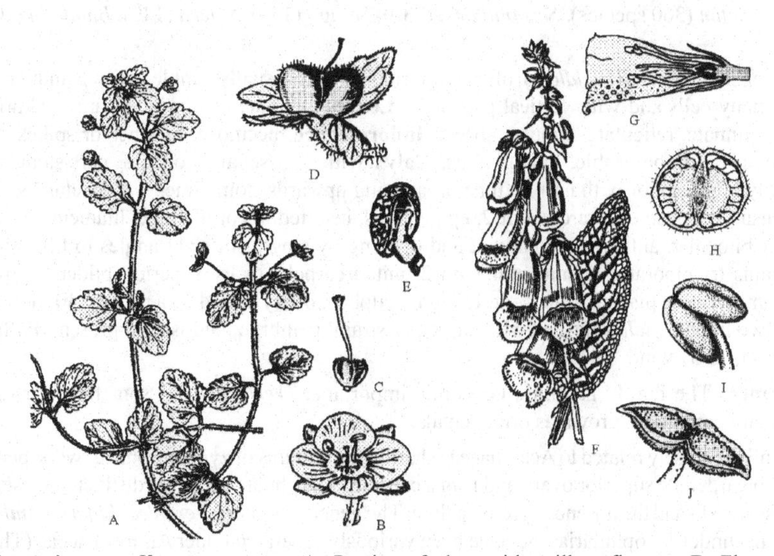

Figure 13.121: Plantaginaceae. *Veronica persica.* **A:** Portion of plant with axillary flowers; **B:** Flower with 4 sepals and petals each and 2 stamens; **C:** Gynoecium; **D:** Fruit with persistent calyx and style; **E:** Seed. *Digitalis purpurea.* **F:** Branch with inflorescence; **G:** Vertical section of flower; **H:** Transverse section of ovary with axile placentation; **I:** Young anther; **J:** Anther dehisced through 2 slits.

Species of *Digitalis,* mainly *D. purpurea* and *D. lanata,* are used for the extraction of drugs digitalin and digoxin used as cardiac stimulants and tonics. The juice of *Limnophila indica* is used in fevers, tonic and as stomachic. Various species of *Veronica* yield glucoside rhinanthis and used for ulcers and burns.

Phylogeny: The family is closely related to Scrophulariaceae and Acanthaceae sharing the features of zygomorphic flowers, pentamerous flowers, stamens less than 5, bicarpellate superior ovary and fruit a capsule. The genera were originally included under Scrophulariaceae from which they differ in having distinct bithecous anthers dehiscing by 2 slits and absence of vertical partitions in heads of glandular hairs. Thorne had earlier (1999, 2000) removed all three *Plantago, Callitriche,* and *Hippuris* to separate families Plantaginaceae, Callitrichaceae and Hippuridaceae, respectively, and used the name Antirrhinaceae for the family. Judd et al. (1999, 2002), APG II and APweb had, however, combined all the four families under Plantaginaceae, a placement also followed subsequently (2003) by Thorne. Monophyly of the family is supported by cpDNA characters. Olmstead (2001) suggested the removal of *Calceolaria* (with highly saccate corolla) and related genera to separate family Calceolariaceae, a change that has been incorporated in APG II, APweb and Thorne (2003, 2007).

The family is commonly divided into 12 tribes: Angelonieae, Antirrhineae, Callitricheae, Cheloneae, Digitalideae, Globularieae, Gratioleae, Hemiphragmeae, Plantagineae, Russelieae, Sibthorpieae and Veroniceae.

Scrophulariaceae A. L. de Jussieu *Figwort Family*

62 genera, 1,830 species

Widely distributed from temperate to tropical regions, especially diverse in Africa.

Placement:

	B & H	Cronquist	Takhtajan	Dahlgren	Thorne	APG IV/(APweb)
Informal Group						Asterids / (Euasterids/ Lamiid/Asterid I)
Division		Magnoliophyta	Magnoliophyta			
Class	Dicotyledons	Magnoliopsida	Magnoliopsida	Magnoliopsida	Magnoliopsida	
Subclass	Gamopetalae	Asteridae	Lamiidae	Magnoliidae	Lamiidae	
Series+/ Superorder	Bicarpellatae+		Lamianae	Lamianae	Lamianae	
Order	Personales	Scrophulariales	Lamiales	Lamiales	Lamiales	Lamiales

Salient features: Leaves alternate or opposite, stipules absent, flowers zygomorphic, anther commonly opening by single slit, carpels 2, ovary superior, 2-chambered, ovules many, fruit a capsule.

Major genera: *Verbascum* (360 species), *Scrophularia* (230), *Selago* (150), *Sutera* (140), *Buddleja* (100), *Manulea* (55) and *Nuxia* (30).

Description: Herbs or small shrubs (*Buddleja*), often with Iridoids, hairs usually simple, when glandular with short discoid head composed of many cells and with vertical partitions. **Leaves** alternate or opposite, rarely whorled, simple, entire or dentate, venation pinnate, reticulate, stipules absent. **Inflorescence** racemose: racemes or spikes. **Flowers** bisexual, zygomorphic, or almost actinomorphic, hypogynous. **Calyx** with 3–5 sepals, connate, persistent. **Corolla** with 4–5 petals, united, usually bilabiate, or with narrow tube broadening upwards, sometimes with nectar sac or spur, imbricate. **Androecium** with usually 5 stamens, rarely 4 or 2, epipetalous, inserted in corolla tube, filaments free, sometimes hairy (*Verbascum*), anther bilocular, anther sac confluent and opening by single slit right angles to filament, anther base not sagittate, pollen grains tricolporate. **Gynoecium** with 2 united carpels, ovary superior, bilocular, ovules several to 1 (*Selago*) in each chamber, axile placentation, style 1, stigma bilobed, ovary seated on a nectary. **Fruit** a septicidal capsule, or schizocarp with two nutlets (*Selago*); seed with curved or straight embryo, endosperm present. Pollination by insects. Seeds or nutlets dispersed by wind.

Economic importance: The family has little economic importance. *Verbascum* is sometimes grown as ornamental. *Buddleja* and *Nuxia* are commonly grown as ornamentals.

Phylogeny: The family is closely related to Acanthaceae sharing the features of zygomorphic flowers, pentamerous flowers, stamens less than 5, bicarpellate superior ovary and fruit a capsule. Scrophulariaceae are distinct in presence of endosperm, anthers opening by single slit and the absence of retinaculum. The genera including *Veronica, Linaria, Antirrhinum, Digitalis,* etc., formerly included under Scrophulariaceae have been variously separated under Antirrhinaceae (Thorne, 1999, 2000) or Plantaginaceae (Judd et al., 2002; APG II and APweb; Thorne, 2006). The Budlejaceae and Selaginaceae have been merged with Scrophulariaceae in these systems. Monophyly of Scrophulariaceae is clearly supported by morphology, rbcL and ndhF sequences (Olmstead and Reeves, 1995). *Verbascum* and *Scrophularia* form a clade characterised by hairy

Figure 13.122: Scrophulariaceae. *Scrophularia elatior*. **A:** Portion of plant with terminal inflorescence; **B:** Flower with long-exserted stamens and style. *Verbascum chinense*. **C:** Lower part of plant with basal and lower cauline leaves; **D:** Upper part of inflorescence; **E:** Flower; **F:** Corolla spread to show epipetalous stamens; **G:** Stamen with glandular hairy filament; **H:** Flower after removal of corolla and one lobe of calyx to show gynoecium; **I:** Capsule with persistent calyx.

filaments, endosperm development and distinctive seeds and are sister to rest of genera. *Selago* and relatives (former Selaginaceae) form a clade based on uniovulate locules and achene-like fruits. *Budleja* is very much paraphyletic, but several lines of evidence place it here (Maldonado de Magnano, 1986b); *Teedia* and *Oftia* have strong support as the sister group to *Budleja* s. l. (Wallick et al., 2001, 2002).

Oxelman et al. (2005) recognized following 8 tribes within Scrophulariaceae: Aptosimeae, Buddlejeae, Hemimerideae, Leucophylleae, Limoselleae, Myoporeae, Scrophularieae and Teedieae, genus regarded *Camptoloma* is sister to other Buddlejeae. Tank et al., while following above scheme found three genera *Androya* (previously under placed separately under Myoporae in APweb, 2018), *Camptoloma* and *Phygelius,* falling outside these tribal limits. Myoporeae and Leucophylleae are clearly sister taxa and the combined clade has synapomorphies, except the problem with *Androya* (APweb, 2018).

Acanthaceae A. L. de Jussieu *Acanthus Family*

210 genera, 4,000 species

Cosmopolitan in distribution, mainly in tropics and warm temperate regions.

Placement:

	B & H	Cronquist	Takhtajan	Dahlgren	Thorne	APG IV/(APweb)
Informal Group						Asterids/ (Euasterids/ Lamiid/Asterid I)
Division		Magnoliophyta	Magnoliophyta			
Class	Dicotyledons	Magnoliopsida	Magnoliopsida	Magnoliopsida	Magnoliopsida	
Subclass	Gamopetalae	Asteridae	Lamiidae	Magnoliidae	Lamiidae	
Series+/ Superorder	Bicarpellatae+		Lamianae	Lamianae	Lamianae	
Order	Personales	Scrophulariales	Lamiales	Lamiales	Lamiales	Lamiales

Salient features: Leaves opposite, stipules absent, flowers zygomorphic, with prominent bracts and bracteoles, stamens 2–4, anther lobes unequal in size, carpels 2, ovary superior, 2-chambered, ovules 4 or more, fruit a capsule, seeds with jaculators.

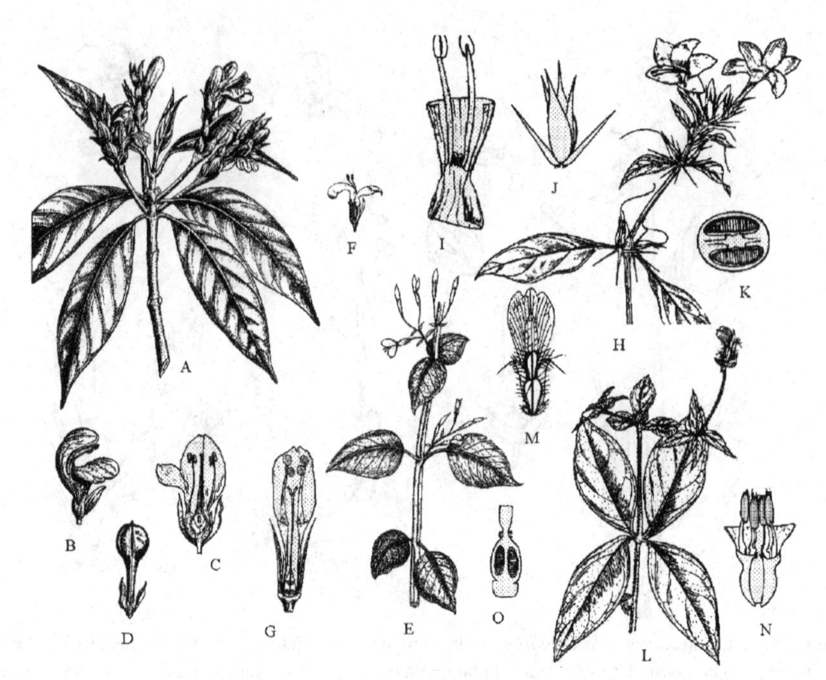

Figure 13.123: Acanthaceae. *Adhatoda vasica.* **A:** Branch with flowers in spikes; **B:** Flower with bilabiate corolla; **C:** Longitudinal section of flower showing 2 epipetalous stamens; **D:** Capsule with persistent calyx. *Peristrophe bicalyculata.* **E:** Branch with flowers; **F:** Flower with bilabiate corolla; **G:** Vertical section of flower. *Barleria prionitis.* **H:** portion of branch with spines at nodes and axillary clusters of flowers; **I:** Corolla tube opened to show stamens, corolla limb cut away; **J:** Spiny calyx and bracteoles; **K:** Transverse section of ovary. *Blepharis maderaspatensis.* **L:** Portion of branch with flowers; **M:** Flower; **N:** Corolla tube opened to show epipetalous stamens, corolla limb partly cut away; **O:** Longitudinal section of gynoecium.

Major genera: *Justicia* (400), *Beloprone* (300), *Barleria* (240), *Strobilanthus* (230), *Ruellia* (190), *Dicliptera* (140), *Thunbergia* (140) and *Adhatoda* (20).

Description: Herbs or shrubs (*Adhatoda*), sometimes small trees (*Strobilanthus*) or lianas (*Thunbergia*), a rarely aquatic herbs (*Cardentha*), sometimes spiny (*Barleria*), usually with anomalous secondary growth, often with Iridoids, alkaloids and diterpenoids, cystoliths often present, nodes unilacunar, vessels with simple end-walls. **Leaves** opposite, rarely alternate (*Nelsonia, Elytraria*), simple, entire or dentate, stipules absent. **Inflorescence** cymose, racemose (usually spike) or of solitary flowers (*Bontia*). **Flowers** bisexual, zygomorphic, hypogynous, with prominent bracts and bracteoles. **Calyx** with 4 (*Acanthus*) to 5 (*Adhatoda*) sepals, free or united. **Corolla** with 5 petals, united, usually bilabiate, sometimes nearly regular (*Acanthus*). **Androecium** with 2 (*Adhatoda*) or 4 (*Acanthus, Ruellia*) didynamous stamens, rarely 5 (*Pentstemonacanthus*), epipetalous, inserted in corolla tube, filaments free, anther lobes unequal in size, sometimes one lobe aborted, dehiscence longitudinal, tapetum glandular, pollen grains 2–8 aperturate or colpate. **Gynoecium** with 2 united carpels, ovary superior, bilocular, ovules 2 in each chamber, axile placentation, style 1, stigma bilobed, ovary seated on a nectary. **Fruit** a loculicidal capsule; seed with jaculator or retinaculum a hook-shaped projection of funiculus, embryo large, curved or straight, endosperm absent.

Economic importance: The family includes many ornamentals such as *Barleria, Thunbergia, Pachystachys, Eranthemum* and *Acanthus*. Extract from *Adhatoda vasica* is component of some cough syrups. An extract from the boiled leaves of *Acanthus ebracteatus* (sea holy) is used as a cough medicine in parts of Malaya, whereas the roots of *A. mollis* (bear's breech) are used to treat diarrhoea in some parts of Europe.

Phylogeny: The family is closely related to Scrophulariaceae sharing the features of zygomorphic flowers, pentamerous flowers, stamens less than 5, bicarpellate superior ovary and fruit a capsule. The Acanthaceae are distinct in absence of endosperm, anthers opening by two slits and the presence of retinaculum. Thorne had earlier (1999, 2000) recognized 5 subfamilies: Nelsonioideae, Thunbergioideae, Mendoncioideae, Acanthoideae and Ruellioideae. The first two include aberrant genera. Nelsonoideae with sometimes alternate leaves, presence of endosperm and absence of retinacula may represent a paraphyletic basal group within the family. Nelsonioideae have often been Placed in Scrophulariaceae s.l. or considered 'intermediate' between Scrophulariaceae and Acanthaceae, but they are placed sister to rest of Acanthaceae s.l. in Hedren et al. (1995). According to Scotland and Vollesen (2000), the absence of retinacula or cystoliths, descending cochlear aestivation (i.e., the adaxial petals overlapping the abaxial petals in bud)—are likely to be plesiomorphies. Acanthoideae are clearly monophyletic (Scotland, 1990) and characterized by the absence of cystoliths, nodes not swollen, colpate pollen and monothecous anthers. In Mendoncioideae, one of the carpels is often aborted, fruit is a drupe and style bifid. Mendoncioideae and Ruellioideae have subsequently been merged under Thunbergioideae and Acanthoideae, respectively (APweb and Thorne, 2007). APweb includes Avicennioideae as fourth subfamily, stressing that the position of Avicenniaceae within Acanthaceae s.l. is fairly well established; it shows a rather weakly supported sister group relationship with Thunbergioideae (Schwarzbach and McDade, 2002). This placement based on molecular evidence is also supported by articulated nodes, inflorescence structure, flowers with bract and 2 bracteoles, a reduction in number of ovules and absence of endosperm (Judd et al., 2002). Thorne (2000, 2003), who earlier treated Avicenniaceae as a distinct family, has finally (2007) relegated it to subfamily level like APweb. APweb (2018) recognized 4 subfamilies Nelsonioideae, Acanthoideae (7 tribes Acantheae, Ruellieae, Justicieae, Barlerieae, Andrographidae, Whitfieldieae, Nemacanthus following McDade et al. (2008)), Thunbergioideae and Avicennioideae. Within Nelsonioideae, Nelsonia and Elytraria are successively sister to the rest of the subfamily (McDade et al., 2008).

Bignoniaceae A. L. de Jussieu ***Trumpet Creeper Family***

113 genera, 800 species (excluding *Paulownia*)

Widely distributed in tropical and subtropical regions, a few species in temperate regions, most diverse in northern South America from temperate to tropical regions, especially diverse in Africa.

Placement:

	B & H	**Cronquist**	**Takhtajan**	**Dahlgren**	**Thorne**	**APG IV/(APweb)**
Informal Group						Asterids/ (Euasterids/ Lamiid/Asterid I)
Division		Magnoliophyta	Magnoliophyta			
Class	Dicotyledons	Magnoliopsida	Magnoliopsida	Magnoliopsida	Magnoliopsida	
Subclass	Gamopetalae	Asteridae	Lamiidae	Magnoliidae	Lamiidae	
Series+/ Superorder	Bicarpellatae+		Lamianae	Lamianae	Lamianae	
Order	Personales	Scrophulariales	Lamiales	Lamiales	Lamiales	Lamiales

Salient features: Usually woody lianas or trees, leaves usually opposite, often compound, sometimes with tendrils, stipules absent, nectaries on leaves, flowers zygomorphic, showy, stamens 4, carpels 2, ovary superior, 2-chambered, ovules many, fruit a woody capsule, seeds often winged.

Major genera: *Tabebuia* (100 species), *Arrabidaea* (70), *Adenocalyma* (45), *Jacaranda* (40), *Spathodia* (20), *Catalpa* (11), *Campsis* (2) and *Kigelia* (1).

Description: Shrubs, trees or lianas (*Bignonia*, *Campsis*), lianas often with characteristic secondary growth resulting in lobed or furrowed xylem cylinder, usually with iridoids and phenolic glycosides. **Leaves** usually opposite or whorled, pinnately or palmately compound, sometimes simple (*Catalpa*), venation pinnate to palmate, reticulate, some leaflets often modified into tendrils, stipules absent but glands often present at the base of petiole. **Inflorescence** cymose, raceme or panicle, rarely solitary. **Androecium** with usually 4 stamens, fifth represented by a staminode, rarely 5 (*Oroxylum*) or 2 (*Catalpa*), epipetalous, inserted in corolla tube, filaments free, anther bithecous, sagittate, dehiscence longitudinal, pollen grains sometimes in tetrads or polyads. **Gynoecium** with 2 united carpels, ovary superior, bilocular with axile placentation, rarely unilocular with free-central placentation, ovules many, anatropous, style short, stigma with unequal lobes. **Fruit** a woody capsule, occasionally a berry **Flowers** bisexual, zygomorphic, hypogynous, usually showy. **Calyx** with 5 sepals, connate. **Corolla** with 5 petals, united, showy, usually bilabiate, sometimes with sac or spur on the lower lip, imbricate or pod; seed winged or fringed with hairs, endosperm absent, cotyledons deeply bilobed. Pollination by insects. Seeds dispersed by wind.

Economic importance: The family contributes several ornamentals such as *Spathodia* (African tulip tree), *Kigelia* (sausage tree), *Tabebuia* (poui, gold tree), *Crescentia* (calabash tree) and *Tecoma*. Common climbers include *Bignonia* (cross vine), *Campsis, Tecomaria* (Cape honeysuckle) and *Pyrostegia* (flame vine). *Tabebuia* and *Catalpa* are exploited as timbers, mostly for fence posts.

Phylogeny: The family is closely related to Scrophulariaceae sharing the features of zygomorphic flowers, pentamerous flowers, stamens less than 5, bicarpellate superior ovary and fruit a capsule. The family is monophyletic as evidenced by morphology. Pinnate compound leaves are considered to be ancestral. The genus *Paulownia* and *Schlegelia* often included

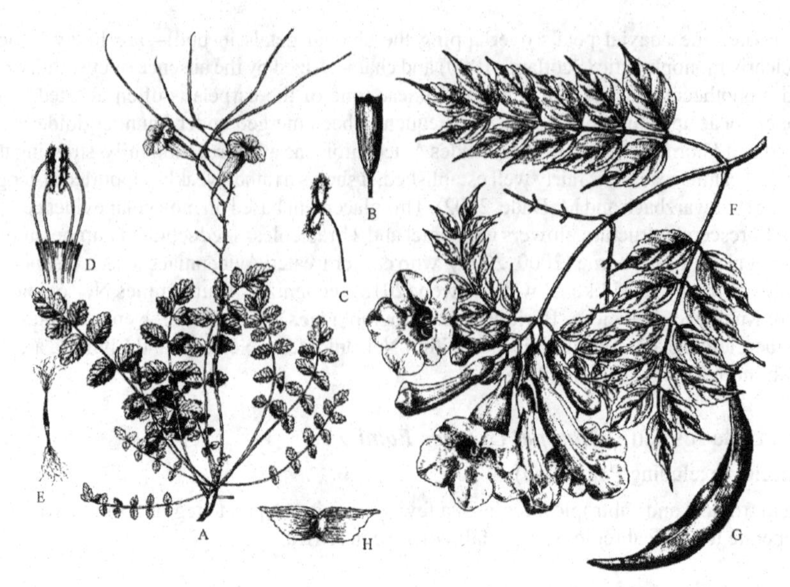

Figure 13.124: Bignoniaceae. *Incarvillea emodi.* **A:** Plant with terminal raceme and long linear capsules; **B:** Calyx with minute lobes; **C:** Stamen with arched filaments and spreading villous anther lobes; **D:** Portion of corolla spread to show stamens; **E:** Seed, linear and fibrillate at both ends. *Campsis radicans.* **F:** Branch with flowers; **G:** Fruit; **H:** Winged seed.

in Bignoniaceae are intermediate between this family and the Scrophulariaceae and as such treated under distinct families Paulowniaceae and Schlegeliaceae respectively by Thorne, APG II, APG IV (2016) and APweb. *Paulownia* is superficially like *Catalpa,* but it has endosperm and lacks the ovary and seed anatomy of Bignoniaceae (Armstrong, 1985; Manning, 2000). Olmstead et al. (2009) divided Bignoniaceae into 8 tribes based on molecular studies: Bignonieae, Catalpeae, Coleeae, Crescentieae, Jacarandeae, Oroxyleae, Tecomeae and Tourrettieae, treatment followed in APweb (2018), though several genera are unassigned. Phylogenetic position of *Argylia* and *Delostoma* is ambiguous. *Delostoma* may be sister to the [Bignonieae [[Catalpeae + Oroxyleae] [Crescentieae + Coleeae]]] clade (Pace et al., 2015).

Verbenaceae Jaume St.-Hilaire ***Verbena Family***

32 genera, 1,000 species

Widely distributed, mainly in tropical regions, also in temperate regions, prominent in new world.

Placement:

	B & H	Cronquist	Takhtajan	Dahlgren	Thorne	APG IV/(APweb)
Informal Group						Asterids/ (Euasterids/ Lamiid/Asterid I)
Division		Magnoliophyta	Magnoliophyta			
Class	Dicotyledons	Magnoliopsida	Magnoliopsida	Magnoliopsida	Magnoliopsida	
Subclass	Gamopetalae	Asteridae	Lamiidae	Magnoliidae	Lamiidae	
Series+/ Superorder	Bicarpellatae+		Lamianae	Lamianae	Lamianae	
Order	Lamiales	Lamiales	Lamiales	Lamiales	Lamiales	Lamiales

Salient features: Plants aromatic, leaves opposite, serrate, stem often angular, nonglandular hairs if present unicellular, flowers zygomorphic, in racemes, spikes or heads, pollen exine thickened near apertures, style simple with bilobed stigma, stigmatic area conspicuously swollen and glandular, ovary with four ovules, ovules attached to the margin of false septa.

Major genera: *Verbena* (200), *Lippia* (180), *Lantana* (140), *Citharexylum* (65), *Glandularia* (55), *Duranta* (28) and *Phyla* (10).

Description: Aromatic herbs (*Lippia*), shrubs (*Lantana*), sometimes trees, rarely lianas, sometimes with prickles or thorns, stem usually 4-angled, often with iridoids and phenolic glycosides, usually with glandular hairs, nonglandular hairs if

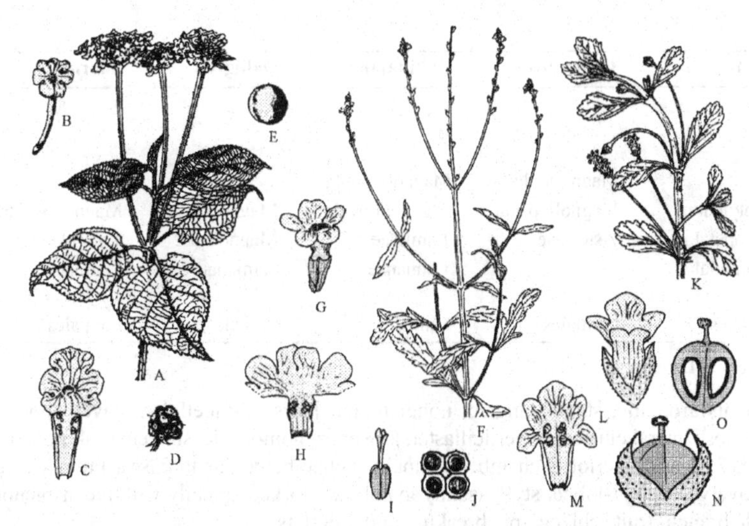

Figure 13.125: Verbenaceae. *Lantana camara.* **A:** Branch with ovoid compact inflorescences on long peduncles; **B:** Flower with long corolla tube and zygomorphic limb; **C:** Corolla spread out to show epipetalous stamens; **D:** Cluster of fruits; **E:** Fruit. *Verbena officinalis.* **F:** Plant with terminal spikes; **G:** Corolla with shorter broader tube and zygomorphic limb; **H:** Corolla spread out to show epipetalous stamens; **I:** Gynoecium; **J:** Transverse section of ovary with 4 one-seeded chambers. *Phyla nodiflora.* **K:** Portion of plant with pedunculate globose inflorescences; **L:** Flower with short broad tube and zygomorphic limb; **M:** Corolla spread out to show epipetalous stamens; **N:** Fruit with persistent calyx; **O:** Longitudinal section of fruit.

present unicellular. **Leaves** opposite, sometimes whorled, simple or sometimes lobed, usually aromatic, entire to serrate, stipules absent. **Inflorescence** racemose: racemes, spikes or heads. **Flowers** bisexual, zygomorphic, hypogynous. **Calyx** with 5 sepals, united, tubular to campanulate, persistent, sometimes enlarged in fruit. **Corolla** with 5 petals, sometimes appearing 4 due to fusion of two posterior petals, united, weakly bilabiate, lobes imbricate. **Androecium** with 4 stamens, epipetalous, didynamous, inserted in corolla tube, filaments free, dehiscence longitudinal, pollen grains tricolpate, exine thickened near apertures. **Gynoecium** with 2 united carpels, ovary superior, bilocular, ovules 2 in each chamber, finally 4-locular due to false septum with 1 ovule in each chamber, unitegmic, axile placentation, ovary not or slightly 4-lobed, style 1, terminal, style simple with bilobed stigma, stigmatic area conspicuously swollen and glandular, ovary seated on a nectary disc. **Fruit** a drupe with 2 or 4 pits, or schizocarp splitting into 2 or 4 nutlets; seed with straight embryo, endosperm absent. Pollination by insects. Dispersal by birds, wind or water.

Economic importance: The family contributes some ornamentals such as *Verbena*, *Lantana*, *Duranta*, and *Glandularia*. *Lippia* (lemon verbena) and *Privea* are used as herbal teas or yield essential oils. *Verbena officinalis* (vervain) is used for a number of herbal remedies including treatment of skin diseases.

Phylogeny: The family is closely related to Lamiaceae. The circumscription of the family has undergone considerable revision with several genera (nearly two-thirds) from older Verbenaceae such as *Clerodendrum*, *Callicarpa*, *Vitex* and *Tectona* transferred to Lamiaceae (Judd et al., 2002; Thorne, 2000, 2003; APG II, APweb). The family is now circumscribed to include only subfamily Verbenoideae. The traditionally delimited Verbenaceae are paraphyletic and Lamiaceae polyphyletic. With narrowly defined Verbenaceae and broadly defined Lamiaceae, both become monophyletic. The family is distinguished from Lamiaceae in racemose inflorescence, ovules attached on margins of false septa, style simple with conspicuous bilobed stigma, pollen exine thickened near apertures, hairs unicellular, weakly bilabiate corolla and usually terminal style. *Phryma* (Phrymaceae) with one carpel aborted and ovary with single basal ovule may be closely related to Verbenaceae (Chadwell et al., 1992). *Avicennia* often included in distinct family or broadly circumscribed Verbenaceae is more appropriately included in Acanthaceae (APweb). Thorne (2003) treats Phrymaceae and Avicenniaceae as distinct families. APWeb (2018) while retaining Phrymaceae, merge Avicenniaceae with Acanthaceae recognizing as subfamily Avicennioideae.

Lamiaceae Martinov *Mint Family*
(= Labiatae A. L. de Jussieu)

241 genera, 7,530 species

Worldwide in distribution, largely concentrated in the Mediterranean Region.

Placement:

	B & H	**Cronquist**	**Takhtajan**	**Dahlgren**	**Thorne**	**APG IV/(APweb)**
Informal Group						Asterids/ (Euasterids/ Lamiid/Asterid I)
Division		Magnoliophyta	Magnoliophyta			
Class	Dicotyledons	Magnoliopsida	Magnoliopsida	Magnoliopsida	Magnoliopsida	
Subclass	Gamopetalae	Asteridae	Lamiidae	Magnoliidae	Lamiidae	
Series+/ Superorder	Bicarpellatae+		Lamianae	Lamianae	Lamianae	
Order	Lamiales	Lamiales	Lamiales	Lamiales	Lamiales	Lamiales

B & H as Labiatae, others as Lamiaceae.

Salient features: Plants aromatic, stem 4-angled, nonglandular hairs multicellular, leaves opposite, stipules absent, inflorescence with cymose lateral clusters or verticillaster, flowers zygomorphic, stamens 2–4, pollen exine not thickened near apertures, carpels 2, ovary superior, 2-chambered, finally 4 chambered due to false septum, ovules 4, attached to the sides of false septa, ovary deeply 4-lobed, style usually gynobasic, forked apically with inconspicuous stigmatic region at the tip of each style branch, fruit schizocarpic breaking into 4 nutlets.

Major genera: *Salvia* (700 species), *Clerodendrum* (400), *Thymus* (340), *Plectranthus* (300), *Scutellaria* (300), *Stachys* (300), *Nepeta* (260), *Teucrium* (200), *Callicarpa* (150), *Ocimum* (150), *Lamium* (50), *Marrubium* (40), *Mentha* (30), *Lavandula* (30) and *Tectona* (3).

Description: Aromatic herbs or shrubs (*Rosmarinus*, *Teucrium*), sometimes small (*Hyptis*) or large (*Tectona*) trees, rarely climbers (*Scutellaria*), stem 4-angled, often with iridoids, phenolic glycosides, sometimes with suckers (*Mentha*) or stolons (*Ajuga*), sometimes green and assimilatory (*Hedeoma*), usually with glandular hairs, nonglandular hairs when present multicellular. **Leaves** opposite (rarely alternate), simple or pinnate compound, usually aromatic, sometimes reduced (*Hedeoma*), stipules absent. **Inflorescence** verticillaster [two opposite whorls (verticels) of cymose clusters initially biparous and subsequently uniparous], arranged in raceme, spike or panicle. **Flowers** bracteate (*Coleus*) or ebracteate (*Salvia*), bisexual, zygomorphic, hypogynous, often bilabiate. **Calyx** with 5 sepals, united, often bilabiate 1/4 (*Ocimum*) or 3/2 (*Salvia*), persistent. **Corolla** with 5 petals, united, usually bilabiate 4/1 (*Ocimum*) or 2/3 (*Salvia*), upper lip sometimes absent (*Ajuga*), rarely corolla 4-lobed (*Pogostemon*). **Androecium** with 2 (*Salvia*) to 4 (*Ocimum*) stamens, epipetalous, usually didynamous, inserted in corolla tube, sometimes with turn-pipe mechanism (lever mechanism) as in *Salvia* (anther lobes separated by a long connective and swinging like a lever, one anther-lobe sterile, another fertile), filaments free, dehiscence longitudinal, pollen grains tricolpate or 6-colpate. **Gynoecium** with 2 united carpels, ovary superior, bilocular, ovules 2 in each chamber, finally 4-locular due to false septa with 1 ovule in each chamber, anatropous, axile placentation, ovules attached to the sides of the false septa, ovary 4-lobed, style 1, gynobasic, rarely terminal (*Ajuga*), forked apically with inconspicuous stigmatic region at the tip of each style branch, ovary seated on a nectary disc. **Fruit** a schizocarp (carcerulus) splitting into 4 nutlets or a drupe or indehiscent 4-seeded pod; seed with straight embryo, endosperm minute or absent. Pollination by insects, lower lip providing landing platform. Dispersal by birds, wind or water.

Economic importance: The family includes several plants used in cooking and flavouring such as spearmint (*Mentha spicata*), peppermint (*M. piperita*) thyme (*Thymus vulgaris*), sweet basil 'niazbo' (*Ocimum basilicum*), pot marjoram (*Origanum vulgare*) and sage (*Salvia officinalis*). The family is also source of popular perfumes such as lavender (*Lavandula angustifolia*) and rosemary (*Rosmarinus officinalis*). Holy basil "Tulsi" (*Ocimum tenuiflorum* (syn: *O. sanctum*)) is sacred in India. Common ornamentals include sage (*Salvia*), horsemint (*Monarda*), *Molucella*, *Clerodendrum* and *Coleus*. The tubers of few species of *Stachys* are edible. Teak (*Tectona grandis*) is valuable timber, known for its hard and durable wood and extensively cultivated in India and Burma.

Phylogeny: The family Lamiaceae is generally considered to be one of the most highly evolved of all dicotyledonous families, and closely related to Vebenaceae. The circumscription of the family has undergone considerable revision with several genera (nearly two-thirds) from older Verbenaceae such as *Clerodendrum*, *Callicarpa*, *Vitex* and *Tectona* transferred to Lamiaceae (Judd et al., 2002; Thorne, 2000, 2003, 2007; APG III, APG IV, APweb). The family Lamiaceae is distinguished from Verbenaceae in cymose lateral whorls, ovules attached on sides of false septa, bilobed style with small stigmatic region, pollen exine not thickened near apertures, hairs multicellular, strongly bilabiate corolla and usually gynobasic style. According to Wagstaff et al., 1998 the following 5 clades (subfamilies) are distinct: Nepetoideae (pollen trinucleate, hexacolpate, style gynobasic; myxocarpy; endosperm absent, embryo investing), Lamioideae (laballenic acid in seed oils, embryo sac with micropylar lobe longer and broader than chalazal lobe, style gynobasic), Pogostemonoideae (stamens 4, about the same length), Scutellarioideae (style bilabiate, with rounded lips; seeds tuberculate), and Teucrioideae

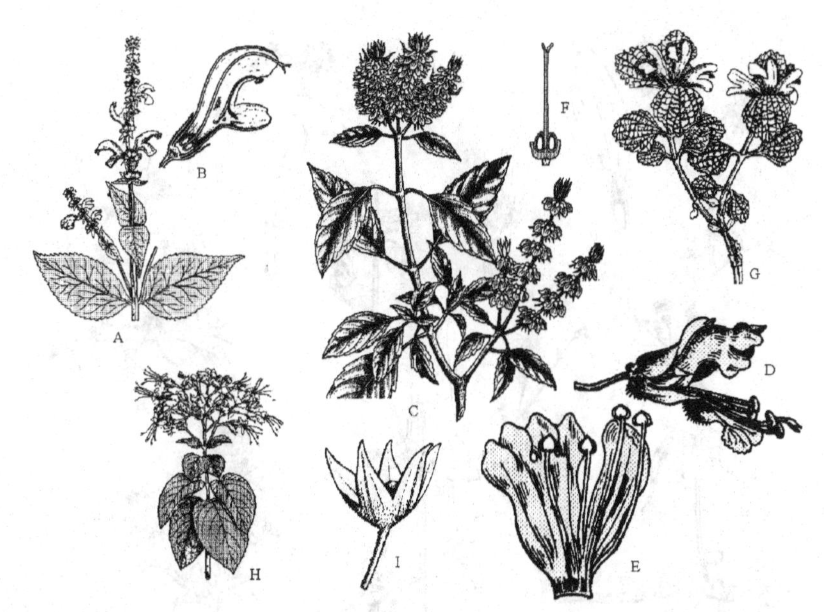

Figure 13.126: Lamiaceae. *Salvia splendens.* **A:** Branch with axillary and terminal inflorescences; **B:** Vertical section of flower to show bilabiate corolla and stamen with turn-pipe mechanism, gynobasic style and nectary below ovary. *Ocimum basilicum.* **C:** Portion of plant with inflorescences; **D:** Flower with bilabiate corolla and 4 didynamous stamens; **E:** Corolla spread out to show epipetalous stamens of 2 sizes; **F:** Gynoecium with gynobasic style and bifid stigma, nectary below the ovary. **G:** *Lamium rhomboideum*, plant with inflorescences. *Clerodendrum viscosum.* **H:** Branch with terminal spreading inflorescence; **I:** Fruit with persistent calyx.

[(inc. Ajugoideae) exine with branched to granular columellae]. Bootstrap support for the family as circumscribed is 100 per cent (Wagstaff et al., 1998). Thorne (2000, 2003, 2007) adds two more subfamilies Symphorematoideae and Prostantheroideae, establishing Ajugoideae (instead of Teucrioideae) thus recognizing a total of seven subfamilies. APWeb (2018) recognizes 12 subfamilies adding Callicarpoideae, Viticoideae, Tectonoideae, Premnoideae, Peronematoideae and Cymaroideae, merging Pogostemonoideae.

Asteraceae Martinov *Sunflower or Aster Family* (= Compositae Giseke)

1,623 genera, 24,700 species (second largest family of flowering plants after Orchidaceae with 28000 species)

Worldwide in distribution mainly in temperate and subtropical climates, mainly in mountain regions, also common in tropics.

Placement:

	B & H	Cronquist	Takhtajan	Dahlgren	Thorne	APG IV/(APweb)
Informal Group						Asterids/ (Campanulid/ Asterid II)
Division		Magnoliophyta	Magnoliophyta			
Class	Dicotyledons	Magnoliopsida	Magnoliopsida	Magnoliopsida	Magnoliopsida	
Subclass	Gamopetalae	Asteridae	Asteridae	Magnoliidae	Asteridae	
Series+/ Superorder	Inferae+		Asteranae	Asteranae	Asteranae	
Order	Asterales	Asterales	Asterales	Asterales	Asterales	Asterales

B & H as Compositae, others as Asteraceae.

Salient features: Usually herbs, lacking Iridoids, leaves usually alternate, stipules absent, inflorescence a capitulum with ray florets and disc florets (one type or both in a head), surrounded by involucre bracts (phyllaries), calyx represented by pappus, anthers united forming a cylinder around style, style with two branches, fruit a cypsela (commonly called achene, although typical achene is formed from single carpel and superior ovary), ovary inferior.

Figure 13.127: Asteraceae. *Helianthus annuus.* **A:** Portion of plant with inflorescences, the capitulum with both ray florets and disc florets (radiate head); **B:** Vertical section of ray floret lacking androecium; **C:** Vertical section of disc floret. *Ageratum houstonianum.* **D:** Portion of plant with capitula in clusters, each with only disc florets (discoid head); **E:** Vertical section of disc floret; **F:** Achene with pappus consisting of 5 scales. **G:** Plant of *Sonchus oleraceous* with auricled leaves and capitula with only ray florets (ligulate head). **H:** Plant of *Launaea nudicaulis* with ligulate heads. *Carthamus lanatus.* **I:** Portion of plant with spinose leaves and discoid heads; **J:** Capitulum with spiny involucre bracts. *Bidens chinensis.* **K:** Lower part of plant with pinnate leaves; **L:** Upper part with one flowering and one fruiting capitulum; **M:** Ray floret with three-toothed corolla; **N:** Disc floret; **O:** Disc floret with corolla partly removed to show androecium; **P:** Achene.

Major genera: *Senecio* (1470 species), *Vernonia* (1050), *Cousinia* (600), *Eupatorium* (590), *Centaurea* (590), *Hieracium* (470), *Helichrysum* (460), *Saussurea* (300), *Cirsium* (270), *Aster* (240), *Bidens* (210), *Chrysanthemum* (200), *Crepis* (200), *Inula* (200), *Gnaphalium* (140), *Solidago* (110), *Helianthus* (100), *Carduus* (90), *Lactuca* (90), *Taraxacum* (80), *Tragopogon* (70), *Sonchus* (50) and *Calendula* (30).

Description: Usually herbs or shrubs, rarely trees (*Vernonia arborea*; *Leucomeris*) or lianas (*Vernonia scandens*), sometimes producing tubers (*Dahlia*, *Helianthus tuberosus*), usually storing inulin, laticifers usually present, rarely lacking, terpenoids usually present, usually sesqueterpene lactones, Iridoids absent. **Leaves** usually alternate and simple, sometimes compound (*Dahlia, Artemisia*), rarely opposite (*Dahlia*) or whorled, stipules absent. **Inflorescence** a capitulum with broad receptacle containing disc florets (discoid head—*Ageratum*, *Vernonia*), ray florets (ligulate head—*Sonchus*, *Launaea*) or both type of florets with latter towards the periphery (radiate head—*Helianthus*, *Aster*), all types of heads having florets surrounded by involucre bracts (phyllaries), rarely capitulum with single floret (*Echinops*) with capitula arranged into globose heads. Flowers bisexual (usually disc florets and ray florets of a ligulate head) or unisexual (commonly ray florets in a ligulate head, which may even be sterile), actinomorphic (usually disc florets) or zygomorphic (usually ray florets), epigynous. **Calyx** absent or represented by pappus in the form of scales (*Helianthus*), bristles (*Bidens*), simple hairs (*Sonchus*) or plumose (*Carduus*). **Corolla** with 5 petals, united, tubular and 5-lobed (disc floret) or ligulate with 3–5 teeth (ray floret: sometimes also bilabiate). **Androecium** with 5 stamens with free filaments and united anthers (syngenesious) forming a tube around the style, epipetalous, anthers bithecous, dehiscence longitudinal. **Gynoecium** with 2 united carpels, unilocular a single ovule, placentation basal, ovary inferior, style with two branches. Fruit a cypsela (often called achene which typically, however, is formed from single carpel with superior ovary) usually with pappus at tip. Seeds 1, embryo straight, endosperm usually absent.

Economic importance: Compared to the number of species included, the family is of lesser economic importance. Common valuable ornamentals include species of *Aster, Dahlia, Chrysanthemum, Gerbera, Helichrysum, Tagetus* and *Zinnia*. A few food plants include *Lactuca* (lettuce), *Cynara* (artichoke), *Helianthus* (sunflower oil), and *Cichorium* (chicory, added to coffee). Safflower a red dye is obtained from *Carthamus tinctorius*. *Chrysanthemum cinerariefolium* is the source of natural insecticide pyrethrum.

Phylogeny: Interestingly Asteraceae in spite of huge size form a well-defined clade, easily recognizable and evidently monophyletic. The family is often considered related to Rubiaceae, Caprifoliaceae, Dipsacaceae, Valerianaceae, Campanulaceae and a few others. The first four are basically cymose and also differ in biochemical features. Stylidiaceae, Goodeniaceae and Brunoniaceae resemble the Asteraceae in being mostly racemose and in possessing inulin but differ in biochemical features of taxonomic significance. Recent molecular studies (Bremer et al., 2002; Lundberg and Bremer, 2002) indicate that Asteraceae, Calyceraceae, Goodeniaceae and their sister group Menyanthaceae form a monophyletic group. All the four families are placed under Asterales by Thorne (1999, 2003). The relationships between the first three families are not very clear. The *rbcL* and *ndhF* (Kårehed et al., 1999) and *ndhF* data (Olmstead et al., 2000) support Asteraceae and Calyceraceae as sister families whereas *rbcL* together with *atpB* and 18S *r*DNA (Soltis et al., 2000) support Goodeniaceae and Calyceraceae as sister taxa. A combination of morphological data, and *rbcL, ndhF* and *atpB* sequences provided a strong support for Calyceraceae and Asteraceae as sister groups (Lundberg and Bremer, 2002). Similar conclusion was also reached by the analysis of six DNA regions (Bremer et al., 2002).

The family Asteraceae is usually divided into three subfamilies: Barnadesioideae (style papillate, stigma lobed; cypsela with spines; lacks chloroplast DNA inversion found in other two subfamilies), Cichorioideae (Latex present, style branches long with inner surface stigmatic, acute; those with ray florets often separated into a distinct subfamily Lactucoideae) and Asteroideae (latex absent, both disc and ray florets). Thorne (2006, 2007) recognizes Carduoideae including Cichorioideae and Lactucoideae. He subdivides these three subfamilies further to include a total of 25 tribes in Asteraceae. Heywood had earlier (1978) recognized 17 tribes under two subfamilies Lactucoideae and Asteroideae. APweb (2003) recognized 11 subfamilies including one undefined 'The Stifftia group', the number increased to 13 and establishing Stifftioideae in 2018 version.

Adoxaceae E. Meyer *Elderberry Family*

5 genera, 225 species (Inc. Sambucaceae Borkh)

Distributed mainly in northern temperate region, a few in montane tropical and subtropics.

Placement:

	B & H	Cronquist	Takhtajan	Dahlgren	Thorne	APG IV/(APweb)
Informal Group						Asterids/ (Campanulid/ Asterid II)
Division		Magnoliophyta	Magnoliophyta			
Class	Dicotyledons	Magnoliopsida	Magnoliopsida	Magnoliopsida	Magnoliopsida	
Subclass	Gamopetalae	Asteridae	Rosidae	Magnoliidae	Asteridae	
Series+/ Superorder	Inferae+		Rhamnanae	Cornanae	Aralianae	
Order	Rubiales	Dipsacales	Dipsacales	Cornales	Dipsacales	Dipsacales

B & H under Caprifoliaceae.

Salient features: Mainly shrubs or lianas, leaves opposite, usually simple, flowers small, sepals with 1 trace, seed single.

Major genera: *Viburnum* (200 species), *Sambucus* (22), *Adoxa* (1), *Sinadoxa* (1), *Tetradoxa* (1).

Description: Usually shrubs or small trees (*Viburnum*), tall herbs (some species of *Sambucus*), or small geophytic herbs with creeping rhizome (*Adoxa*), sometimes with storied cambium and crystal sand (*Sambucus*). **Leaves** opposite or rarely verticillate, in basal rosette and alternate in *Adoxa*, simple, entire to variously toothed or lobed, with or without small stipules (*Viburnum*), or pinnately or bipinnately compound, the leaflets mostly with serrate margins, and large foliaceous stipules (*Sambucus*), stipules sometimes reduced to glandular appendages or absent. **Inflorescence** usually flat-topped, arranged in terminal cymes, corymbs or panicles, or small head (*Adoxa*). **Flowers** small, bisexual or rarely unisexual, occasionally with some marginal flowers neutral and a greatly enlarged corolla, slightly zygomorphic. **Calyx** with 5 sepals, rarely 3 (*Adoxa*; usually interpreted to represent bract + 2 bracteoles, calyx treated as absent) or 4, imbricate. **Corolla** with 5 petals, rarely 3–4, united, rotate to campanulate, occasionally tubular, imbricate, lobes sometimes with nectary (*Adoxa*). **Androecium** with as many stamens as petals, epipetalous, alternating with petals, anthers bithecous, in *Adoxa, Sinadoxa*

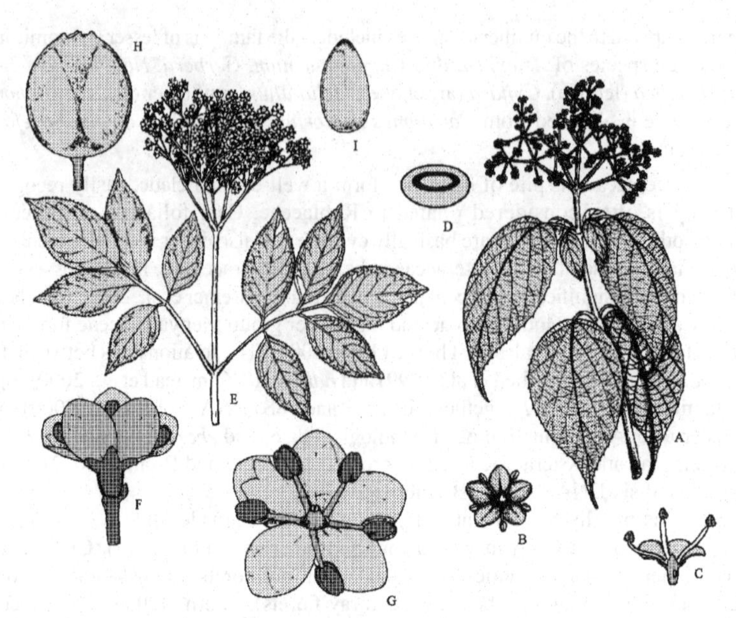

Figure 13.128. Adoxaceae. *Viburnum mullaha.* **A:** Branch with inflorescence; **B:** Flower, top view; **C:** Vertical section of flower; **D:** Transverse section of ovary. *Sambucus nigra.* **E:** Branch with inflorescence; **F:** Flower, side view; **G:** Flower, top view; **H:** Fruit; **I:** Seed (A–B, after Polunin and Stainton, Fl. Himal., 1984).

and *Tetradoxa* the stamens appear to be divided into two half stamens, each with a separate filament and monothecal anther, half anther being peltate in *Tetradoxa*, dehiscence by longitudinal slits. **Gynoecium** with 3–5 united carpels, 1 in *Sinadoxa*, ovary semi-inferior, rarely superior (*Tetradoxa*), 3–5 locular, the style solitary and terminal, divided in *Adoxa*, *Sinadoxa* and *Tetradoxa*, the stigma mostly capitate, the ovule 1 per locule, pendulous, anatropous, unitegmic and tenuinucellar or crassinucellate. **Fruit** a small drupe, or drupaceous berry with 3–5 stones, the embryo minute, straight, endosperm oily. Pollination by insects, musky flowers of *Adoxa* attract flies. Dispersal mostly by birds and animals.

Economic importance: The family is known for its well-known ornamentals belonging to *Sambucus* and *Viburnum*. Black elder (*S. nigra*) is widely cultivated in temperate as well tropical regions, whereas snow ball (*Viburnum opulus*) is mainly grown in temperate climates.

Phylogeny: The genera included here were originally placed in broadly circumscribed Caprifoliaceae, which has now been reduced to mere 5 genera, including well known genus *Lonicera*. The genus *Adoxa* (and recently recognised *Sinadoxa* and *Tetradoxa*) has traditionally been considered distinct from caprifoliaceae in absence of calyx, divided filaments with monothecous anthers, split styles and fruit with more than one stones, but now two major genera *Viburnum* and *Sambucus* have been included in expanded Adoxaceae. Donoghue et al. (2001) studied the taxonomy of Dipsacales especially Adoxaceae using *rbc*L and nuclear ribosomal internal transcribed spacer (ITS) sequences. They concluded that the 5 genera formed a clearly defined clade with *Viburnum* being sister to rest of the four genera, followed by *Sambucus,* which is sister to remaining three. Data established Adoxaceae and Caprifoliaceae as two major clades within the order. Within Adoxaceae *Adoxa*, *Sinadoxa* and *Tetradoxa* form **Adoxina clade** marked by herbaceous habit, reduction in the number of perianth parts, nectaries of multicellular hairs on the perianth, and bifid stamens. Thorne (2006, 2007) recognized two subfamilies Adoxoideae (4 genera) and Opuloideae (*Viburnum*). APweb (2008) also recognized these two clades as **Adoxa** clade and **Viburnum** clade, respectively, giving them formal tribal names Viburneae and Adoxeae in version 14 of APweb (update, 2018).

Araliaceae A. L. de Jussieu **Aralia *Family***

43 genera, 1,650 species

Widely distributed with most species in tropics and subtropics.

Placement:

	B & H	Cronquist	Takhtajan	Dahlgren	Thorne	APG IV/(APweb)
Informal Group						Asterids/ (Campanulid/ Asterid II)
Division		Magnoliophyta	Magnoliophyta			
Class	Dicotyledons	Magnoliopsida	Magnoliopsida	Magnoliopsida	Magnoliopsida	
Subclass	Polypetalae	Rosidae	Asteridae	Magnoliidae	Asteridae	
Series+/ Superorder	Calyciflorae+		Cornanae	Aralianae	Aralianae	
Order	Umbellales	Apiales	Apiales	Araliales	Araliales	Apiales

Salient features: Leaves often large, alternate, compound, leaving large scars on falling, flowers small, actinomorphic, pentamerous, usually in umbels, ovary inferior, each locule with single ovule, fruit a berry.

Major genera: *Schefflera* (650 species), *Polyscias* (150), *Hydrocotyle* (130), *Oreopanax* (85), *Dendropanax* (70), *Aralia* (65), *Osmoxylon* (50), *Trachymene* (45), *Eleutherococcus* (38), and *Hedera* (15).

Description: Herbs, shrubs, trees or lianas with prickly or stellate hairs, sometimes palm-like, a few root-climbers (*Hedera*). **Leaves** alternate, rarely opposite or whorled, large, lobed or more commonly pinnately or palmately compound, rarely simple, petioles often with sheathing base usually formed by membranous stipules, leaving large scar on stem after falling. **Inflorescence** usually umbellate, rarely corymbose, racemose or panicled, spikes or heads. **Flowers** small, usually pedicelled, greenish or whitish, bisexual, rarely unisexual and dioecious, epigynous, rarely hypogynous, bracts very small. **Calyx** with 5 sepals, adnate to ovary, lobes reduced to small teeth or seam-like rim. **Corolla** with 5 petals, rarely 3–10, free, broader at base, arising from disc, valvate, caducous, falling separately or as calyptra-like cap. **Androecium** with 5 stamens, rarely 3–10, as many as petals and alternating them, free, anthers bithecous, dorsifixed, dehiscence longitudinal. **Gynoecium** with usually 5 united carpels, sometimes 2–15, rarely 1, ovary inferior, locules as many as carpels, ovule one in each locule, on apical-axile placentas, anatropous, raphe ventral, styles as many as carpels, free and recurved or connate into a column or cone (stylopodium), rarely absent and stigmas sessile. **Fruit** a berry or drupe, rarely schizocarpic and splitting into pyrenes or mericarps; seed with small embryo at one end, endosperm copious, sometimes ruminate.

Economic importance: The family is known for its ornamental foliage plants such as *Schefflera*, *Fatsia japonica*, *Eleutherococcus*, *Aralia* and *Hedera*. Many cultivars of *Hedera helix* (Ivy) are used as house plants. Roots of Ginseng plant (*Panax ginseng*) from China and Korea yield drug Ginseng used medicinally as stimulant and aphrodisiac. American Ginseng (*P. quinquefolia*) is also being used as substitute for true ginseng. *Aralia cordata* and *A. racemosa* are also used medicinally.

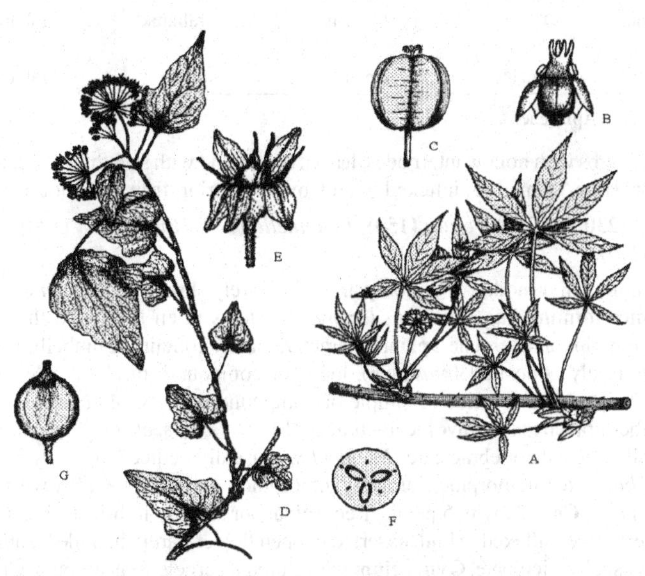

Figure 13.129: **Araliaceae.** *Eleutherococcus cissifolius.* **A:** Portion of plant with inflorescence. **B:** Flower; **C:** Fruit. *Hedera nepalensis.* **D:** Branch with inflorescence; **E:** Flower; **F:** Transverse section of ovary; **G:** Fruit (A–C, after Polunin and Stainton, Fl. Himal., 1984).

Phylogeny: Apiaceae and Araliaceae have been considered as closely related families for a long time, often included in the same order (Bentham and Hooker, Engler and Prantl), a trend continued by almost all recent authors, though Hutchinson (1926, 1973) had separated the two under distinct orders, and even under different groups Lignosae and Herbaceae. This separation was arbitrary and as such in most recent classifications they are placed closer together under Araliales (Dahlgren; Takhtajan, Thorne) or Apiales (Cronquist, APG II, APweb). The genera formerly included in Hydrocotyloideae of Apiaceae (including genera *Hydrocotyle, Centella,* etc.) form a polyphyletic group and as such have been segregated to Araliaceae (*Hydrocotyle*) and Mackinlayaceae (*Centella, Trachymene*, etc.) by Downie et al. (2000) and Chandler and Plunkett (2003, 2004). Stevens (APweb, 2003) points out that sampling must improve to resolve affinities especially with regard to *Hydrocotyle* and *Trachymene*. Thorne (2003, 2006), has shifted *Centella* and 5 other genera to Mackinlayaceae but placed *Hydrocotyle* and *Trachymene* in Araliaceae. Judd et al. (1999, 2002) argued that if Apiaceae and Araliaceae, in close to their traditional circumscriptions, were recognized, they would be poorly characterized morphologically, and certain genera would have no well-supported familial placement. They accordingly merge Araliaceae and Mackinlayaceae with Apiaceae, recognizing three subfamilies Aralioideae, Apioideae and Saniculoideae. Thorne (2003) and APG II (2003) treated all the three families as independent, but Thorne subsequently (2006, 2007) merged Saniculoideae and Mackinlayaceae with Apiaceae and divided Araliaceae into two subfamilies Aralioideae and Hydrocotyloideae. APweb (2003, 2008) follows the same treatment. Basal Araliaceae may well be bicarpellate and have simple leaves. Both these are features of the herbaceous Hydrocotyloideae, sister to the rest of the family (Chandler and Plunkett, 2004; Plunkett et al., 2004a). The position of *Harmsiopanax*, which has fruits that are schizocarpic like those of Hydrocotyloideae, is uncertain (Nicolas and Plunkett, 2009), APweb (2018) as such places *Harmsiopanax* separately between Hydrocotyloideae and Aralilioideae.

Apiaceae Lindley ***Carrot Family***
(= Umbelliferae A. L. de Jussieu)

442 genera, 3,575 species

Mainly distributed in north temperate regions.

Placement:

	B & H	**Cronquist**	**Takhtajan**	**Dahlgren**	**Thorne**	**APG IV/(APweb)**
Informal Group						Asterids/ (Campanulid/ Asterid II)
Division		Magnoliophyta	Magnoliophyta			
Class	Dicotyledons	Magnoliopsida	Magnoliopsida	Magnoliopsida	Magnoliopsida	
Subclass	Polypetalae	Rosidae	Asteridae	Magnoliidae	Asteridae	
Series+/ Superorder	Calyciflorae+		Cornanae	Aralianae	Aralianae	
Order	Umbellales	Apiales	Apiales	Araliales	Araliales	Apiales

B & H as Umbelliferae, others as Apiaceae.

Salient features: Aromatic herbs with hollow internodes, leaves compound with sheathing base, inflorescence umbel, petals incurved in bud, yellow or white, stamens 5, inflexed in bud, ovary inferior, fruit a cremocarp with stylopodium at apex.

Major genera: *Eryngium* (230 species), *Ferula* (150), *Pimpinella* (150), *Bupleurum* (100), *Heracleum* (60), *Sanicula* (40), and *Chaerophyllum* (40).

Description: Herbs with hollow internodes, commonly aromatic, rarely shrubs (*Eryngium giganteum*), or even climbers (*Pseudocarpum*), sometimes forming huge cushions (*Azorella*). Stems often fistular, with secretary canals containing ethereal oils and resins, coumarins, and terpenes, plants characteristically containing umbelliferose, a trisaccharide storage product. **Leaves** alternate, rarely opposite (*Apiastrum*), lobed or compound, rarely simple (*Bupleurum*), petioles with sheathing base, stipules absent. **Inflorescence** of simple or compound umbels, often subtended by involucre of bracts (involucre—bracts of umbel branches and involucel—bracts of flowers; absent in *Foeniculum*), sometimes like a head (*Eryngium*). **Flowers** small, bracteate or ebracteate (*Foeniculum*), usually pedicelled, rarely sessile (*Eryngium*) bisexual, rarely unisexual (*Echinophora*), actinomorphic (rarely zygomorphic), epigynous. **Calyx** with 5 sepals, adnate to ovary, 5-lobed, lobes often very small. **Corolla** with 5 petals, free, valvate or slightly imbricate, incurved in bud, notched at tip. **Androecium** with 5 stamens, free, inflexed in bud, exserted in open flower, rarely included, anthers bithecous, dehiscence longitudinal, pollen grains usually tricolpate. **Gynoecium** with 2 united carpels (syncarpous), with inferior ovary, bilocular with 1 ovule in each chamber, placentation axile, style surrounded at base by bilobed nectary, the basal portion of style

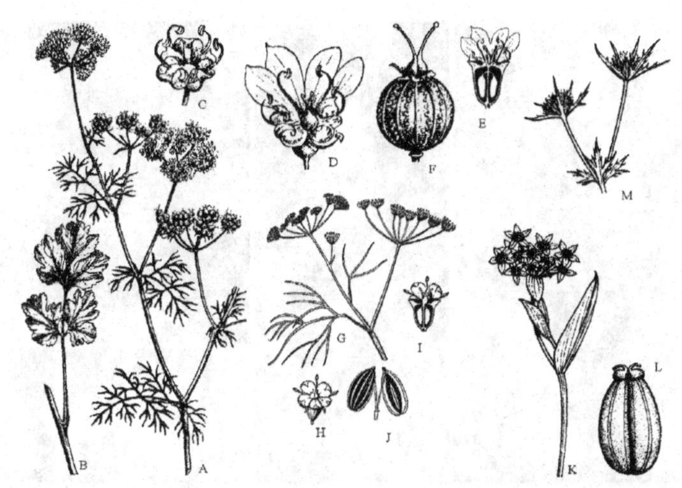

Figure 13.130: Apiaceae. *Coriandrum sativum.* **A:** Upper portion of plant with compound umbels in flower and fruit; **B:** Part of lower leaf with broader segments; **C:** Inner actinomorphic flower; **D:** Outer zygomorphic flower; **E:** Vertical section of flower; **F:** Cremocarp with persistent stylopodium at tip. *Foeniculum vulgare.* **G:** Portion of branch with compound umbels without bracts; **H:** Flower; **I:** Vertical section of flower; **J:** Cremocarp with forked carpophore separating 2 mericarps. *Bupleurum candollii.* **K:** Upper portion of plant with simple entire leaves (rare situation in this family) and umbels; **L:** Cremocarp. **M:** Upper portion of plant of *Eryngium biebersteinianum* with spiny leaves and sessile head-like umbels.

along with nectary persisting in fruit as stylopodium. **Fruit** schizocarpic known as cremocarp splitting at maturity into two mericarps attached by a common stalk carpophore, mericarp containing oil canals called vittae inside. Seeds with small embryo, endosperm oily.

Economic importance: The family is the source of food plants, spices and condiments. Carrot (*Daucus carota*) and parsnip (*Pastinaca sativa*) are important root crops. Important flavouring plants include fennel (*Foeniculum vulgare*), coriander (*Coriandrum sativum*), caraway (*Carum carvi*), anise (*Pimpinella anisum*) and celery (*Apium graveolens*). *Cicuta*, *Conium* (hemlock, which Socrates is said to have used for suicide) and *Oenanthe* include poisonous plants.

Phylogeny: Apiaceae and Araliaceae have been considered as closely related families for a long time, often included in the same order (Bentham and Hooker, Engler and Prantl), a trend continued by almost all recent authors, though Hutchinson (1926, 1973) had separated the two under distinct orders, and even under different groups Lignosae and Herbaceae. This separation was arbitrary and as such in most recent classifications they are placed closer together under Araliales (Dahlgren, Takhtajan, Thorne) or Apiales (Cronquist, APG II, APweb), Monophyly of the family is supported by morphology, secondary metabolites, rbcL and matK sequences (Judd et al., 1994; Plunkett et al., 1997). Earlier studies (Judd et al., 1999) had indicated that Apiaceae are most closely related to Pittosporaceae, but recent data (APweb; Plunkett, 2001) points to Pittosporaceae being sister taxon of the whole group or Pittosporaceae may be embedded in Apiaceae + Araliaceae + other taxa. The family Apiaceae is usually divided into two subfamilies: Saniculoideae (Leaves often broad, with hairy or thorny leaf teeth, stylopodium separated from style by groove, fruit scaly or spiny, vittae often poorly developed) and Apioideae (umbels compound, stylopodium lacking groove, carpophore free, bifid, mericarps attached at apex). Recent molecular studies (Downie et al., 2000a, 2000b) have indicated that traditional division into tribes and genera may undergo substantial rearrangement. The genera formerly included in Hydrocotyloideae (including genera *Hydrocotyle, Centella,* etc.) form a polyphyletic group and as such have been segregated to Araliaceae (*Hydrocotyle*) and Mackinlayaceae (*Centella, Trachymene*, etc.) by Downie et al. (2000) and Chandler and Plunkett (2003, 2004, quoted in APweb). Stevens (APweb, 2003) points out that sampling must improve to resolve affinities especially with regard to *Hydrocotyle* and *Trachymene*. Thorne (2003), has shifted *Centella* and 5 other genera to Mackinlayaceae but placed *Hydrocotyle* and *Trachymene* in Araliaceae, recognizing only two subfamilies—Apioideae and Saniculoideae—under Apiaceae. Judd et al. (1999, 2002) argued that if Apiaceae and Araliaceae, in close to their traditional circumscriptions, were recognized, they would be poorly characterized morphologically, and certain genera would have no well-supported familial placement. They accordingly merge Araliaceae and Mackinlayaceae with Apiaceae, recognizing three subfamilies Aralioideae, Apioideae and Saniculoideae. APweb (2003, 2018), Thorne (2006, 2007) and APG IV (2016) recognize Araliaceae as an independent family, but relegates Mackinlayaceae to subfamily Mackinlayoideae, recognizing additional subfamily Azorelloideae (some former members of Hydrocotyloideae), thus recognizing a total of 4 subfamilies (other two being Apioideae and Saniculoideae).

Figure 13.131: Chenopodiaceae. A: *Chenopodium album*, plant; **B:** Flower cluster. **Amaranthaceae. C:** *Amaranthus caudatus*, plant with inflorescence; **D:** Flower cluster; **E:** *Celosia cristata*, inflorescence. **Caryophyllaceae. F:** *Stellaria media*, plant; **G:** Flower; **H:** *Dianthus barbatus*, plant; **I:** Part of inflorescence. **Polygonaceae. J:** *Rumex hymenosepalus*, plant. **K:** *Eriogonum latifolium*, plant; **L:** Flower cluster; **M:** *Polygonum davisae*, portion of plant.

Figure 13.132: Portulacaceae. A: *Portulacaria afra*, plant. **Cactaceae. B:** *Mammillaria densispina*, plant; **C:** *Echinopsis terscheckii*, plant; **D:** Portion of stem. **Nyctaginaceae. E:** *Bougainvillea glabra*, plant; **F:** flower enclosed in showy bracts; **G:** *Mirabilis jalapa*, plant; **H:** Flower. **Aizoaceae. I:** *Mesembryanthemum criniflorum*, plant; **J:** Flower.

Figure 13.133: Hydrangeaceae. A: *Hydrangea macrophylla*, flowering branch; **B:** Flowers. **Polemoniaceae. C:** *Phlox diffusa*, flowering branch; **D:** Flowers. **Cornaceae. E:** *Cornus capitata*, flowering branch; **F:** Flower. **Primulaceae. G:** *Primula florindae*, inflorescence; **H:** *Dodecantheon meadia*, inflorescence; **I:** Flower.

Figure 13.134: Ericaceae. A: *Colluna vulgaris*, flowering branch; **B:** *Erica blanda*, flowering branch; **C:** *Arbutus unedo*, flowering branch; **D:** Fruit; **E:** *Rhododendron giersonianum*, flower cluster; **F:** *R. occidentale*, flowering branch. **G:** *Phyllodoce breweri*, flowering branch; **Adoxaceae. H:** *Sambucus nigra*, flowering branch; **I:** *Viburnum cotinifolium*, branch with young fruits; **J:** Branch with mature fruits.

Figure 13.135: Rubiaceae. A: *Ixora coccinea*, flowering branch; **B:** Inflorescence **C:** *Hamelia patens*, flowering branch; **D:** Flowers and young fruits; **Apocynaceae. E:** *Plumeria alba*, flowering branch; **F:** *Allamanda catharatica*, branch with flowers; **G:** *Catharanthus roseus*, flowering branch; **H:** Flower; **I:** *Nerium oleander*, flowering branch; **J:** Flower; **K:** *Asclepias syriaca*, flowering plant; **L:** *A. fascicularis*, flowering branch; **M:** Flowers; **N:** *Calotropis procera*, flowering branch.

Figure 13.136: Verbenaceae. A: *Lantana camara*, plant; **B:** Flowers. **Bignoniaceae. C:** *Incarvillea arguta*, flowering branch; **D:** *Tecoma stans*, flowering branch. **Acanthaceae. E:** *Justicia brandegeana*, plant; **F:** *Acanthus spinosus*, flowering branch; **G:** Flowers; **H:** *Adhatoda vasica*, flowering branch; **I:** Flowers; **J:** *Thunbergia grandiflora*, flower. **Scrophulariaceae. K:** *Bowkeria gerardiana*, flowering branch; **L:** Part of inflorescence; **M:** *Vebascum thapsus*, plant.

Figure 13.137: Solanaceae. A: *Cestrum elegans*, flowering branch; **B:** flowers; **C:** *Solanum hispidum,* flowering branch; **D:** *Atropa belladonna*, flowering branch; **E:** *Datura suaveolens*, flowering plant; **F:** *Solanum melanogena,* flowering branch; **G:** Flower; **H:** Fruit; **I:** *Solanum nigrum,* plant in flower; **J:** Young fruits. **Convolvulaceae. K:** *Ipomoea cairica*, plant in flower; **L:** Flower; **M:** *Jacquemontia pentantha,* plant in flower; **N:** Flowers. **Boraginaceae. O:** *Ehretia laevis*, plant; **P:** Flowers.

Figure 13.138: Plantaginaceae. A: *Plantago lanceolata*, plant; **B:** *Mimulus cardinalis*, plant; **C:** *M. guttatus*, flower cluster; **D:** *M. puniceus*, flowering branch; **E:** *Digitalis purpurea*, flowering branch; **F:** Flowers enlarged. **Lamiaceae. G:** *Salvia muelleri*, flower; **H:** *S. mexicana*, flowering branch; **I:** Flower; **J:** *Origanum calcaratum*, flowers; **K:** *Salvia splendens*, plant; **L:** Flower; **M:** *S. scorodonifolia*, flower; **N:** *Lavandula angustifolia*, flowering twig.

Figure 13.139: Asteraceae. A: *Helianthus debilis*, plant; **B:** Capitulum; **C:** *Achillea millefolium*, plant; **D:** *Taraxacum officinale*, plant with capitula; **E:** *Pachystegia insignis*, plant with capitula; **F:** *Haplopappus macrocephalus*, plant with capitula and fruiting heads; **G:** *Centaurea solstitialis*, plant with inflorescence; **H:** *Artemisia pycnocephala*, plant; **I:** *Calendula officinalis*, capitulum; **J:** *Sonchus oleraceous*, plant.

Figure 13.140: Apiaceae. A: *Astrantia major*, plant; **B:** Umbel; **C:** *Eryngium paniculatum*, plant with inflorescences; **D:** *Angelica pachycarpa*, inflorescence; **E:** *Foeniculum vulgare*, plant; **F:** Part of inflorescence; **G:** *Coriandrum sativum*, plant; **H:** Part of inflorescence. **Araliaceae. I:** *Pseudopanax crassifolium*, flowering branch; **J:** Inflorescence; **K:** *Hedera helix*, plant.

References

Abrams, L. (1923–1960). *An Illustrated Flora of the Pacific States.* Vol. 4 by Roxana Ferris. Stanford Univ. Press, Stanford, CA.

Adams, K. L. and J. F. Wendel. (2005). Polyploidy and genome evolution in plants. *Curr. Opin. Plant Biol.* **8:** 135–141.

Adanson, M. (1763). *Familles des plantes*, Paris, 2 vols.

Aguilar-Ortigoza, C. J. and V. Sosa. 2004. The evolution of toxic phenolic compounds in a group of Anacardiaceae genera. *Taxon* **53:** 357–364.

Airy Shaw, H. K. and J. C. Willis. (1973). *A Dictionary of the Flowering Plants and Ferns.* Ed. 8, revised by H. K. Airy Shaw. Cambridge University Press, Cambridge.

Albach, D. C., P. S. Soltis, D. E. Soltis and R. G. Olmstead. (2001). Phylogenetic analysis of Asterids based on sequences of four genes. *Ann. Missouri Bot. Gard.* **88:** 163–212.

Albert, V. A., A. Backlund, K. Bremer, M. W. Chase, J. R. Manhart, B. D. Mishler and K. C. Nixon. (1994). Functional constraints and *rbcL* evidence for land plant phylogeny. *Ann. Missouri Bot. Gard.* **81:** 534–567.

Alrich, P. and W. Higgins. (2008). *Illustrated Dictionary of Orchid Genera.* Comstock Publishing Associates, Cornell University Press, Ithaca.

Al-Shehbaz, I. A., M. A. Beilstein and E. A. Kellogg. (2006). Systematics and phylogeny of the Brassicaceae (Cruciferae): An overview. *Plant Syst. Evol.* **259:** 89–120.

Al-Shehbaz, I. A. (2012). A generic and tribal synopsis of Brassicaceae (Cruciferae). *Taxon* **61:** 931–954.

Alston, R. E. and B. L. Turner. (1963). Natural hybridisation among four species of *Baptisia* (Leguminosae). *Amer. J. Bot.* **50**: 159–173.

Alverson, W. S., B. A. Whitlock, R. Nyffeler and D. A. Baum. (1998). Phylogeny of core Malvales: Evidence from *ndhF* sequence data. *Amer. J. Bot.* **86**(6) Suppl.: 112.

Ambrose, B. A., D. R. Lerner, P. Ciceri, C. M. Padilla, M. F. Yanofsky and R. J. Schmidt. (2000). Molecular and genetic analyses of the Silky1 gene reveal conservation in floral organ specification between eudicots and monocots. *Molecular Cell* **5:** 569–579.

Anderberg, A. A., X. Zhang and M. Källersjö. (2000). Maesaceae, a new primuloid family in the order Ericales s. l. *Taxon* **49:** 183–187.

Anderberg, A. A., C.-I. Peng, I. Trift and M. Källersjö. (2001). The *Stimpsonia* problem; evidence from DNA sequences of plastid genes *atp*B, *ndhF* and *rbcL*. *Bot. Jahrb. Syst.* **123:** 369–376.

Anderson, B. and S. D. Johnson. (2009). Geographical covariation and local convergence of flower depth in a guild of fly-pollinated plants. *New Phytol.* **182:** 533–540.

Anderson, E. (1940). The concept of the genus. II. A survey of modern opinion. *Bull. Torrey Bot. Club* **67**: 363–369.

Anderson, J. M., H. M. Anderson and C. J. Cleal. (2007). Brief history of the gymnosperms: Classification, biodiversity, phytogeography and ecology. *Strelitzia* **20:** 1–280.

APG [= Angiosperm Phylogeny Group]. (1998). An ordinal classification for the families of flowering plants. *Ann. Missouri Bot. Gard.* **85:** 531–553.

APG [= Angiosperm Phylogeny Group] II. (2003). An update of the Angiosperm Phylogeny Group classification for the orders and families of flowering plants: APG II. *Bot. J. Linn. Soc.* **141:** 399–436.

APG [= Angiosperm Phylogeny Group] III. (2009). An update of the Angiosperm Phylogeny Group classification for the orders and families of flowering plants: APG III. *Bot. J. Linn. Soc.* **161:** 105–121.

APG [= Angiosperm Phylogeny Group] IV. (2016). An update of the Angiosperm Phylogeny Group classification for the orders and families of flowering plants: APG IV. *Bot. J. Linn. Soc.* **181**: 1–20.

Applequist, W. L. and R. S. Wallace. (2001). Phylogeny of the Portulacaceous cohort based on *ndhF* sequence data. *Syst. Bot.* **26:** 406–419.

Applequist, W. L., W. L. Wagner, E. A. Zimmer and M. Nepokroeff. (2006). Molecular evidence resolving the systematic position of *Hectorella* (Portulacaceae). *Syst. Bot.* **31:** 310–319.

Arber, E. and J. Parkins. (1907). On the origin of Angiosperms. *Bot. J. Linn. Soc.* **38**: 29–80.

Arber, E. (1934). *The Gramineae: a study of cereal, bamboo, and grass.* Cambridge, England.

Arber, E. (1938). *Herbals: Their Origin and Evolution* (2nd ed.). Cambridge Univ. Press.

Armstrong, J. E. (1985). The delimitation of Bignoniaceae and Scrophulariaceae based on floral anatomy, and the placement of problem genera. *Amer. J. Bot.* **72:** 755–766.

Arnold, C. A. (1948). Classification of gymnosperms from the point of view of Palaeobotany. *Bot. Gaz.* **110:** 2–12.

Ashlock, P. H. (1971). Monophyly and related terms. *Syst. Zool.* **20**: 63–69.

Ashlock, P. H. (1979). An evolutionary Systematits's view of classification. *Syst. Zool.* **28:** 441–450.

Asmussen, C. B., W. J. Baker and Dransfield. (2000). Phylogeny of the palm family (Arecaceae) based on *rps*16 intron and *trnL-trnF* plastid DNA sequences. pp. 525–535. *In*: K. L. Wilson and D. A. Morrison (eds.). *Monocots: Systematics and Evolution*. CSIRO, Collingwood.

Augusto, L., T. J. Davies, S. Delzon and A. de Schrijver. (2014). The enigma of the rise of angiosperms: Can we untie the knot? *Ecol. Lett.* **17:** 1326–1338.

Axelrod, D. I. (1970). Mesozoic paleogeography and early angiosperm history. *Bot. Rev.* **36**: 277–319.

Babcock, E. B. (1947). The genus *Crepis* pt. 1. The taxonomy, phylogeny distribution and evolution of *Crepis*. *Univ. Calif. Publs. Bot.* **21**: 1–197.

Bailey, I. W. (1944). The development of vessels in angiosperms and its significance in morphological research. *Am. J. Bot.* **31**: 421–428.

Bailey, L. H. (1949). *Manual of Cultivated Plants* (rev. Ed.). Macmillan, New York.

Bailey, I. W. and B. G. L. Swamy. (1951). The conduplicate carpel of dicotyledons and its initial trend of specialization. *Amer. J. Bot.* **38:** 373–379.

Barber, H. N. (1970). Hybridization and evolution of plants. *Taxon* **19**: 154–160.

Barber, J., M. Diazgranados and P. C. Hoch. (2008). Molecular phylogeny of *Ludwigia* (Onagraceae) reconstructed from multiple nuclear and cpDNA markers. P. 86, in *Botany 2008. Botany without Borders.* [Botanical Society of America, etc. Abstracts.]

Barnes, R. D. (1989) Diversity of organisms: how much do we know? *Amer. Zool.* **29:** 1075–84.

Barthlott, W. and G. Voit. (1979). Mikromorphologie der Samenschalen und Taxonomie der Cactaceae Ein raster-elektronem-microscopischer uberblick. *Plant Syst. Evol.* **132**: 205–229.

Barthlott, W. (1981). Epidermal and seed surface characters of plants: Systematic applicability and some evolutionary aspects. *Nordic J. Bot.* **1**: 345–355.

Barthlott, W. and D. Froelich. (1983). Mikromorphologie und Orientierungsmuster epicuticularer Wachs-Kristalloide: Ein neues systematisches Merkmal bei Monocotylen. *Pl. Syst. Evol.* **142**: 171–185.

Barthlott, W. (1984). Microstructural features of seed surfaces. *In*: V. H. Heywood and D. M. Moore (eds.). *Current Concepts in Plant Taxonomy.* Systematic Association Special Volume No. **25:** 95–105.

Baskin, C. C. and J. M. Baskin. (2007). Nymphaeaceae: A basal angiosperm family (ANITA grade) with a fully developed embryo. *Seed Sci. Res.* **17:** 293–296.

Bate-Smith, E. C. (1958). Plant phenolics as taxonomic guides. *Proc. Linn. Soc. Lond.* **169**: 198–211.

Bate-Smith, E. C. (1962). The phenolic constituents of plants and their taxonomic significance. *J. Linn. Soc. Bot.* **58**: 95–173.

Bate-Smith, E. C. (1968). The phenolic constituents of plants and their taxonomic significance. *J. Linn. Soc. Bot.* **60**: 325–383.

Bauhin, C. (1596). *Phytopinax seu enumeratic plantarum....* Basel.

Bauhin, C. (1623). *Pinax theatri botanici.* Basel.

Baum, B. R. (1977). *Oats: Wild and Cultivated. A Monograph of the Genus Avena* L. (*Poaceae*). Minister of Supply and Services, Ottawa.

Baum, H. (1949). Der einheitliche Bauplan der Angiospermen gynözeen und die Homologie ihrer fertilen Abschnitte. *Bot. Jahrb. Syst. Pflanzen.* **96:** 64–82.

Baum, H. and W. Leinfellner. (1953). Die Peltationsnomeklatur der Karpelle. *Bot. Jahrb. Syst. Pflanzen.* **100:** 424–426.

Bayer, C., M. F. Fay, A. Y. De Bruijn, V. Savolainen, C. M. Mortan, K. K. Kubitzki, W. S. Alverson and M. W. Chang. (1999). Support for an expanded family concept of Malvaceae within a recircumscribed order Malvales: a combined analysis of plastid *atp*B and *rbc*L DNA sequences. *Bot. J. Linn. Soc.* **129**: 267–303.

Beech, E., M. Rivers, S. Oldfield and P. P. Smith. (2017). GlobalTreeSearch: The first complete global database of tree species and country distributions. *J. Sust. Forestry* http://dx.doi.org/10.1080/10548911.2017.1310049.

Behnke, H. D. (1965). Über das phloem der Dioscoreaceen unter besonderer Berücksichtigung ihrer phloembecken. II. Mitteilung: Elektronenoptische untersuchungen zur feinstruktur des phloembeckens. *Z. Pflanzenphysiol.* **53**: 214–244.

Behnke, H. D. (1976). Ultrastructure of sieve-element plastids in Caryophyllales (Centrospermae); evidence for the delimitation and classification of the order. *Plant Syst. Evol.* **126**: 31–54.

Behnke, H. D. (1977). Transmission electron microscopy and systematics of flowering plants. *In*: K. Kubitzki (ed.). *Flowering Plants Evolution and Classification of Higher Categories. Plant Syst. Evol. Suppl.* **1**: 155–178.

Behnke, H. D. and W. Barthlott. (1983). New evidence from the ultrastructural and micromorphological fields in Angiosperm classification. *Nordic. J. Bot.* **1**: 341–460.

Behnke, H. D. (1997). Sarcobataceae—a new family of Caryophyllales. *Taxon* **46**: 495–507.

Belfod, H. S. and W. F. Thomson. (1979). Single copy DNA homologies and phylogeny of *Atriplex*. Carnegie Inst. Wash Year Book **78**: 217–223.

Belk, E. (1939). *Studies in the anatomy and morphology of the spikelet and flower of the Gramineae*. Thesis (Ph.D.). Cornell Univ.

Bell, G. A. (1971). Comparative biochemistry of non-protein amino acids. pp. 179–206. *In*: J. B. Harborne, D. Boulter and B. L. Turner (eds.). *Chemotaxonomy of Leguminosae.* Academic Press, London.

Bellstedt, D. U., L. van Zyl, E. M. Marais, B. Bytebier, C. A. de Villiers, A. M. Makwarela and L. L. Dreyer. (2008). Phylogenetic relationships, character evolution and biogeography of southern African members of Zygophyllum (Zygophyllaceae) based on three plastid regions. *Molec. Phyl. Evol.* **47:** 932–949.

Bennett, M. D. and I. J. Leitch. (2007). Nuclear DNA amounts in angiosperms: 583 new estimates. *Ann. Bot.* **80:** 169–196.

Bensel, C.R. and B. F. Palser. (1975a). Floral anatomy in the Saxifragaceae *sensu lato*. II. Saxifragoideae and Iteoideae. *Amer. J. Bot.* **62:** 661–675.

Bensel, C.R. and B. F. Palser. (1975b). Floral anatomy in the Saxifragaceae *sensu lato*. III. Kirengeshomoideae, Hydrangeoideae and Escallonioideae. *Amer. J. Bot.* **62:** 676–687.

Benson, L. (1957). *Plant Classification*. Oxford and IBH Co., New Delhi.

Bentham, G. (1858). *Handbook of British Flora* (7th ed., revised by A. B. Rendle in 1930). Ashford, Kent.

Bentham, G. and J. D. Hooker (1862–83). *Genera Plantarum*. London, 3 vols.

Bentham, G. (1863–1878). *Flora Australiensis*. London, 7 volumes.

Berry, P. E., V. Savolainen, K. J. Sytsma, J. C. Hall and M. W. Chase. (2001). *Lissocarpa* is sister to *Diospyros* (Ebenaceae). *Kew Bull.* **56:** 725–729.

Bessey, C. E. (1915). Phylogenetic taxonomy of flowering plants. *Ann. Mo. Bot. Gard.* **2**: 109–164.

Bhandari, M. M. (1978). *Flora of the Indian Desert*. Sc. Publ., Jodhpur.

Bhattacharya, B. and B. M. Johri. (1998). *Flowering Plants: Taxonomy and Phylogeny.* Narosa Publishing House, New Delhi.

Blackith, R. E. and R. A. Reyment. (1971). *Multivariate Morphometrics*. Acad. Press, London.

Blake, S. F. and A. C. Atwood. (1941). *Geographical Guide to the Floras of the World.* Part I. Misc. Publ. 401. U. S. Dept. Agr., Washington, D. C.

Blake, S. F. (1961). *Geographical Guide to the Floras of the World.* Part II. Misc. Publ. 797. U. S. Dept. Agr., Washington, D. C.

Blakeslee, A. F. , A. G. Avery, S. Satina and J. Rietsama. (1959). *The genus Datura*. Ronald Press, New York.

Bogler, D. J. and B. B. Simpson. (1996). Phylogeny of Agavaceae based on ITS rDNA sequence variation. *Amer. J. Bot.* **83:** 1225–1235.

Bonnett, H. T. and E. H. Newcomb. (1965). Polyribosomes and cisternal accumulations in root cells of radish. *J. Cell Biol.* **27**: 423–432.

Bordet, J. (1899). Sur l'agglutination et la dissolution des globules rouges par le serum d'animaux injectes de sang defibriné. *Ann. Inst. Pasteur* **13**: 225–250.

Borsch, T. and P. S. Soltis. (2008). Nymphaeales—the first globally diverse clade? *Taxon* **57:** 1051.

Borsch, T., C. Löhne and J. H. Wiersema. (2008). Phylogeny and evolutionary patterns in Nymphaeales: Integrating genes, genomes and morphology. *Taxon* **57:** 1052–1081.

Boulter, D. (1974). The use of amino acid sequence data in the classification of higher plants. *In*: G. Bendz and J. Santesson (eds.). *Chemistry and Botanical Classification, Nobel Symposium*. Acad. Press, London/New York **25**: 211–216.

Bower, F. O. (1935). *Primitive Land Plants, Also Known as the Archegoniatae*. Macmillan, London.

Bramwell, D. (1972). Endemism in the Flora of Canary Islands. pp. 141–159. *In*: D. H. Valentine (ed.) *Taxonomy, Phytogeography and Evolution*. Academic Press, London.

Bremer, B. and L. Struwe. (1992). Phylogeny of the Rubiaceae and Loganiaceae: Congruence or conflict between morphologican and molecular data? *Amer. J. Bot.* **79:** 1171–1184.

Bremer, B., K. Andreasen and D. Olsson. (1995). Subfamilial and tribal relationships in the Rubiaceae based on rbcL sequence data. *Ann. Missouri Bot. Gard.* **82:** 383–397.

Bremer, B., R. K. Jansen, B. Oxelman, M. Backlund, H. Lantz and K. -J Kim. (1999). More characters or more taxa for a robust phylogeny—case study from the coffee family (Rubiaceae). *Syst. Bio.* **48:** 413–435.

Bremer, B., K. Bremer, N. Heidari, P. Erixon, A. A. Anderberg, R. G. Olmstead, M. Kållersjö and E. Barkhordarian. (2002). Phylogenetics of asterids based on three coding and three non-coding chloroplast DNA markers and the utility of non-coding DNA at higher taxonomic levels. *Molecular Phylogenetics and Evolution* **24:** 274–301.

Bremer, K. and H. -E. Wanntorp. (1978). Phylogenetic systematics in botany. *Taxon* **27**: 317–329.

Bremer, K. and H. -E. Wanntorp. (1981). The cladistic approach to plant classification. *In*: V. A. Funk and D. R. Brroks (eds.). *Advances in Cladistics*. New York Botanical Garden, New York **1**: 87–94.

Bremer, K., A. Backlund, B. Sennblad, U. Swenson, K. Andreasen, M. Hjertson, J. Lundberg, M. Backlund and B. Bremer. (2001). A phylogenetic analysis of 100+ genera and 50+ families of euasterids based on morphological and molecular data with notes on possible higher level morphological synapomorphies. *Plant Syst. Evol.* **229:** 137–169.

Bremer, K. (2002). Gondwanan evolution of the grass alliance of families (Poales). *Evolution* **56:** 1374–1387.

Brendan, B. Larsen, Elizabeth C. Miller, Matthew K. Rhodes and John J. Wiens. (2017). Inordinate Fondness Multiplied and Redistributed: the Number of Species on Earth and the New Pie of Life. The Quarterly Review of Biology 92, no. **3**: 229–265.

Brenner, G. H. and I. Bickoff. (1992). Palynology and age of the Lower Cretaceous basal Kurnub Group from the coastal plain to the northern Negev of Israel. *Palynology* **16**: 137–185.

Brenner, G. J. (1963). Spores and Pollen of Potomac Group of Maryland. *Maryland Department of Geology, Mines, Water Resources Bulletin* **27**: 1–215.

Brenner, G. L. (1996). Evidence for the earliest stage of angiosperm pollen evolution. A paleoequatorial section from Israel. pp. 91–115. *In*: D. W. Taylor and L. J. Hickey (eds.). *Flowering Plant Origin, Evolution and Phylogeny.* Chapman & Hall Inc., New York.

Brown, R. (1827). Character and description of *Kingia*, a new genus of plants found on the southwest coast of New Holland with observations on the structure and its unimpregnated ovulum on female flower of Cycadeae and Coniferae. pp. 534–565. *In*: P. P. King's (ed.). *Narrative of a Survey of Intertropical and Westerns Coasts of Australia, Performed Between 1818 and 1822*, John Murray, London.

Brown, N. E. (1920). New and old species of *Mesembranthemum*, with critical notes. *Journ. Linn. Soc. Bot.* **45:** 53–140.

Brown, R. W. (1956). Palmlike plants from the Delores Formation (Triassic) in southwestern Colorado. *U. S. Geological Survey Professional Paper* **274**: 205–209.

Brummitt, R. K. (1992). *Vascular Plant Families and Genera*. Royal Botanic Gardens, Kew. http://www.rbgkew.org.uk/web.dbs/genlist.html.

Brunfels, O. (1530). *Herbarium vivae eicones*. Argentorati, 3 tomes.

Bruneau, A., F. Forest, P. S. Herendeen, B. B. Klitgaard and G. P. Lewis. (2001). Phylogenetic relationships in the Caesalpinioideae (Leguminosae) as inferred from chloroplast *trnL* intron sequences. *Syst. Bot.* **26:** 487–514.

Buerki, S., F. Forest, P. Acevedo-Rodríguez, M. W. Callmander, J. A. A. Nylander, M. Harrington, I. Sanmartín, F. Küpfer and N. Alvarez. (2009). Plastid and nuclear DNA markers reveal intricate relationships at subfamilial and tribal levels in the soapberry family (Sapindaceae). *Molec. Phyl. Evol.* **51:** 238–258.

Buerki, S., P. P. II Lowry, N. Alvarez, S. G. Razafimandimbison, P. Küpfer and M. W. Callmander. (2010). Phylogeny and circumscription of Sapindaceae revisited: Molecular sequence data, morphology, and biogeography support recognition of a new family, Xanthoceraceae. *Plant Ecol. Evol.* **143:** 148–159.

Burke, J. M. and A. Sanchez. (2011). Revised subfamily classification of Polygonaceae, with tribal classification of Eriogonoideae. *Brittonia* **63**(4): 510–520.

Byng, J. W. (2014). *The Flowering Plants Handbook: A practical Guide to Families and Genera of the World.* Plant Gateway, Hertford.

Byng, J. W. (2015). *The Gymnosperms Handbook: A practical Guide to Families and Genera of the World.* Plant Gateway, Hertford.

Caddick, L. R., C. A. Furness, P. Wilkons and M. W. Chase. (2000). Yams and their allies: Systematics of Dioscoreales. pp. 475–487. *In*: K. L. Wilson and D. A. Morrison (eds.). *Monocots: Systematics and Evolution*. CSIRO, Collingwood.

Caddick, L. R., C. A. Furness, P. Wilkons, T. A. J. Hedderson and M. W. Chase. (2002a). Phylogenetics of Dioscoreales based on combined analyses of morphological and molecular data. *Bot. J. Linnean Soc.* **138:** 123–144.

Caddick, L. R., P. Wilkin, P. J. Rudall, T. A. J. Hedderson and M. W. Chase. (2002b). Yams reclassified: A recircumscription of Dioscoreaceae and Dioscoreales. *Taxon* **51:** 103–114.

Caesalpino, A. (1583). *De plantis libri*. Florentiae.

Cain, A. J. and G. A. Harrison. (1958). An analysis of the taxonomists's judgement of affinity. *Proc. Zool. Soc. Lond.* **131**: 85–98.

Cain, A. J. and G. A. Harrison. (1960). Phyletic weighting. *Proc. Zool. Soc. Lond.* **135**: 1–31.

Cain, S. A. (1944). *Foundations of Plant Geography*. New York & London.

Cameron, K. M. (2002). Intertribal relationships within Orchidaceae as inferred from analyses of five plastid genes. P. 116, in *Botany 2002: Botany in the Curriculum,* Abstracts. [Madison, Wisconsin.].

Cameron, K. M. (2004). Utility of plastid psaB gene sequences for investigating intrafmilial relationships within Orchidaceae. *Mol. Phyl. Evol.* **31:** 1157–1180.

Cameron, K. M. and C. Fu. (2006). A nuclear rDNA phylogeny of Smilax (Smilacaceae). pp. 598–605. *In*: J. T. Columbus, E. A. Friar, J. M. Porter, L. M. Prince and M. G. Simpson (eds.). *Monocots: Comparative Biology and Evolution. Excluding Poales.* Rancho Santa Ana Botanical Garden, Claremont, Ca. [Aliso 22: 598–605.]

Cantino, P. D. and K. de Queiroz. (2006). PhyloCode: A Phylogenetic Code of Biological Nomenclature. [http://www.ohio.edu.phylocode]

Camp, W. H. (1947). Distribution patterns in modern plants and the problems of ancient dispersals. *Ecol. Monogr.* **17:** 159–183.

Cantino, P. D. and R. W. Sanders. (1986). Subfamilial classification of Labiatae. *Syst. Bot.* **11:** 163–185.

Cantino, P. D. (2000). Phylogenetic nomenclature: addressing some concerns. *Taxon* **49**: 85–93.

Carlquist, S. (1987). Presence of vessels in *Sarcanda* (Chloranthaceae); comments on vessel origin in angiosperms. *Amer. J. Bot.* **64**: 1765–1771.

Carlquist, S. (1996). Wood anatomy of primitive Angiosperms: New perspective and syntheses. pp. 68–90. *In*: D. W. Taylor and L. J. Hickey (eds.). *Flowering Plant Origin, Evolution and Phylogeny.* Chapman & Hall Inc., New York.

Carlquist, S. and E. L. Schneider. (2010). Origin and nature of vessels in monocotyledons. 11. Primary xylem microstructure, with examples from Zingiberales. *Internat. J. Plant Sci.* **171:** 258–266.

Carlquist, S. (2012). Monocot xylem revisited: New information, new paradigms. *Bot. Review* **78:** 87–150.

Carpenter, J. M. (2003). Crique of pure folly. *Bot. Rev.* **69(1):** 79–92.

Catalán, P., E. A. Kellogg and R. G. Olmstead. (1997). Phylogeny of Poaceae subfamily Pooideae based on chloroplast *ndhF* gene sequences. *Mol. Phylog. Evol.* **8:** 150–166.

Cavalier-Smith. (1981). Eukaryotic kingdoms, seven or nine? *BioSystems* **14**: 461–481.

Cavalier-Smith. (1998). A revised six-kingdom system of life. *Biol. Rev.* **73**: 203–266.

Cavalier-Smith. (2000). A revised six kingdom system of life. *Biological Reviews.*

Cavalier-Smith. (2004). Only six kingdoms of life. *Proceedings of the Royal Society B: Biological Sciences* **271**: 251–1262.

Cervantes, A., A. Fuentes, J. Gutiérrez, S. Magallón and T. Borsch. (2016). Successive arrivals since the Miocene shaped the diversity of the Caribbean Acalyphoideae (Euphorbiaceae). *J. Biogeog.* **43:** 1773–1785.

Chacón, J., F. Luebert, H. H. Hilger, S. Ovchinnikova, F. Selvi, L. Cecchi, C. M. Guilliams, K. Hasenstab-Lehman, K. Sutorý, M. G. Simpson and M. Weigend. (2016). The borage family (Boraginaceae s. str.): A revised infrafamilial classification based on new phylogenetic evidence, with emphasis on the placement of some enigmatic genera. *Taxon* **65:** 523–546.

Chamberlain, C. J. (1935). *Gymnosperm structure and evolution*. University of Chicago Press, Chicago.

Chandler, G. T. and G. M. Plunkett. (2003). The phylogenetic placement and evolutionary significance of the polyphyletic subfamily Hydrocotyloideae (Apiaceae). P. 75 in *Botany 2003: Aquatic and Wetland Plants: Wet and Wild.* [Mobile, Alabama.]

Chandler, G. T. and G. M. Plunkett. (2004). Evolution in Apiales: Nuclear and chloroplast markers together in (almost) perfect harmony. *Botanical J. Linnean Soc.* (cited in APweb).

Chandra, S. and K. R. Surange. (1976). Cuticular studies of the reproductive organs of Glossopteris, Part I: Dictyopteridium feismanteli sp. Nov. attached on Glossopteris tenuinervis. *Palaeontographia* **B156:** 87–102.
Chapman, M. (1936). Carpel anatomy of Berberidaceae. *Amer. J. Bot.* **23:** 340–348.
Chapman, A. D. (2009). *Numbers of Living Species in Australia and the World* (2nd ed.). Canberra: Australian Biological Resources Study. pp. 1–80. Online pdf at https://www.environment.gov.au/system/files/pages/2ee3f4a1-f130-465b-9c7a-79373680a067/files/nlsaw-2nd-complete.pdf.
Chappill, J. A. (1994). Cladistic analysis of Leguminosae. The development of an explicit hypothesis. pp. 1–9. *In*: M. D. Crisp and J. J. Doyle (eds.). *Advances in Legume Systematics*. part 7. Royal Botanic Gardens, Kew.
Chase, M. W., D. E. Soltis, R. G. Olmstead et al. (1993). Phylogenetics of seed plants: an analysis of nucleotide sequences from the plastid gene *rbc*L. *Ann. Missourie. Bot. Gdn.* **80**: 528–580.
Chase, M. W., M. R. Duvall, H. G. Hills et al. (1995a). Molecular systematics of Lilianae. pp. 109–137. *In*: M. J. Rudall et al. (eds.). *Monocotyledons: Systematics and Evolution*. Royal Botanic Gardens, Kew.
Chase, M. W., D. W. Stevenson, P. Wilkin and P. Rudall. (1995b). Monocot systematics: A combined analysis. pp. 685–730. *In*: M. J. Rudall et al. (eds.). *Monocotyledons: Systematics and Evolution*. Royal Botanic Gardens, Kew.
Chase, M. W., D. E. Soltis, P. G. Rudall, M. F. Fay, W. J. Hahn, S. Sullivan, J. Joseph, M. Molvray, P. J. Kores, T. J. Givnish, K. J. Sytsma and J. C. Pires. (2000). Higher level systematics of monocotyledons: An assessment of current knowledge and a new classification. pp. 3–16. *In*: K. L. Wilson and D. A. Morrison (eds.). *Systematics and Evolution of Monocots. Proceedings of the 2nd International Monocot Symposium*. Melbourne: CSIRO.
Chase, M. W., S. Zmarzty, M. D. Lledó, K. J. Wurdack, S. M. Swensen and M. F. Fay. (2002). When in doubt, put it in Flacourtiaceae: A molecular phylogenetic analysis based on plastid *rbc*L DNA sequences. *Kew Bull.* **57:** 141–181.
Chase, M. W., J. L. Reveal and M. F. Fay. (2009). A subfamilial classification for the expanded asparagalean families Amaryllidaceae, Asparagaceae and Xanthorrhoeaceae. *Bot. J. Linn. Soc.* **161**(2): 132–136. doi:10.1111/j.1095-8339.2009.00999.x.
Chase, M. W. et al. (2015). An updated classification of Orchidaceae. *Bot. J. Linn. Soc.* **177**(2). https://doi.org/10.1111/boj.12234.
Chatterjee, D. (1939). Studies on the endemic flora of India and Burma. *Journ. Roy. As. Soc. Beng. Sci.* **5:** 19–67.
Chatton, E. 1937. *Titres et travaux scientifiques*, Sette, Sottano, Italy.
Cheadle, V. I. (1953). Independent origin of vessels in the monocotyledons and dicotyledons. *Phytomorphology* **3**: 23–44.
Chen, Z. -D., S. R. Manchester and H. - Y. Sun. (1999). Phylogeny and evolution of Betulaceae as inferred from DNA sequences, morphology, and paleobotany. *Amer. J. Bot.* **86:** 1168–1181.
Chen, Z. -D. et al. (several authors). (2016). Tree of life for the genera of Chinese vascular plants. *J. Syst. Evol.* **54:** 277–306.
Christenhusz, M. J. M., J. L. Reveal, A. Farjon, M. F. Gardner, R. R. Mill and M. W. Chase. (2011). A new classification and linear sequence of extant gymnosperms (PDF). *Phytotaxa* **19:** 55–70.
Christenhusz, M. J. M. and J. W. Byng. (2016). The number of known plants species in the world and its annual increase. *Phytotaxa.* Magnolia Press. **261**(3): 201–217. doi:10.11646/phytotaxa.261.3.1.
Christenhusz, M. J. M. and M. W. Chase. (2014). Trends and concepts in fern classification. *Ann. Bot.* **113:** 571–594.
Christenhusz, M. J. M. and M. W. Chase. (2018). PPG recognises too many fern genera. *Taxon* **67:** 481–487.
Clark, P. J. (1952). An extension of the coefficient of divergence for use with multiple characters. *Copeia* **2**: 61–64.
Clark, L. G., W. Zhang and J. F. Wendel. (1995). A phylogeny of grass family (Poaceae) based on *ndhF* sequence data. *Syst. Bot.* **20:** 436–460.
Clark, L. G. and J. K. Triplett. (2006). Phylogeny of the Bambusoideae (Poaceae): An update. P. 212. *In*: Botany 2006—*Looking to the Future, Conserving the Past.* [Abstracts: Botanical Society of America, etc.]
Clarke, C. B. (1898). Sub-subareas of British India. *Journ. Linn. Soc.* **34.**
Clement, W. L. and G. D. Weiblen. (2009). Morphological evolution in the mulberry family (Moraceae). *Syst. Bot.* **34:** 530–552.
Clifford, H. T. (1977). Quantitative studies of inter-relationships amongst the Liliatae. *In*: K. Kubitzki (ed.). *Flowering Plants—Evolution and Classification of Higher Categories. Pl. Syst. Evol. Suppl.* **1**: 77–95.
Cocucci, A. E. (1983). New evidence from embryology in angiosperm classification. *Nordic J. Bot.* **3:** 67–73.
Colless, D. H. (1967). An examination of certain concepts in phenetic taxonomy. *Syst. Zool.* **16**: 6–27.
Collett, H. (1921). *Flora Simlensis*, 2nd ed. Thacker, Spink, Calcutta.
Constance, L. (1964). Systematic Botany—an unending synthesis. *Taxon* **13:** 257–273.
Coode, M. J. E. (1967). Revision of Genus *Valerianella* in Turkey. *Notes Roy. Bot. Gard. Edinb.* **27**: 219–256.
Copeland, H. F. (1938). The kingdoms of organisms. *Quart. Rev. Biol.* **13**: 383–420.
Copeland. (1956). *The Classification of Lower Organisms*, Palo Alto: Pacific Books.
Core, E. L. (1955). *Plant Taxonomy*. Prentice-Hall, Englewood Cliffs.
Corner, E. J. H. (1946). Centrifugal Stamens. *Journ. Arn. Arbor.* **27:** 423.
Corner, E. J. H. (1953). The durian theory extended, Part I. *Phytomorph.* **3:** 465–476, *Phytomorph.* **4:** 152–165 [Part II], 263–274 [Part III].
Cornet, B. (1986). Reproductive structures and leaf venation of Late Triassic angiosperm, *Sanmiguelia lewisii. Evol. Theory* **7**: 231–309.
Cornet, B. (1989). Reproductive morphology and biology of *Sanmiguelia lewisii* and its bearing on angiosperm evolution in Late Triassic. *Evol. Trends Plants* **3:** 25–51.
Cornet, B. (1993). Dicot-like leaf and flowers from the Late Triassic tropical Newark Supergroup rift zone, U. S. A. *Modern Geol.* **19**: 81–99.
Cornet, B. (1996). A New Gnetophyte from the Late Carnian (Late Triassic) of Texas and its bearing on the origin of the angiosperm carpel and stamen. pp. 32–67. *In*: D. W. Taylor and L. J. Hickey (eds.). *Flowering Plant Origin, Evolution and Phylogeny.* Chapman & Hall Inc., New York.

Coulter, J. M. and C. J. Chamberlain. (1913). *Morphology of Angiosperms.* D. Appleton, New York.

Coulter, J. M. and C. J. Chamberlain. (1917). *Morphology of Gymnosperms.* University of Chicago Press, Chicago.

Couper, R. A. (1958). British Mesozoic microspores and pollen grains. *Palaeontographica*, Abt. B, **103**: 75–179.

Crane, P. R., E. M. Friis and K. R. Pedersen. (1995). The origin and early diversification of angiosperms. *Nature* **374**: 27–33.

Crawford, D. J. and E. A. Julian. (1976). Seed protein profiles in the narrow-leaved species of *Chenopodium* of the Western United States: Taxonomic value and comparison with distribution of flavonoid compounds. *Am. J. Bot.* **63**: 302–308.

Crepet, W. L. (1974). Investigations of North American Cycadeoides. The Reproductive biology of *Cycadeoidea Palaeontographia* **148:** 144–169.

Crepet, W. L., K. C. Nixon and M. A. Gandolfo. (2004). Fossil evidence and phylogeny: The age of major angiosperm clades based on mesofossil and macrofossil evidence from Cretaceous deposits. *American J. Bot.* **91**: 1666–1682.

Cronquist, A., A. L. Takhtajan and W. Zimmerman. (1966). On the higher taxa of Embryobionta. *Taxon* **15**: 129–134.

Cronquist, A. (1968). *Evolution and Classification of Flowering Plants.* Houghton Mifflin, New York.

Cronquist, A. (1977). On the taxonomic significance of secondary metabolites in Angiosperms. *Plant Syst. Evol.* Suppl. **1**: 179–189.

Cronquist, A. (1981). *An Integrated System of Classification of Angiosperms.* Columbia Univ. Press, New York.

Cronquist, A. (1988). *Evolution and Classification of Flowering Plants* (2nd ed.). New York Botanical Garden, Bronx, New York.

Cuénoud, P. (2002). Introduction to expanded Caryophyllales. pp. 1–4. *In*: K. Kubitzki (ed.). *The Families and Genera of Vascular Plants. IV. Flowering Plants. Dicotyledons. Malvales, Capparales and Non-betalain Caryophyllales.* Springer, Berlin.

Cuénoud, P., V. Savolainen, L. W. Chatrou, M. Powell, R. J. Grayer and M. W. Chase. (2002). Molecular phylogenetics of Caryophyllales based on nuclear 18S rDNA and plastid *rbc*L, *atp*B, and *mat*K DNA sequences. *American J. Bot.* **89:** 132–144.

Dahlgren, G. (1989). An updated angiosperm classification. *Bot. J. Linn. Soc.* **100**: 197–203.

Dahlgren, G. (1989). The last Dahlgrenogram. System of classification of dicotyledons. pp. 249–260. *In*: K. Tan (ed.). *The Davis and Hedge Festschrift.* Edinburgh Univ. Press, Edinburgh.

Dahlgren, G. (1991). Steps towards a natural system of the dicotyledons: embryological characters. *Aliso* **13**(1): 107–165.

Dahlgren, R. in cooperation with B. Hansen, K. Jakobsen and K. Larsen. (1974). Angiospermernes taxonomy, 1. (2 and 3 in 1975, 4 in 1976). In Danish. Kobenhsvn: Akademisk Forlag.

Dahlgren, R. (1975). A system of classification of angiosperms to be used to demonstrate the distribution of characters. *Bot. Notiser* **128**: 119–147.

Dahlgren, R. (1977). Commentary on a diagrammatic presentation of the Angiosperms. pp. 253–283. *In*: K. Kubitzki (ed.). *Flowering Plants: Evolution and Classification of Higher Categories.* Plant Systematics and Evolution Suppl. 1. Springer-Verlag Wien/New York.

Dahlgren, R. (1980). A revised system of classification of angiosperms. *Bot. J. Linn. Soc.* **80**: 91–124.

Dahlgren, R., S. Rosendal-Jensen and B. J. Nielsen. (1981). A revised classification of the angiosperms with comments on the correlation between chemical and other characters. pp. 149–199. *In*: D. A. Young and D. S. Seigler (eds.). *Phytochemistry and Angiosperm Phylogeny.* Praeger, New York.

Dahlgren, R. (1983). General aspects of angiosperm evolution and macrosystematics. *Nordic. J. Bot.* **3**: 119–149.

Dahlgren, R. and F. N. Rasmussen. (1983). Monocotyledon evolution: characters and phylogenetic estimation. *Evol. Biol.* **16**: 255–395.

Dahlgren, R., H. T. Clifford and P. F. Yeo. (1985). *The Families of Monocotyledons.* Springer-Verlag, Berlin.

Dransfield, J., N. W. Uhl, C. B. Asmussen, W. J. Baker, M. M. Harley and C. E. Lewis. (2008) *Genera Palmarum—The Evolution and Classification of Palms.* Richmond, UK: Royal Botanic Gardens, Kew.

Darlington, C. D. and E. K. Janaki-Ammal. (1945). *Chromosome Atlas of Cultivated Plants.* Allen and Unwin, London.

Darlington, C. D. and A. P. Wylie. (1955). *Chromosome Altas of Flowering Plants.* Allen and Unwin, London.

Darwin, C. (1859). *The Origin of Species.* London.

Daugherty, L. H. (1941). *The Upper Triassic Flora of Arizona with a Discussion on its Geological Occurrence.* Contributions to Paleontology 526, Carnegie Institution of Washington.

Davis, C. C. and H. Schaefer. (2011). Plant evolution: Pulses of extinction and speciation in gymnosperm diversity. *Curr. Biol.* **21:** 995–998.

Davis, J. I. and R. Soreng. (1993). Phylogenetic structure in grass family (Poaceae) as inferred from chloroplast DNA restriction site variation. *Am. J. Bot.* **80:** 1444–1454.

Davis, P. H. (1960). Materials for the Flora of Turkey. IV. Ranunculaceae, II. *Notes Roy. Bot. Gard. Edinburgh* **23**: 103–161.

Davis, P. H. and V. H. Heywood. (1963). *Principles of Angiosperm Taxonomy.* Oliver and Boyd, London.

de Candolle, A. P. (1813). *Theorie elementaire de la botanique.* Paris.

de Candolle, A. P. (1824–73). *Prodromus systematis naturalis regni vegetabilis.* Paris, 17 vols.

de Jussieu, A. L. (1789). *Genera plantarum.* Paris.

de Queiroz, K. and J. Gauthier. (1990). Phylogenetic taxonomy. *Ann. Rev. Ecol. Syst.* **23:** 449–480.

de Queiroz, K. and J. Gauthier. (1992). Phylogeny as central principle in taxonomy: Phylogenetic definitions of taxon names. *Syst. Zool.* **39:** 307–322.

De Smet, Y., C. Granados Mendoza, S. Wamke, P. Goethebeur and M.-S. Samain. (2015). Molecular phylogenetics and a new (infra) generic classification to alleviate polyphyly in tribe Hydrangeeae (Cornales: Hydrangeaceae). *Taxon* **64:** 741–753.

de Soo, C. R. (1975). A review of new classification system of flowering plants (Angiospermatophyta, Magnoliophytina). *Taxon* **24**(5/6): 585–592.

Delavoryas. (1971). Biotic provinces and the Jurassic-Cretaceous floral transition. *Proc. N. Aer. Paleontol.* **L:** 1660–1674.
Dilcher, D. L. (1979). Early angiosperm reproductions: An introductory report. *Rev. Palaeobot. Palynol.* **27:** 291–328.
Dilcher, D. L. and P. L. Crane. (1984). An early Angiosperm from the Western Interior of North America. *Ann. Missouri Bot. Gard.* **71:** 380–388.
Dilcher, D. L. and P. R. Crane. (1984). Archaeanthus: an early angiosperm from the Cenomanian of the Western Interior of North America. *Ann. Mo. Bot. Gard.* **71**: 351–383.
Donoghue, M. J. and J. A. Doyle. (1989). Phylogenetic analysis of angiosperms and the relationships Hamamelidae. pp. 17–45. *In*: Pr. R. Crane and S. Blackmore (eds.). *Evolution, Systematics and Fossil History of the Hamamelidae*. Claredon Press, Oxford.
Donoghue, M. J., R. H. Ree and D. A. Baum. (1998). Phylogeny and the evolution of flower symmetry in the Asteridae. *Trends Plant Sci.* **3:** 311–317.
Donoghue, M. J., T. Eriksson, P. A. Reeves and R. G. Olmstead. (2001). Phylogeny and phylogenetic taxonomy of Dipsacales, with special reference to *Sinadoxa* and *Tetradoxa* (Adoxaceae). *Harvard Papers Bot.* **6:** 459–479.
Douglas, N. A. and P. S. Manos. (2007). Molecular phylogeny of Nyctaginaceae: Taxonomy, biogeography, and characters associated with radiation of xerophytic genera in North America. *American J. Bot.* **95:** 856–872.
Douglas, N. and R. Spellenberg. (2010). A new tribal classification of Nyctaginaceae. *Taxon* **59:** 905–910. [See also Erratum in *Taxon* **60:** 615. 2011.]
Doweld, A. B. (2001). *Tentamen Systematis Plantarum Vascularium (Tracheophytorum)*. Moscow: GEOS.
Downie, S. R., D. S. Katz-Downie and M. F. Watson. (2000a). A phylogeny of the flowering plant family Apiaceae based on chloroplast *rpl16* and *rpoC1* sequences: Towards a suprageneric classification of subfamily Apioideae. *Amer. J. Bot.* **87:** 273–292.
Downie, S. R., M. F. Watson, K. Spalik and D. S. Katz-Downie. (2000b). Molecular systematics of Old World Apioideae (Apiaceae): Relationships among some members of tribe Peucedaneae *sensu lato*, the placement of several island-endemic species, and resolution within the apioid superclade. *Canad. J. Bot.* **78:** 506–528.
Doyle, J. A. (1969). Cretaceous angiosperm pollen of Atlantic Coastal Plain and its evolutionary significance. *J. Arnold Arboretum.* **50**: 1–35.
Doyle, J. A. (1978). Origin of angiosperms. *Ann. Rev. Ecol. Systematics* **9**: 365–392.
Doyle, J. A. (2001). Significance of molecular phylogenetic analyses for paleobotanical investigations on the origin of angiosperms. *Palaeobotanist* **50:** 167–188.
Doyle, J. A. and Donoghue. (1987). The origin of angiosperms: a cladistic approach. pp. 17–49. *In*: E. M. Friis, W. G. Chaloner and P. R. Crane (eds.). *The Origins of Angiosperms and Their Biological Consequences*. Cambridge University Press, U. K.
Doyle, J. A., C. L. Hotton and J. V. Ward. (1990). Early Cretaceous tetrads, zonasulculate pollen, and Winteraceae. I. Taxonomy, morphology, and ultrastructure. *Am. J. Bot.* **77**: 1544–1557.
Doyle, J. A. and Donoghue. (1993). Phylogenies and angiosperm diversification. *Paleobiology* **19**: 141–167.
Doyle, J. A., M. Van Campo and B. Lugardon. (1975). Observations on exine structure of Eucommiidites and lower Cretaceous angiosperm pollen. *Pollen and Spores* **17**: 429–486.
Doyle, J. A. and P. K. Endress. (2000). Morphological phylogenetic analysis of basal angiosperms: comparison and combination with molecular data. *International Journal of Plant Sciences* **161**(6 suppl.): S121–S153.
Doyle, J. A., H. Sauquet, T. Scharaschkin and A. Le Thomas. (2004). Phylogeny, molecular and fossil dating, and biogeographic history of Annonaceae and Myristicaceae (Magnoliales). *Int. J. Plant Sci.* **165**(4 Suppl.): S55–S67.
Doyle, J. J. (1994). Phylogeny of legume family: An approach to understanding the origins of nodulation. *Ann. Rev. Ecol. Syst.* **25:** 325–349.
Doyle, J. J., J. A. Chappill, C. D. Bailey and T. Kajita. (2000). Towards a comprehensive phylogeny of legumes: Evidence from *rbcL* sequences and non-molecular data. pp. 1–20. *In*: P. S. Herendeen and A. Bruneau (eds.). *Advances in Legume Systematics, Part 9*. Royal Botanic Gardens, Kew.
Doyle, J. A., P. K. Endress and G. R. Upchurch. (2008). Early Cretaceous monocots: a phylogenetic evaluation. *Sborník Národního muzea v Praze*, B **64:** 59–87.
Downie, S. R., D. S. Katz-Downie and M. F. Watson. (2000). A phylogeny of the flowering plant family Apiaceae based on chloroplast *rpl*16 and *rpo*C1 sequences: Towards a suprageneric classification of subfamily Apioideae. *American J. Bot.* **87:** 273–292.
Dransfield, J., N. W. Uhl, C. B. Asmussen, W. J. Baker, M. M. Harley and C. E. Lewis. (2008). *Genera palmarum: The Evolution and Classification of Palms*. Kew Publishing, Royal Botanic Gardens, Kew.
Du Rietz. (1930). Fundamental units of biological taxonomy. *Svensk bot. Tidskr.* **24**: 333–428.
Duangjai, S., B. Wallnoeffer, R. Samuel, F. Munzinger and M. W. Chase. (2006a). Phylogenetic relationships and infrafamilial classification of Ebenaceae s.l. based on six plastid markers. pp. 218–219. *In*: *Botany 2006—Looking to the Future—Conserving the Past.* [Abstracts: Botanical Society of America, etc.]
Duangjai, S., B. Wallnoeffer, R. Samuel, F. Munzinger and M. W. Chase. (2006b). Generic delimitation and relationships in Ebenaceae *sensu lato*: Evidence from six plastid DNA regions. *American J. Bot.* **93:** 1808–1827.
Du Toit, A. (1937). *Our Wandering Continents*. Edinburgh and London.
Du, Z. -Y., Q. -F. Wang and China Phylogeny Consortium. (2016). Phylogenetic tree of vascular plants reveals the origins of aquatic angiosperms. *J. Syst. Evol.* **54:** 342–348.
Ducousso, M., G. Béna, C. Bourgeois, B. Buyck, G. Eyssartier, M. Vincelette, R. Rabevohitra, L. Randrihasipara, B. Dreyfus and Y. Prin. (2004). The last common ancestor of Sarcolaenaceae and Asian dipterocarp trees was ectomycorrhizal before the India-Madagascar separation, about 88 million years ago. *Molecular Ecology* **13:** 231.
Dykes, W. R. (1913). *The Genus Iris.*

Eames, A. J. (1961). *Morphology of Angiosperms.* McGraw-Hill Book Co., New York.
Eckenwalder, J. E. (1976). Re-evaluation of Cupressaceae and Taxodiaceae: A proposed merger. *Madroño* **23:** 237–256.
Edwards, D. (2003). Xylem in early tracheophytes. *Plant Cell. Environ.* **26:** 57–72.
Edwards, D. and J. B. Richardson. (2004). Silurian and Lower Devonian plant assemblages from the Anglo-Welsh basin: A palaeobotanical and palynological synthesis. *Geol. J.* **39:** 375–402.
Ehrendorfer, F. (1968). Geographical and ecological aspects of infraspecific differentiation. pp. 261–296. *In*: V. H. Heywood (ed.). *Modern Methods in Plant Taxonomy.* Acad. Press, New York.
Ehrendorfer, F. (1983). Summary Statement. *Nord. J. Bot.* **3**: 151–155.
Eichler, A. W. (1883). *Syllabus der Vorlesungen über Specielle und Medicinisch-Pharmaceutische Botanik.* Leipzig.
Endress, M. E., B. Sennblad, S. Nilsson, L. Civeyrel, M. W. Chase, S. Huysmans, E. Grafröm and B. Bremer. (1996). A phylogenetic analysis of Apocynaceae s. str. and some related taxa in Gentianales: A multidisciplinary approach. *Opera Bot. Belg.* **7:** 59–102.
Endress, M. E., S. Leide-Schumann and U. Meve. (2014). An updated classification for Apocynaceae. *Phytotaxa* 159: 3.2. DOI: http://dx.doi.org/10.11646/phytotaxa.159.3.2.
Endress, P. K. (1977). Evolutionary tends in Hamamelidales-Fagales-Group. *In*: H. Kubitzki (ed.). *Flowering Plants: Evolution and Classification of Higher Categories. Plant. Syst. Evol. Suppl.* **1**: 321–347.
Endress, P. K. (2008). The whole and the parts: Relationships between floral architecture, floral organ shape, and their repercussions on the interpretation of fragmentary floral fossils. *Ann. Missouri Bot. Gard.* **95:** 101–120.
Engler, A. and L. Diels. (1936). *Syllabus der pflanzenfamilien.* 11th ed. Berlin.
Engler, A. and K. Prantl. (1887–1915). *Die naturlichen pflanzenfamilien.* Leipzig, 23 vols.
Engler, A. (1892). *Syllabus der Pflanzenfamilien.* Berlin.
Engler, A. (ed.). (1900–1953). *Das Pflanzenreich.* Regni vegetabilis conspectus Im Auftrage der Preus. Akademie der Wissenschaften, Leransgegeben von A. Engler, Berlin. (after Engler's death subsequent volumes, continuing upto 1953 were edited by other authors).
Engler, A. (H. Melchior and E. Werdermann, eds.). (1954). *Syllabus der pflanzenfamilien.* 12th ed., vol. 1. Gebruder Borntraeger, Berlin.
Engler, A. (H. Melchior, ed.). (1964). *Syllabus der pflanzenfamilien.* 12th ed., vol. 2. Gebruder Borntraeger, Berlin.
Erbar, C., S. Kusma and P. Leins. (1999). Development and interpretation of nectary organs in Ranunculaceae. *Flora* **194:** 317–332.
Erdtman, G. (1948). Did dicotyledonous plants exist in Early Jurrassic time? *Geol. Fören. Stockholm Förh.* **70**: 265–271.
Erdtman, G. (1966). *Pollen Morphology and Plant Taxonomy. Angiosperms. (An Introduction to Palynology. I.).* Hafner Publ. Co., London.
Evans, R. C. and T. A. Dickinson. (2002). How do studies of comparative ontogeny and morphology aid in elucidation of relationships within the Rosaceae? P. 108, *In*: *Botany 2002: Botany in the Curriculum,* Abstracts. [Madison, Wisconsin.]
Evans, R. C., C. Campbell, D. Potter, D. Morgan, T. Eriksson, L. Alice, S. -H Oh, E. Bortiri, F. Gao, J. Smedmark and M. Arsenault. (2002a). A Rosaceae phylogeny. P. 108. *In*: *Botany 2002: Botany in the Curriculum,* Abstracts. [Madison, Wisconsin.]
Evans, R. C., T. A. Dickinson and C. Campbell. (2002b). The origin of the apple subfamily (Maloideae; Rosaceae) is clarified by DNA sequence data from duplicated GBSSI genes. *American J. Bot.* **89:** 1478–1484.
Faden, R. B. and D. R. Hunt. (1991). The classification of the Commelinaceae. *Taxon* **40:** 19–31.
Fairbrothers, D. E. (1983). Evidence from nucleic acid and protein chemistry, in particular serology, in angiosperm classification. *Nordic. J. Bot.* **3**: 35–41.
Farjon, A. (2005). *A Monograph of Cupressaceae and Sciadopitys.* Royal Botanic Gardens, Kew.
Farjon, A. (2008). *A Natural History of Conifers.* Timber Press, Portland, OR.
Farjon, A. (2017). *A Handbook of the World's Conifers.* 2 vols, Ed. 2. Brill, Leiden.
Fan, C. and Q. -Y. Xiang. (2003). Phylogenetic analyses of Cornales based on 26S rRNA and combined 26S rDNA-MATK-RBCL sequence data. *American J. Botany* **90:** 1357–1372.
Farr, E. R., J. A. Leussink and F. A. Stafleu (eds.). (1979). *Index Nominum Genericorum (Plantarum). Regnum Veg.* **100-102**: 1–1896.
Farr, E. R., J. A. Leussink and G. Zijlstra (eds.). (1986). *Index Nominum Genericorum (Plantarum) Supplementum* I. *Regnum Veg.* **113**: 1–126.
Fassett, N. C. (1957). *A Manual of Aquatic Plants.* Univ. Wisconin Press, Madison.
Faust, W. Z. and S. B. Jones. (1973). The systematic value of trichome complements in a North American Group of *Vernonia* (Compositae). *Rhodora* **75**: 517–528.
Fay, M. F., P. J. Rudall, S. Sullivan, K. L. Stobart, A. Y. de Bruijn, G. Reeves, M. Qamaruz-Zaman, W.-P. Hong, J. Joseph, W. J. Hahn, J. G. Conran and M. W. Chase. (2000). Phylogenetic studies of Asparagales based on four plastid DNA regions. pp. 360–371. *In*: K. L. Wilson and D. A. Morrison (eds.). *Monocots: Systematics and Evolution.* CSIRO, Collingwood.
Fay, M. F., B. Bremer, G. T. Prance, M. van der Bank, D. Bridson and M. W. Chase. (2000a). Plastid rbcL sequence data show Dialypetalanthus to be member of Rubiaceae. *Kew Bull.* **55:** 853–864.
Federov, A. A. (ed.). (1969). *Chromosome Numbers of Flowering Plants.* Akad. Nauk SSSR, Leningrad.
Fiori, A. and G. Paoletti. (1896). *Flora analitica d'Italia* **1**: 1–256. Padova.
Fishbein, M., C. Hibsch-Jetter, D. E. Soltis and L. Hufford. (2001). Phylogeny of Saxifragales (Angiosperms, Eudicots): Analysis of a rapid, ancient radiation. *Syst. Biol.* **50:** 817–847.
Fisher, M. J. (1928). Morphology and anatomy of flowers of Salicaceae. I. *Amer. J. Bot.* **15:** 307–326.
Fisher, R. A. (1930). *Genetical Theory of Natural Selection.* Oxford: Claredon Press.

Fischer, E. (2004). Balsaminaceae. pp. 20–25. *In*: K. Kubitzki (ed.). *The Families and Genera of Vascular Plants. VI. Flowering Plants. Dicotyledons. Ceslastrales, Oxalidales, Rosales, Cornales, Ericales.* Springer, Berlin.

Fleming, T. H., C. Geiselman and W. J. Kress. (2009). The evolution of bat pollination: A phylogenetic perspective. *Ann. Bot.* **104:** 1017–1043.

Flicker, B. J. and H. E. Jr. Ballard. (2015). *Afrohybanthus* (Violaceae), a new genus for a distinctive and widely distributed Old World hybanthoid lineage. *Phytotaxa* **230:** 39–53.

Folk, R. A., C. J. Visger, P. S. Soltis, D. E. Soltis and R. P. Guralnick. (2018). Geographic range dynamics drove ancient hybridization in a lineage of angiosperms. *American Naturalist* **192:** 171–187.

Ford, B. A., M. Iranpour, R. F. C. Naczi, J. R. Starr and C. A. Jerome. (2006). Phylogeny of Carex subg. Vignea (Cyperaceae) based on non-coding nrDNA sequence data. *Syst. Bot.* **31:** 70–82.

Forest, F., V. Savolainen, M. W. Chase, R. Lupia, A. Bruneau and P. R. Crane. (2005). Teasing apart molecular—versus fossil-based error estimates when dating phylogenetic trees: A case study in the birch family (Betulaceae). *Syst. Bot.* **30:** 118–133.

Freudenstein, J. V. and M. W. Chase. (2001). Analysis of mitochondrial *nad1*b-c intron sequences in Orchidaceae: Utility and coding of length-change characters. *Syst. Bot.* **26:** 643–657.

Friedrich, H. C. (1956). Studien über die natürliche verwandtschaft der Plumbaginales und Centrospermae. *Phyton* (*Austria*) **6**: 220–263.

Friis, E. M., K. R. Pedersen and P. R. Crane. (2010). Diversity in obscurity: fossil flowers and the early history of angiosperms. *Philos Trans R Soc. Lond. B Biol. Sci.* Feb 12; **365**(1539): 369–382.

Friis, E. M., P. R. Crane and Kaj R. Pedersen. (2011). *Early Flowers and Angiosperm Evolution.* Cambridge University Press, 585 pp.

Frodin, D. G. (1984). *Guide to the Standard Floras of the World.* Cambridge Univ. Press.

Frost, F. H. (1930). Specialization in secondary xylem in dicotyledons. I. Origin of vessels. *Botanical Gazette* **89**: 67–94.

Fuse, S. and M. N. Tamura. (2000). A phylogenetic analysis of the plastid *matK* gene with emphasis on Melanthiaceae *sensu lato. Plant Biology* **2:** 415–427.

Gagnepain, F. and Boureau. (1947). Nouvelles considerations systématische á propos du Sarcopus abberans Gagnepain. *Bull. Soc. Bot. Fr.* **94**: 182–185.

García, M. A., M. Costea, M. Kuzmina and S. Stefanovic. (2014). Phylogeny, character evolution, and biogeography of *Cuscuta* (dodders; Convolvulaceae) inferred from coding plastid and nuclear sequences. *American J. Bot.* **101:** 670–690.

Garcke, A. (1972). *Illustrierte Flore von Deutschland und Angrenzende Gebiete*, 23rd ed. (revised by K. von Weihe ed.). Parey, Berlin.

Garnock-Jones, P. J. and C. J. Webb. (1996). The requirement to cite authors of plant names in botanical journals. *Taxon* **45**: 285–286.

Gaussen, H. (1946). *Les Gymnosperms actuelles et fossiles.* Pt. 3. Travaux du Laboratoire Forestier, Toulouse.

Gershenzon, J. and T. J. Mabry. (1983). Secondary metabolites and the higher classification of angiosperms. *Nordic J. Bot.* **3**: 5–34.

Geuten, K., E. Smets, P. Schols, Y. -M. Yuan, S. Janssens, P. Küpfer and N. Pyck. (2004). Conflicting phylogenies of balsaminoid families and the polytomy in Ericales: Combining data in a Bayesian framework. *Mol. Phyl. Evol.* **31:** 711–729.

Geuten, K., A. Becker, K. Kaufmann, P. Caris, S. Janssens, T. Viaene, G. Theißen and E. Smets. (2006). Petaloidy and petal identity MADS-box genes in the balsaminoid genera. *Impatiens* and *Marcgravia. Plant J.* **47:** 501–518.

Gifford, E. M. and A. S. Foster. (1988). *Morphology and Evolution of Vascular Plants*, 3rd ed. W. H. Freeman, New York.

Givnish, T. J., A. Zuluaga, I. Marques, V. K. Y. Lam, M. S. Gomez, W. J. D. Iles, M. Ames, D. Spalink, J. R. Moeller, B. G. Briggs, S. P. Lyon, D. W. Stevenson, W. Zomlefer and S. W. Graham. (2016). Phylogenomics and historical biogeography of the monocot order Liliales: Out of Australia and through Antarctica. *Cladistics* **32:** 581–605. doi: 10.1111/cla.12153.

Givnish, T. J., K. W. Sparks, S. J. Hunter and A. Pavlovic. (2018). Why are plants carnivorous? Cost/benefit analysis, whole plant growth, and the context-specific advantages of botanical carnivory. pp. 232–255. *In*: A. M. Ellison and L. Adamec (eds.). *Carnivorous Plants. Physiology, Ecology, and Evolution.* Oxford University Press, Oxford.

Gleason, H. A. (1963). *The New Britton and Brown Illustrated Flora.* 3 Vols. Hafner, New York.

Goldblatt, Peter, A. Rodriguez, M. P. Powell, J. T. Davies, J. C. Manning, M. van der Bank and V. Savolainen. (2008). Iridaceae 'Out of Australasia'? Phylogeny, biogeography, and divergence time based on plastid DNA sequences. *Syst. Bot.* **33**(3): 495–508.

Good, R. (1931). A Theory of Plant Geography. *New Phytologist* **30**.

Good, R. (1974). The Geography of Flowering Plants. 4th ed. (Ist in 1947). Longman, London.

Gornall, R. J. 1989. Anatomical evidence and the taxonomic position of *Darmera* (Saxifragaceae). *Bot. J. Linnean Soc.* **100:** 173–182.

Gottsberger, G. (1974). Structure and function of primitive angiosperm flower—A discussion. *Acta Bot. Neerl.* **23**: 461–471.

Gower, J. C. (1966). Some distance properties of latent root and vector methods used in multivariate analysis. *Biometrika* **53**: 325–338.

Graham, S. W. and R. G. Olmstead. (2000a). Evolutionary significance of an unusual chloroplast DNA inversion found in two basal angiosperm lineages. *Curr. Genet.* **37:** 183–188.

Graham, S. W. and R. G. Olmstead. (2000b). Utility of 17 chloroplast genes for inferring the phylogeny of the basal angiosperms. *American J. Bot.* **87:** 1712–1730.

Graham, S. A., J. Hall, K. Sytsma and S. -H. Shi. (2005). Phylogenetic analysis of the Lythraceae based on four gene regions and morphology. *Internat. J. Plant Sci.* **166**: 995–1017.

Grant, V. (1957). The plant species in theory and practice. pp. 39–80. *In*: E. Mayr (ed.). *The Species Problem*. Amer. Assoc. Adv. Sci. Washington, D. C.

Grant, V. (1981). *Plant Speciation* (2nd ed.). Columbia Univ. Press, New York.

Grayum, M. (1987). A summary of evidence and arguments supporting the removal of *Acorus* from Araceae. *Taxon* **36:** 723–729.

Greenberg, A. K. and M. J. Donoghue. (2011). Molecular systematics and character evolution in Caryophyllaceae. *Taxon* **60:** 1637–1652.

Gregory, W. C. (1941). Phylogenetic and cytological studies in the Ranunculaceae. *Trans. Am. Phil. Soc.* **3**: 443–520.

Greuter, W., D. L. Hawksworth, J. Mcneill, M. A. Mayo, A. Minelli, P. H. A. Sneath, B. J. Tindall, P. Trehane and P. Tubbs. (1998). Draft BioCode (1997): the prospective international rules for the scientific names of organisms. *Taxon* **47**: 127–150.

Greuter, W., R. K. Brummitt, E. Farr, N. Kilian, P. M. Kirk and P. C. Silva. (1993). *Names in current use for extant plant genera.* Koeltz, Königstein, Germany. xxvii + 1464 pp. *Regnum veg.* Vol. **129**.

Greuter, W., J. Mcneill, F. R. Barrie, H. M. Burdet, V. Demoulin, T. S. Filgueiras, D. H. Nicolson, P. C. Silva, J. E. Skog, P. Trehane, N. J. Turland and D. L. Hawksworth (editors & compilers). (2000). International code of botanical nomenclature (St. Louis Code) adopted by the Sixteenth International Botanical Congress St. Louis, Missouri, July–August 1999. *Regnum Veg.* Vol. **138**.

Groppo, M. J. A. Kallunki, J. R. Pirani and A. Antonelli. (2012). Chilean Pitavia more closely related to Oceania and Old World Rutaceae than to Neotropical groups: Evidence from two cpDNA non-coding regions, with a new subfamilial classification of the family. *PhytoKeys* 19: 9–29.

Gunderson, A. (1939). Flower buds and phylogeny of dicotyledons. *Bull. Torrey Bot. Club* **66**: 287–295.

Gustafsson, M. H. G. (2002). Phylogeny of Clusiaceae based on *rbcL* sequences. *Int. J. Plant Sci.* **163**(6): 1045–1054.

Haeckel, E. (1866). *Generelle Morphologie der Organismen.* Reimer, Berlin.

Haeckel, E. (1887). Echte Gräser. Engler and Prantl Die natürlichen Pflanzenfamilien, II, 2.

Hahn, W. J. 2002. A molecular phylogenetic study of the Palmae (Arecaceae) based on *atp*B, *rbc*L and 18s nrDNA sequences. *Syst. Biol.* **51:** 92–112.

Hall, D. W. (1981). Microwave: a method to control herbarium insects. *Taxon* **30**: 818–819.

Hammond, P. M. (1992). Species inventory. *In*: *Global Diversity. Status of the Earth's Living Resources* (B. Groombridge, ed.) pp. 17–39. London: Chapman & Hall.

Hanelt, P. and J. Schultze-Motel. (1983). Proposal (715) to conserve *Triticum aestivum* L. (1753) against Triticum *hybernum* L. (1753) (Gramineae). *Taxon* **32**: 492–498.

Hanelt, P. (1990). Taxonomy, evolution and history. pp. 1–26. *In*: H. Rabinowitch and J. L. Brewster (eds.). *Onions and Allied Crops*, vol. 1. Boca Raton, FLA.

Hansen, A. (1920). *Die Pflanzendecke der Erde*. Leipzig.

Harbaugh, D. T., M. Nepokroeff, R. K. Rabeler, J. Mc Neill, E. A. Zimmer and W. L. Wagner. (2010). A new lineage-based tribal classification of the family Caryophyllaceae. *Internat. J. Plant Sci.* **171:** 185–198.

Harborne, J. B. and B. L. Turner. (1984). *Plant Chemosystematics*. Acad. Press, London.

Harris, J. G. and M. W. Harris. (1994). *Plant Identification Terminology: An Illustrated Glossary.* Spring Lake, Publishing, Spring Lake, UT.

Harris, T. M. (1932). The fossil flora of scorseby sound east greenland. Part 3: Caytoniales and Bennettitales. *Meddelelser om Grônland* **85**(5): 1–133.

Harris, D. J., M. F. Newman, M. L. Hollingsworth, M. Möller and A. Clark. (2006). The phylogenetic position of Aulotandra (Zingiberaceae). *Nordic J. Bot.* **23:** 725–734.

Hart, J. A. (1987). A cladistic analysis of conifers: Preliminary results. *J. Arnold Arbor.* **68:** 269–307.

Hartl, D. L. and E. W. Jones. (1998). *Genetics: Principles and Analysis.* 4th ed. Jone and Bartlett Publishers, London.

Hartmann, H. E. K. (1993). Aizoaceae. pp. 37–69. *In*: K. Kubitzki, J. G. Rohwer and V. Bittrich (eds.). *The Families and Genera of Vascular Plants*, Vol. 2, Magnoliid, Hamamelid and Caryophyllid Families. Springer-Verlag, Berlin.

Haszprunar, G. (1987). Vetigastropoda and systematics of Streptpneuros Gastropoda (Mollusca). *Journal of Zoology* **211:** 747–770.

Hawksworth, D.L. (1995). Steps along the road to a harmonized bionomenclature. *Taxon* **44**: 447–456.

Hedge, I. C. and J. M. Lamond. (1972). Umbelliferae. Multi-access key to the Turkish genera. pp. 171–177. *In*: P. H. Davis (ed.). *Flora of Turkey*. Edinburgh Univ. Press, Edinburgh, vol. 4.

Hedrén, M., M. W. Chase and R. G. Olmstead. (1995). Relationships in the Acanthaceae and related families as suggested by cladistic analysis of *rbc*L nucleotide sequences. *Plant Syst. Evol.* **194:** 93–109.

Hegi, G. (1906–1931). *Illustrierte Flora Von Mitteleuropa*. Ed. I, Munchen.

Heimerl, A. (1934). Nyctaginaceae in Engler and Prantl, Die naturlichen pflanzenfamilien, ed. 2, Bd. 16c: 86–134.

Henderson, D. M. (1983). *International Directory of Botanical Gardens IV*. Koeltz, Koenigstein.

Hennig, W. (1950). *Grundzüge einer Theorie der phylogenetischen Systematik.* Deutscher Zentralverlag, Berlin.

Hennig, W. (1957). Systematik und Phylogenese. *Ber. Hundertjahrfeier Deutsch. Entomol. Ges.*, pp. 50–70.

Hennig, W. (1966). *Phylogenetic Systematics*. Translated by D. D. Davies and R. Zangerl. Univ. Illinois Press, Urbana.

Hermsen, E. J., T. N. Taylor, E. L. Taylor and D. W. Stevenson. (2006). Cataphylls of the middle Triassic cycad *Antarcticycas schopfii* and new insights into cycad evolution. *American J. Bot.* **93:** 724–738.

Heslop-Harrison, J. (1952). A reconsideration of plant teratology. *Phyton* **4**: 19–34.

Heslop-Harrison, J. (1958). The unisexual flower—a reply to criticism. *Phytomorphology* **8**: 177–184.

Heyduk, K., M. R. McKain, F. Lalani and J. Leebens-Mack. (2016). Evolution of CAM anatomy predates the origin of crassulacean acid metabolism in the Agavoideae (Asparagaceae). *Molec. Phyl. Evol.* **105:** 102–113.
Heywood, V. H. (ed.). 1978. *Flowering Plants of the World.* Oxford University Press, London.
Hibbett, D. and M. J. Donoghue. (1998). Integrating phylogenetic analysis and classification in fungi. *Mycologia* **90:** 347–356.
Hickey, L. J. and J. A. Doyle. (1977). Early Cretaceous fossil evidence for angiosperm evolution. *Bot. Rev.* **43:** 1–104.
Hickey, L. J. and D. W. Taylor. (1992). Paleobiology of early angiosperms: evidence from sedimentological associations in Early Cretaceous Potomac Group of eastern U. S. A. *Paleontological Soc. Spec. Publ.* **6**: 128.
Hickey, L. J. and D. W. Taylor. (1996). Origin of angiosperm flower. pp. 176–231. *In*: D. W. Taylor and L. J. Hickey (eds.). *Flowering Plant Origin, Evolution and Phylogeny.* Chapman & Hall Inc., New York.
Hidalgo, O., Jaume Pellicer Maarten J. M. Christenhusz, Harald Schneider and Ilia J. Leitch. (2017). Genomic gigantism in the whisk-fern family (Psilotaceae): *Tmesipteris obliqua* challenges record holder *Paris japonica*. *Bot. J. Linn. Soc.* **183**(4): 509–514.
Hill, S. R. (1983). Microwave and the herbarium specimens: potential dangers. *Taxon* **32**: 614–615.
Hilu, K., T. Borsch, K. Muller, D. E. Soltis, P. S. Soltis, V. Savolainen, M. W. Chase, M. P. Powell, L. A. Alice, R. Evans, H. Sauquet, C. Neinhuis, T. A. B. Slotta, J. G. Rohwer, C. S. Campbell and L. W. Chatrou. (2003). Angiosperm phylogeny based on *mat*K sequence information. *American J. Bot.* **90:** 1758–1766.
Hjelmquist, H. (1948). Studies on the floral morphology and phylogeny of the Amentiferae. *Bot. Notiser,* Suppl. **2:** 1–171.
Hofmann, U. and V. Bittrich. (2016). Caprifoliaceae (with Zabelia incert. sed.), Morinaceae. pp. 117–129, 275–280. *In*: J. W. Kadereit and V. Bittrich. (eds.). *The Families and Genera of Vascular Plants, Volume 14*: Flowering Plants: Eudicots—Aquifoliales, Boraginales, Bruniales, Dipsacales, Escalloniales, Garryales, Paracryphiales, Solanales (except Convolvulaceae), Icacinaceae, Metteniusaceae, Vahliaceae. Springer.
Holmgren, P. K., N. H. Holmgren and L. C. Barnett. (1990). Index herbariorum. Part I: The Herbaria of the World (8th ed.). *Regnum Veg.* **120.**
Hooker, J. D. (1870). *Student's Flora of British Isles.* Macmillan and Co., London. (3rd ed. in 1884).
Hooker, J. D. (1872–97). *Flora of British India.* L. Reeve and Co., London, 7 vols.
Hooker, J. D. (1907). Botany. *Imperial Gazetter of India*, **1**.
Hoot, S. B. (1995). Phylogeny of Ranunculaceae based on epidermal preliminary *atpB*, *rbcL* and *18S* nuclear ribosomal DNA sequence data. *Plant Syst. Evol. Suppl.* **9:** 241–251.
Hoot, S. B., J. W. Kadereit, F. R. Blattner, K. B. Jork, A. E. Schwarzbach and P. R. Crane. (1998). Data congruence and phylogeny of the Papaveraceae *s.l.* based on four data sets: *atpB* and *rbcL* sequences, *trnK* restriction sites, and morphological characters. *Syst. Bot.* **22:** 575–590.
Horn, J. W. (2009). Phylogenetics of Dilleniaceae using sequence data from four plastid loci (rbcL, infA, rps4, rpl16 Intron). *International Journal of Plant Sciences* **170**(6): 794–813.
Horn, J. W., K. J. Wurdack and L. J. Dorr. (2016). Phylogeny and diversification of Malvales. P. 157. *In*: *Botany 2016. Celebrating our History, Conserving our Future. Savannah, Georgia.* [Abstracts.]
Hsiao, C., N. J. Chatterton, K. H. Asay and K. B. Jensen. (1994). Molecular phylogeny of Pooideae (Poaceae) based on nuclear rDNA (ITS) sequences. *Theor. Appl. Genet.* **90:** 389–398.
Huber, H. (1991). *Angiospermen. Leitfaden durch die Ordnungen und Familien der Bedektsamer.* Gustav Fischer, Stuttgart.
Hufford, L. (1997). A phylogenetic analysis of Hydrangeaceae based on morphological data. *Int. J. Plant Sci.* **158:** 652–672.
Hufford, L. (2001). Ontogeny and morphology of the fertile flowers of *Hydrangea* and allied genera of tribe Hydrangeeae (Hydrangeaceae). *Bot. J. Linn. Soc.* **137:** 139–187.
Hughes, N. F. (1961). Further interpretation of *Eucommiidites* Erdtman, 1948. *Palaeontology* **4**: 292–299.
Humboldt, A. Von. (1817). *De distributione geographica plantarum*. Paris.
Hutchinson, J. and J. M. Dalzeil. (1927–1929). *Flora of West Tropical Africa.* London, 2 vols.
Hutchinson, J. (1946). *A Botanist in South Africa.* London.
Hutchinson, J. (1948). *British Flowering Plants*. London.
Hutchinson, J. (1964–67). *The Genera of Flowering Plants.* Claredon, Oxford, 2 vols.
Hutchinson, J. (1968). *Key to the Families of Flowering Plants of the World.* Clarendon, Oxford, 117 pp.
Hutchinson, J. (1969). *Evolution and Phylogeny of Flowering Plants.* Acad. Press, London.
Hutchinson, J. (1973). *The Families of Flowering Plants.* (3rd ed.). Oxford Univ. Press. (2nd ed. 1959; Ist ed. 1926, 1934).
Index Kewensis Plantarum Phanerogamarum. (1893–95), 2 vols. Oxford. 16 *supplements* up to 1971.
Jaccard, P. (1908). Nouvelles recherches sur la distribution florale. *Bull. Soc. Vaud. Sci. Nat.* **44**: 223–270.
Janesen, R. K. and J. D. Palmer. (1987). A chloroplast DNA inversion marks an ancient evolutionary split in sunflower family (Asteraceae). *Proc. Nat. Acad. Sc., USA* **84**: 5818–5822.
Janssens, S., K. Geuten, Y. M. Yuan, Y. Song, P. Küpfer and E. Smets. (2006). Phylogenetics of *Impatiens* and *Hydrocera* (Balsaminaceae) using chloroplast *atp*B-*rbc*L spacer sequences. *Syst. Bot.* **31:** 171–180.
Jardine, N. and R. Sibson. (1971). *Mathematical Taxonomy.* Wiley, London.
Ji, Q., L. M. Bowe, Y. Liu and D. W. Taylor. (2004). Early Cretaceous *Archaefructus eoflora* sp. nov., with bisexual flowers from Beipiao, *Western Liaoning, China. Acta Geol. Sinica* **78**: 883–896.
Jian, S., P. S. Soltis, A. Dhingra, R. Li, Y. -L. Qiu, M. -J. Yoo, C. Bell and D. E. Soltis. (2006). Phylogenetic relationships and diversification within Saxifragales based on molecular data. P. 229, *In*: *Botany 2006—Looking to the Future—Conserving the Past.* [Abstracts: Botanical Society of America, etc.]

Johnson, B. L. (1972). Seed protein profiles and the origin of the hexaploid wheats. *Amer. J. Bot.* **59**: 952–960.

Johnson, L. A. S. (1959). The Families of Cycads and Zamiaceae of Australia. *Proc. Linn. Soc.* **84:** 64–117.

Johnson, L. A. S. and K. L. Wilson. (1993). Casuarinaceae. pp. 237–242. *In*: K. Kubitzki, J. G. Rohwer and V. Bittrich (eds.). *The Families and Genera of Vascular Plants. II. Flowering Plants: Dicotyledons, Magnoliid, Hamamelid and Caryophyllid Families.* Springer, Berlin.

Jones, S. B. Jr. and A. E. Luchsinger. (1986). *Plant Systematics*, 2nd ed. McGraw-Hill Book Co., New York.

Jordan, A. (1873). Remarques sur le fait de l'existence en société à l'état sauvage des espéces végétales affines. *Bull. Ass. Fr. Avanc. Sci.* **2**, session Lyon.

Judd, W. S., R. W. Sanders and M. J. Donoghue. (1994). Angiosperm family pairs: Preliminary cladistic analyses. *Harvard Pap. Bot.* No. **5:** 1–51.

Judd, W. S., C. S. Cambell, E. A. Kellogg, P. F. Stevens and M. J. Donoghue. (2002). *Plant Systematics: A Phylogenetic Approach.* 2nd ed. Sinauer Associates, Inc., USA (Ist ed., 1999).

Kallersjo, M., G. Bergqvist and A. A. Anderberg. (2000). Generic realignment in primuloid families of the Ericales s.l. (Angiosperms): A phylogenetic analysis based on DNA sequences of *rbc*L and *ndh*F. *American J. Bot.* **87:** 1325–1341.

Kårehed, J., J. Lundberg, B. Bremer and K. Bremer. (1999). Evolution of the Australian Australasian families Alseuosmiaceae, Argophyllaceae and Phellinaceae. *Systematic Botany* **24:** 660–682.

Karp, G. (2002). *Cell and Molecular Biology: Concepts and Experiments.* 3rd ed. John Wiley and Sons, New York.

Keating, R. C. (2003a). *Anatomy of the Monocotyledons. IX. Acoraceae and Araceae* (ed. Gregory, M., & Cutler, D. F.). Oxford University Press, Oxford.

Keating, R. C. (2003b). Vegetative anatomical data and its relationship to a revised classification of the genera of Araceae. *Missouri Bot. Gard. Monogr. Syst. Bot.* [in press; cited in APweb].

Keating, R. C. (2004a). Vegetative anatomical data and its relationship to a revised classification of the genera of Araceae. *Ann. Missouri Bot. Gard.* **91:** 485–494.

Keating, R. C. (2004b). Systematic occurrence of raphide crystals in Araceae. *Ann. Missouri Bot. Gard.* **91:** 495–504.

Keller, R. (1996). *Identification of Tropical Woody Plants in the Absence of Flowers and Fruits: A Field Guide.* Birkauser, Basel.

Keller, R. A., R. N. Boyd and Q. D. Wheeler. (2003). Illogical basis of Phylogenetic nomenclature. *Bot. Rev.* **69(1):** 93–110.

Kellogg, E. A. (2000). The grasses: A case study in macroevolution. *Ann. Rev. Ecol. Syst.* **31:** 217–38.

Kerguélen, M. (1980). Proposal (68) on article 57.2 to correct the *Triticum* example. *Taxon* **29**: 516–517.

Kimura, M. (1968). Evolutionary rate at the molecular level. *Nature* **217**: 624–626.

Kimura, M. (1983). *The Neutral Theory of Molecular Evolution.* Cambridge University Press, Cambridge.

Klak, C., A. Khunou, G. Reeves and T. A. J. Hedderson. (2003). A phylogenetic hypothesis for the Aizoaceae (Caryophyllales) based on four plastid DNA regions. *American J. Bot.* **90:** 1433–1445.

Klak, C., P. V. Bruyns and T. A. J. Hedderson. (2007). A phylogeny and new classification for Mesembryanthemoideae (Aizoaceae). *Taxon* **56:** 737–756.

Klak, C., P. Hanácek and P. V. Bruyns. (2017). Disentangling the Aizooideae: New generic concepts and a new subfamily in Aizoaceae. *Taxon* **66:** 1147–1170.

Kluge, A. G. and J. S. Farris. (1969). Quantitative phyletics and the evolution of anurans. *Syst. Zool.* **18**: 1–32.

Komarov, V. L. and B. K. Shishkin. (1934–1964). *Flora SSSR.* AN SSSR Press, Moscow/Leningrad, 30 vols.

Köppen, W. P. and R. Gieger. (1930). *Handbuch der Klimatologie*. Vol. 1, part A. Berlin.

Kosakai, H., M. F. Moseley and V. I. Cheadle. (1970). Morphological studies in the Nymphaeaceae. V. Does *Nelumbo* have vessels? *Amer. J. Bot.* **57**: 487–494.

Kramer, K. U. and P. S. Green (eds.). (1990). Vol. I, Pteridophytes and Gymnosperms. pp. 284–391. *In*: K. Kubitzki (ed.). *Families and Genera of Vascular Plants.* Springer-Verlag, Berlin.

Krassilov, V. A. (1977). Contributions to the knowledge of the Caytoniales. *Rev. Paleobot. Palynology* **24**: 155–178.

Kraus, R. (1897). Über Specifishe Reactionin in Keimfreien Filtraten aus Cholera, Typhus und Pestbouillon Culturen erzeugt durch homologes Serum. *Weiner Klin. Wechenschr.* **10**: 136–138.

Kron, K. A. (1996). Phylogenetic relationships of Empetraceae, Epacridaceae, Ericaceae, Monotropaceae and Pyrolaceae. Evidence from nucleotide ribosomal 18S sequence data. *Ann. Bot.* **77:** 293–303.

Kubitzki, K. (ed.) (1993). *The Families and Genera of Vascular Plants, Vol. II. Flowering Plants, Dicotyledons: Magnoliid, Hamamelid and Caryophyllid Families.* Springer-Verlag, New York.

Kubitzki, K. and M. W. Chase. (2002). Introduction to Malvales. pp. 12–16. *In*: K. Kubitzki (ed.). *The Families and Genera of Vascular Plants. IV. Flowering Plants. Dicotyledons. Malvales, Capparales and Non-betalain Caryophallales.* Springer, Berlin.

Lam, H. J. (1961). Reflections on angiosperm phylogeny. I and II. Facts and theories. *Koninklijke Akademie van Wetenschappen te Amsterdam, Afdruunken Natuurkunde, Procesverbaal* **64:** 251–276.

Lamarck, J. B. P. (1778). *Flore Francaise*. Imprimerie Royale, Paris, 3 vols.

Lamarck, J. B. P. (1809). *Philosophie Zoologique*. Paris.

Lamb Frye, A. S. and K. A. Kron. (2003). *Rbc*L phylogeny and character evolution in Polygonaceae. *Syst. Bot.* **28:** 326–332.

Lance, G. N. and W. T. Williams. (1967). A general theory of classificatory sorting strategies. 1. Hierarchical systems. *Computer J.* **9**: 373–380.

Lagoudakis, C. H., C. B. Lund, N. E. Iwanycki, O. Seberg, G. Petersen, A. K. Jäger and N. Rønstedb. (2015). Identification of common horsetail (Equisetum arvense L.; Equisetaceae) using thin layer chromatography versus DNA barcoding. *Sci Rep.* **5**: 11942.
Lapage, S. P., P. H. A. Sneath, E. F. Lessel, V. B. D. Skerman, H. P. R. Seeliger and W. A. Clark (eds.). (1992). *International Code of Nomenclature of Bacteria* (Bacteriological Code 1990 Revision). *Amer. Soc. Microbiol.*, Washington, D.C. xlii + 189 pp.
Laubengayer, R. A. (1937). Studies in the anatomy and morphology of the polygonaceous flower. *Amer. J. Bot.* **24:** 329–343.
Lawrence, G. H. M. (1951). *Taxonomy of Vascular Plants.* Macmillan, New York.
Lee, T. B. (1979). *Illustrated Flora of Korea.* Hyangmunsa, Seoul.
Lee, Y. S. (1981). Serological investigations in *Ambrosia* (Compositae: Ambrosieae) and relatives. *Syst. Bot.* **6**: 113–125.
Lemesle, R. (1946). Les divers types de fibres a ponctuations areolees chez les dicotyledones apocarpiques les plus archaiques et leur role dans la phylogenie. *Ann. Sci. Nat. Bot. et Biol. Vegetal* **7**: 19–40.
Les, D. H., D. K. Garvin and C. F. Wimpee. (1991). Molecular evolutionary history of ancient aquatic angiosperms. *Proc. Nat. Acad. Sc. USA* **88:** 10119–10123.
Les, D. H. and R. R. Haynes. (1995). Systematics of subclass Alismatidae: A synthesis of approaches. pp. 353–377. *In*: P. J. Rudall, P. J. Cribb, D. F. Cutler and C. J. Humphries (eds.). *Monocotyledons: Systematics and Evolution*, vol. 2. Royal Botanic Gardens, Kew.
Les, D. H., M. A. Cleland and M. Waycott. (1997). Phylogenetic studies in Alismatidae, II: Evolution of marine angiosperms (seagrasses) and hydrophily. *Syst. Bot.* **22:** 443–463.
Les, D. H. and N. P. Tippery. (2013). In time and with water... The systematics of alismatid monocotyledons. pp. 118–164. *In*: P. Wilkin and S. J. Mayo (eds.). *Early Events in Monocot Evolution*, Cambridge University Press, Cambridge. [Systematics Association Special Volume 83.]
Levin, R. A., W. L. Wagner, P. C. Hoch, M. Nepokroeff, J. C. Pires, E. A. Zimmer and K. J. Sytsma. (2003). Family-level relationships of Onagraceae based on rbcL and ndhF data. *Amer. J. Bot.* **90(1):** 107–115.
Levin, R. A., W. L. Wagner, P. C. Hoch, W. J. Hahn, A. Rodriguez, D. A. Baum, L. Katinas, E. A. Zimmer and K. J. Sytsma. (2004). Paraphyly in Tribe Onagreae: Insights into phylogenetic relationships of onagraceae based on nuclear and chloroplast sequence data. *Systematic Botany* **29(1):** 147–164.
Linnaeus, C. (1730). *Hortus uplandicus.* Stockholm.
Linnaeus, C. (1735). *Systema naturae* (2nd ed.). Lugduni Batavorum. Stockholm.
Linnaeus, C. (1737). *Critica botanica.* Leyden.
Linnaeus, C. (1737). *Flora Lapponica.* Amsterdam.
Linnaeus, C. (1737). *Genera* plantarum. Lugduni Batavorum.
Linnaeus, C. (1737). *Hortus Cliffortianus.* Amsterdam.
Linnaeus, C. (1751). *Philosophica botanica.* Stockholm, 362 pp.
Linnaeus, C. (1753). *Species* plantarum. Stockholm, 2 vols.
Linnaeus, C. (1762). *Fundamenta fructificationis.* Stockholm.
Li, L. -F., Y. -M. Tuan Häkkinen, G. Hao and X. -J. Ge. (2010). Molecular systematics and phylogeny of the banana family (Musaceae) inferred from multiple nuclear and chloroplast DNA fragments, with a special reference to the genus Musa. *Mol. Phyl. Evol.* **57:** 1–10.
Li, R. -Q., Z. -D. Chen, A. -M. Lu, D. E. Soltis, P. S. Soltis and P. S. Manos. (2004). Phylogenetic relationships in Fagales based on DNA sequences from three genomes. *Internat. J. Plant Sci.* **165:** 311–324.
Lin, R. -Z., J. Zeng and Z. -D. Chen. (2010). Organogenesis of reproductive structures in Betula alnoides (Betulaceae). *Internat. J. Plant Sci.* **171:** 586–594.
Liu, A. -Z., W. J. Kress and D. -Z. Li. (2010). Phylogenetic analyses of the banana family (Musaceae) based on nuclear ribosomal (ITS) and chloroplast (trnL-F) evidence. *Taxon* **59:** 20–28.
Liu, S. -H., P. C. Hoch, M. Diazgranados, P. H. Raven and J. C. Barber. (2017). Multi-locus phylogeny of *Ludwigia* (Onagraceae): Insights on infrageneric relationships and the current classification of the genus. *Taxon* **66:** 1112–1127.
Loconte, H. (1993). Berberidaceae. pp. 147–152. *In*: K. Kubitzki, J. G. Rohwer and V. Bittrich (eds.). *The Families and Genera of Vascular Plants, vol. 2, Magnoliid, Hamamelid and Caryophyllid Families.* Springer-Verlag, Berlin.
Loconte, H. (1996). Comparison of alternative hypotheses for the origin of Angiosperms. pp. 267–285. *In*: D. W. Taylor and L. J. Hickey (eds.). *Flowering Plant Origin, Evolution and Phylogeny.* Chapman and Hall, Inc., New York.
Loconte, H. and D. W. Stevenson. (1991). Cladistics of Magnoliidae. *Cladistics* **7**: 267–296.
Löve, A., D. Löve and R. E. G. Pichi-Sermolli. (1977). *Cytotaxonomic Atlas of Pteridophytes.* Cramer, Koenigstein.
LPWG (The Legume Phylogeny Working Group). (2017). A new subfamily classification of the Leguminosae based on a taxonomically comprehensive phylogeny. *Taxon* **66**(1): 44–77.
Lundberg, J. (2001). A well resolved and supported phylogeny of Euasterids II based on a Bayesian inference, with special emphasis on Escalloniaceae and other incertae sedis. Chapter V. *In*: J. Lundberg (ed.). *Phylogenetic Studies in the Euasterids II with Particular Reference to Asterales and Escalloniaceae.* Acta Universitatis Upsaliensis, Uppsala.
Lundberg, J. and K. Bremer. (2002). A phylogenetic study of the order Asterales using one large morphological and three molecular data sets. *International Journal of Plant Sciences.*
Maas, P. J. M. and L. Y. T. Westra. (1993). Neotropical Plant Families. Koeltz, Koenigstein.
Mabberley, D. J. (1997). *The Plant Book.* Ed. 2. Cambridge University Press, Cambridge.
Mabry, T. J. (1976). Pigment dichotomy and DNA-RNA hybridization data for Centrospermous families. *Pl. Syst. Evol.* **126**: 79–94.

Maheshwari, J. K. (1963). *Flora of Delhi*. CSIR, New Delhi.

Maheshwari, P. (1964). Embryology in relation to taxonomy. pp. 55–97. *In*: W. B. Turril (ed.). *Vistas in Botany*. Pergamon Press, London, vol. **4**.

Malcomber, S. T. and E. A. Kellogg. (2005). SEPALLATA gene diversification: Brave new whorls. *Trends Plant Sci.* **10:** 427–435.

Malcomber, S. T., J. C. Preston, R. Reinheimer, J. Kossuth and E. A. Kellogg. (2006). Developmental gene evolution and the origin of grass inflorescence diversity. *Adv. Bot. Res.* **44:** 425–481.

Maldonado de Magnano, S. (1986). Estudios embriologicos en *Buddleja* (Buddlejaceae) I: Endosperma y episperma. *Darwiniana* **27:** 225–236.

Meng, Shao-Wu, Zhi-Duan Chen, De-Zhu Li and Han-Xing Liang. (2002). Phylogeny of Saururaceae based on mitochondrial *matR* gene sequence Data. *J. Plant Res.* **115**: 71–76.

Manning, S. D. (2000). The genera of Bignoniaceae in the Southeastern United States. *Harvard Papers Bot.* **5:** 1–77.

Manos, P. S., K. C. Nixon and J. J. Doyle. (1993). Cladistic analysis of restriction site variation within the chloroplast DNA inverted repeat region of selected Hamamelididae. *Syst. Bot.* **18:** 551–562.

Manos, P. S. and K. P. Steele. (1997). Phylogenetic analysis of "higher" hamamelididae based on plastid sequence data. *Syst. Bot.* **84:** 1407–1419.

Manos, P. S., Z.-K. Zhou and C. H. Cannon. (2001). Systematics of Fagaceae: Phylogenetic tests of reproductive trait evolution. *Int. J. Plant Sci.* **162:** 1361–1379.

Marchant, C. J. (1968). Evolution in *Spartina* (Gramineae) III. Species chromosome numbers and their taxonomic significance. *Bot. J. Linn. Soc. (London)* **60**: 411–417.

Markham, K. R., L. J. Porter, E. O. Cambell, J. Chopin and M. L. Bouillant. (1976). Phytochemical support for the existence of two species in the genus *Hymenophyton*. *Phytochemistry* **15**: 1517–1521.

Martin, W., D. Lydiate, H. Brinkmann, G. Forkmann, H. Saedler and R. Cerff. (1993). Molecular phylogenies in angiosperm evolution. *Mol. Biol. Evol.* **10**: 140–162.

Mason, H. L. (1936). The principles of geographic distribution as applied to floral analysis. *Madrono* **12:** 161–169.

Mason-Gamer, R. J. and E. A. Kellogg. (1996). Chloroplast DNA analysis of the monogenomic Triticeae: Phylogenetic implications and genome-specific markers. pp. 301–325. *In*: P. Jauhaur (ed.). *Methods of Genome Analysis in Plants: Their Merits and Pitfalls*. CRC Press, Boca Raton, FL.

Mason-Gamer, R. J., C. F. Weil and E. A. Kellogg. (1998). Granule bound starch synthase: Structure, function and phylogenetic utility. *Mol. Biol. Evol.* **15:** 1658–1673.

Mathew, K. M. (1983). *The Flora of the Tamil Nadu Carnatic*. The Rapinat Herbarium, St. Joseph's College, Tirucherapalli, India, 3 vols.

Mathews, S. and R. A. Sharrock. (1996). The phytochrome gene family in grasses (Poaceae): A phylogeny and evidence that grasses have a subset of loci found in dicot angiosperms. *Mol. Biol. Evol.* **13:** 1141–1150.

Matthews, M. L. and P. K. Endress. (2002). Comparative floral morphology and systematics in Oxalidales (Oxalidaceae, Connaraceae, Brunelliaceae, Cephalotaceae, Cunoniaceae, Elaeocarpaceae, Tremandraceae). *Bot. J. Linn. Soc.* **140:** 321–381.

Mauseth, J. D. (1998). *Botany: An Introduction to Plant Biology.* 2/e multimedia enhanced edition. Jones and Bartlett Publishers, Massachusetts.

Mayo, S. J., L. Cabrera, G. Salazar and M. Chase. (2003). Aroids and their watery beginnings. Ms. (cited in APweb)

Mayr, E. (1942). *Systematics and the Origin of Species.* Columbia Univ. Press, New York.

Mayr, E. (1957). *The Species Problem.* Amer. Assoc. Adv. Sci. Pub. No. 50.

Mayr, E. (1963). *Animal Species and Evolution.* Belknap Press, Harvard Univ. Press, Cambridge.

Mayr, E. (1966). The proper spelling of taxonomy. *Systematic Zool.* **15**: 88.

Mayr, E. (1969). *Principles of Systematic Zoology*. McGraw-Hill, New York.

Mayr, E. and P. D. Ashlock. (1991). *Principles of Systematic Zoology*. 2nd. ed. McGraw-Hill, New York.

Mayr, E. (1998). Two empires or three? *Proceedings of the National Academy of Science* **95**: 9720–9723.

McDade, L. A. (1992). Pollinator relationships, biogeography, and phylogenetics. *Bioscience* **42:** 21–26.

McDade, L. A., T. F. Daniel and C. A. Kiel. (2008). Toward a comprehensive understanding of phylogenetic relationships among lineages of Acanthaceae s.l. (Lamiales). *American J. Bot.* **95:** 1136–1152.

McDade, L. A., T. F. Daniel, C. A. Kiel and A. J. Borg. (2012). Phylogenetic placement, delimitation, and relationships among genera of the enigmatic Nelsonioideae (Lamiales: Acanthaceae). *Taxon* **61:** 637–651.

McLay, T. G. B. and M. J. Bayly. (2016). A new family placement for Australian blue squill, Chamaescilla: Xanthorrhoeaceae (Menerocallidoideae), not Asparagaceae. *Phytotaxa* **275:** 97–111.

McMillan, C., T. J. Mabry and P. I. Chavez. (1976). Experimental hybridization of *Xanthium strumarium* (Compositae) from Asia and America, II. Sesquiterpene lactones of F1 hybrids. *Amer. J. Bot.* **63**: 317–323.

McMinn, H. E. and E. Maino. (1946). *Illustrated Manual of Pacific Coast Trees,* 2nd Ed. Univ. of Calif. Press, Berkeley, CA.

McNeill, J., F. R. Barrie et al. (eds.). (2006). International Code of Botanical Nomenclature (Vienna Code) 2006. Adopted by the Seventeenth International Botanical Congress, Vienna, Austria, July 2005. Series: Regnum Vegetabile Vol. **146**. 568 pages. Gantner Verlag.

Meeuse, A. J. D. (1963). The multiple origins of the angiosperms. *Advancing Frontiers of Plant Sciences* **1**: 105–127.

Meeuse, A. J. D. (1972). Facts and fiction in floral morphology with special reference to the Polycarpicae. *Acta Bot. Neerl.* **21**: 113–127, 235–252, 351–365.
Meeuse, A. J. D. (1990). *All about Angiosperms.* Eburon, Delft.
Meglitsch, P. A. (1954). On the nature of species. *Zyst. Zool.* **3**: 49–65.
Melville, R. (1962). A new theory of the angiosperm flower, I. The gynoecium. *Kew Bull.* **16**: 1–50.
Melville, R. (1963). A new theory of the angiosperm flower, II. *Kew Bull.* **17**: 1–63.
Melville, R. (1983). Glossopteridae, Angiospermidae and the evidence of angiosperm origin. *Bot. J. Linn. Soc.* **86**: 279–323.
Melville, R. and H. M. Heybroek. (1971). The Elms of the Himalayas. *Kew Bull.* **26**(1): 5–28.
Mennes, C. B., E. F. Smets, S. N. Moses and V. S. Merckx. (2013). New insights into the long-debated evolutionary history of Triuridaceae (Pandanales). *Molec. Phyl. Evol.* **69:** 994–1004.
Mérat, F. V. (1821). *Nouvelle flore des environs de Paris* ed. 2, 2 Paris, 107 pp.
Merckx, V., L. W. Chatrou, B. Lemaire, M. N. Sainge, S. Huysmans and E. F. Smets. (2008). Diversification of myco-heterotrophic angiosperms: Evidence from Burmanniaceae. *BMC Evol. Biol.* **8:** 178. http://www.biomedcentral.com/1471-2148/8/178.
Metcalfe, C. R. and L. Chalk. (eds.) (1983). *Anatomy of Dicotyledons* (2nd ed.). Claredon Press, Oxford. (Takhtajan's classification, vol., 2 pp. 258–300).
Miao, M. -Z., H. -F. Liu, Y. -F. Kuang, P. Zou and J. P. Liao. (2014a). Floral vasculature and ontogeny in *Canna indica. Nordic J. Bot.* **32:** 485–492.
Miao, M. -Z., H. -F. Liu, Y. -F. Kuang, P. Zou, Y. -K. He and J. P. Liao. (2014b). Floral vasculature anatomy of *Canna glauca* 'Erebus' (Cannaceae). *J. Trop. Subtrop. Bot.* **22:** 344–350.
Michener, C. D. and R. R. Sokal. (1957). A quantitative approach to a problem in classification. *Evolution* **11**: 130–162.
Miki, S. (1937). The origin of *Najas* and *Potamogeton. Bot. Mag. Tokyo* **51:** 290–480.
Mirov, N. T. (1961). *Composition of Gum Terpentine's of Pines.* U. S. Dept. Agric. Tech. Bull. 1239.
Mirov, N. T. (1967). *The Genus Pinus.* Ronald Press, New York.
Mitchell, J. D., D. C. Daly, S. K. Pell and A. Randrianasolo. (2006). *Poupartiopsis* gen. nov. and its context in Anacardiaceae classification. *Syst. Bot.* **31:** 337–348.
Moore, G., K. M., Z. Devos, Z. Wang and M. D. Gale. (1995). *Current Opinion Genet. Devel.* **5:** 737.
Moore M. J., C. D. Bell, P. S. Soltis and D. E. Soltis. (2007). Using plastid genome-scale data to resolve enigmatic relationships among basal angiosperms. *Proc. Nat. Acad. Sc., USA* **104:** 19363–19368.
Moore M. J., S. Soltis, C. D. Bell, J. G. Burleigh and D. E. Soltis. (2010). Phylogenetic analysis of 83 plastid genes further resolves the early diversification of eudicots. *Proc. Nat. Acad. Sci., USA* **107:** 4623–4628.
Mora, Camilo, Derek P. Tittensor, Sina Adl and Alastair G. B. Simpson. (2011). How many species are there on earth and in the ocean? *PLoS Biology* 9(8): e1001127. DOI: 10.1371/journal.pbio.1001127.
Muasya, A. M., D. A. Simpson, M. W. Chase and A. Culham. (1998). An assessment of suprageneric phylogeny in Cyperaceae using *rbc*L DNA sequences. *Plant Syst. Evol.* **211:** 257–271.
Muhammad, A. F. and R. Sattler. (1982). Vessel structure of *Gnetum* and the origin of angiosperms. *Amer. J. Bot.* **69**: 1004–1021.
Müller, K. and T. Borsch. (2005a). Phylogenetics of Amaranthaceae based on *mat*K/*trn*K sequence data—evidence from parsimony, likelihood, and Bayesian analysis. *Ann. Missouri Bot. Gard.* **92:** 66–102.
Müller, K. and T. Borsch. (2005b). Multiple origins of a unique pollen feature: Stellate pore ornamentation in Amaranthaceae. *Grana* **44:** 266–281.
Nadaf, A. and R. Zanan. (2012). *Indian Pandanaceae—An Overview*. Springer India, Heidelberg.
Naik, V. N. (1984). *Taxonomy of Angiosperms.* Tata McGraw Hill, New Delhi.
Nandi, O. I., M. W. Chase and P. K. Endress. (1998). A combined cladistic analysis of angiosperms using *rbc*L and non-molecular data sets. *Ann. Missouri Bot. Gard.* **85:** 137–212.
Neubig, K., F. Herrera, S. Manchester and J. R. Abbott. (2012). Fossils, biogeography and dates in an expanded phylogeny of Ulmaceae. P. 143. *In*: *Botany 2012: The Next Generation. July 7–11—Columbus, Ohio. Abstracts.*
Neumayer, H. (1924). Die Geschichte der Blûte. *Abhandlung Zoologischen Botanische Gesellschaft* **14**: 1–110.
Nickrent, D. L. (2002). Orígenes filogenéticos de las plantas parásitas. pp. 29–56. *In*: J. A. López-Sáez, Catalán and L. Sáez (eds.). *Plantas Parásitas de la Península Ibérica e Islas Baleares.* Mundi-Prensa, Madrid.
Nicolas, A. N. and G. M. Plunkett. (2009). The demise of subfamily Hydrocotyloideae (Apiaceae) and the re-alignment of its genera across the whole order Apiales. *Molec. Phyl. Evol.* **53:** 134–151.
Nicolson, D. H. (1974). Paratautonym, a comment on proposal 146. *Taxon* **24**: 389–390.
Nixon, K. C. and J. M. Carpenter. (2000). On the other "Phylogenetic Systematics". *Cladistics* **16**: 298–318.
Nixon, K. C., J. M. Carpenter and D. W. Stevenson. (2003). The PhyloCode is fatally flawed, and the "Linnaean" System can be easily fixed. *Bot. Rev.* **69(1):** 111–120.
Nyananyo, B. L. and S. I. Mensah. (2004). Distribution and origins of members of the family Portulaceae (Centrospermae). *Journal of Applied Sciences & Environmental Management* **8**(2): 59–62.
Nyffeler, R. 2002. Phylogenetic relationships in the cactus family (Cactaceae) based on evidence from *trn*K/*mat*K and *trn*L-*trn*F sequences. *American J. Bot.* **89:** 312–326.
Ogundipe, O. T. and M. Chase. (2009). Phylogenetic analyses of Amaranthaceae based on matK DNA sequence data with emphasis on West African species. *Turkish J. Bot.* **33**: 153–161.

Ohba, H. and S. Akiyama. (2016). Generic segregation of some sections and subsections of the genus *Hydrangea* (Hydrangeaceae). *J. Japanese Bot.* **91:** 345–350.

Olmstead, R. G. and P. A. Reeves. (1995). Evidence for polyphyly of Scrophulariaceae based on chloroplast *rbcL* and *ndhF* sequences. *Ann. Missouri Bot. Gard.* **82:** 176–193.

Olmstead, R. G., J. A. Sweere, R. E. Spangler, L. Bohs and J. D. Palmer. (1999). Phylogeny and provisional classification of the Solanaceae based on chloroplast DNA. pp. 111–137. *In*: M. Nee, D. Symon, R. N. Lester and J. P. Jessop (eds.). *Solanaceae IV: Advances in Biology and Utilization.* Royal Botanic Gardens, Kew.

Olmstead, R. G., K. -J Kim, R. K. Jansen and S. J. Wagstaff. (2000). The phylogeny of the Asteridae *sensu lato* based on chloroplast *ndhf* gene sequences. *Molecular Phylog. Evol.* **16:** 96–112.

Olmstead, R. G., C. W. dePamphilis, A. D. Wolfe, N. D. Young, W. J. Elisens and P. A. Reeves. (2001). Disintegration of the Scrophulariaceae. *American J. Bot.* **88:** 348–361.

Olson, A. R. (1991). Postfertilization changes in ovules of *Monotropa uniflora* L. (Monotropaceae). *American J. Bot.* **78:** 99–107.

Owen, R. (1848). Report on the archetype and homologies of vertebrate skeleton. *Rep. 16th Meeting Brit. Assoc. Adv. Sci.* 169–340.

Owenby, M. (1950). Natural hybridisation and amphiploidy in the genus *Tragopogon. Amer. J. Bot.* **37**(10): 487–499.

Pace, M. R., L. G. Lohmann, R. G. Olmstead and V. Angyalossy. (2015). Wood anatomy of major Bignoniaceae clades. *Plant Syst. Evol.* **301**: 967–995.

Page, C. N. (1979). The herbarium preservation of Conifer specimens. *Taxon* **28**: 375–379.

Page, C. N. (1990). Pinata. pp. 290–361. *In*: K. Kubitzki (ed.). *The Families and Genera of Vascular Plants. Volume 1. Pteridophytes and Gymnosperms.* Springer, Berlin.

Pant, D. D. (1957). Classification of gymnospermous plants. *Palaeobotanist* **6:** 65–70.

Pant, D. D. and P. F. Kidwai. (1964). On the diversity in the development and organisation of stomata in *Phyla nodiflora* Michx. *Curr. Sci.* **33**: 653–654.

Pappas, S. (2016). *There Might Be 1 Trillion Species on Earth.* https://www.livescience.com/54660-1-trillion-species-on-earth.html.

Patchell, M. J., E. H. Roalson and J. C. Hall. (2014). Resolved phylogeny of Cleomaceae based on all three genomes. *Taxon* **63:** 315–328.

Paton, A., N. Brummitt, R. Govaerts, K. Harman, S. Hinchcliffe, B. Allkin and E. Lughadha. (2008). Towards target 1 of the global strategy for plant conservation: a working list of all known plant species—progress and prospects. *Taxon* **57**: 602–611.

Pax, F. and K. Hoffmann. (1934). Aizoaceae in Engler and Prantl, *Die naturlichen pflanzenfamilien*, 2ed, Bd. 16c: 179–233.

Pederneiras, L. C., J. P. P. Carauta, S. Romaniuc-Neto and V. de F. Mansano. (2015a). An overview of the infrageneric nomenclature of *Ficus* (Moraceae). *Taxon* **64:** 589–594.

Pederneiras, L. C., S. Romaniuc-Neto and V. de F. Mansano. (2015b). Molecular phylogenetics of *Ficus* section *Pharmacosycea* and the description of *Ficus* subsection *Carautaea* (Moraceae). *Syst. Bot.* **40:** 504–509.

Peery, R., K. R. Wilcox, C. Morton and L. A. Raubeson. (2008). Exploring the utility of the nuclear XDH gene for gymonsperm phylogenetics. pp. 196–197. *In*: *Botany 2008. Botany without Borders.* Botanical Society of America.

Pell, S. K. (2004). *Molecular Systematics of the Cashew Family (Anacardiaceae).* Ph.D. Thesis, Louisiana State University.

Pell, S. K. and J. D. Mitchell. (2007). Evolutionary trends in Anacardiaceae inferred from nuclear and plastid molecular data and morphological evidence. P. 178. *In*: *Plant Biology and Botany 2007. Program and Abstract Book.* Chicago.

Pell, S. K., J. D. Mitchell, A. J. Miller and T. A. Lobova. (2011). Anacardiaceae. pp. 7–50. *In*: K. Kubitzki (ed.). *The Families and Genera of Flowering Plants. X. Flowering Plants: Eudicots. Sapindales, Cucurbitales, Myrtaceae.* Springer, Berlin.

Pennington, T. D. (1991). *The Genera of Sapotaceae*. Royal Botanic Gardens, Kew and New York Botanic Gardens, Bronx.

Pérez-Gutiérrez, M. A., A. T. Romero-García, M. C. Fernández, G. Blanca, M. J. Salinas-Bonillo and V. N. Suárez-Santiago. (2015). Evolutionary history of fumitories (subfamily Fumarioideae, Papaveraceae): An old story shaped by the main geological and climatic events in the Northern Hemisphere. *Molec. Phyl. Evol.* **88:** 75–92.

Peruzzi, L. (2015). A new intrafamilial taxonomic setting for Liliaceae, with a key to genera and tribes. *Plant Biosystems* doi: 10.1080/11263504.2015.1115435.

Petersen, G. and O. Seberg. (1997). Phylogenetic analysis of Triticeae (Poaceae) based on *rpoA* sequence data. *Mol. Phylog. Evol.* **7:** 217–230.

Pichon, H. (1946). Sur les Alismatacées et les Butomacées [includes *Albidella*, gen. nov., key to genera of redefined Alismaceae]. *Not. Syst.* [Paris] **12:** 170–183.

Plunkett, G. M., D. E. Soltis and P. S. Soltis. (1995). Phylogenetic relationships between Juncaceae and Cyperaceae: Insights from *rbcL* sequence data. *American J. Bot.* **82:** 520–525.

Plunkett, G. M., D. E. Soltis and P. S. Soltis. (1997). Clarification of the relationship between Apiaceae and Araliaceae based on *matK* and *rbcL* sequence data. *Amer. J. Bot.* **84:** 567–580.

Plunkett, G. M. (2001). Relationship of the order Apiales to subclass Asteridae: A re-evaluation of morphological characters based on insights from molecular data. *Edinburgh J. Bot.* **8:** 183–200.

Plunkett, G. M., J. Wen and P. P. Lowry (2004). Infrafamilial classifications and characters in Araliaceae: Insights from the phylogenetic analysis of nuclear (ITS) and plastid (*trn*L-*trn*F) sequence data. *Plant Syst. Evol.* **245:** 1–39.

Porter, C. L. (1959). *Taxonomy of Flowering Plants.* W. H. Freeman, San Francisco.

Porter, E. A. and L. A. Johnson. (1998). Phylogenetic relationships of Polemoniaceae: Inferences from mitochondrial *nad1b* intron sequences. *Aliso* **17:** 157–188.

Porter, E. A., E. Nic Lughadha and M. S. J. Simmonds. (2000). Taxonomic significance of polyhydroxyalkaloids in the Myrtaceae. *Kew Bull.* **55:** 615–632.

Potgeiter, J. and V. A. Albert. (2001). Phylogenetic relationships within Apocynaceae s.l. based on *trn*L intron and *trn*L-F spacer sequences and propagule characters. *Ann. Missouri Bot. Gard.* **88:** 523–549.

Potter, D., F. Gao, P. E. Bortiri, S. –H. Oh and S. Baggett. (2002). Phylogenetic relationships in Rosaceae inferred from chloroplast *matK* and *trnL-trnF* nucleotide sequence data. *Plant Syst. evol.* **231:** 77–89.

PPG I (Pteridophyte Phylogeny Group). (2016). A community-derived classification for extant lycophytes and ferns. *Journal of Systematics and Evolution* **54**(6): 563–603. https://doi.org/10.1111/jse.12229. Wiley Online Library.

Prat, W. (1960). Vers une classification naturelles des Graminées. *Bull. Soc. Bot. Fr.* **107**: 32–79.

Prather, C. A., C. J. Ferguson and R. K. Jansen. (2000). Polemoniaceae phylogeny and classification: Implications of sequence data from the chloroplast gene *ndhF. Amer. J. Bot.* **87:** 1300–1308.

Pratt, D. B., L. G. Clark and R. S. Wallace. (2001). A tale of two families: Phylogeny of the Chenopodiaceae-Amaranthaceae. P. 135, in *Botany 2001: Plants and People*, Abstracts. [Albuquerque.].

Preston, J. C. and E. A. Kellogg. (2006). Reconstructing the evolutionary history of paralogous APETALA1? FRUITFULL-like genes in Grasses (Poaceae).

Price, R. A. (2003). Generic and familial relationships of the Taxaceae from *rbc*L and *mat*K sequence comparisons. pp. 235–238. *In*: R. R. Mill (ed.). *IV International Conifer Conference.* [ISIS Acta Horticulturae 615.]

Puri, G. S. (1960). *Indian Forest Ecology*. Oxford Book & Stationary Co., New Delhi. 2 Vols.

Qui, Y. -L., M. W. Chase, D. H. Les and C. R. Parks. (1993). Molecular phylogenetics of the Magnoliidae: Cladistic analyses of nucleotide sequences of plastid gene *rbcL. Ann. Missouri Bot. Gard.* **80:** 587–606.

Qiu, Y. -L., M. W. Chase, S. B. Hoot, E. Conti, P. R. Crane, K. J. Sytsma and C. R. Parks. (1998). Phylogenetics of the Hamamelidae and their allies: Parsimony analyses of nucleotide sequences of the plastid gene *rbcL. Int. J. Plant Sci.* **159:** 891–905.

Qiu, Y. -L., L. Li, B. Wang, Z. Chen, V. Knoop, M. Groth-Malonek, O. Dombrovska, J. Lee, L. Kent, J. S. Rest, G. F. Estabrook, T. A. Hendry, D. W. Taylor, C. M. Testa, M. Ambros, B. Crandall-Stotler, R. J. Duff, M. Stech, W. Frey, D. Quandt and C. C. Davis. (2006). The deepest divergences in land plants inferred from phylogenomic evidence. *Proc. National Acad. Sci. U.S.A.* **103:** 15511–15516.

Qi, Z., P. Li, Y. Zhao, K. Cameron and C. Fu. (2012). Molecular phylogeny and biogeography of Smilacaceae (Liliales), a cosmopolitan family of monocots. P. 197. *In*: *Botany 2012: The Next Generation. July 7–11—Columbus, Ohio. Abstracts.*

Qui, Y. -L., J. Lee, F. Bernasconi-Quadroni, D. E. Soltis, P. S. Soltis, M. Zanis, E. A. Zimmer, Z. Chen, V. Savolainen and M. W. Chase. (1999). The earliest angiosperms: evidence from mitochondrial, plastid and nuclear genomes. *Nature* **402:** 404–407.

Qui, Y. -L., J. Lee, F. Bernasconi-Quadroni, D. E. Soltis, P. S. Soltis, M. Zanis, E. A. Zimmer, Z. Chen, V. Savolainen and M. W. Chase. (2000). Phylogeny of basal angiosperms: Analyses of five genes from three genomes. *Int. J. Plant Sci.* **161**(6: suppl.): S3–S27.

Qui, Z., P. Li and C. Fu. (2013). New combinations and a new name in Smilax for species of Heterosmilax in Eastern and Southeast Asian Smilacaceae (Liliales). *Phytotaxa* **117**(2): 58–60. DOI: 10.11646/phytotaxa.117.2.4.

Quinn, C. J., R. A. Price and P. A. Gadek. (2002). Familial concepts and relationships in the conifers based on *rbc*L and *mat*K sequence comparisons. *Kew Bull.* **57:** 513–531.

Radford, A. E., W. C. Dickison, J. R. Massey and C. R. Bell. (1974). *Vascular Plant Systematics*. Harper and Row, New York.

Radford, A. E. (1986). *Fundamentals of Plant Systematics.* Harper and Row, New York.

Rai, H. S., P. A. Reeves, R. Peakall, R. G. Olmstead and S. W. Graham. (2008). Inference of higher-order conifer relationships from a multi-locus plastid data set. *Botany* **86:** 658–669.

Raimann, R. (1893). Onagraceae. in Engler and Prantl, *Die naturlichen Pflanzenfamilien* **III(7):** 199–223.

Ram, Manasi. (1959). Morphological and embryological studies in the family Santalaceae II. *Exocarpus*, with a discussion on its systematic position. *Phytomorphology* **8**: 4–19.

Raunkiaer, C. (ed. A. G. Tansley). (1934). *The Life Forms of Plants and Statistical Plant Geography*. Oxford.

Raven, P. H. (1975). The bases of angiosperm phylogeny: cytology. *Ann. Miss. Bot. Gard.* **62**: 725–764.

Ray, J. (1682). *Methodus plantarum nova.* London, 3 vols.

Ray, J. (1686–1704). *Historia plantarum.* London, 3 vols.

Rechinger, K. H. (1963). *Flora Iranica.*

Reeves, R. G. (1972). *Flora of Central Texas*. Prestige Press, Ft. Worth, TX, 320 pp.

Rehder, A. (1940). *Manual of Cultivated Trees and Shrubs Hardy in North America* (2nd ed.). Macmillan, New York.

Rendle, A. B. (1904). *Classification of Flowering Plants.* Cambridge, England. Vol. 2 1925; 2nd ed. Vol. 1 1930.

Renner, S. S., A. Weerasooriya and M. E. Olson. (2002). Phylogeny of Cucurbitaceae inferred from multiple chloroplast loci. P. 169. *In*: *Botany 2002: Botany in the Curriculum*, Abstracts. [Madison, Wisconsin.]

Renner, S. S. and H. Schaefer. (2016). Phylogeny and evolution of the Cucurbitaceae. *In*: R. Grumet, N. Katzir and J. Garcia-Mas (eds.). *Genetics and Genomics of Cucurbitaceae.* Springer International, Switzerland. [*Plant Genetics and Genomics: Crops and Models.* Vol. **19**.]

Retallack, G. and D. L. Dilcher. (1981). A coastal hypothesis for the dispersal and rise of dominance of flowering plants. pp. 27–77. *In*: K. J. Niklas (ed.). *Paleobotany, Paleoecology and Evolution.* Praeger, New York, Vol. 2.

Richardson, J. E., F. M. Weitz, M. F. Fay, Q. C. Cronk, H. P. Linder, G. Reeves and M. W. Chase. (2001). Phylogenetic analysis of Phylica L. (Rhamnaceae) with an emphasis on island species: Evidence from plastid trnL-F and nuclear internal transcribed spacer (ribosomal) DNA sequences. *Taxon* **50:** 405–427.

Richardson, J. E., B. A. Whitlock, A. W. Meerow and S. Madriñ&n. (2015). The age of chocolate: A diversification history of Theobroma and Malvaceae. *Front. Ecol. Evol.* **3:** 120.

Rieppel, Olivier. (2006). The PhyloCode: a critical discussion of its theoretical foundation. https://onlinelibrary.wiley.com/doi/full/10.1111/j.1096-0031.2006.00097.x.

Riesberg, L. H., B. Sinervo, C. R. Linder, M. Ungerer and D. M. Arias. (1996). Role of gene interactions in hybrid speciation: Evidence from ancient and experimental hybrids. *Science* **272:** 741–745.

Renner, S. S. and H. Schaefer. (2016). Phylogeny and evolution of the Cucurbitaceae. *In*: R. Grumet, N. Katzir and J. Garcia-Mas (eds.). *Genetics and Genomics of Cucurbitaceae. Springer International, Switzerland. [Plant Genetics and Genomics: Crops and Models. Vol. 19.]*

Rise, K. A., M. J. Donoghue and R. G. Olmstead. (1997). Analyzing large data sets: *rbcL* 500 revisited. *Syst. Biol.* **46:** 554–563.

Robbrecht, E and J. F. Manen. (2006). The major evolutionary lineages of the coffee family (Rubiaceae, angiosperms). Combined analysis (nDNA and cpDNA) to infer the position of *Coptosapelta* and *Luculia*, and supertree construction based on *rbcL*, *rps16*, *trnL-trnF* and *atpB-rbcL* data. A new classification in two subfamilies, Cinchonoideae and Rubioideae. *Systematic Geography of Plants* **76**: 85–146.

Rodman, J. E., R. A. Price, K. Karol, E. Conti, K. J. Sytsma and J. D. Palmer. (1993). Nucleotide sequences of *rbcL* gene indicate monophyly of mustard oil plants. *Ann. Missouri Bot. Gard.* **80:** 686–699.

Rogers, D. J. (1963). Taximetrics, new name, old concept. *Brittonia* **15**: 285–290.

Rollins, R. C. (1953). Cytogenetical approaches to the study of genera. *Chronica Botanica* **14**(3): 133–139.

Rose, J. P., T. J. Kleist, S. D. Löfstrand, B. T. Drew, J. Schönenberger and K. J. Sytsma. (2018). Phylogeny, historical biogeography, and diversification of angiosperm order Ericales suggest ancient Neotropical and East Asian connections. *Molec. Phyl. Evol.* **122:** 59–79.

Ross, K. L. (2004). All Living Things: In Seven Kingdoms. *The Proceedings of the Friesian School, Fourth Series.* http://www.friesian.com/life-1.htm#one.

Rousi, A. (1973). Studies on the cytotaxonomy and mode of reproduction of *Leontodon* (Compositae). *Ann. Bot. Fenn.* **10**: 201–215.

Rova, J. H. E., P. G. Delprete, L. Andersson and V. A. Albert. (2002). A *trnL*-F cpDNA sequence study of the Condamineeae-Rondeletieae-Sipaneeae complex with implications on the phylogeny of the Rubiaceae. *Amer. J. Bot.* **89:** 145–159.

Rudall, P., C. A. Furness, M. W. Chase and M. F. Fay. (1997). Microsporogenesis and pollen sulcus type in Asparagales (Lilianae). *Canad. J. Bot.* **75:** 408–430.

Rudall, P. (2001). Floral morphology of Asparagales: unique structures and iterative evolutionary themes. P. 16. *In*: *Botany 2001: Plants and People*, Abstracts. [Albuquerque.]

Ruggiero, M. A., D. P. Gordon, T. M. Orrell, N. Bailly, T. Bourgoin, R.C. Brusca, T. Cavalier-Smith, M.D. Guiry and P.M. Kirk. (2015). A higher-level classification of all living organisms. *PLoS ONE* **10:** e0119248.

Ruhfel, B. R., C. P. Bove, C. T. Philbrick and C. C. Davis. (2016). Dispersal largely explains the Gondwanan distribution of the ancient tropical clusioid plant clade. *American J. Bot.* **103:** 1117–1128.

Rydin, C., M. Källersjö and E. M. Friis. (2002). Seed plant relationships and the systematic position of Gnetales based on nuclear and chloroplast data: Conflicting data, rooting problems, and the monophyly of conifers. *Internat. J. Plant Sci.* **163:** 197–214.

Rydin, C., N. Wikström and B. Bremer. (2017). Conflicting results from mitochondrial genomic data challenge current views of Rubiaceae phylogeny. *American J. Bot.* **104:** 1522–1532.

Sahasrabudhe, S. and C. A. Stace. (1974). Developmental and structural variation in the trichomes and stomata of some Gesneriaceae. *New Botanist* **1**: 46–62.

Sahni, B. (1925). Ontogeny of vascular plants and theory of recapitulation. *J. Indian Bot. Soc.* **4**: 202–216.

Sahni, B. (1948). Pentoxyleae: A new group of Jurassic gymnosperms from the Rajmahal Hills of India. *Bot. Gaz.* **110:** 47–80.

Salywon, A., N. Snow and L. R. Landrum. (2002). Phylogenetic relationships in the berry-fruited Myrtaceae as inferred from ITS sequences. P. 149. *In*: *Botany 2002: Botany in the Curriculum*, Abstracts. [Madison, Wisconsin.]

Sanchez, A., T. M. Schuster and K. A. Kron. (2009). A large-scale phylogeny of Polygonaceae based on molecular data. *Internat. J. Plant Sci.* **170:** 1044–1055.

Sauquet, H., L. Carrive, N. Poullain, J. Sannier, C. Damerval and S. Nadot. (2015). Zygomorphy evolved from disymmetry in Fumarioideae (Papaveraceae, Ranunculales): New evidence from an expanded molecular phylogenetic framework. *Ann. Bot.* **115:** 895–914.

Savolainen, V., M. W. Chase, S. B. Hoot, C. M. Morton, D. E. Soltis, C. Bayer, M. F. Fay, A. Y. de Bruijn, S. Sulllivan and Y. -L. Qiu. (2000). Phylogenetics of flowering plants based on combined analysis of plastid *atp*B and *rbc*L sequences. *Syst. Biol.* **49:** 306–362.

Scamardella, J. M. (1999). Not plants or animals: a brief history of the origin of Kingdoms Protozoa, Protista and Protoctista. *International Microbiology* **2**: 207–221.

Schaefer, H. and S. S. Renner. (2011). Phylogenetic relationships in the order Cucurbitales and a new classification of the gourd family (Cucurbitaceae). *Taxon* **60:** 122–138.

Scharaschkin, T. and J. A. Doyle. (2005). Phylogeny and historical biogeography of *Anaxagorea* (Annonaceae) using morphology and non-coding chloroplast sequence data. *Syst. Bot.* **30:** 712–735.

Schlueter, J. A., P. Dixon, C. Granger, D. Grant, L. Clark, J. J. Doyle and R. C. Schoemaker. (2004). Mining EST databases to resolve evolutionary events in major crop species. *Genome* **47:** 868–876.

Schneider, H. A. W. and W. Liedgens. (1981). An evolutionary tree based on monoclonal antibody-recognised surface features of plastid enzume (5-aminolevulinate dehydratase). *Z. Naturforsch.* **36**(c): 44–50.

Schneider, H., S. J. Russell, C. J. Cox, F. Bakker, S. Henderson, M. Gibby and J. C. Vogel. (2004). Chloroplast phylogeny of asplenioid ferns based on *rbcL* and *trnL-F* spacer sequences (Polypodiidae, Aspleniaceae) and its implications for the biogeography. *Syst. Bot.* **29:** 260–274.

Schneider, H., A. R. Smith, R. Cranfill, T. E. Hildebrand, C. H. Haufler and T. A. Ranker. (2004b). Unraveling the phylogeny of polygrammoid ferns (Polypodiaceae and Grammitidaceae): exploring aspects of the diversification of epiphytic plants. *Molec. Phylog. Evol.* **31:** 1041–1063.

Schubert, I., H. Ohle and P. Hanelt. (1983). Phylogenetic conclusions from Geisma banding and NOR staining in Top Onions (Liliaceae). *Pl. Syst. Evol.* **143**: 245–256.

Schulz, O. E. (1936). *Cruciferae. In*: E. Engler, Die Naturlichen Pflanzenfamilien, ed. 2, **17B**: 227–658.

Schuster, T. M., J. L. Reveal, M. J. Bayly and K. A. Kron. (2015). An updated molecular phylogeny of Polygonoideae (Polygonaceae): Relationships of *Oxygonum*, *Pteroxygonum*, and *Rumex*, and a new circumscription of *Koenigia. Taxon* **64:** 1188–1208.

Schwarzbach, A. E. and L. A. McDade. (2002). Phylogenetic relationships of the mangrove family Avicenniaceae based on chloroplast and nuclear ribosomal DNA sequences. *Syst. Bot.* **27:** 84–98.

Scotland, R. W. (!990). *Palynology and systematics of Acanthaceae.* Ph.D. Thesis, University of Reading, England.

Scotland, R. W. and K. Volleson. (2000). Classification of Acanthaceae. *Kew Bull.* **55:** 513–589.

Sennblad, B. and B. Bremer. (2002). Classification of Apocynaceae s.l. according to a new approach combining Linnaean and phylogenetic taxonomy. *Syst. Biol.* **51:** 389–409.

Seward, A. C. (1925). Arctic vegetation past and present. *J. Hort. Soc.* **50**, i.

Shen, H., D. Jin, J. P. Shu, X. L. Zhou, M. Lei, R. Wei, H. Shang, H. J. Wei, R. Zhang, L. Liu, Y. F. Gu, X. C. Zhang and Y. H. Yan. (2018). Large-scale phylogenomic analysis resolves a backbonephylogeny in ferns. *GigaScience* **7:** 1–11.

Shouw, J. F. (1923). *Grunfzüge einer allgemeinen Pflanzengeographie*. Berlin.

Shukla, P. and S. P. Misra. (1979). *An introduction to Taxonomy of Angiosperms.* Vikas Publishing House, New Delhi.

Simpson, G. G. (1961). *Principles of Animal Taxonomy.* New York/London.

Simpson, D. A., C. A. Furness, T. R. Hodkinson, A. M. Muasya and M. W. Chase. (2003). Phylogenetic relationships in Cyperaceae subfamily Mapanioideae inferred from pollen and plastid DNA sequence data. *American J. Bot.* **90:** 1071–1086.

Simpson, M. G. (2006). *Plant Systematics.* 590 pp. Elsevier Academic Press, San Diego, CA.

Singh, G., Bimal Misri and P. Kachroo. (1972). Achene morphology: An aid to the taxonomy of Indian Plants. 1. Compositae, Liguliflorae. *J. Indian Bot. Soc.* **51**(3-4): 235–242.

Singh, G. (1999). *Plant Systematics.* Science Publishers, New York.

Singh, G. (1999). *Plant Systematics—Theory and Practice*. Oxford & IBH Publishing Co. Pvt. Ltd., New Delhi.

Sinnot, E. W. and I. W. Bailey. (1914). Investigations on the phylogeny of angiosperms. Part 3. *Amer. J. Bot.* **1**: 441–453.

Sinnott, Q. P. (1983). A solar thermoconvective plant drier. *Taxon* **32**: 611–613.

Sivarajan, V. V. (1984). *Introduction to Principles of Plant Taxonomy.* Oxford & IBH, New Delhi.

Smith, A. C. (1970). *The Pacific as a Key to Flowering Plant History.* Harold L. Lyon Arboretum Lecture Number 1.

Smith, A. R., M. K. Pryer, E. Schuettpelz, P. Korall, Harald Schneider and P. G. Wolf. (2006). Classification of extant Ferns. *Taxon* **55**(3): 705–731.

Smith, P. M. (1972). Serology and species relationship in annual bromes (*Bromus* L. sect. *Bromus*). *Ann. Bot.* **36**: 1–30.

Smith, P. M. (1983). Protein, mimicry and microevolution in grasses. pp. 311–323. *In*: U. Jensen and D. E. Fairbrothers (eds.). *Proteins and Nucleic Acids in Plant Systematics.* Springer-Verlag, Berlin.

Sneath, P. H. A. (1957). The application of computers to taxonomy. *J. Gen. Microbiol.* **17**: 201–226.

Sneath, P. H. A. and R. R. Sokal. (1973). *Numerical Taxonomy.* W. H. Freeman and Company, San Francisco.

Snustad, D. P. and M. J. Simmons. (2000). *Principles of Genetics.* 2nd ed. John Wiley and Sons, New York.

Sokal, R. R. and C. D. Michener. (1958). A statistical method for evaluating systematic relationships. *Univ. Kansas Sci. Bull.* **44**: 467–507.

Sokal, R. R. (1961). Distance as a measure of taxonomic similarity. *Systematic Zool.* **10**: 70–79.

Sokal, R. R. and P. H. A. Sneath. (1963). *Principles of Numerical Taxonomy.* W. H. Freeman and Company, San Francisco.

Solsbrig, O. T. (1970). *Principles and Methods of Plant Biosystematics.* Macmillan, London.

Soltis, D. E., D. R. Morgan, A. Grable, P. S. Soltis and R. Kuzoff. (1993). Molecular systematics of Saxifragaceae *sensu stricto*. *Amer. J. Bot.* **80:** 1056–1081.

Soltis, D. E., P. S. Soltis, D. L. Nickrent, L. A. Johnson, W. J. Hahn, S. B. Hoot, J. A. Sweere, R. K. Kuzoff, K. A. Kron, M. W. Chase, S. M. Swensen, E. A. Zimmer, S. -M. Chaw, L. J. Gillespie, W. J. Kresss and K. J. Sytsma. (1997). Angiosperm phylogeny inferred from 18S ribosomal DNA sequences. *Ann. Missouri Bot. Garden* **84:** 1–49.

Soltis, D. E., P. S. Soltis, M. E. Mort, M. W. Chase, V. Savolainen, S. B. Hoot and C. M. Morton. (1998). Inferring complex phylogenies using parsimony: An empirical approach using three large DNA data sets for angiosperms. *Syst. Biol.* **47:** 32–42.

Soltis, D. E., P. S. Soltis, M. W. Chase, M. E. Mort, D. C. Albach, M. Zanis, V. Savolainen, W. H. Hahn, S. B. Hoot, M. F. Fay, M. Axtell, S. M. Swensen, L. M. Prince, W. J. Kress, K. C. Nixon and J. A. Farris. (2000). Angiosperm phylogeny inferred from 18S *r*DNA, *rbcL,* and *atpB* sequences. *Bot. Journ. Linn. Soc.* **133:** 381–461.

Soltis, D. E., R. K. Kuzoff, M. E. Mort, M. Zanis, M. Fishbein, L. Hufford, J. Koontz and M. Arroyo. (2001). Elucidating deep-level phylogenetic relationships in Saxifragaceae using sequences for six chloroplastic and nuclear DNA regions. *Ann. Missouri Bot. Gard.* **88:** 669–693.

Soltis, D. E., P. S. Soltis, P. K. Endress and M. W. Chase. (2005). *Phylogeny and Evolution of Angiosperms.* 370 pp. Sinauer Associates, Sunderland, MA.

Soltis, D. E., M. E. Mort, M. Latvis, E. V. Mavrodiev, B. C. O'Meara, P. S. Soltis, J. G. Burleigh and R. R. de Casas. (2013). Phylogenetic relations and character evolution analysis of Saxifragales using a supermatrix approach. *American J. Bot.* **100:** 916–929.

Soltis, P. S., D. E. Soltis and M. W. Chase. (1999). Angiosperm phylogeny inferred from multiple genes as a tool for comparative biology. *Nature* **90:** 461–470.

Soltis, P. S., D. E. Soltis, M. J. Zanis and S. Kim. (2000). Basal lineages of angiosperms: Relationships and implications for floral evolution. *International Journal of Plant Sciences* **161**(6, suppl.): S97–S107.

Soltis, P. S., D. E. Soltis, M. W. Chase, M. E. Mort, D. C. Albach, M. Zanis, V. Savolainen, W. H. Hahn, W. H. Hoot, M. F. Fay, M. Axtell, S. M. Swensen, L. M. Prince, W. J. Cress, K. C. Nixon and J. A. Farris. (2000a). Angiosperm phylogeny inferred from 18S rDNA, *rbcL*, and *atpB* sequences. *Bot. J. Linn. Soc.* **133:** 381–461.

Soltis, S. A. Smith, N. Cellinese, K. J. Wurdack, D. C. Tank, S. F. Brockington, N. F. Refulio-Rodriguez, J. B. Walker, M. J. Moore, B. S. Carlsward, C. D. Bell, M. Latvis, S. Crawley, C. Black, D. Diouf, Z. Xi, C. A. Rushworth, M. A. Gitzendanner, K. J. Sytsma, Y. L. Qiu, K. W. Hilu, C. C. Davis, M. J. Sanderson, R. S. Beaman, R. G. Olmstead, W. S. Judd, M. J. Donoghue and P. S. Soltis. (2011). Angiosperm phylogeny: 17 genes, 640 taxa. *American J. Bot.* **98:** 704–730.

Soreng, R. J. and J. I. Davis. (1998). Phylogenetics and character evolution in the grass family (Poaceae): Simultaneous analysis of morphological and chloroplast DNA restriction site character sets. *Bot. Rev.* **64:** 1–85.

Soros, C. L. and D. H. Les. (2002). Phylogenetic relationships in the Alismataceae. P. 152. *In*: *Botany 2002: Botany in the Curriculum,* Abstracts. [Madison, Wisconsin.]

Sosef, M. S. M. (1997). Hierarchical models, reticulate evolution and the inevitability of paraphyletic supraspecific taxa. *Taxon* **46**: 75–85.

Speta, F. (1979). Weitere untersuchungen über proteinkörper in Zellkernen und ichre taxonomische Bedentung. *Plant Syst. Evol.* **132**: 1–126.

Sporne, K. R. (1971). *The mysterious origin of flowering plants.* Oxford Biology Readers 3, F. F. Head and O. E. Lowenstein (eds.). Oxford Univ. Press, Oxford.

Sporne, K. R. (1974). *Morphology of Angiosperms.* Hutchinson Univ. Library, London.

Sporne, K. R. (1976). Character correlation among angiosperms and the importance of fossil evidence in assessing their significance. *In*: C. B. Beck (ed.). *Origin and Early Evolution of Angiosperms.* Columbia Univ. Press, New York.

Stace, C. A. (1973). Chromosome numbers in British Species of *Calystegia* and *Convolvulus. Watsonia* **9**: 363–367.

Stace, C. A. (1980). *Plant Taxonomy and Biosystematics.* Edward Arnold, London.

Stace, C. A. (1989). *Plant Taxonomy and Biosystematics.* (2nd ed.) Edward Arnold, London.

Stafleu, F. A. and E. A. Mennega. (1997). Taxonomic literature, 2nd ed., suppl. 4 (Ce-Cz). *Regnum Veg.* 134. (suppl. 3 (Br-Ca) *Regnum Veg.* 132 publ. 1995; suppl. 1 (Aa-Ba) *Regnum Veg.* 125 publ. 1992).

Stanier, R. Y. and C. B. van Niel. (1962). The concept of a bacterium. *Arch. Microbiol.* **42**: 17–35.

Starr, J. R., S. A. Harris and D. A. Simpson. (2004). Phylogeny of the unispicate taxa in Cyperaceae tribe Cariceae I: Generic relationships and evolutionary scenarios. *Syst. Bot.* **29:** 528–544.

Starr, J. R., V. Teoh, E. Roalson, A. M. Muasya and D. A. Simpson. (2006). Towards a phylogenetic classification of sedges (Cyperaceae): chloroplast (rbcL, matK, NdhF) and nuclear (ADC) data. pp. 258–259. *In: Botany 2006—Looking to the Future, Conserving the Past.* [Abstracts: Botanical Society of America, etc.].

Stebbins, G. L. (1950). *Variation and Evolution in Plants.* Columbia Univ. Press, NY.

Stebbins, G. L. (1974). *Flowering Plants; Evolution above the Species Level.* The Belknap Press, Harvard Univ. Press, Cambridge.

Steenis, C. G. G. J. van (ed.). (1948). *Flora Malesiana.* Series I: Spermatophyta. Groninger, Jakarta.

Stefanovic, S., L. Krueger and R. G. Olmstead. (2002). Monophyly of the Convolvulaceae and circumscription of their major lineages based on DNA sequences of multiple chloroplast loci. *American J. Botany* **89:** 1510–1522.

Stefanovic, S., D. F. Austin and R. G. Olmstead. (2003). Classification of Convolvulaceae: A phylogenetic approach. *Syst. Bot.* **28:** 791–806.

Stevens, P. F. (1994). *The Development of Biological Systematics.* Columbia University Press, New York.

Stevens, P. F. (1998). What kind of classification should the practicing taxonomist use to be saved? pp. 295–319. *In*: J. Dransfield, M. J. E. Coode and D. A. Simpson (eds.). *Plant Diversity in Malesia III,* Royal Botanical Gardens, Kew.

Stevens, P. F. (2001 onwards). Angiosperm Phylogeny Website. Version 14, July 2017 [and more or less continuously updated since]. http://www.mobot.org/MOBOT/research/APweb. Website developed and maintained by Hilary Davis (last updation incorcorated 25/09/2018).

Stevenson, D. W. (1981). Observations on ptyxis, phenology and trichomes in the Cycadales and their systematic implications. *Am. J. Bot.* **68:** 1104–1114.

Stevenson, D. W. and H. Loconte. (1995). Cladistic analysis on monocot families. pp. 543–578. *In*: M. J. Rudall et al. (eds.). *Monocotyledons: Systematics and Evolution.* Royal Botanic Gardens, Kew.

Stewart, W. N. and G. W. Rothwell. (1993). *Paleobotany and the Evolution of Plants.* 2nd ed. Cambridge University press, Cambridge.

Steyermark, J. A. (1963). *Flora of Missouri.* Iowa State Univ. Press, Ames, IA, 1725 pp.

Stork, N. E. and M. J. D. Brendel (1990). Variation in the insect fauna of Sulawesi trees with season, altitude and forest type. pp. 173–90. *In*: W. J. Knight and J. D. Holloway (eds.). *Insects and the Rain Forests of South East Asia (Wallacea).* London: Royal Entomological Society of London.

Strahler, A. N. and A. H. Strahler. (1977). *Geography and Man's Environment.* John Wiley and Sons, New York.

Stuessy, T. F. (1990). *Plant Taxonomy.* Columbia Univ. Press, New York.

Sun, B. -Y., M. H. Kim, C. H. Kim and C. -W. Park. (2001). *Mankyua* (Ophioglossaceae): a new fern genus from Cheju Island, Korea. *Taxon* **50:** 1019–1024.

Sun, B. -Y. (2002). Characteristics of fern flora of Korea with emphasis on the endemic genus *Mankyua* (Ophioglossaceae) from Cheju Island, Korea. *First Korean Academy of Science and Technology/Hungary Academy of Science Bilateral Symposium Proceedings* **1:** 62–68.

Sun, G., G. L. Dilcher, S. Zheng and Z. Zhou. (1998). In search of the first flower: A jurassic angiosperm, archaefructus, from northeast china. *Science* **282:** 1692–1695.

Sun, G., Q. Ji, D. L. Dilcher, S. Zheng, K. C. Nixon and X. Wang. (2002). Archaefru-ctaceae, a new basal angiosperm family. *Science* **296:** 899–904.

Sun, G., D. L. Dilcher, H. Wang and Z. Chen. (2011). A eudicot from the early Cretaceous of China. *Nature* **471**: 625–628.

Sun, M., R. Naeem, J. -X. Su, Z. -Y. Cao, G. J. Burleigh, P. S. Soltis, D. E. Soltis and Z. -D. Chen. (2016). Phylogeny of the Rosidae: A dense taxon sampling analysis. *J. Syst. Evol.* **54**(4): 363–391. doi:10.1111/jse.12211.

Surange, K. R. and S. Chandra. (1975). Morphology of gymnospermous fructifications of the Glossopteris flora and their relationships. *Palaeontographia* **149:** 153–180.

Sutter, D. and P. K. Endress. (1995). Aspects of gynoecial structure and macrosystematics in Euphorbiaceae. *Bot. Jahrb. Syst.* **116:** 517–536.

Swain, T. (1977). Secondary compounds as protective agents. *Ann. Rev. Pl. Physiol.* **28**: 479–501.

Swenson, K. M. and N. El-Mabrouk. (2012). Gene trees and species trees: irreconcilable differences. *BMC Bioinformatics* **13** (Suppl 19).

Swenson, U. and A. A. Anderberg. (2005). Phylogeny, character evolution, and classification of Sapotaceae (Ericales). *Cladistics* **21:** 101–130.

Sytsma, K. J. and D. A. Baum. (1996). Molecular phylogenies and the diversification of the angiosperms. pp. 314–340. *In*: D. W. Taylor and L. J. Hickey (eds.). *Flowering Plant Origin, Evolution and Phylogeny.* Chapman & Hall Inc., New York.

Sytsma, K. J., J. Morawetz, J. C. Pires, M. Nepokroeff, E. Conti, M. Zjhra, J. C. Hall and M. W. Chase. (2002). Urticalean rosids: Circumscription, rosid ancestry, and phylogenetics based on *rbc*L, *trn*L-F, and *ndh*F sequences. *American J. Bot.* **89:** 1531–1546.

Takahashi, A. (1988). Morphology and ontogeny of stem xylem elements in *Sarcandra glabra* (Thunb.) Nakai (Chloranthaceae): additional evidence for the occurrence of vessels. *Bot. Mag. (Tokyo)* **101**: 387–395.

Takhtajan, A. L. (1958). *Origin of Angiospermous Plants.* Amer. Inst. Biol. Sci. (Translation Russian edition of 1954).

Takhtajan, A. L. (1959). *Die Evolution der A*ngiospermen. Gustav Fischer Verlag, Jena.

Takhtajan, A. L. (1966). *Systema et Phylogenia Magnoliophytorum.* Soviet Publishing Institution, Nauka.

Takhtajan, A. L. (1969). *Flowering Plants—Origin and Dispersal* (English translation by C. Jeffrey). Smithsonian Institution Press, Washington.

Takhtajan, A. L. (1980). Outline of classification of flowering plants (Magnoliophyta). *Bot. Rev.* **46**: 255–369.

Takhtajan, A. L. (1986). *Floristic Regions of the World.* Berkeley.

Takhtajan, A. L. (1987). *Systema Magnoliophytorum.* Nauka, Leningrad.

Takhtajan, A. L. (1991). *Evolutionary Trends in Flowering Plants.* Columbia Univ. Press, New York.

Takhtajan, A. L. (1997). *Diversity and Classification of Flowering Plants.* Columbia Univ. Press, New York, 642 pp.

Takhtajan, A. L. (2009). *Flowering Plants.* Second Edition. Springer, 918 pp.

Tamboli, A. S., B. Pradnya, P. B. Yadav, A. A. Gothe, S. R. Yadav and S. P. Govindwar. (2018). Molecular phylogeny and genetic diversity of genus Capparis (Capparaceae) based on plastid DNA sequences and ISSR markers. *Plant Systematics and Evolution* **304**(2): 205–217.

Tanaka, N., H. Setoguchi and J. Murata. (1997). Phylogeny of the family Hydrocharitaceae inferred from *rbc*L and *mat*K gene sequence data. *J. Plant Res.* **110:** 329–337.

Taylor, D. W. (1981). *Paleobotany: An Introduction to Fossil Plant Biology.* McGraw Hill, New York.

Taylor, D. W. and L. J. Hickey. (1992). Phylogenetic evidence for herbaceous origin of angiosperms. *Plant Systematics and Evolution* **180**: 137–156.

Taylor, D. W. and G. Kirchner. (1996). Origin and evolution of angiosperm carpel. pp. 116–140. *In*: D. W. Taylor and L. J. Hickey (eds.). *Flowering Plant Origin, Evolution and Phylogeny.* Chapman & Hall Inc., New York.

Taylor, D. W. and L. J. Hickey. (1996). Evidence for and implications of an Herbaceous origin of Angiosperms. pp. 232–266. *In*: D. W. Taylor and L. J. Hickey (eds.). *Flowering Plant Origin, Evolution and Phylogeny.* Chapman & Hall Inc., New York.

Taylor, P. E., G. Card, J. House, M. H. Dickinson and R. C. Flagan. (2006). High-speed pollen release in the white mulberry, *Morus alba* L. *Sex. Plant Reprod.* **19:** 19–24.

Terrell, E. (1983). Proposal (695) to conserve the name of the tomato as *Lycopersicon esculentum* P. Miller and reject the combination *Lycopersicon lycopersicum* (L.) Karsten (Solanaceae). *Taxon* **32**: 310–313.

Theophrastus. (1916). *Enquiry into plants.* Translated by A. Hort. W. Heinemann, London, 2 vols.

Theophrastus. (1927). *De causis plantarum* Translated by R. E. Dengler. Philadelphia.

Thorne, R. F. (1968). Synopsis of a putative phylogenetic classification of flowering plants. *Aliso* **6**(4): 57–66.

Thorne, R. F. (1974). A phylogenetic classification of Annoniflorae. *Aliso* **8**: 147–209.
Thorne, R. F. (1976). A phylogenetic classification of angiosperms. *Evol. Biol.* **9**: 35–106.
Thorne, R. F. (1981). Phytochemistry and angiosperm phylogeny: A summary statement. pp. 233–295. *In*: D. A. Young and D. S. Seigler (eds.). *Phytochemistry and Angiosperm Phylogeny.* Praeger, New York.
Thorne, R. F. (1983). Proposed new realignments in angiosperms. *Nordic J. Bot.* **3**: 85–117.
Thorne, R. F. (1992). An updated classification of the flowering plants. *Aliso* **13**: 365–389.
Thorne, R. F. (1992b). Classification and geography of the flowering plants. *Bot. Rev.* **58**: 225–348.
Thorne, R. F. (1996). The least specialized angiosperms. pp. 286–313. *In*: D. W. Taylor and L. J. Hickey (eds.). *Flowering Plant Origin, Evolution and Phylogeny.* Chapman & Hall Inc., New York.
Thorne, R. F. (1999). *An Updated Classification of the Class Angiospermae*. http://www.inform.umd.edu/PBIO/fam/thorneangiosp99.html. Website maintained (with nomenclatural additions) by Dr. J. L. Reveal of University of Maryland.
Thorne, R. F. (2000). The classification and geography of the monocotyledon subclasses Alismatidae, Liliidae, and Comelinidae. pp. 75–124. *In*: B. Nordenstam, G. El Ghazaly and M. Kassas (eds.). *Plant Systematics for the 21st Century*. Portland, Oregon.
Thorne, R. F. (2000). The classification and geography of flowering plants: Dicotyledons of the class Angiospermae (subclasses Magnoliidae, Ranunculidae, Caryophyllidae, Dilleniidae, Rosidae, Asteridae, and Lamiidae). *Bot. Rev.* **66:** 441–647.
Thorne, R. F. (2003). *An Updated Classification of the Class Angiospermae*. www.rsabg.org/publications/angiosp.htm.
Tournefort, J. P. de. (1696). *Elements de botanique*. Paris, 3 vols.
Tournefort, J. P. de. (1700). *Institutiones rei herbariae*. Imprimerie Royale, Paris, 3 vols.
Trehane, P. et al. (1995). International code of nomenclature for cultivated plants. *Regnum Veg.* **133**.
Treutlein, J., G. F. Smith, B. -E. van Wyk and M. Wink. (2003). Phylogenetic relationships in Asphodelaceae (subfamily Alooideae) inferred from chloroplast DNA sequences (*rbc*L, *mat*K) and from genomic finger-printing (ISSR). *Taxon* **52:** 193–207.
Troitsky, A. V., Y. F. Melekhovets, G. M. Rakhimova, V. K. Bobrova, K. M. Valiegoroman and A. S. Antonov. (1991). Angiosperm origin and early stages of seed plant evolution deduced from rRNA sequence comparison. *J. Molecular Evolution* **32**: 253–261.
Tsutsumi, C. and M. Kato. (2006) Evolution of epiphytes in Davalliaceae and related ferns. *Bot. J. Linn. Soc.*
Tucker, S. C. and W. Douglas. (1996). Floral structure, development and relationships of Paleoherbs: Saruma, Cabomba, Lactoris, and selected Piperales. pp. 141–175. *In*: D. W. Taylor and L. J. Hickey (eds.). *Flowering Plant Origin, Evolution and Phylogeny.* Chapman and Hall, Inc., New York.
Tucker, S. C. (2002). Floral ontogeny of *Cercis* (Leguminosae: Caesalpinioideae: Cercideae): Does it show convergence with Papilionoids? *Int. J. Plant Sci.* **163:** 75–87.
Turrill, W. B. (1939). Principles of Plant Geography. *Kew Bull.* 208–237.
Turrill, W. B. (1959). On the Flora of St. Helena. *Kew Bull.* 358–362.
Turrill, W. B. (1959). Plant geography. pp. 171–229. *In*: W. B. Turrill (ed.). *Vistas in Botany*. Pergamon Press, London.
Turrill, W. B. (1964). Plant taxonomy, phytogeography and ecology. *In*: W. B. Turrill (ed.). *Vistas in Botany.* Pergamon Press, London **4**: 187–224.
Tutin, T. G. et al. (ed.). (1964–1980). *Flora Europaea.* Cambridge University Press, Cambridge, 5 vols.
Uhl, N. W. (1947). Studies in the floral morphology and anatomy of certain members of Helobiae. Ph.D. Thesis (Cited by Lawrence, 1951).
Uhl, N. W. and J. Dransfield. (1987). *Genera palmarum: a classification of palms based on the work of Harold E. Moore, Jr.* Allen Press, Lawrence, Kansas.
Uhl, N. W., J. Dransfield, J. I. Davis, M. A. Luckow, K. H. Nansen and J. J. Doyle. (1995). Phylogenetic relationships among palms: Cladistic analysis of morphological and chloroplast DNA restriction site variation. *In*: P. J. Rudall, P. J. Cribb, D. F. Cutter and C. J. Humphries (eds.). *Monocotyledons: Systematics and Evolution.* 623–661.
Upchurch, G. R. Jr. and J. A. Wolfe. (1987). Mid-cretaceous to early tertiary vegetation and climate: evidence from fossil leaves and woods. pp. 75–105. *In*: E. M. Friis, W. G. Chaloner and P. R. Crane (eds.). *The Origin of Angiosperms and their Biological Consequences.* Cambridge Univ. Press, Cambridge.
Valentine, D. H. and A. Love. (1958). Taxonomic and biosystematic categories. *Brittonia* **10**: 153–166.
Vasconcelos, T. N. C., C. E. B. Proença, B. Ahmad, D. S. Aguilar, R. Aguilar, B. S. Amorim, K. Campbell, I. R. Costa, P. S. De-Carvalho, J. E. Q. Faria, A. Giaretta, P. W. Kooij, D. F. Lima, F. F. Mazine, B. Peguero, G. Prenner, M. F. Santos, J. Soewarto, A. Wingler and E. J. Lucas. (2017). Myrteae phylogeny, calibration, biogeography and diversification patterns: Increased understanding in the most species rich tribe of Myrtaceae. *Molec. Phyl. Evol.* **109**: 113–137. doi: http://dx.doi.org/10.1016/j.ympev.2017.01.002.
Vegter, I. H. (1988). Index herbariorum: a guide to the location and contents of the world's public herbaria. Part 2(7). Collectors T-Z. *Regnum Veg.* **117**. (Part 2(6). Collectors S. *Regnum Veg.* **114** publ. 1986).
Wahlert, G. A., T. Marcussen, J. de Paula-Souza, M. Feng and H. E. Jr. Ballard. (2014). A phylogeny of the Violaceae (Malpighiales) inferred from plastid DNA sequences: Implications for generic diversity and intrafamilial classification. *Syst. Bot.* **39**: 239–252.
Walker, J. W. and A. G. Walker. (1984). Ultrastructure of Lower Cretaceous angiosperm pollen and the origin and early evolution of flowering plants. *Ann. Mo. Bot. Gard.* **71**: 464–521.
Wallace, R. S. and A. C. Gibson. (2002). Evolution and systematics. pp. 1–21. *In*: P. S. Nobel (ed.). *Cacti: Biology and Uses*. University of California Press, Berkeley.
Wallick, K., W. Elisens, P. Kores and M. Molvray. (2001). Phylogenetic analysis of *trn*L-F sequence variation indicates a monophyletic Buddlejaceae and a paraphyletic *Buddleia*. pp. 148–149. *In*: *Botany 2001: Plants and People,* Abstracts. [Albuquerque.]

Wallick, K., W. Elisens, P. Kores and M. Molvray. (2002). Phylogenetic analysis of *trnL*-F sequence variation indicates a monophyletic Buddlejaceae and a paraphyletic *Buddleia*. pp. 156–157. *In*: *Botany 2002: Botany in the Curriculum,* Abstracts. [Madison, Wisconsin.]

Walters, D. R. and D. J. Keil. (1995). *Vascular Plant Taxonomy,* 4th ed. Kendall? Hunt, Dubuque, IA.

Wang, N., H. A. McAllister, P. R. Bartlett and R. J. Buggs. (2016). Molecular phylogeny and genome size evolution of the genus Betula (Betulaceae). *Ann. Bot.* **117:** 1023–1035.

Wang, W., A. -M. Lu, Y. Ren, M. E. Endress and Z. -D. Chen. (2009). Phylogeny and classification of Ranunculales: Evidence from four molecular loci and morphological data. *Persp. Plant Ecol. Evol. Syst.* **11:** 81–110.

Wannan, B. S. (2006). Analysis of generic relationships in Anacardiaceae. *Blumea* **51:** 165–195.

Watson, L. and M. J. Dallwitz. (2000). *The Families of Flowering Plants: Descriptions and Illustrations*. Website: http://biodiversity.uno.edu/delta/.

Webster, G. L. (1967). The genera of Euphorbiaceae in the southeastern United States. *J. Arnold Arbor.* **48:** 303–430.

Webster, G. L. (1994). Classification of the Euphorbiaceae. *Ann. Missouri Bot. Gard.* **81:** 3–32.

Weeks, A., F. Zapata, S. K. Pell, D. C. Daly, J. Mitchell and P. V. A. Fine. (2014). To move or evolve: Contrasting patterns of intercontinental connectivity and climatic niche evolution in "Terebinthaceae" (Anacardiaceae and Burseraceae). *Front. Genet.* **5:** 409. doi:10.3389/fgene.2014.00409.

Wegener, A. (1915). *Die Ensttehung der Kontinente und Oceane.* In German.

Wegener, A. (trans. Skerl, J. G. A.). (1924). *The Origin of Continents and Oceans.* London.

Wendel, J. F., A. Schnabel and T. Seelanan. (1995). An unusual ribosomal DNA sequence from Gossypium gossypioides reveals ancient, cryptic, intergenomic, introgression. *Mol. Phylog. Evol.* **4:** 298–313.

Wettstein, R. R. von. (1907). *Handbuch der systematischen Botanik* (2nd ed.) Franz Deuticke, Leipzig.

Williams, S. E., V. A. Albert and M. W. Chase. (1994). Relationships of Droseraceae: A cladistic analysis of *rbc*L sequence and morphological data. *American J. Bot.* **81**: 1027–1037.

Whitlock, B. A., J. Lee, O. Dombrovska and Y. L. Qiu. (2002). Effects of rate heterogeneity on estimates of the age of angiosperms. P. 158. *In*: *Botany 2002: Botany in the Curriculum*, Abstracts. [Madison, Wisconsin.]

Whittaker, R. H. (1969). New concepts of kingdoms of organisms. *Science* **163**: 150–160.

Wieland, G. R. (1906). *American Fossil Cycads.* Carnegie Institute, Washington, D. C.

Wieland, G. R. (1916). *American Fossil Cycads* vol. II. Carnegie Institute, Washington, D. C.

Wiley, E. O. (1978). The evolutionary species concept reconsidered. *Syst. Zool.* **27**: 17–26.

Wiley, E. O. (1981). *Phylogenetics—The Theory and Practice of Phylogenetic Systematics.* John Wiley and Sons, New York.

William, W. T., J. M. Lambert and G. N. Lance. (1966). Multivariate methods in plant ecology. V. Similarity analyses and information-analysis. *J. Ecol.* **54**: 427–445.

Willis, J. C. (1922). *Age and Area.* Cambridge.

Willis, J. C. (1973). *A. Dictionary of Flowering Plants and Ferns,* 8th ed. (revised by H. K. Airy-Shaw). Cambridge Univ. Press, Cambridge, 1245 pp.

Wilson, O. E. (1992). *The Diversity of Life.* Harvard University Press, Cambridge.

Wilson, P. G., M. M. O'Brien, P. A. Gadek and C. J. Quinn (2001). Myrtaceae revisited: A reassessment of infrafamilial groups. *American J. Bot.* **88:** 2013–2025.

Wilson, P. G. (2011). Myrtaceae. pp. 212–271. *In*: K. Kubitzki (ed.). *The Families and Genera of Flowering Plants. X. Flowering Plants: Eudicots. Sapindales, Cucurbitales, Myrtaceae.* Springer, Berlin.

Woese, C. R., W. E. Balch, L. J. Magrum, G. E. Fox and R. S. Wolfe. (1977). An ancient divergence among the bacteria. *Journal of Molecular Evolution* **9**: 305–311.

Woese, C. R., O. Kandler and M. L. Wheelis. (1990). Towards a natural system of organisms: proposal for the domains archaea, bacteria, and eucarya. *Proceedings of the National Academy of Sciences* **87**: 4576–4579.

Wojciechowski, M. F., M. Lavin and M. J. Sanderson. (2003). A phylogeny of legumes based on sequences of the plastid *mat*K gene, P. 99. *In*: *Botany 2003: Aquatic and Wetland Plants: Wet and Wild.* Abstracts. [Mobile, Alabama.]

Wolfe, K. H., M. Gouy, Y. -W. yang, P. M. Sharp and W. -H. Li. (1989). Date of monocot-dicot divergence estimated from Choloroplast DNA sequence data. *Proc. Nat. Acad. Sci. USA* **86**: 6201–6205.

Woodland, D. W. (1991). *Contemporary Plant Systematics.* Prentice Hall. New Jersey.

Worsdell, W. C. (1908). The affinities of *Paeonia. J. Bot. (London)* **46:** 114.

Wu, H. -C., H. -J. Su and J. -M. Hu. (2007). The identification of A-, B-, C-, and E-class MADS-box genes and implications for perianth evolution in the basal eudicot *Trochodendron aralioides* (Trochodendraceae). *Internat. J. Plant Sci.* **168:** 775–799.

Wu, Z. -Y., A. M. Lu and Y. C. Tang. (1998a). A comprehensive study of "Magnoliidae" *sensu lato*—with special consideration on possibility and necessity for proposing a new "polyphyletic-polychronic-polytopic" system of angiosperms. pp. 269–334. *In*: A. I. Zhang and S. G. Wu (eds.). *Floristic Characteristics and Diversity of East Asian Plants.* Beijing & Berlin: China Higher Education Press & Springer Verlag.

Wu, Z. -Y., Y. C. Tang, A. M. Lu et al. (1998b). On primary subdivisions of the Magnoliophyta—towards a new scheme for an eight-class system of classification of the angiosperms. *Acta. Phytotax Sinica.* **36:** 385–402.

Wu, Z. -Y., A. M. Lu, Y. -C. Tang, Z. -D. Chen and D. -Z. Li. (2002). Synopsis of a new 'polyphyletic-polychronic-polytopic' system of the angiosperms. *Acta Phytotax. Sinica* **40:** 289–322.

Wu, Z. -Y., A. M. Lu, Y. -C. Tang, Z.- D. Chen and D. -Z. Li. (2003). *The Families and Genera of Angiosperms in China: A Comparative Analysis.* Science Press, Beijing.

Wu, Z. -Y., A. K. Monro, R. I. Milne, H. Wang, J. Liu and D. -Z. Li. (2013). Molecular phylogeny of the nettle family (Urticaceae) inferred from multiple loci of three genomes and extensive generic sampling. *Molecular Phylogenetics and Evolution* **69**(3): 814–827. doi:10.1016/j.ympev.2013.06.022. PMID 23850510.

Wu, Z. -Y., R. I. Milne, C. -J. Chen, J. Liu, H. Wang and D. -Z. Li. (2015). Ancestral state reconstruction reveals rampant homoplasy of diagnostic morphological characters in Urticaceae, conflicting with current classification schemes. *PLoS ONE* **10**(11): e0141821. doi:10.1371/journal.pone.0141821.

Wurdack, K. J. and M. W. Chase. (2002). Phylogenetics of Euphorbiaceae s. str. using plastid (*rbcL* and *trnL*-F) sequences. P. 160. *In*: *Botany 2002: Botany in the Curriculum,* Abstracts. [Madison, Wisconsin.]

Xiang, Q. -Y., D. E. Soltis and P. S. Soltis. (1998). Phylogenetic relationships of Cornaceae and close relatives inferred from matK and rbcL sequences. *Amer. J. Bot.* **85:** 285–297.

Xiang, Y., C. H. Huang, Y. Hu, J. Wen, S. Li, T. Yi, H. Chen, J. Xiang and H. Ma. (2016). Evolution of Rosaceae fruit types based on nuclear phylogeny in the context of geological times and genome duplication. *Molec. Biol. Evol.* **34:** 262–281. doi: https://doi.org/10.1093/molbev/msw242.

Yuan, Y. -M., Y. Song, K. Geuten, E. Rahelivololona, S. Wohlhauser, E. Fischer, E. Smets and P. Küpfer. (2004). Phylogeny and biogeography of Balsaminaceae inferred from ITS sequences. *Taxon* **53:** 391–403.

Yoo, K. -O. and J. Wen. (2002). Phylogeny and biogeography of *Carpinus* and subfamily Coryloideae (Betulaceae). *Int. J. Plant Sci.* **163:** 641–650.

Young, D. A. (1981). Are the angiosperms primitively vesselless? *Syst. Bot.* **6**: 313–330.

Young, D. J. and L. Watson. (1970). The classification of the dicotyledons: A study of the upper levels of hierarchy. *Aust. J. Bot.* **8**: 387–433.

Yu, Sheng-Xiang, S. B. Jansenns, Xiang-Yun Zhu, M. Liden, Tiang-Gang Gao and W. Wang. (2015). Phylogeny of *Impatiens* (Balsaminaceae): integrating molecular and morphological evidence into a new classification. *Cladistics* **32**(2): 179–197.

Zahn, L. M., J. H. Leebens-Mack, C. W. dePamphilis, H. Ma and G. Theissen. (2005). To B or not to B a flower: The role of DEFICIENS and GLOBOSA orthologs in the evolution of the angiosperms. *J. Hered.* **96:** 225–240.

Zanis, M. J., D. E. Soltis, P. S. Soltis, Y. -L. Qiu, S. Mathews and M. J. Donoghue. (2002). The root of the angiosperms revisited. *Proc. Nat. Acad. Sci.* **99:** 6848–6853.

Zanis, M. J., D. E. Soltis, P. S. Soltis, Y. -L. Qiu and E. A. Zimmer. (2003). Phylogenetic analyses and perianth evolution in basal angiosperms. *Ann. Missouri Bot. Gard.*

Záveská Drábková, L. and C. Vlcek. (2009). DNA variation within Juncaceae: Comparison of impact of organelle regions on phylogeny. *Plant Syst. Evol.* **278:** 169–186.

Záveská Drábková, L. and J. Kirschner. (2013). Oreojuncus, a new genus in the Juncaceae. *Preslia* **85:** 483–503.

Zgurski, J. M., H. S. Rai, Q. M. Fai, D. J. Bogler and J. Francisco-Ortega. (2008). How well do we understand the overall backbone of cycad phylogeny? New insights from a large, multigene plastid data set. *Mol. Phyl. Evol.* **47:** 1232–1237.

Zhang, N., L. Zeng, H. Shan and H. Ma. (2012). Highly conserved low-copy nuclear genes as effective markers for phylogenetic analyses in angiosperms. *New Phytologist* **195:** 923–937. doi: 10.1111/j.1469-8137.2012.04212.x.

Zhou, Z.K., P. M. Barrett and J. Hilton. (2003). An exceptionally preserved Lower Cretaceous ecosystem. *Nature* **421**: 807–814.

Zomlefer, W. B. (1994). *Guide to Flowering Plant Families.* University of Carolina Press, Chapel Hill.

Index

C

D

F

I

J

K

L

M

N

O

P

Q

R

S

T

U

V

W

X

Y

Z

About the Author

The author, Dr. Gurcharan Singh, is a practicing taxonomist having worked extensively with the floristics and phytosociology of N. W. Himalayas and the general nomenclatural problems. He has nearly 40 years of experience in teaching taxonomy, ecology and environment. He has retired from Sri Guru Tegh Bahadur Khalsa College, University of Delhi, India. He has published nearly 42 research papers including several new species and nomenclatural changes. He has also authored 14 books. He is presently engaged in developing online database, helping online identifications and providing guidance in various online forums. He also has accumulated a fair knowledge of Floristics of SW United States, mainly California plants.

Printed in the United States
by Baker & Taylor Publisher Services

Printed in the United States
by Baker & Taylor Publisher Services